Geophysical Monograph Series

Including
Maurice Ewing Volumes
Mineral Physics Volumes

GEOPHYSICAL MONOGRAPH SERIES

Geophysical Monograph Volumes

1 **Antarctica in the International Geophysical Year** *A. P. Crary, L. M. Gould, E. O. Hulburt, Hugh Odishaw, and Waldo E. Smith (Eds.)*
2 **Geophysics and the IGY** *Hugh Odishaw and Stanley Ruttenberg (Eds.)*
3 **Atmospheric Chemistry of Chlorine and Sulfur Compounds** *James P. Lodge, Jr. (Ed.)*
4 **Contemporary Geodesy** *Charles A. Whitten and Kenneth H. Drummond (Eds.)*
5 **Physics of Precipitation** *Helmut Weickmann (Ed.)*
6 **The Crust of the Pacific Basin** *Gordon A. Macdonald and Hisashi Kuno (Eds.)*
7 **Antarctica Research: The Matthew Fontaine Maury Memorial Symposium** *H. Wexler, M. J. Rubin, and J. E. Caskey, Jr. (Eds.)*
8 **Terrestrial Heat Flow** *William H. K. Lee (Ed.)*
9 **Gravity Anomalies: Unsurveyed Areas** *Hyman Orlin (Ed.)*
10 **The Earth Beneath the Continents: A Volume of Geophysical Studies in Honor of Merle A. Tuve** *John S. Steinhart and T. Jefferson Smith (Eds.)*
11 **Isotope Techniques in the Hydrologic Cycle** *Glenn E. Stout (Ed.)*
12 **The Crust and Upper Mantle of the Pacific Area** *Leon Knopoff, Charles L. Drake, and Pembroke J. Hart (Eds.)*
13 **The Earth's Crust and Upper Mantle** *Pembroke J. Hart (Ed.)*
14 **The Structure and Physical Properties of the Earth's Crust** *John G. Heacock (Ed.)*
15 **The Use of Artificial Satellites for Geodesy** *Soren W. Henriksen, Armando Mancini, and Bernard H. Chovitz (Eds.)*
16 **Flow and Fracture of Rocks** *H. C. Heard, I. Y. Borg, N. L. Carter, and C. B. Raleigh (Eds.)*
17 **Man-Made Lakes: Their Problems and Environmental Effects** *William C. Ackermann, Gilbert F. White, and E. B. Worthington (Eds.)*
18 **The Upper Atmosphere in Motion: A Selection of Papers With Annotation** *C. O. Hines and Colleagues*
19 **The Geophysics of the Pacific Ocean Basin and Its Margin: A Volume in Honor of George P. Woollard** *George H. Sutton, Murli H. Manghnani, and Ralph Moberly (Eds.)*
20 **The Earth's Crust: Its Nature and Physical Properties** *John G. Heacock (Ed.)*
21 **Quantitative Modeling of Magnetospheric Processes** *W. P. Olson (Ed.)*
22 **Derivation, Meaning, and Use of Geomagnetic Indices** *P. N. Mayaud*
23 **The Tectonic and Geologic Evolution of Southeast Asian Seas and Islands** *Dennis E. Hayes (Ed.)*
24 **Mechanical Behavior of Crustal Rocks: The Handin Volume** *N. L. Carter, M. Friedman, J. M. Logan, and D. W. Stearns (Eds.)*
25 **Physics of Auroral Arc Formation** *S.-I. Akasofu and J. R. Kan (Eds.)*
26 **Heterogeneous Atmospheric Chemistry** *David R. Schryer (Ed.)*
27 **The Tectonic and Geologic Evolution of Southeast Asian Seas and Islands: Part 2** *Dennis E. Hayes (Ed.)*
28 **Magnetospheric Currents** *Thomas A. Potemra (Ed.)*
29 **Climate Processes and Climate Sensitivity (Maurice Ewing Volume 5)** *James E. Hansen and Taro Takahashi (Eds.)*
30 **Magnetic Reconnection in Space and Laboratory Plasmas** *Edward W. Hones, Jr. (Ed.)*
31 **Point Defects in Minerals (Mineral Physics Volume 1)** *Robert N. Schock (Ed.)*
32 **The Carbon Cycle and Atmospheric CO_2: Natural Variations Archean to Present** *E. T. Sundquist and W. S. Broecker (Eds.)*
33 **Greenland Ice Core: Geophysics, Geochemistry, and the Environment** *C. C. Langway, Jr., H. Oeschger, and W. Dansgaard (Eds.)*
34 **Collisionless Shocks in the Heliosphere: A Tutorial Review** *Robert G. Stone and Bruce T. Tsurutani (Eds.)*
35 **Collisionless Shocks in the Heliosphere: Reviews of Current Research** *Bruce T. Tsurutani and Robert G. Stone (Eds.)*
36 **Mineral and Rock Deformation: Laboratory Studies—The Paterson Volume** *B. E. Hobbs and H. C. Heard (Eds.)*
37 **Earthquake Source Mechanics (Maurice Ewing Volume 6)** *Shamita Das, John Boatwright, and Christopher H. Scholz (Eds.)*
38 **Ion Acceleration in the Magnetosphere and Ionosphere** *Tom Chang (Ed.)*
39 **High Pressure Research in Mineral Physics (Mineral Physics Volume 2)** *Murli H. Manghnani and Yasuhiko Syono (Eds.)*
40 **Gondwana Six: Structure, Tectonics, and Geophysics** *Garry D. McKenzie (Ed.)*
41 **Gondwana Six: Stratigraphy, Sedimentology, and Paleontology** *Garry D. McKenzie (Ed.)*
42 **Flow and Transport Through Unsaturated Fractured Rock** *Daniel D. Evans and Thomas J. Nicholson (Eds.)*
43 **Seamounts, Islands, and Atolls** *Barbara H. Keating, Patricia Fryer, Rodey Batiza, and George W. Boehlert (Eds.)*

Maurice Ewing Volumes

1 **Island Arcs, Deep Sea Trenches, and Back-Arc Basins** *Manik Talwani and Walter C. Pitman III (Eds.)*
2 **Deep Drilling Results in the Atlantic Ocean: Ocean Crust** *Manik Talwani, Christopher G. Harrison, and Dennis E. Hayes (Eds.)*
3 **Deep Drilling Results in the Atlantic Ocean: Continental Margins and Paleoenvironment** *Manik Talwani, William Hay, and William B. F. Ryan (Eds.)*
4 **Earthquake Prediction—An International Review** *David W. Simpson and Paul G. Richards (Eds.)*
5 **Climate Processes and Climate Sensitivity** *James E. Hansen and Taro Takahashi (Eds.)*
6 **Earthquake Source Mechanics** *Shamita Das, John Boatwright, and Christopher H. Scholz (Eds.)*

Mineral Physics Volumes

1 **Point Defects in Minerals** *Robert N. Schock (Ed.)*
2 **High Pressure Research in Mineral Physics** *Murli H. Manghnani and Yasuhiko Syono (Eds.)*

Geophysical Monograph 44

T.M

Modeling Magnetospheric Plasma

T. E. Moore
J. H. Waite, Jr.
Editors

T. W. Moorehead
Technical Editor

W. B. Hanson
Editorial Advisor

American Geophysical Union, Washington, D.C.
1988

Published under the aegis of AGU Geophysical Monograph Board.

Library of Congress Cataloging-in-Publication Data

Modeling magnetospheric plasma.

(Geophysical monograph ; 44)
"Published under the aegis of AGU Geophysical Monograph Board"—T.p. verso.

1. Space plasmas—Mathematical models. 2. Magnetosphere—Mathematical models. 3. Ionosphere—Mathematical models. I. Moore, T. E. II. Waite, J. H. (John H.) III. AGU Geophysical Monograph Board. IV. Series.

QC809.P5M63 1988 551.5'14 88-10398
ISBN 0-87590-070-4

Printed in the United States of America.

CONTENTS

PREFACE

Existing models of the plasma distribution and dynamics in magnetosphere/ionosphere systems form a patchwork quilt of different techniques and boundaries chosen to define tractable problems. With increasing sophistication in both observational and modeling techniques has come the desire to overcome these limitations and strive for a more unified description of these systems. On the observational side, we have recently acquired routine access to diagnostic information on the lowest energy bulk plasma, completing our view of the plasma and making possible comparisons with magnetohydrodynamic calculations of plasma moments. On the theoretical side, rising computational capabilities and shrewdly designed computational techniques have permitted the first attacks on the global structure of the magnetosphere. Similar advances in the modeling of neutral atmospheric circulation suggest an emergent capability to globally treat the coupling between plasma and neutral gases. Simultaneously, computer simulation has proven to be a very useful tool for understanding magnetospheric behaviors on smaller space and time scales.

We now possess enough observational data to begin forming empirical models of the magnetospheric plasma analogous to those models of the magnetospheric magnetic field which have been available for some time. As more data are acquired, we may move toward a well specified knowledge of the systematic variations of the plasma conditions that are produced by variations in the solar radiation field and solar wind. Comparison of such empirical models with the behavior of theoretical models injects reality into the selection of initial conditions and background states and provides the basis for learning about the detailed dynamics of the systems modeled. Modeling and observations on smaller scales provide information on kinetic and transport processes that will influence the behavior of the global models.

In order to address the current state of these modeling efforts and the prospects for synthesizing the various models into a unified and predictive description of magnetosphere/ionosphere systems, the First Huntsville Workshop on Magnetosphere/Ionosphere Plasma Models was held October 14-16, 1987, at Guntersville State Park Lodge, a facility of the State of Alabama. The purpose of the workshop was to explore the relationships between the various models (e.g., MHD, kinetic, particle simulation, empirical) using observations as a guide.

The workshop attracted over 90 scientists from several nations including the U.S., UK, France, Sweden, Norway, The Netherlands, Germany, and Canada. A number of graduate students attended the workshop with support from the sponsors, and several of them presented poster papers describing their contributions in this area of research.

The format of the workshop was based on invited review talks with extensive periods of time reserved for open discussion of issues of interest. Evening sessions were devoted to poster presentation of contributed papers and general discussion among the workshop participants. The resulting schedule was very intense, packing a great deal into the brief 3-day period of the workshop.

The idea of organizing a topical workshop on models of the magnetosphere and ionosphere grew out of an effort at Marshall Space Flight Center, spearheaded by C. R. Chappell, to develop an empirical model of bulk plasma parameters in the magnetosphere from spacecraft data. The extremely wide range of magnetospheric plasma particle energies and the consequent need to assemble bulk parameters from a variety of instrumentation made this a highly challenging undertaking. It was clear that the required effort would benefit from an overall assessment of the current state of theoretical knowledge as well as spacecraft observations.

Our idea then was to convene a workshop to consider how to best pursue the goal of more global treatments of the magnetospheric plasma. We chose a cloistered environment where the participants could interact intensively for a 3-day period. In consultation with our program committee, it was decided to pursue publication of the presentated papers in hopes that the volume would spark interest and be of help to workers wishing to pursue global modeling.

The contents of this monograph represent most of the materials presented at the workshop which, in turn, reflect the emphases established by the program committee in designing the program. The program committee members were:

M. Ashour-Abdalla, UCLA
J. B. Cladis, Lockheed Palo Alto Research Laboratory
A. J. Dessler, Rice Univ.
J. A. Fedder, Naval Research Laboratory

T. E. Holzer, National Center for Atmospheric Research
R. J. Moffett, Univ. of Sheffield
A. F. Nagy, Univ. of Michigan
P. J. Palmadesso, Naval Research Laboratory
B.U.Ö. Sonnerup, Dartmouth College
R. W. Schunk, Utah State Univ.
T. W. Speiser, Univ. of Colorado
D. G. Torr, Univ. of Alabama in Huntsville

We are greatly indebted to these individuals for the efforts to define the program and to secure the participation of the invited speakers.

In accordance with AGU policy for its monograph series, each submitted manuscript underwent peer review. The resulting papers have been grouped under headings that follow essentially the organization of the workshop itself. In general, a roughly equal mix of observational and theoretical perspectives were presented for each topic.

A historical perspective and a view of the future of global modeling has been provided as an introduction to this volume by C. R. Chappell. The current status of global modeling is addressed by J. L. Horwitz; R. G. Roble; V. M. Vasyliunas; and R. J. Walker and T. Ogino. The status of research on the structure of the extended ionosphere and plasmasphere is discussed by D. L. Gallagher and P. D. Craven; P. G. Richards and D. G. Torr; R. Schmidt and A. Pedersen; and N. Singh. The special characteristics of the auroral zone and boundary layer region plasmas are discussed in papers by J. L. Burch and C. Gurgiolo; W. Lotko and C. G. Schultz; P. J. Palmadesso, S. B. Ganguli, and H. G. Mitchell, Jr.; and W. K. Peterson. The polar magnetosphere and topside ionosphere are the subject of papers by R. W. Schunk; and A. W. Yau and M. Lockwood. The plasma characteristics of the plasma sheet and ring current regions are addressed by J. Birn and K. Schindler; L. A. Frank; and T. W. Speiser. Finally, the prospects for a synthesis of local models into a more global view are discussed by T. G. Forbes; and R. L. Lysak. These invited papers along with several important contributed papers form the basis for the published material.

In addition to these review and commentary papers, similar but unpublished presentations were made by D. A. Gurnett, C. F. Kennel, R. R. Anderson, P.M.E. Decreau, R. A. Heelis, J. D. Winningham, T. I. Gombosi, G. Gloeckler, R. A. Wolf, J. Lyon, and T. L. Killeen. The workshop benefitted greatly from these instructive lectures and we are grateful for their contribution. We also thank R. Lundin for preparing a presentation which he was unfortunately unable to deliver.

We are likewise grateful to our effective session chairpersons who found themselves in the role of discussion leaders as well. They were: M. Ashour-Abdalla, R. J. Moffett, L. R. Lyons, T. E. Holzer, J. B. Cladis, and J. A. Fedder. Special thanks are also due to Professor W. B. Hanson for serving as Editorial Advisor.

We wish to acknowledge the organizational efforts of Sharon Chunn of MSFC and Ms. Linda Rabbitt as well as other staff members of the University of Alabama Division of Continuing Education who skillfully arranged for everything to happen. Technical arrangements for the workshop were efficiently handled by Paul Craven and a team which included Barbara Giles, Victoria Newman, and Craig Pollock.

Special thanks are due to Jeanne Nevin and Judy C. Holoviak, who together piloted our proposal for an AGU monograph through to endorsement. P. Rayner and S. Mansberg have ably guided the proper preparation of manuscripts. We are also most appreciative of the efforts of Sharon Chunn, to whom fell the job of tracking the progress of the manuscripts through the peer review and production process. Special thanks also go to Shelby Morris and Susan Burrer who were responsible for the typing of the monograph. The efforts of the authors in producing their manuscripts and the referees in providing constructive criticism have, of course, been crucial.

This workshop was sponsored by the NASA/Marshall Space Flight Center, Space Science Laboratory and the University of Alabama in Huntsville. We wish to acknowledge financial support from the National Aeronautics and Space Administration, Space Plasma Physics Branch; the National Science Foundation, Atmospheric Sciences Division; and the sponsoring agencies of all of the participants.

T. E. Moore
J. H. Waite, Jr.
T. W. Moorehead
Space Science Laboratory
NASA Marshall Space Flight Center
Huntsville, Alabama
1987

INTRODUCTION

Our picture of the magnetosphere-ionosphere-atmosphere portion of the solar-terrestrial system is coming progressively more into focus as a result of a series of satellite missions and ground-based observations over the past 25 years. During this same time period, we have also seen the initial steps taken toward modeling different pieces of this closely coupled system. This foreword assesses the state of knowledge of each of the different regions of the magnetosphere, ionosphere, and upper atmosphere from both an experimental and modeling perspective. In the course of the discussions, the areas of uncertainty in our current understanding have been identified. The effective merger of the current magnetospheric, ionospheric, and atmospheric models, combined with the continuous increase in their fidelity and agreement with observations, represent the next fundamentally important challenge toward our eventual goal of understanding and predicting future changes in the solar-terrestrial environmental system.

Our goals as solar-terrestrial scientists involve attaining an understanding of the sun-earth environment and gaining the ability to predict the changes at earth that are caused by variations in the solar output, both in terms of electromagnetic radiation as well as particles and magnetic fields. Our ability to make the desired predictions is dependent on the development of mathematical models which are derived both from a numerical fit to the observations and from a detailed knowledge of the physical processes involved. It is a fundamental requirement, therefore, that all the different regions of the solar-terrestrial coupled system be measured thoroughly and in a correlated way, and that models be developed that predict the behavior of these different regions as they are driven both internally and by other portions of the system.

Within the solar-terrestrial system the magnetosphere plays the primary role in coupling the energy of the variable solar wind into the near earth regions of the ionosphere and atmosphere. The magnetosphere is also responsible for the deflection of the solar wind particles and the control of access of these particles to the earth's immediate neighborhood. The magnetosphere is a very complex system which is broad in spatial extent and made up of many interconnected regions of plasma and energization phenomena. As a key link in the coupling of solar variability to the earth's environment, its character and dynamic processes must be modeled in an intercorrelated and self-consistent way. This is the challenge which faces the magnetospheric physics community and which has brought together the scientists who have contributed to this volume.

It is very important for magnetospheric scientists to begin to push toward a unified model of magnetospheric plasmas and processes at this time. The best model that we have today derives from a loose collection of many pieces, both observationally and mathematically. As we begin to bring these pieces together, it will be important to establish boundary conditions and assumptions that will allow the different pieces to mesh. For example, an agreement on the levels of Kp, AE, or Dst which represent quiet, moderate, and high magnetic conditions is of fundamental importance for combining the results of models of individual magnetospheric regions or processes. In addition, the adoption of a certain convection flow pattern, for example, for ionospheric modeling should be consistently applied to atmospheric dynamic models and magnetospheric plasma flow models alike. Agreement on topics such as these is imperative and should be reached at an early stage of model development. The discussions during this workshop can lead to the required definition and agreement on these boundary conditions and assumptions.

The invited review papers given in this volume contain excellent detail in reviewing and presenting both observational data and models for each of the magnetospheric regions. Therefore, I will not go into detail, but will try to give a top level view which lays out a framework into which these different individual papers can fit and which identifies the spectrum of understanding that we currently have of magnetospheric plasmas and processes in terms of both observations and models. It is my goal to have this overview bring out the specific areas which need particular attention and effort as we bring the unified model together over the coming months and years.

The Pieces of the Puzzle

Figure 1 shows an artist's portrayal of the earth's magnetospheric environment as it is influenced by the sun and solar wind. We are working to fill in the specific pieces of this diffuse environment with the goal of bringing it more into focus. We have historically thought of the plasma source as the external solar wind as pictured in Figure 1, but more and more the magnitude of the of the internal ionospheric source has been recognized. We now find the key focus shifting to the understanding of the mixing of these two plasma sources.

The magnetospheric magnetic field provides the coupling between the various plasma regimes and is the framework within which the plasmas are guided. Our approach will be to begin with an assessment of the understanding of this framework and then to extend the assessment as we fill in the different plasma regions by following the circulation of the plasmas from their sources to their sinks as best we currently understand it. By necessity, this approach requires a discussion of both plasma regions and plasma processes. To give an overview of the status of both observations and models we have adapted a classification scheme which gives a first order assessment of their completeness for each of the regions. This classification scheme is color-coded and is shown as the key in Figure 2.

The Skeleton

The spatial character and motion of particles in the magnetosphere are for the most part dominated by the earth's magnetic field. The shape of the magnetosphere is determined by its interaction with the external solar wind and the internal currents that interconnect different magnetospheric regions. Observations of the solar wind/magnetosphere interface are in progress both for the magnetic merging in flux transfer events (Russell and Elphic, 1979) and for the more general case of magnetopause energy transfer (Haerendel et al., 1978). Initial steps toward modeling of this interaction have been taken for example by Cowley (1980) with a more global modeling approach done by Walker and Ogino (1987). The overall energy coupling of the solar wind to the magnetosphere has been studied by Burton et al. (1975) and recently modeled empirically by Akasofu (1981).

The resulting shape of the magnetosphere has been measured extensively beginning with the initial measurements of the magnetic field by Ness et al. (1964). Observations from many magnetospheric spacecraft have built an extensive data base on the configuration of the field and its changes with magnetic activity (Ness, 1965; Russell and Brody, 1967; Fairfield, 1971; Sugiura and Poros, 1973; Tsurutani et al., 1984). Development of models of the magnetic field shape and dynamics is in progress (Mead and Fairfield, 1975; Olson and Pfitzer, 1974; Luhmann and Friesen, 1979; Tsyganenko, 1976) and is converging on a working predictor of magnetospheric configuration and dynamic response to solar wind changes.

The Sources

Given the magnetic framework of the magnetosphere we turn to the sources of its plasma. Historically the external solar wind has been considered as the dominant source. Its access through the polar cusp and plasma mantle has been measured rather thoroughly by the HEOS spacecraft (Paschmann et al., 1976) and at lower altitudes by the Injun, ISEE, Prognoz, and Dynamics Explorer satellites among others (Frank, 1971; Burch et al., 1980; Heikkila and Winningham, 1971). Modeling of the cusp plasma source is in its initial phases with particle transport into the plasma mantle calculated by Pilipp and Morfill (1978), Cowley (1980), and others.

A second potential source of magnetospheric plasmas comes from the entry of magnetosheath plasma through the flanks of the magnetotail and into the plasma sheet. Acquisition of data on the extent of this source is underway (Eastman et al., 1985), but the magnitude of the source has been diffucult to determine. Models of the specific plasma access to the tail are very incomplete at this time adding to the current uncertainty regarding the relative contribution of the external solar wind and internal ionosphere to magnetospheric plasmas.

To look at the internal ionospheric source one begins with the character of the neutral atmosphere. Of particular interest to the origin of magnetospheric plasmas is the upper portion of the earth's atmosphere, the thermosphere. Data acquisition on this region is in progress, both in terms of its composition and its dynamics (Carignan et al., 1982; Spencer et al., 1982; Killeen et al., 1982). Models for this region are reaching an advanced state and can be used to predict densities, composition, and neutral wind patterns, both as a driver of ionospheric motions and in response to changing magnetospheric conditions (Hedin et al., 1977; Roble et al., 1982; Fuller-Rowell and Rees, 1980).

Given the neutral atmosphere and its motion, one can begin to generate the observed ionosphere through the action of solar electromagnetic radiation and energetic particle precipitation. To characterize the ionosphere one must consider both chemistry and dynamics. There is an extensive observational data base on the ionosphere. The character of the incident solar spectrum has been observed by Hinteregger et al. (1981), and the particle input measured by a host of spacecraft. Models of the atmospheric response are in development and include predictions of atmospheric emissions as measured by Torr and Torr (1984) and others as well as iono-

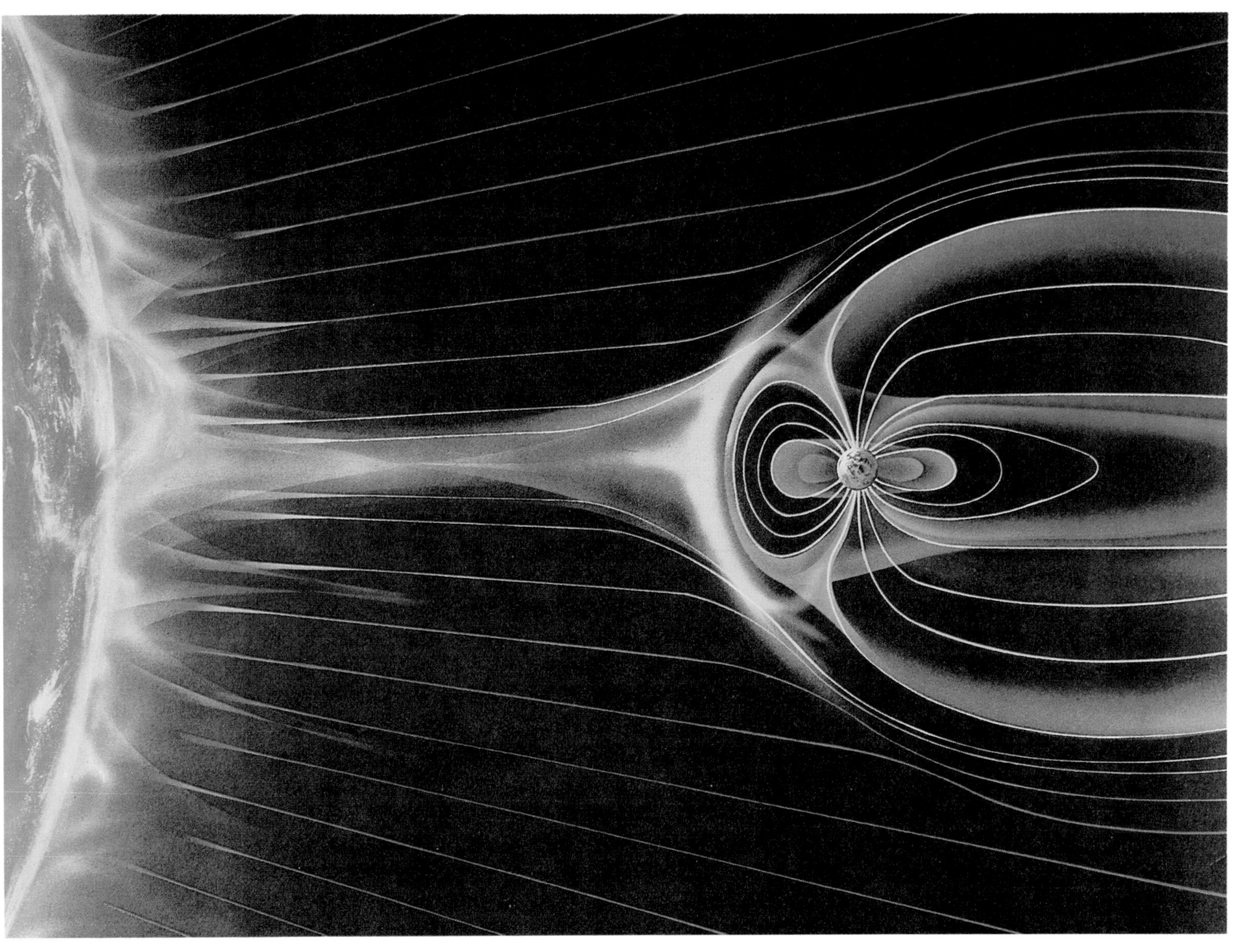

Fig. 1. The earth's magnetospheric environment as influenced by the solar wind.

spheric plasma generation (Banks et al., 1974; Rees, 1963; Sojka and Schunk, 1986).

An intrinsic part of the formation of the earth's ionosphere is the motion of the ions during and after their creation. The flowing motion of the ions which is driven by both the winds in the neutral atmosphere and the electric fields from the magnetosphere affects the character of the resulting ionosphere, particularly the F region. There have been extensive observations of ionospheric dynamics both for motions up and down the magnetic field lines (Hoffman and Dodson, 1980; Lockwood et al., 1985; Shelley et al., 1982) and in the horizontal direction generally perpendicular to the magnetic field (Heelis, 1984; Foster and Doupnik, 1984; Heppner, 1972), and a general pattern of ionospheric circulation is beginning to emerge. Models of the ionospheric evolution and dynamics are in development with results showing an encouraging agreement with observations (Sojka et al., 1981).

Auroral process. Given the presence and magnitude of the ionospheric plasma source at the feet of the magnetospheric field lines, we must next turn to processes which can affect the transport of this plasma into the magnetosphere and its concurrent energization. There are two categories of this transport and energization - the low-energy polar wind mechanism which will be treated below and those energization and transport phenomena associated with the aurora. As we look to the aurora, we begin to look at things on a much more detailed scale and have to treat a set of coupled phenomena on almost a single particle basis. Involved in the auroral process are particle acceleration and precipitation, large- and small-scale electric fields and currents, and a host of plasma phenomena including wave particle interactions which exchange energy within the different plasma populations. All of these auroral processes have been measured extensively; however, the requirements for measurement resolution to very small spatial scales combined with the great variety of auroral phenomena as seen around the auroral oval have made complete data acquisition difficult. The measurements continue, particularly using the multiple spacecraft approach. The modeling of auroral processes is also underway, although present activity tends to address only small individual pieces of the auroral picture with a synthesis of the multiple models likely to be a more distant goal. Numerical simulation techniques are now being employed and should contribute to our abilities to describe aurora phenomena mathematically.

It is in the auroral region that the energy of the solar wind is coupled most directly into the ionosphere and it is here that the ionospheric particles are altered significantly in pitch angle and energy by all of the auroral processes. The result is an energization of a significant portion of the ionospheric plasma and its transport into the magnetosphere. Observations of the particle transport are continuing (Yau et al., 1985) with models of the energization and transport in an early stage of development (Horwitz, 1984; Cowley, 1980; Sauvaud and Delcourt, 1987; Cladis and Francis, 1985). The auroral region acts as an energizer and an initiator of particle transport to inject particles of ionospheric origin into the different plasma regions of the magnetosphere.

The polar wind. The auroral processes result in concentrated energization and injection of ionospheric plasma into the magnetosphere producing high fluxes and energies. The polar wind mechanism on the other hand injects more modest fluxes of particles at an energy of a few electron volts. However, this injection is thought to take place over a very extensive range of latitudes resulting in a total flux into the magnetosphere of very large magnitude (Chappell et al., 1987). Observations of the polar wind source are very limited (Hoffman and Dodson, 1980; Nagai et al., 1984) and are in fact being led by theory and modeling (Raitt and Schunk, 1983). Global statistics on the polar wind transport mechanism for ionospheric plasma are badly needed to determine the true strength of the ionospheric source of magnetospheric plasmas.

Magnetospheric Plasma Regions

Inner and outer plasmasphere. Given the ionospheric sources discussed above, we now turn to the different plasma regions of the magnetosphere beginning with the innermost regions and working outward. The upward flow of the polar wind from the ionosphere is thought to populate directly the inner and outer plasmasphere. Observations of density, temperature, ion composition, and flow are underway (Carpenter, 1963; Park, 1970; Chappell, 1970a; Gringauz et al., 1960; Comfort et al., 1985), and models are in development at the present time (Richards and Torr, 1986; Khazanov et al., 1984; Murphy et al., 1984; Wolf and Harel, 1979; Singh et al., 1986). One can expect that these models will begin to portray the mutual coupling between ionosphere and the plasmasphere.

Plasma trough. Outside of the plasmasphere region which is dominated by corotation is the plasma trough where plasma is fed upward from the ionosphere via the polar wind and lost through convection to the dayside magnetopause. Although hampered by the limitations of spacecraft charging effects, the measurements of density, temperature, composition and flow are in progress (Chappell, 1970b; Higel and Lei, 1984; Young, 1983), but the model development is just in its earliest steps (Lemaire, 1986). Within the nightside region, the plasma trough and the plasma sheet and ring current are colocated. The latter two regions are treated below.

Polar magnetosphere. In the polar regions of the magnetosphere the upward flow of particles from the cleft region, the auroral zone, and the polar cap is carried through the polar region and through the tail lobes either to be captured and energized in the plasma sheet or vented out the tail in the anti-sunward direction. Because of positive spacecraft potentials, relatively little is known about the particle distributions in the polar region particularly at the low energies which are expected to be present there. Measurements are in progress (Yau et al., 1985; Waite et al., 1985), and some modeling is underway (Horwitz, 1984; Cladis and Francis, 1985).

The plasma sheet. Probably the most difficult magnetospheric region to measure thoroughly and to model is the plasma sheet, which is thought to be a region of extensive mixing of ionospheric and solar wind plasmas. All three major sources for the magnetosphere mentioned above are probably operative to some degree in the plasma sheet and are subject to energization, transport, mixing, and evolution in charge state. The plasma sheet has been measured by many spacecraft (e.g., Eastman et al., 1985; Williams, 1981; Lennartsson and Shelley, 1986), although the low-energy plasma component has probably not been adequately observed because of spacecraft charging phenomena. First-order models of the plasma sheet have been developed (Cowley, 1980; Jaggi and Wolf, 1973) but are only in the initial stages.

The energization processes acting in the plasma sheet have been measured indirectly in terms of interpreting changing plasma distribution characteristics (Eastman et al., 1985; Frank et al., 1976), but the specific magnetic merging mechanisms are difficult to observe. Initial models of the energization (Cowley, 1980; Lyons and Speiser, 1982; Cladis and Francis, 1985) have been developed but require much more work because of the complexity of plasma sheet processes. We have only the initial data on plasma sheet composition (Lennartsson and Shelley, 1986; Gloeckler et al., 1985) with no models to describe its evolution in charge state.

The dynamic changes in the plasma sheet that drive the substorm phenomenon have been measured for many years (Hones, 1979) but have not led to a convergence of understanding, probably because of the need for multiple spacecraft to measure the extensive spatial size of the plasma sheet and its differing dynamic characteristics in differing regions such as inner edge transport versus far tail merging phenomena. The plasma sheet dynamics are intimately connected to the ionosphere through the particle flow and currents mentioned above in the auroral processes section. Models of the plasma sheet dynamics are in their initial steps (Hill, 1974; Walker and Ogino, 1987; Wolf and Harel, 1979; Hill and Reiff, 1980).

Ring current. At the earthward edge of the plasma sheet and extending around the earth is the ring current. Extensive observations of the ring current have been made over the past two decades; however, it was only recently that the particle composition could be determined at the energies responsible for the bulk of the ring current energy density (Gloeckler et al., 1985). Models of the ring current are in a conceptual stage, with the element of compositional changes having been addressed to only a minor extent (Jaggi and Wolf, 1973).

Radiation belts. The most energetic particles in the magnetosphere are found in the radiation belts. Because of their high energies, these particles were assessible to early measurement from spacecraft using Geiger tubes and solid state detectors. The more stable configuration of radiation belt particles with changing magnetospheric conditions and the long history of measurement have led to a very extensive observational data base (Van Allen et al., 1958; Williams and Mead, 1965; Vampola et al., 1977) as well as predictive models (Spjeldvik and Fritz, 1978; Singley and Vette, 1972). This region is probably the best understood and theoretically characterized of the magnetospheric plasma populations, although work still needs to be done on its evolving particle composition and on the mechanisms responsible for precipitative loss of radiation belt particles.

Magnetospheric Plasma Composition

A strong conclusion that becomes obvious immediately when scanning the state of magnetospheric models is the lack of information on the ion composition of the different plasma regions. Except for obscuration by spacecraft charging effects at low energy, the ability to determine composition in the very low-energy (~1 eV) and very high-energy (~1 MeV) plasmas has existed for some time. However, the composition of the intermediate energy regimes remained unmeasured until the last decade. As a result of this late measurement as well as the difficulty of including charge state as a parameter on top of the other changing characteristics, composition models are almost non-existent. Yet the evolving states of plasma composition and charge state become the key distinguishing features in determining the origin of magnetospheric plasmas and their subsequent circulation through and energization in the magnetosphere. It is only through an ability to model the evolution of the multiple charge states of oxygen and helium ions, for example, that we will be able to understand the relative contributions of the solar wind and ionosphere to the magnetospheric plasmas.

Because of the importance of plasma composi-

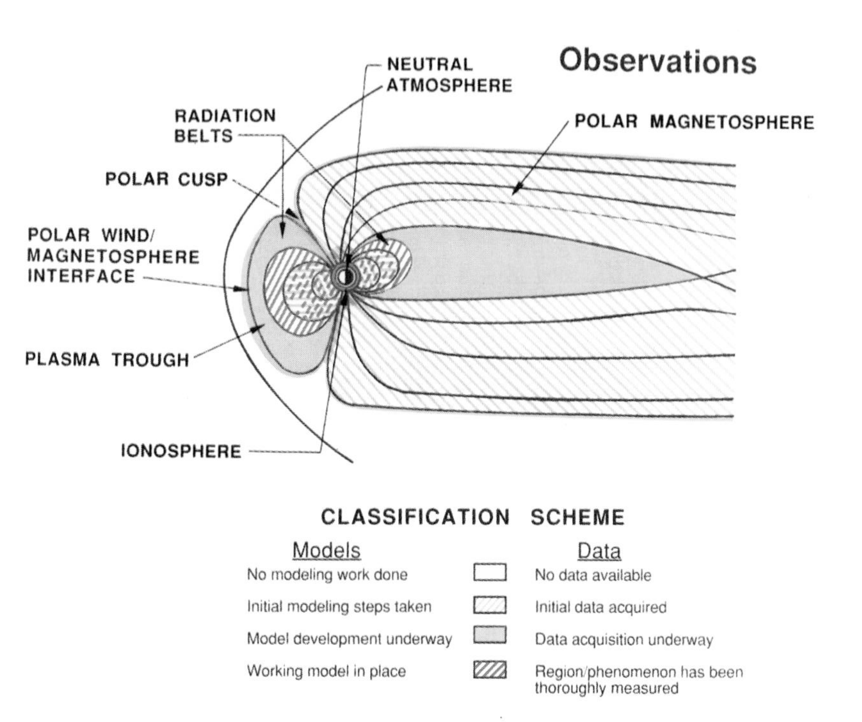

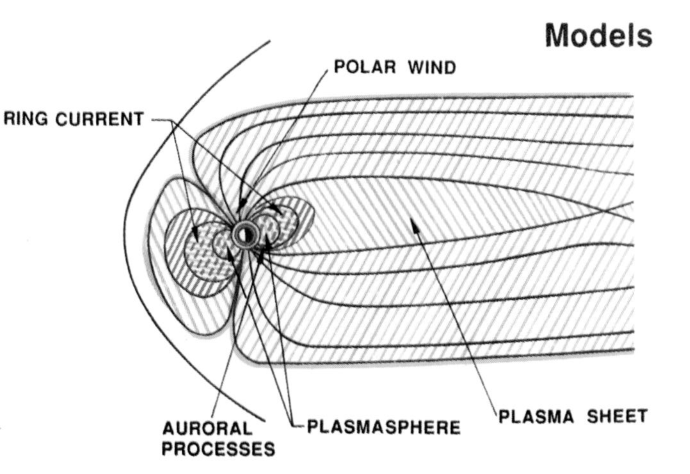

Fig. 2. A summary status of observations and models of magnetospheric plasmas and fields. The color classification scheme is shown in the center of the figure.

Composition

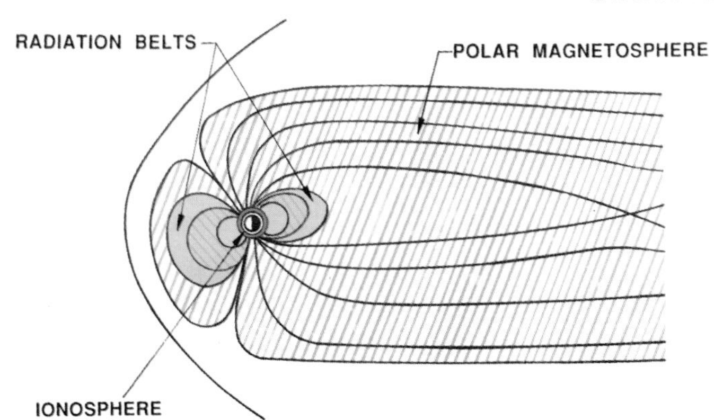

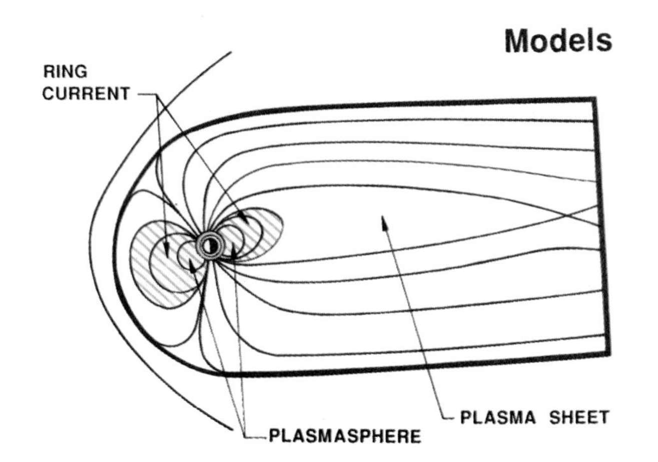

Fig. 3. A summary status of magnetospheric composition, both observations and models. The color classification scheme is the same as that shown in Figure 2.

tion we have shown a separate assessment of its observational and modeling state in Figure 3. Data on the ionospheric, plasmaspheric, and radiation belt composition have been relatively abundant (Hanson, 1975; Hoffman, 1970; Taylor et al., 1968; Chappell, 1972; Young, 1986), but only recently has information on the plasma sheet and ring current become available (Lennartsson and Shelley, 1986; Gloeckler et al., 1985). Only the most limited modeling has been done on the evolution of composition and charge state in the magnetosphere (Spjeldvik and Fritz, 1978), and it has been limited to a very well-behaved region of the inner magnetosphere. Ionospheric composition models are under development with the first steps toward plasmasphere composition having been taken (Sojka et al., 1981; Richards and Torr, 1986; Chandler et al., 1987). The large, uncolored areas in Figure 3 illustrate the significant lack of models which address magnetospheric composition and its changes during changing magnetospheric conditions.

Summary

An immediate conclusion that can be drawn from this overview is the breadth of the job that lies ahead in consolidating the many pieces of the model puzzle into a unified picture. There is a spectrum of understanding both in observations and models which ranges from the relatively sophisticated atmosphere and radiation models to the almost non-existent magnetospheric composition models. In the model section of Figure 2, for example, our goal of a completely "red" classification is far from being accomplished.

Attention and effort must be given to those areas which are lagging in order to move toward convergence in a unified model. For example, we are beginning to see evidence for the circulation of particles within the earth's environment from neturals in the atmosphere to ionospheric ions which are transported into the magnetosphere and energized in the plasma sheet and ring current where they are convected into the inner magnetosphere to be charge exchanged back into energetic neutrals where they are either returned to the atmosphere by precipitation or ejected out of the magnetosphere into the interplanetary medium. Circulation systems of this sort must be modeled if the magnetospheric system is to be predictable in the future.

Finally, it is clear that because of the patchwork nature of the current models, we need to decide on boundary conditions and guiding assumptions that will allow the pieces to mesh easily in the future. Common assumptions regarding magnetic activity levels and convection models were mentioned above. There are many other assumptions of this nature which should be discussed by the many modeling groups and agreed upon in advance. Of equal importance is the coordination between different interactive pieces of the model in which the output of one model becomes the input for another because of the interactive nature of the magnetosphere. It is clear that some overall management of this effort is called for, perhaps by a steering committee, that can lead to the interaction and cooperation that will be required to attain the final product.

Should we push toward a unified model of magnetospheric plasmas? Is it possible to accomplish such a task? I believe that the answer to both questions is yes and it is time to organize ourselves to attack this challenge. This workshop has been a first step toward this goal.

References

Akasofu, S.-I., Energy coupling between the solar wind and the magnetosphere, Space Sci. Rev., 28, 121, 1981.

Banks, P. M., C. R. Chappell, and A. F. Nagy, A new model for the interaction of auroral electrons with the atmosphere: Spectral degradation, backscatter, optical emission, and ionization, J. Geophys. Res., 79, 1459, 1974.

Burch, J. L., P. A. Reiff, R. W. Spiro, R. A. Heelis, and S. A. Fields, Cusp region particle precipitation and ion convection for northward interplanetary magnetic field, Geophys. Res. Lett., 7, 393, 1980.

Burton, R. K., R. L. McPherron, and C. T. Russell, An empirical relationship between interplanetary conditions and Dst, J. Geophys. Res., 80, 4204, 1975.

Carignan, G. R., T. Dachev, A. E. Hedin, C. A. Reber, and N. W. Spencer, Neutral composition in the polar thermosphere: Observations made on Dynamics Explorer, Geophys. Res. Lett., 9, 949, 1982.

Carpenter, D. L., Whistler evidence of a 'knee' in the magnetospheric ionization density profile, J. Geophys. Res., 68, 1675, 1963.

Chandler, M. O., J. J. Ponthieu, T. E. Cravens, A. F. Nagy, and P. G. Richards, Model calculations of minor ion populations in the plasmasphere, J. Geophys. Res., 92, 5885, 1987.

Chappell, C. R., Observations of the plasmapause from OGO 5, J. Geophys. Res., 75, 219, 1970a.

Chappell, C. R., The morphology of the bulge region of the plasmasphere, J. Geophys. Res., 75, 3848, 1970b.

Chappell, C. R., Recent satellite measurements of the morphology and dynamics of the plasmasphere, Rev. Geophys. Space Phys., 10, 951, 1972.

Chappell, C. R., T. E. Moore, and J. H. Waite, Jr., The ionosphere as a fully adequate source of plasma for the earth's magnetosphere, J. Geophys. Res., 92, 5896, 1987.

Cladis, J. B., and W. E. Francis, The polar ionosphere as a source of the storm time ring current, J. Geophys. Res., 90, 3465, 1985.

Comfort, R. H., J. H. Waite, Jr., and C. R. Chappell, Thermal ion temperatures from the

Retarding Ion Mass Spectrometer on DE 1, J. Geophys. Res., 90, 3475, 1985.
Cowley, S.W.H., Plasma populations in a simple open model magnetosphere, Space Sci. Rev., 26, 217, 1980.
Eastman, T. E., L. A. Frank, and C. Y. Hwang, The boundary layers as the primary transport regions of the earth's magnetotail, J. Geophys. Res., 90, 9541, 1985.
Fairfield, D. H., Average and unusual locations of the earth's magnetopause and bow shock, J. Geophys. Res., 76, 6700, 1971.
Frank, L. A., Plasma in the earth's polar magnetosphere, J. Geophys. Res., 76, 5202, 1971.
Frank, L. A., K. L. Ackerson, and R. P. Lepping, On hot tenuous plasmas, fireballs, and boundary layers in the earth's magnetotail, J. Geophys. Res., 81, 5859, 1976.
Foster, J. C., and J. R. Doupnik, Plasma convection in the vicinity of the dayside cleft, J. Geophys. Res., 89, 9107, 1984.
Fuller-Rowell, T. J., and D. Rees, A three-dimensional, time-dependent, global model of the thermosphere, J. Atmos. Sci., 37, 2545, 1980.
Gloeckler, G., B. Wilken, W. Stüdemann, F. M. Ipavich, D. Hovestadt, D. C. Hamilton, and G. Kremser, First composition measurement of the bulk of the storm-time ring current (1 to 300 keV/e) with AMPTE-CCE, Geophys. Res. Lett., 12, 325, 1985.
Gringauz, K. I., V. G. Kurt, V. I. Moroz, and I. S. Shklovskii, Results of observations of charged particles observed out to R = 100,000 km with the aid of charged particle traps on Soviet space rockets, Astron. Zh., 37, 716, 1960.
Haerendel, G., G. Paschmann, N. Sckopke, H. Rosenbauer, and P. C. Hedgecock, The frontside boundary layer of the magnetosphere and the problem of reconnection, J. Geophys. Res., 83, 3195, 1978.
Hanson, W. B., Earth's dynamic thermosphere, Astronaut. Aeronaut., 13, 16, 1975.
Hedin, A. E., C. A. Reber, G. P. Newton, N. W. Spencer, H. C. Brinton, H. G. Mayr, and W. E. Potter, A global thermospheric model based on mass spectrometer and incoherent scatter data MSIS 2. Composition, J. Geophys. Res., 82, 2148, 1977.
Heelis, R. A., The effects of interplanetary magnetic field orientation on dayside high latitude ionospheric convection, J. Geophys. Res., 89, 2873, 1984.
Heikkila, W. J., and J. D. Winningham, Penetration of magnetosheath plasma to low altitudes through the dayside magnetospheric cusps, J. Geophys. Res., 76, 883, 1971.
Heppner, J. P., Polar-cap electric field distributions related to the interplanetary magnetic field direction, J. Geophys. Res., 77, 4877, 1972.
Higel, B., and W. Lei, Electron density and plasmapause characteristics at 6.6 R_E: A statistical study of the GEOS 2 relaxation sounder data, J. Geophys. Res., 89, 1583, 1984.
Hill, T. W., Origin of the plasma sheet, Rev. Geophys. Space Phys., 12, 379, 1974.
Hill, T. W., and P. H. Rieff, Plasma sheet dynamics and magnetospheric substorms, Planet. Space Sci., 28, 363, 1980.
Hinteregger, H. E., K. Fukui, and B. R. Gilson, Observational, reference and model data on solar EUV, from measurements on AE-E, Geophys. Res. Lett., 8, 1147, 1981.
Hoffman, J. H., Studies of the composition of the ionosphere with a magnetic deflection mass spectrometer, Int. J. Mass Spectrom. Ion Phys., 4, 315, 1970.
Hoffman, J. H., and W. H. Dodson, Light ion concentrations and fluxes in the polar regions during magnetically quiet times, J. Geophys. Res., 85, 626, 1980.
Hones, E. W., Jr., Plasma flow in the magnetotail and its implications for substorm theories, in Dynamics of the Magnetosphere, edited by S.-I. Aksaofu, P. 545, D. Reidel Publ. Co., Dordrecht, Holland, 1979.
Horwitz, J. L., Features of ion trajectories in the polar magnetosphere, Geophys. Res. Lett., 11, 1111, 1984.
Jaggi, R. K., and R. A. Wolf, Self-consistent calculation of the motion of a sheet of ions in the magnetosphere, J. Geophys. Res., 78, 2852, 1973.
Khazanov, G. V., M. A. Koen, Y. V. Konikov, and I. M. Sidorov, Simulation of ionosphere-plasmasphere coupling taking into account ion inertia and temperature anisotropy, Planet. Space Sci., 32, 585, 1984.
Killeen, T. L., P. B. Hays, N. W. Spencer, and L. E. Wharton, Neutral winds in the polar thermosphere as measured from Dynamics Explorer, Geophys. Res. Lett., 9, 957, 1982.
Lemaire, J., Plasma transport in the plasmasphere, Adv. Space Res., 6, 157, 1986.
Lennartsson, W., and E. G. Shelley, Survey of 0.1- to 16-keV/e plasma sheet ion composition, J. Geophys. Res., 91, 3061, 1986.
Lockwood, M., M. O. Chandler, J. L. Horwitz, J. H. Waite, Jr., T. E. Moore, and C. R. Chappell, The cleft ion fountain, J. Geophys. Res., 90, 9736, 1985.
Luhmann, J. G., and L. M. Friesen, A simple model of the magnetosphere, J. Geophys. Res., 84, 4405, 1979.
Lyons, L. R., and T. W. Speiser, Evidence for current sheet acceleration in the geomagnetic tail, J. Geophys. Res., 87, 2276, 1982.
Mead, G. D., and D. H. Fairfield, A quantitative magnetospheric model derived from spacecraft magnetometer data, J. Geophys. Res., 80, 523, 1975.
Murphy, J. A., A. A. Nagmoosh, G. J. Bailey, and R. J. Moffett, A theoretical study of the time-dependent behavior of O^{2+} in the mid-latitude plasmasphere, Planet. Space Sci., 32, 1591, 1984.

Nagai, T., J. H. Waite, Jr., J. L. Green, C. R. Chappell, R. C. Olsen, and R. H. Comfort, First measurements of supersonic polar wind in the polar magnetosphere, Geophys. Res. Lett., 11, 669, 1984.

Ness, N. F., The earth's magnetic tail, J. Geophys. Res., 70, 2989, 1965.

Ness, N. F., C. S. Scearce, and J. B. Seek, Initial results of Imp 1 magnetic field experiment, J. Geophys. Res., 69, 3531, 1964.

Olson, W. P., and K. A. Pfitzer, A quantitative model fo the magnetospheric magnetic field, J. Geophys. Res., 79, 3739, 1974.

Park, C. G., Whistler observations of the interchange of ionization between the ionosphere and the protonosphere, J. Geophys. Res., 75, 4249, 1970.

Paschmann, G., G. Haerendel, N. Sckopke, H. Rosenbauer, and P. C. Hedgecock, Plasma and magnetic field characteristics of the distant polar cusp near local noon: The entry layer, J. Geophys. Res., 81, 2883, 1976.

Pilipp, W., and G. Morfill, The formation of the plasma sheet resulting from plasma mantle dynamics, J. Geophys. Res., 83, 5670, 1978.

Raitt, W. J., and R. W. Schunk, Composition and characteristics of the polar wind, in Energetic Ion Composition in the Earth's Magnetosphere, edited by R. G. Johnson, p. 99, Terra Scientific Publ. Co., Tokyo, 1983.

Rees, M. H., Auroral ionization and excitation by incident energetic electrons, Planet. Space Sci., 11, 1209, 1963.

Richards, P. G., and D. G. Torr, Thermal coupling of conjugate ionospheres and the tilt of the earth's magnetic field, J. Geophys. Res., 91, 9017, 1986.

Roble, R. G., R. E. Dickinson, and E. C. Ridley, Global circulation and temperature structure of the thermosphere with high-latitude plasma convection, J. Geophys. Res., 87, 1599, 1982.

Russell, C. T., and K. I. Brody, Some remarks on the position and shape of the neutral sheet, J. Geophys. Res., 72, 6104, 1967.

Russell, C. T., and R. C. Elphic, ISEE observations of flux transfer events at the dayside magnetopause, Geophys. Res. Lett., 6, 33, 1979.

Sauvaud, J. A., and D. Delcourt, A numerical study of suprathermal ionospheric ion trajectories in three-dimensional electric and magnetic field models, J. Geophys. Res., 92, 5873, 1987.

Shelley, E. G., W. K. Peterson, A. G. Ghielmetti, and J. Geiss, The polar ionosphere as a source of energetic magnetospheric plasma, Geophys. Res. Lett., 9, 941, 1982.

Singh, N., R. W. Schunk, and H. Thiemann, Temporal features of the refilling of a plasmaspheric flux tube, J. Geophys. Res., 91, 13,433, 1986.

Singley, G. W., and J. I. Vette, A model environment for outer zone electrons, NASA Report NSSDC-72-13, 1972.

Sojka, J. J., and R. W. Schunk, A theoretical study of the production and decay of localized electron density enhancements in the polar ionosphere, J. Geophys. Res., 91, 3245, 1986.

Sojka, J. J., W. J. Raitt, and R. W. Schunk, Theoretical predictions for ion composition in the high latitude winter F-region for solar minimum and low magnetic activity, J. Geophys. Res., 86, 2206, 1981.

Spencer, N. W., L. E. Wharton, G. R. Carignan, and J. C. Maurer, Thermosphere zonal winds, vertical motions and temperature as measured from Dynamics Explorer, Geophys. Res. Lett., 9, 953, 1982.

Spjeldvik, W. N., and T. A. Fritz, Energetic ionized helium in the quiet time radiation belts: Theory and comparison with observations, J. Geophys. Res., 83, 654, 1978.

Sugiura, M., and D. J. Poros, A magnetospheric field model incorporating the Ogo 3 and 5 magnetic field observations, Planet. Space Sci., 21, 1763, 1973.

Taylor, H. A., Jr., H. C. Brinton, and M. W. Pharo III, Contraction of the plasmasphere during geomagnetically disturbed periods, J. Geophys. Res., 73, 961, 1968.

Torr, M. R., and D. G. Torr, Atmospheric spectral imaging, Science, 225, 169, 1984.

Tsurutani, B. T., J. A. Slavin, E. J. Smith, R. Okida, and D. E. Jones, Magnetic structure of the distant geotail from -60 to -220 R_e: ISEE-3, Geophys. Res. Lett., 11, 1, 1984.

Tsyganenko, H. A., A model of the cis-lunar magnetospheric field, Ann. Geophys., 32, 1, 1976.

Vampola, A. L., J. B. Blake, and G. A. Paulikas, A new study of the magnetospheric electron environment, J. Spacecr. Roc., 14, 690, 1977.

Van Allen, J. A., G. H. Ludwig, E. C. Ray, and C. E. McIlwain, Observation of high intensity radiation by satellites, Jet Propul., 28, 588, 1958.

Waite, J. H., Jr., T. Najai, J.F.E. Johnson, C. R. Chappell, J. L. Burch, T. L. Killeen, P. B. Hays, G. R. Carignan, W. K. Peterson, and E. G. Shelley, Escape of suprathermal O^+ ions in the polar cap, J. Geophys. Res., 90, 1619, 1985.

Walker, R. J., and T. Ogino, Field-aligned current and magnetospheric convection - A comparison between MHD simulations and observations, this volume, 1987

Williams, D. J., Energetic ion beams at the edge of the plasma sheet: ISEE 1 observations plus a simple model, J. Geophys. Res., 86, 5507, 1981.

Williams, D. J., and G. D. Mead, Nightside magnetosphere configuration as obtained from trapped electrons at 1100 kilometers, J. Geophys. Res., 70, 3017, 1965.

Wolf, R. A., and M. Harel, Dynamics of the magnetospheric plasmas, in Dynamics of the Magnetosphere, edited by S.-I. Akasofu, p. 143, D. Reidel Publ. Co., Dordrecht, Holland, 1979.

Yau, A. W., E. G. Shelley, W. K. Peterson, and L. Lenchyshyn, Energetic auroral and polar ion outflow at DE 1 altitudes: Magnitude, composi-

tion, magnetic activity dependence, and long-term variations, J. Geophys. Res., 90, 8417, 1985.
Young, D. T., Near-equatorial magnetospheric particles from ~1 eV to ~1 Mev, Rev. Geophys. Space Phys., 21, 402, 1983.
Young, D. T., Experimental aspects of ion acceleration in the Earth's magnetosphere, in Ion Acceleration in the Magnetosphere and Ionosphere, Geophys. Monogr. Ser., vol. 38, edited by T. Chang, p. 17, AGU, Washington, D.C., 1986.

C. R. Chappell
Space Science Laboratory
NASA Marshall Space Flight Center
Huntsville, Alabama

THE KINETIC APPROACH IN MAGNETOSPHERIC PLASMA TRANSPORT MODELING

J. L. Horwitz

Department of Physics, The University of Alabama in Huntsville, Huntsville, Alabama 35899

Abstract. The need for a kinetic approach in magnetospheric plasma transport problems is reviewed, as are the trends in its recent applications. The need for kinetic modeling is particularly obvious when confronted with the astonishing variety of magnetospheric particle measurements that display compelling energy and pitch angle-related spatial and/or temporal dispersion, and various types of highly non-Maxwellian features in the distribution functions. Global problems in which the kinetic approach has recently been applied include solar wind plasma injection and dispersion over the cusp, substorm particle injection near synchronous orbit, synergistic energization of ionospheric ions into ring current populations by waves and induced electric field-driven convection, and ionospheric outflow from restricted source regions into the magnetosphere. Kinetic modeling can include efforts ranging from test-particle techniques to particle-in-cell studies, and this range is considered here. There are some areas where fluid and kinetic approaches have been combined or patched together, and these will be briefly discussed.

Introduction

For most parts of the magnetosphere, say generally above about 2000 km altitude, the plasma may be regarded as classically collisionless in the sense that a characteristic mean free path is longer than the determining scale sizes of the system. For instance, for an H^+ plasma with a density of about 10^3 H^+ cm^{-3} and a characteristic energy 1 eV, the mean free path is about 10^4 km. This is almost 2 R_E, which is greater than, for example, the scale of the diverging magnetic field. Indeed, only in the ionosphere and most of the plasmasphere is it common for the classical mean free path to be smaller than the relevant scale lengths (chiefly the scale of the magnetic field, gravitational scale height, or temperature gradient). In certain regions of the magnetosphere, such as the central plasma sheet, the magnetosheath, and the magnetopause boundary layer, the distributions often appear not to deviate from Maxwellians as much as might be expected for such classically collisionless plasmas, and it is common to presume that electric and/or magnetic field fluctuations perform the role of randomizing scattering centers to bring about such near-Maxwellian characteristics. R. W. Schunk (private communication, 1986) has pointed out that even if such wave-particle interactions do take place and the resulting distribution appears to be near-Maxwellian, it would still be invalid, for example, to use the classical heat conduction terms in the energy transport equations, since those collision terms have temperature dependences that are specifically appropriate only for Coulomb collisions (in the case of a fully ionized plasma, for instance). Some recent hydrodynamic models utilize progressively more sophisticated expansions and approaches for regions where the plasma is classically collisionless. These may include the use of higher order moments in the transport equations. Generally these are billed as thirteen- or sixteen-moment expansions (e.g., Schunk and Watkins, 1982), sometimes expanding about bi-Maxellian zero-order distributions, though actually aximuthal symmetry about the magnetic field direction typically reduces the number of real moments to be solved to five or six. In these treatments, the question of closure of the equations to be solved typically involves making some assumption relating the higher moments to the lower ones. In this area, some new steps have been taken by Palmadesso and colleagues (e.g., Palmadesso et al., 1987) to incorporate anomalous transport coefficients (such as in frictional interactions between streaming O^+ and H^+), where the transport coefficients are calculated or distilled from some form of a kinetic treatment of microscopic plasma processes. The value of fluid-type treatments clearly lies in their facility in dealing with coupled, large-scale systems, and, as such, they can often yield valuable insights on the characteristic behavior of the plasma systems.

However, as will be reviewed in the next section, observed magnetospheric particle distribu-

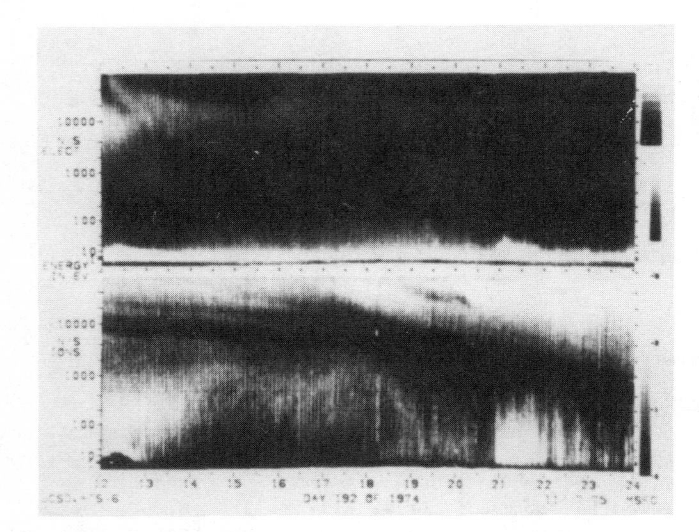

Fig. 1. Typical energy-time spectrogram of gray-scale coded particle fluxes, from the UCSD auroral particles experiment on the synchronous orbiter ATS 6.

tion functions frequently depart so far from even the more modern descriptions that higher order moment approahces can provide that it is clear that fluid approaches must be augmented by kinetic models that attempt more directly to calculate the distribution functions. Often only these kinetic models contain enough flexibility to model the complex magnetospheric distribution functions observed. In this paper dealing with the kinetic approach in plasma transport modeling, we will exclude consideration of work in which fluid-type moments are the basis of the plasma description, even the higher order moment calculations. Thus, we are considering work that elicits the actual details of the velocity distribution function. Since we are chiefly interested in the plasma description, we will also exclude work that deals specifically and primarily with the instability and propagation of waves in plasmas, although these are often based on kinetic descriptions and the waves considered may ultimately have important effects on the plasma itself (usually in nonlinear treatments).

We consider the kinetic approach to include modeling ranging from test-particle calculations of particles in pre-arranged or independently given fields (steady state or time-dependent), through efforts where assumptions about constants of motion (e.g., energy and first adiabatic invariant) may allow self-consistent calculation of particle distributions and fields, to particle-in-cell (PIC) numerical simulations. In the next section, we briefly review some representative magnetospheric observations that fairly obviously imply the need for kinetic models of these phenomena. In the following section, we review the work that has transpired in the past 6 years using kinetic modeling approaches of one form or another. We conclude by summarizing the recent trends in kinetic modeling and commenting briefly on the areas of future progress and where there might be some patching or blending of fluid and kinetic models.

Observations of Kinetic Effects

There is, or was, a San Diego-based contractor company called MAYA, as would ordinarily refer to the peoples of Yucatan prior to the European settlement of the Americas. But in this case, the company name is an acronym for Most Anything You Ask. The magnetosphere can often seem to behave this way and produce just about any plasma phenomenon we might seek. This is certainly the case for the particle distribution functions within the magnetosphere, where several new geometrical velocity space configurations seem to be discovered each year. To stress some points on which the fluid approaches are clearly deficient and clarify the need for kinetic approaches, we wish to review briefly some magnetospheric observations in which the kinetic effects are easily evident.

One region of the magnetosphere where kinetic effects have long been observed is geosynchronous orbit. Figure 1 shows a typical particle energy-time spectrogram for a 24-hour period, in this case from the University of California at San Diego (UCSD) auroral particles experiment on ATS 6 (Mauk and McIlwain, 1975). The gray-scale coded particle fluxes display numerous striation features, with such flux striations starting at the highest energies for various times winding downward in energy with increasing time. As has been known since DeForest and McIlwain (1971) first described these phenomena, the particle behavior is readily understood in terms of time- and space-localized injection and subsequent dispersion of plasma clouds during magnetospheric substorms. At a single time, a cut of the spectrogram could be cast in terms of a distribution function containing many slope changes and perhaps bumps and dents versus energy. Fluid parameters could be calculated, but it is difficult to see how these would be of much assistance in understanding the phenomenon itself; rather, this is a well-known, large-scale magnetospheric process in which kinetic, in fact single particle, considerations are the superior basis for its understanding.

Another region of large-scale dispersive particle dynamics is found in the plasma sheet boundary layer (PSBL). Figure 2, from Forbes et al. (1981), presents four successive ion velocity distributions in the vicinity of the PSBL, as observed by ISEE 2. In the first upper left panel distribution function, earthward streaming ions are observed coming from the deep tail. In panels b and c, we begin to see new, lower earthward velocity ions appearing, as well as ions now streaming in the opposite, tailward direction.

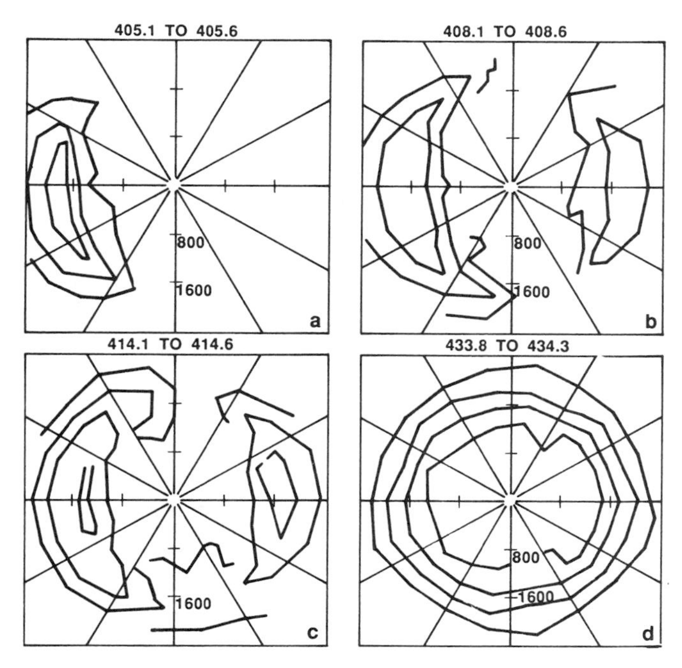

Fig. 2. Four successive two-dimensional ion velocity distributions in the vicinity of the plasma sheet boundary layer, as observed by ISEE 2 (from Forbes et al., 1981).

In the last panel, a merged, nearly isotropic distribution is seen. Although it is now known that some of the ions streaming tailward are of ionospheric origin in these PSBL events (Eastman et al., 1984), the evolution of this situation is basically understood in terms of the energy dispersion and mirrored return (from the stronger magnetic field region on the earthward side) of mainly free-streaming ions. But it would be misleading to attempt a (single) fluid picture, as the resulting bulk flow velocity could be close to zero for the distributions in b and c, for instance, and the pressure/temperature moments around this bulk velocity would be much larger than those for each of the separate peaks.

The examples above illustrate relatively large-scale aspects of kinetic effects in the magnetosphere. But, particularly in regions such as the auroral magnetosphere, the details of the increasingly intriguing types of auroral distribution functions reveal much about the acceleration and transport processes that can probably best be understood in the particle, as opposed to fluid, picture. Figure 3, from Klumpar et al. (1984), displays a new type of observed ion distribution function, the ion bowl distribution. The upper panel of Figure 3 displays the actual observed velocity space distribution function seen by Klumpar et al., as contours versus parallel and perpendicular velocity, with the resulting bowl-shaped contours of a conic at high speeds and a minimum $v_{\parallel}$ for zero perpendicular velocity. Klumpar et al. (1984) modeled these distributions as transversely accelerated O^+ ions (bottom panel) which have moved through the diverging magnetic field (middle panel) and then been accelerated by a parallel electric field to produce the features indicated by the top panel of observations. Horwitz (1986a) and Temerin (1986) have pointed out that ion bowl distributions can also result from velocity filter effects and extended transverse heating, respectively. The point is not which of these explanations is the best for a particular event, but that the explanations motivated by this observation are all from a kinetic picture. Other recently observed auroral distribution functions include ring or toroidal distribution functions in O^+ (Moore et al., 1986) and various bumps, holes, and conical features of auroral electron distribution functions (e.g., Burch and Gurgiolo, 1987). All of these observations seem to be best understood in terms of ensembles of particle trajectories in wave fields, diverging magnetic field geometry, parallel electric fields, etc., i.e., kinetic considerations.

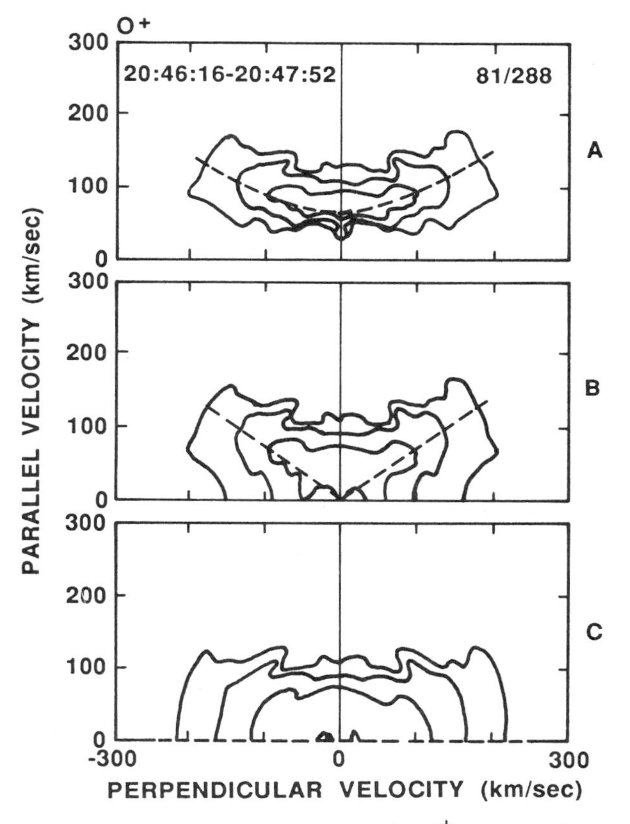

Fig. 3. Ion bowl distribution in O^+ observed by Klumpar et al. (1984). Top panel shows observed O^+ distribution function. Middle panel adiabatically deforms distribution back through parallel potential drop, and bottom panel deforms back through the varying magnetic field.

Recent Kinetic Modeling Work

In this section, we want to examine the directions that the kinetic modeling efforts have been going in the past few years. In the reference section, we have listed articles from the period January 1981 through August 1986 that appeared in the Journal of Geophysical Research or Geophysical Research Letters in which some form of kinetic modeling was employed. By restricting ourselves only to these journals and this period, we will not really be attempting a fully comprehensive review of all the kinetic modeling studies that are potentially relevant. However, our purpose is more to obtain the general trends and areas that are being studied. We have attempted to organize the discussion according to the magnetospheric region or phenomenon of principal application. These are: (1) the ring current/synchronous orbit region, (2) the plasma sheet, plasma sheet boundary layer regions, (3) ion outflow in the polar cap and cleft, (4) solar wind/magnetosphere interactions, (5) tail neutral sheet/current sheet, (6) radiation belts, (7) auroral processes, and (8) general/miscellaneous. Again, we note that instability analyses and other works focused on wave features are not included.

Ring Current/Synchronous Orbit Region

Between January 1981 and August 1986, there were 24 articles published in the American Geophysical Union journals that applied kinetic modeling techniques to understand the behavior of particles in the ring current/synchronous orbit region. Several of these papers presented improvements in the modeling of the time-dependent injection, acceleration, and dispersin of particles into this middle magnetospheric region during substorms (e.g., Mauk and Meng, 1983a,b; Strangeway and Johnson, 1983, 1984; Chiu and Kishi, 1984; Cladis and Francis, 1985; Baker et al., 1982; Greenspan et al., 1985; Quinn and Southwood, 1982; Quinn and Johnson; 1985). All of these works were of the test-particle variety, in which particles were assumed to be fed in, generally at a double-spiral injection boundary in the equatorial magnetosphere, and allowed to drift and disperse in presumed geoelectric and geomagnetic fields. One of the interesting touches included by Cladis and Francis (1985) was to incorporate presumed electrostatic waves which could stochastically energize ions, since these ions could be convected earthward and accelerated by the betatron process, particularly during short-lived induced electric field events, to synergistically accelerate the ions to ring current energies. Other related studies were by Horwitz (1984a,b), who derived relatively compact functional relationships between the electric field and electron energy dispersion in the dusk sector.

A second research theme applied to this region was the study of the local evolution of particle distributions, generally under the influence of various types of waves (Singh et al., 1982a; Southwood and Kivelson, 1982; Lin and Parks, 1982; Mauk, 1982; Kivelson and Southwood, 1983, 1985; Roux et al., 1984; Curtis, 1985; Berchem and Gendrin, 1985; Tanaka, 1985; Omura et al., 1985), although Lyons and Moore (1981) examined the formation of loss cones in ion distributions due to charge exchange. Most studies examined the evolution of paticles and their distributions using essentially test-particle approaches, but Tanaka (1985) and Omura et al. (1985) employed PIC simulations to investigate the interaction of hot, anisotropic plasmas with cold plasma populations.

Plasma Sheet/Plasma Sheet Boundary Layer

For the plasma sheet and plasma sheet boundary layer, there were six papers (see also tail neutral/current sheet section), with most of them dealing with the test-particle dispersion of ions injected at or near the PSBL (Williams, 1981; Andrews et al., 1981; Williams and Speiser, 1984; Lyons, 1984; Green and Horwitz, 1986), and with two additional self-consistent simulations by Swift (1982, 1983).

Ion Outflow in the Polar Cap and Polar Cleft

Two different kinetic approaches for ion outflow in the polar cap and cleft have received the principal attention in recent years. In one approach, a test-particle method is used to elicit the ion trajectories, distribution functions, and bulk parameters in a two-dimensional geometry for the flow of ions from the cleft topside ionosphere into the polar cap magnetosphere (Horwitz, 1984c, 1986b; Horwitz et al., 1985; Horwitz and Lockwood, 1985; Lockwood et al., 1985). Also, though the article appeared beyond our nominal review cutoff date, we wish to note the recent contribution of Cladis (1986) on what might be referred to as centrifugal acceleration of ions flowing into the magnetosphere. A separate direction is taken by Barakat and Schunk (1983, 1984), in which a self-consistent, one-dimensional, steady state adiabatic model, where the ions are taken to be kinetic and the electrons obey a Boltzmann relation, is used to investigate the ambi-polar outward electric propulsion of heavy ionospheric ions due to coupling with hot thermal electrons and the polar rain. An additional approach was developed by Singh and Schunk (1982a) to study the initial expansion of the polar wind into a vacuum, wherein electron forerunners electrically pull the lagging ions out of the ionosphere.

Solar Wind-Magnetosphere Interactions

Kinetic studies associated with this topic have examined, again mostly from a test-particle perspective, ions getting across the magnetopause, between the magnetosheath and magnetosphere (Speiser et al., 1981; Whipple and Silevitch, 1982; Speiser and Williams, 1982; Burch et al., 1982; Tsurutani and Thorne, 1982; Olson and Pfitzer, 1984), while Fairfield and Scudder (1985) looked at the transport of the polar rain electrons. The only PIC simulation study of the interaction of a simulated solar wind carrying a southward magnetic field with a vacuum dipole field was performed by Wagner et al. (1986). This pioneering simulation study was able to produce current sheets, filaments, and magnetic X and O lines that were previously available mostly in MHD investigations.

Tail Neutral Sheet/Current Sheet

Papers dealing with particles in the neutral or current sheet have mainly focused on the complex ion orbits and energization in the vicinity of idealized magnetic neutral sheets and points (Lyons and Speiser, 1982, 1985; Gray and Lee, 1982; Speiser and Lyons, 1984; Chapman and Cowley, 1984; Chen and Palmadesso, 1986). Sato et al. (1982) blended MHD and kinetic techniques to some extent by first calculating the time- and space-varying electric and magnetic fields in an MHD simulation of X-type magnetic reconnection and then numerically integrating particle trajectories in these fields.

Radiation Belts

Studies of the pitch angle and radial diffusion of radiation belt particles under natural and artificial conditions remain active. In the former category are works by Jentsch (1982, 1984), West et al. (1981), Spjeldvik and Fritz (1981), Fritz and Spjeldvik (1982), and Westphalen and Spjeldvik (1982). The effects of injected wave fields on the radiation belts have been investigated by Chang and Inan (1983, 1985), Matsumoto and Omura (1983), Bell (1984), Matthews et al. (1984), and Inan et al. (1985).

Auroral Processes

The area of auroral processes is by far the most active one of investigation for kinetic modelers. Several papers have examined the interrelationship of the particle distribution functions and the self-consistent parallel electric potential distributions (Greenspan et al., 1981; Yeh and Hill, 1981; Chiu et al., 1981; Silevitch, 1981; Stern, 1981; Cornwall and Chiu, 1982; Kan and Okuda, 1983; Serizawa and Sato, 1984; Yamamoto and Kan, 1985; Bruning and Goertz, 1985; Chiu, 1986; Newman et al., 1986; Bujarbarua and Goswami, 1985). Other investigations have sought to model ions as test particles in various types of transversely accelerating wave fields (Singh et al., 1981, 1982b, 1983a; Chang and Coppi, 1981; Retterer et al., 1983b; Singh and Schunk, 1984; Chang et al., 1986), while more self-consistent approaches to such acceleration have been taken by Ashour-Abdalla et al. (1981), Okuda and Ashour-Abdalla (1981, 1983), Nishikawa and Okuda (1985), and Nishikawa et al. (1983, 1985). Test-particle analysis of the effects of two-dimensional structure in the elecric potential distribution on the generation of ion conics has been performed by Borovsky (1984) and Greenspan (1984), while Gorney et al. (1985) have studied the enhanced transverse acceleration of ions in diverging magnetic field geometry and downward electric fields.

Other types of stochastic acceleration or pitch angle diffusion of particle distributions in auroral processes have been studied by Melrose and White (1981), Dusenbery and Lyons (1981), Retterer et al. (1983a), Belmont et al. (1983), Lotko (1986), Davidson (1986a,b), Temerin et al. (1986), Roth and Hudson (1985), and Kaufmann et al. (1986). Klumpar et al. (1984) and Horwitz (1986a) have used relatively simple adiabatic analyses to investigate the formation of the ion bowl distributions discovered by Klumpar et al. (1984). Simulations focusing on the electrodynamics of auroral double layers have been presented by Sato and Okuda (1981), Singh and Schunk (1982b,c,d), Singh et al. (1985), Hudson et al. 1983), Borovsky and Joyce (1983), Wagner et al. (1983), and Smith (1986).

General/Miscellaneous

Somewhat general studies involving kinetic treatments have been conducted by Brinca (1984), Dungey (1985), Whipple et al. (1986), and Goldstein et al. (1986). Whipple et al. (1986), for example, sought to develop new adiabatic variants or constants of motion appropriate to regions of strong gradients in the magnetic and electric fields, while Goldstein et al. (1986) utilized MHD methods to obtain magnetic islands and bubbles in reconnection regions to obtain intense, short-lived electric fields to estimate the amount of stochastic acceleration possible for test particles.

Recent Trends in Kinetic Modeling

In terms of the general trends which can be sifted out of the above cited investigations, we might separate these into two categories: studies not using PIC techniques, and the PIC studies. The trends for the non-PIC studies are: much emphasis on particle acceleration and pitch angle scattering in electric and magnetic wave fields either given or calculated by separate techniques, such as MHD methods; also studies of complex particle motions in regions of strong field gradients (e.g., magnetic neutral sheets); mostly numerical but a few attempts at new invariants; studies of large-scale particle transport from presumed source regions in assumed electric and magnetic fields, chiefly in two-dimensional geometries; some attempts to calculate electric potential self-consistently in auroral processes and polar ion outflow situations, mostly one-dimensional, steady state treatments; very few attempts at self-consistent calculation of magnetic field; and relatively popular usage of single-particle type modeling in direct comparisons with particle observations, perhaps due to flexibility in fitting complexities of observed particle distribution functions. An additional trend for PIC simulations which is worth noting is that, in addition to the microscopic works, there are now some that might be referred to as mesoscale or perhaps even large-scale simulations in the areas of auroral arcs, magnetotail dynamics, and solar wind-magnetosphere interactions.

Conclusions

As noted in the introduction, kinetic modeling approaches are an essential component of the techniques we need to use in addressing plasma modeling in the magnetosphere, particularly where particle energy and pitch angle dispersion or non-Maxwellian aspects are of paramount interest. In some of the large-scale transport problems, there is a need to move from one- and two-dimensional models to fully three-dimensional models and to do a better job of estimating the self-consistent fields.

In terms of combining kinetic and fluid approaches in addressing specific phenomena, at least two directions are being taken. One of these involves the calculation of fields (magnetic and/or electric) through fluid (e.g., MHD) techniques and then test-particle modeling of the particle acceleration and transport in such fields. Another direction is to utilize basically fluid moments and hierarchy of equations, but to incorporate certain kinetic effects in the form of anomalous transport coefficients whose parametric dependences are distilled or separately calculated from microscopic studies. From the discussions at this conference, there appears to be significant encouragement and enthusiasm for this approach, exemplified here perhaps best by Palmadesso et al. (1987), tempered by the realization that at this stage there is a great deal of work to be done on how to parametrically represent the (non-Maxwellian) particle distributions and especially on the transport coefficients that might be calculated from the microscopic analyses or simulations.

Acknowledgments. This work was supported in part by NSF grant ATM-8503102 and NASA grant NAG8-058.

References

Andrews, M. K., P. W. Daly, and E. Keppler, Ion jetting at the plasma sheet boundary: Simultaneous observations of incident and reflected particles, Geophys. Res. Lett., 8, 987, 1981.

Ashour-Adballa, M., H. Okuda, and C. Z. Cheng, Acceleration of heavy ions on auroral field lines, Geophys. Res. Lett., 8, 795, 1981.

Baker, D. N., T. A. Fritz, B. Wilken, P. R. Higbie, S. M. Kaye, M. G. Kivelson, T. E. Moore, W. Studemann, A. J. Masely, P. H. Smith, and A. L. Vampola, Observation and modeling of energetic particles at synchronous orbit on July 29, 1977, J. Geophys. Res., 87, 5917, 1982.

Barakat, A. R., and R. W. Schunk, O^+ ions in the polar wind, J. Geophys. Res., 88, 7887, 1983.

Barakat, A. R., and R. W. Schunk, Effect of hot electrons on the polar wind, J. Geophys. Res., 89, 9771, 1984.

Bell, T. F., The nonlinear gyroresonance interaction between energetic electrons and coherent VLF waves propagating at an arbitrary angle with respect to the earth's magnetic field, J. Geophys. Res., 89, 905, 1984.

Belmont, G., D. Fontaine, and P. Canu, Are equatorial electron cyclotron waves responsible for diffuse auroral electron precipitation?, J. Geophys. Res., 88, 9163, 1983.

Berchem, J., and R. Gendrin, Nonresonant interaction of heavy ions with electromagnetic ion cyclotron waves, J. Geophys. Res., 90, 10,945, 1985.

Borovsky, J. E., and G. Joyce, Numerically simulated two-dimensional double layers, J. Geophys. Res., 88, 3116, 1983.

Borovsky, J. E., The production of ion conics by oblique double layers, J. Geophys. Res., 89, 2251, 1984.

Brinca, A. L., On the coupling of test ions to magnetoplasma flows through turbulence, J. Geophys. Res., 89, 115, 1984.

Bruning, K., and C. K. Goertz, Influence of the electron source distribution on field-aligned currents, Geophys. Res. Lett., 12, 53, 1985.

Bujarbarua, S., and K. S. Goswami, Weak ion acoustic double lyers and solitary waves on the auroral field lines, J. Geophys. Res., 90, 7611, 1985.

Burch, J. L., and C. Gurgiolo, A model of auroral potential structure based on Dynamics Explorer plasma data, this volume, 1987.

Burch, J. L., P. H. Reiff, R. A. Heelis, J. D. Winningham, W. B. Hanson, C. Gurgiolo, J. D. Menietti, R. A. Hoffman, and J. N. Barfield, Plasma injection and transport in the mid-altitude polar cusp, Geophys. Res. Lett., 9, 921, 1982.

Chang, H. C., and U. S. Inan, Quasi-relativistic electron precipitation due to interactions with coherent VLF waves in the magnetosphere, J. Geophys. Res., 88, 318, 1983.

Chang, H. C., and U. S. Inan, Test particle modeling of wave-induced energetic electron precipitation, J. Geophys. Res., 90, 6409, 1985.

Chang, T., and B. Coppi, Lower hybrid acceleration and ion evolution in the suprauroral region, Geophys. Res. Lett., 8, 795, 1981.

Chang, T., G. B. Crew, N. Hershowitz, J. R. Jasperse, J. M. Retterer, and J. D. Winningham, Transverse acceleration of oxygen ions by

electromagnetic ion cyclotron resonance with broad band left-hand polarized waves, Geophys. Res. Lett., 13, 636, 1986.
Chapman, S. C., and S.W.H. Cowley, Acceleration of lithium test ions in the quiet time geomagnetic tail, J. Geophys. Res., 89, 7357, 1984.
Chen, J., and P. Palmadesso, Chaos and nonlinear dynamics of single-particle orbits in a magnetotail-like magnetic field, J. Geophys. Res., 91, 1499, 1986.
Chiu, Y. T., A simple kinetic theory of auroral arc scales, J. Geophys. Res., 91, 204, 1986.
Chiu, Y. T., and A. M. Kishi, Kinetic model of auroral plasma formation by magnetospheric convection and injection, 1. Electrons, J. Geophys. Res., 89, 5531, 1984.
Chiu, Y. T., A. L. Newman, and J. M. Cornwall, On the structures and mapping of auroral electrostatic potentials, J. Geophys. Res., 86, 10,029, 1981.
Cladis, J. B., Parallel acceleration and transport of ions from polar ionosphere to plasma sheet, Geophys. Res. Lett., 13, 893, 1986.
Cladis, J. B., and W. E. Francis, The polar ionosphere as a source of the stormtime ring current, J. Geophys. Res., 90, 3465, 1985.
Cornwall, J. M., and Y. T. Chiu, Ion distribution effects of turbulence on a kinetic auroral arc model, J. Geophys. Res., 87, 1517, 1982.
Curtis, S. A., Equatorial trapped plasmasphere ion distributions and transverse stochastic acceleration, J. Geophys. Res., 90, 1765, 1985.
Davidson, G. T., Pitch angle diffusion in morningside aurorae, 1. The role of the loss cone in the formation of impulsive bursts of precipitation, J. Geophys. Res., 91, 4413, 1986a.
Davidson, G. T., Pitch angle diffusion in morningside aurorae, 2. The formation of repetitive auroral pulsations, J. Geophys. Res., 91, 4429, 1986b.
DeForest, S. E., and C. E. McIlwain, Plasma clouds in the magnetosphere, J. Geophys. Res., 76, 3587, 1971.
Dungey, J. W., Perturbations in the velocity distribution in a collisionless plasma, J. Geophys. Res., 90, 370, 1985.
Dusenbery, P. B., and L. R. Lyons, Generation of ion-conic distribution by upgoing ionospheric electrons, J. Geophys. Res., 86, 7627, 1981.
Eastman, T. E., L. A. Frank, W. K. Peterson, and W. Lennartsson, The plasma sheet boundary layers, J. Geophys. Res., 89, 1553, 1984.
Fairfield, D. W., and J. D. Scudder, Polar rain: Solar coronal electrons in the earth's magnetosphere, J. Geophys. Res., 90, 4055, 1985.
Forbes, T. G., E. W. Hones, S. J. Bame, J. R. Asbridge, G. Paschmann, N. Sckopke, and C. T. Russell, Evidence for the tailward retreat of a magnetic neutral line in the magnetotail during substorm recovery, Geophys. Res. Lett., 8, 261, 1981.
Fritz, T. A., and W. A. Spjeldvik, Pitch angle distributions of geomagnetically trapped MeV helium ions during quiet times, J. Geophys. Res., 87, 5095, 1982.
Goldstein, M. L., W. H. Mattheus, and J. J. Ambrosiano, Acceleration of charged particles in magnetic reconnection: Solar flares, the magnetosphere, and solar wind, Geophys. Res. Lett., 13, 205, 1098.
Gorney, D. J., Y. T. Chiu, and D. R. Croley, Jr., Trapping of ion conics by downward parallel electric fields, J. Geophys. Res., 90, 4205, 1985.
Gray, P. C., and L. C. Lee, Particle pitch angle diffusion due to non-adiabatic effects in the plasma sheet, J. Geophys. Res., 87, 6089, 1982.
Green, J. L., and J. L. Horwitz, Destiny of earthward streaming plasma in the plasma sheet boundary layer, Geophys. Res. Lett., 13, 76, 1986.
Greenspan, M., Effects of oblique double layers on upgoing ion pitch angle and gyrophase, J. Geophys. Res., 89, 2842, 1984.
Greenspan, M. E., M. B. Silevitch, and E. C. Whipple, Jr., On the use of electron data to infer the structure of parallel electric fields, J. Geophys. Res., 86, 2175, 1981.
Greenspan, M. E., D. J. Williams, B. H. Mauk, and C.-I. Meng, Ion and electron energy dispersion features detected by ISEE-1, J. Geophys. Res., 90, 4079, 1985.
Horwitz, J. L., Relationship of dusk sector radial electric field to energy dispersion at the inner edge of the electron plasma sheet, J. Geophys. Res., 89, 3011, 1984a.
Horwitz, J. L., Relationship of dusk-sector electric field to energy dispersion at the inner edge of the electron plasma sheet for non-equatorially mirroring electrons, J. Geophys. Res., 89, 10,865, 1984b.
Horwitz, J. L., Features of ion trajectories in the polar magnetosphere, J. Geophys. Res., 89, 1111, 1984c.
Horwitz, J. L., Velocity filter mechanism for ion bowl distributions (bimodal conics), J. Geophys. Res., 91, 4513, 1986a.
Horwitz, J. L., The tail lobe ion spectrometer, J. Geophys. Res., 91, 5689, 1986b.
Horwitz, J. L., and M. Lockwood, The cleft ion fountain: A two-dimensional kinetic model, J. Geophys. Res., 90, 9749, 1985.
Horwitz, J. L., J. H. Waite, Jr., and T. E. Moore, Supersonic ion outflows in the polar magnetosphere, Geophys. Res. Lett., 12, 757, 1985.
Hudson, M. K., W. Lotko, I. Roth, and E. Witt, Solitary waves and double layers on auroral field lines, J. Geophys. Res., 88, 916, 1983.
Inan, U. S., H. C. Chang, R. A. Helliwell, W. L. Imhof, J. B. Reagan, and M. Walt, Precipitation of radiation belt electrons by man-made waves: A comparison between theory and measurement, J. Geophys. Res., 90, 359, 1985.
Jentsch, V., On the role of external and internal source in generating energy and pitch angle distributions of inner zone protons, J. Geophys. Res., 86, 701, 1981.

Jentsch, V., The radial distribution of radiation belt protons: Approximate solution of the steady state transport equation at arbitrary pitch angle, J. Geophys. Res., 89, 1527, 1984.

Kan, J. R., and H. Okuda, Generation of auroral arc elements in an inverted-V arc due to ion cyclotron turbulence, J. Geophys. Res., 88, 6339, 1983.

Kaufmann, R. L., G. R. Ludlow, H. L. Collin, W. K. Peterson, and J. L. Burch, Interaction of upgoing auroral H^+ and O^+ beams, J. Geophys. Res., 91, 10,080, 1986.

Kivelson, M. G., and D. J. Southwood, Charged particle behavior in low-frequency geomagnetic pulsations, 3. Spin phase dependence, J. Geophys. Res., 88, 174, 1983.

Kivelson, M. G., and D. J. Southwood, Charged particle behavior in low-frequency geomagnetic pulsations, 4. Compressional waves, J. Geophys. Res., 90, 1486, 1985.

Klumpar, D. M., W. K. Peterson, and E. G. Shelley, Direct evidence for two-stage (bimodal) acceleration of ionospheric ions, J. Geophys. Res., 89, 10,779, 1984.

Lin, C. S., and G. K. Parks, Modulation of energetic particle fluxes by a mixed mode of transverse and compressional waves, J. Geophys. Res., 87, 5102, 1982.

Lockwood, M., M. O. Chandler, J. L. Horwitz, J. H. Waite, Jr., T. E. Moore, and C. R. Chappell, The cleft ion fountain, J. Geophys. Res., 90, 9736, 1985.

Lotko, W., Diffusive acceleration of auroral primaries, J. Geophys. Res., 91, 191, 1986.

Lyons, L. R., Electron energization in the geomagnetic tail current sheet, J. Geophys. Res., 89, 5479, 1984.

Lyons, L. R., and T. E. Moore, Effects of charge exchange on the distribution of ionospheric ions trapped in the radiation belts near synchronous orbit, J. Geophys. Res., 86, 5885, 1981.

Lyons, L. R., and T. W. Speiser, Evidence for current sheet acceleration in the geomagnetic tail, J. Geophys. Res., 87, 2276, 1982.

Lyons, L. R., and T. W. Speiser, Ohm's law for a current sheet, J. Geophys. Res., 90, 8543, 1985.

Matsumoto, H., and Y. Omura, Computer simulation studies of VLF triggered emissions deformation of distribution function by trapping and detrapping, Geophys. Res. Lett., 10, 607, 1983.

Matthews, J. P., Y. Omura, and H. Matsumoto, A study of particle trapping by whistler mode waves in the geomagnetic field: The early development of the VLF quiet band, J. Geophys. Res., 89, 2275, 1984.

Mauk, B. H., Electromagnetic wave energization of heavy ions by the electric "phase bunching" process, Geophys. Res. Lett., 9, 1163, 1982.

Mauk, B. H., and C. E. McIlwain, ATS-6 UCSD auroral particles experiment, IEEE Trans. Aerosp. Electron. Syst., 11, 1125, 1975.

Mauk, B. H., and C.-I. Meng, Characterization of geostationary particle signatures based on the "injection boundary" model, J. Geophys. Res., 88, 3055, 1983a.

Mauk, B. H., and C.-I. Meng, Dynamical injections as the source of near geostationary quiet time particle spatial boundaries, J. Geophys. Res., 88, 10,011, 1983b.

Melrose, D. B., and S. M. White, Precipitation from an asymmetric magnetic flux tube, J. Geophys. Res., 86, 2183, 1981.

Moore, T. E., J. H. Waite, Jr., M. Lockwood, and C. R. Chappell, Observations of coherent transverse ion acceleration, in Ion Acceleration in the Magnetosphere and Ionosphere, Geophys. Monogr. Ser., vol. 38, edited by J. Chang, pp. 50-55, AGU, Washington, D.C., 1986.

Newman, A. L., Y. T. Chiu, and J. M. Cornwall, Two-dimensional quasi-neutral description of particles and fields above discrete auroral arcs, J. Geophys. Res., 91, 3167, 1986.

Nishikawa, K.-I., and H. Okuda, Heating of light ions in the presence of a large-amplitude heavy ion cyclotron wave, J. Geophys. Res., 90, 2921, 1985.

Nishikawa, K.-I., H. Okuda, and A. Hasegawa, Heating of heavy ions on auroral field lines, Geophys. Res. Lett., 10, 553, 1983.

Nishikawa, K.-I., H. Okuda, and A. Hasegawa, Heating of heavy ions on auroral field lines in the presence of a large-amplitude hydrogen cyclotron wave, J. Geophys. Res., 90, 419, 1985.

Okuda, H., and M. Ashour-Abdalla, Formation of a conical distribution and intense ion heating in the presence of hydrogen cyclotron waves, Geophys. Res. Lett., 8, 811, 1981.

Okuda, H., and M. Ashour-Abdalla, Acceleration of hydrogen ions and conic formation along auroral field lines, J. Geophys. Res., 88, 899, 1983.

Olson, W. P., and K. P. Pfitzer, The entry of ampte lithium ions into a magnetically closed magnetosphere, J. Geophys. Res., 89, 7347, 1984.

Omura, Y., M. Ashour-Abdalla, R. Gendrin, and K. Quest, Heating of thermal helium in the equatorial magnetosphere: A simulation study, J. Geophys. Res., 90, 8281, 1985.

Palmadesso, P. J., S. B. Ganguli, and H. G. Mitchell, Jr., Multimoment fluid simulations of transport processes in the auroral zones, this volume, 1987.

Quinn, J. M., and D. J. Southwood, Observations of parallel ion energization in the equatorial region, J. Geophys. Res., 87, 10,536, 1982.

Quinn, J. M., and R. G. Johnson, Observation of ionospheric source cone enhancements at the substorm injection boundary, J. Geophys. Res., 90, 4211, 1985.

Retterer, J. M., J. R. Jasperse, and T. S. Chang, A new approach to pitch angle scattering in the magnetosphere, J. Geophys. Res., 88, 201, 1983a.

Retterer, J. M., T. Chang, and J. R. Jasperse,

Ion acceleration in the suprauroral region: A Monte Carlo model, J. Geophys. Res., 88, 583, 1983b.
Roth, I., and M. K. Hudson, Lower hybrid heating of ionospheric ions due to ion ring distributions in the cusp, J. Geophys. Res., 90, 4191, 1985.
Roux, A., N. Cournilleau-Wehrlin, and J. L. Rauch, Acceleration of thermal electrons by ICW's propagating in a multicomponent magnetospheric plasma, J. Geophys. Res., 89, 2267, 1984.
Sato, T., and H. Okuda, Numerical simulations on ion acoustic double layers, J. Geophys. Res., 86, 3357, 1981.
Sato, T., H. Matsumoto, and K. Nagai, Particle acceleration in time developing magnetic reconnection process, J. Geophys. Res., 87,6089,1982.
Schunk, R. W., and D. S. Watkins, Proton temperature anisotropy in the polar wind, J. Geophys. Res., 87, 171, 1982.
Serizawa, Y., and T. Sato, Generation of large scale potential difference by currentless plasma jets along the mirror field, Geophys. Res. Lett., 11, 595, 1984.
Silevitch, M. B., On a theory of temporal fluctuations in the electrostataic potential structures associated with auroral arcs, J. Geophys. Res., 86, 3573, 1981.
Singh, N., and R. W. Schunk, Numerical calculations relevant to the initial expansion of the polar wind, J. Geophys. Res., 87, 9154, 1982a.
Singh, N., and R. W. Schunk, Current carrying properties of double layers and low frequency auroral fluctuations, Geophys. Res. Lett., 9, 446, 1982b.
Singh, N., and R. W. Schunk, Current-driven double layers and the auroral plasma, Geophys. Res. Lett., 9, 1345, 1982c.
Singh, N., and R. W. Schunk, Dynamical features of moving double layers, J. Geophys. Res., 87, 3561, 1982d.
Singh, N., and R. W. Schunk, Energization of ions in the auroral plasma by broadband waves: Generation of ion conics, J. Geophys. Res., 89, 5538, 1984.
Singh, N., R. W. Schunk, and J. J. Sojka, Energization of ionospheric ions by electrostatic hydrogen cyclotron waves, Geophys. Res. Lett., 8, 1249, 1981.
Singh, N., W. J. Raitt, and F. Yasuhara, Low-energy ion distribution functions on a magnetically quiet day at geostationary altitude (L = 7), J. Geophys. Res., 87, 681, 1982a.
Singh, N., R. W. Schunk, and J. J. Sojka, Cyclotron resonance effects on stochastic acceleration of light ionospheric ions, Geophys. Res. Lett., 9, 1053, 1982b.
Singh, N., R. W. Schunk, and J. J. Sojka, Preferential perpendicular acceleration of heavy ionospheric ions by interactions with electrostatic hydrogen cyclotron waves, J. Geophys. Res., 88, 4055, 1983a.
Singh, N., H. Thiemann, and R. W. Schunk, Simulation of auroral current sheet equilibria and associated V-shaped potential structures, Geophys. Res. Lett., 10, 745, 1983b.
Singh, N., H. Thiemann, and R. W. Schunk, Dynamical features and electric field strengths of double layers driven by currents, J. Geophys. Res., 90, 5173, 1985.
Smith, R. A., Simulation of double layers in a model auroral circuit with nonlinear impedance, Geophys. Res. Lett., 13, 809, 1986.
Southwood, D. J., and M. G. Kivelson, Charged particle behavior in low-frequency geomagnetic pulsations, 2. Graphical approach, J. Geophys. Res., 87, 1707, 1982.
Speiser, T. W., and D. J. Williams, Magnetopause modeling: Flux transfer events and magnetosheath quasi-trapped distributions, J. Geophys. Res., 87, 2177, 1982.
Speiser, T. W., and L. R. Lyons, Comparison of an analytical approximation for particle motion in a current sheet with precise numerical calculations, J. Geophys. Res., 89, 147, 1984.
Speiser, T. W., D. J. Williams, and H. A. Garcia, Magnetospherically trapped ions as a source of magnetosheath energeticions, J. Geophys. Res., 86, 723, 1981.
Spjeldvik, W. N., and T. A. Fritz, Energetic heavy ions with nuclear charge $Z \geq 4$ in the equatorial radiation belts of the earth: Magnetic storms, J. Geophys. Res., 86, 2349, 1981.
Stern, D. P., One-dimensional models of quasi-neutral parallel electric fields, J. Geophys. Res., 86, 5839, 1981.
Strangeway, R. J., and R. G. Johnson, On the injection boundary model and dispersion ion signatures at near-geosynchronous altitudes, J. Geophys. Res., 88, 549, 1983.
Strangeway, R. J., and R. G. Johnson, Energetic ion mass composition as observed at near-geosynchronous and low altitudes during the storm period of February 21 and 22, 1979, J. Geophys. Res., 89, 8919, 1984.
Swift, D. W., Numerical simulation of the interaction of the plasma sheet with the lobes of the earth's magnetotail, J. Geophys. Res., 87, 2287, 1982.
Swift, D. W., A. Two-dimensional simulation of the interaction of the plasma sheet with the lobes of the earth's magnetotail, J. Geophys. Res., 88, 125, 1983.
Tanaka, M., Simulations of heavy ion heating by electromagnetic ion cyclotron waves driven by proton temperature anisotropies, J. Geophys. Res., 90, 6459, 1985.
Temerin, M., Evidence for a large bulk ion conic heating region, Geophys. Res. Lett., 13, 1059, 1986.
Temerin, M., J. McFadden, M. Boehm, C. W. Carlson, and W. Lotko, Production of flickering aurora and field-aligned electron flux by electromagnetic ion cyclotron waves, J. Geophys. Res., 91, 5769, 1986.
Tsurutani, B. T., and R. M. Thorne, Diffusion processes in the magnetopause boundary layer, Geophys. Res. Lett., 9, 1247, 1982.

Wagner, J. S., R. D. Sydora, T. Tajima, T. Hallinan, L. C. Lee, and S.-I. Akasofu, Small-scale auroral arc deformations, J. Geophys. Res., 88, 8013, 1983.

Wagner, J. S., P. C. Gray, T. Tajima, and S.-I. Akasofu, A plasma simulation study of the transformation from a closed to an open magnetic configuration, J. Geophys. Res., 91, 1491, 1986.

West, H. I., Jr., R. M. Buck, and G. T. Davidson, The dynamics of energetic electrons in the earth's outer radiation belt during 1968 as observed by the Lawrence Livermore National Laboratory's Spectrometer on OGO-5, J. Geophys. Res., 86, 2111, 1981.

Westphalen, H., and W. N. Spjeldvik, On the energy dependence of the radial diffusion coefficient and spectra of inner radiation belt particles: Analytic solutions and comparison with numerical results, J. Geophys. Res., 87, 8321, 1982.

Whipple, E. C., and M. B. Silevitch, Ion current in a magnetic neutral region: Generation of an incipient magnetopause, J. Geophys. Res., 87, 651, 1982.

Whipple, E. C., T. G. Northrop, and T. J. Birmingham, Adiabatic theory in regions of strong field gradients, J. Geophys. Res., 91, 4149, 1986.

Williams, D. J., Energetic ion bems at the edge of the plasma sheet: ISEE-1 observations plus a simple explanatory model, J. Geophys. Res., 86, 5507, 1981.

Williams, D. J., and T. W. Speiser, Sources for energetic ions at the plasma sheet boundary: Time-varying or steady state?, J. Geophys. Res., 89, 8877, 1984.

Yamamoto, T., and J. R. Kan, The field-aligned scale length of one-dimensional double layers, J. Geophys. Res., 90, 1553, 1985.

Yeh, H.-C., and T. W. Hill, Mechanism of parallel electric fields inferred from observations, J. Geophys. Res., 86, 6706, 1981.

THERMOSPHERIC GENERAL CIRCULATION AND RESPONSE TO MAGNETOSPHERIC FORCINGS

R. G. Roble

High Altitude Observatory, National Center for Atmospheric Research*
P. O. Box 3000, Boulder, Colorado 80307

Abstract. The earth's thermosphere is driven primarily by the absorption of solar ultraviolet radiation at wavelengths less than 200 nm. In the upper thermosphere, solar heating drives a diurnal circulation from the dayside to the nightside of the earth. In the lower thermosphere, a semidiurnal circulation dominates. It is driven by semidiurnal in situ heating components and by upward propagating tides from the lower atmosphere. Thermospheric winds calculated by the National Center for Atmospheric Research (NCAR) thermospheric general circulation model (TGCM) have been used with a global ionospheric dynamo model to calculate the global distribution of electric currents and fields for equinox conditions during solar minimum. At high latitudes, the solar-driven circulation is modified greatly by magnetospheric convection and auroral particle precipitation. Time-dependent thermospheric responses to auroral energy and momentum sources calculated by the TGCM display a wide variety of dynamic phenomena that depends upon the nature and time history of the forcings. Results from two TGCM simulations, November 21-22, 1981, and March 22, 1979, are described.

Introduction

Measurements from the Dynamics Explorer (DE) satellites and from ground-based optical and radar observatories have provided considerable information on global thermospheric dynamics and its response to various magnetospheric forcings. There has also been considerable progress in our ability to model many of the observed features of the global circulation, temperature, and compositional structure of the thermosphere. A brief review of this progress is presented.

*Sponsored by the National Science Foundation.

Progress in Modeling the Global Thermospheric Structure

Solar EUV and UV-Driven Circulation

Solar energy in the EUV ($5 \leq \lambda \leq 103$ nm) and UV ($103 \leq \lambda \leq 200$ nm) portions of the solar spectrum is absorbed primarily by the major constituents of the thermosphere. In the optically thin upper thermosphere, above about 200 km, the solar heating distribution is rather uniform over the dayside, whereas below 200 km the optical depth increases and solar heating varies with the solar zenith angle. The solar heating distributions in the upper and lower thermosphere for equinox conditions at solar cycle minimum are shown in Figure 1 (Dickinson et al., 1984). Solar EUV radiation also ionizes the constituents of the thermosphere and produces the ionosphere. Photochemical equilibrium exists in the lower ionosphere and produces an electron density distribution that varies with the solar zenith angle as shown in Figure 2b. In the upper ionosphere, near 300 km, the ion gyrofrequency exceeds the ion-neutral collision frequency and the plasma is controlled by the geomagnetic field line configuration which produces the complex structure illustrated in Figure 2a. The ionospheric plasma, above 150 km, moves under the combined influence of the earth's geomagnetic field and any superimposed electric field. Its motion is largely independent of the neutral gas motion and collisions between the ions and neutrals exert an ion drag on the neutral gas motions. This ion drag provides either a resistance to the neutral flow or it can accelerate it. The ion drag is proportional to the electron number density and the relative velocity difference between the ionospheric plasma and neutral gas (e.g., Roble et al., 1982; Killeen and Roble, 1984). At low- and mid-latitudes, the thermospheric flow above 150 km, to a first approximation, is governed by a balance between the pressure force, generated

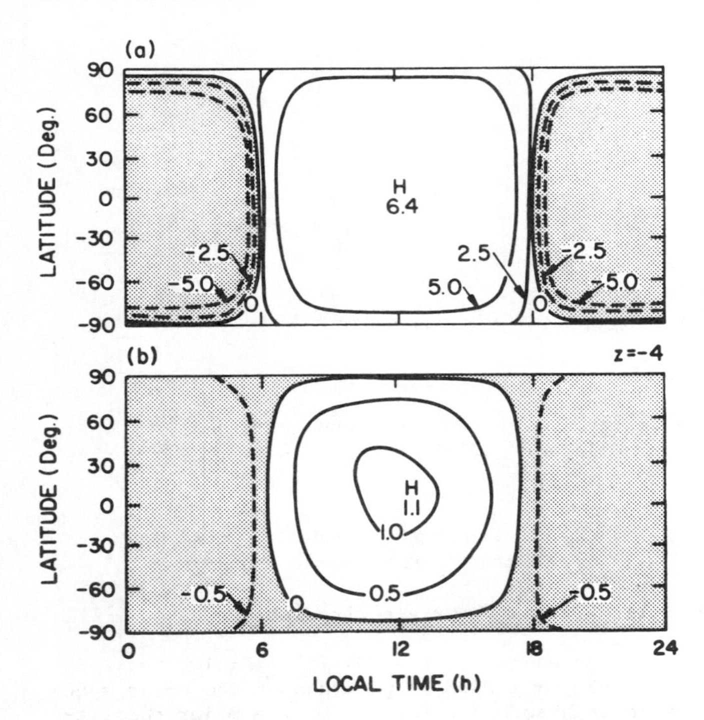

Fig. 1. Contours of calculated neutral gas heating rate (J kg^{-1} s^{-1}) for equinox conditions during solar cycle minimum along (a) the model Z = +2 constant pressure surface level near 300 km and (b) the Z = -4 level near 120 km (Dickinson et al., 1984).

to solar heating, and the ion drag force. Below 150 km, the coriolis force becomes important. Using the solar heating and electron density distributions shown in Figures 1 and 2 for equinox conditions during solar minimum the NCAR thermospheric general circulation model has been used to calculate the global circulation, temperature, and composition structure for the idealized case where auroral processes are neglected (Dickinson et al., 1981, 1984). The calculated global circulation and perturbation temperature structure, from a global mean value, driven by in situ solar heating only, is shown in Figure 3.

In the upper thermosphere, the distribution of solar heating drives a largely diurnal circulation from the high temperature and high pressure region on the dayside to the low temperature and low pressure region on the nightside (Figure 3a). The magnitude of the wind speed is less on the dayside because of increased electron density, and hence ion drag, than on the nightside. In the lower thermosphere, there are larger semidiurnal in situ heating components that give rise to a semidiurnal circulation and temperature structure as shown in Figure 3b.

The solar-driven circulation of the thermosphere is also influenced by upward propagating tides that are excited by absorption of solar energy by ozone and water vapor in the middle and lower atmosphere. Fesen et al. (1986) have specified the tidal structure at the lower boundary of the NCAR-TGCM and calculated their influence on the global thermospheric structure as shown in Figure 4. By comparing Figures 3 and 4 it can be seen that tides primarily affect the structure of the lower thermosphere and they are gradually dissipated by viscosity, thermal conductivity, and ion drag as they reach the upper thermosphere. The longer vertical wavelength 2,2 tidal mode does, however, reach the upper thermosphere and produces the midnight temperature bulge that has been observed at equatorial latitudes.

Global Dynamo

Richmond and Roble (1987) have developed a global dynamo model that calculates the ionospheric electric fields and currents, and the associated ground magnetic variations, that are generated by the dynamo actions of winds simu-

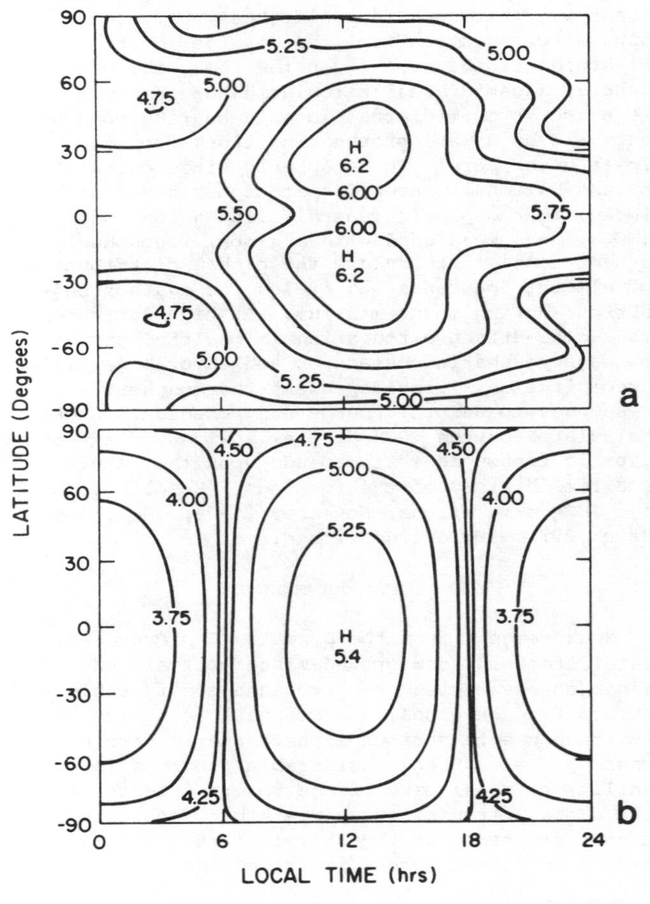

Fig. 2. Contours of the electron number density [$\log_{10}(n_e$; $cm^{-3})$] obtained from the Chiu (1975) empirical model for equinox conditions during solar minimum (a) along the 300 km constant height surface and (b) along the 120 km constant height surface.

lated with the NCAR-TGCM. The calculated electrodynamic structure, using the winds shown in Figure 4, is shown in Figure 5. The dynamo model uses a tilted dipole geomagnetic field and allows for field-aligned current flow between conjugate points, but no magnetospheric dynamo effects are included. Richmond and Roble (1987) have shown that without tidal forcing the TGCM winds produce grond magnetic variations that have the general pattern of Sq variations but that are only about half as strong. The addition of tidal forcing improves the agreement between calculated and observed magnetic variations and electric fields. The calculated electric potential, field-aligned current distribution, total horizontal, and equivalent current function at 11 UT for equinox conditions during solar cycle minimum are shown in Figure 5. The calculated potential distribution in Figure 5 is in reasonable agreement with empirical model predictions by Richmond et al. (1980). Field-aligned currents between southern

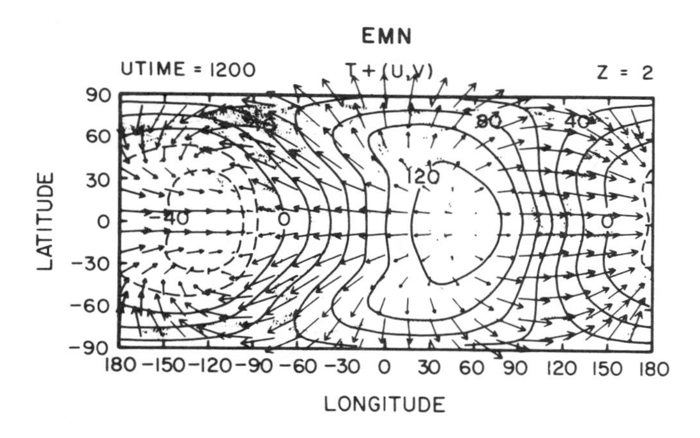

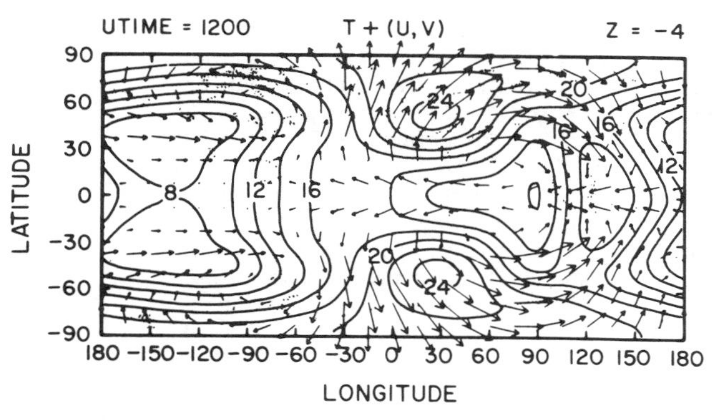

Fig. 3. Perturbation temperatures from a constant global mean value (solid curves) and winds (arrows) at fixed pressure levels for equinox conditions during solar minimum calculated by the TGCM without tides (Fesen et al., 1986). The pressure levels Z = +2 and -4 correspond to approximate altitudes of 300 and 135 km. The maximum wind arrows are 136 and 24 ms^{-1} on the Z = +2 and -4 pressure levels, respectively.

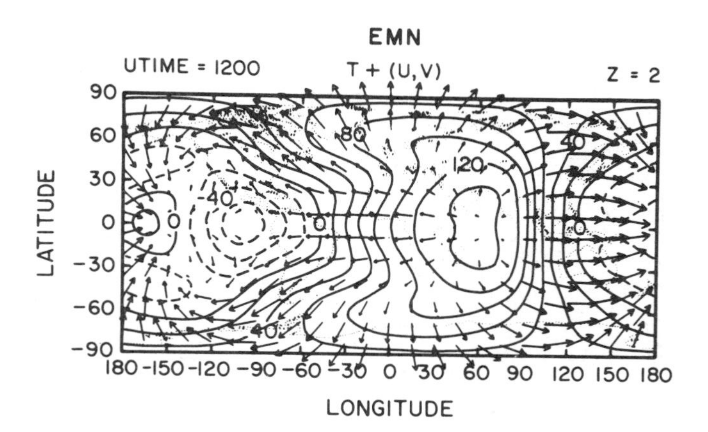

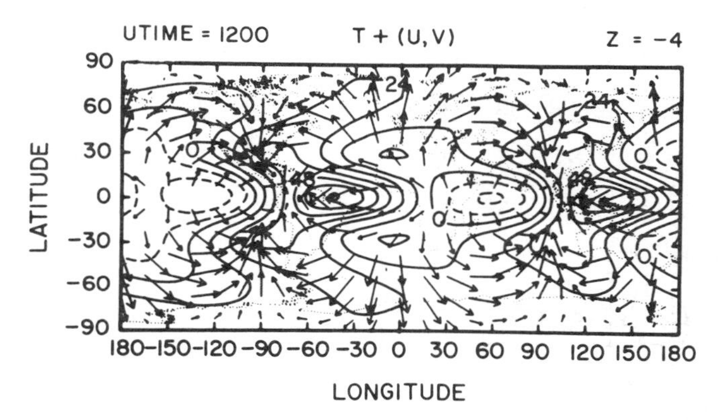

Fig. 4. Same captions as for Figure 3 except with inclusion of tides in the TGCM (Fesen et al., 1986). The maximum wind arrows are 166 and 71 ms^{-1} on the Z = +2 and -4 pressure levels, respectively.

and northern magnetic hemisphere are substantial even at equinox, with about 290 kA flowing at 11 UT. The field-aligned current distribution is considerably more sensitive to the wind and conductivity patterns than is the horizontal ionospheric current distribution. Other details of the dynamo simulation are discussed in detail by Richmond and Roble (1987).

High-Latitude Thermospheric Dynamics

At high latitudes, the solar-driven circulation is modified greatly by magnetospheric convection and auroral particle precipitation. The ion drag momentum source of magnetospheric plasma convection drives a largely rotational, nondivergent, double vortex wind system at F region altitudes that essentially follows the two cell pattern of magnetospheric convection. Wind velocities vary with the magnitude of the cross-polar cap potential drop and magnitudes greater than 500 ms^{-1} are common at high latitudes during moderate levels of geomagnetic activity. The circulation and perturbation temperature struc-

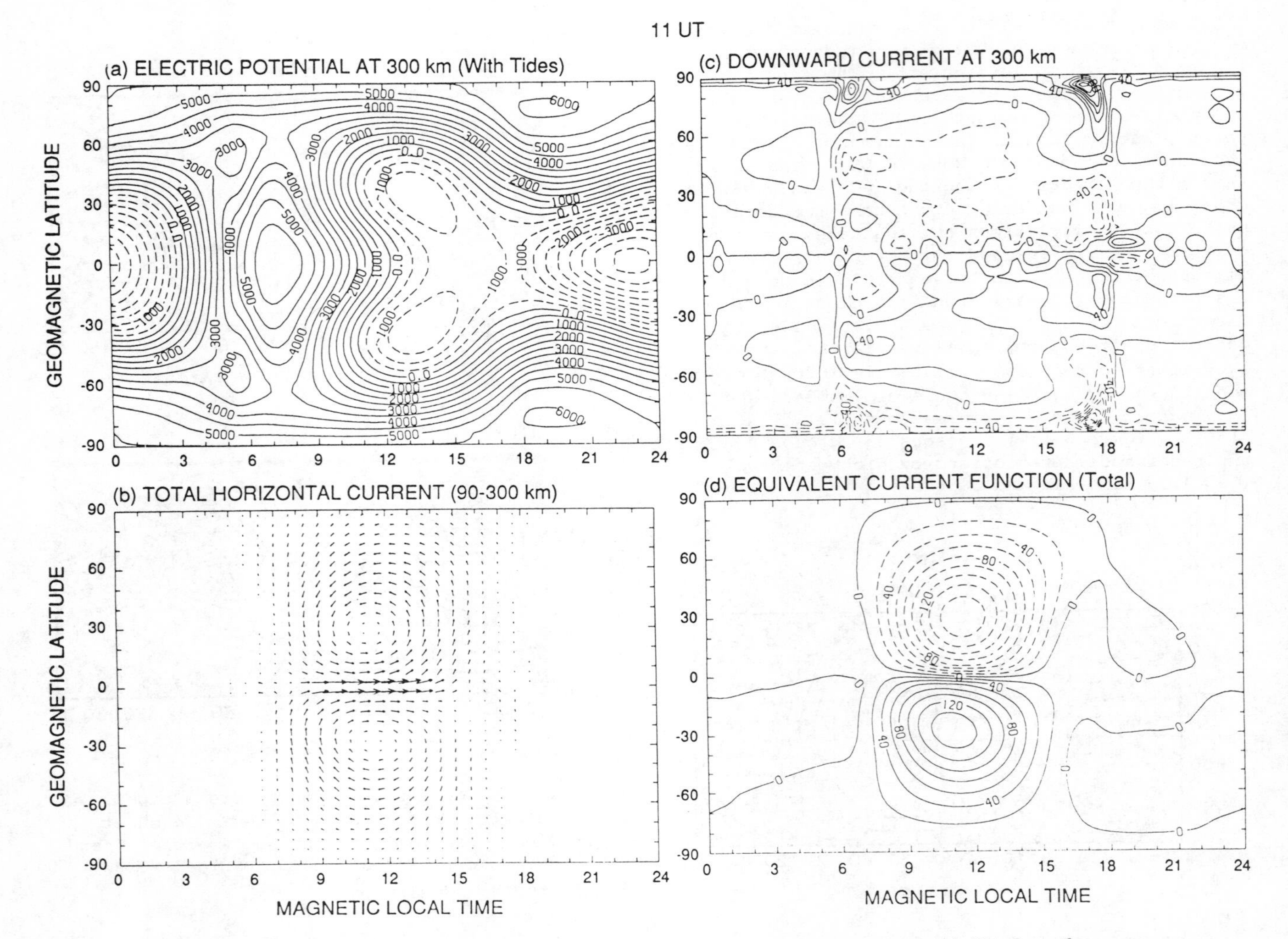

Fig. 5. (a) Electrostatic potential at 300-km altitude computed at 11 UT for the case including semidiurnal tidal forcing. The contour interval is 1 kV. (b) Total horizontal current flowing between 90 and 300 km at 11 UT for the case including tidal forcing. A vector equal in length to 1 hour of magnetic local time represents 0.1 A m^{-1}. (c) Downward component of field-aligned current at 300 km computed for 11 UT for the case including tidal forcing. The contour interval is 2 x 10^{-9} A m^{-2}. Downward currents have solid contours; upward currents have dashed contours. (d) Equivalent current function for the entire three-dimensional ionospheric-magnetospheric current system at 11 UT for the case including tidal forcing. The contour interval is 20 kA. Equivalent current flows counterclockwise around the dashed contours and clockwise around the solid contours (Richmond and Roble, 1987).

ture over the summer hemisphere (southern for December solstice) polar cap for three different cases is shown in Figure 6 (Roble et al., 1983). The solar only driven circulation, shown in Figure 6a, has a wind speed of about 100 ms^{-1} and it flows from the dayside to the nightside over the polar cap. With magnetospheric convection ion drag forcing, with a cross-polar cap potential drop of 20 kV, the wind speeds in the polar cap increase to 200 ms^{-1} with the two cell pattern of magnetospheric convection being superimposed on the solar-driven flow. At 60 kV cross-polar cap potential drop, the two-cell pattern clearly stands out in the neutral wind pattern with speeds approaching 400 ms^{-1}. Joule heating raises the perturbation temperature increasing from 200 K to 300 K to 450 K for the three cases, respectively.

The vorticies generated by magnetospheric convection extend downward into the lower thermosphere, as shown in Figure 7, with a cold low pressure cyclonic circulation near the dawn terminator and a warm high pressure anti-cyclonic circulation in the evening sector. The strength of the vorticies increases with the cross-polar cap potential drop as shown in the figure.

The DE satellites have made major contributions to our overall understanding of thermospheric dynamics at high latitudes and its response to various magnetospheric forcings. High-latitude neutral winds have been shown to be driven mainly by and generally follow the basic two-cell pattern of magnetospheric convection (Killeen et al., 1982, 1983, 1984, 1985, 1986, 1987; Hays et al., 1984; Rees et al., 1983, 1985a,b; Roble et al., 1983, 1984). Neutral wind measurements have shown relationships dependent upon the B_z (Killeen et al., 1985) and the B_y component directions of the interplanetary magnetic field (IMF) (McCormac et al., 1985). In addition, satellite data have shown considerable variability and small-scale structure in both the ion drift velocity and neutral wind vectors.

A number of the observed large-scale features

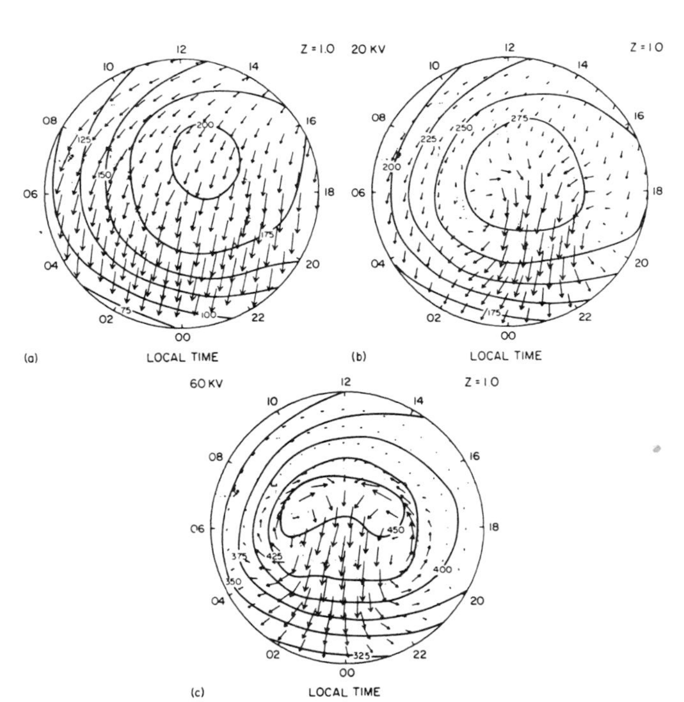

Fig. 6. Polar plots giving the direction and magnitude of the calculated southern hemisphere (summer) high-latitude circulation and contours of perturbations temperature (K) along the Z = +1 (~300 km) constant pressure surface at December solstice for (a) solar heating only, (b) solar heating plus magnetospheric convection with a cross-polar cap potential drop of 20 kV, and (c) same as (b) except for 60 kV. The wind speed associated with the maximum arrow is 110 ms^{-1} in (a), 200 ms^{-1} in (b), and 380 ms^{-1} in (c), respectively (Roble et al., 1983).

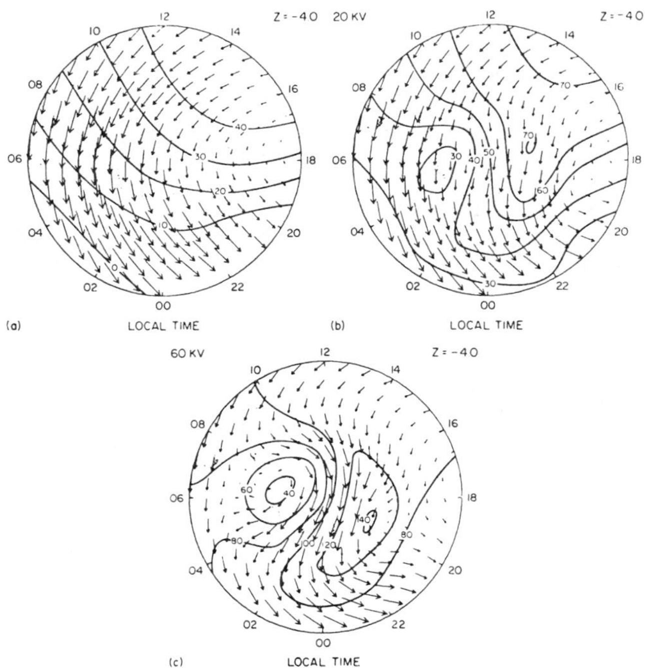

Fig. 7. Same as in Figure 6 except along the Z = -4(~130 km) constant pressure surface. The wind speed associated with the maximum arrow is 75 ms^{-1} in (a), 77 ms^{-1} in (b), and 89 ms^{-1} in (c), respectively (Roble et al., 1983).

of high-latitude dynamics have been well modeled by TGCMs. These include the strong anti-sunward flow over the polar caps (Rees et al., 1983, 1985a,; Roble et al., 1983, 1984; Hays et al., 1984; Larsen and Mikkelsen, 1983), a strong sunward-directed flow observed in the evening auroral oval (Fuller-Rowell and Rees, 1984; Rees et al., 1985b; Roble et al., 1987a), the thermospheric response to the changing ion drift pattern associated with the B_y component of the IMF (Rees et al., 1986), and the general temperature and compositional structure over the polar (Roble et al., 1984; Rees et al., 1985a).

Time-Dependent TGCM Simulations

The NCAR-TGCM has been modified recently to accommodate time-dependent auroral forcings and it has been used to calculate the time-dependent thermospheric circulation, temperature, and compositional patterns for specific events or days. An empirical magnetospheric convection pattern (Heelis et al., 1982) is used to specify the ion drift and it has been modified to mimic the observed B_y interplanetary magnetic field-induced shifts. The time-dependent cross-polar cap potential drop is determined using measured solar wind parameters and an empirical relationship developed by Reiff et al. (1985).

Auroral oval parameters are based on the

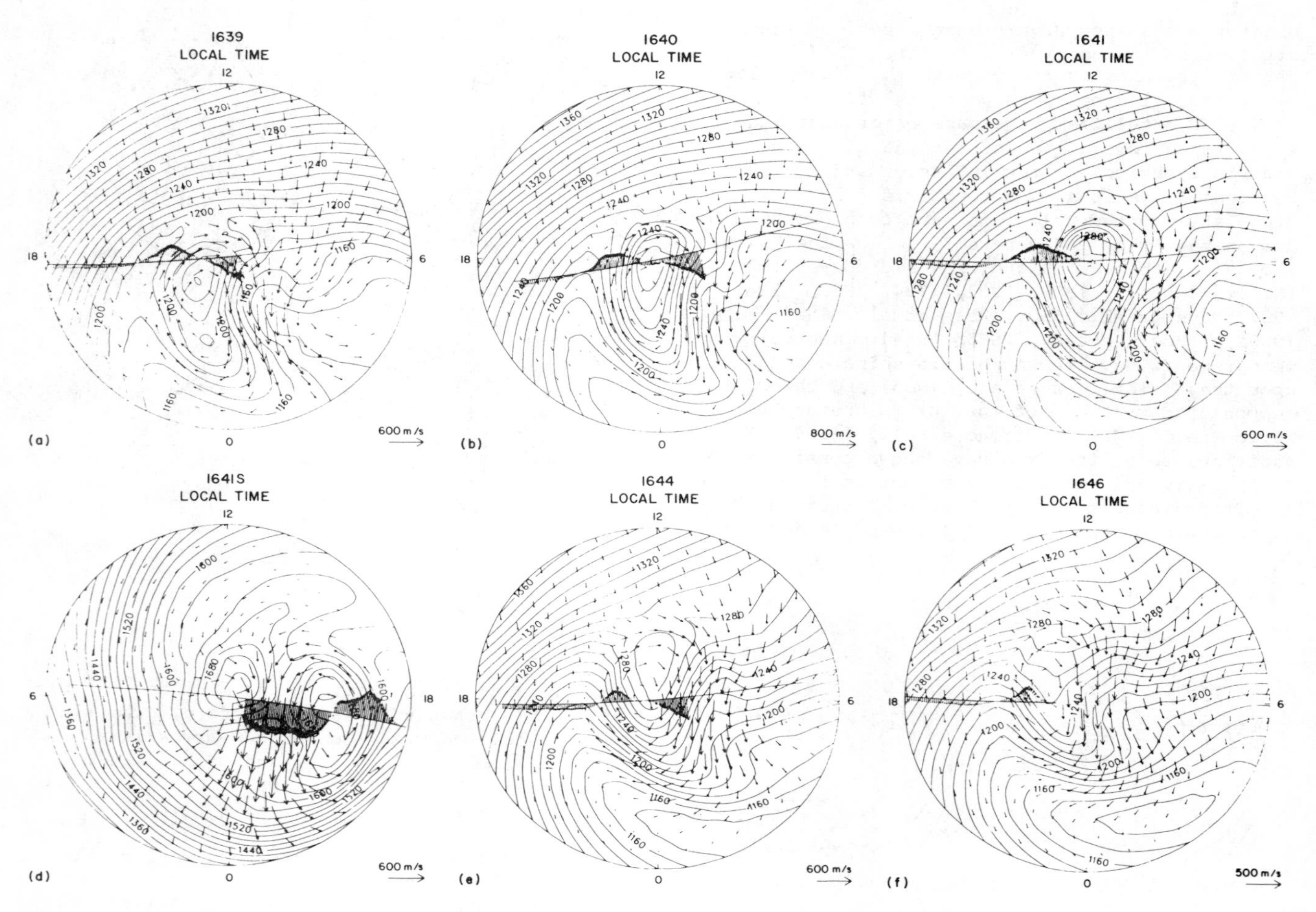

Fig. 8. Measured meridional neutral wind by the wind and temperature spectrometer instrument along the DE 2 satellite track over northern and southern hemispherie polar caps for six orbits: (a) 1639, NH; (b) 1640, NH; (c) 1641, NH; (d) 1641, SH; (e) 1644, NH; and (f) 1646, NH, on November 22, 1981. Also shown are TGCM predicted neutral temperature pattern (K) and neutral wind vectors along the model Z = +2 constant pressure surface (~400 km) over the polar caps at the times of the DE 2 satellite passes. The outer circle represents a latitude of either ±37.5 latitude depending upon the hemisphere (Roble et al., 1987a).

empirical model developed by Fuller-Rowell and Evans (1987) using NOAA satellite data of aurora particle precipitation. The oval characteristics are specified by a single parameter, the estimated hemispherical power input from which particle energy flux, characteristic energy, and auroral boundaries are obtained through empirical relationships. A fast parameterization that allows the ionization rate and the ion and electron density to be calculated self-consistently within the NCAR-TGCM has been described by Roble and Ridley (1987). The enhanced electron density from auroral processes is appropriately added to the electron density obtained from the Chiu (1975) empirical model and the sum is used to derive the high-latitude ion drag parameters for time-dependent TGCM simulations.

The parameterized auroral forcings were used in the TGCM to investigate thermospheric dynamics during November 21-22, 1981. Geomagnetic activity was relatively steady and a detailed comparison of the TGCM predicted high-latitude winds was made with wind measurements made by the DE satellite on an orbit-by-orbit basis during the period (Roble et al., 1987a). DE 2 measurements of electric fields and auroral particle precipitation also enabled a detailed comparison to be made between the TGCM auroral parameterizations and measurements. TGCM predictions and DE 2 measurements for six different passes over the northern and southern hemisphere polar caps are shown in Figure 8. In general, this study (Roble et al., 1987a) shows that at least for this period of steady forcing during moderate geo-

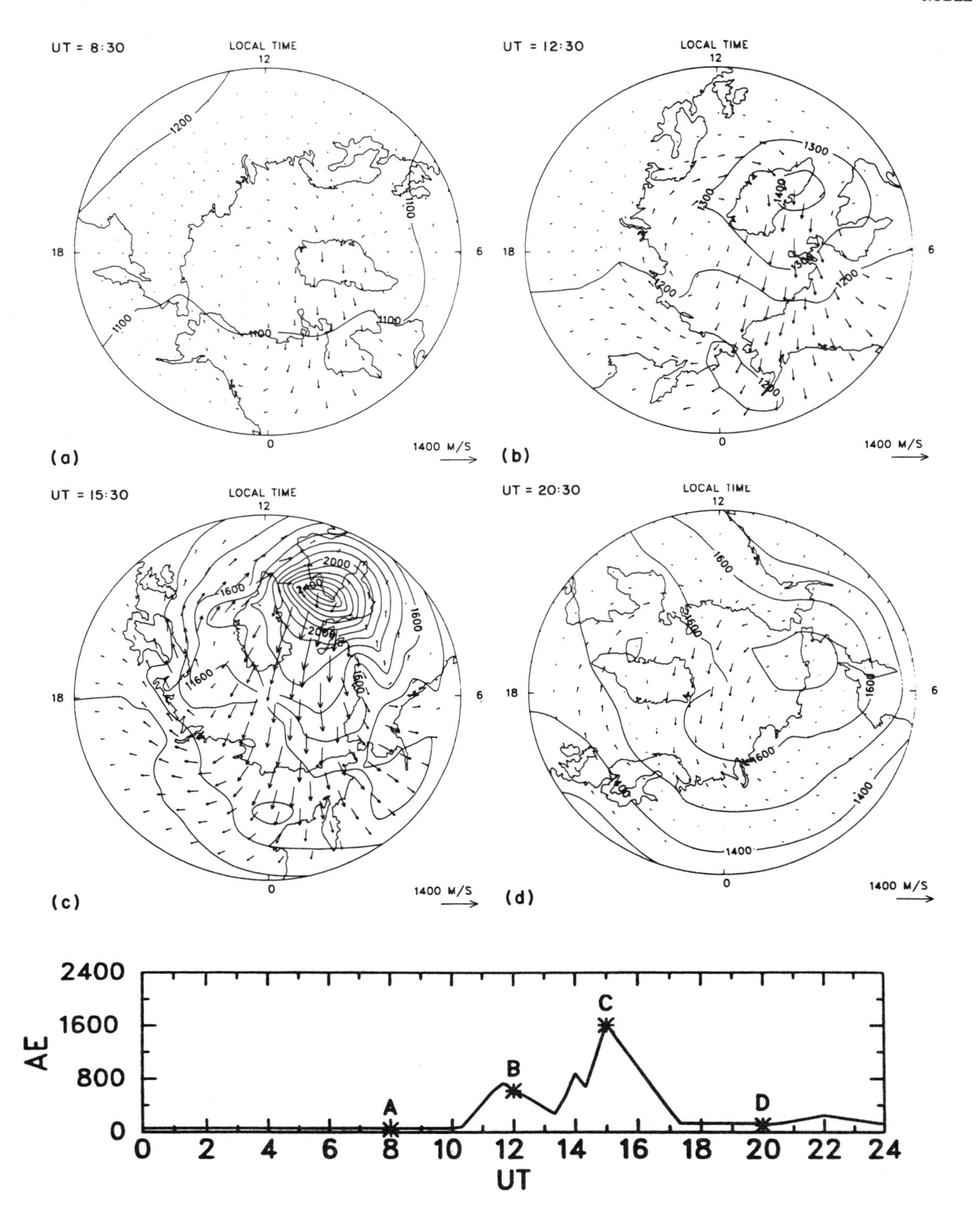

Fig. 9. Contours of TGCM calculated neutral temperature (K) and neutral wind vectors (ms^{-1}) on the Z = +1 constant pressure surface near 300 km over the northern hemisphere polar cap for four different times on March 22, 1979: (a) 0830 UT, (b) 1230 UT, (c) 1530 UT, and (d) 2030 UT. The length of the maximum arrow is 1400 ms^{-1} and is shown for reference (Roble et al., 1987b).

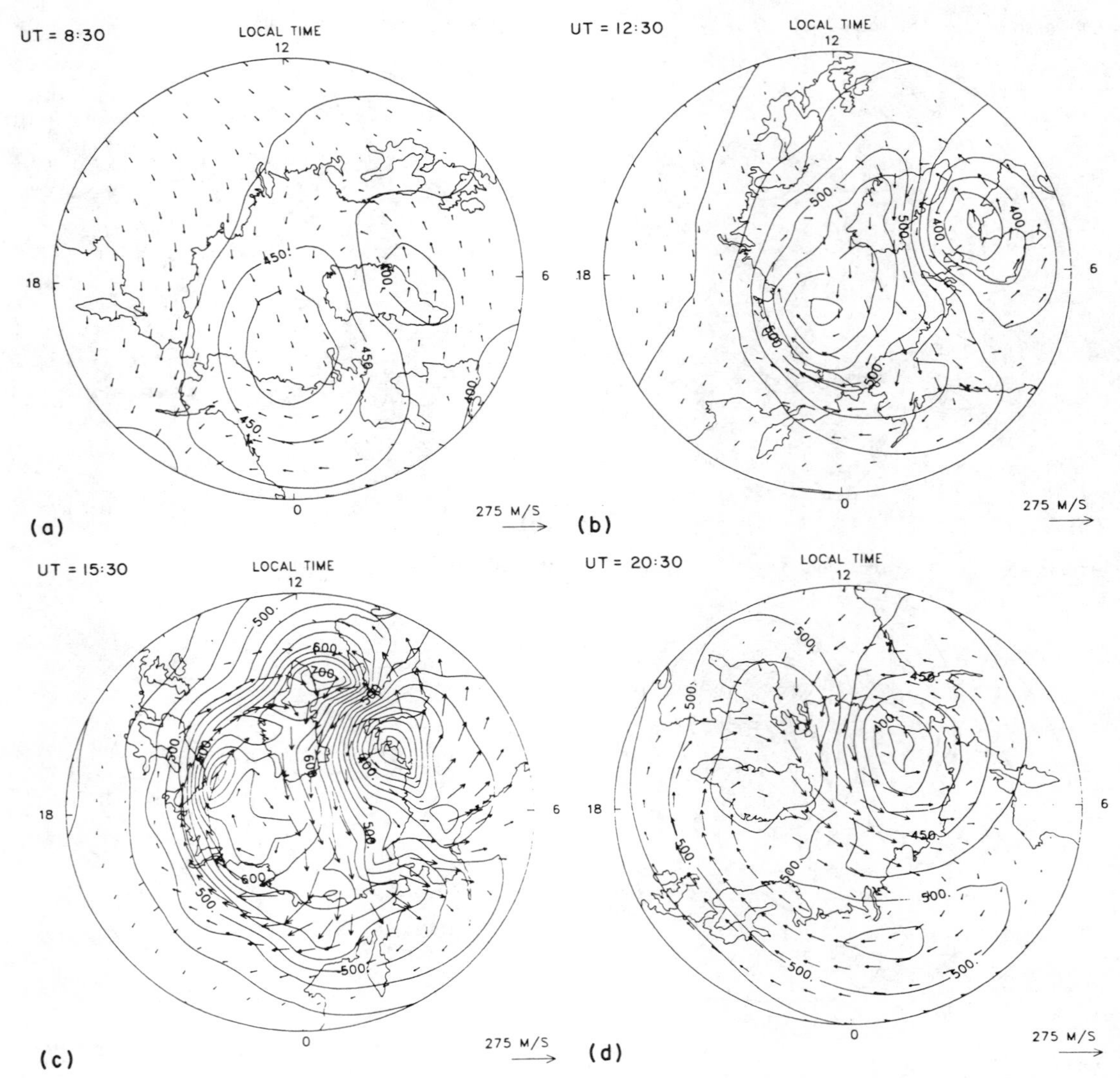

Fig. 10. Same as in Figure 9 except for the Z = -4 constant pressure surface near 120 km. The length of the maximum arrow is 275 ms^{-1} and all arrows are scaled relative to this maximum value (Roble et al., 1987b).

magnetic activity there is good general agreement between model predictions and observations of the large-scale structure on the evening side of the auroral oval and polar cap. There is, however, considerable small-scale structure observed by the DE 2 satellite that is not predicted by the TGCM.

Roble et al. (1987b) also used the TGCM to study the thermospheric response to the geomagnetic storm of March 22, 1979. The magnetospheric and ionospheric characteristics of this storm have been studied intensively by the CDAW-6 coordinated workshop (McPherron and Manka, 1985). The time-dependent auroral forcings were estimated from the studies done by CDAW-6 and used as inputs for the TGCM. The calculated thermospheric temperature and wind structure on a constant pressure surface near 300 km at various times during the storm are shown in Figure 9. Geomagnetic activity was low for the period preceding the storm and the wind velocities were generally small with a uniform temperature over the polar cap as shown in Figure 9a. At 1230 UT the storm intensified and both the wind velocities and temperature increased as shown in Figure 9b. By 1530 UT, near the peak of the storm when the auroral electrojet index AE was near 1600, the wind velocities approached 1 km s^{-1} with temperatures exceeding 2000 K near the throat of the convection pattern as shown in Figure 9c. By 2030 UT (Figure 9d) both the wind velocities and temperatures decreased but there was still resid-

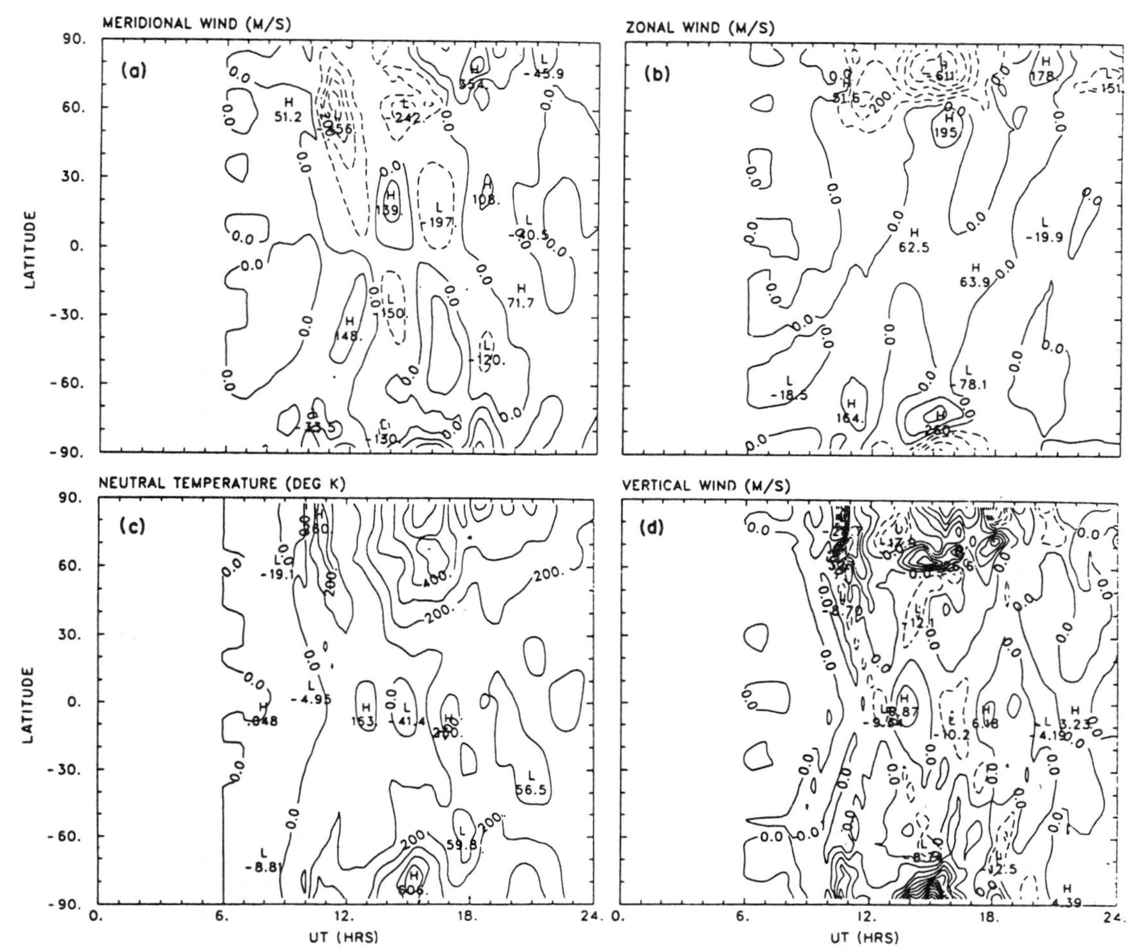

Fig. 11. Contours of the time variation of difference fields (storm simulation minus quiet variation). (a) Meridional wind (ms^{-1}), (b) Zonal wind (ms^{-1}), (c) Temperature (K), and (d) Vertical wind (ms^{-1}) along the Z = +1 constant pressure surface near 330 km. Contour intervals are: (a) and (b) 100 ms^{-1}, (c) 100 K, and (d) 5 ms^{-1} (Roble et al., 1987b).

ual heating and dynamics lingering after the intense storm period. In the lower thermosphere, near 120 km, the dawn and dusk vortices were spun up during the storm period as shown in Figure 10. Once the vorticies were spun up, they decayed slowly with time with some residual dynamics remaining at 2400 UT, 9 hours after the peak of the event.

Because of the impulsive nature of the storm, large-scale thermospheric disturbances were generated at F region heights that propagated equatorward from both auroral zones. To illustrate the stormtime response two TGCM simulations were made and difference fields were constructed: TGCM storm simulation minus a TGCM simulation where the auroral forcing remained constant at its prestorm values. The aurora-generated disturbances in magnetic conjugate hemispheres propagated equatorward and both reached the equator in about 2 hours, as shown in Figure 11, where they interacted. Converging waves are associated with sinking motion and adiabatic heating, and subsequent diverging motion is associated with upwelling and adiabatic cooling. As a result, the temperature at the equator first increases by about 160 K and then decreases by about 50 K. These interacting waves were shown by Fesen et al. (1987) to have an important influence on the equatorial ionosphere especially in the midnight sector where impulses travel effectively toward lower latitudes. Finally, comparisons of temperature difference fields of the storm response calculated by the TGCM and MSIS-83 empirical model are shown in Figure 12. There is general agreement in the large-scale features indicating that the overall magnitude of the storm is well modeled; however, there is considerably more structure associated with local regions of

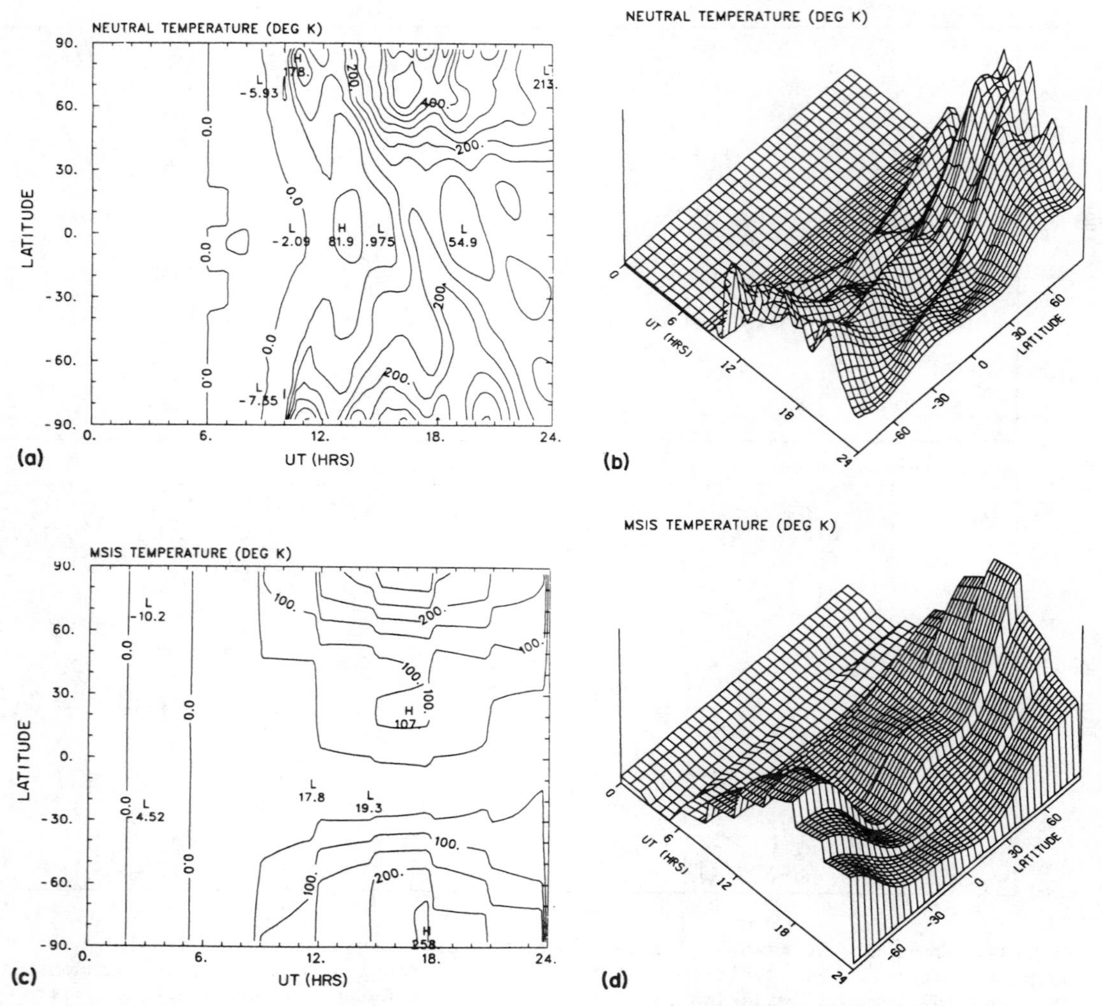

Fig. 12. Contours of the time variation of the zonally averaged temperature difference fields (K) (storm simulation minus quiet variation) calcualted by the TGCM and MSIS-83 models, respectively, illustrating the response to the March 22, 1979, substorms: (a) TGCM-calculated temperature difference along the 325-km constant height surface, (b) a perspective illustration of that temperature response, (c) MSIS-83 calculated temperature difference along the 325-km constant height surface, and (d) a perspective illustration of that temperature response. Contour intervals are 50 K. (Roble et al., 1987b).

intense heating and wave propagation in the TGCM fields that is not shown in the MSIS-83 predictions. It appears that MSIS-83 gives a good climatology or mean representation of the storm response, whereas the TGCM accounts for time-dependent weather processes with a much higher time resolution.

References

Chiu, Y. T., An improved phenomenological model of ionospheric density, J. Atmos. Terr. Phys., 37, 1563, 1975.

Dickinson, R. E., E. C. Ridley, and R. G. Roble, A three-dimensional, time-dependent general circulation model of the thermosphere, J. Geophys. Res., 86, 1499, 1981.

Dickinson, R. E., E. C. Ridley, and R. G. Roble, Thermospheric general circulation with coupled dynamics and composition, J. Atmos. Sci., 41, 205, 1984.

Fesen, C. G., R. E. Dickinson, and R. G. Roble, Simulation of thermospheric tides at equinox with the NCAR thermospheric general circulation model, J. Geophys. Res., 91, 4471, 1986.

Fesen, C. G., R. G. Roble, and G. Crowley, Ionospheric effects at low latitudes during the March 22, 1979 geomagnetic storm, J. Geophys. Res., in press, 1987.

Fuller-Rowell, T. J., and D. Rees, Interpretation of an anticipated long-lived vortex in the lower thermosphere following simulation of an isolated substorm, Planet. Space Sci., 32, 69, 1984.

Fuller-Rowell, T. J., and D. S. Evans, Height integrated Pedersen and Hall conductivity patterns inferred from the TIROS-NOAA satellite data, J. Geophys. Res., 92, 7606, 1987.

Hays, P. B., T. L. Killeen, N. W. Spencer, L. E. Wharton, R. G. Roble, B. E. Emery T. J. Fuller-Rowell, D. Rees, L. A. Frank, and J. D. Craven, Observations of the dynamics of the polar thermosphere, J. Geophys. Res., 89, 5597, 1984.

Heelis, R. A., J. K. Lowell, and R. W. Spiro, A model of the high-latitude ionospheric convection pattern, J. Geophys. Res., 87, 6339, 1982.

Killeen, T. L., and R. G. Roble, An analysis of the high latitude thermospheric wind pattern calculated by a thermospheric general circulation model, 1, Momentum forcing, J. Geophys. Res., 89, 7509, 1984.

Killeen, T. L., P. B. Hays, N. W. Spencer, and L. E. Wharton, Neutral winds in the polar thermosphere as measured from Dynamics Explorer, Geophys. Res. Lett., 9, 957, 1982.

Killeen, T. L., P. B. Hays, N. W. Spencer, and L. E. Wharton, Neutral winds in the polar thermosphere as measured from Dynamics Explorer, Adv. Space Res., 2(10), 133, 1983.

Killeen, T. L., P. B. Hays, G. R. Carignan, R. A. Heelis, W. B. Hanson, N. W. Spencer, and L. H. Brace, Ion-neutral coupling in the high latitude F-region: Evaluation of ion heating terms from Dynamics Explorer 2, J. Geophys. Res., 89, 7495, 1984.

Killeen, T. L., R. A. Heelis, P. B. Hays, N. W. Spencer, and W. B. Hanson, Neutral motions in the polar thermosphere for northward interplanetary magnetic field, Geophys. Res. Lett., 12, 159, 1985.

Killeen, T. L., R. W. Smith, N. W. Spencer, J. W. Meriwether, D. Rees, G. Hernandez, P. B. Hays, L. L. Cogger, D. P. Sipler, M. A. Biondi, and C. A. Tepley, Mean neutral circulation in the winter polar F-region, J. Geophys. Res., 91, 1633, 1986.

Killeen, T. L., J. D. Craven, L. A. Frank, J.-J. Ponthieu, N. W. Spencer, R. A. Hellis, L. H. Brace, R. G. Roble, P. B. Hays, and G. R. Carignan, On the relationship between the dynamics of the polar thermosphere and the morphology of the aurora: Global-scale observations from Dynamics Explorers 1 and 2, J. Geophys. Res., in press, 1987.

Larsen, M. F., and I. S. Mikkelsen, The dynamic response of the high-latitude thermosphere and geostrophic adjustment, J. Geophys. Res., 88, 3158, 1983.

McCormac, F. G., T. L. Killeen, E. Gombosi, P. B. Hays, and N. W. Spencer, Configuration of the high-latitude neutral thermosphere circulation for IMF By negative and positive, Geophys. Res. Lett., 12, 155, 1985.

McPherron, R. L., and R. H. Manka, Dynamics of the 1054 UT March 22, 1979, substorm event: CDAW 6, J. Geophys. Res., 90, 1175, 1985.

Rees, D., T. J. Fuller-Rowell, R. Gordon, T. L. Killeen, P. B. Hays, L. E. Wharton, and N. W. Spencer, A comparison of wind observations of the upper thermosphere from the Dynamics Explorer satellite with the predictions of a global time-dependent model, Planet. Space Sci., 31, 1299, 1983.

Rees, D., R. Gordon, T. J. Fuller-Rowell, M. Smith, G. R. Carignan, T. L. Killeen, P. B. Hays, and N. W. Spencer, The composition, structure, temperature, and dynamics of the upper thermosphere in the polar regions during October to December 1981, Planet. Space Sci., 33, 617, 1985a.

Rees, D., T. J. Fuller-Rowell, M. F. Smith, R. Gordon, T. L. Killeen, P. B. Hays, N. W. Spencer, L. E. Wharton, and N. C. Maynard, The westward thermospheric jet-stream of the evening auroral oval, Planet. Space Sci., 33, 425, 1985b.

Rees, D., T. J. Fuller-Rowell, R. Gordon, J. P. Heppner, N. C. Maynard, N. W. Spencer, L. E. Wharton, P. B. Hays, and T. L. Killeen, A theoretical and empirical study of the response of the high-latitude thermosphere to the sense of the "Y" component of the interplanetary magnetic field, Planet. Space Sci., 34, 1, 1986.

Reiff, P. H., R. W. Spiro, R. A. Wolf, Y. Kamide, and J. H. King, Comparison of polar cap potential drops estimated from solar wind and ground magnetometer data, J. Geophys. Res., 90, 1318, 1985.

Richmond, A. D., and R. G. Roble, Electrodynamic effects of thermospheric winds from the NCAR thermospheric general circulation model, J. Geophys. Res., in press, 1987.

Richmond, A. D., M. Blanc, B. A. Emery, R. H. and, B. G. Fejer, R. F. Woodman, S. Ganguly, P. Amayenc, R. A. Behnke, C. Calderon, and J. V. Evans, An empirical model of quiet-day ionospheric electric fields at middle and low latitudes, J. Geophys. Res., 85, 4658, 1980.

Roble, R. G., and E. C. Ridley, An auroral model for the NCAR thermospheric general circulation model, Annales Geophysicae, in press, 1987.

Roble, R. G., R. E. Dickinson, and E. C. Ridley, Global circulation and temperature structure of the thermosphere with high-latitude plasma convection, J. Geophys. Res., 87, 1599, 1982.

Roble, R. G., R. E. Dickinson, E. C. Ridley, B. A. Emery, P. B. Hays, T. L. Killeen, and N. W. Spencer, The high latitude circulation and temperature structure of the thermosphere near solstice, Planet. Space Sci., 31, 1479, 1983.

Roble, R. G., B. A. Emery, R. E. Dickinson, E. C.

Ridley, T. L. Killeen, P. B. Hays, G. R. Carignan, and N. W. Spencer, Thermospheric circulation, temperature, and compositional structure of the Southern Hemisphere polar cap during October-November 1981, J. Geophys. Res., 89, 9057, 1984.

Roble, R. G., T. L. Killeen, G. R. Carignan, N. W. Spencer, R. A. Heelis, P. H. Reiff, J. D. Winningham, and D. S. Evans, Thermospheric dynamics during 21/22 November 1981; Dynamics Explorer measurements and TGCM predictions, J. Geophys. Res., in press, 1987a.

Roble, R. G., J. M. Forbes, and F. A. Marcos, Thermospheric dynamics during the March 22, 1979 magnetic storm (a) Model simulations, J. Geophys. Res., 92, 6045, 1987b.

LARGE-SCALE MODELS OF THE IONOSPHERE/MAGNETOSPHERE/SOLAR WIND SYSTEM - MHD AS A UNIFYING PRINCIPLE

V. M. Vasyliunas

Max-Planck-Institut für Aeronomie, D-3411 Katlenburg-Lindau, Federal Republic of Germany

Abstract. The magnetohydrodynamic approximation provides the basic framework for most models of large-scale plasma processes in the interaction between the solar wind and the magnetosphere. It may be applied directly, with the magnetohydrodynamic coupling between the plasma flow and the magnetic field being built into the model, or additional well-defined terms in the generalized Ohm's law may be introduced to permit departures from ideal magnetohydrodynamics. In models of the magnetosheath (1), the magnetic field is determined from the flow which is given by a hydrodynamic calculation. In models of the inner magnetosphere (2) and its coupling to the ionosphere, the flow is constrained by the given magnetic field, and the finite ionospheric resistivity plays a crucial role. A self-consistent treatment of plasma dynamics and the magnetic field is introduced in models of the magnetotail (3) but generally only for simplified geometries and special symmetries. Most unsolved problems lie at the interfaces between these categories; (2)-(3) requires primarily improved approximation and solution techniques, but (1)-(2) also involves unsolved questions of physics of magnetic merging and plasma entry.

Introduction

Large-scale models of plasma phenomena in the ionosphere/magnetosphere/solar wind system are, for the most part, constructed on the basis of magnetohydrodynamics (MHD) or its extensions and modifications. This paper is intended as a brief general review of these models, emphasizing the assumptions, approximations, and boundary conditions invoked in their construction and identifying unsolved problems as well as areas of potential improvement or extension, noting in the latter cases whether the primary need is merely improved approximation and solution techniques or new physical understanding. After some general remarks on the nature and role of MHD, discussions are given on models for the various distinct regions sketched in Figure 1 (magnetosheath, inner magnetosphere and ring current, magnetotail) and then the interfaces between these regions (inner magnetosphere/magnetotail, magnetosheath/dayside magnetosphere, magnetosheath/magnetotail).

Role of Magnetohydrodynamics

The meaning and the basis of the MHD approximation as applied in contemporary space plasma physics have been described in numerous publications (e.g., Rossi and Olbert, 1970; Vasyliunas, 1975, 1980; Siscoe, 1983). The starting point is the so-called generalized Ohm's law, which relates the electric and magnetic fields **E** and **B** and the plasma bulk velocity **V** to various electromagnetic and mechanical quantities; for our purposes here it suffices to write it in the form

$$c\mathbf{E} + \mathbf{V} \times \mathbf{B} = O(\lambda_{ch}/L), \qquad (1)$$

where the right-hand side stands for various terms (given in detail in the references cited) which are all, compared to the left-hand side, of order of magnitude λ_{ch}/L, with λ_{ch} any one of various characteristic lengths (gyroradius, collisionless skin depth, etc.) and L the gradient length scale of macroscopic quantities. The MHD approximation comes from the recognition that in many applications $L >> \lambda_{ch}$; therefore, (1) can be approximated as

$$c\mathbf{E} + \mathbf{V} \times \mathbf{B} \approx 0 . \qquad (2)$$

The time rate of change of magnetic flux Φ_M through a loop moving with the plasma is given rigorously by

$$d\Phi_M/dt = \int d\boldsymbol{\ell} \cdot (\mathbf{E} + \mathbf{V} \times \mathbf{B}/c) , \qquad (3)$$

and evaluating the integrand from the generalized Ohm's law (1) yields, in order of magnitude,

$$d\Phi_M/dt = O(\lambda_{ch}/L)BA/\tau , \qquad (4)$$

where A is the area of the loop and B and τ the typical field strength and convective time scale,

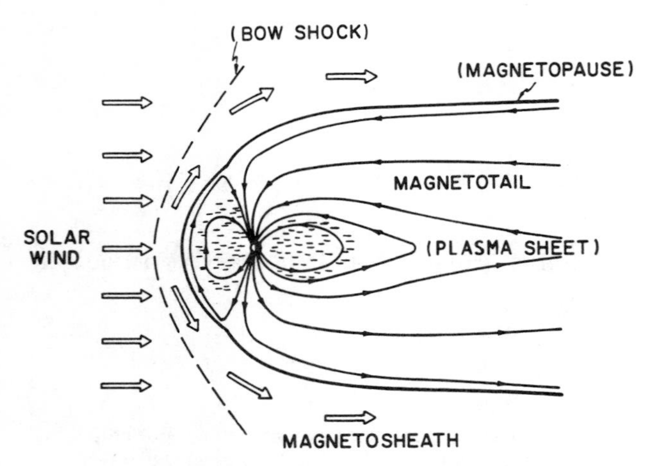

Fig. 1. Schematic view of the magnetosphere and adjacent regions, in the noon-midnight meridian plane. Magnetic field lines (solid lines with arrows) inside the magnetosphere and plasma flow (open arrows) outside are shown. The shaded region is the inner magnetosphere and ring current. The region between the bow shock and the magnetopause is the magnetosheath.

respectively. The MHD approximation then implies $O(\lambda_{ch}/L) \approx 0$, or

$$d\Phi_M/dt \approx 0 \; ; \tag{5}$$

i.e., the plasma flow is flux-preserving and therefore also line-preserving (i.e., Stern, 1966; Vasyliunas, 1972). The significance of MHD for large-scale modeling lies primarily in this close coupling of the flow and the magnetic field which allows either one to be calculated (or at least strongly constrained) if the other one is given.

Note that the validity (or otherwise) of the MHD approximation is distinct from and independent of any assertions about plasma dynamics embodied in the momentum equation

$$\begin{aligned}\rho(d\mathbf{V}/dt - g) + \nabla \cdot \underset{\approx}{P} &= \mathbf{J} \times \mathbf{B}/c \\ &= (\nabla \times B) \times B/4\pi \; .\end{aligned} \tag{6}$$

The relative order of magnitude of various terms in (6) determines whether the flow dominates the field or vice versa, but it does not by itself imply anything about whether the flow and the field are coupled. The characteristic lengths λ_{ch} do not appear at all in the one-fluid momentum equation (6), which therefore says nothing about the ratio λ_{ch}/L.

Significant deviations from the MHD approximation are expected in small-scale systems for which $L \sim \lambda_{ch}$ and therefore the right-hand side of (1) cannot be neglected. On large scales, MHD turbulence may sometimes lead to equations for mean quantities that deviate from ideal MHD: if **B**, **V**, and **E** obey equation (2) but each is split into a mean and a fluctuating part,

$$\mathbf{B} = \langle \mathbf{B} \rangle + \delta\mathbf{B} \; , \text{ etc.,}$$

then the mean of (2) yields

$$\langle c\mathbf{E} \rangle + \langle \mathbf{V} \rangle \times \langle \mathbf{B} \rangle = -\langle \delta\mathbf{V} \times \delta\mathbf{B} \rangle \; , \tag{7}$$

and there is no general reason why the right-hand side of (7) should be zero.

Models for Various Regions

Magnetosheath

Over much of the magnetosheath region, the magnetic field is weak in the sense

$$B^2/8\pi \ll \rho V^2 \ll P \; ; \tag{8}$$

therefore, the **J** x **B** force in the momentum equation (6) can be neglected and the flow calculated from a purely hydrodynamic model, as flow around an obstacle having the shape of the magnetosphere (e.g., Spreiter et al., 1966, 1968). MHD then allows the determination of **B** given the flow; with the use of the equation derived from the curl of (2),

$$\partial\mathbf{B}/\partial t = \nabla \times (\mathbf{V} \times \mathbf{B}) \; , \tag{9}$$

the calculation is (at least in principle) straightforward and requires as a boundary condition only **B** in the upstream solar wind region. The magnetic field of the magnetosheath calculated by this method (Alksne, 1967) is draped around the magnetospheric obstacle and in this respect agrees well with observations (Fairfield, 1968). Near the magnetopause **B** may become large enough so that the **J** x **B** force may no longer be negligible, and some models have attempted to include **B** to a limited extent in the dynamics (e.g., Zwan and Wolf, 1976).

An important consequence of MHD, independent of the inequalities (8), is that a stagnation point flow around the magnetosphere requires zero normal component of **B** at the magnetopause. Consider a loop moving with the plasma flow on the surface of the obstacle (Figure 2a): the area of the loop is zero when passing through the stagnation point; hence the magnetic flux through it is zero and (by virtue of the MHD approximation) remains zero, implying $B_n = 0$. A non-zero B_n implies that the flow has a stagnation line (Figure 2b) rather than a stagnation point; this line is to be identified with the X line or reconnection line or, more precisely, the dayside merging line (Vasyliunas, 1984) of the open magnetic field topology.

The draping of the magnetosheath magnetic field around the magnetosphere is a simple consequence of $B_n = 0$ and therefore of the stagnation point flow in the hydrodynamic model.

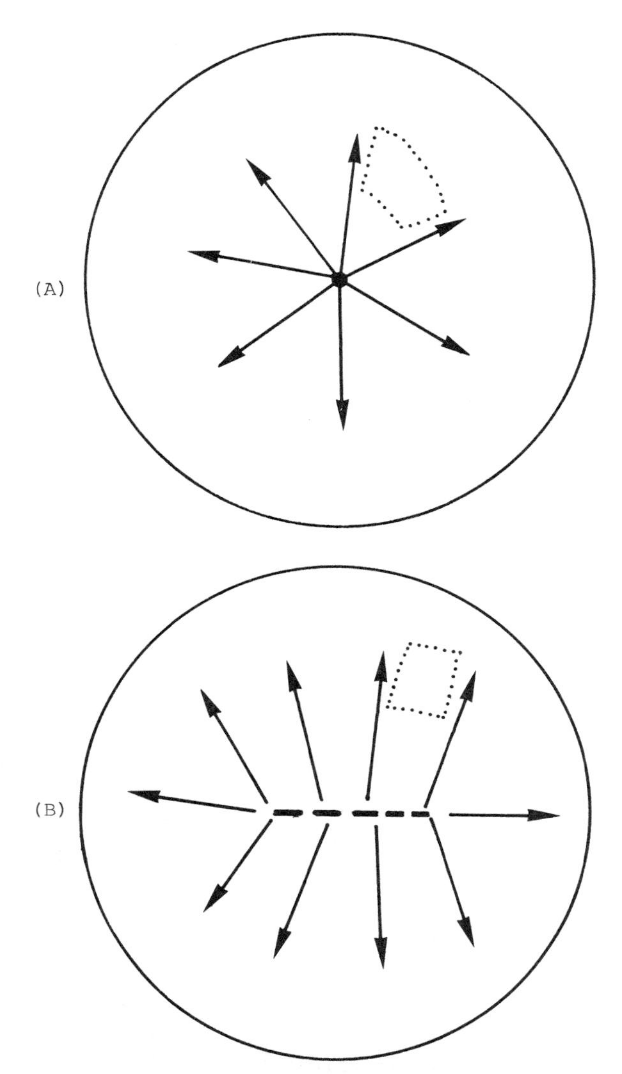

Fig. 2. Sketch of plasma flow lines on or just outside the magnetopause, viewed from the sun. A loop moving with the plasma is shown by dotted lines. (a) Stagnation point flow. (b) Stagnation line flow.

The agreement of the calculated and the observed fields implies that the actual normal component and the deviation of the flow from a stagnation point pattern are (in some suitable sense) small. The large-scale models of the magnetosheath are thus insensitive to the detailed boundary conditions at the magnetopause and, by the same token, provide no information on them.

Inner Magnetosphere and Ring Current

Within the inner magnetosphere, defined here as the region bounded roughly by the geocentric radial distance 7-8 R_E, the following inequalities hold,

$$B^2/8\pi \gg P \gg \rho V^2 \quad , \tag{10}$$

and the magnetic field is therefore strong and dominates over the flow. To first approximation, the magnetic field in this region may be taken as known; it is determined largely by the dipole field of the earth (for quantitative models, see e.g., Walker, 1979 and Olson et al., 1979). MHD then constrains the plasma flow to preserve the field lines of the known magnetic field. If the plasma bulk flow perpendicular to **B**, or the equivalent electric field according to (2), is known at one point along a magnetic field line, it can immediately be calculated along the entire field line merely from the geometry of **B**.

The main consequence of MHD in the inner magnetosphere is thus to allow the determination of the plasma flow pattern from the electric field distribution at, say, the base of all the field lines threading the region. Since this electric field is also imposed on the ionosphere, the finite electrical resistivity of the ionosphere assumes a crucial role. Adding the requirement of continuity between electric currents in the magnetosphere, with $\mathbf{J}_\perp$ fixed by the momentum equation (6), and in the ionosphere, one obtains the scheme of self-consistent electrodynamic coupling between the magnetosphere and the ionosphere illustrated in Figure 3 (from Vasyliunas, 1979) which has been the basis of quantitative models for large-scale plasma dynamics in the inner magnetosphere and ring current region (see, e.g., review by Wolf, 1983, and references therein).

The boundary conditions needed by these models are apparent from Figure 3: either the electric potential or the field-aligned currents on the high-latitude boundary (the driving field or driving current, respectively, of Figure 3), and the distribution of plasma sources from the solar wind and from the ionosphere.

Various extensions of the self-consistent magnetosphere-ionosphere coupling scheme are possible at least in principle; among them are inclusion of the effects of small-scale non-MHD parallel electric fields, modification of ionospheric conductivities by particle precipitation associated with field-aligned currents, dynamo effects of thermospheric neutral winds, ion-drag influences on neutral winds, and modification of the magnetic field by currents in the magnetosphere and ionosphere. Several of these extensions have been studied or are being implemented (e.g., Wolf, 1987, and references therein). Their inclusion in quantitative models is limited by computer size and power as well as knowledge of the relevant physical processes and parameters.

Magnetotail

The central region of the magnetotail, the plasma sheet, is characterized by the fact that $B^2/8\pi$ and P are of the same order of magnitude,

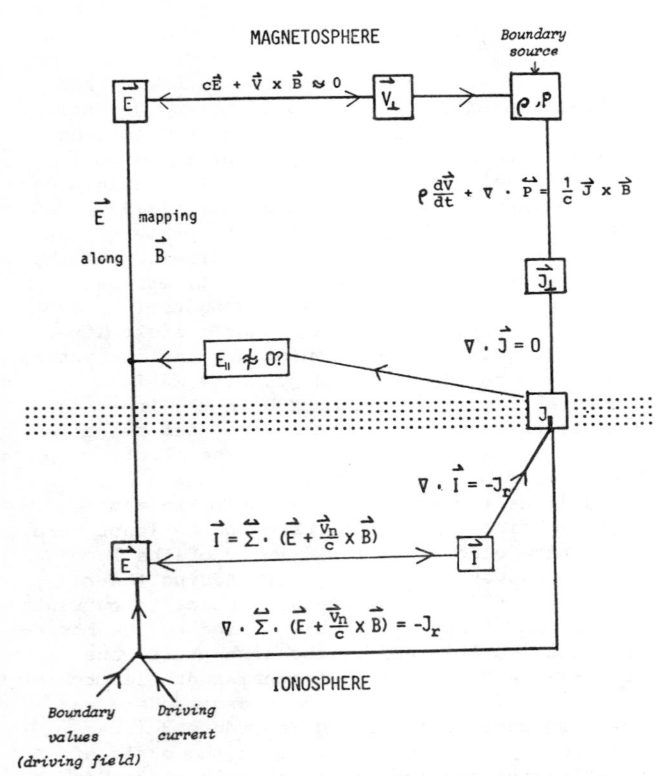

Fig. 3. Calculation scheme of self-consistent electrodynamic coupling between the magnetosphere and the ionosphere (from Vasyliunas, 1979). Quantities being calculated are enclosed in boxes and the equations relating them are indicated along the lines joining the boxes. Italics denote boundary conditions. The dotted band in the middle separates magnetosphere and ionosphere.

while usually, but not always, $\rho V^2 \gg P$. It is thus no longer possible, not even as a first approximation, to compute plasma flow ignoring **B**, as in the magnetosheath, or to fix the magnetic field more or less independently of the plasma, as in the inner magnetosphere. The magnetic field and the plasma are now of equal importance and must be determined together self-consistently from the full MHD equations (2) and (6). However, there are now geometrical approximations that can be invoked to simplify the equations: the length scale for variations along the magnetotail axis x is generally much longer than the length scale for variations along the north-south direction z, so that the scaling

$$\partial/\partial x \ll \partial/\partial z \tag{11}$$

may be consistently applied, and for many purposes the variations in the dawn-dusk direction y may be ignored and the problem treated as two dimensional.

Models of the magnetotail, in a sequence of increasing sophistication, have been developed both analytically, based on these geometrical simplifications, and by purely numerical solution of the MHD equations (see, e.g., Schindler and Birn, 1987, and references therein).

Needed as boundary conditions for models of the magnetotail are the electric field at its north and south lobe boundaries and the plasma content of magnetic flux tubes; these are in essence the same boundary conditions as for the inner magnetosphere, only applied at a boundary pushed much farther out. In addition, boundary conditions are needed in principle at the boundary of the magnetotail toward the inner magnetosphere, but it is still not firmly settled what form these should take, particularly as far as field-aligned electric currents (Birkeland currents) and field-aligned plasma flows are concerned.

Interfaces Between Regions

Within each of the regions of the ionosphere/magnetosphere/solar wind system described previously, quantitative models for large-scale phenomena have been developed on the basis of approximations appropriate to the particular region. These approximations, however, necessarily fail at the interface between two regions. It is thus not surprising that our understanding of the interfaces is far less developed than our understanding of the individual regions. The following will be confined largely to listing and describing unsolved problems for each interface.

Inner Magnetosphere/Magnetotail

For this interface, the problems are primarily mathematical: the matching of two geometrically different approximation regimes. The magnetotail has a quasi-two-dimensional plasma geometry; whereas in the inner magnetosphere, the magnetic field tends toward axial symmetry. A unified treatment of both regions or a successful way of mating both geometries would automatically solve the previously mentioned problem of specifying boundary conditions, particularly field-aligned currents, at the earthward face of the magnetotail.

Magnetosheath/Magnetosphere (Dayside)

Between the magnetosheath and the dayside magnetosphere lies a relatively narrow interface region containing the magnetopause and the various magnetospheric boundary layers. The plasma and magnetic field structure of this interface has been reasonably well surveyed observationally but is far from being understood theoretically, let alone reproduced by quantitative models. Here the unsolved problems are primarily physical rather than mathematical and include the global configuration of the magnetic field (in particular the location and possible motion of singular lines mentioned in connection with Figure 2 and the distribution of **B** normal to the magnetopause), the mechanism of plasma entry from the magnetosheath into the magnetosphere, and the nature and role of both large- and small-scale instabilities and fluctuations.

Magnetosheath/Magnetotail

The boundary region between the magnetosheath and the magnetotail is generally thought to be intimately associated with the so-called solar wind/magnetosphere dynamo. Successful modeling of this region would be tantamount to explaining and predicting the configuration of polar cap electric fields and associated Birkeland currents, the mechanism and rate of energy input from the solar wind into the magnetosphere, and the formation of the plasma sheet; it would thus provide the boundary conditions needed by models of the inner magnetosphere and the magnetotail (in fundamental terms rather than, as is now the case, purely empirically). We are, however, very far from being able to develop such models. Lacking is an adequate physical and mathematical understanding of how to handle MHD in highly inhomogeneous systems involving significant small-scale effects (such as magnetic field line merging and particle acceleration near magnetic X lines).

Conclusion

The ionosphere/magnetosphere/solar wind system contains regions of vastly different parameter regimes, but the principles of MHD, judiciously used, have shown themselves sufficiently flexible to serve as a unified underlying basis for large-scale models of the various regions. Major remaining problems occur primarily at the interfaces between the various regions and their regimes of approximation; they involve some of the fundamental unsolved aspects of the interaction between the solar wind and the magnetosphere.

References

Alksne, A. Y., The steady-state magnetic field in the transition region between the magnetosphere and the bow shock, Planet. Space Sci., 15, 239, 1967.

Olson, W. P., K. A. Pfitzer, and G. J. Mroz, Modeling the magnetospheric magnetic field, in Quantitative Modeling of Magnetospheric Processes, Geophys. Monogr. Ser., vol. 21, edited by W. P. Olson, pp. 77-85, AGU, Washington, D.C., 1979.

Rossi, B., and S. Olbert, Introduction to the Physics of Space, pp. 372-389, McGraw-Hill, New York, 1970.

Schindler, K., and J. Birn, Magnetotail theory, Space Sci. Rev., 44, 307, 1987.

Siscoe, G. L., Solar system magnetohydrodynamics, in Solar-Terrestrial Physics, edited by R. L. Carovillano and J. M. Forbes, pp. 11-100, D. Reidel Publ. Co., Dordrecht, Holland, 1983.

Spreiter, J. R., A. L. Summers, and A. Y. Alksne, Hydromagnetic flow around the magnetosphere, Planet. Space Sci., 14, 223, 1966.

Spreiter, J. R., A. Y. Alksne, and A. L. Summers, External aerodynamics of the magnetosphere, in Physics of the Magnetosphere, edited by R. L. Carovillano, J. F. McClay, and H. R. Radoski, pp. 301-375, D. Reidel Publ. Co., Dordrecht, Holland, 1968.

Stern, D. P., The motion of magnetic field lines, Space Sci. Rev., 6, 147, 1966.

Vasyliunas, V. M., Nonuniqueness of magnetic field line motion, J. Geophys. Res., 77, 6271, 1972.

Vasyliunas, V. M., Theoretical models of magnetic field line merging, 1, Rev. Geophys. Space Phys., 13, 303, 1975.

Vasyliunas, V. M., Interaction between the magnetospheric boundary layers and the ionosphere, in Proceedings of Magnetospheric Boundary Layers Conference, Eur. Space Agency Spec. Publ., ESA SP-148, edited by B. Battrick, pp. 387-393, Noordwijk, The Netherlands, 1979.

Vasyliunas, V. M., Plasma sheet dynamics: Effects on, and feedback from, the polar ionosphere, in Exploration of the Polar Upper Atmosphere, edited by C. S. Deehr and J. A. Holtet, pp. 229-244, D. Reidel Publ. Co., Dordrecht, Holland, 1980.

Vasyliunas, V. M., Steady state aspects of magnetic field line merging, in Magnetic Reconnection in Space and Laboratory Plasmas, Geophys. Monogr. Ser., vol. 30, edited by E. W. Hones, Jr., pp. 25-31, AGU, Washington, D.C., 1984.

Walker, R. J., Quantitative modeling of planetary magnetospheric magnetic fields, in Quantitative Modeling of Magnetospheric Processes, Geophys. Monogr. Ser., vol. 21, edited by W. P. Olson, pp. 9-34, AGU, Washington, D.C., 1979.

Wolf, R. A., The quasi-static (slow-flow) region of the magnetosphere, in Solar-Terrestrial Physics, edited by R. L. Carovillano and J. M. Forbes, pp. 303-368, D. Reidel Publ. Co., Dordrecht, Holland, 1983.

Wolf, R. A., Prospects for a unified theoretical model of large-scale dynamics in the magnetosphere-ionosphere system, this volume, 1987.

Zwan, B. J., and R. A. Wolf, Depletion of solar wind plasma near a planetary boundary, J. Geophys. Res., 81, 1636, 1976.

FIELD-ALIGNED CURRENTS AND MAGNETOSPHERIC CONVECTION - A COMPARISON BETWEEN MHD SIMULATIONS AND OBSERVATIONS

Raymond J. Walker

Institute of Geophysics and Planetary Physics
University of California, Los Angeles, California 90024

Tatsuki Ogino

Research Institute of Atmospherics, Nagoya University, Toyokawa, Japan

Abstract. Recently we have simulated the interaction between the solar wind and the earth's magnetosphere by using a time-dependent three-dimensional magnetohydrodynamic (MHD) model. We have used this model to investigate the magnetospheric configuration as a function of the interplanetary magnetic field (IMF) direction when it was in the y-z plane in geocentric solar magnetospheric coordinates. The model results show four types of convection cells. They are the large global convection cells, the tail lobe cells, the high-latitude polar cap cells, and the low-latitude cells. There are also four main field-aligned current systems: two systems at subpolar cap latitudes, the region 1 and 2 currents, and two current systems in the polar cap region, the tail region 1 currents and the polar cap currents. The polar cap currents evolve into the polar cusp currents when we rotate the IMF from northward to southward. The polar cap convection pattern and the distribution of field-aligned currents are determined by the location of magnetopause reconnection, which is determined by the IMF direction. Recent polar cap observations are consistent with the model results. For northward IMF, we have made direct comparisons between MAGSAT observations and the MHD simulation and found good agreement between the model current distributions and the observations. We also found good agreement between the current patterns from the simulation and polar cap currents observed on S3-2 for southward IMF. Finally we compared the polar cap convection patterns observed on Atmosphere Explorer C and Dynamics Explorer 2 for both northward and southward IMF with the model and again found reasonable agreement.

Introduction

It is difficult to understand a magnetospheric system from the limited measurements provided by a spacecraft. This is because the observations are a time series limited to the spacecraft trajectory. In order to understand a magnetospheric system, the experimenter is challenged to interpret the limited single point measurements in terms of a large-scale and highly dynamic system. During the past few years a new technique, magnetohydrodynamic (MHD) simulation, has been developed to help solve this problem and to provide a self-consistent picture of the solar wind-magnetosphere-ionosphere system.

In an MHD simulation, the plasma is modeled as a conducting fluid and the time-dependent equations for fluid flow plus Maxwell's equations are solved numerically. The resulting models give each parameter at every point in the model magnetosphere as a function of time.

Over the past several years, several of these three-dimensional global MHD simulation codes have been developed to model the interaction of the solar wind with the magnetosphere (Leboeuf et al., 1981; Wu et al., 1981; Brecht et al., 1981, 1982; Ogino and Walker, 1984; Ogino et al., 1985, 1986; Ogino, 1986; Wu, 1983, 1984). These reproduced many of the observed magnetospheric features such as the plasma sheet and the cross tail and the magnetopause current systems (Wu et al., 1981). The models have been used to study the effects of the east-west component of the interplanetary magnetic field (IMF) on the magnetospheric configuration (Brecht et al., 1981; Ogino et al., 1985, 1986) and substorm-like events associated with magnetic reconnection in the magnetotail were modeled when the IMF had a southward component (Brecht et al., 1982).

In this investigation, we have tried to extend

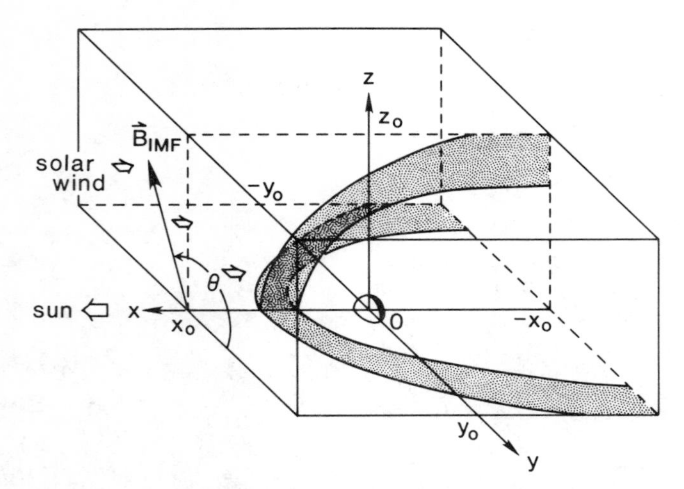

Fig. 1. Coordinate system used in the simulation. The angle θ gives the direction of the interplanetary magnetic field (from Ogino et al., 1985).

these simulation studies by linking the model results to observations. As a first step in this process, we have asked how well the models reproduce the observations in the magnetosphere. We have done this for two reasons. First we need to calibrate the models. We need to learn the limitations of the calculations and to learn what new physics must be added to the models to properly model the magnetosphere. Second, we would like to start using the model as a tool to interpret the spacecraft observations. We hope eventually to use the simulations to aid in the planning for future missions as well as in the interpretation of data from current satellites.

In this study, we will compare the field-aligned currents generated in the model magnetosphere and the model magnetospheric convection system with observations from low-altitude polar orbiting satellites. We use data from the low-altitude polar satellites because these spacecraft move rapidly in latitude and provide the closest thing available to a photograph of the magnetospheric configuration.

The Model

Our simulation model has been described in detail elsewhere (Ogino and Walker, 1984; Ogino et al., 1985, 1986; Ogino, 1986) so we will describe only the main features here. We have solved the MHD and Maxwell's equations as an initial value problem by using the two step Lax-Wendroff scheme (see Potter, 1970). The normalized resistive MHD equations which we solved are written as follows:

$$\frac{\partial \rho}{\partial t} = -\nabla \cdot (\mathbf{v}\rho) + D\nabla^2 \rho \qquad (1a)$$

$$\frac{\partial \mathbf{v}}{\partial t} = -(\mathbf{v} \cdot \nabla)\mathbf{v} - \frac{1}{\rho}\nabla p + \frac{1}{\rho}\mathbf{J} \times \mathbf{B} + \mathbf{g} + \frac{1}{\rho}\boldsymbol{\phi} \qquad (1b)$$

$$\frac{\partial p}{\partial t} = -(\mathbf{v} \cdot \nabla)p - \gamma p \nabla \cdot \mathbf{v} + D_p \nabla^2 p \qquad (1c)$$

$$\frac{\partial \mathbf{B}}{\partial t} = \nabla \times (\mathbf{v} \times \mathbf{B}) + \eta \nabla^2 \mathbf{B} \qquad (1d)$$

$$\mathbf{J} = \nabla \times (\mathbf{B} - \mathbf{B}_d) \ , \qquad (1e)$$

where ρ is the plasma density, $\mathbf{v}$ the flow velocity, p the plasma pressure, $\mathbf{B}$ the magnetic field, $\mathbf{B}_d$ the internal magnetic field of the earth, $\mathbf{J}$ the current density, $\mathbf{g}$ the gravity force, $\boldsymbol{\phi} \equiv \mu\nabla^2 v$ the viscosity, $\gamma = 5/3$ the ratio of the specific heats, $\eta = \eta_o(T/T_o)^{-3/2}$ the resistivity, and T/T_o is the temperature normalized by its value in the ionosphere. The units, for distance, velocity, and time, are the earth's radius, $R_E = 6.37 \times 10^6$ m, the Alfvén speed at one earth radius on the equator, $v_A = 6.80 \times 10^6$ m s^{-1}, and the Alfvén transit time, $t_A = 1\ R_E/v_A = 0.937$ s. The numerical values for η, μ, D, and D_p are $\eta_o = 0.01$ and $\mu/\rho_{sw} = D = D_p = 0.005$, where ρ_{sw} is the density of the solar wind. The magnetic Reynold number, which is the magnetic diffusion time divided by the Alfvén transit time, is $S = \tau_\eta/\tau_A = 100$-$1000$.

The diffusion and viscosity terms in (1a)-(1c) were added to suppress MHD fluctuations which come from unbalanced forces at the start of the calculation. These numerical oscillations tend to occur in front of the bow shock and have a scale which is the same as the grid spacing. This does not affect the magnetospheric configuration much because the forces between neighboring meshes tend to cancel each other. The very small viscous and diffusion coefficients used in the calculations are sufficient to suppress these oscillations. Ogino (1986) investigated the changes in the magnetospheric configuration between the case presented here and the case with $\mu/\rho_{sw} = D = D_p = 0$. He found that the maximum pressure at the bow shock changed by less than 10% while the changes elsewhere in the magnetosphere were negligible.

The solar magnetospheric coordinate system used in the calculation has been sketched in Figure 1. A uniform solar wind $n_{sw} = 5\ cm^{-3}$, $v_{sw} = 300\ km\ s^{-1}$, and $T_{sw} = 2 \times 10^5$ K flows into a simulation box of dimension $-x_o \le x \le x_o$, $-y_o \le y \le y_o$, and $0 \le z \le z_o$ at $x = x_o$, where $x_o = y_o = z_o = 24.5\ R_E$. The uniform IMF is initially given by $\mathbf{B}_{IMF} = (0, B_{IMF}\cos\theta, B_{IMF}\sin\theta)$ in the whole simulation region where the angle θ is determined by rotating in a counterclockwise direction from the y-axis so that θ = 90° is northward and θ = 270° is southward. All of these model parameters (Ψ) are held constant (Ψ = constant) at $x = x_o$. We used forms of free boundary conditions through which plasmas and

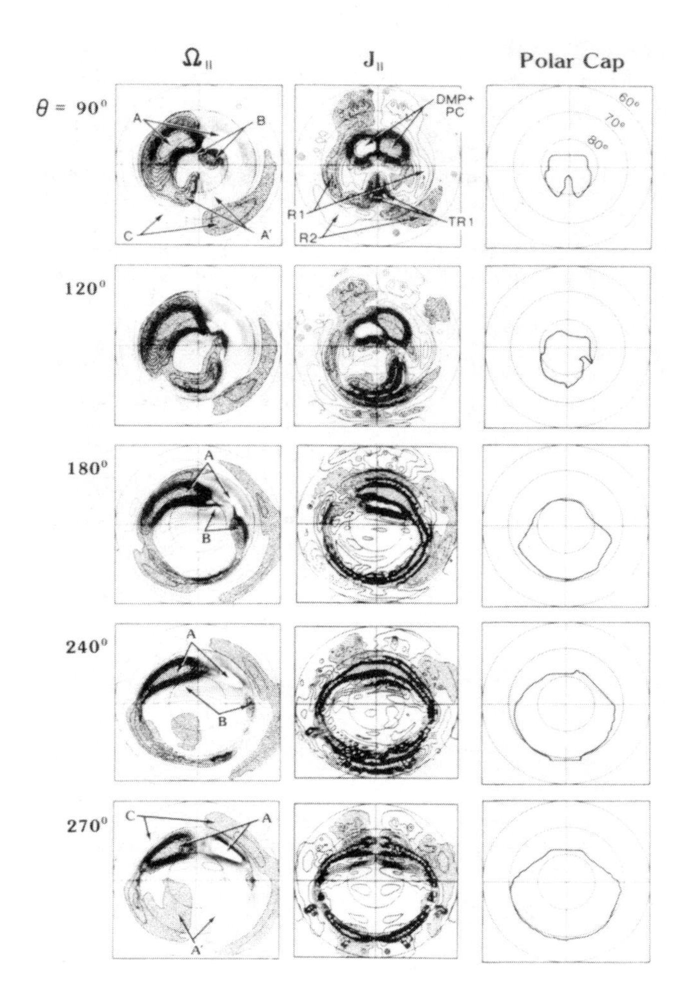

Fig. 2. North polar projections of the field-aligned vorticity, $\Omega_{\parallel}$, the field-aligned current density, $J_{\parallel}$, and a plot of the polar cap for five IMF orientations. Positive parallel vorticity is shaded as is negative parallel current density (upward currents). The model polar cap is delimited by the heavy line.

waves can freely enter or exit the system at $x = -x_o$, $y = \pm y_o$, and $z = z_o$. At $x = -x_o$, we set $\partial\Psi/\partial x = 0$, while at $y = \pm y_o$ and $z = z_o$, the free boundary is at 45° to the x axis. A form of the mirror boundary conditions was used at z = 0 such that quantities map from the northern dawn (dusk) side to the southern dusk (dawn) side according to

$$\frac{\partial\rho}{\partial z} = \frac{\partial p}{\partial z} = \frac{\partial v_x}{\partial z} = \frac{\partial B_z}{\partial z} = \frac{\partial B_y}{\partial z} = 0$$

$$v_y(y,-z) = -v_y(-y,z) \; v_z(y,-z) \qquad (2)$$

$$= -v_z(-y,z) \; B_x(y,-z) = -B_x(-y,z).$$

The ionospheric boundary condition imposed near the earth was determined by requiring a static equilibrium (Ogino, 1986). We kept all of the parameters ($p, \mathbf{v}, \rho, \mathbf{B}$) fixed for $\xi \equiv (x^2 + y^2 + z^2)^{1/2} < \xi_a$ where $\xi_a = 3.5\ R_E$. All perturbations were damped out by using a smoothing function near the ionosphere ($\xi < 5.5\ R_E$). Quantities outside of the smoothing region (Ψ_{ex}) are related to internal quantities (Ψ_{in}) by

$$\Psi = f\Psi_{ex} + (1 - f)\Psi_{in} ,$$

where $f \equiv a_o h/(a_o h + 1)$ with $a_o = 100$, and $h = (\xi/\xi_a)^2 - 1$ for $\xi \geq \xi_a$ and $h = 0$ for $\xi < \xi_a$. The resulting ionosphere is passive to the extent that all perturbations go to zero. This includes field-aligned currents which therefore close in the magnetosphere not in the ionosphere.

Within the ionosphere, all of the parameters are maintained at their initial values, which are:

Density

$$\rho_o = \xi^{-3} \qquad \rho_o \geq 0.2\ \rho_{sw}$$

$$\rho_o = 0.2\ \rho_{sw} \qquad \rho_o < 0.2\ \rho_{sw}$$

Plasma pressure

$$p_o = p_{oo}\xi^{-2} \qquad p_o \geq p_{sw}$$

$$p_o = p_{sw} \qquad p_o < p_{sw}$$

where $p_{oo} = 5.4 \times 10^{-7}$.

We used a dipole for the magnetic field near the earth [$B_d = 1/\xi^5(-3xz, -3yz, x^2 + y^2 - 2z^2)$].

The MHD equations were solved on a 48 x 48 x 24 point grid not including the boundary grid points. The mesh size was $\Delta x = \Delta y = \Delta z = 1\ R_E$ and the time step was $\Delta t = 3.8$ s in order to assure that the numerical stability criterion, $v_g^{max}\,\Delta t/\Delta x < 1$, where v_g^{max} is the maximum group velocity in the calculation domain, was met.

Simulation Results

The code was run until a quasi-steady state magnetospheric configuration resulted. This required 512 time steps which corresponded to about 32 min in real time. The model is intrinsically time dependent, however, we have analyzed the results at only one instant in time. All of the results used in this study are a snapshot taken at the end of the run.

We evaluate each of the model parameters at each grid point at each time step. This results in a large amount of data. Figure 2 shows one way of representing three-dimensional data in a two-dimensional format. In the two left-hand columns the parallel vorticity ($\Omega_{\parallel} = \mathbf{B} \cdot \nabla \times \mathbf{v}/|\mathbf{B}|$) and the field-aligned current ($J_{\parallel}$) have been mapped along magnetic field lines into the northern polar cap. The mapped values were determined by calculating $\int(f/|\mathbf{B}|)d\ell/\int(1/|\mathbf{B}|)d\ell$ along the magnetic field lines, where f stands for each of the parameters and $|\mathbf{B}|$ is the field magnitude. Only that part of open field lines within the magnetosphere was included in the integral. The results are plotted on a polar plot with noon at the top. The circular rings indicate 10° steps in latitude. The results in each row are for a constant IMF direction. From top to bottom, the IMF orientation changes from 90° (northward) through 180° (dawnward) to 270° (southward).

We have identified four major convection systems from the parallel vorticity plots. In this column parallel vorticity parallel to **B** has been shaded. The four convection systems are the large viscous cells (A), tail lobe cells (A'), high-latitude merging cells (B), and low-latitude cells (C). The largest system is the global system (A). The flow corresponding to this system is clockwise at dusk (i.e., anti-sunward over the polar cap and sunward at lower latitudes) while it is counterclockwise in the cell at dawn. This system has a large viscous contribution since it is found even when $|\mathbf{B}_{IMF}| = 0$ (Ogino, 1986). However, there is also a contribution from reconnection as can be seen from the changes which occur when the IMF orientation changes.

The nightside extension of the A system is called the A' system. For northward IMF the A' system is associated with a localized region of the plasma sheet which extends into the polar cap (Ogino and Walker, 1984). This can be seen in the right-hand column where the open field line region is poleward of the heavy dark line. Here the extension of the plasma sheet into the polar cap can be seen near midnight. As the IMF is rotated toward dawn this feature moves toward dawn (Ogino et al., 1985). For southward IMF there also is convection in the tail. As we will see later, in both cases the A' convection is related to reconnection but the reconnection is in different locations. The low-latitude C cells are found in the inner magnetosphere and are primarily a viscous effect.

The final convection system, the polar cap or high-latitude system (B), is most readily seen when $B_{y\ IMF} \neq 0$. The flow in this system has the sense opposite to that of the A cells. On the dusk side the flow is tailward at lower latitudes and sunward at the higher latitudes. When the IMF has a component toward dawn this negative parallel vorticity dominates the polar cap. It is interesting to note that this counterclockwise flow dominates for both northward and southward IMF provided $B_{y\ IMF}$ is toward dawn (Ogino et al., 1986).

There are also four types of field-aligned currents: region 1 currents (R1), region 2 currents (R2), tail region 1 currents (TR1), and high-latitude currents composed of the dayside magnetopause currents and polar cap currents (PC). In Figure 2, regions with current directed away from the ionosphere have been shaded. It is easiest to recognize the current systems when the IMF points either entirely northward (θ = 90°) or southward (θ = 270°). At high latitudes, on closed field lines, the currents have the same sense, as the region 1 currents observed by Iijima and Potemra (1976). These currents are away from the ionosphere on the dusk side and toward the ionosphere on the dawn side. At lower latitudes, the currents reverse sign and are toward the earth at dusk and away at dawn. These currents have the same sign as the observed region 2 currents. The dayside region 1 currents are closely correlated with the A convection cells and are driven by that flow (Ogino and Walker, 1984; Ogino et al., 1985). The region 2 currents, while associated with the C convection cells, are mainly generated by pressure gradients in the inner magnetosphere (Ogino et al., 1986).

We call the currents with the region 1 sense on the nightside the tail region 1 currents. For northward IMF, region 1 sense currents are associated with the localized thickening of the plasma sheet. The tail region 1 currents move toward dawn along with the plasma sheet thickening when $B_{y\ IMF} < 0$ (θ = 120°). Region 1 sense currents are also found in the tail for southward IMF (θ = 270°). When $B_{z\ IMF} < 0$ and $B_{y\ IMF} < 0$ (θ = 240°), this current undergoes a major change and becomes very structured and almost filamentary. An additional current system appears in the tail at lower latitudes for $B_{y\ IMF} < 0$. This current has a region 2 sense and can be seen in the θ = 240° panel and in Figure 1 of Walker et al. (1987). In subsequent runs this current was found for purely southward IMF as well but at slightly later times.

For northward IMF, the polar cap is dominated by currents with the region 2 sense (i.e., toward the earth at dusk and away at dawn). These currents can be seen in Figure 1 just poleward of the region 1 currents. As we rotate the IMF toward dawn, the earthward current which was on the dusk side for θ = 90° moves toward the center of the polar cap while the corresponding outward current moves toward dusk and toward the equator. By θ = 180° (dawnward IMF) this upward current has joined with the dusk side region 1 current to form an arc of upward current at auroral latitudes. Some of this region 1 current is on open field lines and some is on closed field lines. The largest current density in the arc of upward current is associated with the boundary between open and closed field lines. The polar cap continues to be dominated by earthward currents even for southward IMF as long as $B_{y\ IMF} \neq 0$ (θ = 240°). For purely southward IMF (θ = 270°) the currents in the polar cap region again have a region 2 sense just as in the purely northward IMF (θ = 90°) case. However, the currents for

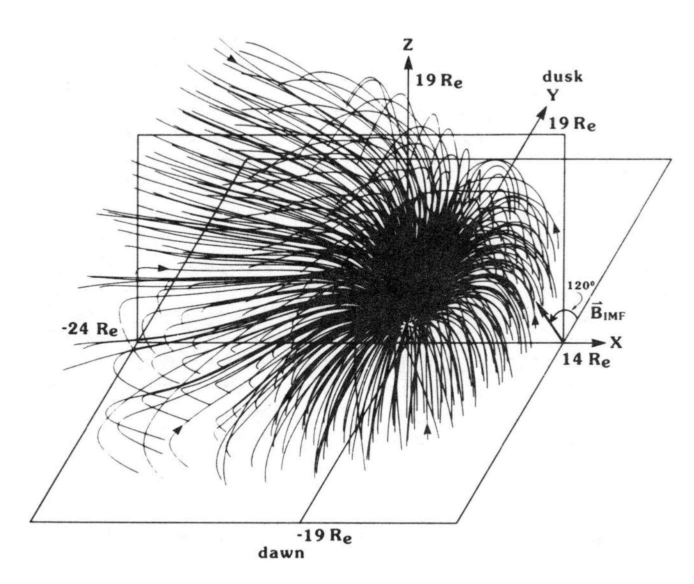

Fig. 3. Three-dimensional plot of the magnetic field line configuration. The IMF had both northward and dawnward components (θ = 120°) (from Ogino et al., 1985).

θ = 270° are equatorward of those for θ = 90°. Note that the region of open field lines also has expanded equatorward.

For Figure 3 we have plotted magnetic field lines for a case with $B_{z\ IMF} > 0$ and $B_{y\ IMF} < 0$ (θ = 120°). Note the field lines on the dawn side which have a larger normal component than their neighbors. These are plasma sheet field lines in the local thickening of the plasma sheet. These field lines occur beneath a region in which localized reconnection is occurring on the nightside magnetopause (see Figure 5). The field lines with a large normal component result when the plasma sheet expands into the region vacated by the reconnected lobe field (Ogino and Walker, 1984). Field lines when the IMF direction was between θ = 180° and 255° are plotted in Figure 4. For θ = 180° reconnection occurs on the dawn magnetopause. (This is difficult to see in this projection.) When the IMF has a southward component the reconnection moves onto the dayside magnetopause and into the magnetotail. In the panels at θ = 210°, 240°, and 255° the formation of neutral lines in the tail can be clearly seen. We have used dashed lines to indicate some of the neutral lines at θ = 240° and

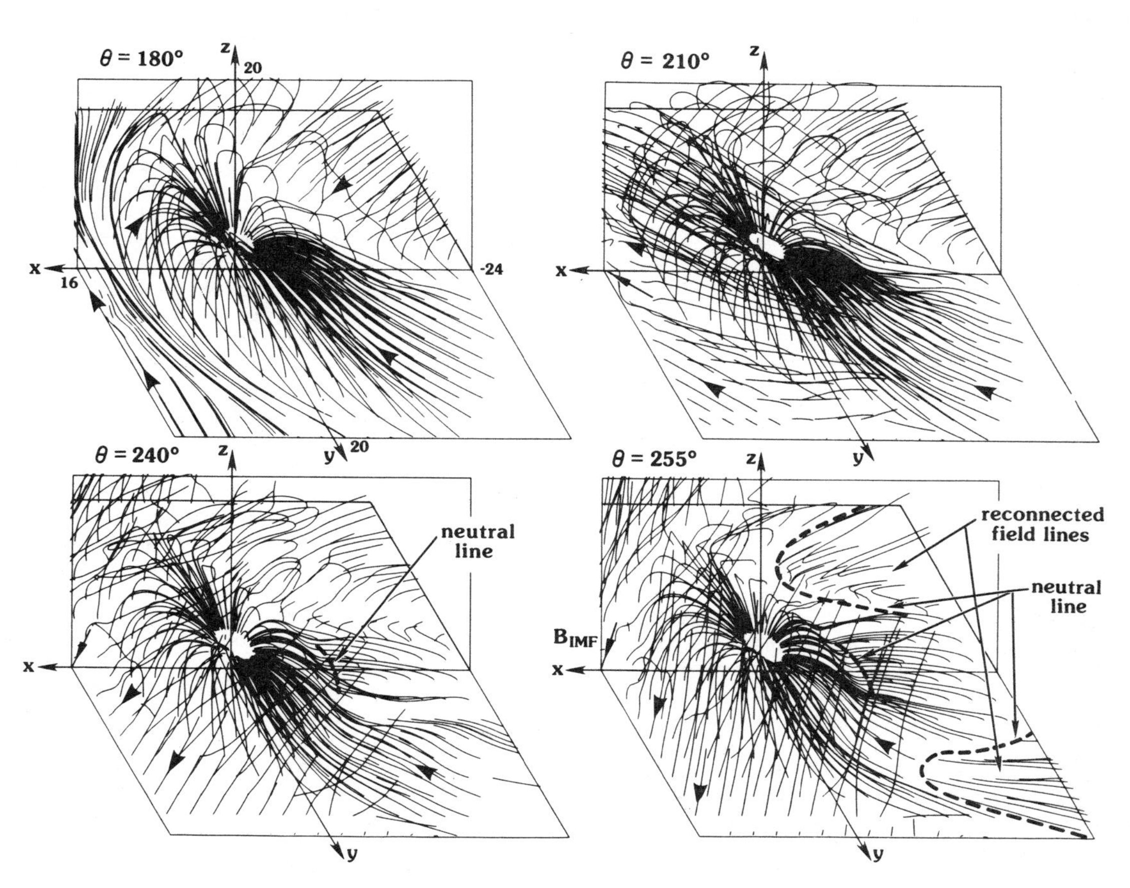

Fig. 4. Three-dimensional plots of the magnetic field configuration for 180° ≤ θ ≤ 255°. The dashed lines show magnetic neutral lines (from Ogino et al., 1986).

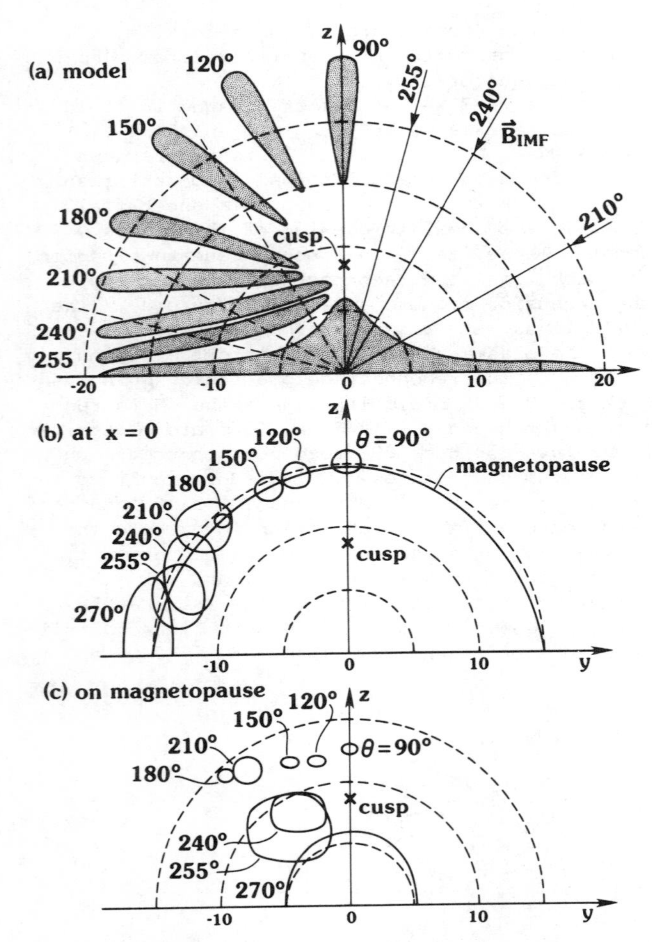

Fig. 5. Areas of dayside magnetic reconnection in a view from the sun for IMF directions between θ = 90° and 270°. Figure 5a shows the predicted anti-parallel merging regions. Figures 5b and 5c are both from the simulation. Figure 5b shows anti-parallel magnetic field regions for $|\mathbf{B}| < 0.3\ B_{IMF}$ in the plane x = 0, while Figure 5c shows the anti-parallel magnetic field regions for $|\mathbf{B}| < 0.3\ B_{IMF}$ on the dayside magnetopause (from Ogino et al., 1986).

θ = 255°. For $B_{y\ IMF} \neq 0$, the reconnection in the tail is localized and patchy. Another view of the location of the tail neutral lines showing the region where $|\mathbf{B}| \sim 0$ can be seen in Figure 6 of Ogino et al. (1986). For θ = 270° a single neutral line forms in the tail.

We have plotted predicted areas of magnetopause reconnection for IMF directions between θ = 90° and 270° in Figure 5a. The potential magnetic merging sites were modeled by using a dipole magnetic field for the magnetospheric field and a uniform IMF of 5 nT. The shaded regions are anti-parallel field areas for which $[(B_y^2 + B_x^2)^{1/2}] < 0.1\ B_{IMF}$. If anti-parallel merging is important in the simulation, we would expect it to occur on the magnetopause in that part of the shaded regions nearest the subsolar point. Thus, we would expect reconnection on the magnetopause tailward of the polar cusp for θ = 90° and for the location of the reconnection to move along the tail magnetopause until θ = 180°. For θ > 180° (i.e., southward IMF), we would expect the reconnection region to move across the dayside magnetopause toward the subsolar point. Anti-parallel magnetic field regions from the simulations for which $|\mathbf{B}| < 0.3\ B_{IMF}$ in the plane x = 0 are shown in Figure 5b while Figure 5c shows the anti-parallel magnetic field regions for $|\mathbf{B}| < 0.3\ B_{IMF}$ on the dayside magnetopause. The reconnection region from the simulation starts tailward of the polar cusp for θ = 90°. It moves dawnward as θ increases to 180° and approaches the subsolar point as θ increases from 210° to 270°. Thus, the regions of reconnection on the magnetopause from the simulation are consistent with those expected from anti-parallel merging.

As noted above, when the magnetopause reconnection moves from the tail magnetopause to the dayside magnetopause for θ > 180°, it also moves into the tail and the multiple neutral lines in Figure 4 occur (see also Walker et al., 1987).

The Physical Model

There are two types of field-aligned current generation mechanisms that are thought to be dominant in the magnetosphere. One is associated with field-aligned vorticity and the other with pressure gradients (see Sato, 1982, and Vasyliunas, 1984, for a more detailed discussion). The former can be further subdivided into two subgroups: cases in which the currents are caused by magnetospheric convection on closed field lines and cases in which the currents are generated by the twisting of reconnected open field lines. The region 1 currents in the simulation mainly come from the field-aligned vorticity of the A cells. The model polar cap currents on the other hand are generated by the twisting of reconnected field lines. The region 2 currents are primarily generated by pressure gradients in the inner magnetosphere since the low-latitude convection is weak (Ogino et al., 1985). Pressure gradients in the plasma sheet are also responsible for the tail region 2 currents (Ogino et al., 1986; Walker et al., 1987). The tail region 1 currents are found at the interface between open and closed field lines for both northward and southward IMF. For northward IMF, these currents are generated by the convection established by reconnection on the tail magnetopause (Ogino and Walker, 1984), while for southward IMF, they are generated by reconnection within the tail (Ogino et al., 1986; Walker et al., 1987).

The polar cap convection cells are found in the open field line region and are associated with high-latitude nightside or dayside recon-

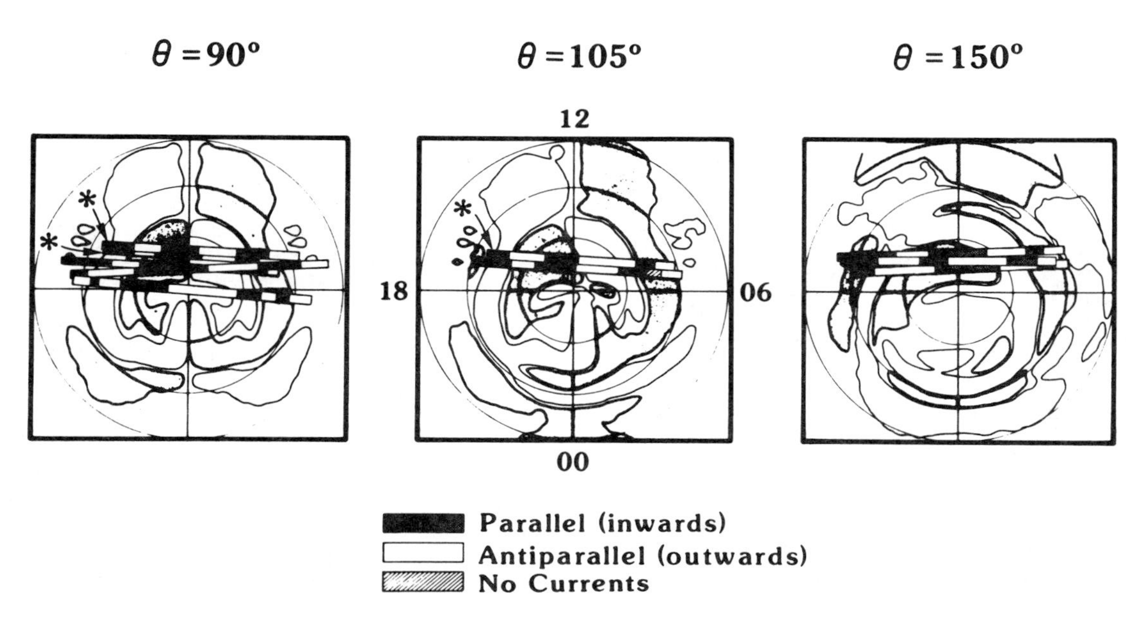

Fig. 6. MAGSAT trajectories superimposed on the simulated current distribution for three IMF orientations. We have used shading to indicate the direction of the simulated currents and the directions of the currents inferred from magnetic field observations. The shaded regions contain earthward currents. The circles indicate 80°, 70°, and 60° northern latitude.

nection. For $B_{y\ IMF} < 0$ field lines which reconnect on the dawn magnetopause are convected across the polar cap. The exact pattern depends on the location of reconnection.

Similarly, the polar cap currents evolve from currents which resemble the northward B_z currents (Iijima et al., 1984) for θ = 90° to currents which resemble the polar cusp currents for θ = 270° (Iijima and Potemra, 1976, 1978). The exact pattern of the currents changes as the location of reconnection changes on the magnetopause.

Comparison with Observations

The field-aligned current patterns projected onto the northern polar cap have been reproduced in Figure 6. Note that in this plot the region with earthward currents has been shaded. Superimposed on the model results are the trajectories of the MAGSAT satellite. Regions of earthward currents from the observations of Iijima et al. (1984) also have been shaded. The observations are from intervals when the B_y and B_z components of the IMF were within ±7.5° of the model value. All of the observations are from the southern hemisphere while the model calculations were carried out in the northern hemisphere. Therefore, we used the boundary conditions in (2) to map the observations to the northern hemisphere. The IMF was northward for all of the cases plotted, and we have included every published case for which the IMF remained nearly constant. Since the model does not have a $B_{x\ IMF}$ dependence, orbits for which the observed $B_x < 2$ nT have been indicated with an asterisk.

There is qualitative agreement between the simulation and the observations in 9 of 10 cases. In both the simulation and observations we find region 2 sense currents over the polar cap when the IMF is northward, and in all but one case the dominant current is inward when $B_{y\ IMF} < 0$. In general, the polar cap current region for the model is larger than is observed and the observed current region is shifted slightly dawnward. Two factors which may account for some of the differences are the dipole tilt and the $B_{x\ IMF}$ component. Both can cause north-south asymmetries in the currents which are not included in the simulations. Also, the simulation has only been run for a single value of the solar wind dynamic pressure.

The published MAGSAT orbits only provide information about the current structure along a dawn-dusk trajectory and for northward IMF. There is also considerable structure in the calculated current distribution in the direction

Field Aligned Currents: Model and Observations

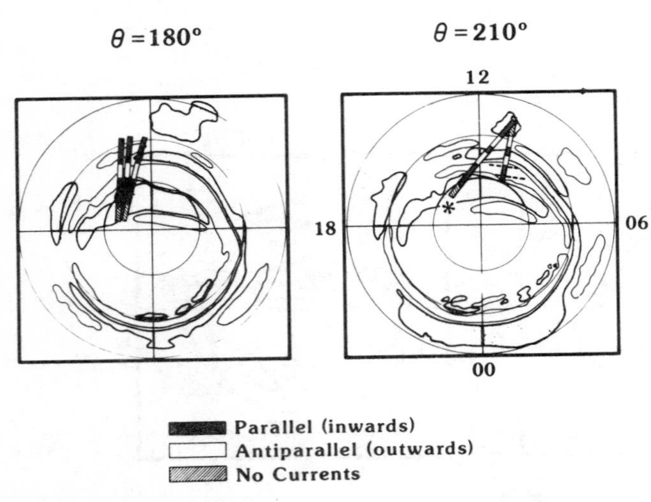

Fig. 7. S3-2 trajectories superimposed on the simulated current distributions for two IMF orientations. The format is the same as in Figure 6. Again, the shaded regions contain earthward currents. The crosshatching indicates regions where no evidence of field-aligned currents could be found in the magnetic field observations.

parallel to the noon-midnight meridian. Some orbits from S3-2, which have a large component of the motion parallel to the noon-midnight meridian, have been superimposed on the model current pattern in Figure 7. We have indicated current directions deduced from S3-2 magnetic field observations by Doyle et al. (1981) by shading. The IMF had a southward component for all of the published S3-2 orbits. We have reproduced all of the cases for which IMF data were available. For three of the orbits, the southward component was very small so we have plotted them on the θ = 180° panel. The model polar cap currents agree with those observed. The observations suggest that the dawnside earthward (region 1) current (see between 70° and 75°) may extend to later local times than in the model. The observed polar cap currents are displaced slightly equatorward of the model currents. At θ = 210°, there is again qualitative agreement. All of the currents, including the polar cap currents, seem displaced equatorward. To emphasize this, we have indicated the model polar cap boundary (poleward dashed line) and the trapping boundary for 100-keV protons (equatorward dashed line). All of these orbits occurred when there was a large negative dipole tilt and all but one had a large $B_{x\ IMF}$ component. Both the tilt and $B_{x\ IMF}$ can change the latitude of the currents.

Another magnetospheric diagnostic with which we have compared the simulations is the flow. Our convection diagnostic is the field line-averaged parallel vorticity. For closed field lines, the contours in Figure 2 mainly reflect the contributions from the equatorial regions. For open field lines, the contours reflect the contributions of processes near the boundaries. To the extent that field lines in the magnetosphere are contours of constant vorticity, the plotted contours should approximate the flow pattern in the ionosphere.

In Figure 8, we have superimposed velocity flow vectors observed on Atmosphere Explorer C (AE-C) (Heelis, 1984) on our parallel vorticity contours. Here we have plotted two orbits with a southward and dawnward IMF (θ = 210°). The agreement between the model and observations for the case on the left is excellent. The flow around the A cell which has been distorted by the presence of a $B_{y\ IMF}$ component can be clearly seen in the observations. Note the tailward and dawnward flow between the two A cells and the flow around the sunward part of the B or polar cap cell. The agreement is not as good in the case on the right. There would be much better agreement if the observations were displaced a few degrees poleward with respect to the model. However, the observed flow pattern seems to be consistent with the overall patterns from the model. The flow vectors for three other southward IMF cases are plotted in Figure 9. Again, the flow patterns are similar at θ = 180° and = 240° with better agreement at θ = 255°. As

Model Parallel Vorticity and Flow Observations

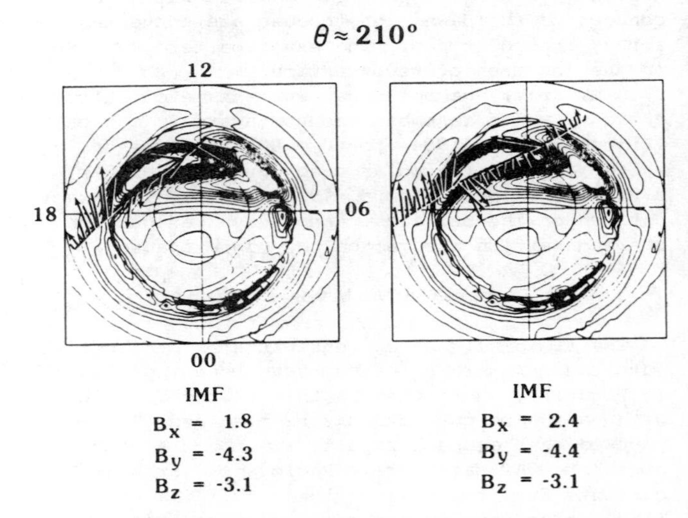

Fig. 8. Ionospheric convection along the AE-C trajectory superimposed on parallel vorticity contours from the simulation for θ = 210°. The IMF is given at the bottom. The format of the plot is the same as that used for the vorticity panels in Figure 2.

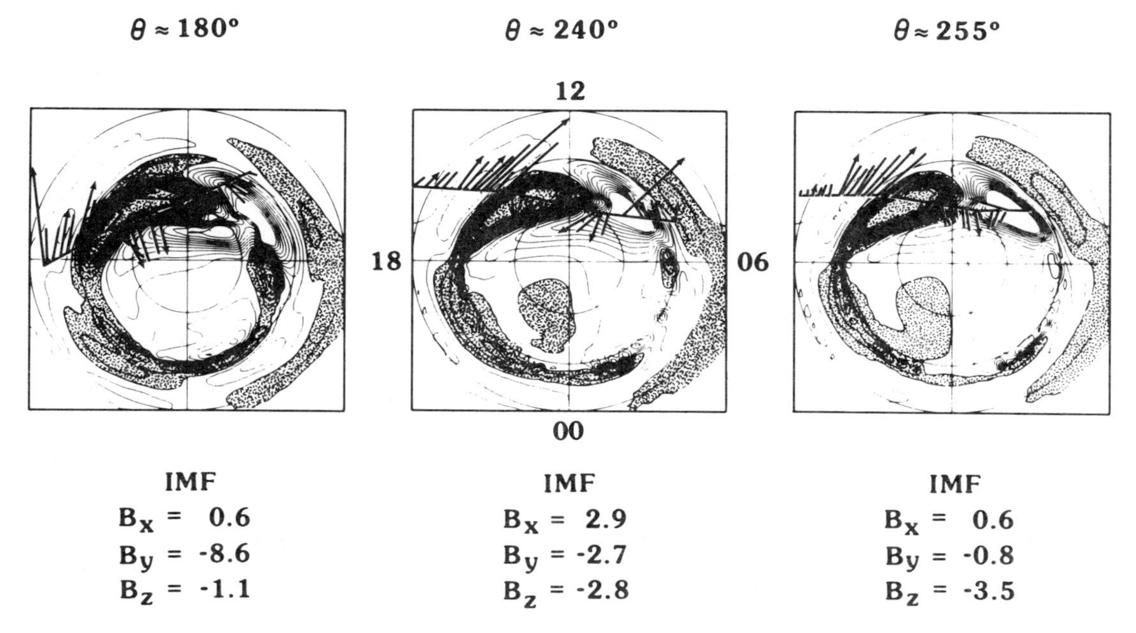

Fig. 9. Same as Figure 8 for three more IMF orientations.

before, the observations are displaced slightly equatorward of the corresponding feature in the model.

Flow for the case when the IMF is northward has been plotted in Figure 10. These observations are from the Dynamics Explorer 2 (DE 2) spacecraft (Heelis et al., 1986). The agreement between the simulation and observations on the left is excellent. Most importantly, the observations on this pass show strong evidence for the B cells. The agreement in the right-hand case is good also. In particular these observations also show evidence for the B cells. There is disagreement on the right side near the A cell where the observations become variable. We found good agreement in four of the six published cases with $B_{z\ IMF} > 0$ and $B_{y\ IMF} < 0$.

When the IMF is northward, the potential difference across the A cell is about 70 kV with the electric field pointing from dawn to dusk (Ogino et al., 1986; Walker and Ogino, 1987). That this is somewhat higher than the 12 to 60 kV determined from polar orbiting satellites (Reiff et al., 1981; Wygant et al., 1983; Doyle and Burke, 1983) indicates that the model is too viscous. For the case when the IMF is entirely southward, the flow driven by reconnection adds to that of the viscous cells (see Figure 2). Then the potential difference is about 180 kV. The difference between the northward IMF case and the southward case (110 kV) is caused by reconnection. This potential difference is consistent with the 40 to 170 kV inferred from the observations.

Summary and Conclusions

We have identified four major convection systems in our global magnetospheric simulation. They are the: global convection cells (A), tail lobe cells (A'), high-latitude polar cap cells (B), and low-latitude cells (C). There are also

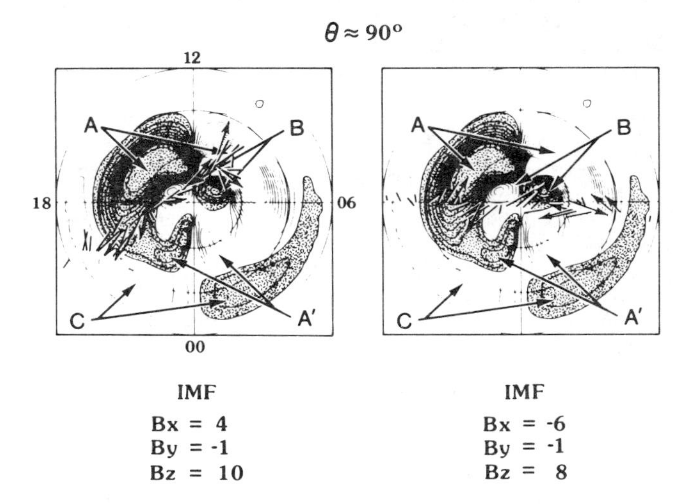

Fig. 10. Same as Figure 8, but using DE 2 observations.

four types of field-aligned currents: region 1 currents (R1), region 2 currents (R2), tail region 1 currents (TR1), and high-latitude polar cap currents (PC).

The IMF B_y component introduces a dawn-dusk asymmetry into the magnetospheric configuration. This can be seen most readily in the distribution of polar cap field-aligned currents. The polar cap currents evolve into the polar cusp currents when we rotate the IMF from northward to southward. The polar cap current distributions depend on the location of reconnection on the magnetopause. For northward IMF, the reconnection occurs on the tail magnetopause and moves onto the dayside magnetopause for southward IMF.

For a dawnward $B_{y\ IMF}$, the dominant polar cap current is earthward. The companion upward current is found equatorward and duskward of the earthward current. This upward current becomes indistinquishable from the duskside region 1 current. Thus, we would expect some instances when region 1 currents are observed on open field lines and some instances when the region 1 currents are on closed field lines. In these latter cases, the largest dayside currents in the simulation are observed at the boundary between open and closed field lines (see Figure 2).

The model does well in reproducing the large-scale features of the dayside parallel current distribution and convection pattern. The model reproduces the structure of the polar cap currents very well. We believe that the simulation will do an even better job of reproducing the observations when the $B_{x\ IMF}$ component and dipole tilt are included. The observations are consistent with the model results that the currents are driven by localized reconnection on the magnetopause and the resulting convection, and that the pattern changes as the region of reconnection moves.

Our treatment of the ionospheric boundary condition is overly simple. In addition, the ionosphere is passive to the extent that there is no feedback from the ionosphere to the magnetosphere. We expect the ionosphere to be important in magnetospheric dynamics. The good agreement between the simulation and observations suggests that the details of the ionosphere-magnetosphere interaction are not important on the very large scale considered in this comparison. It appears that in order to study the effects of the ionosphere on the coupled system, we will need to reexamine the problem with finer resolution and examine time-dependent effects.

Our comparison between observations and the simulations has been primarily confined to the dayside of the magnetosphere. The next step is to consider the nightside. We are currently working on this comparison and hope to be able to report those results in future publications.

Acknowledgments. We would like to acknowledge helpful conversations with Maha Ashour-Abdalla. This work was supported by NASA Solar Terrestrial Theory Program grant NAGW-78.

References

Brecht, S. H., J. G. Lyon, J. A. Fedder, and K. Hain, A simulation study of east-west IMF effects on the magnetosphere, Geophys. Res. Lett., 8, 397, 1981.

Brecht, S. H., J. G. Lyon, J. A. Fedder, and K. Hain, A time dependent three-dimensional simulation of the earth's magnetosphere: Reconnection events, J. Geophys. Res., 87, 6098, 1982.

Doyle, M. A. and W. J. Burke, S3-2 measurements of the polar cap potential, J. Geophys. Res., 89, 9125, 1983.

Doyle, M. A., F. J. Rich, W. J. Burke, and M. Smiddy, Field-aligned currents and electric fields observed in the region of the dayside cusp, J. Geophys. Res., 86, 5656, 1981.

Heelis, R. A., The effects of interplanetary magnetic field orientation and dayside high-latitude ionospheric convection, J. Geophys. Res., 89, 2873, 1984.

Heelis, R. A., P. H. Reiff, J. D. Winningham, and W. B. Hanson, Ionospheric convection signatures observed by DE 2 during northward interplanetary magnetic field, J. Geophys. Res., 91, 5817, 1986.

Iijima, T., and T. A. Potemra, Field-aligned current in the dayside cusp observed by Triad, J. Geophys. Res., 81, 5971, 1976.

Iijima, T., and T. A. Potemra, Large-scale characteristics of field-aligned currents associated with substorms, J. Geophys. Res., 83, 599, 1978.

Iijima, T., T. A. Potemra, L. J. Zanetti, and P. F. Bythrow, Large-scale Birkeland currents in the dayside polar region during strongly northward IMF: A new Birkeland current system, J. Geophys. Res., 89, 7441, 1984.

Leboeuf, J. N., T. Tajima, C. F. Kennel, and J. M. Dawson, Global simulations of the three-dimensional magnetosphere, Geophys. Res. Lett., 8, 257, 1981.

Ogino, T., A three-dimensional MHD simulation of the interaction of the solar wind with the earth's magnetosphere: The generation of field-aligned currents, J. Geophys. Res., 91, 6791, 1986.

Ogino, T., and R. J. Walker, A magnetohydrodynamic simulation of the bifurcation of tail lobes during intervals with a northward interplanetary magnetic field, Geophys. Res. Lett., 11, 1018, 1984.

Ogino, T., R. J. Walker, M. Ashour-Abdalla, and J. M. Dawson, An MHD simulation of B_y-dependent magnetospheric convection and field-aligned currents during northward IMF, J. Geophys. Res., 90, 10835, 1985.

Ogino, T., R. J. Walker, M. Ashour-Abdalla, and J. M. Dawson, An MHD simulation of the effects of the interplanetary magnetic field B_y component on the interaction of the solar wind with the earth's magnetosphere during southward interplanetary magnetic field, J. Geophys. Res., 91, 10,029, 1986.

Potter, D., in Computational Physics, p. 304, John Wiley, Rochester, New York, 1970.
Reiff, P. H., R. W. Spiro, and T. W. Hill, Dependence of polar cap potential drop on interplanetary parameters, J. Geophys. Res., 86, 7639, 1981.
Sato, T., Auroral physics, in Magnetospheric Plasma Physics, edited by A. Nishida, p. 197, D. Reidel Publ. Co., Hingham, Massachusetts, 1982.
Vasyliunas, V. M., Fundamentals of current description, in Magnetospheric Currents, Geophys. Monogr. Ser., vol. 28, edited by T. A. Potemra, p. 63, AGU, Washington, D.C. 1984.
Walker, R. J., T. Ogino, and M. Ashour-Abdalla, A magnetohydrodynamic simulation of reconnection in the magnetotail during intervals with southward interplanetary magnetic field, in Magnetotail Physics, edited by A. T. Lui, p. 183, The Johns Hopkins University Press, Baltimore, Maryland, 1987.
Wu, C. C., Shape of the magnetosphere, Geophys. Res. Lett., 10, 545, 1983.
Wu, C. C., The effects of the dipole tilt on the structure of the magnetosphere, J. Geophys. Res., 89, 11,048, 1984.
Wu, C. C., R. J. Walker, and J. M. Dawson, A three-dimensional MHD model of the earth's magnetosphere, Geophys. Res. Lett., 8, 523, 1981.
Wygant, J. R., R. B. Torbett, and F. S. Mozer, Comparison of S3-3 polar cap potential drops with the interplanetary magnetic field and models of magnetopause reconnection, J. Geophys. Res., 88, 5727, 1983.

THE MAGNETIC MIRROR FORCE IN PLASMA FLUID MODELS

R. H. Comfort

Physics Department, The University of Alabama in Huntsville, Huntsville, Alabama 35899

Abstract. In the past decade, there have been several attempts to include the magnetic mirror force in the equation of motion for a plasma in a fluid formalism. In the process, some confusion has been evident regarding when and how this should be done. This problem has been addressed in the literature, but these treatments appear to have been forgotten or misunderstood. Here we summarize the mathematical arguments so that the physical consequences are readily perceived. It is shown that for an isotropic plasma fluid, in the direction parallel or anti-parallel to a magnetic field, the forces associated with a diverging magnetic field cancel out. Only for anisotropies in the fluid properties does the diverging field influence the plasma dynamics.

Introduction

The magnetic mirror force is well known from treatments of single particle motions, where it plays a significant role in particle dynamics. The transition from microscopic variables, used in kinetic models, to macroscopic variables, used in fluid theories, is not always intuitively straightforward (e.g., Chen, 1984). The details of this transition, including the mirror force, have been discussed at length in several scientific papers and monographs (e.g., Spitzer, 1952; Lehnert, 1964). In the last decade, however, there have appeared at least three papers pointing to the need for explicitly introducing this force in fluid models for certain magnetized plasma regimes (Fahr et al., 1976; Chiu et al., 1979; Li et al., 1983; see also Fahr et al., 1977). It is argued that for plasmas which are intermediate between the collisionless and collision-dominated extremes, such that collision times are long compared to gyroperiods, but sufficiently short to maintain quasi-isotropic Maxwellian distributions, the mirror force should be significant for parallel plasma flows. Intuitively, the argument appears plausible, and it raises the counter question: if this is not the case, why not? Conversations with colleagues have indicated some ambivalence and confusion in this regard, and, in general, the topic has not been addressed directly.

The purpose of this brief review is to demonstrate why the mirror force need not be explicitly included in a fluid model for isotropic plasmas. The mathematical summaries are adapted from those of Lehnert (1964, chapters 3 and 5). They are not intended to be rigorous (see that work for the details), but rather to indicate where the terms in the equations originate.

Kinetic Theory Approach

The first approach is to derive the fluid equation of motion from kinetic theory, beginning with the Vlasov or collisionless Boltzmann equation for a charged particle species

$$\frac{\partial f}{\partial t} + \mathbf{w} \cdot \nabla f + \frac{1}{m} (\mathbf{F} + q\mathbf{w} \times \mathbf{B}) \cdot \nabla_w f = 0 , \quad (1)$$

where $\mathbf{F}$ represents all non-magnetic external forces. This provides the basis for a consistent development of fluid theory from kinetic theory, giving a reference result for comparison with that of a more intuitive physical approach. Define the average of a variable $\chi(\mathbf{r},\mathbf{w},t)$ over the velocity distribution $f(\mathbf{r},\mathbf{w},t)$ as

$$\overline{\chi} \equiv \frac{1}{n(\mathbf{r},t)} \iiint \chi \, f \, d^3w \quad ,$$

where

$$n \equiv \iiint f(\mathbf{r},\mathbf{w},t) d^3w \ .$$

Let $\chi = mw_k$, where w_k is the velocity component in the k direction. We obtain the momentum conservation equation by multiplying (1) by χ and integrating over all velocity space. The result is

$$\frac{\partial}{\partial t} (nmv_k) + \nabla \cdot (nm\overline{w_k \mathbf{w}}) - n(\mathbf{F} + q\mathbf{v} \times \mathbf{B})_k = 0, \quad (2)$$

where $\mathbf{v} \equiv \overline{\mathbf{w}}$. Define a pressure tensor

$$\overline{\overline{\pi}} = (\pi_{jk}), \ \pi_{jk} \equiv nm\overline{\tilde{w}_j \tilde{w}_k} \ ,$$

where $\tilde{\mathbf{w}} = \mathbf{w} - \mathbf{v}$. We can then put (2) into the form

$$nm[\frac{\partial \mathbf{v}}{\partial t} + (\mathbf{v} \cdot \nabla)\mathbf{v}] = n(\mathbf{F} + q\mathbf{v} \times B) - \nabla \cdot \bar{\bar{\pi}} . \quad (3)$$

If off-diagonal elements of the pressure tensor can be neglected and if pressure is assumed to be anisotropic due to the magnetic field, for a 'strong' magnetic field we can write

$$\pi_{jk} = p_\parallel \hat{B}_j \hat{B}_k + p_\perp(\delta_{jk} - \hat{B}_j \hat{B}_k) ,$$

so that $\nabla \cdot \bar{\bar{\pi}}$ can be placed into the form

$$\nabla \cdot \bar{\bar{\pi}} = \nabla_\parallel p_\parallel - (\frac{p_\parallel - p_\perp}{B})\nabla_\parallel B + \nabla_\perp p_\perp + (p_\parallel - p_\perp)(\hat{B} \cdot \nabla)\hat{B} , \quad (4)$$

where

$$\nabla_\parallel = \hat{B}(\hat{B} \cdot \nabla) \text{ and } \nabla_\perp = \nabla - \nabla_\parallel .$$

With these results, the parallel component of (3) can then be written as:

$$nm[\frac{\partial \mathbf{v}}{\partial t} + (\mathbf{v} \cdot \nabla)\mathbf{v}]_\parallel = nF_\parallel - \nabla_\parallel p_\parallel + (\frac{p_\parallel - p_\perp}{B})\nabla_\parallel B . \quad (5)$$

Clearly, if the plasma is isotropic, so that $p_\parallel = p_\perp$, the last term vanishes, leaving the usual parallel equation of motion for a fluid comprised of charged particles of a given species.

Physical Derivation

Now we rederive the parallel equation of motion from orbit theory using physical arguments. This more intuitive approach gives a clearer picture of how these terms originate. We have the result in (5) to assure that nothing is added or left out of this treatment.

Consider the volume element in Figure 1. In the guiding center treatment, the volume element Sds will be acted on by volume forces with parallel components given by

$$n\mathbf{F} \cdot \mathbf{B} \text{ and } n\bar{\mu}(\hat{B} \cdot \nabla)B ,$$

where the latter term is the magnetic mirror force and $\mathbf{F}$ represents all non-magnetic forces. Here $\bar{\mu}$ represents the magnetic moment averaged over the particles in the volume element:

$$\bar{\mu} \equiv \frac{1}{n}\iiint \mu f d^3w = \frac{1}{n}\iiint \frac{\frac{1}{2}mw_\perp^2}{B} f d^3w = \frac{1}{2}\frac{m}{B}\overline{w_\perp^2} \equiv \frac{p_\perp}{nB} \quad (6)$$

In addition to the volume forces, there is a parallel pressure force $p_\parallel S$ on each of the cross-sectional surfaces of the volume element, giving a net pressure force in the parallel direction of $-\partial/\partial s(p_\parallel S)ds$. The sum of these forces is balanced by the inertial force of the total mass flow of the volume element along $\mathbf{B}$. The force balance equation is then given by

$$nm[\frac{\partial \mathbf{v}}{\partial t} + (\mathbf{v} \cdot \nabla)\mathbf{v}]_\parallel Sds = n[\mathbf{F} \cdot \hat{B} - \bar{\mu}(\hat{B} \cdot \nabla B)]Sds - \frac{\partial}{\partial s}(p_\parallel S)ds . \quad (7)$$

Noting that $\nabla_\parallel = \partial/\partial s$ and using (6), we can write the last two terms on the right side of equation (7) as

$$-[n\bar{\mu}(\hat{B} \cdot \nabla)B]Sds - \frac{\partial}{\partial s}(p_\parallel S)ds = \frac{p_\perp}{B}(\frac{\partial B}{\partial s})Sds - [S\frac{\partial p_\parallel}{\partial s} + p_\parallel \frac{\partial S}{\partial s}]ds . \quad (8)$$

Since $\Phi = BS$ = constant along the flux tube, we can write $S = \Phi/B$ and

$$\frac{\partial S}{\partial s} = \Phi \frac{\partial}{\partial s}(\frac{1}{B}) = \frac{-S}{B}\frac{\partial B}{\partial s} . \quad (9)$$

Hence, equation (8) becomes

$$-[n\bar{\mu}(\hat{B} \cdot \nabla)B]Sds - \frac{\partial}{\partial s}(p_\parallel S)ds = \frac{-p_\perp}{B}(\frac{\partial B}{\partial s})Sds - \frac{\partial p_\parallel}{\partial s}Sds + \frac{p_\parallel}{B}\frac{\partial B}{\partial s}Sds . \quad (10)$$

Substituting (10) into (7) and dividing through by the volume element Sds gives the parallel equation of motion

$$nm[\frac{\partial \mathbf{v}}{\partial t} + (\mathbf{v} \cdot \nabla)\mathbf{v}]_\parallel = n\mathbf{F} \cdot \hat{B} - \nabla_\parallel p_\parallel + (\frac{p_\parallel - p_\perp}{B})\nabla_\parallel B , \quad (11)$$

which is identical to equation (5) derived from kinetic theory. Once again, for an isotropic plasma, the effect of the diverging magnetic field vanishes.

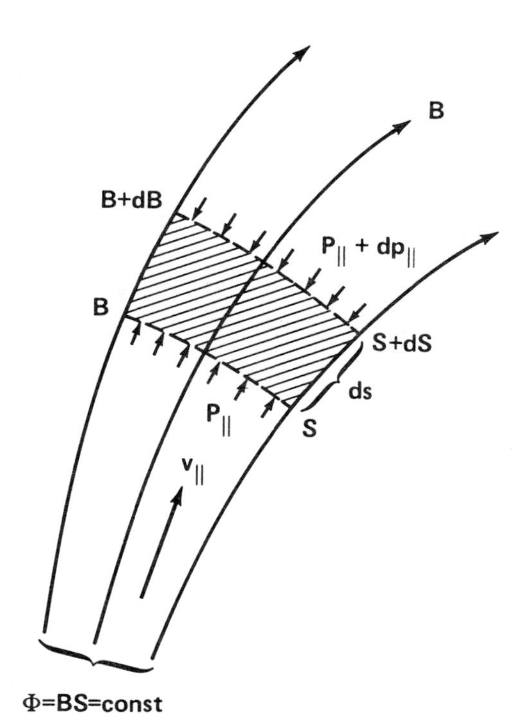

Fig. 1. Volume element of thickness ds along the field, bounded by surfaces of areas S and S + dS perpendicular to the field (after Lehnert, 1964, Figure 3.5).

Discussion and Conclusions

From equations (8)-(11), we see that the magnetic mirror force corresponds to the $p_\perp$ part of the last term in (11), while the $p_\parallel$ part of that term comes from the pressure force associated with the changing area of the flux tube. These two forces are oppositely directed. Hence, in an isotropic plasma, the magnetic mirror force is exactly balanced by that part of the pressure force which results from the diverging magnetic field. Therefore, the lack of a term corresponding to the magnetic mirror force in the fluid equations for an isotropic plasma comes about not because that force has been neglected, or because it was lost in an approximation, or because it is unimportant--it is missing because it is exactly cancelled out by the change in the pressure gradient force that must also occur for a diverging magnetic field.

Although this brief review has been concerned with studies in which the mirror force was not included correctly in plasma fluid treatments, there are in fact a number of such studies in which it has been treated properly. Among examples of investigations of polar wind, auroral, or plasmaspheric plasmas which have correctly included the mirror force in the context of plasma fluid simulations are those by Holzer et al. (1971), Schunk and Watkins (1982), Khazanov et al. (1984), and Demars and Schunk (1986). A recent comprehensive review of the mathematical formulaton of the fluid transport equations for anisotropic plasmas has also been given by Barakat and Schunk (1982). These should provide ample guidance for future studies.

Acknowledgments. This research has been supported in part by NASA grant NAG8-058 and NSF grant ATM-8506642 to The University of Alabama in Huntsville.

References

Barakat, A. R., and R. W. Schunk, Transport equations for multicomponent anisotropic space plasmas: A review, Plasma Phys., 24, 389, 1982.
Chen, F. F., Introduction to Plasma Physics and Controlled Fusion, vol. 1, 2nd ed., chapter 3, Plenum Press, New York, 1984.
Chiu, Y. T., J. G. Luhmann, B. K. Ching, and D. J. Boucher, Jr., An equilibrium model of plasmaspheric composition and density, J. Geophys. Res., 84, 909, 1979.
Demars, H. G., and R. W. Schunk, Solutions to bi-Maxwellian transport equations for SAR-arc conditions, Planet. Space Sci., 34, 1335, 1986.
Fahr, H. J., H. W. Ripken, and M. K. Bird, Effects of diverging coronal fields on the solar wind expansion, Astrophys. Space Sci., 43, 19, 1976.
Fahr, H. J., H. W. Ripken, and M. K. Bird, The effects of diverging coronal fields on the solar wind expansion, Astrophys. Space Sci., 46, L11, 1977.
Holzer, T. E., J. E. Fedder, and P. M. Banks, A comparison of kinetic and hydrodynamic models of an expanding ion-exosphere, J. Geophys. Res., 76, 2453, 1971.
Khazanov, G. V., M. A. Koen, Y. V. Konikov, and I. M. Sidorov, Simulation of ionosphere-plasmasphere coupling taking into account ion inertia and temperature anisotropy, Planet. Space Sci., 32, 585, 1984.
Lehnert, B. (Ed.), Dynamics of Charged Particles, North-Holland Publ. Co., Amsterdam, 1964.
Li, W., J. J. Sojka, and W. J. Raitt, A study of plasmaspheric density distributions for diffusive equilibrium conditions, Planet. Space Sci., 31, 1315, 1983.
Schunk, R. W., and D. S. Watkins, Proton temperature anisotropy in the polar wind, J. Geophys. Res., 87, 171, 1982.
Spitzer, L., Jr., Equations of motion for an ideal plasma, Astrophys. J., 116, 299, 1952.

SELF-CONSISTENT NEUTRAL POINT CURRENT AND FIELDS FROM SINGLE PARTICLE DYNAMICS

R. F. Martin, Jr.

Cooperative Institute for Research in Environmental Science, University of Colorado
Boulder, Colorado 80302

and

Department of Physics, Colorado State University, Ft. Collins, Colorado 80523

Abstract. In order to begin to build a global model of the magnetotail-auroral region interaction, it is of interest to understand the role of neutral points as potential centers of particle energization in the tail. In this paper, the single particle current is calculated near a magnetic neutral point with magnetotail properties. This is balanced with the Ampere's law current producing the magnetic field to obtain the self-consistent electric field for the problem. Also calculated is the current-electric field relationship and, in the regime where this relation is linear, an effective conductivity. Results for these macroscopic quantities are surprisingly similar to the values calculated for a constant normal field current sheet geometry. Application to magnetotail modeling is discussed.

Introduction

In attempting to understand the physics of particle acceleration in the geomagnetic tail, one has several possible tools. Magnetohydrodynamics (MHD) is capable of describing large-scale phenomena, but it is not clear whether this approximation contains all the relevant physics. For example, if particle distributions are strongly non-Maxwellian or if velocity space instabilities are important, MHD theory is less likely to give physically realistic results. On the other hand, velocity space Vlasov theory is probably realistic enough, but is valid only locally in space, and spatially inhomogeneous kinetic theory presents a nearly intractable problem. In this light, the time-honored method of single particle dynamics can lend a hand. A good approximation in low-density plasma, test particle theory can handle non-adiabatic kinetic effects, which MHD cannot, and it is relatively easy to solve over macroscopic distance scales. Thus, breakdowns in the MHD approximation can be determined, and first approximations to a kinetic treatment can be obtained by statistical calculations with many orbits.

In this paper we present results of a preliminary study of quasi-self-consistent currents and fields near an X-type magnetic neutral line. Such a field geometry has been considered for many years as a natural result of MHD-style reconnection or of a collisionless tearing mode instability. Recently, Coroniti (1985) used single particle orbits in a simple current sheet as a basis for his model of explosive reconnection in the tail. One of the goals of the present study is to determine the differences between current sheet motion and motion near a neutral point in order to determine whether the more complicated neutral point motion is required for a complete understanding of tail dynamics. It is already known that neutral point dynamics can be chaotic for realistic tail parameters (Martin, 1986a). While current sheet dynamics can also exhibit asymptotic chaos (Chen and Palmadesso, 1985), realistic tail parameters yield orbits which appear regular over current sheet trapping time scales (Speiser and Lyons, 1984). Moreover, particle acceleration is more energy dependent for the neutral line fields (Martin, 1986b) although similar in magnitude to the current sheet acceleration.

We will determine the relationship between the average current density and the electric field for neutral point motion and, when well defined, the effective conductivity near the neutral point. These will be compared to similar quantities derived for current sheet motion. Although the problem can be partially solved analytically if ergodicity is assumed, the fact that particles can leave the region of interest at random times precludes a simple analytic result. Thus, the results in this paper are numerical.

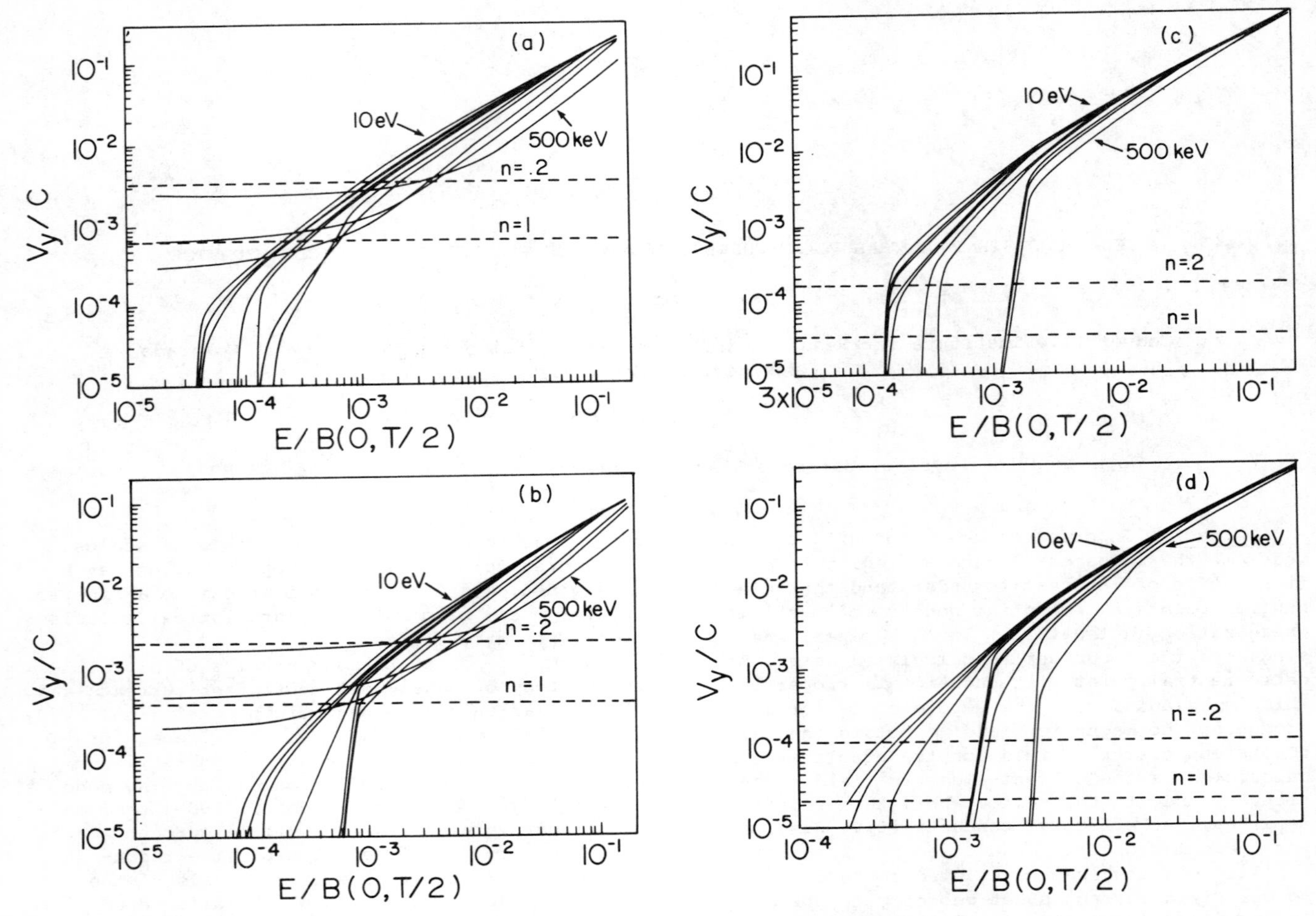

Fig. 1. Average y-velocity (proportional to current density) as a function of electric field for fixed input energies of 10, 50, 100, and 500 eV and 1, 5, 100, and 500 keV. The intersection of the curves with the horizontal dashed lines determines the self-consistent condition of equation (3) for the densities labeled in the figure. The following parameters are used: (a) T = 1000 km, δ = 0.03; (b) T = 1000 km, δ = 0.3; (c) T = 2 x 10^5 km, δ = 0.03, (d) T = 2 x 10^5 km, δ = 0.3.

The Model

Consider a magnetic neutral point of the form

$$\mathbf{B} = \frac{B_o}{D}(z\hat{x} + \delta x\hat{z}),$$

where B_o and D are reference field strength and distance scale, respectively. The neutral point is centered in a rectangle of height T and length $L = T/\sqrt{\delta}$ which defines the extent of usefulness of the field model in the geomagnetic tail. T can be thought of as the current sheet thickness.

From Ampere's law, the current producing this field is

$$J_y = \frac{cB_o}{4\pi D}(1 - \delta) \tag{1}$$

For self-consistency, we calculate single particle orbits in the above magnetic field and a uniform electric field in the y-direction and obtain the average ion current within the T x L rectangle

$$J_y^{\text{numerical}} = n_i e\bar{v}_y^{\text{ numerical}}, \tag{2}$$

where the bar denotes an average over the region. Adjusting the value of the cross tail electric field until equations (1) and (2) match gives the self-consistent electric field. The matching condition for protons reads

$$\frac{\bar{v}_y^{\text{ numerical}}}{c} = \frac{0.633(1-\delta)}{n_i(\text{cm}^{-3})T(\text{km})}. \tag{3}$$

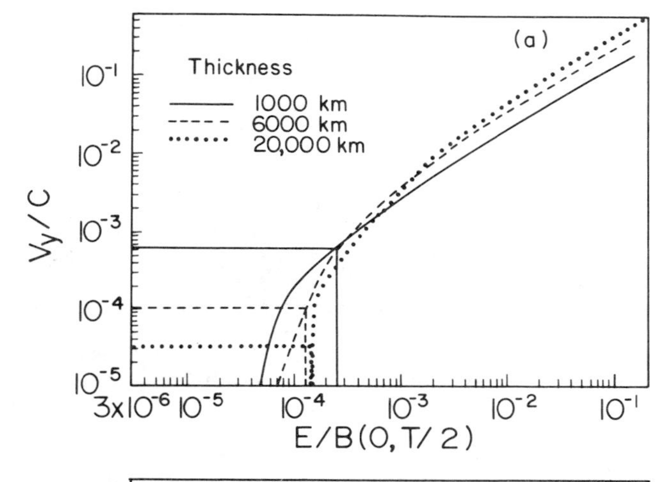

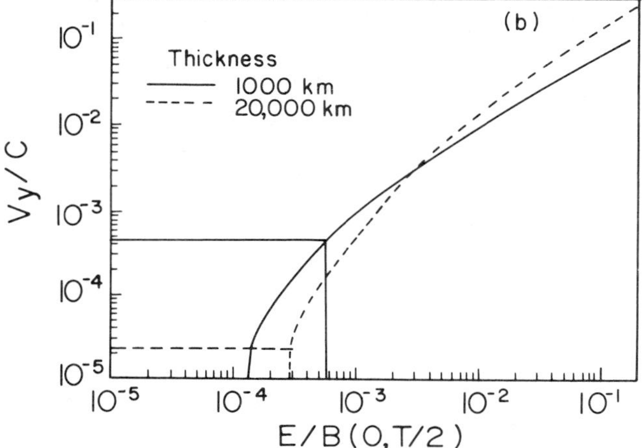

Fig. 2. Average y-velocity as a function of electric field resulting from the input energy distribution of equation (4) for (a) $\delta = 0.03$ and (b) $\delta = 0.3$. The linear portions of the curves are used to determine the effective conductivity of the neutral point region. The self-consistent values of electric field are also shown for the $n = 1$ cm^{-3} case.

In the process, we also calculate the relation between J and E (an Ohm's law) and an effective conductivity. We ignore the contribution of the electrons to lowest order. While electrons are strongly accelerated very near the neutral point, we are averaging over the much longer length scale of the current sheet thickness (defined essentially by ion oscillation amplitudes). Over this length scale, the electrons are mainly adiabatic, producing little effect on the average current. This was verified by sample numerical calculations which included electrons.

Initial conditions are chosen at $z = T/2$, where the magnetic field is taken to be 20 nT. Initial x-position, phase angle, and pitch angle are sampled uniformly, while initial y-position is taken to be zero. Fixed input energies from 10 eV to 500 keV are used. At least 1000 initial conditions are used for each input energy in order to get good statistics in this chaotic system (Martin, 1986a). Finally, an input energy distribution of the form

$$f(K) =$$

$$5 \times 10^{-13} \exp\left[-\frac{(\sqrt{K(\mathrm{keV})} - \sqrt{0.052})^2}{0.862}\right], \quad K < 2.5 \text{ keV} \tag{4}$$

$$9.34 \times 10^{-13} [K(\mathrm{keV})]^{-3}, \quad K > 2.5 \text{ keV}$$

is used (from Lyons and Speiser, 1982). This Maxwellian with a power law tail is based on observations of mantle plasma.

Results: J-E Relation

Figure 1 shows plots of numerical $\overline{v}_y$ as a function of $E/B(x = 0, z = T/2)$ for various input energies and current sheet thicknesses T. The self-consistent values of $\overline{v}_y$ are also shown, for densities of 1 cm^{-3} (typical plasma sheet) and 0.2 cm^{-3} (typical mantle).

The curves that drop abruptly pass through zero and become negative [due to gradient drift (Martin, 1986b)] for smaller electric fields. However, for T = 1000 km and T = 6000 km there are curves (for high input energy) which reach a limiting value of $\overline{v}_y$. For these cases there is an ion current with no electric field. This is because the orbits in this regime are of similar character to Speiser orbits (Speiser, 1965) in a current sheet geometry and obtain the y-motion due to gyromotion about the normal magnetic field. As the thickness T increases, the motion becomes more chaotic and such behavior does not occur.

For no case is a simple linear J-E relation valid for smaller electric fields. For larger electric fields the slope of the $\overline{v}_y$-E curves does approach a limit. The resulting conductivity

$$\sigma(\mathrm{mho/m}) = 8 \times 10^{-6} n(\mathrm{cm}^{-3}) \times (\text{slope of graph})$$

in nearly independent of input energy but increases with sheet thickness. This is consistent with the idea that the chaotic motion in thicker current sheets can increase the lifetime near the neutral point and therefore increase the conductivity.

Figure 2 shows the results when the energy distribution (4) is used as input (note that two different separatrix angles are used). The increase in slope with T is now apparent in these linear plots, as is the decrease of the conductivity as δ is increased. Values of the ef-

TABLE 1. Conductivity of the neutral point region estimated from the linear part of the curves in Figure 2, using a number density of 1 cm^{-3}.

	T(km)	σ(mho/m)
$\delta = 0.03$	1000	7.5×10^{-6}
	6000	1.6×10^{-5}
	20,000	2.0×10^{-5}
$\delta = 0.3$	1000	4.4×10^{-6}
	20,000	9.4×10^{-6}

fective conductivity for $n = 1$ cm^{-3} calculated from this slope are given in Table 1.

Lyons and Speiser (1985) give a value of about

$$\sigma(\text{mho/m}) = \frac{1 \times 10^{-4}\ n(\text{cm}^{-3})}{B_z(\text{nT})}$$

for a current sheet magnetic field with a constant normal component B_z. If such a region were attached to the neutral point region at $x = L/2$, the normal field would be 3.4 nT. This gives a conductivity of about 3×10^{-5} mho m^{-1} for $\delta = 0.03$, which is somewhat larger than the neutral point conductivities.

On the other hand, Rusbridge (1971) estimated the conductivity near a neutral line with perpendicular separatrices to be

$$\sigma(\text{mho/m}) = 8 \times 10^{-7}\ n(\text{cm}^{-3})$$

which is smaller than the present calculation gives. Note, however, that σ decreases as the separatrix angle increases, so our results may approach Rusbridge's for larger angles.

Martin (1986a) estimated a chaotic conductivity based on the Lyapunov exponent time scale, giving about 2×10^{-4} mho m^{-1} for $\delta = 0.03$. The discrepancy between this value and the values calculated here could mean the Lyapunov exponent is not the correct time scale, or that other effects such as Speiser's (1970) inertial and gyromotion contributions are acting to decrease the conductivity over the stochastic value.

Results - Self-Consistent E

Figure 2 also shows self-consistency results for the case with the above input energy distribution. The self-consistent values of electric field for this distribution are remarkably independent of region thickness, since the J-E curves tend to drop rapidly near this point. For all cases the self-consistent electric field is about $E = 2 \times 10^{-4}$ B(0,T/2) (in Gaussian units). For a magnetic field at the edge of the current sheet of 20 nT and $n = 1$ cm^{-3}, this electric field is about 1.2×10^{-3} V m^{-1}.

Lyons and Speiser (1985) give a self-consistency condition of $E = 3.06 \times 10^{-3}$ B_z(nT) for the constant normal field and $n = 1$ cm^{-3}. For the normal field $B_z = 3.4$ nT calculated earlier this is about 1×10^{-3} V m^{-1}, very similar to the neutral point result.

A nominal value for the actual cross tail electric field is somewhat smaller than both these estimates. By mapping observed potential drops from the polar regions to the tail one gets a field of about 3×10^{-4} V m^{-1}, assuming the potential drop occurs across the whole width of the tail. There is some evidence from observations and MHD simulations, however, that a neutral line may extend only part way across the tail, in which case our calculated values of E could still produce the observed potential drop. Another possibility is that field-aligned potential drops exist in the plasma sheet boundary layer for example, so polar potential drops do not map directly into the plasma sheet.

It is further interesting to note that because of the limiting behavior at low electric field for thin sheets, an energy distribution peaked at several hundred kiloelectron volts would not be able to support such a magnetic field; i.e., no self-consistent current can be obtained. Thus, it is possible that steady reconnection could not occur in this case.

Summary

We have calculated J-E relations, conductivities, and self-consistent electric fields for a model plasma sheet neutral point region. The numerical results are summarized below for a number density of 1 cm^{-3}, with $\delta = 0.03$ and 0.3:

Conductivity

1. Due to non-Ohmic nature of J-E curve, the conductivity is not well defined for $E/B(0,T/2) \lesssim 10$.
2. When defined, the conductivity ranges from about 10^{-6} to 10^{-5} mho m^{-1}, somewhat greater than the Rusbridge (1971) value and somewhat smaller than Lyons and Speiser (1985).
3. The conductivity increases as sheet thickness increases, although this depends on input energy distribution.

Self-Consistent Electric Field

1. No self-consistent value is possible for high-energy particles in the thinner current sheets, due to the limiting behavior of J at small E.
2. For the input energy distribution (4) the value of the electric field is about 1.2×10^{-3} V m^{-1}. This is similar to the value obtained by Lyons and Speiser (1985) for the constant normal field current sheet.
3. Considering only high input energy

particles in a thick current sheet, one can get electric fields which are an order of magntiude larger than the above value.

We interpret the results as follows. For a thick plasma sheet, the fact that the particle spends more time undergoing adiabatic motion than it spends near the neutral point, where the electric field is strong, results in the rapid drop in current density as electric field is decreased. This represents a signficant deviation from a simple linear current-electric field behavior. Another deviation occurs for thin sheets at small electric field, where the orbits become Speiser-like and the current becomes independent of electric fields. It is important to note that the electric fields for which the linear Ohm's law fails are realistic tail fields, based on mapping polar potential drops into the plasma sheet. The related question of which terms in the generalized Ohm's law might be causing this deviation requires a more sophisticated calculation of spatially inhomogeneous distribution functions, which is beyond the scope of this paper. One perhaps surprising result from the above analysis is that the macroscopic properties of conductivity and electric field for the neutral point field are quite close to those for the dynamically simpler current sheet field. This is in spite of the fact that the microscopic dynamics is significantly different in the two cases. One interpretation of this is that for the tail acceleration mechanism, there is no great need for a neutral point region, especially in light of the fact that Lyons and Speiser (1982) were able to obtain reasonable plasma sheet boundary layer distributions based solely on current sheet acceleration (note, however, the possible non-uniqueness of such distribution function matching). If the current sheet is unstable, as in a model like Coroniti's, and a neutral point is formed, our results indicate that the electric field is then no stronger, implying that the instability may not be effective in converting magnetic energy into particle energy via an intermediary electric field. However, such conclusions are still preliminary, as the microscopic dynamics may yet be important. For example, in the tearing mode theory of Chen and Lee (1985), non-Maxwellian features in the distribution function can strongly affect growth rates of the instability. The fact that phase space is broken up into regions with differing behavior (e.g., different degrees of chaos) can result in non-Maxwellian distributions. Thus, further calculations are needed to fully determine the effect of chaos on possible instabilities.

Acknowledgments. This research was partially supported by the National Oceanic and Atmospheric Administration, Space Environment Laboratory, Boulder, Colorado. The author wishes to thank T. W. Speiser for many helpful discussions on magnetotail physics.

References

Chen, J., and Y. C. Lee, Collisionless tearing instability in a non-Maxwellian neutral sheet: an integro-differential formulation, Phys. Fluids, 28, 2137, 1985.

Chen, J., and P. J. Palmadesso, Chaos and nonlinear dynamics of single-particle orbits in a magnetotail-like magnetic field, J. Geophys. Res., 91, 1499, 1985.

Coroniti, F. V., Explosive tail reconnection: the growth and expansion phases of magnetospheric substorms, J. Geophys. Res., 90, 7427, 1985.

Lyons, L. R., and T. W. Speiser, Evidence for current sheet acceleration in the geomagnetic tail, J. Geophys. Res., 87, 2276, 1982.

Lyons, L. R., and T. W. Speiser, Ohm's law for a current sheet, J. Geophys. Res., 90, 8543, 1985.

Martin, R. F., Jr., Chaotic particle dynamics near a two-dimensional neutral point, with application to the geomagnetic tail, J. Geophys. Res., 91, 11,985, 1986a.

Martin, R. F., Jr., The effect of plasma sheet thickness on ion acceleration near a magnetic neutral point, in Ion Acceleration in the Magnetosphere and Ionosphere, Geophys. Monogr. Ser., vol. 38, edited by T. Chang, p. 141, AGU, Washington, D.C., 1986b.

Rusbridge, M. G., Shielded supports in toroidal multipoles, Plasma Phys., 13, 33, 1971.

Speiser, T. W., Particle trajectories in model current sheets, 1, analytical solutions, J. Geophys. Res., 70, 4219, 1965.

Speiser, T. W., Conductivity without collisions or noise, Planet. Space Sci., 18, 613, 1970.

Speiser, T. W., and L. R. Lyons, Comparison of analytical approximation for particle motion in a current sheet with precise numerical calculations, J. Geophys. Res., 89, 147, 1984.

INITIAL DEVELOPMENT OF A NEW EMPIRICAL MODEL OF THE EARTH'S INNER MAGNETOSPHERE FOR DENSITY, TEMPERATURE, AND COMPOSITION

D. L. Gallagher and P. D. Craven

NASA Marshall Space Flight Center, Huntsville, Alabama 35812

Abstract. The analytical representation of plasma characteristics in the near earth environment is a valuable tool for studying wave propagation, for new instrument and spacecraft design, and for developing a better theoretical understanding of plasmaspheric processes. There are no empirical models currently available that encompass the near earth environment and include the core or low-energy plasma characteristic of that region. The initial steps to constructing a new empirical model of plasmaspheric density, temperature, and composition are discussed. A limited set of density measurements from the retarding ion mass spectrometer and the plasma wave instrument on the Dynamics Explorer 1 spacecraft is used to demonstrate features of the proposed analytical formalism.

Introduction

Empirical models of the magnetospheric plasma are needed to facilitate numerical modeling of global magnetospheric properties and processes in a realistic environment. Models are also needed to anticipate spacecraft charging and other effects associated with spacecraft design, to complete the overall picture of ionospheric outflow, to stimulate studies toward physical understanding, for comparison to theoretical models of the magnetosphere, and for modeling of other planetary bodies. The list is of course incomplete.

The development of theoretical or physical models of magnetospheric properties continues to depend, in part, upon the availability of corresponding empirical models. In addition, some modeling studies, such as wave propagation ray tracing, depend upon analytical descriptions of magnetopheric properties. Although the inner magnetosphere or plasmasphere has been studied for many years, few analytical models of the plasma population in this region exist that are based entirely upon observation.

Garrett and DeForest (1979) modeled the plasma population at geosynchronous orbit using the first four moments of the ion and electron population. Density, number flux, pressure, and energy flux were expressed in an analytical formula as a function of local time and magnetic activity (Ap). Two free parameters were used to linearly fit variations in magnetic activity. Three free parameters were used in a truncated Fourier series expansion on local time. The University of California at San Diego particle detectors on the ATS 5 spacecraft were used to measure ions and electrons in the energy range from 51.6 eV to 51.6 keV. The lower limit effectively eliminated much of the core plasma (as defined by Horwitz, 1987) from their study.

Brace and Theis (1981) used the Atmosphere Explorer-C, ISIS 1, and ISIS 2 spacecraft measurements to model electron temperature at 300, 400, 1400, and 3000 km altitude. The ISIS 1 observations from the 2000 to 3600 km altitude range were used to model the temperature at 3000 km. The temperature at 1400 km altitude was obtained from the ISIS 2 satellite, which was in a circular orbit at that altitude. Atmosphere Explorer-C measured electron temperatures from circular orbits first at 300 km and later at 400 km altitude. In the Brace and Theis model, electron temperature was expressed as a function of Dip latitude and local time in a spherical harmonic expansion.

Kohnlein (1986) has most recently produced an empirical model for electron and ion temperatures in the altitude range from 50 to 4000 km. Measurements from six satellites (AE-C, AE-D, AE-E, ISIS 1, ISIS 2, and OGO 6) were used. Five incoherent scatter radar stations (Arecibo, Chatanika, Jicamarca, Millstone Hill, and St. Santin) and rocket measurements were also incorporated. Temperatures are expressed as a sum of spherical harmonics and Fourier series functions (for the horizontal structure) and spline functions (for the vertical structure). Although only measurements from quiet geophysical conditions are included, annual or seasonal effects are treated.

Each of these studiess provide analytical expressions for determining various plasma characteristics with differing, but substantial, res-

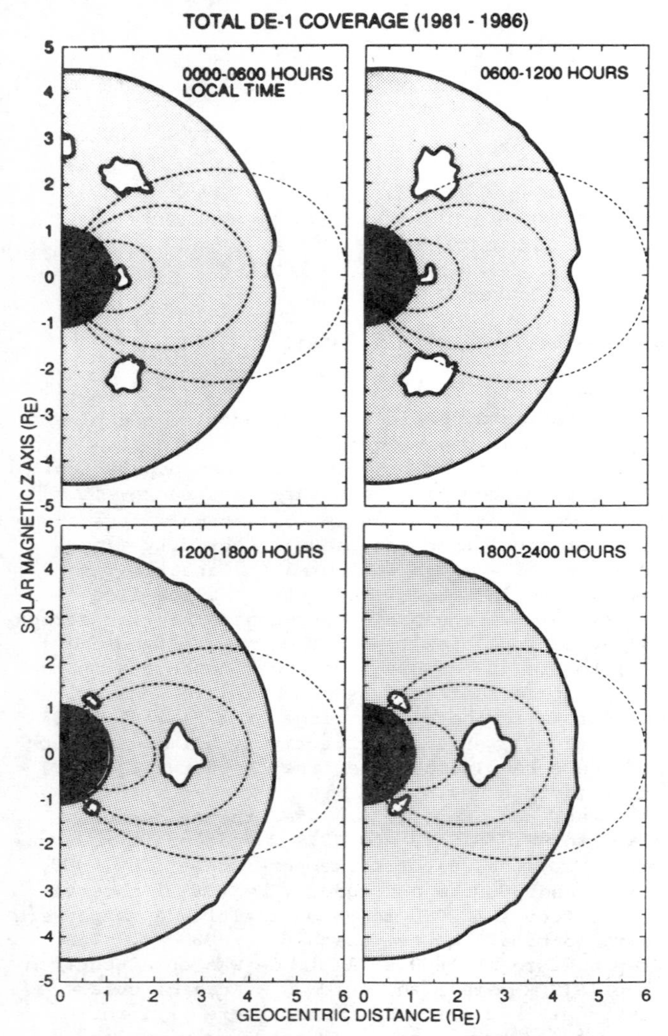

Fig. 1. Available orbital coverage of the DE 1 spacecraft from 1981 through 1986 in four magnetic local time intervals. The shaded areas indicate locations reached by the spacecraft. The earth's origin is located in the center left of each panel with ±5 R_E along the vertical axis and 0-5 R_E along the horizontal axis in a meridian projection including the z-axis in solar magnetic coordinates.

trictions on applicable spatial region, ion species, or geophysical conditions. In addition, particle measurements of both the ions and electrons that are part of the core or low-energy plasmaspheric plasma have not been included in these empirical models.

Numerical programs exist that solve the transport equations including reaction rates for the different species. These programs determine the density and temperature of a plasma at a given location assuming an initial set of conditions. The programs are generally computer intensive and dependent on the initial conditions and on the degree to which all the dynamics and interactions have been included. Empirical models cannot compete with numerical methods in describing small-scale temporal changes. However, a qualitative understanding of processes can be obtained with good empirical models, without requiring excessive computer memory or execution times.

We report here our first steps in constructing a comprehensive empirical model of the plasmasphere. The modeling framework is first outlined and the analytical format for representing ion densities and temperatures is described. The generalized expression consists of a basic global description of the plasmasphere plus corrections. Example fits of the proposed analytical expression for the model are also shown. Finally, significant aspects important to producing an empirical model and using statistical measurements of the plasmasphere are discussed.

New Empirical Study

With the launch of the Dynamics Explorer 1 (DE 1) spacecraft, a new opportunity was created for obtaining multi-species ion measurements of the core plasmaspheric plasma. On DE 1, the retarding ion mass spectrometer (RIMS) measures ion energies from 0-100 eV and ion species from 1-32 AMU. The RIMS instrument consists of one detector in the spacecraft spin plane and two along the spin axis (see Chappell et al., 1981 for futher details). Using the 5 years of RIMS observations of density, temperature, and composition of core plasma and measurements from other instruments on DE 1 and other spacecraft, a more complete empirical model of the near earth environment can now be obtained. Figure 1 shows a representative sketch of the orbital coverage of DE 1, which provides nearly complete spatial sampling of the near earth environment between 1 R_E and 4.6 R_E geocentric distances.

A study has been initiated to collect near earth plasma measurements of density, temperature, and composition for the purpose of producing new analytical expressions for these quantities, based entirely upon observation. The intent is to produce a standard reference plasmasphere for core plasma, not unlike that now available for the ionosphere (Bilitza, 1986). With this new empirical model plasma density, temperature, and composition at arbitrary locations in the near earth environment and for varying geophysical conditions can easily be calculated. When complete, this model may be used to facilitate new spacecraft and instrument design, modeling of magnetospheric processes, and as a guide for theoretical models.

Modeling Framework

The analytical framework for the empirical model will follow McIlwain (1972, 1974). McIlwain

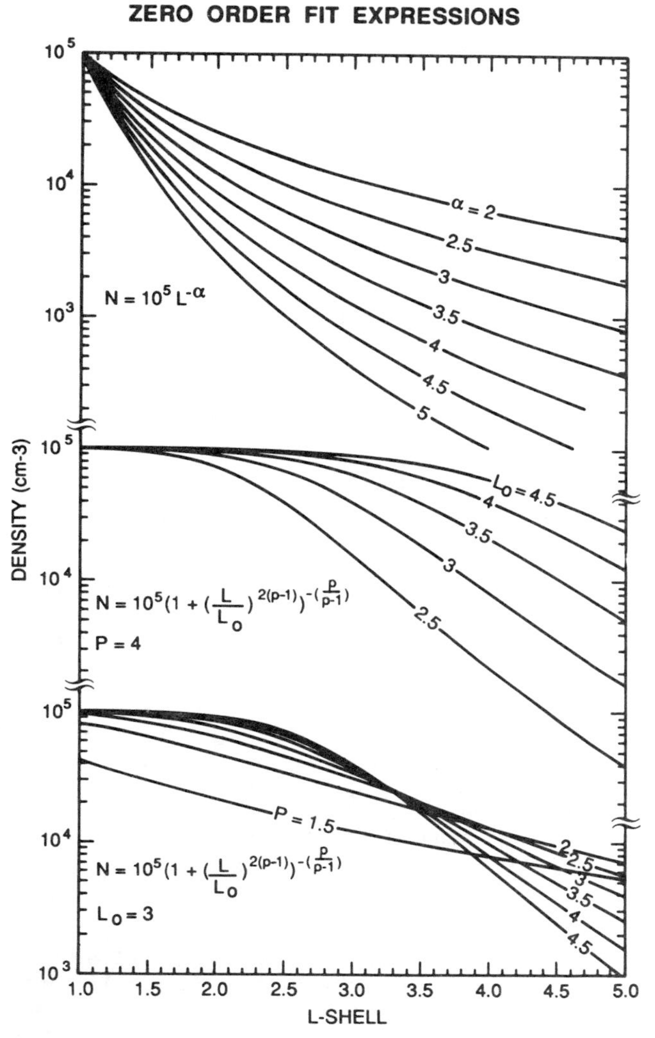

Fig. 2. Three families of curves for two possible analytical expressions of the first term (or zero-order term) in equation (1). The upper set of curves represents a power law in L shell for the density. The power law in L is varied from $\alpha = 2$ to $\alpha = 5$ and encompasses likely values for the plasmasphere. The center and bottom sets of curves are for a modified Lorentzian equation in L shell. The free parameters L_0 and p are varied in the middle and lower panels, respectively, to display various shapes available with this expression. A density of 10^5 cm^{-3} at L = 1 is arbitrarily chosen for displaying all curves.

constructed an expression for the electric field such that calculated values of energy versus local time and charge times energy versus local time matched the values observed with ATS 5 (in geosynchronous orbit) during periods of quite geomagnetic activity. In these works, McIlwain expresses the electric field as the sum of three terms [McIlwain, 1972, his equation (3)]. The first term is a constant term that is just an offset. The second term is one that gives a corotation field for high values of the magnetic field (**B**). The third term is the sum of 120 functions which are products of fit parameters and exponentials. Following this example in the present study, ion densities and temperatures will each be expressed as

$$A(\varepsilon_1,\varepsilon_2,\varepsilon_3) = B_0(\varepsilon_1,\varepsilon_2,\varepsilon_3) + \Sigma_\ell B_\ell \exp[-(\varepsilon_1-\varepsilon'_{1\ell})^2/\sigma_1 - (\varepsilon_2-\varepsilon'_{2\ell})^2/\sigma_2 - (\varepsilon_3-\varepsilon'_{3\ell})^2/\sigma_3] \quad (1)$$

where $\sigma_1,\sigma_2,\sigma_3$ are constants dependent upon the separation between fit terms, B_0 and B_ℓ are the fit parameters, $(\varepsilon'_1,\varepsilon'_2,\varepsilon'_3)_\ell$ are the fixed coordinates of the fit terms, and $(\varepsilon_1,\varepsilon_2,\varepsilon_3)$ are free spatial coordinates.

The zero-order fit to observed values is found in the term B_0. B_0 is a function of $(\varepsilon_1,\varepsilon_2,\varepsilon_3)$ and its form is chosen to generally follow the observed variation in ion density or temperature. In the case of fitting to density, B_0 may be expressed as a power law in L shell,

$$B_0(\varepsilon_1,\varepsilon_2,\varepsilon_3) = C \cdot L^{-\alpha} , \quad (2)$$

where $(\varepsilon_1,\varepsilon_2,\varepsilon_3)$ is the location in solar magnetic coordinates, while C and α are expressed as first- or second-order polynomials in local time. When fitting temperature, B_0 may be chosen to be constant or increase with altitude along an L shell, as observed by Comfort (1986).

Alternatively, B_0 may take other forms as needed. We have chosen here to include the plasmapause in the zero-order fit to density by using a modified Lorentzian function,

$$B_0(\varepsilon_1,\varepsilon_2,\varepsilon_3) = C \cdot \left[1 + \left(\frac{L}{L_0}\right)^{2(p-1)}\right]^{-\left(\frac{p}{p-1}\right)} \quad (3)$$

where C, L_0, and p become the fit parameters. Figure 2 shows a family of curves for varying fit parameter values for both equations (2) and (3). The upper panel shows characteristic examples of equation (2), while the middle and bottom panels show families of curves for equation (3), where first L_0 and then p take on different values.

The second term in equation (1) accommodates localized variation in density or temperature away from the zero-order term B_0. Each term contributes to the sum in a limited spatial region centered at $(\varepsilon'_1,\varepsilon'_2,\varepsilon'_3)_\ell$ where ℓ designates locations uniformly distributed throughout the plasmasphere. The constant divisor of each exponential (σ_1, σ_2, and σ_3) is chosen to insure a smooth variation between terms in the summation

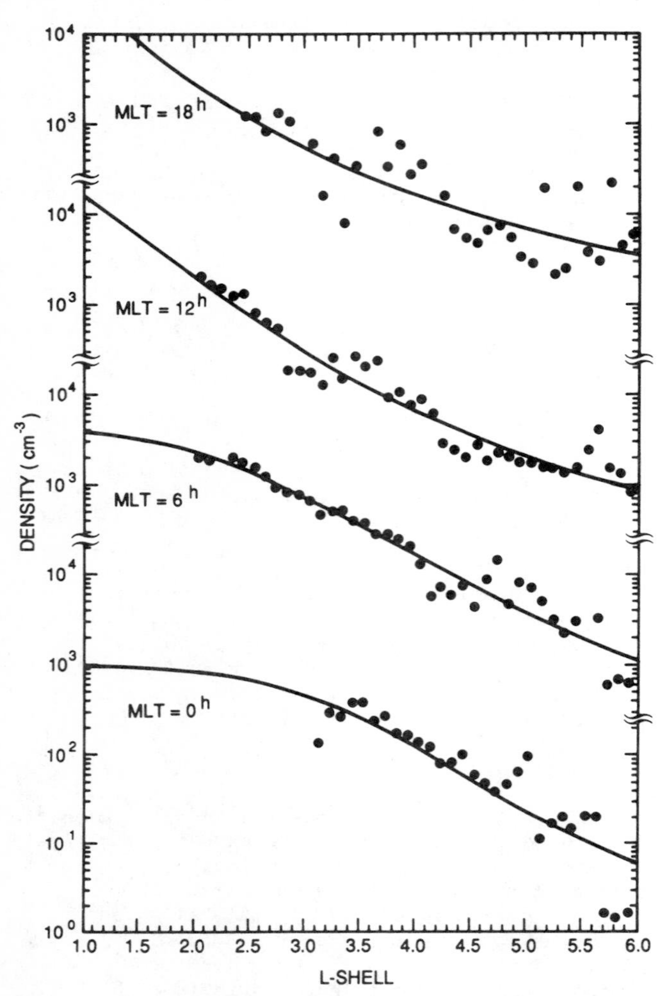

Fig. 3. Fits of the modified Lorentzian expression to a limited set of RIMS and PWI measurements shown as solid curves in four (2-hour wide) magnetic local time intervals. The measurements, shown as filled circles, are averaged along L shells and into the four local time intervals. The modified Lorentzian expression successfully follows the changing shape of the plasmaspheric density from midnight into the evening buldge region.

and is related to the full width at half maximum of the exponentials. Each multiplier is given by the following

$$\sigma_j = \frac{(\Delta\varepsilon'_j)^2}{2}, \quad (4)$$

where $\Delta\varepsilon'_j$ is the separation between the locations of the exponential terms along each coordinate direction (j = 1,2,3). The parameters B_ℓ are linearly fit (using a proven least square fitting technique from Bevington, 1969) to measurements in the localized regions corresponding to each term in the summation.

Initial Observations

The initial set of observations for this new empirical model of the plasmasphere is being compiled from the RIMS and the plasma wave instrument (PWI) on DE 1. The RIMS detects low-energy ions (0-100 eV) and is capable of measuring density and energy of the major and minor ions (H^+, He^+, O^+, N^+, O^{++}, He^{++}, N^{++}). Plasma density and temperature are derived using the thin sheath approximation of Comfort et al. (1982, 1985). The PWI step frequency receiver measures electric and magnetic field strengths between 1.8 Hz and 409 kHz in 136 frequency bands (see Shawhan et al., 1981). Measurement of the upper hybrid emission frequency by PWI is used together with the local magnetic field strength to determine the total electron density. Where both RIMS and PWI measurements exist, the PWI-derived total density is used to calibrate the RIMS-derived composition.

Figure 3 shows example fits to a limited set of RIMS and PWI measurements using the modified Lorentzian function at four magnetic local times. Density versus magnetic L shell is plotted. Each curve represents an average over 2 hours in magnetic local time for low geomagnetic activity ($Kp \leq 3+$). The modified Lorentzian function is shown to reproduce the observations for varying plasmapause shapes, e.g., the more sharply defined density decrease at the plasmapause at midnight and the more gradual falloff but higher densities in the dusk bulge region.

Discussion

The empirical modeling framework outlined herein is designed both to facilitate its implementation and its interpretation. The first term (B_0) in equation (1) allows a relatively simple expression to be used to describe the general characteristics of the plasmasphere. The overall density, plasmapause location, and systematic variations with local time and latitude can be included in that term. The relatively few free parameters in the first term simplify the fitting procedure. Once these parameters are determined, the first term may provide some insight into the global characteristics of plasmaspheric temperature and composition. For example, regions of diffusive equilibrium or heating may be suggested by inspection of the first term alone. When B_0 is combined with the second term in equation (1), localized variation away from the generalized global characteristics of the plasmasphere can be included in the empirical model. The spacial decoupling of each fit term in the summation by exponential multipliers greatly simplifies the fitting process. Only a limited number of fit parameters B_ℓ need to be included when fitting equation (1) in a given region of the plasmaphere.

The dynamic nature of the near earth environment threatens to obscure important features of plasma characteristics in any statistical study. Plasmapause location and shape at all local times, along with heating in the outer plasmasphere, are strongly influenced by the level and history of magnetic activity in the magnetosphere. Simple averaging of plasma characteristics may produce an unrealistic picture of plasmaspheric properties. Averaging density for various L shells, for example, without regard to the level of magnetic activity would smooth any sharp variations characteristic of plamaspheric density, like that at the plasmapause. It is our intent to limit the consequence of averaging measurements obtained over long periods of time by evaluating equation (1) within limited ranges of magnetic activity (such as defined by Kp). Sets of fit parameters will be obtained for specific ranges of geophysical activity or fit parameters will be allowed to vary continuously.

The DE 1 spacecraft offers an excellent source of plasmaspheric measurements from the RIMS and PWI instruments. Plasma bulk density and temperature measurements of at least H^+, He^+, and O^+ will be included in the final empirical model. Measurements of other ions are also available and will be included if they are sufficiently well sampled. Although the procedure for obtaining density and temperature from RIMS is complex (Comfort et al., 1982), it has been automated and, together with PWI, will be used to provide plasma characteristics from the last quarter of 1981 through 1986.

In this initial study, only measurements within L = 6 will be considered. The interpretation of measurements from RIMS under conditions of low density and spacecraft charging are significantly more difficult and will not be included. Also not initially included are measurements from other excellent instruments on DE 1 and other spacecraft which have widely sampled the near earth environment. As this study progresses, it is hoped that a broader collection of near earth plasma measurements can be included in this empirical model.

The first steps have now been taken toward the generation of a new empirical model of the earth's plasmasphere. As efforts continue, a more complete discription of ion composition and temperatures will emerge than has been available in the past. The model is designed to be both usable and easily updated as more or better measurements become available. When complete, every effort will be made to insure the widest possible availability of this new empirical model.

Acknowledgments. The authors are grateful to D. A. Gurnett for the important addition of plasma wave-derived densities to this study from the University of Iowa plasma wave instrument on the Dynamics Explorer 1 spacecraft. A rather large programming effort has been and continues to be required for this study. Most, if not all, of the programming for this effort has been undertaken by Richard West of Boeing Computer Support Services. The authors would like to acknowledge his efforts and capable work. We also would like to thank those others who are contributing to the development of the empirical model, including Hugh Comfort, Barbara Giles, Tom Six, and Eric Peterson.

References

Bevington, P. R., Data Reduction and Error Analysis for the Physical Sciences, McGraw-Hill Book Company, New York, 1969.

Bilitza, Dieter, International Reference Ionosphere: Recent developments, Radio Science, 21, 343, 1986.

Brace L. H., and R. F. Theis, Global empirical models of ionospheric electron temperature in the upper F-region and plasmasphere based on in situ measurements from the Atmosphere Explorer-C, ISIS-1, and ISIS-2 satellites, J. Atmos. Terr. Phys., 43, 1317, 1981.

Chappell, C. R., S. A. Fields, C. R. Baugher, J. H. Hoffman, W. B. Hanson, W. W. Wright, H. D. Hammack, G. R. Carignan, and A. F. Nagy, The retarding ion mass spectrometer on Dynamics Explorer-A, Space Sci. Instrum., 5, 477, 1981.

Comfort, R. H., Plasmasphere thermal structure as measured by ISEE-1 and DE-1, Adv. Space Res., 6, 31, 1986.

Comfort R. H., C. R. Baugher, and C. R. Chappell, Use of the thin sheath approximation for obtaining ion temperatures from the ISEE 1 limited aperture RPA, J. Geophys. Res., 87, 5109, 1982.

Comfort, R. H., J. H. Waite, Jr., and C. R. Chappell, Thermal ion temperatures from the retarding ion mass spectrometer, J. Geophys. Res., 90, 3475, 1985.

Garrett, H. B., and S. D. DeForest, An analytical simulation of the geosynchronous plasma environment, Planet. Space Sci., 27, 1101, 1979.

Horwitz, J. H., Core plasma in the magnetosphere, Rev. Geophys., in press, 1987.

Kohnlein, W., A model of the electron and ion temperatures in the ionosphere, Planet. Space Sci., 34, 609, 1986.

McIlwain, C. E., Plasma convection in the vicinity of the geosynchronous orbit, in Earth's Magnetospheric Processes, edited by B. M. McCormac, pp. 268-279, D. Reidel Publ. Co., Dordrecht-Holland, 1972.

McIlwain, C. E., Substorm injection boundaries, in Magnetospherics Physics, edited by B. M. McCormac, pp. 143-154, D. Reidel Publ. Co., Dordrecht-Holland, 1974.

Shawhan, S. D., D. A. Gurnett, D. L. Odom, R. A. Helliwell, and C. G. Park, The plasma wave and quasi-static electric field instrument (PWI) for Dynamics Explorer-A, Space Sci. Instrum., 5, 535, 1981.

HYDRODYNAMIC MODELS OF THE PLASMASPHERE

P. G. Richards and D. G. Torr

The University of Alabama in Huntsville, Huntsville, Alabama 35899

Abstract. A brief history of the application of hydrodynamic models to the understanding of the plasmaspheric density distribution is presented. Then results from recent modeling are discussed in more detail and areas of agreement between theory and measurement are established. Finally, some calculations are presented which indicate important discrepancies between theory and measurement.

Introduction

During the early 1950's, whistler measurements led to the discovery of surprisingly large electron densities at altitudes of several earth radii (Storey, 1953). The source of this plasma was identified as the reaction $O^+ + H \rightarrow O + H^+$ in the topside ionosphere below approximately 1000 km (Dungey, 1955; Johnson, 1960). This is an accidentally, energetically resonant reaction and is therefore fast in both directions. The reverse reaction is the main chemical sink for H^+.

In the early 1960's, Hanson and Ortenburger (1961) and Hanson and Patterson (1963) investigated ionosphere-plasmasphere coupling. They showed that both the upward and downward H^+ flows are limited by the O^+ diffusive barrier and that, consequently, the F region and protonosphere are only weakly coupled. Subsequently, Geisler (1967) derived an analytical expression for the limiting upward H^+ flux.

A simple picture of ionosphere-plasmasphere coupling is depicted in Figure 1 (Banks and Kockarts, 1973). The plasma is treated as being confined to magnetic flux tubes with only flow along the flux tubes being important. Ideally, in equilibrium, the plasma pressure along the flux tube would be such that there would be no flow (Figure 1a). Although there may be no net flow of ions, in reality, there must always be some flow due to the nonuniform nature of the production and loss mechanisms. Young et al. (1979) showed that in the steady state, significant flows occur with O^+ streaming to high altitudes where it charge exchanges to form H^+ which flows back into the ionosphere.

Even within the inner plasmasphere the plasma is rarely, if ever, in a state of steady equilibrium, mainly because of diurnal variations in the neutral constituents, ion production, and plasma heating. At night, when solar EUV production ceases, the plasma pressure in the equatorial region may exceed that in the topside ionosphere and H^+ flows into the topside ionosphere where it may charge exchange with O to provide a nocturnal source of O^+ (Figure 1b). Conversely, during the day, ion production and thermal breathing in the topside ionosphere generally causes an excess of plasma pressure in the topside and plasma flow will be upward (Figure 1c).

The simple picture depicted in Figure 1 applies at best to a very limited region of the earth's environment within the inner plasmasphere, a torus-shaped region bounded by magnetic field lines that cross the equatorial plane at altitudes of 1-2 earth radii (L = 2-3). Periodically, beyond L = 2 or 3, plasma is rapidly swept away by processes associated with solar activity. When this happens, the plasmaspheric densities are of the order of a few ions cm^{-3}. Once the magnetic activity quietens, the flux tubes begin to refill from the ionosphere, but it takes several days to completely refill an empty flux tube to the stage where there is a daily ebb and flow. By that time, a new magnetic storm usually occurs to once again sweep away the plasma.

In the early stages of flux tube refilling, the flows may be supersonic, collisionless, and anisotropic, and on very long flux tubes, this state, termed the polar wind (Banks and Holzer, 1969), may persist between magnetic storms. This supersonic state has been studied extensively but is outside the scope of the present review which deals with the gentler subsonic flows that occur within the plasmasphere (see Raitt and Schunk (1983) for a comprehensive review of earlier work on the composition and characteristics of the

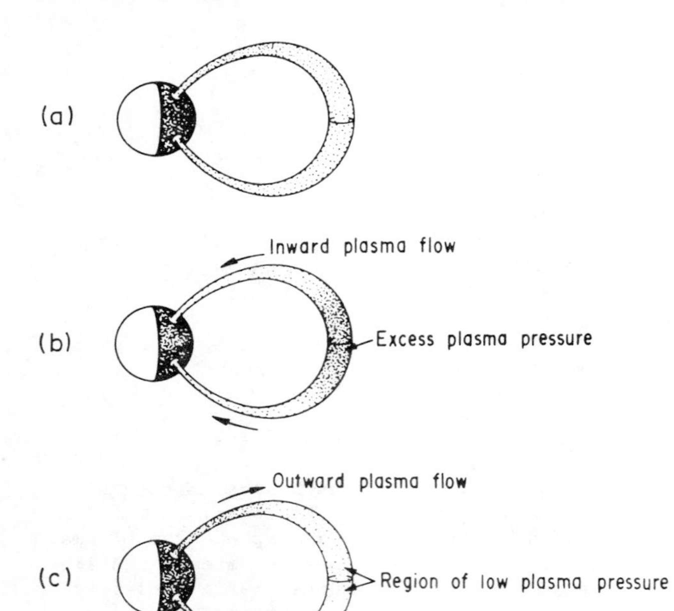

Fig. 1. Schematic illustration of the three basic states of motion for the topside ionosphere. (a) Diffusive equilibrium: static distribution of plasma with small inward flow at low altitudes to accommodate ion losses in the F_2 region. (b) Inward diffusive flow: plasma moves inward and is lost in the F_2 region. Speed of flow may become supersonic if excess plasma pressure is sufficiently large. (c) Outward diffusive flow: plasma moves outward to equalize plasma pressure. Speed of flow may become supersonic if pressure deficit is large (from Banks and Kockarts, 1973).

polar wind). More recent developments in this area are discussed in this volume by Schunk (1987), Gombosi (1987), and Singh (1987).

Fluxes

Figure 2, from Raitt et al. (1975), illustrates the changes in the density and velocity profiles induced by different velocity boundary conditions at 3000 km. Increasing the upper boundary velocity has the same effect as decreasing the high-altitude plasma pressure. As the upper boundary velocity is increased, the density changes from one close to diffusive equilibrium, typical of the inner plasmasphere (curve a), to one appropriate for the outer plasmasphere 2 < L < 6 where the flux tubes are continuously refilling (curves b-f). Curves g and h are more representative of high latitudes where the flows may be supersonic.

For the conditions represented by curve a, the topside O^+ density follows a diffusive equilibrium profile while H^+ undergoes a transition from chemical equilibrium in the region below 1000 km, where H^+ is a minor ion, to diffusive equilibrium where it is the major ion. If H^+ were the major ion, there would be a peak in the H^+ distribution below 1000 km where the diffusive lifetime is approximately equal to the chemical lifetime in a manner analogous to the F_2 layer peak in O^+ density. There is no peak in the equilibrium case at the altitude where the diffusion and chemical lifetimes are equal because the chemical and diffusive scale heights are similar (Hanson and Ortenburger, 1961). The similarity in the chemical and diffusive scale heights is brought about by the influence of the polarization electric field set up by the major O^+ ion. There is a

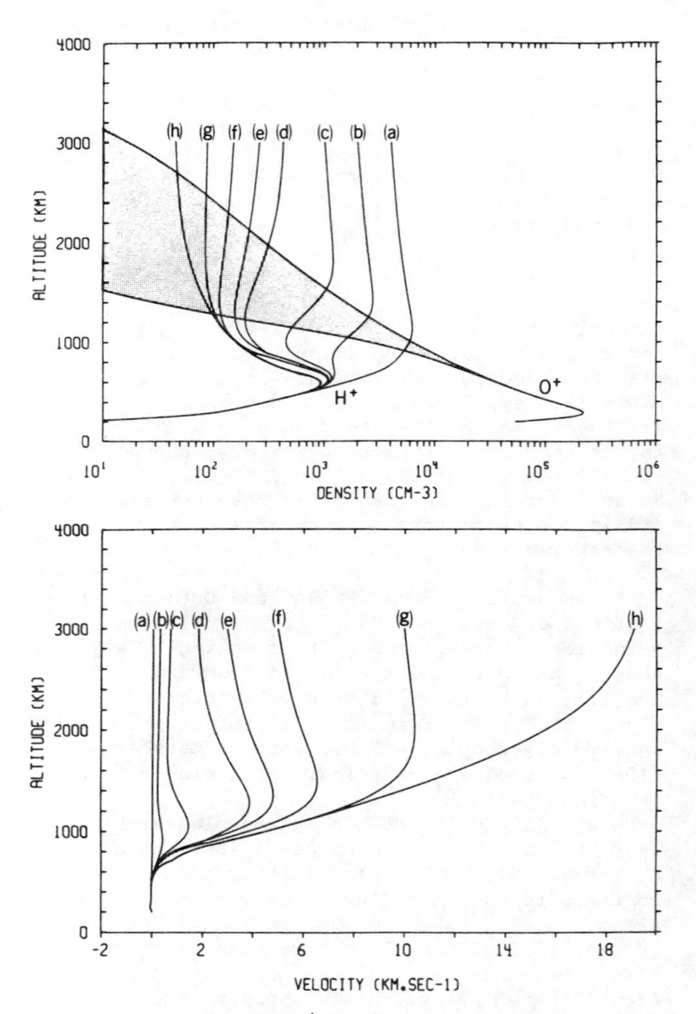

Fig. 2. Theoretical H^+ density and field-aligned drift velocity profiles for the earth's daytime high-latitude ionosphere. The different curves correspond to different H^+ escape velocities at 3000 km: (a) 0.06, (b) 0.34, (c) 0.75, (d) 2.0, (e) 3.0, (f) 5.0, (g) 10.0, (h) 20.0 km s^{-1}. The shaded region shows the range of O^+ densities (from Raitt et al., 1975).

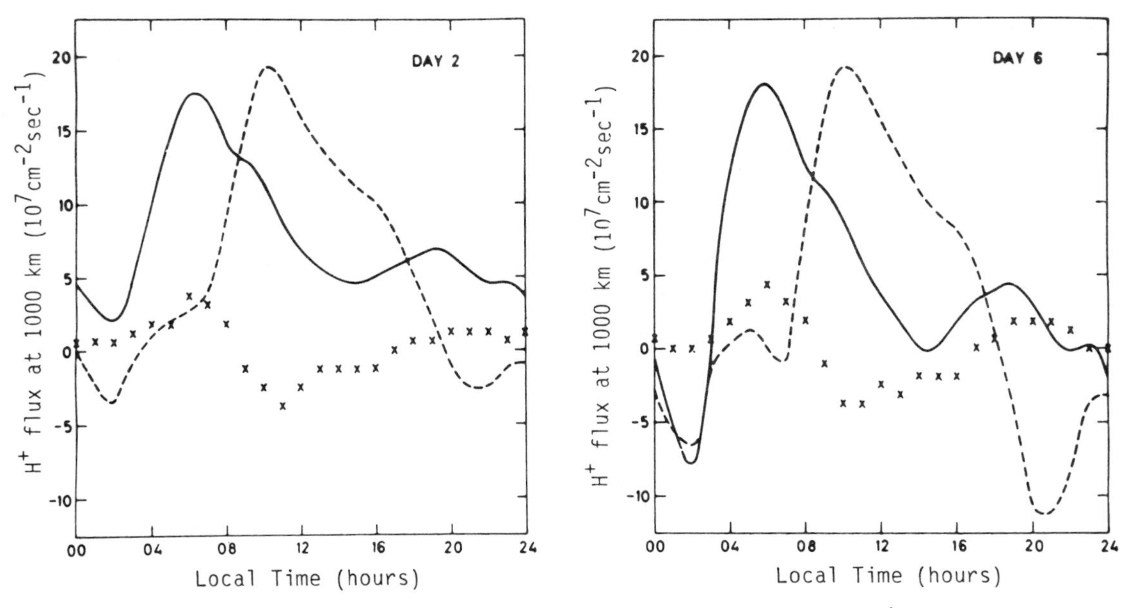

Fig. 3. Calculated daily behavior of the magnetic field-aligned H^+ flux at 1000 km and at the equator for day 2 and day 6: ______, summer hemisphere; -----, winter hemisphere; xxxxx, equator (Bailey et al., 1978).

peak in the H^+ profile above 1000 km where it is the major ion and less influenced by O^+. In contrast, the transition from chemical to diffusive control is evident from the peak in the topside ionosphere in curves b-h where the plasmaspheric density is reduced. In the region where H^+ is major and collisions with O^+ are not important, H^+ still follows a diffusive equilibrium profile. For depleted flux tubes, the magnitude of the upward H^+ flow is determined within one or two scale heights of the altitude for transition from chemical to diffusive control, although the ultimate driving force is the low plasma density at great altitudes.

An interesting feature of Figure 2 is the rapidity with which the H^+ flux at 3000 km approaches saturation as the H^+ density at 3000 km is reduced. Both curves g and h indicate a limiting H^+ flux of about 8 x 10^7 cm^{-2} s^{-1} but the flux for curve b is 6.8 x 10^7 cm^{-2} s^{-1}, which is within 15% of the limiting value. The flux corresponding to curve a is a moderate 2.6 x 10^7 cm^{-2} s^{-1}. Thus, this figure shows that when the plasmaspheric density is reduced by a factor of 2 from the equilibrium value, the upward H^+ flux is already close to its limiting value. This means that the rate of refilling of flux tubes is probably near maximum most of the time, at least during the day.

In order to calculate the plasmaspheric density distribution, it is necessary to solve the equations of continuity and momentum for the major species (O^+, H^+). These are coupled nonlinear, partial differential equations that require numerical solutions. Early attempts at numerical solution for these equations were beset by instability problems caused by the large diffusion coefficients in the plasmasphere and the long distances that have to be covered. Moffett and Murphy (1973) developed a searching procedure which proved to be fast and stable. Young et al. (1980a,b) developed a flux preserving approach and used a Newton iterative method to improve the accuracy and stability of the solutions. The use of these techniques also eliminated the need to linearize the equations. Subsequently, the increase in the size and accuracy of computers has overcome the problems experienced earlier.

The English group of Moffett, Murphy, and Bailey have made extensive investigations of the coupling of the ionosphere and plasmasphere. Figure 3, from Bailey et al. (1978), shows the diurnal variation of the H^+ flux at the 1000-km level for days 2 and 6 of a refilling flux tube and for both summer and winter hemispheres. Note that the daytime variations are very similar for both days. In both hemispheres, the H^+ flux peaks soon after sunrise and then decreases, but in the summer hemisphere there is a minimum in the afternoon that is absent from the winter curves. Although Bailey et al. (1978) do not comment on this behavior, the minimum is probably caused in part by a minimum in the H density at that time. Other factors that could affect the upward flux in this way include: (1) the poleward neutral wind that is maximum in the afternoon in the model, and (2) a decrease in the topside plasma temperatures. The neutral wind affects the flux by lowering the F_2 layer and reducing the topside O^+ density. At night the flows are

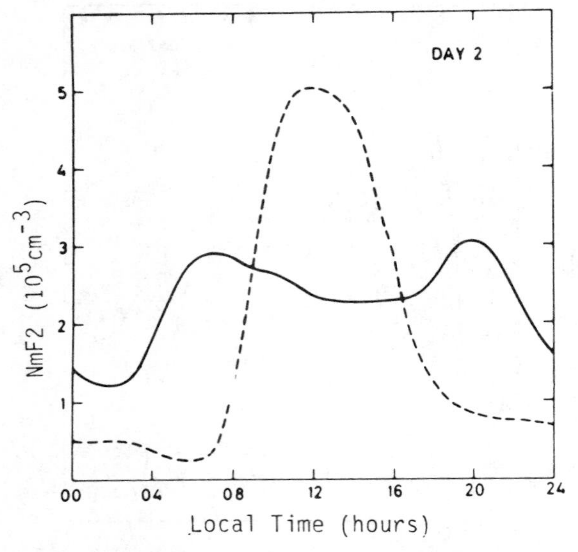

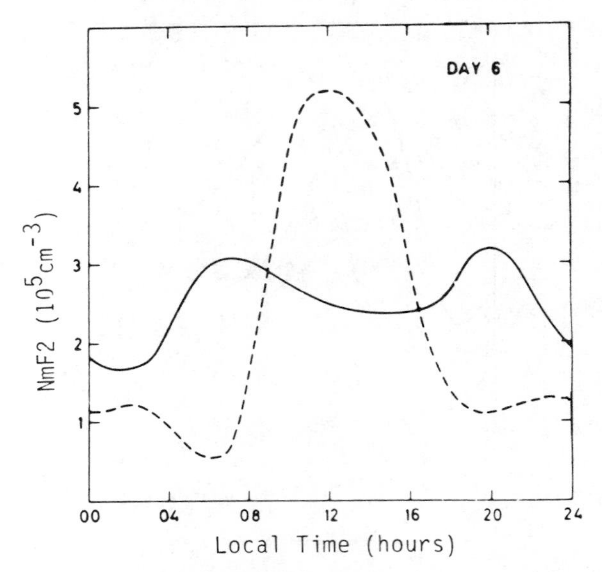

Fig. 4. Calculated daily behavior of NmF_2 for day 2 and day 6: ______, summer hemisphere; -----, winter hemisphere.

generally small and downward except in the summer hemisphere on day 2 where the flow is upward at all times. The interhemisphere flux (multiplied by 10) which is also shown in Figure 3 shows that during the day the flow is from the winter to the summer hemisphere.

Since the plasmasphere acts as both a source and sink for ionospheric plasma, it is of interest to see how the state of replenishment of the plasmasphere affects the ionosphere. In particular, there is the perennial question of whether or not the plasmasphere helps to support the nocturnal F_2 layer. Figure 4, also from Bailey et al. (1978), shows the behavior of the F_2 region peak density (N_mF_2) for the two days depicted in Figure 3 where the plasmasphere was more depleted on day 2 than on day 6. The effects are small during the day, but at night on day 6 the winter N_mF_2 is maintained a factor of 2 higher than for day 2 purely as a result of the increased night flow of H^+ into the ionosphere as the plasmaspheric flux tube fills. Moreover, this figure also reveals that the nighttime F region can be maintained without the assistance of a plasma flow from the plasmasphere. On day 2 the summer H^+ fluxes are upward throughout the day and night, yet the F region density is maintained above 10^5 cm^{-3} during that time.

Richards and Torr (1985) have studied the seasonal and solar cyclical variations of the H^+ flux on depleted flux tubes. The seasonal variation is only of the order of 50% with a maximum in winter and minimum in summer (Figure 5). The seasonal variation is mainly due to the seasonal variation in the H density. The H^+ flux is most sensitive to the topside H density and O^+ scale height. The theoretical variation of the H^+ flux as a function of solar activity index $F_{10.7}$ is displayed in Figure 6. There is a factor of 3 variation on the H^+ flux with solar cycle. The flux tends to follow the H variation, but a number of factors, including a reduction in topside plasma temperature, help to reduce the variation

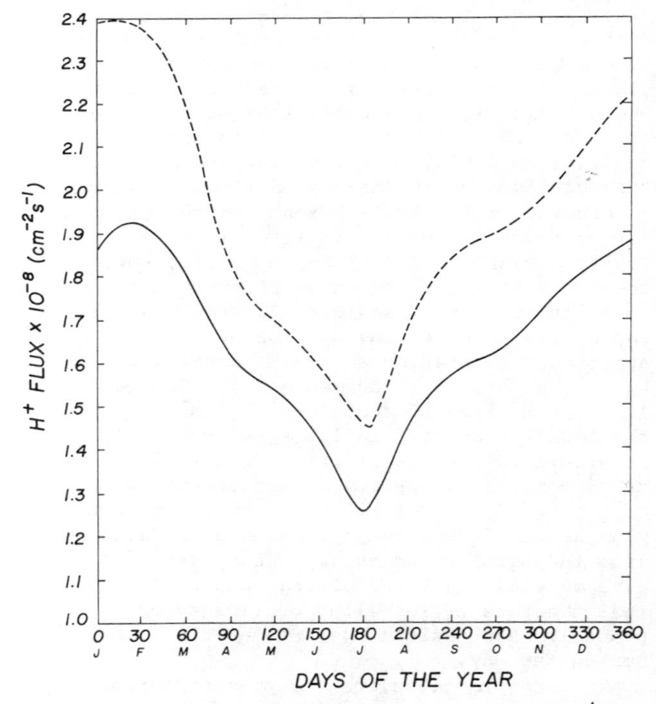

Fig. 5. Seasonal variation in the limiting H^+ flux at solar minimum. The solid line is the flux from the full model and the dashed line is the flux predicted by the analytical formula. The $F_{10.7}$ index was 70 and the day number was 120 (Richards and Torr, 1985).

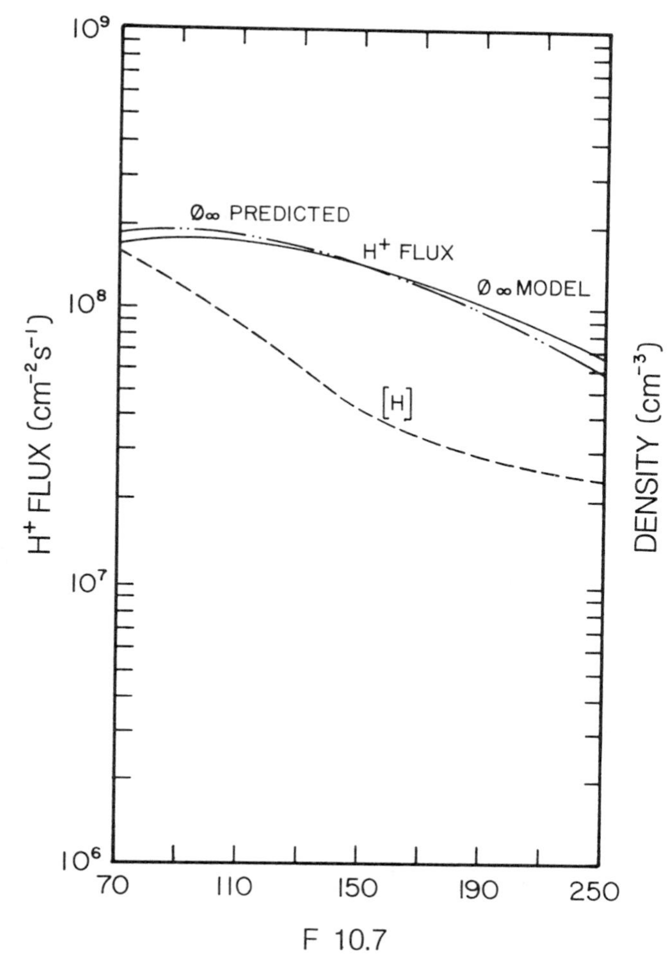

Fig. 6. Diurnal variation in the limiting H^+ flux at solar maximum. The dashed line is the flux from the full model and the solid line is the flux predicted by the analytical formula. The day number is 120 (Richards and Torr, 1985).

in H^+ flux below that of H (Richards and Torr, 1985).

Interhemispheric Fluxes

The sensitivity of the H^+ flow to the topside O^+ density scale height produces some interesting effects related to conjugate photoelectron heating. The diurnal calculation of the H^+ flux shown in Figure 7, from Richards and Torr (1986), was done for a tilted dipole magnetic field geometry in which the southern (winter) end of the flux tube was 1.6 hours behind the northern (summer) end in local time. When plotted as a function of local time, the winter (dark) hemisphere H^+ flux shows a pronounced peak 1.6 hours in local time before sunrise in the summer hemisphere and several hours before local sunrise. This phenomenon results from the strong thermal coupling of the conjugate hemispheres. Photoelectrons produced in the sunlit hemisphere escape into the plasmasphere where they heat the ambient thermal electrons via Coulomb collisions. This heat is then conducted down to both hemispheres where it is lost to the ions and neutrals. Direct heating also occurs as conjugate photoelectrons precipitate into the ionosphere. The heating causes a sharp rise in the plasma temperature and a consequent increase in the O^+ scale height. As a result, H^+, which has settled into the topside ionosphere during the period of darkness, is expelled into the plasmasphere. Once the initial H^+ has been expelled, there is insufficient production to maintain the large upward flux, and there is a gradual relaxation to near normal nighttime values until at local sunrise there is sufficient production to generate a large upward flux.

The effects of conjugate photoelectron heating on protonospheric content, slab thickness, and shape factor have been studied by Sethia et al. (1985). These effects are shown in Figure 8 for shape factor (defined as the electron density weighted average of the gyrofrequency along the ray path) and protonospheric content. The curves labeled A include conjugate heating, while the curves labeled D do not. The C curves are for the case where the heating rates of case A were reduced by a factor of 2. Observed variations are also shown. It is clear that neglect of conjugate photoelectron heating reduces the protonospheric content (N_p) and thereby increases the shape factor (F).

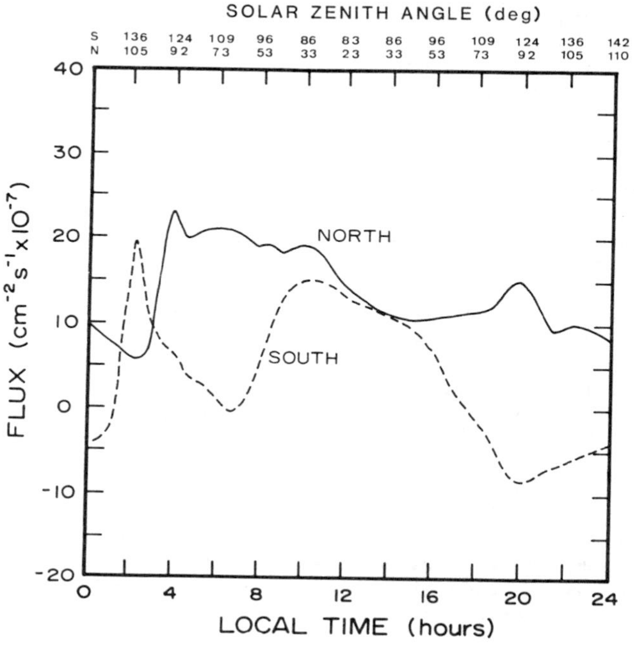

Fig. 7. Diurnal variation of the H^+ fluxes through the 3000-km level as a function of local time in the northern hemisphere at L = 3 at the June solstices (Richards and Torr, 1986).

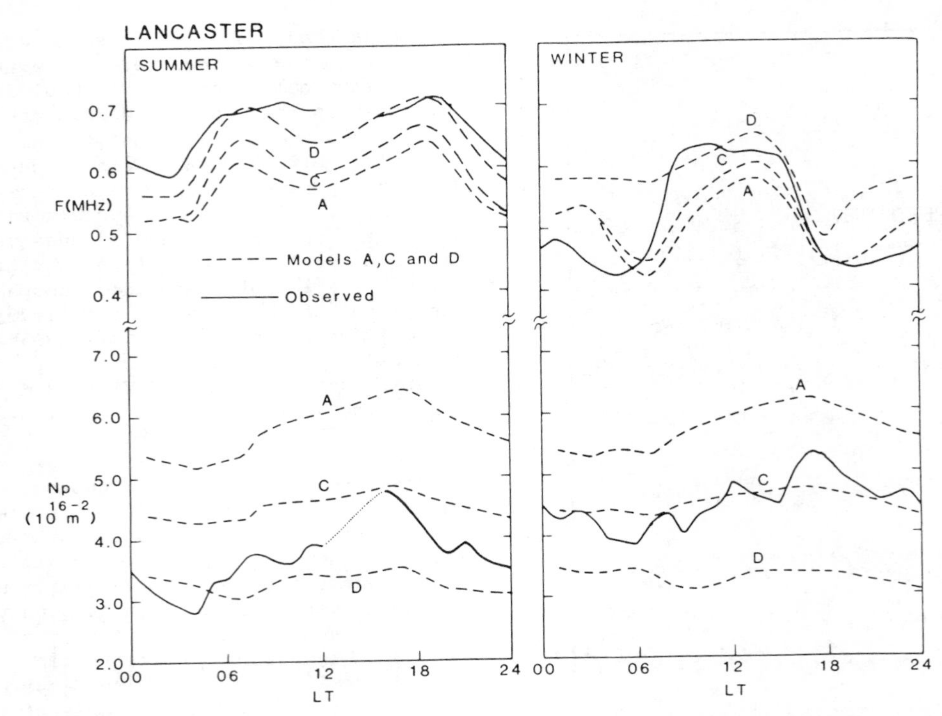

Fig. 8. Effects of plasma heating by conjugate photoelectrons on shape factor (F) and protonospheric content (N_p) (Sethia et al., 1985).

Temperatures

The results discussed above illustrate the importance of the thermal structure of the plasmasphere, especially the thermal coupling of the conjugate ionospheres. Yet, until recently, very little attention was paid to the thermal structure. This lack of consideration for thermal effects may have resulted from the difficulty of calculating appropriate heating rates or because early measurements of the temperature at high altitudes appeared to indicate much higher temperature gradients along the field lines than would be expected theoretically. Furthermore, there were large uncertainties in the photoelectron flux due to uncertainties in the solar EUV flux and electron impact cross sections.

Recent measurements of the plasma temperatures within the plasmasphere are more in line with theoretical expectations. The measurements of Rich et al. (1979) showed that the temperature gradient along the field line is very small at high altitudes and that the temperature is of the order of 6000 K. With the improvements in both theory and measurement, we are now in a position to make more detailed comparisons of calculated plasmaspheric temperatures with measured temperatures.

Figure 9, from Brace and Theis (1981), shows electron temperature measurements at 1400 km in the equatorial plane that were taken by the ISIS 2 satellite during equinox in 1971-1973. They also presented 3000-km data from ISIS 1 for 1969-1970 and 400-km data from Atmosphere Explorer C in 1975-1978. These data are ideal for comparison with theoretical models because they were taken in a plasmaspheric region that should be well behaved. That is, the flux tubes are near full and there is thought to be little interaction with high-energy particles and convection electric fields. Richards and Torr (1985) have developed an interhemispheric model which not only solves the continuity and momentum equations for O^+ and H^+ and the energy equations for T_e and T_i but also solves the two-stream equation for the photoelectron flux. The model takes only the neutral atmosphere and solar EUV as inputs and calculates the densities and temperatures of the species along complete field lines.

In Figure 10, the data of Brace and Theis from Figure 9 are compared with the calculated diurnal variation from the model. In this low-altitude equatorial region, the agreement between the theory and data is good. The model temperature slightly exceeds the measured temperatures, but the difference is well within the limits of the accuracy of the model inputs.

At higher altitudes the agreement is not so good and the model temperatures are usually lower than the measured temperatures. Figure 11 compares the model temperatures with the 3000-km ISIS 1 equatorial data presented by Brace and

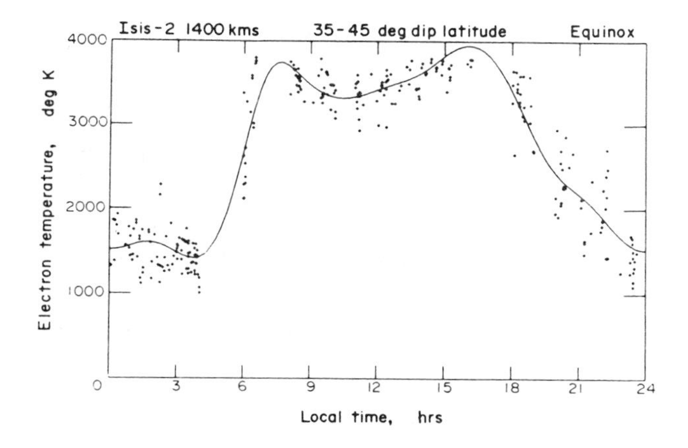

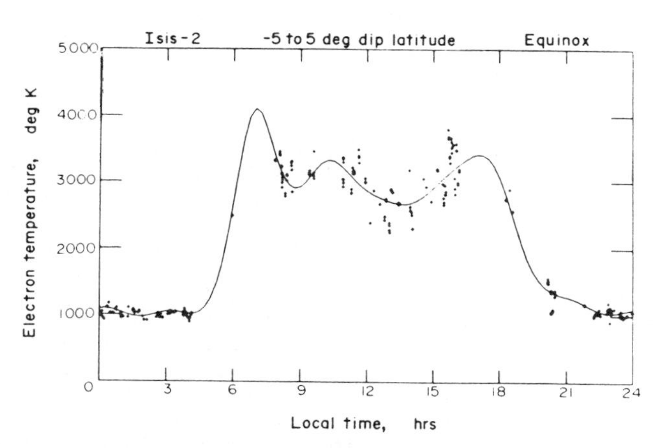

Fig. 9. Electron temperature at 1400 km from the ISIS 2 satellite during 1971-1973. Upper panel: L = 2.2 to 3.0. Lower panel: L = 1.2 (Brace and Theis, 1981).

Theis (1981). The two peaks at sunrise and sunset are not well reproduced by the theory. There is a broad morning peak in the theoretical curve which results from a peak in the photoelectron flux in the morning. There is also a hint of a peak near sunset for similar reasons, but neither peak is as pronounced as those in the measurements. The difference between the theory and data is significant because the plasmaspheric heating rate would have to be doubled to achieve the measured temperatures at sunrise and sunset. Another discrepancy between the model and measurement is evident at night. The L = 1.5 flux tube does not have a large enough reservoir of heat to maintain the electron temperature above the neutral temperature at night, and the model temperature quickly decays to the neutral temperature after sunset. However, the measured temperature remains well in excess of the neutral temperatures. The ISIS 1 data have recently been

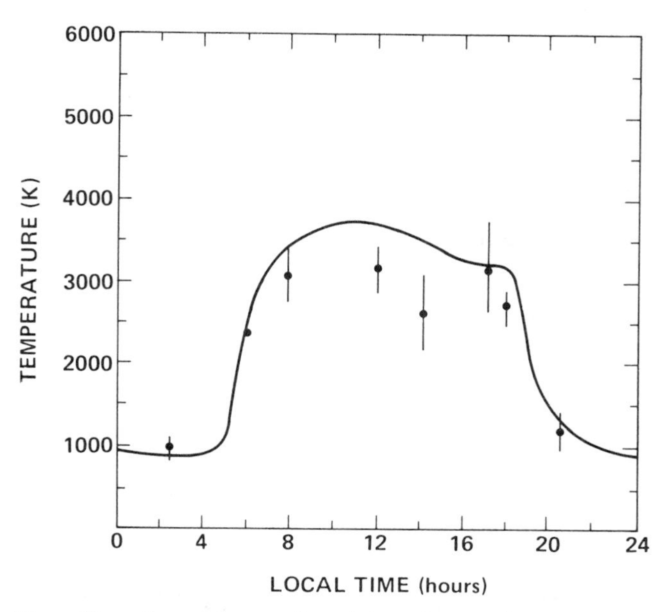

Fig. 10. Comparison of model values of T_e at 1400 km at equatorial latitudes with the data from Brace and Theis (1981). The bars indicate the spread in the data.

reexamined by Kohnlein (1986) who found that the ISIS 1 temperatures were always too high when compared to Jicamarca incoherent scatter data at 600 km and also when compared to other satellite measurements. The renormalized ISIS 1 data, shown in Figure 11, are in excellent agreement with the theory except for the noon minimum that

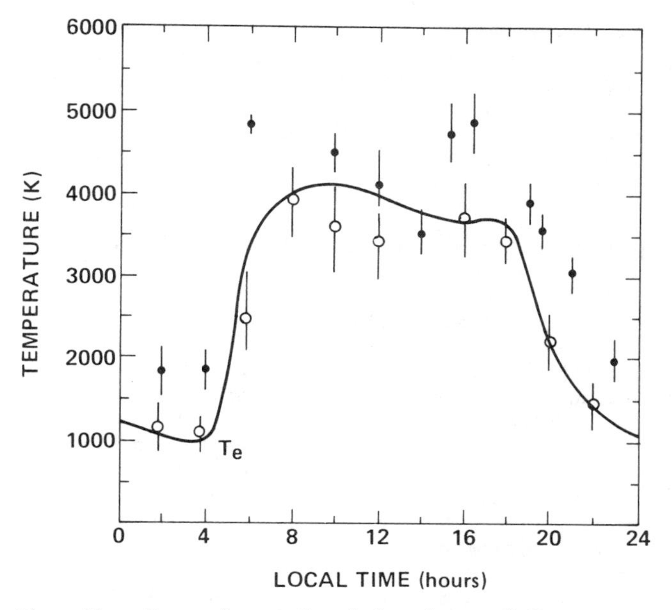

Fig. 11. Comparison of model values of T_e at 3000-km altitude at equatorial latitudes compared with the observations of Brace and Theis (1981).

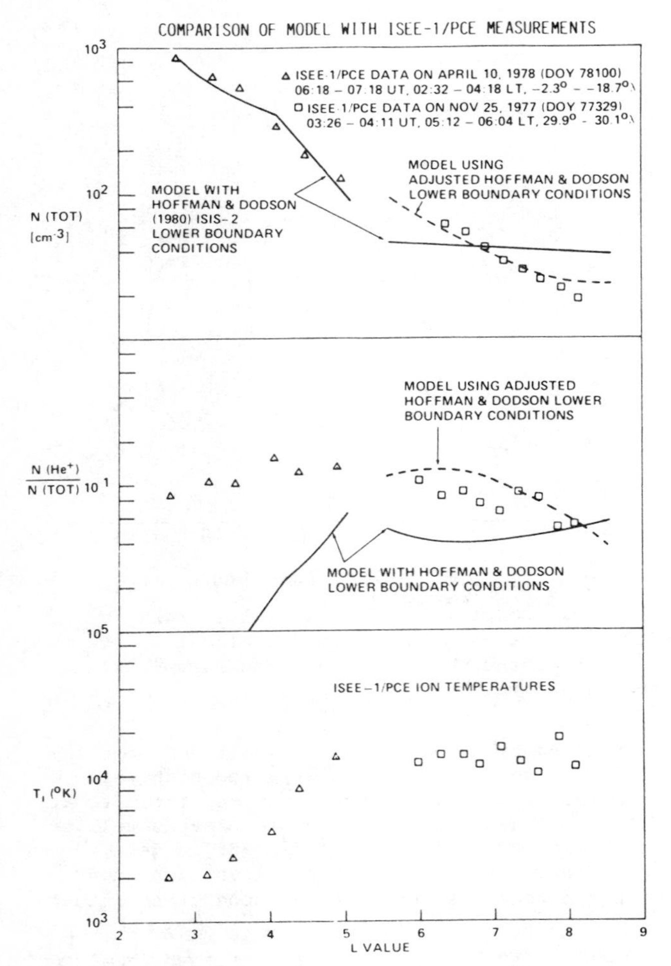

Fig. 12. Comparison of model predictions for total densities and He^+: total ion ratio with ISEE 1 PCE data for two passes, plotted versus L value. The bottom panel gives the ion temperatures as determined by the ISEE 1 PCE for the two passes (Waite et al., 1984).

appears in the data but not in the theoretical calculations.

If photoelectrons were the only plasmaspheric heat source, there would be little change in equatorial T_e and T_i with increasing L value. This expectation comes from the fact that the heating rate is proportional to the photoelectron flux and the total electron content of the plasmasphere, and neither are expected to have large latitudinal gradients. The observations of Comfort et al. (1985) do show a constant T_i up to L = 2.8 but in the region of the plasmapause, there is a prominent peak in T_i. Similar signatures of the plasmapause have been regularly observed in the topside T_e beginning with Brace et al. (1967). This enhancement has recently been examined in great detail by Kozyra et al. (1986, 1987). They attribute the elevated temperature to transfer of energy from the ring current ions to the thermal electrons. Horwitz et al. (1986a) showed that the enhancement in T_i often coincides with enhancements in the high-altitude O^+ and O^{++} and Chandler et al. (1987), using the model developed by Richards and Torr (1985), were able to show that the high electron temperatures would lead to increased O^+ and O^{++} at high altitudes.

Similar double peaked T_e behavior has been observed at 800 km in the topside ionosphere over Millstone Hill (L = 3.2) at equinox (Roble, 1975). By using the T_e measurements as upper boundary conditions in his ionospheric model, Roble deduced the required heat flow through the 800-km level. The heat flow was steady during the day but just prior to sunset there was a factor of 2 increase in heat flow that decreased again after sunset.

Electric Fields

Convective electric fields have profound effects on the structure of the ionosphere and plasmasphere. Brief discussions with references can be found in reviews by Horwitz et al. (1986b) and Richards et al. (1986).

Early work on the effects of large-scale plasmaspheric convection was done by Chen and Wolf (1972) who showed that a time-varying electric field can cause the plasmasphere to develop a long tail that gradually wraps itself around the main plasmasphere. Periodic gusts in a spatially uniform convection field produce extremely complicated fine structure. Chen and Wolf (1972) used a simple model for the refilling of flux tubes, but Marubashi (1979) solved the continuity and momentum equations between 500 and 3000 km in the presence of convective electric fields in order to examine the dynamic behavior of the coupled ionosphere-protonosphere system. He found that the expansion of magnetic flux tubes on the dusk side causes enhancement of the upward H^+ flow; whereas the contraction of the dawn side causes enhancement of the downward flow.

By calculating the drift paths of plasma density irregularities taking into account plasma interchange motion, Lemaire and Kowalkowski (1981) showed that plasma interchange motion plays a determining role in the formation of the equatorial plasmapause.

Bailey (1983) has investigated the effect of $\mathbf{E} \times \mathbf{B}$ drifts on the ionosphere at L = 1.4. The effect of $\mathbf{E} \times \mathbf{B}$ drift is to reduce the calculated daytime electron temperature at 1000 km by several hundred degrees. The effects of cross-L drifts on the behavior of the thermal plasma at mid-latitudes was studied by Murphy et al. (1980). They found that the large changes in H^+ tube content at L = 3.2 that occur as the protonosphere is replenished can be explained in terms of electromagnetic drifts. The combined effect of diurnal drifts and frequency of magnetic storms can produce multiple peaks in the diurnal behavior of the H^+ tube content.

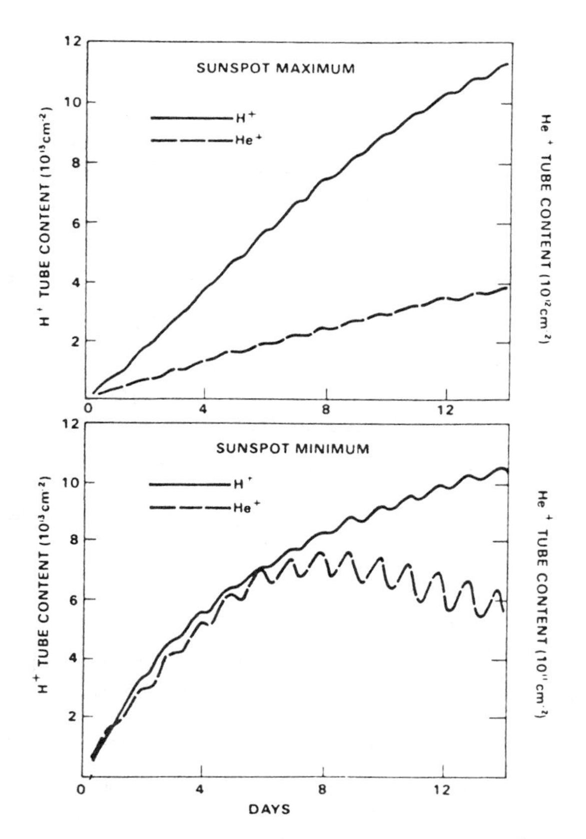

Fig. 13. Time-dependent behavior of the H^+ and He^+ flux tube contents over the integration period (Naghmoosh and Murphy, 1983). Note that a different scale is used for He^+ on the right vertical axes.

He^+

H^+ is the dominant species in the plasmasphere but there are times when both He^+ and O^+ are comparable with H^+ (Horwitz et al., 1984). Indeed, according to Horwitz et al. (1986b) the He^+/H^+ concentration ratio appears to be about 0.2 most of the time, independent of large differences in the overall plasmaspheric density and ionospheric characteristics. By using a diffusive equilibrium model Waite et al. (1984) found that ionospheric composition and temperature distribution along flux tubes are important factors controlling equatorial He^+ composition, through the plasma scale height and thermal diffusion effects. Some of the calculations of Waite et al. (1984) are shown in Figure 12. This figure shows the total ion (upper panel) and the proportion of He^+ (middle panel) plotted versus L for 2 ISEE 1 plasma composition experiment (PCE) passes. The bottom panel of Figure 12 shows the ion temperatures. Waite et al. (1984) found that direct comparison of plasmaspheric H^+ and He^+ observations indicates good agreement with total densities but comparisons with the $He^+:H^+$ ratio show mixed results. While a steady state diffusive equilibrium model helps to illuminate the relevant processes, the long time constants for refilling plasmaspheric flux tubes require that the time-dependent problem be solved.

A time-dependent study has been made by Naghmoosh and Murphy (1983) at sunspot minimum and sunspot maximum. Figure 13 compares the time-dependent behavior of He^+ and H^+ total tube content for both solar maximum and minimum. The time-dependent behavior is different for the two periods of the solar cycle. The sunspot maximum variation in the top panel shows a continuing divergence of the flux tube contents over a 10-day simulation. Because most of the volume of a flux tube is in the equatorial plane, the He^+/H^+ ratio is approximately the same as or less than the ratio of the total tube contents. On day 8, the ratio of 0.03 is an order of magnitude lower than the experimental data of Horwitz et al. (1986b). Note that a different scale is used for He^+ on the right vertical axes. The sunspot minimum behavior differs from the maximum behavior in that the ratio of the total tube contents begins to decrease after 7 days. This is attributed to the increasing downward H^+ flux dragging the He^+ to low altitudes where the chemical loss rate is high.

Thus, the measurements which indicate He^+/H^+ concentration ratios of the order of 0.2 demand order of magnitude larger upward He^+ fluxes than are currently predicted by the theory. Alternatively, the calculated H^+ fluxes may be too high, but this appears unlikely because many theoretical models are in agreement and are also in agreement with measurements of Hoffman and Dodson (1980).

Conclusion

Our knowledge of the plasmasphere has increased steadily since its existence was first inferred in the early 1950's. The basic plasma processes in the well-behaved subsonic region in the inner plasmasphere appear to be well understood and the numerical models are well developed. A number of problems still remain to be solved. These problems include the relationship between the ionospheric and plasmaspheric signatures of the plasmapause, the details of the convection effects, and the anomalously high He^+-to-H^+ ratios that seem ubiquitous in the plasmasphere.

Acknowledgments. This work was supported by NSF grants ATM-8545227, ATM-8603264, and ATM-8714461 and NASA grants NAGW-922 and NAGW-996 at The University of Alabama in Huntsville.

References

Bailey, G. J., The effect of a meridional E x B drift on the thermal plasma at L = 1.4, Planet. Space Sci., 31, 4, 389, 1983.
Bailey, G. J., R. J. Moffett, and J. A. Murphy, Interhemispheric flow of thermal plasma in a

closed magnetic flux tube at mid-latitudes under sunspot minimum conditions, Planet. Space Sci., 26, 753, 1978.

Banks, P. M., and T. E. Holzer, High latitude plasma transport: The polar wind, J. Geophys. Res., 74, 6317, 1969.

Banks, P. M., and G. Kockarts, Aeronomy, Academic Press, New York, 1973.

Brace, L. H., B. M. Reddy, and H. G. Mayr, Global behavior of the ionosphere at 1000-kilometer altitude, J. Geophys. Res., 72, 265, 1967.

Brace, L. H., and R. F. Theis, Global empirical models of ionospheric electron temperature in the upper F-region and plasmasphere based on in situ measurements from the Atmosphere Explorer-C, ISIS-1 and ISIS-2 Satellites, J. Atmos. Terr. Phys., 43, 1317, 1981

Chandler, M. O., J. U. Kozyra, J. L. Horwitz, R. H. Comfort, and L. H. Brace, Modeling of the thermal plasma in the outer plasmasphere: A magnetospheric heat source, this volume, 1987.

Chen, A. J., and R. A. Wolf, Effects on the plasmasphere of a time-varying convection electric field, Planet. Space Sci., 20, 483, 1972.

Comfort, R. H., J. H. Waite, Jr., and C. R. Chappell, Thermal ion temperatures from the retarding ion mass spectrometer on DE 1, J. Geophys. Res., 90, 3475, 1985.

Dungey, J. W., The physics of the ionosphere, Physical Society, p. 406, London, 1955.

Geisler, J. E., On the limiting daytime flux of ionization into the protonosphere, J. Geophys. Res., 72, 81, 1967.

Gombosi, T. I., Time-dependent polar wind models, presented First Huntsville Workshop on Magnetosphere Ionosphere Plasma Models, Guntersville State Park, Alabama, 1986.

Hanson, W. B., and I. B. Ortenburger, The coupling between the protonosphere and the normal F region, J. Geophys. Res., 66, 1425, 1961.

Hanson, W. B., and T.N.L. Patterson, Diurnal variation of the hydrogen concentration in the exosphere, Planet. Space Sci., 11, 1035, 1963.

Hoffman, J. H., and W. H. Dodson, Light ion concentrations and fluxes in the polar regions during magnetically quiet times, J. Geophys. Res., 85, 626, 1980.

Horwitz, J. L., R. H. Comfort, and C. R. Chappell, Thermal ion composition measurements of the formation of the new outer plasmasphere and double plasmapause during storm time recovery phase, Geophys. Res. Lett., 11, 701, 1984.

Horwitz, J. L., R. H. Comfort, and C. R. Chappell, Plasmasphere and plasmapause region characteristics as measured by DE-1, Adv. Space Res., 6, 21, 1986b.

Horwitz, J. L., L. H. Brace, R. H. Comfort, and C. R. Chappell, Dual-spacecraft measurements of plasmasphere-ionosphere coupling, J. Geophys. Res., 91, 11,203, 1986a.

Johnson, F. S., The ion distribution above the F_2 maximum, J. Geophys. Res., 65, 577, 1960.

Kohnlein, W., A model of the electron and ion temperatures in the ionosphere, Planet. Space Sci., 34, 609, 1986.

Kozyra, J. U., L. H. Brace, T. E. Cravens, and A. F. Nagy, A statistical study of the subauroral electron temperature enhancement using Dynamics Explorer 2 Langmuir probe observations, J. Geophys. Res., 91, 11,270, 1986.

Kozyra, J. U., W. K. Peterson, R. H. Comfort, L. H. Brace, T. E. Cravens, and A. F. Nagy, The role of ring current O^+ in the formation of stable auroral red arcs, J. Geophys. Res., in press, 1987.

Lemaire, J., and L. Kowalkowski, The role of plasma interchange motion for the formation of a plasmapause, Planet. Space Sci., 29, 469, 1981.

Marubashi, K., Effects of convection electric field on the thermal plasma flow between the ionosphere and the protonosphere, Planet. Space Sci., 27, 603, 1979.

Moffett, R. J., and J. A. Murphy, Coupling between the F-region and protonosphere: Numerical solution of the time-dependent equations, Planet. Space Sci., 21, 43, 1973.

Murphy, J. A., G. J. Bailey, and R. J. Moffett, A theoretical study of the effects of quiet-time electromagnetic drifts on the behavior of thermal plasma at mid-latitudes, J. Geophys. Res., 85, 1979, 1980.

Naghmoosh, A. A., and J. A. Murphy, A comparative study of H^+ and He^+ at sunspot minimum and sunspot maximum, J. Atmos. Terr. Phys., 38, 673, 1983.

Raitt, W. J., and R. W. Schunk, Composition and characteristics of the polar wind, in Energetic Ion Composition in the Earth's Magnetosphere, edited by R. G. Johnson, pp. 99-141, Terra Scientific Publ. Co., Tokyo, 1983.

Raitt, W. J., R. W. Schunk, and P. M. Banks, A comparison of the temperature and density structure in high and low speed thermal proton flows, Planet. Space Sci., 23, 1103, 1975.

Rich, F. J., R. C. Sagalyn, and P.J.L. Wildman, Electron temperature profiles measured up to 8000 km by S3-3 in the late afternoon sector, J. Geophys. Res., 84, 1328, 1979.

Richards, P. G., and D. G. Torr, Seasonal, diurnal, and solar cyclical variations of the limiting H^+ flux in the Earth's topside ionosphere, J. Geophys. Res., 90, 5261, 1985.

Richards, P. G., and D. G. Torr, Thermal coupling of conjugate ionospheres and the tilt of the Earth's magnetic field, J. Geophys. Res., 91, 9017, 1986.

Richards, P. G., D. G. Torr, J. L. Horwitz, and M. R. Torr, Models of the plasmaspheric thermal plasma distribution, Adv. Space Res., 6, 141, 1986.

Roble, R. G., The calculated and observed diurnal variation of the ionosphere over Millstone Hill on 23-24 March 1970, Planet. Space Sci., 23, 1017, 1975.

Schunk, R. W., The polar wind, this volume, 1987.

Sethia, G. C., G. J. Bailey, R. J. Moffett, and

J. K. Hargreaves, Mid-latitude electron content modelling: The role of interhemispheric coupling, Planet. Space Sci., 33, 321, 1985.

Singh, N., Plasma processes in the refilling of a plasmaspheric flux tube - A review, this volume, 1987.

Storey, L.R.O., An investigation of whistling atmospherics, Phil. Trans. Roy., A246, 113, 1953.

Waite, J. H., J. L. Horwitz, and R. H. Comfort, Diffusive equilibrium distributions of He^+ in the plasmasphere, Planet. Space Sci., 32, 611, 1984.

Young, E. R., D. G. Torr, and P. G. Richards, Counterstreaming of O^+ and H^+ ions in the plasmasphere, Geophys. Res. Lett., 6, 925, 1979.

Young, E. R., P. G. Richards, and D. G. Torr, A flux preserving method of coupling first and second order equations to simulate the flow of plasma between the protonosphere and the ionosphere, J. Comp. Phys., 38, 141, 1980a.

Young, E. R., D. G. Torr, P. Richards, and A. F. Nagy, A computer simulation of the mid-latitude plasmasphere and ionosphere, Planet. Space Sci., 28, 881, 1980b.

PLASMASPHERE AND RING CURRENT ELECTRIC FIELDS OBSERVED BY GEOS 2

R. Schmidt and A. Pedersen

Space Science Department of ESA/ESTEC, Noordwijk, The Netherlands

Abstract. The electric field double probe data from GEOS 2 have been statistically examined to study the consecutive passage of the afternoon plasmaspheric bulge and the trough at the geostationary orbit. It was found that the average location of the bulge depends on the magnetic activity and was encountered at earlier local times for higher magnetospheric activity. Within the bulge the electric field showed very frequently a typical directional change from dawnward outside to duskward inside the bulge. The magnitude of the magnetic field was frequently much smaller near the outbound crossing of the plasmaspheric bulge than is expected from a long-term average. The $\mathbf{E} \times \mathbf{B}/B^2$ drift pointed azimuthally eastward prior tc the encounter of the bulge and rotated into the sunward direction within the bulge. Following its passage through the dense, cold plasma in the bulge, GEOS 2 encountered a hot and tenuous plasma sheet-type plasma in the trough that occasionally corrupted the electric field measurements. Generally, the electric field in the trough is much smaller than in the bulge. A possible cause of the sunward plasma flow within the bulge is discussed on the basis of these data.

Introduction

Electric field measurements in the plasmasphere are sparse. Pedersen and Grard (1979) reported for some selected intervals that their double probe electric field data from GEOS 1 and GEOS 2 revealed the existence of large-amplitude electric fields (>3 mV m^{-1}) between $5 < L < 7.5$ in the local afternoon sector. The orientation of the electric fields is predominantly from dawn to dusk and the inferred $(\mathbf{E} \times \mathbf{B})/B^2$ drift therefore sunward. Data obtained inside L = 6 by the cylindrical probes on ISEE 1 have been reported by Maynard et al. (1983). Their Figure 6 contains the equatorial projection of a 1-year data sample. Their averaged electric fields are smaller than the individual cases reported by Pedersen and Grard (1979). However, similar tendencies toward larger than average electric fields in the local afternoon to evening sector are evident also inside L = 6.

Plasma instruments also detected sunward flows corresponding to electric fields pointing in the duskward direction. Chappell (1974) explained the drift of detached regions of high-density, cold plasma in the afternoon-dusk sector as the peeling off of the plasmasphere's outer layers in the dusk bulge region and the subsequent convection of these regions to higher L shell values in a sunward direction. Chappell attributed the detachment of the plasma to enhanced magnetospheric convection electric fields which exist during substorms. He concluded that the occurrence of detached plasma regions exhibits a preference for moderate and disturbed magnetospheric conditions ($3 < Kp < 5$), based on his correlation between the number of observations of detached plasma and Kp. Convection would then transport these detached plasma regions in the sunward direction. Further observations of isolated plasma regions were reported by Maynard and Chen (1975) using data from the equatorially orbiting Explorer 45. Its apogee at 5.2 R_E limited their observations of isolated regions of enhanced cold plasma to times when the plasmapause occurs at low L shells (magnetic storms). The authors found that the occurrence of such blobs inside L = 5.2 has a strong correlation with a negative Dst which rather pionts toward changes in the large-scale convection structure and a triggering by magnetic storms.

Lennartsson and Reasoner (1978) reported low-energy plasma observations from the geosynchronous spacecraft ATS 6. Their Figure 8 summarizes the flow direction of low-energy ions. It must be pointed out that the only plasma flow in the equatorial plane which can be directly measured by the ATS 6 particle detectors is the eastward or westward flow within 20° of the tangent of the orbit. From the observed east-west asymmetries in the measured ion flow patterns (see Figure 8 in Lennartsson and Reasoner, 1978), Lennartsson and Reasoner inferred a dominant low-energy ion flow pattern in the sunward direction after 1400 LT (see their Figure 10). Nagai et al. (1983) calculated the flow direction of the non-field-aligned ions (perpendicular to the magnetic field) "from the flux peak direction relative to the sun direction." Although Nagai et al. (1983) did not correct the data for satellite motion, the ISEE 1 data clearly indicate a consistent, sunward, anti-corotational flow in the 1200-2000 LT interval (see Figure 5 in Nagai et al., 1983) which is consistent with the sunward, anti-corotational, low-energy ion flows observed by Lennartsson and Reasoner (1978; see their Figures 8 and 10). These flows occurred almost exclusively in the noon to dusk sector at $L \simeq 6$ to 9. Nagai et al. (1985) presented density profiles of some ISEE 1 plasmaspheric bulge crossings and reported the frequent occurrence of sunward convective flows of low-energy plasma in the plasmasphere between $L = 6$ and $L = 8$.

Spiro et al. (1981) simulated the evolution of the plasmapause during magnetic substorms. Their results are relevant to this study because the simulations predict growth and motion of the bulge region which are consistent with the observed increase in magnitude and the direction of the electric field with increasing magnetic activity in the afternoon sector. The theory predicts that filamentary structures are drawn out from the dusk plasmasphere during magnetic substorms and subsequently drift toward the daytime magnetopause. The evolution of these electric field structures is consistent with the origin and drift pattern of the enhanced plasma regions in the afternoon sector (Chappell, 1974; Maynard and Chen, 1975).

In this paper GEOS 2 electric field data are presented which contains clear signatures of the adjacent bulge and trough regions in quiet to moderately disturbed magnetospheric conditions. These regions are located in the local afternoon to early night sector. At times of higher magnetospheric activity the bulge and trough regions move to the post-noon sector. The events were selected from the large amount of data from GEOS 2 accummulated due to almost continuous operation between August 1978 and December 1983 with a 6-month gap in 1980. We looked at the electric and magnetic field signatures during the bulge traversal and the subsequent encounter of the hot and tenuous plasma sheet-like plasma in the trough region. The electric field data in combination with flux-gate magnetometer data were used to derive the $\mathbf{E} \times \mathbf{B}/B^2$ drifts during periods when $|\mathbf{E}|$ exceeded about 0.5 mV m^{-1}. In view of the large number of analyzed events we were able to derive an average picture of the plasma drift in the afternoon to pre-midnight sector with special emphasis on the bulge/trough traversal.

Selection of Events and Data Processing

The electric field measured by the double probe is the spin-averaged, spin plane component of the ambient electric field. The spin plane of GEOS 2 is identical with the earth's equatorial plane. GEOS 2 was for most of its lifetime placed in a geostationary orbit. Its longitudinal position was magnetically conjugated with Northern Scandinavia, and at this position the spacecraft is about 4° south of the magnetic equator. The spin period is about 6 s. The full **E** field vector can be reconstituted for variations slower than the spin period under the condition that $E_\perp \gg E_\parallel$. With only one antenna any variation faster than 6 s cannot be resolved in direction and magnitude.

The double probe technique is sensitive to photo-electron emission on sunlit surfaces and to spurious electric fields produced by the satellite. In a tenuous plasma, where the drift velocity is higher than the thermal velocity, the body of the spacecraft creates a wake. The wake is a product of the downstream reduction of the ion density, which causes a negative space charge. This space charge, together with the positively charged satellite, produces a spurious electric field in the $\mathbf{E} \times B/B^2$ direction. The photo-electron and wake effects are functions of the spacecraft potential and can thus be corrected for. A detailed description of the photo-electron correction is given in Schmidt and Pedersen (1987) and of the wake correction in Pedersen et al. (1984). After subtraction of the spurious signals it is possible to determine the spin plane component of the **E** field with an accuracy of $\geq$0.5 mV m^{-1} in the GSE-X direction and somewhat better than that in the GSE-Y direction. The 0.5 mV m^{-1} error is predominantly due to the not perfect correction for the photo-electron and wake effects. Both corrections are based on statistical studies and comparisons of limited samples of data with data from other instruments onboard GEOS 2. Hence, the corrections apply rather to average data than to individual data points. The error in E_x is larger as this component is fully affected by photo-electron induced electric fields.

The electric field data are presented throughout the paper in an inertial frame of reference. In such a coordinate system a corotating plasma shows up as a small azimuthal flow with a flow speed of less than 3 km s^{-1}.

The measurement technique used on GEOS 2 provided as valuable by-product the potential difference, V_{ps}, between one probe and the floating spacecraft. The probe floating potential is forced by electronic means to a fairly stable potential 0.5 to 1 V above the ambient plasma potential. V_{ps} is therefore the negative value of the spacecraft potential determined with an accuracy of about 1 V or 10-20% of typical GEOS 2 spacecraft potentials. The probe can be considered as a reference against which the spacecraft floating potential can be measured. In the geostationary orbit, the latter potential is a function of the photo-electron characteristics (temperature and current density) and the ambient plasma electrons (temperature and number density). The photo-electron characteristics are fairly well understood from theoretical and laboratory investigations. With suitable instrumentation on the spacecraft, the density of the ambient plasma electrons and their mean temperature can be obtained. Using the GEOS 2 "active" plasma experiments, Knott et al. (1983) were able to derive a relation between V_{ps} and the ambient plasma electron density N_e for a few selected cases. Schmidt and Pedersen (1987) performed a statistical study of a much larger data set and obtained a similar result. The relation between V_{ps} and N_e provides a continuous monitoring of N_e, makes it possible to decide whether GEOS 2 was embedded in a more tenuous or more dense plasma, and is helpful for the determination of the plasma bulge crossing. The different plasma densities near noon and in the early afternoon (electron density $N_e \approx 10$ cm^{-3}), the bulge $N_e \gg 10$ cm^{-3}) and the subsequent trough ($N_e < 1$ cm^{-3}), cause pronounced changes of V_{ps} that can be accurately attributed to these regions. V_{ps} usually drops step like during the outbound transition from the bulge into the trough. The inbound bulge encounter, however, is not always so significant because the post-noon plasma density is occasionally high or, as sometimes observed, it gradually increases until GEOS 2 is in the bulge.

The plasma density has been measured on GEOS 2 by the relaxation sounding instrument and reported, for instance, by Higel and Lei (1984). This method provides reliable plasma density information but does not give continuous data. The data constitute a reference against which our approach for the identification of the bulge traversal using V_{ps} can be verified. We found good agreement between relaxation sounder data from the bulge and trough and signatures in the spacecraft potential. This supports our assumption that V_{ps} is a reliable tool to localize the bulge/trough traversal.

There might be a bias in the data set as we have selected bulge crossings that were clearly visible in the GEOS 2 data base. Obviously, crossings during quiet magnetospehric conditions are much easier to identify; hence they constitute the majority of the cases in our work.

Signatures of the Bulge Traversal in $|\mathbf{E}|$ and $|\mathbf{B}|$

The data set is based on data from quiet to moderately disturbed magnetospheric conditions (Kiruna derived Kp $\leq$ 3). The typically small **E** fields in the early afternoon, pre-bulge sector have a dominant radially inward component. Consequently, the inferred $\mathbf{E} \times \mathbf{B}/B^2$ plasma drift points eastward along the orbit. With increasing local times the orientations change so as to maintain the azimuthally eastward pointing $\mathbf{E} \times \mathbf{B}/B^2$ drift. GEOS 2 enters the bulge at local times that depend on the level of the magnetospheric activity. The relation between the bulge location in local time and magnetospheric activity is depicted in Figure 1. In this plot, the half-width of the bulge in local time (i.e., the

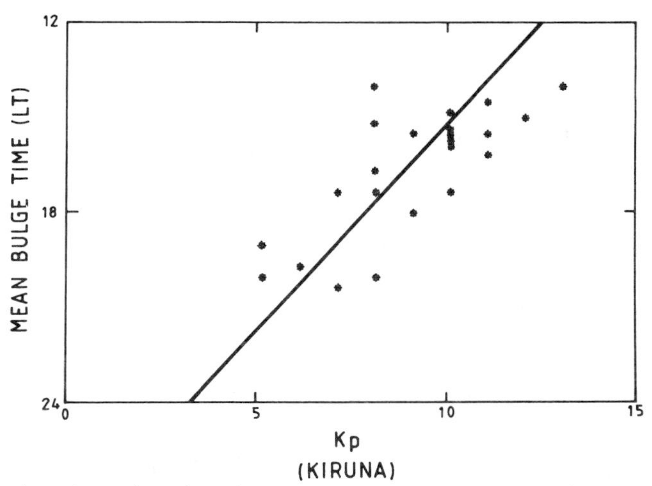

Fig. 1. Distribution of the mean bulge local times as function of the Kiruna Kp values. Note that 2400 LT is at the origin of the y-axis. The equation of the least squares straight line is LT = -1.3 Kp + 28.13. The squared correlation coefficient, r^2, is 0.73.

mean of the outbound and inbound crossing time of all events) is plotted on the ordinate. The Kp value, which is plotted on the abscissa, represents the sum over a 9-hour interval which includes the bulge crossing. The Kp, derived at Kiruna at the time of the bulge crossing, has been added to the Kp values for the two preceding 3-hour intervals. During times of higher magnetic activity (Kp > 10 during the 9-hour interval), the bulge tends to occur at earlier local times (LT < 1500). If Kp is small (Kp < 8) the bulge is encountered at later local times, even as late as 2100 LT. The linear relation between the location of the bulge in LT with increasing geomagnetic activity is apparent in Figure 1. The squared correlation coefficient was found as $r^2 = 0.73$. It should be noted that the local time at the origin of the y-axis is 2400 LT.

An example for a quiet period is given in Figure 2 that contains the bulge traversal on February 11, 1979, at 1910-2400 UT. The Kiruna Kp value over the 9-hour interval is 3. The four panels in Figure 2 are 4-hour plots of E_x, E_y, $|\mathbf{B}|$, and V_{ps}, respectively, where E_x and E_y are the GSE components of the electric field and $|\mathbf{B}|$ is the magnitude of the magnetic field measured by the onboard magnetometer. GEOS 2 enters the bulge at about 1910 UT which is reflected in the significant increase of V_{ps}. No immediate response is found in the electric field or in $|\mathbf{B}|$. However, after a delay of a few minutes, $|\mathbf{B}|$ begins to decrease rapidly, E_x becomes more negative, and E_y tends to become less negative. At 1945 UT, E_y abruptly increases by about 0.6 mV m^{-1} and remains at that level until GEOS 2 leaves the bulge (2040 UT). E_x fluctuates between small positive and negative values but remains positive

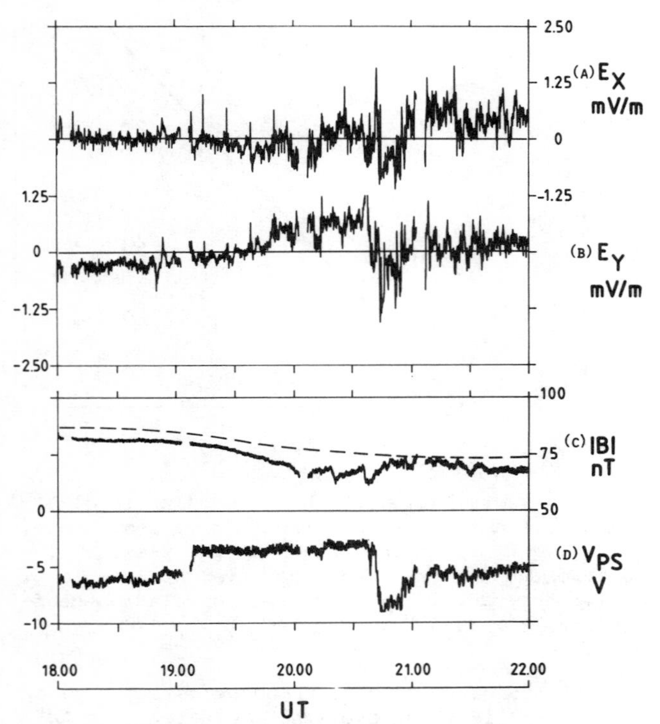

Fig. 2. Data interval from February 11, 1979. Panels a-d represent, respectively, the electric field in E_x, E_y, the total magnetic field $|\mathbf{B}|$, and the measured spacecraft potential V_{ps}. The bulge crossing is clearly visible by the increased spacecraft potential V_{ps}. The broken line in panel c stands for a long-term average of $|\mathbf{B}|$. E_y increases by about 0.6 mV m^{-1} within the bulge at 1945 UT. $|\mathbf{B}|$ is decreased in parts of the bulge (2000-2040 UT). The trough extends from 2040 to 2100 UT.

after leaving the bulge. The broken line in panel c represents the long-term average of magnetic field data measured by the onboard fluxgate magnetometer. This mean $|\mathbf{B}|$ field was obtained from a sample containing data for 180 days from which only intervals with Kp $\leq$ 2 were selected. Between 1945 and 2040 UT, $|\mathbf{B}|$ was considerably smaller than the average. Both the directional change and the absolute decrease in $|\mathbf{E}|$ and $|\mathbf{B}|$ are very typical for bulge crossings and can even be used to identify the bulge whenever the signature in V_{ps} is not so pronounced.

Another characteristic of the region just outside the bulge is the high positive spacecraft potential (V_{ps} negative) that occurs in the trough region (2040-2053 UT). This suggests the presence of a hot and tenuous plasma sheet-type plasma in the vicinity of GEOS 2. Such a period may last several hours; in one case a maximum period of 11 hours has been identified. Occasionally the spacecraft potential was so high that it corrupted the electric field measurements by saturating the input amplifiers so they could only cope with $|V_{ps}| < 9$ V on GEOS 2.

Figure 3 (December 4, 1978), showing the bulge at 1430-1730 UT and the trough at 1730-1930 UT, can be basically discussed along the same line was Figure 2. The Kp value over the 9-hour interval ending at 1800 UT was 7. The higher magnetospheric activity is reflected in the stronger than average frontside compression (1000-1300 UT) of the magnetosphere; the magnetic activity ceased near midnight which is nicely demonstrated by $|\mathbf{B}|$ approaching the long-term average (panel c, broken line). Near the outbound crossing of the bulge (1600-1700 UT) $|\mathbf{B}|$ dropped to about 40% of the long-term average. In the trough (1730-1930 UT) GEOS 2 encountered a tenuous and hot plasma population that became denser near 2030 UT due to the injection processes by a substorm (Knott et al., 1983). The dense, hot plasma led

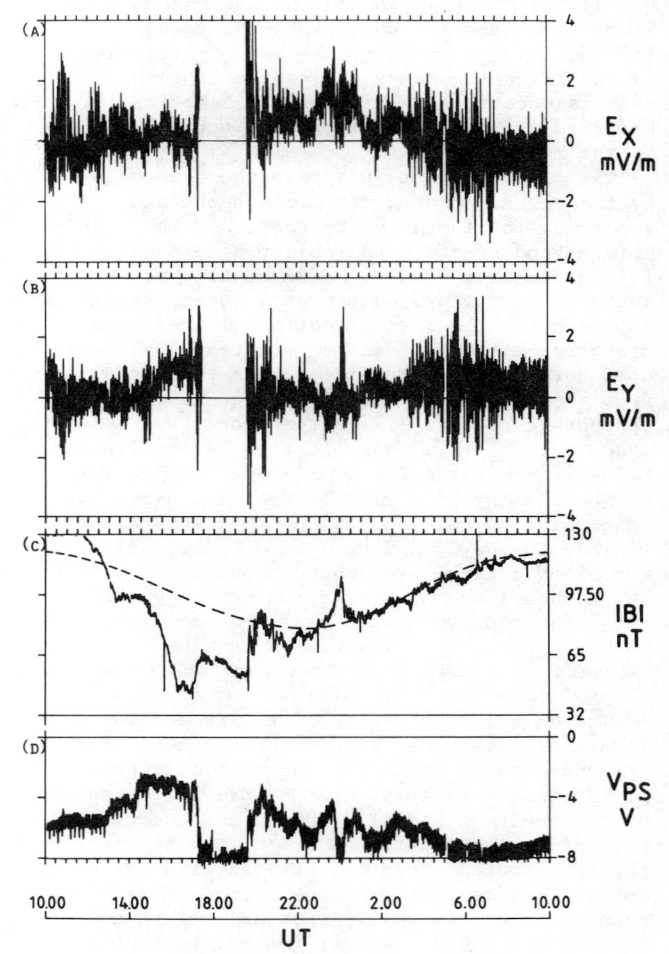

Fig. 3. Similar to Figure 2, however, showing highly compressed data from December 4, 1978. The bulge is located at 1430-1730 UT and the trough at 1730-2030 UT. Signatures in **E** and **B** are very similar to Figure 2.

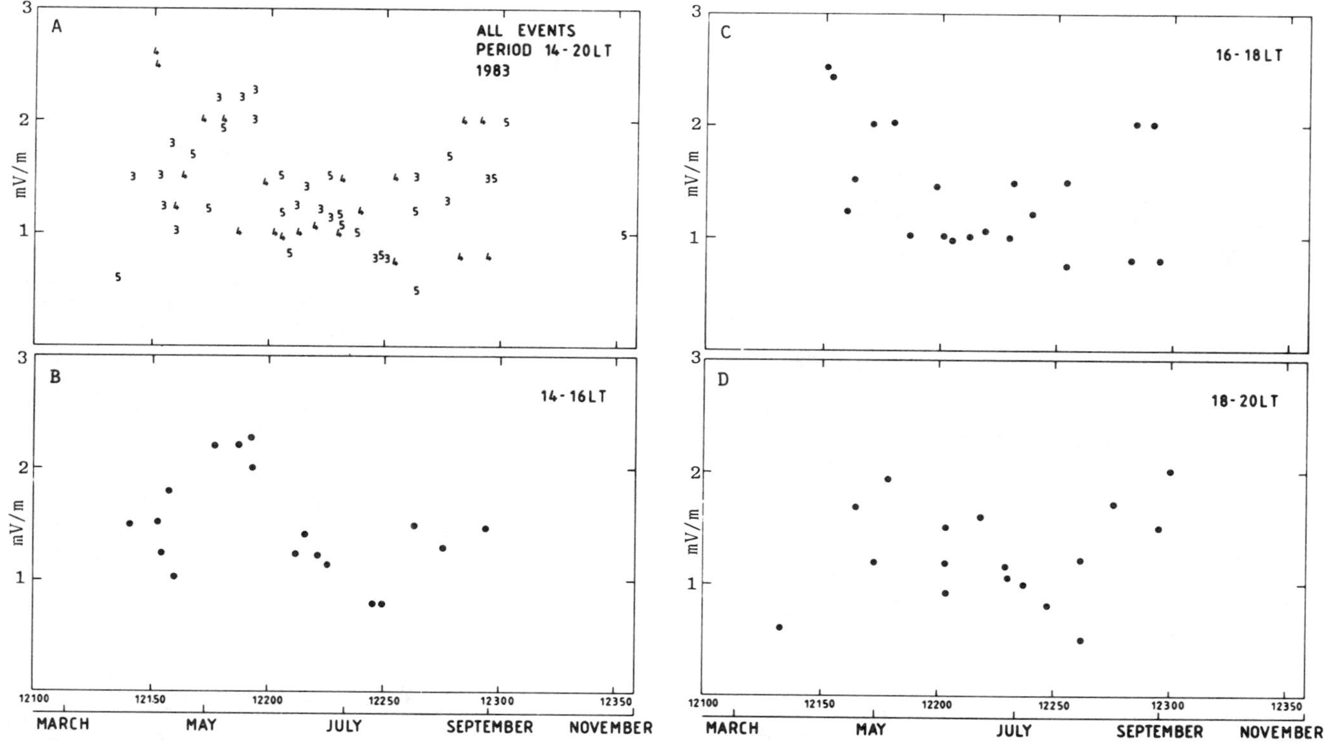

Fig. 4. Panel a summarizes the increases of the electric field component ΔE_y of all bulge crossings between March and October 1983. The numbers identify local time bins as indicated. Panels b-d contain electric field increases of the bins at 1400-1600 LT, 1600-1800 LT, and 1800-2000 LT, respectively. A seasonal variation is apparent in panels b and d.

to a decrease of the spacecraft potential from +8 V to +4 V (V_{ps} from -8 V to -4 V). The spike-like intensity variation of $|\mathbf{B}|$ at 2400 UT is due to another substorm. Figures 2 and 3 as typical examples of bulge and trough traversals yield the impression that E_y has almost always a positive maximum near, but inside, the boundary between the bulge and the trough.

Looking through the entire GEOS 2 data set leads to the impression that the characteristic increase of E_y within the bulge, henceforth called ΔE_y, depends on the time of the year. This seasonal dependence is examined in greater detail by selecting the most obvious bulge crossings in the interval March to October 1983, complemented with a few data from 1978. The actual ΔE_y during the bulge traversal is plotted as a function of the time of the year in Figure 4 (panel a). Each event is coded with a number between 1 and 6 depending on the time at which ΔE_y occurred. These numbers identify local time bins of 2 hours and are chosen such that bin 1 stands for the 1000-1200 LT, bin 2 for 1200-1400 LT, etc. Panels b-d contain the ΔE_y of the individual bins. Panel b (1400-1600 LT) indicates a tendency toward larger ΔE_y in the spring (and probably in the autumn) than in the summer. The same tendency is revealed in panel d (1800-2000 LT) but less clear in panel c (1600-1800 LT).

Plasma Flow in the Plasmaspheric Bulge

In the following we choose to describe the electric field data in terms of $\mathbf{E} \times \mathbf{B}/B^2$ velocities. The period of February 11, 1979, was chosen (see also Figure 2) to demonstrate the changes in the plasma drift direction and velocity. The magnetic field data were provided by the onboard flux-gate magnetometer. Figure 5 presents in panels a-c, respectively, the drift direction, drift velocity, and, as reference, the spacecraft potential V_{ps}. Until 1900 UT, the drift is anti-sunward (±180° in panel a); after 1915 UT, inside the bulge, the drift rotates through the duskward direction (90°) into the sunward direction (0°). After 2100 UT, when GEOS 2 had passed the trough, the drift pointed dawnward (-90°).

Typically the drift velocity in a bulge traversal increases by a factor of 3 to 4. In Figure 5, the drift velocity outside the bulge (until 1900 UT) is 2-3 km s^{-1} and 8-10 km s^{-1} inside the bulge (2000-2040 UT).

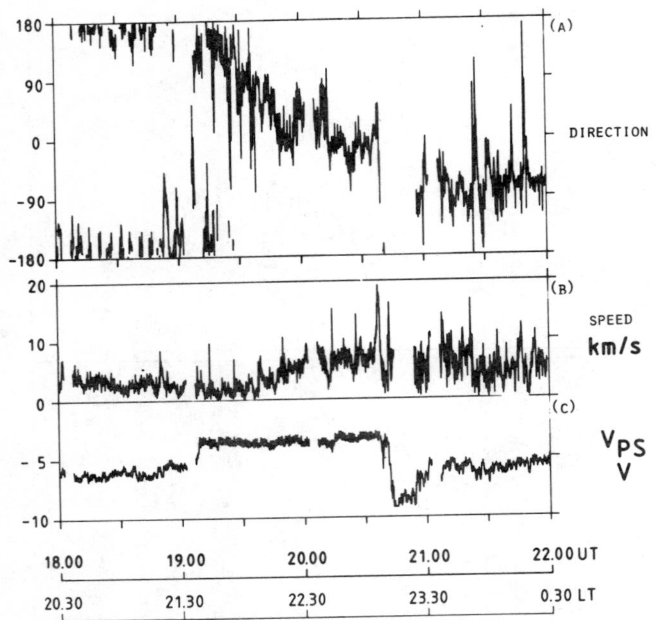

Fig. 5. Interval of February 11, 1979 (1800-2200 UT). The lower time scale reflects the local time of GEOS 2. Panels a and b contain the **E** x **B**/B^2 direction and velocity. Panel c shows the spacecraft potential for identification of the bulge. The drift direction points tailward prior to the bulge (±180°). Within the bulge, at about 1910 UT, it rotates via duskward pointing (90°) in the sunward direction (0°). After the trough it points in dawnward direction (-90°). Accounting for the local time, the drift direction outside the bulge is azimuthally eastward.

Three different typical bulge traversals at 6.6 R_E are shown in Figure 6. The innermost circle marks the local time. Events 1 (March 5, 1979) and 2 (December 4, 1978) are radially displaced as they took place at about the same local time. Case 3 (February 11, 1979) is an event during a quiet magnetosphere. The arrows stand for the **E** x **B**/B^2 drift and indicate both direction and drift speed. The meaning of the bold line enveloping the arrows is to stress the local time extension of the bulge. It does not represent the radial geometry of the bulge. The broken line that follows the outbound crossing marks the sector of the hot and tenuous plasma sheet-like plasma that does not allow for accurate electric field measurements. Corotational plasma flows prior and after the plasma bulge traversal are evident from the tangential orientations of the drift vectors in case 3. The magnitude of **E** due to corotation is about 0.3 mV m^{-1} (see also Figure 2, 1800-1930 UT and after 2100 UT) and is comparable with the measurement accuracy of the instrument.

The sunward flow is very pronounced in all three cases. The way the drift changes during the inbound crossing of the boundary at 6.6 R_E suggests that the sunward drift appears only within the bulge while in the boundary layer it remains azimuthally eastward. This is particularly obvious in event 3 shown with a high time resolution in Figure 5.

In other cases, not discussed here, it was found that narrowly confined plasma flows were superimposed on the general sunward flow. These streamers were also associated with increased flow velocities. The radial plasma streamers are consistent with the theoretical filamentary electric field structures which develop as tail-like appendages to the bulge during magnetic substorms (Spiro et al., 1981). A frequently occurring spiky signature in V_{PS} at the transition from the bulge into the trough could be interpreted as multiple outbound crossings of the bulge and be indicative of such filamentary structures.

Discussion

The electric fields encountered in the consecutive traversal of the plasmaspheric bulge and the trough are very characteristic in the sense

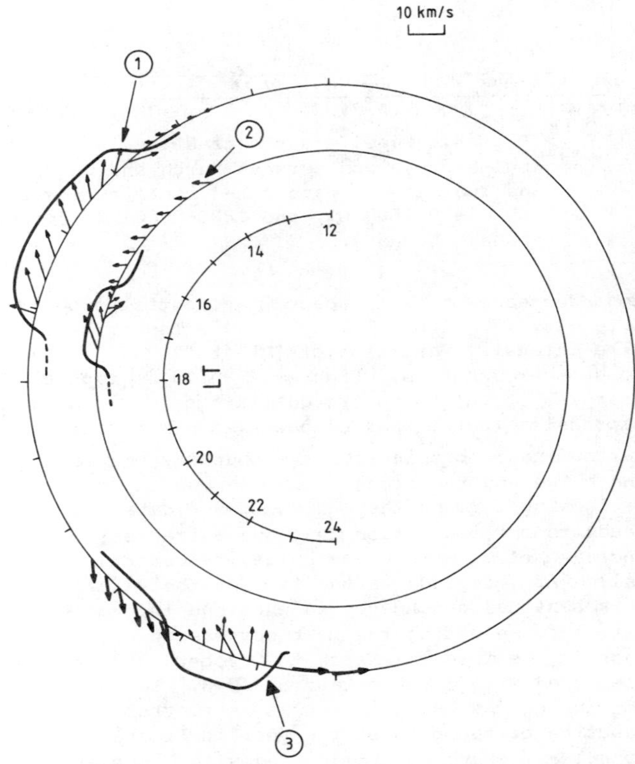

Fig. 6. Three bulge crossings at 6.6 R_E shown in local time. The bold line reflects only the local time extension (not the radial extension). The trough is frequently associated with corrupted **E** field measurements (dotted line). Events 1 (March 5, 1979) and 2 (December 4, 1978) are displaced as they occur at similar local times. Event 3 took place on February 11, 1979. In all cases a prominent sunward plasma drift is apparent.

that the E_y component becomes more positive in all investigated cases. The trough crossing is characterized by a hot and tenuous plasma that on occasion corrupts the electric field measurements. Examples can be found in Figure 2 (2042 UT) and Figure 3 (1730-1930 UT). The **E** x **B**/B^2 plasma drift has a very clear pattern; inside the bulge the GEOS 2 data reveal a flow reversal that rotates the drift from a dusk/tailward orientation (depending on local time) into a clear sunward flow.

The magnetic field in the bulge and usually also in the trough is locally reduced to much smaller values than expected for quiet to moderately disturbed periods. There are only few **E** field measurements available from the trough due to the tenuous plasma; nevertheless it was possible to determine on several occasions that the electric field was very small (<1 mV m^{-1}). In the cases of short trough traversals, e.g., 2040-2053 UT in Figure 2, it was apparent that ΔE_y had decayed after the passage from the bulge to the trough and E_y acquired values similar to the prebulge values. This suggests that the occurrence of ΔE_y is confined to the bulge only. In terms of local times, the trough can extend from several tens of minutes to a few hours. No conclusion could be drawn about a possible relation of its local time extension to magnetospheric activity. It seems that a higher magnetic activity might imply a more extended trough. An extended trough, however, tends to get increasingly interrupted by substorms that generate a hot and somewhat denser plasma environment leading to a less positive spacecraft potential (V_{ps} less negative). These fluctuations of V_{ps} made it difficult to decide whether GEOS 2 was still in the trough.

The decrease of $|\mathbf{B}|$ in the bulge and partly also in the trough indicates that a significant current is flowing outside of the GEOS 2 orbit in the westward direction. Obviously a considerable part of the ring current is frequently outside the geostationary orbit in a narrow local time sector in the afternoon to early night. The results are inconistent with the **B** field measurements on OGO 3 and 5 (Sugiura et al., 1971) that suggest that the ring current has a pronounced maximum near 4 R_E in the dayside and within 6 R_E in the nightside. Recordings of the ground magnetic field at Kiruna on December 4, 1978 (see also Figure 3), indicated that the drop of the equatorial magnetic field (1600-1730 UT) is associated with a simultaneous increase of the horizontal x-component of the high-latitude ground magnetic field (see panel a, Figure 7). Figure 7 depicts the x (north), y (east), and z (vertical) components of the magnetic field, respectively. The bold horizontal line identifies the period of the very small equatorial magnetic field, while the broken branch stands for the encounter of the plasma sheet-like plasma. The x-component peaked at about 200 nT during the trough traversal, while y and z remained essentially unchanged in the early phase but showed a small dip-like fluctuation that coincided with the maximum depression of the equatorial **B** field after 1630 UT in Figure 3. The more positive B_x is indicative of an eastward ionospheric current (i.e., toward later local times) and is likely to be connected to a partial equatorial ring current via Birkeland currents. The direction of the equatorial ring current is sunward and consistent with the drop of $|\mathbf{B}|$ seen in the equatorial plane.

The sunward direction of the plasma flowing in the bulge can be interpreted as the result of momentum transfer from the energetic ions gyrating along field lines from the magnetotail in the sunward direction. These energetic ions have large gyro radii; for example, a proton with $E_\perp \approx 100$ keV gyrates with a radius of ≈500 km. The trajectory of these particles is dominated by the ΔB drift, which drives them in a westward direction, in the Alfvén layer. The bulk of the plasma near the Alfvén layer has energies below 1 keV and only experiences a negligible westward ΔB drift. Consequently, no observable induced electric field is expected in this region. The depletion of the magnetic field that frequently occurs during the bulge/trough traversal of GEOS 2 pushes the Alfvén layer earthward, reaching the geostationary orbit at times of strongly reduced magnetic fields. Figure 3 indicates that this condition might be fulfilled between 1620-1700 UT. Assuming that the low number density and the high energy of these particles yield still a considerable energy flux, we propose an energy and momentum transfer from the energetic flux into the cold plasmaspheric plasma by viscous interaction. The momentum transferred into the cold plasmaspheric plasma causes and maintains

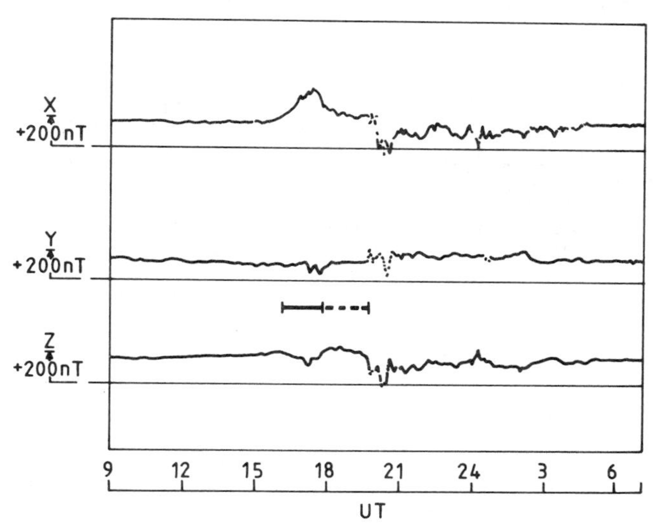

Fig. 7. Ground recording of the magnetic field of December 4, 1978, at Kiruna. x, y, and z stand for the north, east, and vertical component, respectively. The bold line identifies the period of the small equatorial **B** field; the broken branch identifies the trough traversal by GEOS 2.

the sunward convection in the bulge. Unfortunately, there is hardly any overlapping of events seen on GEOS 2 with data from ISEE to conjugate regions. It is therefore impossible to infer what ultimately happens to the sunward drifting plasma. It is likely this sunward drifting plasma is the sunward convective flow of cold plasmaspheric plasma in the dusk bulge region, as reported by Nagai et al. (1985). It also seems feasible that the plasma finally becomes detached from the plasmasphere and forms the sunward convecting enhanced plasma regions reported by Chappell (1974) or Maynard and Chen (1975). Assuming that the current flowing outside the GEOS 2 orbit is very localized, it seems quite feasible that the magnetic field depletions such as the one at 1630 UT on December 4, 1978 (see Figure 3), are associated with a localized magnetic x-line outside the location of GEOS 2, but still within the plasmasphere which could decouple parts of the sunward flowing plasma from the plasmasphere. A multi-spacecraft study of that region will probably be capable of shedding more light on this subject.

Acknowledgment. The raw magnetic field data were provided by ESOC on the experimenter's tapes. They were acquired by the S-331 experiment of the Institute for Physics of the Interplanetary Space of the National Research Council, Frascati, Italy. The authors acknowledge constructive suggestions by the referee.

References

Chappell, C. R., Detached plasma regions in the magnetosphere, J. Geophys. Res., 79, 1861, 1974.

Higel, B., and W. Lei, Electron density and plasmapause characteristics at 6.6 R_E: A statistical study of GEOS 2 relaxation sounder data, J. Geophys. Res., 89, 1583, 1984.

Knott, E., A. Korth, P. Decreau, A. Pedersen, and G. Wrenn, Observations of the GEOS equilibrium potential and its relation to the ambient electron energy distribution, in Spacecraft/Plasma Interactions and Their Influence on Field and Particle Measurements, Eur. Space Agency Spec. Publ. ESA, SP-198, p. 19, 1983.

Lennartsson, W., and D. L. Reasoner, Low-energy plasma observations at synchronous orbit, J. Geophys. Res., 83, 2145, 1978.

Maynard, N. C., T. L. Aggson, and J. P. Heppner, The plasmaspheric electric field as measured by ISEE 1, J. Geophys. Res., 88, 3991, 1983.

Maynard, N. C., and A. J. Chen, Isolated cold plasma regions: Observations and their relation to possible production mechanisms, J. Geophys. Res., 80, 1009, 1975.

Nagai, T., J.F.E. Johnson, and C. R. Chappell, Low-energy (<100 eV) ion pitch angle distribution in the magnetosphere by ISEE 1, J. Geophys. Res., 88, 6944, 1983.

Nagai, T., J. L. Horwitz, R. R. Anderson, and C. R. Chappell, Structure of the plasmapause from ISEE 1 low-energy ion and plasma wave observations, J. Geophys. Res., 90, 6622, 1985.

Pedersen, A., and R.J.L. Grard, Quasistatic electric field measurements on the GEOS-1 and GEOS-2 satellites, in Quantitative Modeling of Magnetospheric Processes, Geophys. Monogr. Ser., vol. 21, edited by W. P. Olson, p. 281, AGU, Washington, D.C., 1979.

Pedersen, A., C. A. Cattell, C.-G. Fälthammar, V. Formisano, P.-A. Lindqvist, F. Mozer, and R. Torbert, Quasistatic electric field measurements with spherical double probes on the GEOS and ISEE satellites, Space Sci. Rev., 37, 269, 1984.

Schmidt, R., and A. Pedersen, Long-term behavior of photo-electron emission from the spherical double probe sensors on GEOS 2, Planet. Space Sci., 35, 61, 1987.

Spiro, R. W., M. Harel, R. A. Wolf, and P. H. Reiff, Quantitative simulation of a magnetospheric substorm, 3. Plasmaspheric electric fields and evolution of the plasmapause, J. Geophys. Res., 86, 2261, 1981.

Sugiura, M., B. G. Ledley, T. L. Skillman, and J. P. Heppner, Magnetospheric distortions observed by OGO 3 and 5, J. Geophys. Res., 76, 7552, 1971.

REFILLING OF A PLASMASPHERIC FLUX TUBE - MICROSCOPIC PLASMA PROCESSES

N. Singh

Department of Electrical and Computer Engineering
The University of Alabama in Huntsville, Huntsville, Alabama 35899

Abstract. Microscopic plasma processes, which facilitate the refilling of the outer plasmaspheric flux tubes after geomagnetic storms by thermalizing and/or trapping the ions in the flux tubes, are studied. The formation of electrostatic shocks in the equatorial region, where the plasma streams originating from the conjugate ionospheres collide, is examined by computer simulations. The mechanism and the conditions for the shock formation are given. A shock pair forms when $T_e > 3 T_i$ and the stream velocity V_b lies in the range $1.3 V_{ti} < V_b < 2.3 C_o$, where T_e and T_i are the electron and ion temperatures, C_o is the ion-acoustic speed, and V_{ti} is the ion thermal velocity. When $V_b > 2.3 C_o$, counterstreaming is expected to continue. Starting with $T_e \sim T_i$ in the ionosphere, shock formation requires a preferential heating of electrons in the equatorial region. In the shocked plasma, electrons are found to be non-Maxwellian. It is shown that a plasma model consisting of two-fluid hydrodynamic treatment for the ion streams originating in the congugate ionospheres and a non-Boltzmann distribution consisting of trapped and free electrons can be successful in including the features of electrostatic shocks in the refilling. The effect of perpendicular ion heating by an extended plasma turbulence along the field lines on the reilling is examined suggesting that an extremely low level of the turbulence with a power spectral density $\lesssim 10^{-11}$ V^2 m^{-2} Hz^{-1} near the ion-cyclotron frequency can be effective in trapping the ions in the flux tubes. A localized perpendicular ion heating in the equatorial region produces a potential barrier for the ionospheric plasma streams. The potential barrier can stop the interhemispheric flow.

1. Introduction

The refilling of the outer plasmaspheric flux tubes after geomagnetic storms presents an interesting situation, where both macroscopic (large-scale) and microscopic (small-scale) plasma processes are operative affecting each other. The plasma for the refilling is supplied by the conjugate ionospheres. During the early stage of the refilling when the densities in the flux tubes are very low (~1 cm^{-3}), the outflow of the plasma from the ionosphere is supersonic. Recent satellite observations clearly demonstrate the presence of such supersonic flows from the conjugate ionospheres when the magnetospheric flux tubes are highly depleted (Sojka et al., 1983). In a recent review Horwitz (1987) has given a detailed account of the observational aspects of the refilling.

The plasma flows from the ionosphere along the outer plasmaspheric flux tubes are large-scale phenomena, but the fundamental problems dealing with the refilling are the small-scale plasma processes (Figure 1) which modify the large-scale flows and thereby trap the plasma in the flux tube and/or thermalize the initially large-velocity (subsonic or supersonic) flow enhancing the plasma density from ~1 cm^{-3} to ~100 cm^{-3}.

One of the major problems associated with the refilling is illustrated in Figure 2 in which two plasma sources occupying the regions $d/2 > |X| > L/2$ (see Figure 2a) correspond to the conjugate ionospheres. In between the two sources at some initial time, there is a region of low-density plasma or a vacuum corresponding to the depletions of the flux tubes at high altitudes during magnetic storms. Because of the pressure gradients, the plasmas from the sources expand into the low-density region creating plasma streams (Singh and Schunk, 1982, 1985, 1986) in which most of the kinetic energy lies in the ion beams. When the plasma streams collide at X = 0 as shown in Figure 2b, the question arises regarding the collisionless coupling between the ion beams. Does the coupling slow down the streams completely, partially, or not at all? In the latter case the streams continue to counterstream, as shown in Figure 2c, and there is a total lack of coupling.

Banks et al. (1971) postulated that the coupling occurs when the plasma streams collide at the equator and it leads to the formation of a pair of electrostatic shocks (Figure 1). However, the mechanism for the shock formation was not explored. Schulz and Koons (1972) suggested that the thermalization occurs in a two-step process: first the ions in the plasma streams are trapped in the equatorial region when they are pitch-angle scattered by some wave-particle interactions in which the waves are excited by the counterstreaming ion beams themselves. After the trapped ion density becomes sufficienty large, the Coulomb collisions become effective in thermalizing the trapped plasma. Schulz and Koons (1972) suggested that the formation of an electrostatic shock pair is unlikely. They argued that the individual plasma streams do not excite waves when they interact with the background plasma and, therefore, no dissipation mechanism exists for the shock formation. However, they did not consider the dissipation mechanism provided by the coupling of the colliding plasma streams.

Recent small-scale simulations by Singh and Schunk (1983) and Singh et al. (1986a,b) have clarified the processes which lead to the formation of electrostatic shocks when counterstreaming flows collide. In the initial study (Singh and Schunk, 1983), it was found that in such a colliding situation an electric potential barrier for the ions develops in association with the plasma density enhancements in the region of the interpenetration of the streams, that the ion streams were slowed down only slightly by the potential barrier, and that eventually a counterstreaming flow is set up without any thermalization of the initially highly supersonic ion beams. In this study it was assumed that $T_e \sim T_i$, where T_e and T_i are the electron and ion temperatures.

In further simulations (Singh et al., 1986a,b), it was found that the enhanced density and the electrical potential barrier evolve into a shock pair provided $T_e > 3\ T_i$. The density enhancement upon the collision of the streams was found to be the seed for the shock formation; the compressive density perturbation steepens into a shock pair when the above temperature criterion is satisfied. The steepening involves generation of short scale lengths, which are Landau damped if $T_e < 3\ T_i$. Singh et al. (1986a,b) also found that the electron dynamics plays a crucial role in the shock formation. The kinetic energy of the ion streams was found to power the electrostatic shocks which accelerate the electrons entering the shocked plasma, and the local electrons in the shocked plasma are trapped between the shocks.

Besides the shocks, the additional plasma processes such as the ion trapping by pitch angle scattering (Schulz and Koons, 1972; Singh et al., 1982) and/or by perpendicular ion heating (see Figure 1) by wave-particle interactions can be effective in the refilling. Satellite observations (e.g., see Horwitz et al., 1982) indicate the presence of low-energy (<100 eV) ion conics along the outer plasmaspheric field lines. The ion conics are generated by a perpendicular heating of the ions. Such heating, even if it is weak, can provide an effective mechanism for the trapping of the upflowing ionospheric ions in the flux tubes; thereby, it can significantly contribute to the refilling.

The purpose of this paper is to present a review of the microscopic processes which are likely to play significant roles in the refilling of the plasmaspheric flux tubes. In section 2 is a detailed discussion on recent small-scale numerical simulations of the formation of an electrostatic shock pair in countersteaming plasma expansions. In section 3 the inadequacy of the hydrodynamic plasma models for the handling of the shock formation in counterstreaming flows is discussed with suggested modifications to the commonly used Boltzmann distributions for electrons in hydrodynamic plasma models to include the effects of the shock formations. In section 4 the role of wave-particle interactions in refilling is discussed. In particular, the role of extended perpendicular ion heating by extremely low level plasma turbulence in trapping the ions in the flux tubes is highlighted (Singh and Hwang, 1987). It is also shown that an intense localized equatorial heating of ions to several hundred electron volts, as observed from satellites (Olsen, 1981; Olsen et al., 1987), can set up downward electric fields supported by the difference between the anisotropies of the local electrons and the heated ions. The potential drop associated with such fields is found to be large enough to stop the interhemispheric flows by reflecting back the ion streams into each hemisphere. The main conclusions of the paper are presented in section 5, which also indicates the directions for future studies dealing with the refilling.

2. Electrostatic Shocks

Banks et al. (1971) postulated that after flux tube depletion the resulting supersonic field-aligned plasma flows, emerging from the conjugate ionospheres, interact at the equator, forming a pair of collisionless electrostatic shocks. Behind the shocks is a relatively high-density, high-temperature plasma (see Figure 1). As time passes, the volume between the two shocks fills with plasma, and the shocks move down (Figure 1) the flux tube toward the conjugate ionospheres at a relatively low speed. However, Banks et al. did not examine the mechanism for the formation of the electrostatic shocks.

2.1 Colliding Plasma Streams

We start our discussion on shock formation by distinguishing between colliding (Figure 2b) and counterstreaming (Figure 2c) plasma steams. The essential difference between the states of the plasma shown in Figures 2b and 2c is that in the situation of Figure 2c the plasma density is uniform (homogeneous), while in Figure 2b for the colliding plasma streams the plasma density approximately doubles in the region of the col-

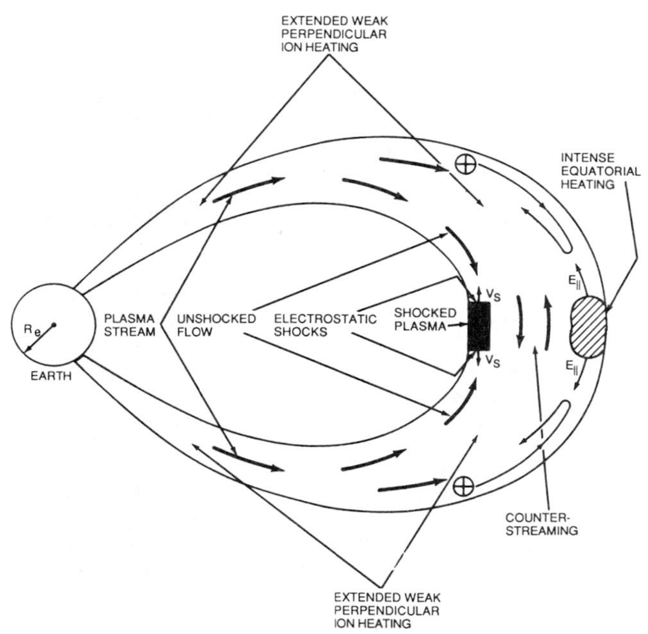

Fig. 1. Schematic diagram showing a flux tube and some plasma processes which affect the refilling. The processes shown are: (i) extended perpendicular heating of the ions in the plasma streams shown by the thick arrows, (ii) formation of a pair of electrostatic shocks at the equator when the plasma streams collide at the equator; the shocks move away from the equator at a velocity V_s leaving behind a dense thermalized plasma, and (iii) equatorial ion heating and the resultant downward parallel electric fields which reflect back the plasma streams coming from the conjugate ionospheres.

lision, $X \simeq 0$, making the plasma density distribution inhomogeneous. The enhanced plasma density in the vicinity of $X \simeq 0$ is a compressional perturbation and it can launch an electrostatic shock pair under suitable conditions. This was shown by Forslund and Shonk (1971) by means of computer simulations, in which the plasma state shown in Figure 2b was studied. Recently, Singh et al. (1986a,b) showed the formation of an electrostatic shock pair in numerical simulations of counterstreaming plasma expansions in which accelerated plasma streams collide. The counterstreaming plasma expansion closely approximates the physical situation of the colliding interhemispheric flows in space. We have chosen to describe here simulations based on a Vlasov code rather than a particle-in-cell code, which was used in previous papers (Singh et al., 1986a,b).

2.2 Small-Scale Simulations

In order to study the electrostatic shock formation in counterstreaming plasma expansion, we assume two plasma sources as shown in Figure 2a. These sources act like plasma reservoirs from which plasma expands into the initially vacuum region $-L/2 < X < L/2$. Here we present results from small-scale simulations in which both the electrons and ions are treated kinetically. The temporal and spatial evolutions of the expanding plasmas are followed by solving the time-dependent Vlasov equations for the distribution functions $f_\alpha(X,V_x)$ of the electrons (α = e) and ions (α = i) along with the Poisson equation for the electrostatic field, given by $E = -\nabla\Phi$, where Φ is the electric potential. These coupled equations are solved numerically (Singh, 1980). In the plasma reservoirs, the plasma density and the particle distribution functions are kept constant. The boundary conditions on the electric potential Φ are $\Phi(X = -L/2) = \Phi(X = L/2) = 0$.

We have used the following definitions and normalizations: distance $\tilde{X} = X/\lambda_{do\ell}$, velocity $\tilde{V} = V/V_{teo}$, time $\tilde{t} = t\omega_{po}$, density $\tilde{N} = N/N_o$, $\tilde{\Phi} = e\Phi/k_B T_e$, where ω_{po} is the electron-plasma frequency in the plasma reservoirs, $V_{teo} =$

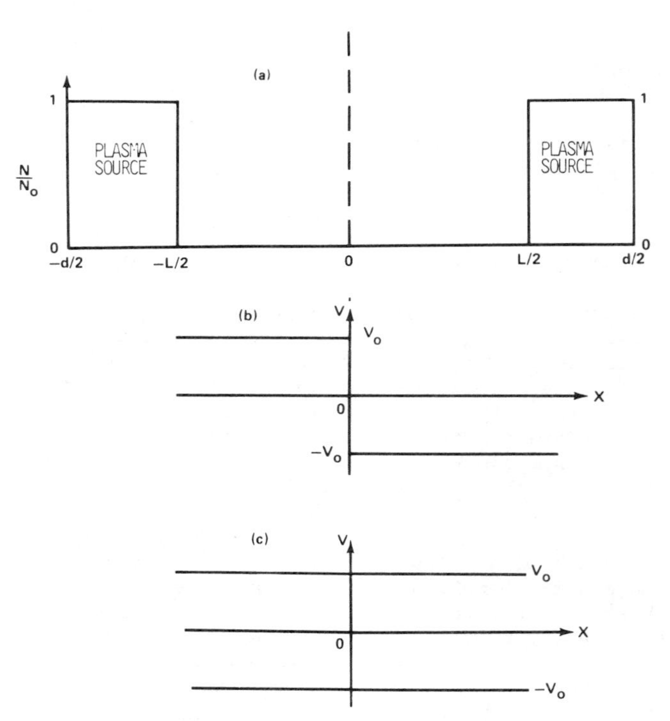

Fig. 2. (a) Plasmas expanding from the two sources occupying the regions $-d/2 < X < -L/2$ and $L/2 < X < d/2$ produce plasma streams in the initially empty region $-L/2 < X < L/2$. In the source regions density is maintained to be constant. The geometry is used for the numerical simulations of the colliding and counterstreaming plasma streams. (b) Colliding plasma streams; for $X < 0$ beam velocity is V_b, while for $X > 0$ it is $-V_b$. (c) Counterstreaming plasma streams with beam velocities $\pm V_b$.

$(k_B T_o/M_e)$, $\lambda_{do} = V_{teo}/\omega_{po}$, and N_o is the plasma density in the plasma reservoirs. T_o is chosen to be a reference tmeperature. Most of the discussion below is based on a simulation in which $T_e = 10\ T_o$ and $T_i = T_o$. We have used an artificial ion mass $M_i = 64\ M_e$ for the purpose of computer economy.

2.3 Numerical Results

The temporal evolution of the plasma from a simulation with $L = 500\ \lambda_{do}$ is shown in Figure 3. The plasma-space density plots are shown for the electrons on the left and for the ions on the right. In these plots contours of constant value of $f_\alpha(X,V_x)$ are plotted in the X-V_x plane.

The ion and elecltron phase-space plots at early times ($\tilde{t} < 200$) show the expansion of the plasmas from the two sources. After $\tilde{t} \approx 200$, the plasma streams from the two sources begin to overlap in space; i.e., counterstreaming has set up; the electron streams have merged while the ion streams are separated in velocity space and their velocity distribution functions appear like beams. It is important to note that at $\tilde{t} \approx 400$, the counterstreaming has set in only for the forerunner ions, which precede the bulk of the ions headed by the expansion fronts as shon by the arrows at $\tilde{t} = 400$.

When the expansion fronts collide ($\tilde{t} > 400$), an instability develops in the mid-plane ($X \sim 0$) of the simulation plasma. In the phase-space plots the main attributes of this instability are: (1) trapping of electrons as shown by the closed loops in the electron phase-space plots for $\tilde{t} \geq 600$, (2) counterstreaming of energized electrons which are colocated with the trapped ones, and (3) slowing down of the ion beams colocated with the trapped electrons.

In order to further illustrate the nature of the instability, the temporal evolution of the density and electric potential profiles for $\tilde{t} > 600$ is shown in Figure 4. This figure shows that when the expansion fronts collide at about $\tilde{t} \approx 660$, a density bump forms near the mid-plane (broken curves), which is accompanied by the formation of a potential hill (solid curves). The density bump grows and steepens into a pair of steep density fronts as it can be seen from the density profiles for $\tilde{t} \geq 900$. These density fronts are electrostatic shocks and propagate away from the mid-plane. The propagation velocity is given by $V_s = dX/dt \approx 0.27\ V_{teo} = 0.7\ C_o$, where C_o is the ion-acoustic speed in the unshocked plasma.

The solid curves in Figure 4 show that, along with the steepening of the density profiles, the potential hill also steepens into a pair of electrostatic shocks. The typical jump conditions for the shocks are

$$N_1/N_o \approx 2, \ \Delta\Phi \approx 2\ (k_B T_e/e)\ , \tag{1}$$

where N_o and N_1 are the low and high densities across the shocks, respectively, and $\Delta\Phi$ is the potential jump across it. The shock width is found to be $\Delta X \approx 30\ \lambda_{do}$.

The closed loops in the electron phase-space plots in Figure 3 are the result of the trapping of the electrons between the two shock fronts. On the other hand, the electrons coming into the shocked plasma are accelerated by the potential jumps across the shocks. The temporal evolution of the electron velocity distribution function in the shocked plasma at $X = 0$ is shown in Figure 5 (right-hand panels). The noteworthy feature of the elelctron velocity distribution function is that it has a composite feature of a flat top distribution, which is a common characteristic of trapped particles, and of beams with positive slopes. Thus, the electron distributions cannot be described by a Maxwellian. However, the width of the distribution can be used to define an effective temperature (T_{eff}) and it is approximately $T_1 \approx T_{eff} \approx 3\ T_e$. Thus, the acoustic speed in the shocked plasma is $C_1 = 1.7\ C_o$ and the shock speed in terms of this speed is about $0.4\ C_1$.

The ion phase-space plots clearly demonstrate the slowing down of the ion beams in the shocked plasma. As a matter of fact, ion kinetic energy powers the shock. This point will be examined further.

Figure 5 (left-hand panels) shows the temporal evolution of the ion velocity distribution function at $X = 0$ from an instant of time just before the onset of the formation of the electrostatic shock. Counterstreaming ion beams and their subsequent slowing down are clearly seen. The beam velocity prior to the onset of the shock formation ($\tilde{t} \sim 550$) is about $V_b \approx 0.8\ V_{teo}$, which, in terms of the ion-acoustic velocity, is $V_b \approx 2\ C_o$.

At this time it is useful to briefly review the linear instability criteria for ion waves. In an ion-beam plasma system ion-ion (I-I) and ion-electron (I-E) interactions cause instabilities. The I-E modes do not provide an effective momentum coupling mechanism between the counterstreaming ion beams; therefore, they do not significantly slow down the beams (McKee, 1970). On the other hand, I-I instability does provide such a couplilng. However, the linear instability criterion for such an instability is given by (McKee, 1970).

$$1.3\ V_{ti} \leq V_b \leq C_o, \quad T_e > 3\ T_i\ , \tag{2}$$

where V_{ti} is the ion thermal velocity. McKee (1970) has shown that this instability condition does not depend on ion mass; therefore, it is applicable to the artificial ions used in the simulations.

As we noted earlier, $V_b \approx 2\ C_o$ in the simulation. Thus, according to the linear instability criterion the beams should not couple. But we see from Figure 3 that they do indeed couple. In this connection the effects of the inhomogeneity of the colliding beams, which are not included in

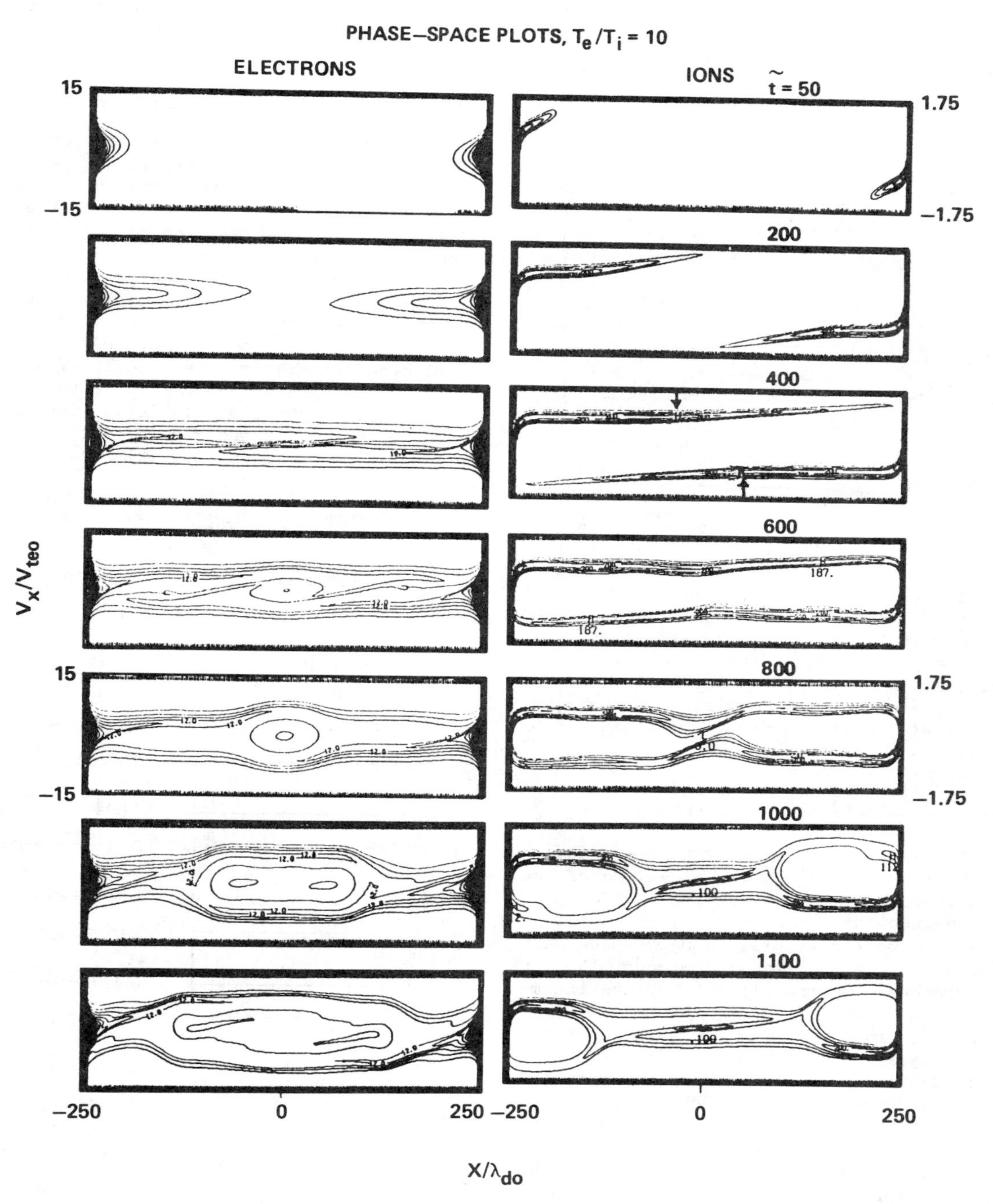

Fig. 3. Phase-space (X-V_x) plots for $T_e/T_i = 10$ showing the contours of constant values of the electron (left-hand panels) and ion (right-hand panels) velocity distribution functions $f_\alpha(X,V_x)$.

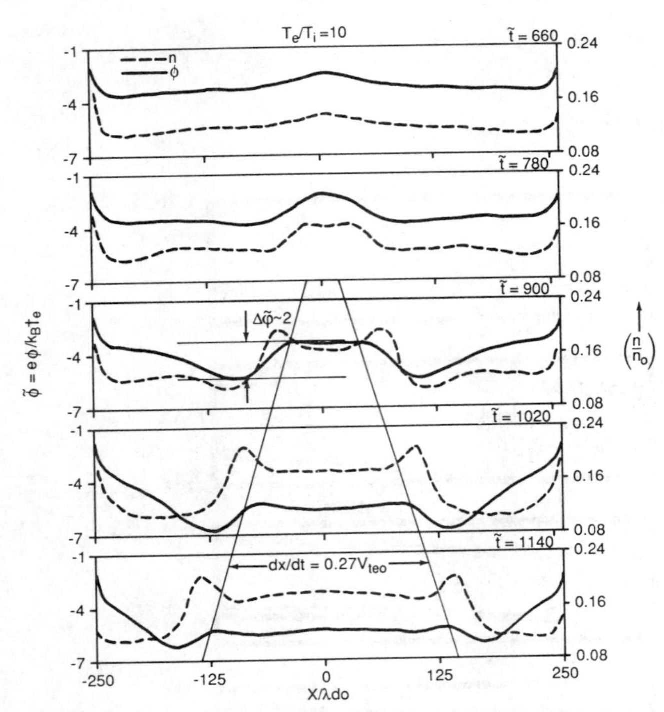

Fig. 4. Ion density distribution $N(x)/N_o$ (broken line curves) from $\tilde{t}$ = 660 to $\tilde{t}$ = 1140 showing the formation of a pair of electrostatic shocks. The propagation velocity of the shocks is shown. Note that the shock evolves from a density enhancement at about t = 660. The solid line curves show the electric potential distribution $\Phi(X)$ corresponding to the density distribution.

the linear instability analysis, become important.

The development of the density bump and the potential hill (Figure 4) in the mid-plane of the simulation region at $\tilde{t} \sim 559$ are the manifestations of the inhomogeneity caused by the colliding beams. The potential hill tends to slow down the ion beams. When the beams slow down, the ion density increases. On the other hand, the incoming electrons get accelerated by the potential hill, therefore, their density decreases. Thus, the charge imbalance caused by the increasing ion density and decreasing electron density makes the potential grow (Singh et al., 1986b). This is the mian cause of the instability, and the inhomogeneity provides the seed for it.

When the beams are sufficiently slowed down by the above process, I-I instability sets in and the slowing down of the beams continues. The ion kinetic energy powers the electrostatic shock, and the electric fields of the shocks accelerate the electrons. The kinetic energy of the ion beam ($V_b \simeq 0.8\ V_{teo}$) is

$$W_b \simeq \frac{1}{2} MV_b^2 = 20\ k_B T_o \approx 2\ k_B T_e \ . \tag{3}$$

As can be seen from Figure 4, the potential jump across the shock is about $\Delta\Phi \simeq W_b/e \approx 2(k_B T_e/e)$. The effective electron temperature in the shocked plasma is about 3 T_e, implying that, on the average, electrons have gained an energy of about 2 $k_B T_e$.

2.4 Equation of State for the Electrons

We noted earlier that the electrons in the shocked plasma are non-Maxwellian. Forslund and Shonk (1971) showed this carrying out simulations with the initial condition shown in Figure 2b. These authors used real ion masses, unlike in the present simulation, in which we used $M_i/M_e = 64$ for the purpose of the computer time economy.

In connection with electrostatic sheath formation, Morse (1965) derived a steady state relation between electron density and the potential including both the trapped and the free electrons. Forslund and Shonk (1971) verified this relation from their simulations. The same relation holds good in the region of the shocked plasma in our simulation. This relation is:

$$\frac{N_1}{N_o} = \tilde{N}_{trap} + \tilde{N}_{free} \simeq (2/\sqrt{\pi})\ \Delta\tilde{\Phi}^{1/2} + \exp(\Delta\tilde{\Phi})\ \mathrm{erfc}(\Delta\tilde{\Phi}^{1/2}),\ \Delta\Phi > 0\ , \tag{4}$$

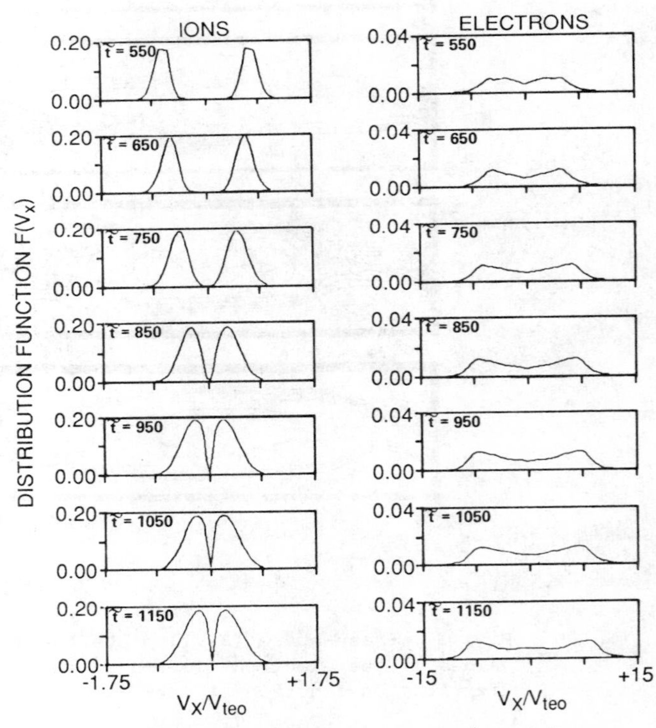

Fig. 5. Electron (right-hand panels) and ion (left-hand panels) velocity distribution functions at X = 0 shown at several times.

where $\tilde{N}_{trap}$ and $\tilde{N}_{free}$ are the relative densities of the trapped and free electrons, respectively, and N_o is the density of the unshocked ion beams. $\Delta\tilde{\Phi}$ is the potential difference in units of $(k_B T_e/e)$.

An interesting consequence of (4) follows from the fact that when the streams overlap the plasma density doubles, $N_1/N_o \simeq 2$, for which the maximum potential jump is

$$\Delta\Phi_{max} \simeq 2.25(k_B T_e/e) \ . \tag{5}$$

Such a potential can slow down the ion beams to velocities $\leq C_o$ if the unshocked beam velocity

$$V_b \leq 2.3\ C_o. \tag{6}$$

Thus, the shock formation discussed here cannot occur for arbitrarily large beam velocities. It is limited to a maximum Mach number $M = V_b/C_o \sim 2.3$. When the streams are faster than this, they do not sufficiently slow down upon their collision and, therefore, do not couple but continue to counterstream.

As mentioned earlier, the above discussion is based on a simulation with $T_e/T_i = 10$ in the plasma reservoir. Similar behavior of the plasma was found when $T_e > 3\ T_i$. On the other hand, when $T_e < 3\ T_i$, the propagating shock pair does not form. However, locally around $X = 0$ the ion streams merge and there are trapped electrons; the merging occurs because of the potential hill setup when the density doubles at $X = 0$ upon the collision of the expanding plasma fronts. In the absence of the launching of a shock pair, the fast ions in the stream cross the hill while slow ones are reflected back.

2.5 Shocks in the Equatorial Region of the Outer Plasmasphere

In view of the above discussion, if shocks do form in the equatorial region of the outer plasmasphere, the electrons must be hotter than the ions. Starting with $T_e \sim T_i$ in the ionosphere, the shock formation requires a preferential heating of electrons or preferential cooling of ions. It is known that in expanding plasmas, the ion temperature decreases. Furthermore, in such situations the ion velocity distribution functions develop elongated tails in the direction of the flow. This reduces ion-Landau damping and relaxes the condition $T_e > 3\ T_i$. This mechanism was suggested to play a role in the excitation of ion waves in the work of Schulz and Koons (1972). However, simulations show that these features are masked away in counterstreaming plasmas, especially in the region where the plasma streams merge. This is primarily due to the formation of the electric potential hill even if the shocks do not form (Singh and Schunk, 1983).

Singh et al. (1986b) suggested that the electron heating in the equatorial region can be facilitated by wave-particle interactions. A promising candidate for the waves is the electromagnetic ion-cyclotron waves; experimental observations (Norris et al., 1983) and theoretical work (Roux et al., 1984) do show that such an equatorial electron heating occurs in the region of low plasma density in the outer plasmasphere. Thus, in the presence of the equatorial electron heating and if the plasma streams are supersonic with respect to the ion thermal speed, but not too fast [see equation (6)], the launching of a shock pair is likely in the equatorial region of the outer plasmasphere.

3. Plasma Models for Plasmaspheric Refilling

While constructing models for space plasmas, there is always the problem of choosing an appropriate description of the plasma so that all the relevant physical processes are properly included. This problem is peculiar to space plasma because there are multitudes of scale lengths involved ranging from those of the plasma, such as the Debye length and the Larmor radius, to the geophysical lengths. The former scale lengths are properly described by kinetic models. However, to solve a complex nonlinear problem at geophysical scale lengths using kinetic plasma models is impossible even by using modern supercomputers. Therefore, fluid or hydrodynamic models are attractive, but then the small-scale and kinetic effects are lost.

In the refilling problem, we have already seen that the small-scale phenomenon of the formation of electrostatic shocks can be an important aspect of the refilling. The crucial question is how to incorporate this small-scale feature into a large-scale hydrodynamic model of the refilling. Furthermore, we notice from the small-scale Vlasov simulations that the process of shock formation involves multistreaming of electrons and ions (see Figure 3). Such multistreaming is inherent to kinetic plasma models, which are based on velocity distribution functions, but not to hydrodynamic models. In the following subsections the weaknesses of hydrodynamic models will be discussed, along with remedies for some of them.

3.1 Single-Fluid Hydrodynamic Model

Time-dependent hydrodynamic models of the refilling of the plasmaspheric flux tube were used by Khazanov et al. (1984) and Singh et al. (1986b). Singh et al. (1986b) solved time-dependent continuity and momentum equations for the hydrogen ions along closed magnetic field lines (Figure 6) and assumed that the proton gas remains isothermal. They further assumed that the electron gas obeys the Boltzmann law and quasi-neutrality prevails. The development of plasma flow was followed in response to an intial evacuation of plasma in the flux tubes. The initial plasma density profiles were analytically modeled to represent the state of the low-density

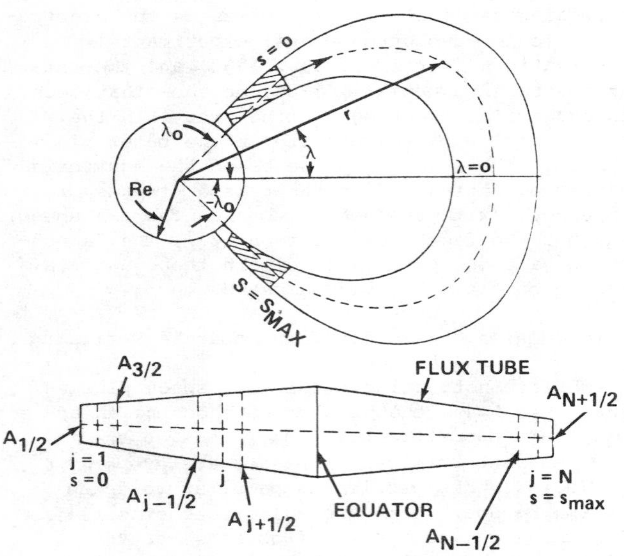

Fig. 6. Geometry of a closed geometric flux tube. The symbols λ and r represent the latitudinal angle and the geocentric distance, respectively. The ionosphere is indicated by the hatched portion; s = 0 and s = s_{max} show the bases in the conjugate ionospheres. The distance along the flux tube from s = 0 is denoted by s.

plasma after geomagnetic storms. It is worth mentioning that in this treatment the plasma flow was treated as one fluid; the plasma streams coming from the two hemispheres were not distinguished from each other when they overlapped.

Figures 7a-7c show the counterstreaming expansion of the plasmas from the conjugate ionospheric sources in response to the initial density profile as shown by the curve at t_o in Figure 7a. The initial velocity is chosen to be zero everywhere (Figure 7b). The curves at t_1, t_2, and t_3 in Figures 7a and 7b show features of expanding plasmas headed by density fronts as marked by F on the density profiles. This feature of the early flow is qualitatively similar to that of expanding plasmas along open field lines as discussed by Singh and Schunk (1985, 1986). We note from these curves a considerable refilling of the flux tube behind the density fronts. We also note a considerable acceleration of the ions behind these fronts. Shortly after t_3, which is about 22 min after t = 0, the two density fronts collide causing the formation of a large density bump at the equator as shown at t_4.

At a time (t = 2.4 hours) later than those shown in Figures 7a and 7b the plasma density and the flow velocity profiles in the flux tube are given in Figure 7c. For the purpose of our present discussion, the important fact to note from Figures 7b and 7c is that upon the collision of the plasma streams, originating from the conjugate ionopshere, the flow velocity in the equatorial region nearly vanishes and a large density buildup occurs.

It is interesting to note that the vanishing of the flow velocity and plasma buildup in the equatorial region are qualitatively analogous to what was seen in the previous section using the small-scale kinetic simulation. However, from the latter simulations we were able to delineate the physical mechanism responsible for the slowing down of the beams. In the absence of this mechanism being operative in the single-fluid

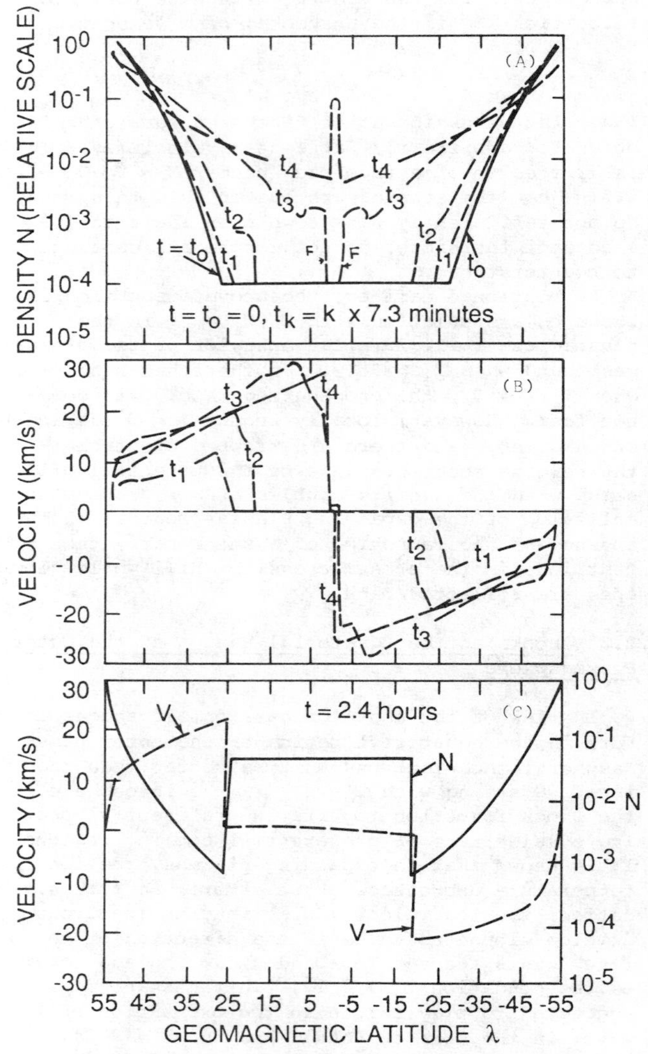

Fig. 7. (a) Density and (b) velocity profiles along the length of the flux tube (L = 4) at several times t_n, n = 1 to 4. the time interval between successive profiles is 7.3 min. (c) Density and velocity profiles at t = 2.4 hours. The horizontal axis is marked in terms of latitudinal angle λ. In top and bottom panels the density is in the units of N_b; the plasma density at the bases s = 0 and s = s_{max}.

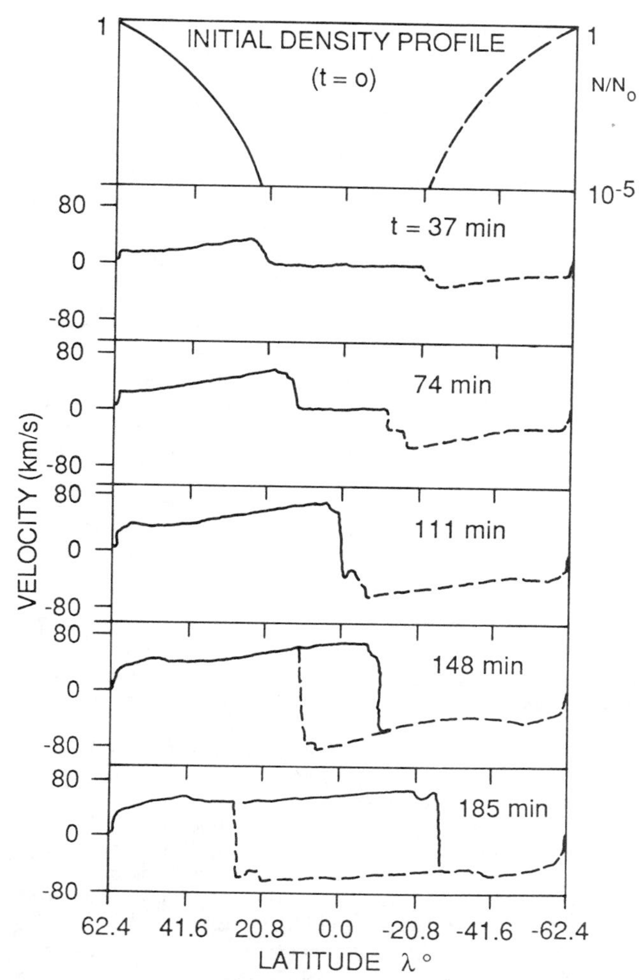

Fig. 8. Initial density profiles (top panel) in an empty flux tube (L = 6). Subsequent panels show the temporal evolution of the flow velocity.

hydrodynamic model, one finds that the flow velocity vanishes simply because in this model the single-valued flow velocity is identically zero when the flows from the conjugate hemispheres overlap in the equatorial region. In the next section this issue, using a two-fluid hydrodynamic model, will be examined.

3.2 Two-Fluid Hydrodynamic Model

In this model we assume that the plasma streams originating at s = 0 and $s = s_{max}$ (Figure 6) are two different fluids; thereby, their identities are kept separate when they overlap in the equatorial region. The top panel in Figure 8 shows the initial density profile in an empty flux tube (L = 6). The rest of the panels show the temporal evolution of the flow velocities of the two fluids. The solid and broken line curves show the flow properties of the fluids coming from the northern ($\lambda > 0$) and southern ($\lambda < 0$) hemispheres. In order to make the flow faster and thereby reduce the computer run time, $T_e = 10\ T_i$ was used in this run.

The early velocity profiles at t < 111 min show the plasma flow headed by the fronts, one in each hemisphere. The profiles at t = 111 min show the fronts crossing each other. At t = 148 min, the plasma fronts have crossed the equator and each front advances without being significantly affected by the presence of the other plasma stream. After about t ≃ 111 min, the plasma fronts cross the equator and the streams begin to overlap in space and counterstreaming is set up. There is no sign of the slowing down of the plasma streams and they do not couple with each other.

When the densities and flow velocities are averaged over the two streams, the total density and the average flow velocity profiles appear similar to those for the single-fluid model shown in Figures 7a-7c. This shows that the single-fluid model just mimicks the slowing down of the streams. As shown by the two-fluid model, the streams do not slow down. Furthermore, the calculation based on the two-fluid model shows that the shocks propagating away from the equator after the collision of the plasma streams are the shocks or the plasma expansion fronts originating from the conjugate hemisphere. In fluid models, there is no formation of an electrostatic shock pair when the plasma streams collide. This is in contrast to the kinetic simulation described earlier.

3.3 Why do the Streams not Couple in the Two-Fluid Hydrodynamic Model?

It is important to note that the ion-ion instability, which plays an important role in the momentum coupling of the counterstreaming ion beams, is essentially a fluid instability because its essential features can be obtained from fluid equations, except for the lower limit on the beam velocity in equation (2) dictated by the ion-Landau damping. Therefore, we need to examine why this instability did not occur in the above fluid calculations. From the instability criterion given by (2) and the modified criterion for the colliding beams given by (6), it is clear that no ion-ion instability can be excited because the flow velocity in the counterstreaming region is highly supersonic; in Figure 8 flow velocity is about 80 km s^{-1} which is 4.7 times the ion-acoustic speed $C_o \simeq 17$ km s^{-1}.

Now, one may ask, if the expansion velocity is reduced somehow, do the plasma streams couple and lead to the formation of electrostatic shocks? In order to answer this question we repeated the two-fluid hydrodynamic calculation with a mild initial depletion of the flux tube; the minimum density in the equatorial region was 10% of the base density N_b at s = 0 and S_{max}, in contrast to $10^{-5}\ N_b$ in the previous calculation (see Figure

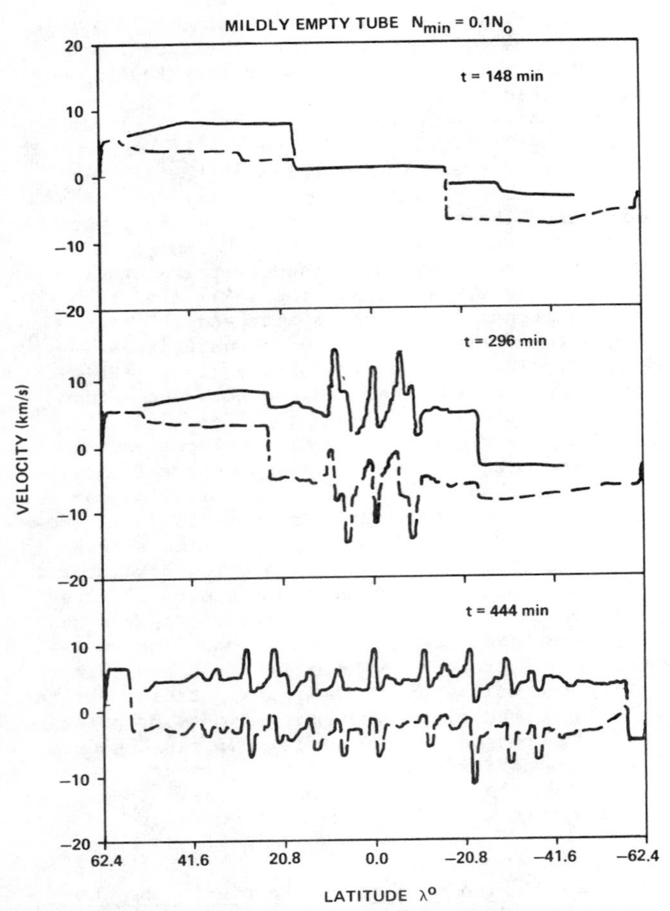

Fig. 9. Same as Figure 8, but the flux tube was depleted very mildly. When the plasma streams overlap, some waves are excited by ion-ion interactions.

8). The resulting plasma expansion is shown in Figure 9 which shows the flow velocity at some selected times. At t = 148 min, we observe the expanding plasma fronts from both hemispheres. At t = 296 min the fronts have crossed the equator and counterstreaming has set up, but the interesting feature to note is that oscillations have broken out in the flow velocity. These oscillations are the manifestation of the ion-ion instability. In the counterstreaming region, the average flow velocity in each stream is $V_b \simeq 10$ km s^{-1}, which is less than the ion-acoustic velocity $C_o = 17$ km s^{-1}. Thus, according to the linear instability criterion for the hydrodynamic model ($V_b < C_o$), the instability is possible and the calculation does show it. At later times, as shown at t = 444 min, the instability spreads over most of the flux tube along with the counterstreaming of the flow.

As can be seen now the I-I instability does occur in the fluid model, but the formation of an electrostatic shock does not occur, in contrast to the Vlasov simulations. This leads us to further scrutinize the physical factors that are lacking in the hydrodynamic model but which are present in the Vlasov model.

3.4 Electron Dynamics

The Vlasov simulation showed that so far as ions are concerned their essential behavior involving slowing down of the plasma streams is fluid-like and a two-fluid model may suffice for their description. However, the electron dynamics is quite complex; it involves a self-consistent generation of multistreaming electrons consisting of trapped (circulating) and accelerated (free) electrons. On the other hand, in the hydrodynamic model we assumed that electrons obey the Boltzmann law. This appears to be an over-simplification of the physical phenomena occurring in colliding plasma streams. The Boltzmann relation relates the density and the electric potential

$$N_1 = N_o \exp(e\Delta\Phi/k_B T_e) , \tag{7}$$

where $\Delta\Phi$ is the potential difference between the points where densities are N_1 and N_o and the potentials are Φ_1 and Φ_o, $\Delta\Phi = \Phi_1 - \Phi_o$.

We recall that in the region of the colliding plasma streams, the electron density from the Vlasov model is given by (4). In Figure 10, the electron density variations are compared with the electric potential $\Delta\tilde{\Phi}$ as given by the Boltzmann law in (7) and the non-Boltzmann relation (4).

The most noteworthy feature of Figure 10 is that when the plasma density doubles ($N_1/N_o = 2$) as expected when the streams collide, the potential difference for the Boltzmann (B) distribu-

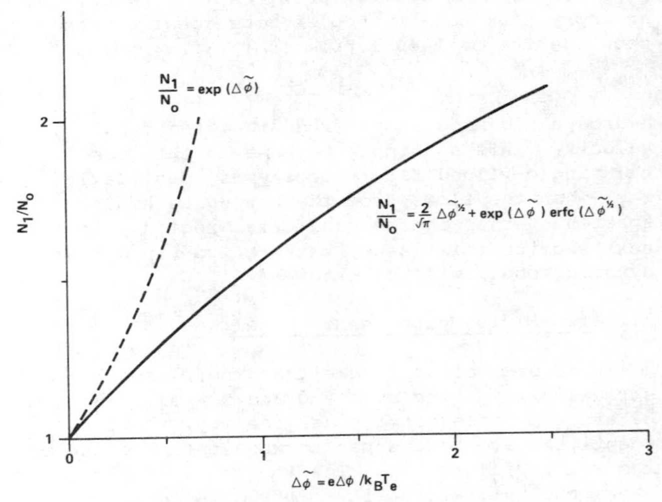

Fig. 10. Plasma density (N_1/N_o) as functions of the electric potential for the Boltzmann and non-Boltzmann distributions; the analytical expressions for the distributions are given.

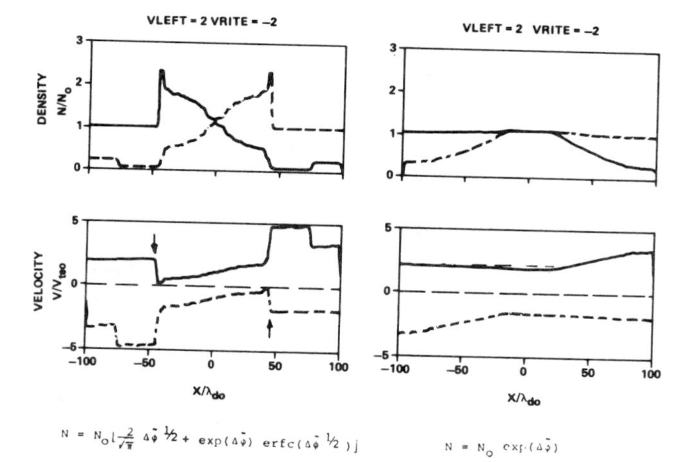

Fig. 11. State of the plasma (left panels) at a time $t = 37.5\ \omega_{pi}^{-1}$ when the colliding plasma streams evolved according to the non-Boltzmann distribution for the electrons. Right panels show the same when the evolution occurred using the Boltzmann distribution. Top and bottom panels show the density and velocity profiles, respectively. Solid and broken curves are for the plasma streams coming from left and right, respectively.

tion $\Delta\tilde{\Phi}_B = e\Delta\Phi_B/k_BT_e = 0.69$. On the other hand, for the non-Boltzmann (NB) distribution $\Delta\tilde{\Phi}_{NB} \approx 2.25$. Thus, the assumption of the Boltzmann distribution unphysically inhibits the development of the elecltric potential in the interaction region. The relatively large potential for the non-Boltzmann distribution is effective in slowing down the streams.

In order to demonstrate the difference between the Boltzmann and non-Boltzmann distributions for the electrons, we followed the temporal evolution of the colliding plasma streams with the initial plasma conditions as follows (Figure 2b). We assumed that at time $t = 0$ there were two plasma streams, one for $X \leq 0$ with a density $N/N_o = 1$ and flow velocity $V = 2\ V_{ti}$, and the other for $X \geq 0$ with the same density and an opposite flow velocity $V = -2\ V_{ti}$. The ion streams were treated as separate fluids.

Figure 11 shows the state of the plasma at a later time $t = 37.5\ \omega_{pi}^{-1}$, where ω_{pi} is the ion-plasma frequency of the individual ion streams. The left-hand panels show the density and flow veocity when the plasma was allowed to evolve according to the non-Boltzmann electron distribution (4). On the other hand, the right-hand panels show the plasma state at the same time when it evolved according to the Boltzmann distribution (7). The solid and broken curves in the figure show the plasma streams coming from the left and right, respectively.

The right-hand bottom panel shows that for the Boltzmann distribution, the ions slow down only slightly, and the essential feature of the density and velocity profiles for this distribution is that the two streams interpenetrate without significantly affecting each other. The interpenetration of the streams is headed by plasma expansion as characterized by the accelerated ions in the region of the decreasing stream densities.

In contrast to the state of plasma in the right-hand panels, we note from the left-hand panels that the incoming streams completely slow down (marked by the arrows) where the individual stream densities jump roughly to 2 N_o. These features are similar to those obtained from the Vlasov treatments for both the electrons and ions. Thus, we conclude that by using hydrodynamic models for the ions but using improved equations of state for the electron it is possible to simulate the physical processes which lead to the formation of electrostatic shocks in the equatorial region of the plasmasphere where the interhemispheric flows collide. However, we note that the calculations for the entire flux tube including the improved equation of state for the electrons have not been carried out so far.

4. Wave-Particle Interactions and Refilling

Schulz and Koons (1972) suggested that the waves excited by the ion strams can cause pitch angle scattering of the ions and trap them in the equatorial region. Singh et al. (1982) showed that field-aligned flows can be transformed into trapped ion populations when the ions interact with the electromagnetic ion-cyclotron waves; the wave intensity required for the trapping ($\sim 10^{-3}\ \gamma^2$/Hz) is readily available (Perraut et al., 1982). However, as yet, a self-consistent study which includes the effect of ion trapping on the refilling has not been performed.

Recently Singh and Hwang (1987) examined the role of perpendicular ion heating in trapping the upflowing ionospheric ions in the outer plasmaspheric flux tubes. They highlighted that the heating by a low level background electrical noise distributed along the flux tubes can be effective in trapping the ions. It was found that a noise level with a spectral power density $\Psi \approx 10^{-11}\ V^2\ m^{-2}\ Hz^{-1}$ asociated with the fluctuations in the perpendicular electric fields near the ion-cyclotron frequencies can scatter the field-aligned ions up to 20° pitch angle keeping the ion energy to a few electron volts.

The ions having such pitch angles are trapped in the flux tubes and when their density becomes sufficiently large they are thermalized by Coulomb collisions producing the bulk of the plasmaspheric plasma population with characteristic energy 1 eV. The spectral power density $\sim 10^{-11}\ V^2\ m^{-2}\ Hz^{-1}$ is near the background noise level as indicated by space observations (Kintner and Gurnett, 1977; Olsen et al., 1987).

When the spectral power density Ψ becomes relatively large $\sim 10^{-9}\ V^2\ m^{-2}\ Hz^{-1}$, the heating leads to the generation of ion conics having superthermal energies and pitch angles in the range 20°-40° (Horwitz et al., 1982; Singh and Hwang, 1987).

Localized equatorial heating is found to be very effective in trapping ions by producing highly anisotropic pitch angle distributions with maximum flux around 90° (Olsen, 1981; Olsen et al., 1987). Singh and Hwang (1987) have shown that the difference in the pitch angle anisotropies of the electrons and the locally heated ions near the equator can set up downward parallel electric fields, which can reflect the upgoing plasma streams (Figure 1). Thus, the interhemispheric flow can be stopped and then each hemisphere fills with its own plasma. The magnitude of the parallel electric potential drop $\Delta\Phi_{\parallel}$ associated with the downward electric field in each hemisphere is approximately given by

$$\Delta\Phi_{\parallel} \simeq 4.5 \left(\frac{k_B T_{\perp}}{e}\right) \lambda_H^{2} , \tag{8}$$

where $T_{\perp}$ is the temperature of the heated ions and λ_H (radians) is the latitudinal width of the heated ions. The observations indicate that $\lambda_H \simeq 5°$ and T can range up to several tens of electron volts (Olsen, 1981; Olsen et al., 1987). For the typical energy (~1 eV) of the ionospheric plasma streams, an equatorial heating leading to $T_{\perp} > 30$ eV is expected to decouple the two hemispheres.

5. Conclusion

The main purpose of this paper has been to highlight the microscopic plasma processes, such as the electrostatic shocks, pitch angle scattering of the ions, and ion perpendicular heating, which help in trapping and/or thermalizing the plasma in the flux tubes.

The mechanism and conditions for the formation of electrostatic shocks, which have been debated since the postulate of Banks et al. (1971), have been discussed. The basic idea behind the shock formation is that the plasma streams must undergo a momentum coupling when they collide. The compressive density perturbation and the associated potential barrier for the ions near the collision points help in coupling the ion streams; when $T_e > 3\ T_i$ and the ion stream velocity lies in the range $1.3\ V_{ti} < V_b < 2.3\ C_o$, where C_o is the ion-acoustic speed, the perturbation is capable of slowing down the streams to C_o, after which the ion-ion instability sets in slowing the streams to C_o, after which the ion-ion instability sets in slowing the streams completely, and a pair of shocks forms. The shocks are powered by the kinetic energy of the ion streams. The electric fields near the shock fronts accelerate the incoming electron and trap the electrons in the shocked plasma. Thus, an energized electron population in the shocked plasma is created. The energy for this is supplied by the ion streams. When the streams are highly supersonic, $V_b > 2.3\ C_o$, the coupling, and hence the shock formation, is not expected. In such a situation the ion streams continue to counterstream after the collision.

It is shown that the assumption of the Boltzmann law for the electrons is not valid in the region of colliding plasma streams. Striking differences between the temporal evolutions of an assumed initial plasma state (simulating colliding plasma streams) using the Boltzmann and non-Boltzmann distributions for the electrons are shown. The use of such a non-Boltzmann description of electrons in large-scale hydrodynamic plasma models for the ions can facilitate constructing theoretical models for the refilling of the flux tubes, including the feature of electrostatic shock formation. However, as yet, such models have not been worked out.

It is highlighted that the extended low level background plasma turbulence at frequencies $\omega \sim \Omega_i$ can cause a weak perpendicular heating of the ions imparting to them an energy of ~1 eV (Singh and Hwang, 1987). Such heating can be effective in trapping the ions in the flux tube. Such trappings followed by Coulomb collisions (Schulz and Koons, 1972) can be efficient processes in refilling the flux tubes. When the plasma streams are too fast or too slow [see equations (2) and (6)] and $T_e \sim T_i$, the shock formation is unlikely, and the extended ion heating may be an effective mechanism for the refilling.

It is shown that the equatorial ion heating has an interesting consequence of stopping the interhemispheric flow by setting up a potential barrier at the equator. The barrier is caused by the electric fields supported by the difference between the pitch angle anisotropies of the equatorially heated ions and the ambient electrons.

It is quite likely that all the above processes must be contributing to the refilling. However, a self-consistent model for the refilling, including all these processes, remains to be studied. Such a study will be able to determine the relative importance of shock formation, pitch angle scattering, and the perpendicular heating of ions by the ambient electromagnetic turbulence in thermalizing the plasma flow from the ionosphere. Observationally, it is important to establish whether or not the extended weak electromagnetic turbulence exists along the field lines. If it does, the trapping of the ions in the flux tube and their subsequent thermalization by Coulomb collisions when the density becomes sufficiently large is easily understandable.

Acknowledgment. This work was supported by NASA contract NAG8-058.

References

Banks, P. M., A. F. Nagy, and W. I. Axford, Dynamical behavior of thermal protons in the mid-latitude ionosphere and magnetosphere, Planet. Space Sci., 19, 1053, 1971.

Forslund, D. W., and C. R. Shonk, Formation of electrostatic collisionless shocks, Phys. Rev. Lett., 25, 1699, 1971.

Horwitz, J. L., Core plasma in the magnetosphere, Rev. Geophys., in press, 1987.

Horwitz, J. L., C. R. Baugher, C. R. Chappell, E. G. Shelley, and D. T. Young, Conical pitch angle distributions of very low-energy ion fluxes observed by ISEE-1, J. Geophys. Res., 87, 2311, 1982.

Kintner, P. M., and D. A. Gurnett, Observations of ion cyclotron waves within the plasmasphere by Hawkeye I, J. Geophys. Res., 82, 2314, 1977.

Khazanov, G. V., M. A. Koen, Y. V. Konikov, and I. M. Sidorov, Simulation of ionosphere-plasmasphere coupling taking into account ion inertia and temperature anisotropy, Planet. Space Sci., 32, 585, 1984.

McKee, C. F., Simulation of counterstreaming plasmas with application to collisionless electrostatic shocks, Phys. Rev. Lett., 24, 990, 1970.

Morse, R. L., Adiabatic time development of plasma sheaths, Phys. Fluids, 8, 308, 1965.

Norris, A. J., J.F.E. Johnson, J. J. Sojka, G. L. Wrenn, N. Cornilleau-Wehrlin, S. Perraut, and A. Roux, Experimental evidence for the acceleration of thermal electrons by ion cyclotron waves in the magnetosphere, J. Geophys. Res., 88, 889, 1983.

Olsen, C. R., Equatorially trapped plasma populations, J. Geophys. Res., 86, 11,235, 1981.

Olsen, C. R., S. D. Shawhan, D. L. Gallagher, J. L. Green, C. R. Chappell, and R. R. Anderson, Plasma observation at the earth's magnetic equator, J. Geophys. Res., 92, 2385, 1987.

Perraut, S., A. Roux, P. Robert, R. Gendrin, and A. Korth, A systematic study of ULF waves above FHT from GEOS 1 and 2 measurements and their relationships with proton ring distributions, J. Geophys. Res., 87, 6219, 1982.

Roux, A., N. Cornilleau-Wehrlin, and J. L. Rauch, Acceleration of thermal electrons by ICW's propagating in a multi-component magnetospheric plasma, J. Geophys. Res., 89, 2267, 1984.

Schulz, M., and H. C. Koons, Thermalization of colliding ion streams beyond the plasmapause, J. Geophys. Res., 77, 248, 1972.

Singh, N., Computer experiments on the formation and dynamics of electric double layer, Plasma Phys., 22, 1, 1980.

Singh, N., and K. S. Hwang, Perpendicular ion heating effects on the refilling of the outer plasmaspheric flux tubes, J. Geophys. Res., in press, 1987.

Singh, N., and R. W. Schunk, Numerical calculations relevant to the initial expansion of the polar wind, J. Geophys. Res., 87, 9154, 1982.

Singh, N., and R. W. Schunk, Numerical simulation of counterstreaming plasmas and their relevance to interhemispheric flows, J. Geophys. Res., 88, 7867, 1983.

Singh, N., and R. W. Schunk, Temporal behavior of density perturbations in the polar wind, J. Geophys. Res., 90, 6487, 1985.

Singh, N., and R. W. Schunk, Ion acceleration in expanding ionospheric plasma, in Ion Acceleration in the Magnetosphere and Ionosphere, Geophys. Monogr. Ser., vol. 38, edited by T. Chang, p. 362, AGU, Washington, D.C., 1986.

Singh, N., W. J. Raitt, and F. Yasuhara, Low energy ion distribution functions on a magnetically quiet day at geostationary altitude (L = 7), J. Geophys. Res., 87, 681, 1982.

Singh, N., H. Thiemann, and R. W. Schunk, Studies on counterstraming plasma expansion, Physica Scripta, 33, 355, 1986a.

Singh, N., R. W. Schunk, and H. Thiemann, Temporal features of the refilling of a plasmaspheric flux tube, J. Geophys. Res., 91, 13,433, 1986b.

Sojka, J. J., R. W. Schunk, J.F.E. Johnson, J. H. Waite, and C. R. Chappell, Characteristics of thermal and suprathermal ions associated with the dayside plasma trough as measured by the Dynamics Explorer retarding ion mass spectrometer, J. Geophys. Res., 88, 7895, 1983.

MODELING OF THE THERMAL PLASMA IN THE OUTER PLASMASPHERE - A MAGNETOSPHERIC HEAT SOURCE

M. O. Chandler,[1] J. U. Kozyra,[2] J. L. Horwitz,[3] R. H. Comfort,[3] and L. H. Brace[4]

Abstract. A case study has been carried out using data from the Dynamics Explorer 1 and 2 spaceraft to study the effect of Coulomb interactions between ring current and suprathermal O^+ and thermal protons on the plasmasphere. Results from a one-dimensional plasmaspheric model suggest that heating due to Coulomb collisions may be sufficient to raise the ion and electron temperatures to observed values. The resultant high temperatures produced enhancements in the model O^+ and O^{++} densities in agreement with observations.

Introduction

There exist several interesting phenomena in the region of the plasmapause that suggest a strong interaction between the magnetosphere, plasmasphere, and underlying ionosphere. These include high ion and electron temperatures (at times exceeding 10,000 K), enhanced heavy ion (O^{++} and O^+) densities at high altitudes, and electron temperature enhancements with associated SAR arc emissions in the ionosphere.

These phenomena have been linked by several studies (e.g., Brace et al., 1967; Brace and Theis, 1974; Titheridge, 1976; Burke et al., 1979; Büchner et al., 1983; Kozyra et al., 1986; Chappell et al., 1971; Horwitz et al., 1986; W. T. Roberts et al., unpublished manuscript, 1987).

These observations suggest the presence of a high-altitude heat source in the outer plasmasphere. This heat source has been associated with the overlap of the energetic ring current and cold plasmaspheric populations. Although the exact method of energy transfer between these populations remains open to question, heating of thermal electrons in the outer plasmasphere via Coulomb collisions with energetic ring current ions (Cole, 1965; Kozyra et al., 1987) and wave-particle interactions between thermal electrons and ion cyclotron waves (Cornwall et al., 1971) have been proposed as mechanisms.

The results of Kozyra et al. (1987) indicate that sufficient energy could be supplied to the thermal electrons to support the subauroral electron temperature enhancement and associated SAR arc, but that the source of this energy is tens of kiloelectron volt O^+ ions and not energetic protons as envisioned by Cole (1965).

Model

A time-dependent hydrodynamic model has been used for the simulations. The model includes solutions to the coupled continuity and momentum equations for O^+ and H^+, the energy and heat flow equations for ions and electrons, and the photoelectron transport equations. These portions of the model have been presented previously (e.g., Young et al., 1980; Chandler et al., 1983). In addition to these equations the minor ion diffusion equation of St.-Maurice and Schunk (1977) has been added for O^{++} (Chandler et al., 1987). The neutral atmosphere parameters were derived from the MSIS model of Hedin (1983). The equations were solved using geophysical inputs appropriate to the time of the observations on October 23, 1981. These inputs include an Ap of 27 and an $F_{10.7}$-cm flux of 208. The solar EUV fluxes are the 1974 values taken from Torr et al. (1979). These have been multiplied by a factor of 2 to simulate the near solar maximum conditions of 1981. Good agreement with observed values of the electron density in the ionosphere suggests that this is a reasonable approximation and that the ionospheric source for the plasmasphere is of the proper magnitude.

The heating rate for the high-altitude outer plasmasphere has been calculated according the methods of Kozyra et al. (1987). The calculated altitude dependence of this heating was approximated by a Gaussian function with a maximum at the magnetic equator and a half width of 17.5° in magnetic latitude.

The heating was assumed to commence at 15 to

[1]Space Science Laboratory, NASA Marshall Space Flight Center, Huntsville, Alabama 35812.
[2]Space Physics Research Laboratory, The University of Michigan, Ann Arbor, Michigan 48105.
[3]Physics Department, The University of Alabama in Huntsville, Huntsville, Alabama 35899.
[4]Planetary Atmospheres Branch, NASA Goddard Space Flight Center, Greenbelt, Maryland 20771.

TABLE 1. Satellite Positions During Observations

	L	Altitude (km)	LT	UT	Λ
DE 1					
1	2.2	6600	20.8	10:47	48°
2	2.5	7460	20.8	10:42	51°
3	3.0	8500	20.8	10:38	55°
DE 2					
1	2.2	895	20.9	10:30	48°
2	2.5	905	20.9	10:29	51°
3	3.0	920	20.9	10:28	55°

16 hours local time. At that time the heating rate was taken as 1% of the calculated rate and it was assumed to rise to the calculated rate over a period of 2 hours. From previous work (Chandler et al., 1987) it was found that the O^{++} and O^{+} densities reached at ~21 LT using such a scheme are near those obtained after a full 24 hours of constant heating. A full 24-hour simulation was carried out for the L = 3 case with similar results.

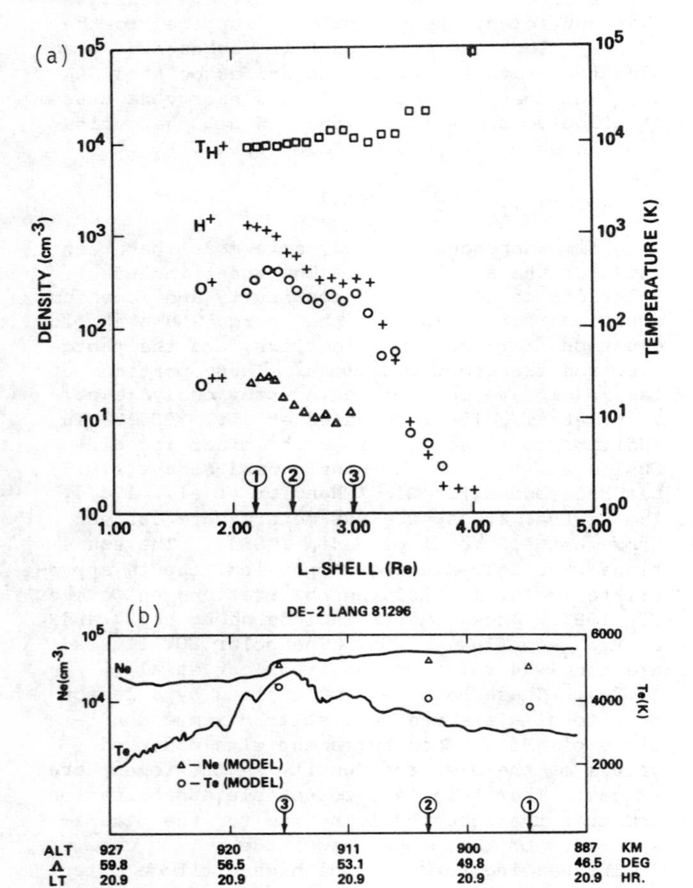

Fig. 1. Plasma observations for October 23, 1981. Top panel: ion densities and the H^{+} temperature from DE 1 as a function of L value; bottom panel: electron density and temperature from DE 2. The numbered arrows mark the regions modeled.

TABLE 2. Heating Rates and Resulting Enhancements

	Max. Heating Rates ($erg\ cm^{-3}\ s^{-1}$)		Heat Flux Through 1000 km ($erg\ cm^{-2}\ s^{-1}$)		Total Heat Input ($erg\ s^{-1}$)		6300 Å Airglow		Density Enhancements		Temperature Enhancements	
L	Ion	Electron	Ion	Electron	Ion	Electron	Col. Emiss.	Energy ($erg\ s^{-1}$)	O^{++}	O^{+}	Ion	Electron
2.2	1.4×10^{-12}	2.8×10^{-12}	3.0×10^{-2}	1.8×10^{-2}	1.4×10^{-3}	2.8×10^{-3}	65	1.5×10^{-5}	7×10^{2}	2×10^{5}	5.1	2.9
2.5	3.8×10^{-13}	2.4×10^{-12}	1.9×10^{-3}	2.4×10^{-2}	4.7×10^{-4}	2.9×10^{-3}	250	3.9×10^{-5}	6×10^{2}	5×10^{5}	6.0	3.0
3.0	1.4×10^{-13}	1.2×10^{-12}	2.1×10^{-3}	2.9×10^{-2}	2.2×10^{-4}	1.8×10^{-3}	630	5.3×10^{-5}	1×10^{3}	5×10^{6}	5.1	4.0

NOTE: All calculations involving flux tube area, used tubes with 1 cm^2 area in the equatorial plane. Factors to convert fluxes to tubes with 1 cm^2 area at 1000 km are: 0.094, 0.062, 0.033, for L = 2.2, 2.5, 3.0, respectively.

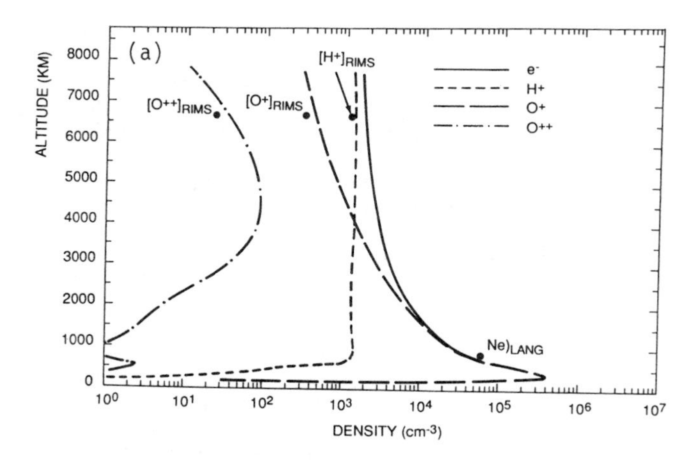

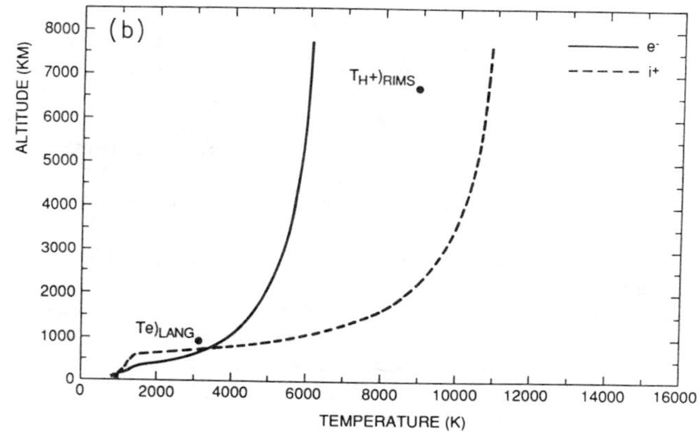

Fig. 2. Model results for L = 2.2. (a) ion densities; (b) ion and electron temperature. DE observations are also shown.

Data

The data for this study were assembled from several instruments on the dual spacecraft DE mission by Kozyra et al. (1987). They included ion composition, density, temperature, and energy measurements at high altitudes from the DE 1 retarding ion mass spectrometer and energetic ion composition spectrometer experiment. At low altitudes, they included electron density and temperature measurements from the DE 2 Langmuir probe. Data were taken for L shell values of 3.0, 2.5, and 2.2 on October 23, 1981. The times and locations of the two satellites for these values are listed in Table 1.

The DE 1 observations (Figure 1a) show a region in the outer plasmasphere where the ion temperature (characterized here by the H^+ temperature) is near 10,000 K. The sharp drop in H^+ density just outside of L = 3 would typically be labeled the plasmapause and is colocated with this region of high ion temperature. In this region the O^+ density is comparable to the H^+ density, while the O^{++} density is about a factor of 20 lower than the O^+ density but still enhanced relative to its ionospheric ratio with O^+ ($\sim 10^{-3}$ cm^{-3}). Such increases in O^+ density have been associated with enhanced ionospheric temperatures (Horwitz et al., 1986). Indeed, at low altitudes the DE 2 (Figure 1b) results show a drop in the electron density near the latitude where the plasmapause was observed at high altitudes (L ~ 3) along with an electron temperature maximum which corresponds, in L, to the O^+ maximum at high altitude. The ring current observations have been discussed in detail by Kozyra et al. (1987). The calculated ion and electron heating rates are given in Table 2.

The plasmaspheric densities and temperatures obtained from the model using these heating rates are shown in Figures 2-4. In all cases the initial H^+ profiles were chosen to match the observed H^+ densities at the altitude of DE 1. From the previous study (Chandler et al., 1987) it was determined that the plasmaspheric H^+ density was weakly affected by the imposition of the

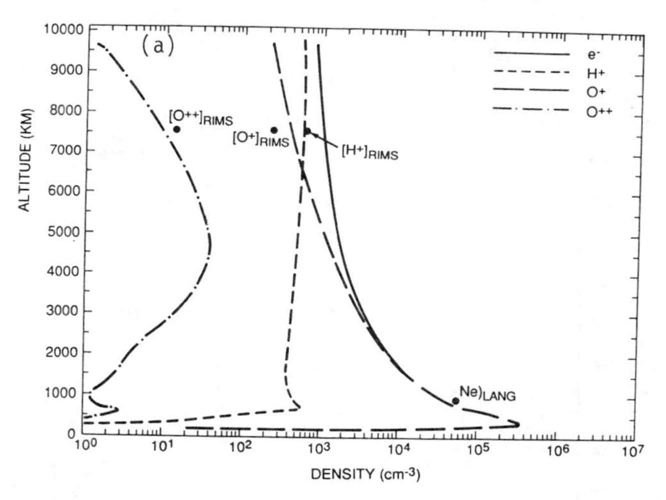

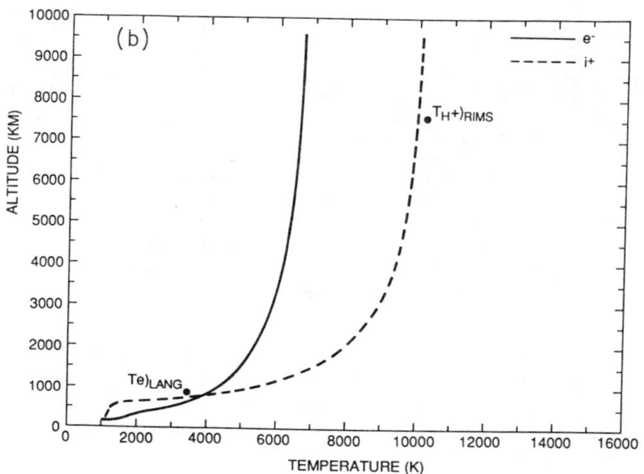

Fig. 3. Model results for L = 2.5. (a) ion densities; (b) ion and electron temperature. DE observations are also shown.

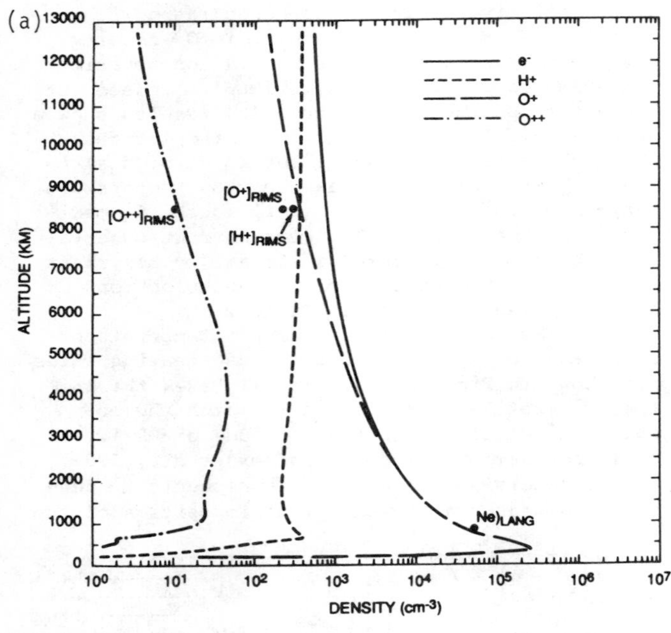

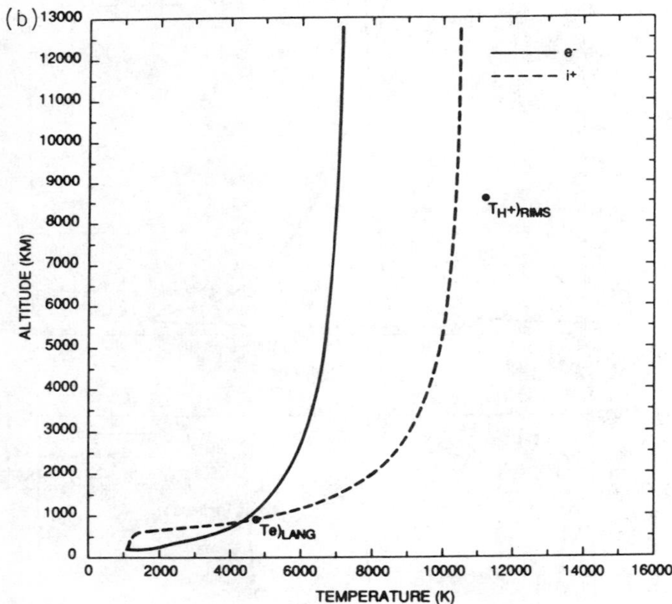

Fig. 4. Model results for L = 3.0. (a) ion densities; (b) ion and electron temperature. DE observations are also shown.

external heating. The final O^{++} and O^{+} densities obtained from the simulations are in good agreement with the observations. The calculated ion temperature was within 20% of the observed H^{+} temperature in all cases.

Discussion and Conclusions

The observations and simulations presented here illustrate one example of the strong interaction between the magnetosphere, plasmasphere, and ionosphere. Several observed phenomena - high plasmaspheric temperatures and heavy ion (O^{++} and O^{+}) densities at high altitudes along with electron temperature enhancements and SAR arc emissions (at 6300 Å) in the ionosphere - are brought together and postulated to be the result of the interaction, through Coulomb collisions, of the ring current and the plasmaspheric particles at lower energies. The energy transferred into the plasmasphere through this process initially heats the plasma at high altitudes. A large fraction of this heat is then conducted into the ionosphere where it heats the ionospheric plasma. This heating in the ionosphere provides the energy necessary to excite the neutral atmosphere, producing airglow, and also supplies energy for the upward expansion and transport of heavy ions into the plasmasphere.

While this heating has important effects in the ionosphere and plasmasphere, it represents a small fraction of the available energy in the ring current. Cole (1965) estimated the total energy in the ring current following a magnetic storm with a Dst decrease of 100 γ at 3×10^{22} erg. The estimated total energy loss rate in this case is $\sim 3 \times 10^{15}$ erg s^{-1}. It is possible that the Coulomb mechanism could be responsible for the rapid depletion of energetic O^{+} observed by AMPTE (Krimigis et al., 1985).

The calculated 6300 Å column emission rates were 65, 250, and 630 Rayleighs for L = 2.2, 2.5, and 3.0, respectively, consistent with observations (D. W. Slater, private communication, 1986).

While this study provides tantalizing hints at the role of Coulomb interactions in magnetosphere-plasmasphere coupling, much additional work is called for. Subsequent studies must include more accurate calculations of the ionospheric effects of this heating such as the enhanced loss of O^{+} due to vibrational excitation of N_2 and changes in the neutral gas temperature and density. The ring current and suprathermal particle distribution functions and their variation with altitude and latitude must be studied in more detail in order to remove uncertainties in the magnitudes of the heating rates. Detailed calculations and simulations of other SAR arc events are currently underway to answer some of these outstanding questions.

Acknowledgments. The majority of this work was conducted while M.O.C. was with The University of Alabama in Huntsville (UAH). The work at UAH was supported by NASA contract NAS8-33982 and grants NAG8-054 and NAG8-058 and NSF grant ATM-8506642. The work at the University of Michigan was supported by NSF grant ATM-8508753 and NASA grants NAG5-472 and NGR-23-005-015.

References

Brace, L. H., and R. F. Theis, The behavior of the plasmasphere at mid-latitudes: Isis 1 Langmuir probe measurements, J. Geophys. Res., 79, 1871, 1974.

Brace, L. H., B. M. Reddy, and H. G. Mayr, Global behavior of the ionosphere at 1000-kilometer altitude, J. Geophys. Res., 72, 265, 1967.

Büchner, J., H.-R. Lehmann, and J. Rendtel, Properties of the subauroral electron temperature peak observed by Langmuir-probe measurements on board intercosmos-18, Gerlands Beitr. Geophys., 92, 368, 1983.

Burke, W. J., H. J. Braun, J. W. Münch, and R. C. Sagalyn, Observations concerning the relationship between the quiet-time ring current and electron temperatures at trough latitudes, Planet. Space Sci., 27, 1175, 1979.

Chandler, M. O., R. A. Behnke, A. F. Nagy, E. G. Fontheim, P. G. Richards, and D. G. Torr, Comparison of measured and calculated low-latitude ionospheric properties, J. Geophys. Res., 88, 9187, 1983.

Chandler, M. O., J. J. Ponthieu, T. E. Cravens, A. F. Nagy, and P. G. Richards, Model calculations of minor ion populations in the plasmasphere, J. Geophys. Res., 92, 5885, 1987.

Chappell, C. R., K. K. Harris, and G. W. Sharp, OGO-V measurements of the plasmasphere during observations of stable auroral red arcs, J. Geophys. Res., 76, 2357, 1971.

Cole, K. D., Stable auroral red arcs, sinks for energy of Dst main phase, J. Geophys. Res., 70, 1689, 1965.

Cornwall, J. M., F. V. Coroniti, and R. M. Thorne, Unified theory of SAR arc formation at the plasmapause, J. Geophys. Res., 76, 4428, 1971.

Hedin, A. E., A revised thermospheric model based on mass spectrometer and incoherent scatter data: MSIS-83, J. Geophys. Res., 88, 10,170, 1983.

Horwitz, J. L., L. H. Brace, R. H. Comfort, and C. R. Chappell, Dual-spacecraft measurements of plasmasphere-ionosphere coupling, J. Geophys. Res., 91, 11,203, 1986.

Kozyra, J. U., L. H. Brace, T. E. Cravens, and A. F. Nagy, A Statistical study of the subauroral electron temperature enhancement using Dynamics Explorer 2 Langmuir probe observations, J. Geophys. Res., 91, 11,270, 1986.

Kozyra, J. U., E. G. Shelley, R. H. Comfort, L. H. Brace, T. E. Cravens, and A. F. Nagy, The role of ring current O^+ in the formation of stable auroral red arcs, J. Geophys. Res., in press, 1987.

Krimigis, S. M., G. Gloeckler, R. W. McEnire, T. A. Potemra, F. L. Scarf, and E. G. Shelley, Magnetic storm of September 4, 1984: a synthesis of ring current spectra and energy densities measured with AMPTE/CCE, Geophys. Res. Lett., 12, 329, 1985.

St.-Maurice, J.-P. and R. W. Schunk, Diffusion and heat flow equations for the mid-latitude topside ionosphere, Planet. Space Sci., 25, 907, 1977.

Titheridge, J. E., Plasmapause effects in the topside ionosphere, J. Geophys. Res., 81, 3227, 1976.

Torr, M. R., D. G. Torr, R. A. Ong, and H. E. Hinteregger, Ionization frequencies for major thermospheric constituents as a function of solar cycle 21, Geophys. Res. Lett., 6, 771, 1979.

Young, E. R., D. G. Torr, P. Richards, and A. F. Nagy, A computer simulation of the midlatitude plasmasphere and ionosphere, Planet. Space Sci., 28, 881, 1980.

PRELIMINARY STATISTICAL SURVEY OF PLASMASPHERIC ION PROPERTIES FROM OBSERVATIONS BY DE 1/RIMS

R. H. Comfort and I. T. Newberry

Physics Department, The University of Alabama in Huntsville, Huntsville, Alabama 35899

C. R. Chappell

NASA Marshall Space Flight Center, Huntsville, Alabama 35812

Abstract. Theoretical modeling of the plasmasphere must be guided by the observed behavior of the plasma properties as they vary in space and time in response to changing conditions. Here we present mean temperatures and densities of plasmaspheric H^+ ions obtained from observations by the retarding ion mass spectrometer on Dynamics Explorer 1. These averages represent more than 100 transits of the plasmasphere in two local time regions: morning side (7-11 hours) and evening side (18-22 hours). They are presented in terms of profiles of temperature or density versus L shell, in which each data point represents an average over a bin of width $\Delta L = 0.2$. In order to see variations of interest, the data have been subdivided in two ways. First, the local time regions noted above have been kept distinct, and second, data above and below (in altitude) the L shell of closest approach are kept separate. Composition ratios are also presented for He^+, O^+, He^{++}, and O^{++}. Lest modelers take these mean profiles too seriously, several representative, but differing, profiles are presented to demonstrate the range of variations that are commonly observed.

Introduction

An essential adjunct to any model is a set of observations of the object being modeled, either to verify a theoretical model or to provide the necessary parameter values for an empirical model. In the past, models of the plasmasphere have had relatively few detailed observations, particularly at high altitudes, to serve these needs. However, observations by the retarding ion mass spectrometer (RIMS) on Dynamics Explorer 1 (DE 1) can provide the data set to meet these objectives. Toward these ends, we present a preliminary set of analyzed data from more than 100 plasmasphere transits by DE 1 during October and November of 1981, shortly after launch. This was a period of high solar activity and moderate, but variable, magnetic activity. Figure 1 shows a DE 1 orbit typical of this time period. Since the orbit is almost polar, it is nearly coplanar with magnetic field lines, and local time variations over an orbit are small, except for the 12-hour shift from morning to evening side. Note that a given plasmasphere pass can be broken into low-altitude and high-altitude segments, divided by the L shell of closest approach, to which the spacecraft is moving tangent. Unless otherwise indicated, all observations presented come from the high-altitude segments of the orbits.

All temperatures and densities are for the H^+ ion. Composition is displayed in terms of ratios of densities of the various ion species to those of H^+. Spatial variations are presented in terms of L shell profiles, with separate profiles for local morning and evening sides. All levels of magnetic activity are included in the profiles.

Observations

Operation of the RIMS system has been described in detail by Chappell et al. (1981). Ion temperatures and densities are obtained from data averaged over 60-s intervals by techniques discussed by Comfort et al. (1982) and Comfort et al. (1985). Typical profiles of these parameters are displayed in Figure 2. Several differences will be noted between these two sets of profiles. Upper and lower panels correspond to different local times and different levels of magnetic activity. Profiles in the lower panel are characteristic of long periods of low magnetic activity. The upper panel shows data from both high- and low-altitude orbital segments (see Figure 1); the lower panel does not because RIMS was not operating during the low-altitude portion of this pass. The instrument was usually turned off near perigee to protect it from high ionospheric fluxes, and this cutoff frequently extended through most of the low-altitude segment of the pass.

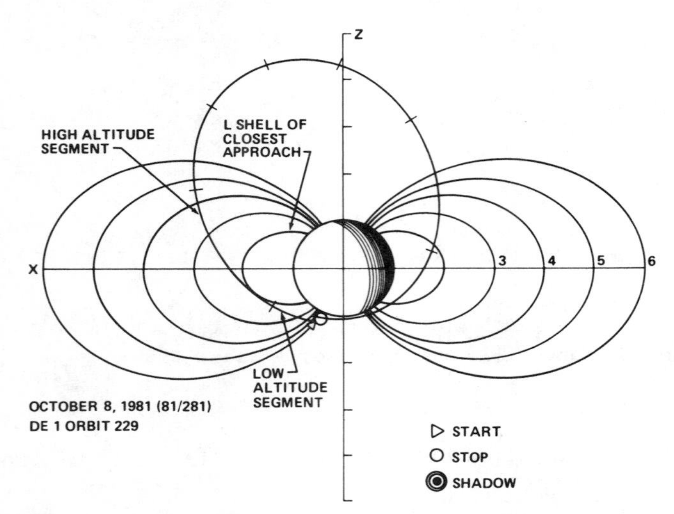

Fig. 1. DE 1 orbit representative of the period covered by the observations presented.

Statistical averages presented below correspond to all observations occurring in a bin of width $\Delta L = 0.2$ centered on the indicated L value. Error bars shown represent one standard deviation in the distribution of values in the bins; i.e., they represent only the statistical spread, not experimental uncertainties of the detectors nor of the analysis procedure.

Temperature Structure

Profiles of H^+ temperature versus L shell are presented in Figures 3a and 3c. Figures 3b and 3d display the distributions of observations among the various L bins. These distributions hold for the density plots presented below as well as for these temperature profiles. The number of observations at high L shell (>4) decreases rapidly because either the count rates were too low to display adequate retarding curves for analysis or the fluxes showed clear indications of being anisotropic.

When the morning side and evening side profiles are laid over one another, it is seen that beyond L ~ 2.5, these profiles are virtually the same. Inside L ~ 2.5, evening temperatures are lower, probably due to cooling to the underlying ionosphere in the absence of photoelectron heating. Inside L ~ 2.5, daytime profile standard deviations are quite small. This indicates that day-to-day temperature variations in this region are small and likewise for variations over the local time period of this sample (~7 to 11 geomagnetic local time). Large standard deviations between L ~ 3 and 4 on the dayside probably reflect the variable location of the enhanced temperature gradient marking the transition between the cool inner plasmasphere and a warmer outer plasmasphere (see the upper panel of Figure 2).

Some indication of altitude variations is gained by comparing the average temperatures from the low- and high-altitude segments of plasmasphere transits. Figure 4 shows such comparisons, the high-altitude profiles being those presented in Figure 3. Since the orbit is almost polar, observations for the high and low orbital segments for the same L shell lie approximately on the same field line. Morning side profiles suggest a division between inner and outer plasmasphere near L ~ 3. Thermal equilibrium

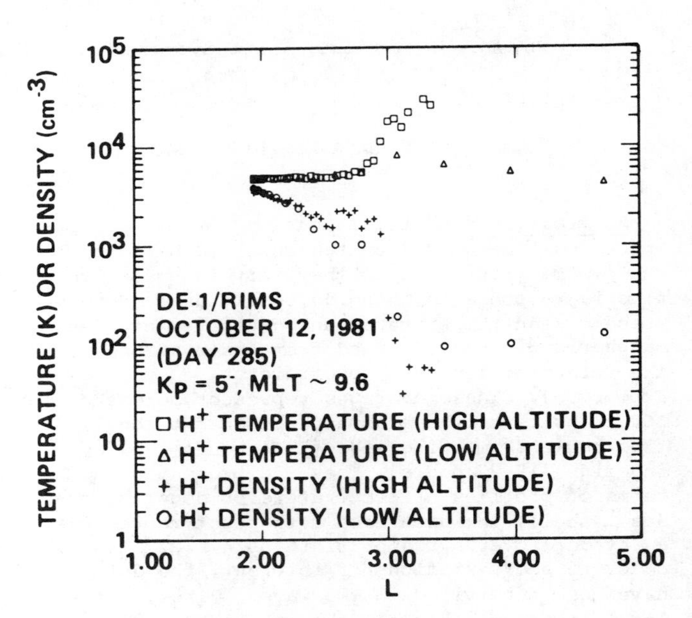

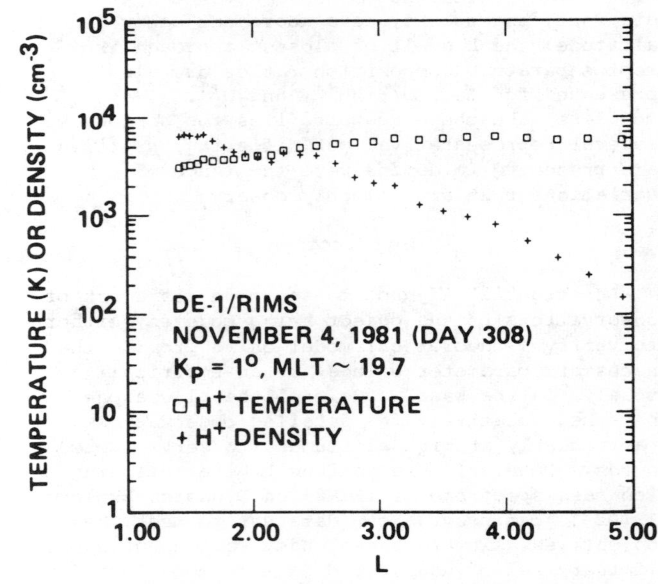

Fig. 2. Example temperature and density versus L shell profiles for H^+ ions. Note that both temperature and density are read from the same scale.

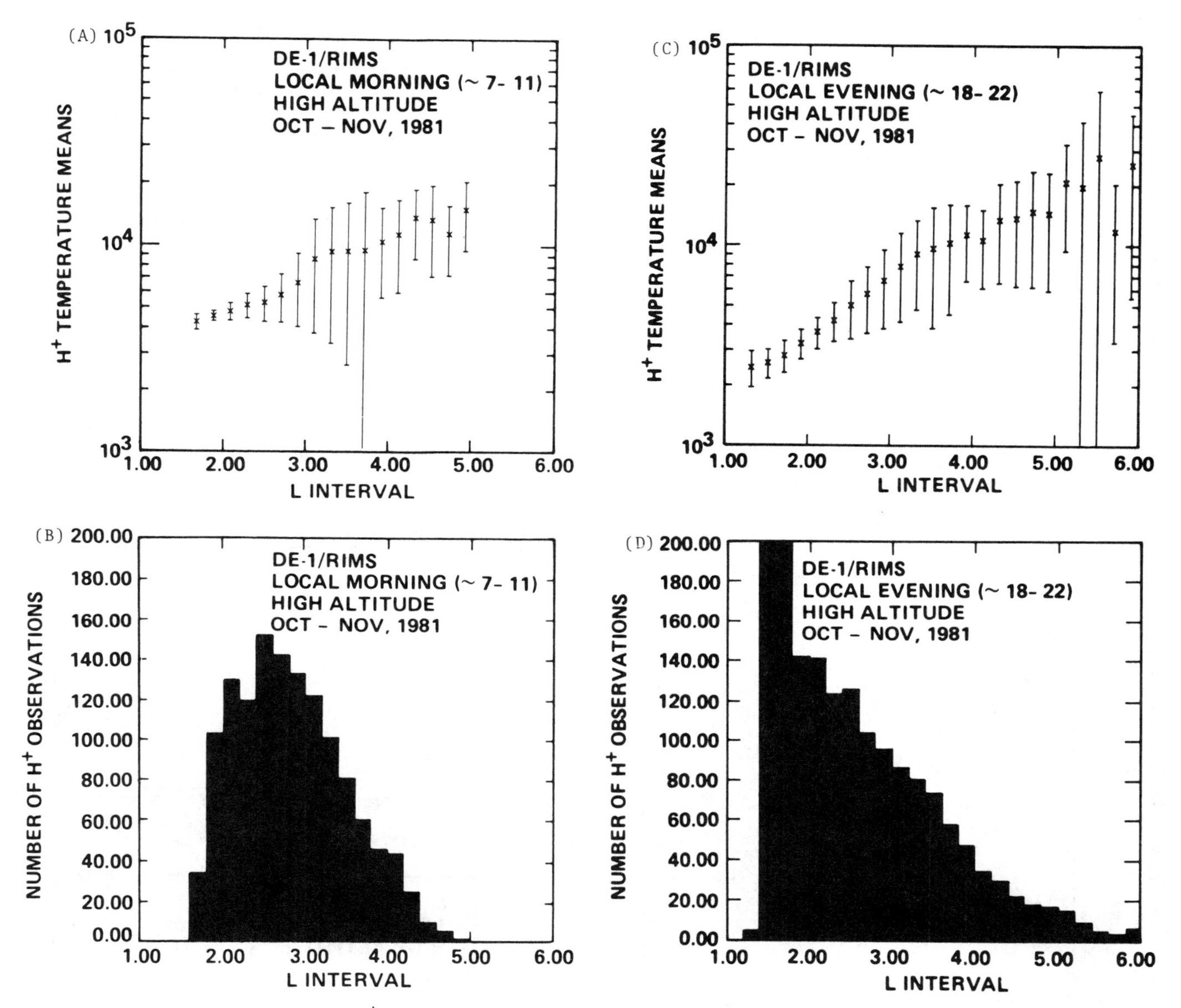

Fig. 3. Mean H^+ temperature profiles together with the distribution of observations which produced them. (a,b) Morning side, (c,d) Evening side.

prevails along field lines in the inner region, while field-aligned temperature gradients develop in the outer region. This is suggestive of a high-altitude, equatorial heat source in the outer region (cf. Chandler et al., 1987). Evening side profiles indicate cooling to the ionosphere over all L shells compared (inner plasmasphere), resulting in field-aligned temperature gradients.

Another property of interest in plasmasphere thermal structure is the degree of thermal equilibrium among the ion species. Figure 5 shows the morning side and evening side profiles of mean He^+ to H^+ temperature ratios. Clearly thermal equilibrium prevails between them to within 10% out to $L \sim 4$ on both morning and evening sides. Evening side standard deviations are generally larger in the outer plasmasphere, with an indication that He^+ may be slightly cooler than H^+ in this region.

Density Structure

Typical H^+ density profiles are presented in Figure 2 above. The lack of a well-defined plasmapause density gradient in the lower panel

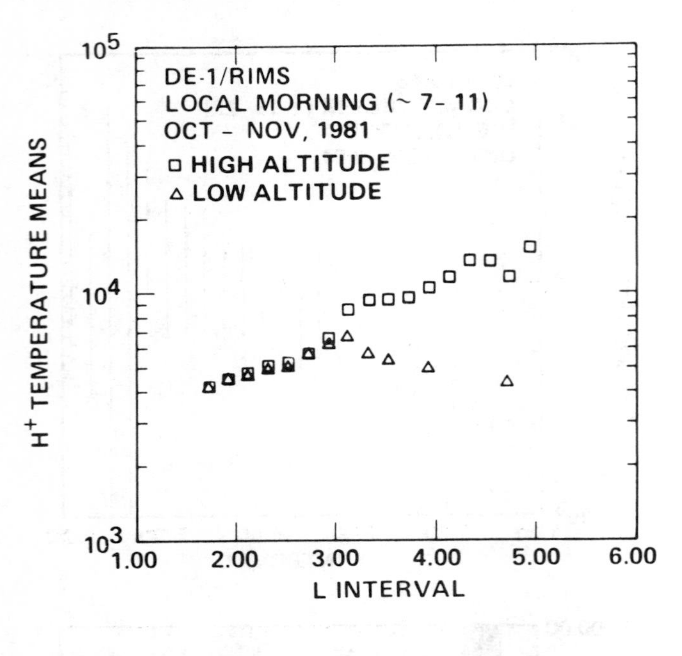

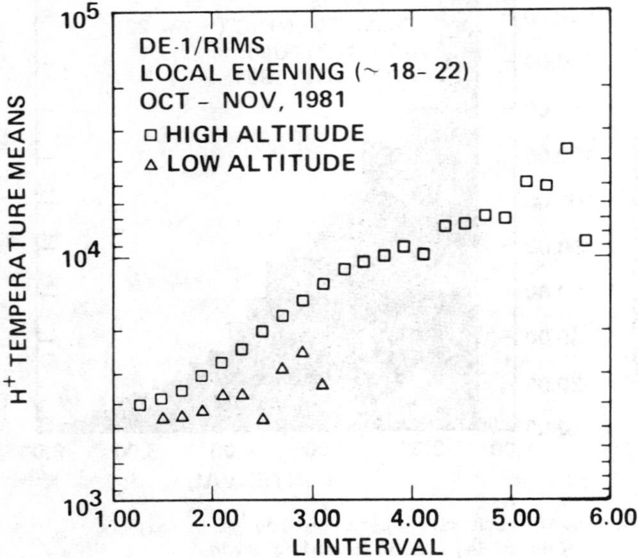

Fig. 4. Mean H^+ temperature profiles for high- and low-altitude segments of the plasmasphere transit (see Figure 1) (upper panel: morning side; lower panel: evening side). Subsequent figures follow the same format.

of that figure is characteristic of long periods of low magnetic activity. As noted previously, the distribution of observations presented in Figures 3a and 3b pertains to densities as well.

Mean density profiles are shown in Figure 6. Morning and evening density profiles are nearly the same out to L ~ 4.5. This suggests that in this region, diurnal variations are small relative to other variations whose effects persist for periods that are long compared to a day. Beyond this region an evening bulge is suggested, but there is also considerable scatter in this region (from the standard deviations) and a relatively small sample (from Figure 3). The H^+ density decrease with decreasing L below L ~ 2 appears to be an altitude effect.

Altitude comparisons of density, similar to those for temperature (Figure 4), are presented in Figure 7. Recall that altitude increases with increasing L for the high-altitude observations and decreases with L for the low-altitude set (see Figure 1). Evening side low-altitude densi-

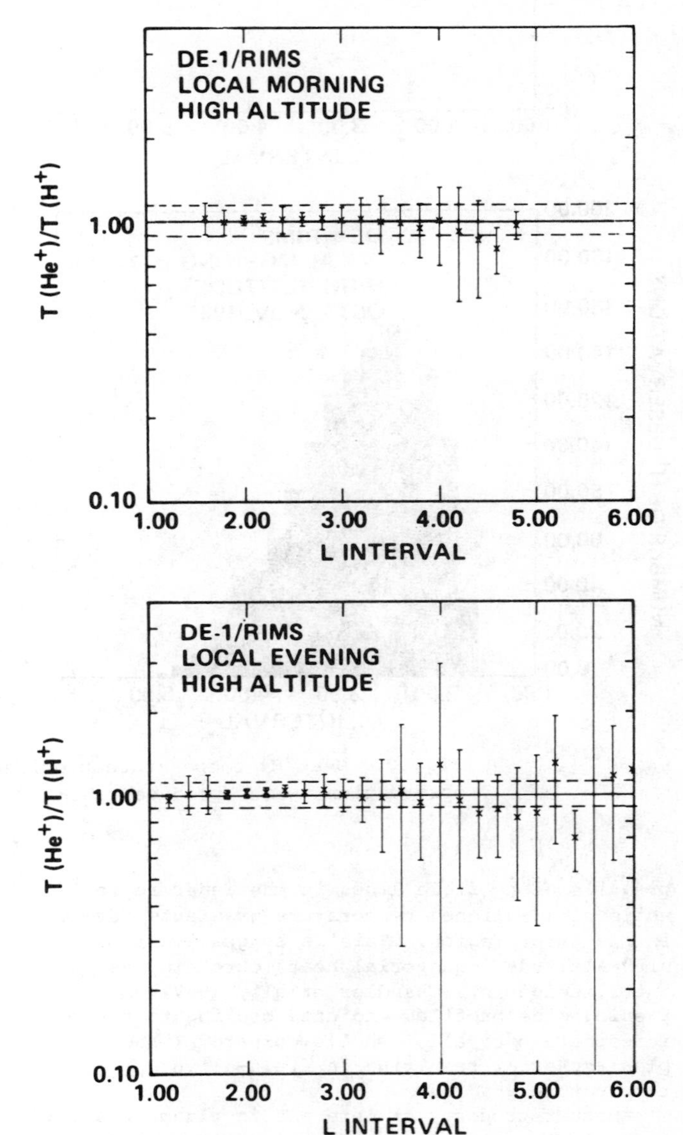

Fig. 5. Profiles of mean ratios of He^+ to H^+ temperatures.

ties are below the H^+ peak and decrease with L at about the same rate as the high-altitude densities. Morning side densities at low altitudes decrease with L more rapidly than those at high altitudes. Because the relationship between L and altitude differs for the DE 1 orbit on morning and evening sides, no firm conclusions can be drawn from these differences.

Composition

Ion composition is represented in terms of ratios of densities of the various ion species to

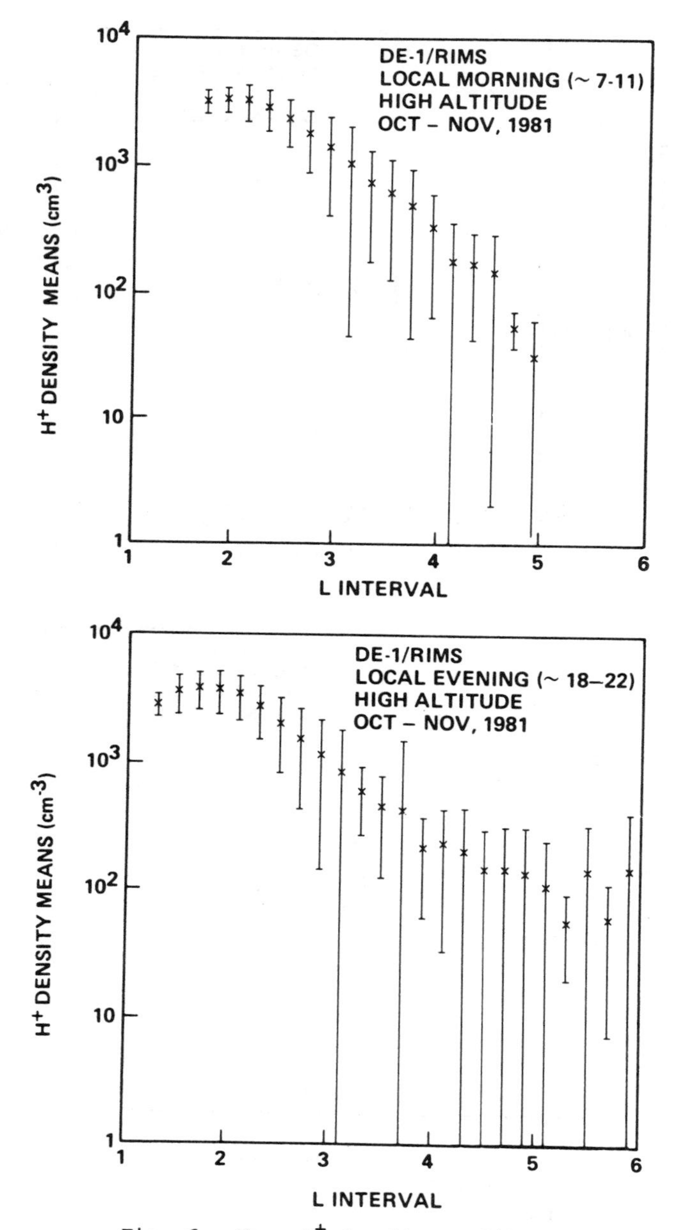

Fig. 6. Mean H^+ density profiles.

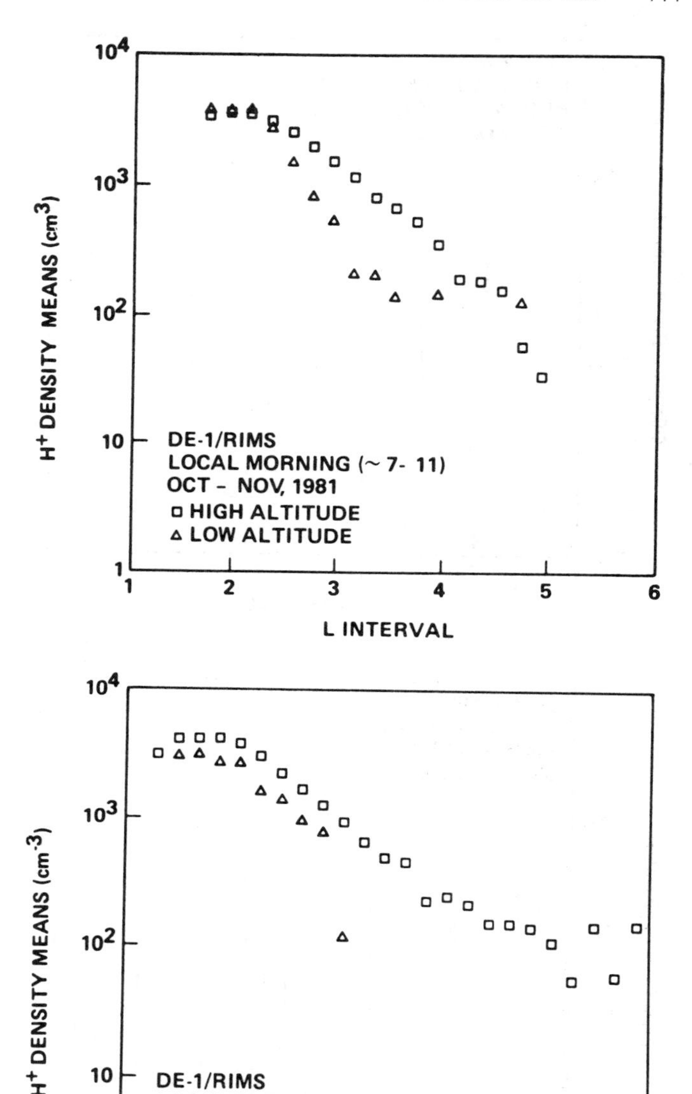

Fig. 7. Mean H^+ density profiles for high- and low-altitude segments of the plasmasphere transit.

those of H^+. Profiles in L are presented for morning and evening sides. For He^+ the distribution of observations is approximately that given for H^+ in Figure 3, since these two species are measured simultaneously. However, the other species are not sampled nearly as well, both because the observation mode of RIMS during the sampled time frame was not always one that could observe other ions and because even when it was in the

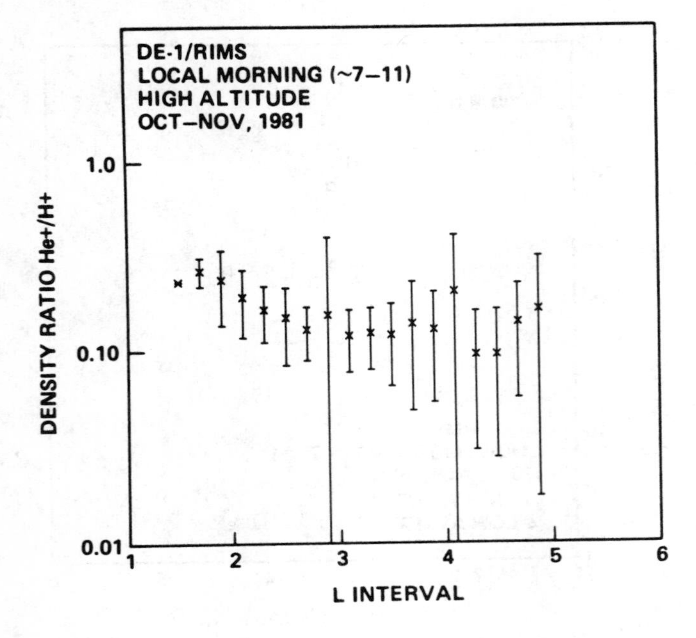

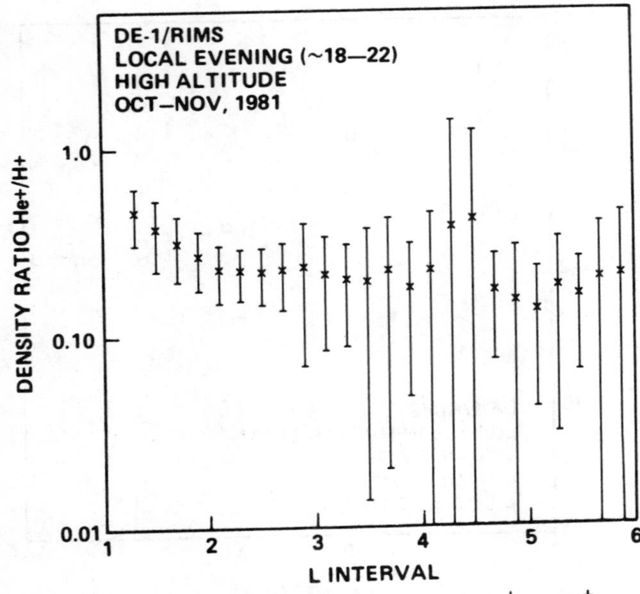

Fig. 8. Mean profiles of ratios of He^+ to H^+ densities.

proper mode, other ions were not always seen.

Profiles for the light ions, H^+ and M/Z = 2 [He^{++} or D^+ (deuterium ions)], are given in Figures 8 and 9. He^+ ratios decrease with L out to L ~ 3 on the morning side and to L ~ 2 on the evening side, at which points they level off and decrease little, if at all, farther out, despite the fact that the H^+ densities are decreasing by more than an order of magnitude in these regions. These relatively stable ratios are about 0.20 and 0.12, respectively. For the M/Z = 2 ions, the ratios are fairly stable at about 0.0017 on the morning side and 0.0022 on the evening side, although they appear to drop off beyond L ~ 3.2.

Heavy ion ratios for O^+ and O^{++} are presented in Figures 10 and 11. These profiles behave similar to one another, as did those of the light ions, but the behaviors of the light and heavy ions are clearly very different. The heavy ions show an enhancement in the outer plasmasphere, beyond minimums near L ~ 2. This characteristic of the heavy ions has been examined in some

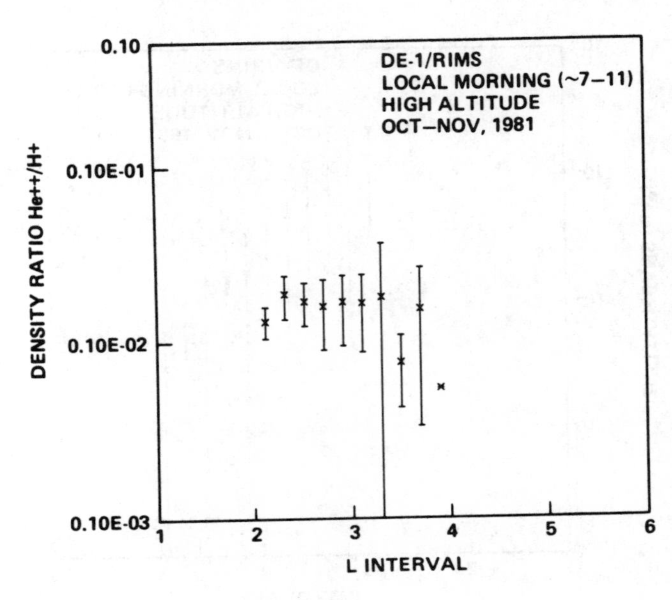

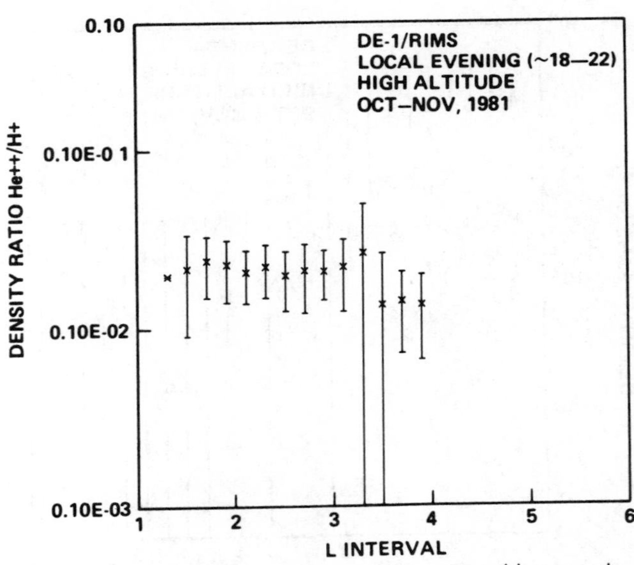

Fig. 9. Mean profiles of ratios of He^{++} (or D^+) to H^+ densities.

detail by Roberts et al. (unpublished manuscript, 1987). Concentrations of the heavy ions in the enhancement region are substantially larger on the evening side than on the morning side. On the evening side, O^+ density ratios are as high as 0.1, while those of O^{++} are about an order of magnitude lower. Standard deviations are rather large in almost all regions for both ion species. The presence of heavy ions in the outer plasmasphere could have significant effects on ring current interactions with the cold plasma and on wave-particle interactions in this region. This enhancement of the heavy ions may be a result of heating in this region, with the heat being con-

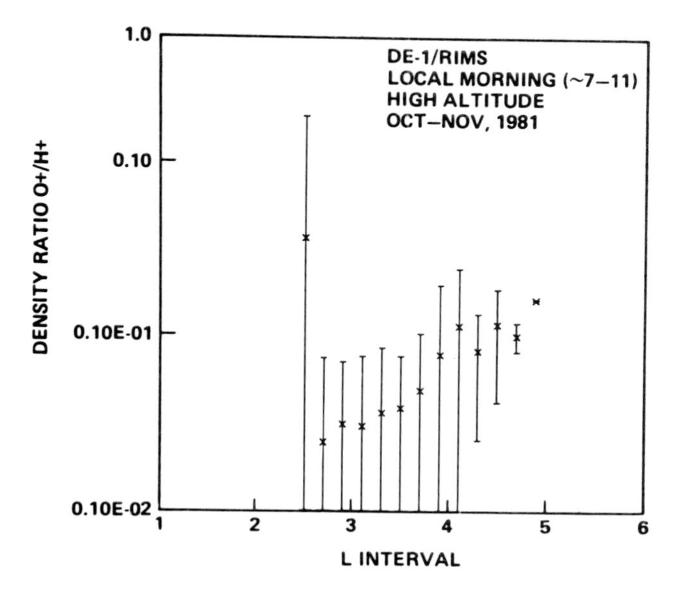

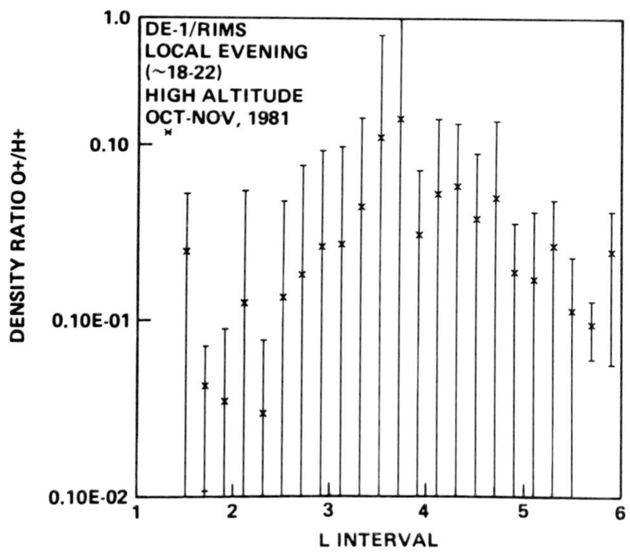

Fig. 10. Mean profiles of ratios of O^+ to H^+ densities.

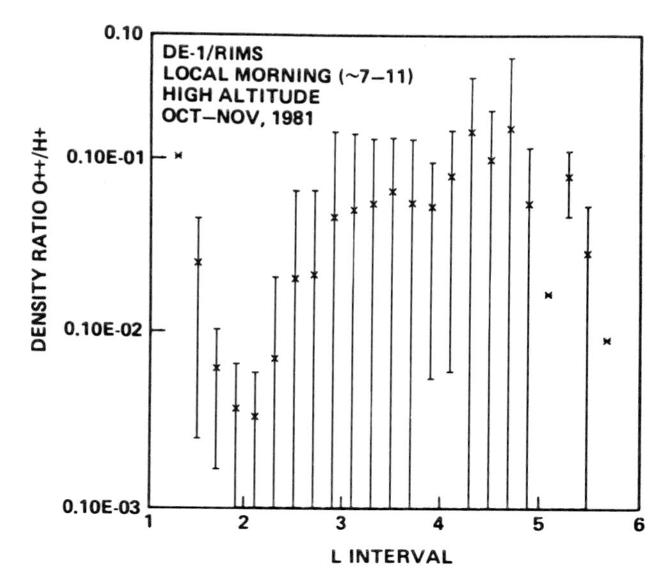

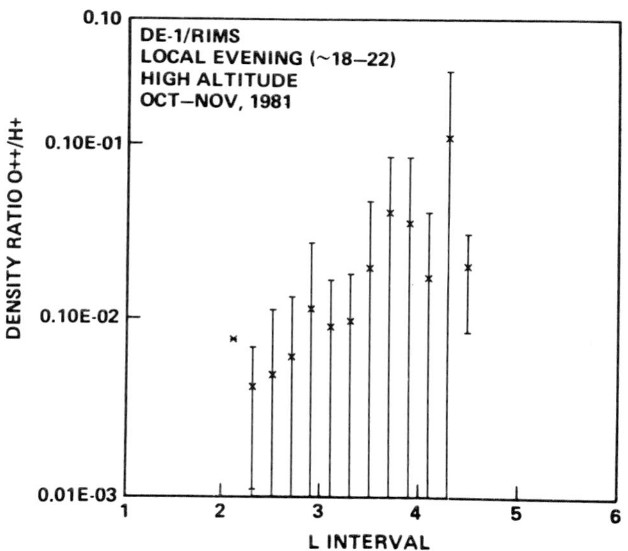

Fig. 11. Mean profiles of ratios of O^{++} to H^+ densities.

ducted to the ionosphere, increasing the scale heights for these outer L shells (cf. Roberts et al., unpublished manuscript, 1987; see also Chandler et al., 1987).

Acknowledgments. This research was supported in part by NASA grant NAG8-058 and NASA contract NAS8-33982 with The University of Alabama in Huntsville.

References

Chandler, M. O., J. U. Kozyra, J. L. Horwitz, R. H. Comfort, W. K. Peterson, and L. H. Brace, Modeling of the thermal plasma in the outer

plasmasphere: A magnetospheric heat source, this volume, 1987.
Chappell, C. R., S. A. Fields, C. R. Baugher, J. H. Hoffman, W. B. Hanson, W. W. Wright, H. D. Hammack, G. R. Carignan, and A. F. Nagy, The retarding ion mass spectrometer on Dynamics Explorer-A, Space Sci. Instrum., 5, 477, 1981.
Comfort, R. H., C. R. Baugher, and C. R. Chappell, Use of the thin sheath approximation for obtaining ion temperatures from the ISEE-1 limited aperture RPA, J. Geophys. Res., 87, 5109, 1982.
Comfort, R. H., J. H. Waite, Jr., and C. R. Chappell, Thermal ion temperatures from the Retarding Ion Mass Spectrometer on DE 1, J. Geophys. Res., 90, 3475, 1985.

A MODEL OF AURORAL POTENTIAL STRUCTURES BASED ON DYNAMICS EXPLORER PLASMA DATA

J. L. Burch and C. Gurgiolo

Southwest Research Institute, San Antonio, Texas 78284

Abstract. Dynamics Explorer 1 hot plasma data are used to investigate three major features of the mid-altitude auroral electron distribution--the hole, the bump (at large pitch angles), and the loss cone. The results of a computer simulation show that these features can be approximately reproduced by a model involving an initial plasma sheet electron distribution and two potential drop regions that are widely separated in altitude. The observed distributions could not be reproduced, even qualitatively, by any of a wide range of distributed potential drops involving parallel electric fields at the point of observation.

Introduction

Knight (1973), using conservation of energy and magnetic moment, showed that, under an assumption of gyrotropy, the two-dimensional velocity space distributions of particles originating in the ionosphere or the plasma sheet can be partitioned by various conic sections into volumes of accessibility which depend on the field-aligned potential difference that may exist between the two regions. Croley et al. (1978) showed good agreement of S3-3 electron and ion measurements with the predicted population boundaries. However, there was one major disagreement involving the existence of electrons in regions of velocity space which are predicted to be forbidden to both plasma sheet and ionospheric electrons. Whipple (1977) had suggested earlier that electrons trapped between the magnetic mirror point and an electrostatic potential above the satellite at its formation might fill the forbidden region. Hence, the forbidden region is often referred to as the trapped electron region. Omidi et al. (1984) referred to the electron distribution in the forbidden region as the bump and identified it as a possible source of free energy for auroral Z mode radiation.

Other prominent features of the auroral electron distribution at altitudes of ≥ 1 R_E are the hole and the loss cone. Croley et al. (1978) identified the hole as an electrostatic reflection of the loss cone, while Omidi et al. (1984) and Lin et al. (1986) identified the hole distribution as a potentially important source of free energy for Z mode radiation.

The purpose of the present study was to investigate, with data from the Dynamics Explorer 1 (DE 1) high altitude plasma instrument (HAPI), the appearance of the hole, bump, and loss cone distributions and to use a computer simulation model to attempt to explain the observations.

Experimental Results

Figure 1 shows an electron spectrogram taken from a high-latitude pass of the DE 1 satellite on October 19, 1981 (day 81292). The middle panel indicates the electron pitch angles, which are approximately 0° for this compressed spectrogram format. The bottom panel shows the electron energy flux integrated over the downward hemisphere. The regions in which potential drops are assumed to exist are clearly indicated by the several inverted-V structures that were encountered following the spacecraft's exit from the polar cap at about 1011:30 UT.

In Figure 2 we take a closer look at the first inverted-V structure, which is centered at about 1015 UT. In the spectrogram format used in Figure 2, all data from one of the five electron/ion sensors of the HAPI are plotted. As in Figure 1, the upper panel is an energy-time spectrogram of electron differential energy flux, and the middle panel gives the pitch angles of the measured particles. In the bottom panel a positive ion spectrogram is plotted.

Several typical features of the auroral particle distributions at altitudes ≥ 1 R_E are clearly seen in Figure 2. From ~1013:00 to ~1013:50 UT, low-energy (E < 100 eV) counterstreaming electron beams are observed, along with a narrow loss cone at high energies (≥ 1 keV). The counterstreaming electrons are illustrative of the type 1 populations identified by Lin et al. (1982).

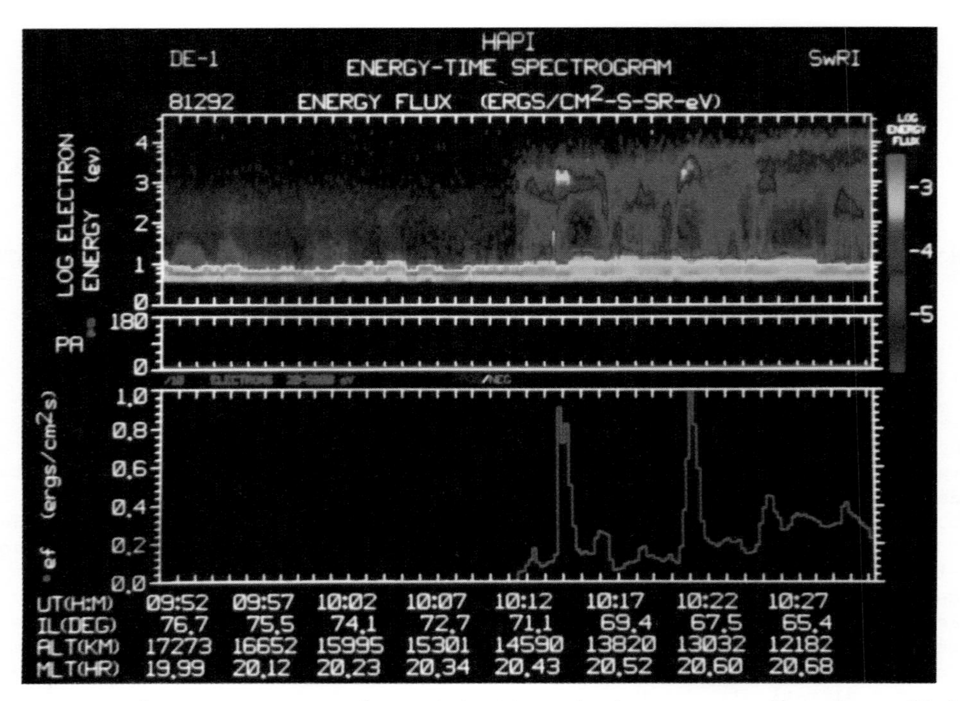

Fig. 1. Energy-time spectrogram of precipitating electron energy flux for a DE 1 pass from the polar cap into the auroral oval (top panel). Electron pitch angles (~0° for this plot) are shown in the middle panel. The bottom panel contains integrated electron energy flux over the downward hemisphere.

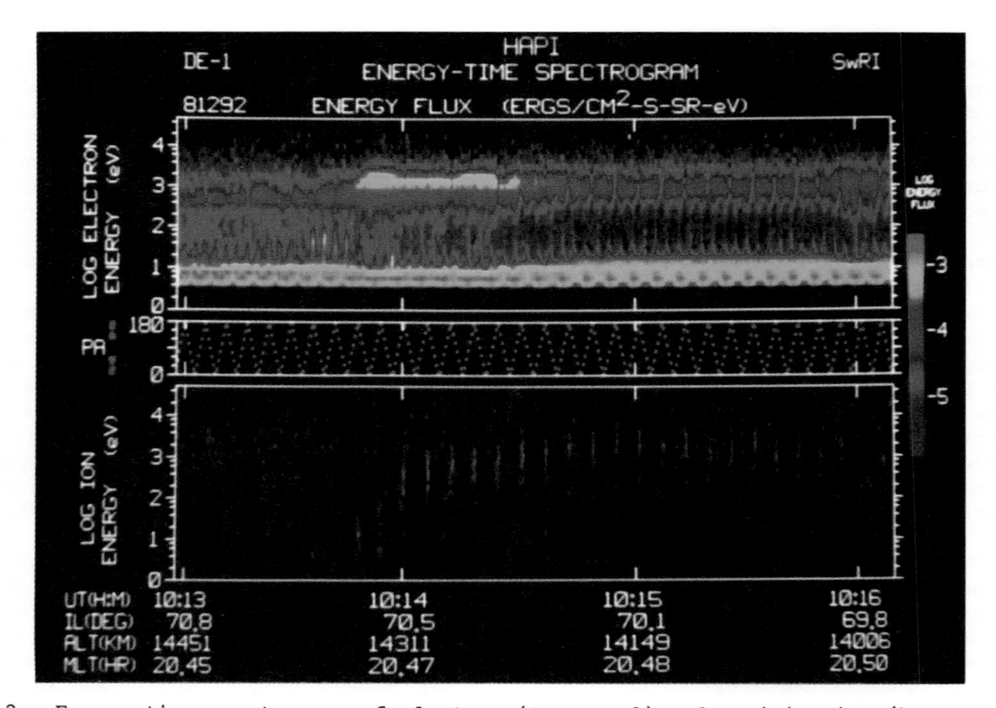

Fig. 2. Energy-time spectrogram of electron (top panel) and positive ion (bottom panel) energy flux for a pass of DE 1 through the most poleward inverted-V region shown in Figure 1. Particle pitch angles are plotted in the middle panel.

Entry of the satellite into the core of the inverted-V structure was accompanied by the appearance of upward-accelerated ion beams with energies increasing toward the center of the event, indicating a large potential drop below the satellite. The most intense electron fluxes were located near the edge of the event at the observing altitude, suggesting a typical V-shaped potential structure. At lower altitudes the event would appear most intense near its center, but at these altitudes DE 1 is passing above most of the potential drop in the middle of the event. In the core of the inverted-V structure (from ~1014:30 to ~1016:00 UT) the hole, bump, and loss cone distributions can all be identified in the electron spectrogram. The loss cone has widened and deepened in this period, as compared to the time period before ~1013:45 UT. The evidence for this expansion of the loss cone is the fact that it is observed on every spin of the spacecraft instead of every other spin. The forbidden (or trapped electron) region is observed as the general void between about 20 and 200 eV. However, it is clearly populated with electrons at large pitch angles, giving the distribution at higher energies for each spin the appearance of bicuspid teeth with double roots extending toward lower energies, which are faintly connected to a more intense sawtooth pattern at lower energies ($\lesssim$40 eV). The hole is located between the two roots of each higher-energy distribution at energies of a few hundred electron volts.

A typical electron velocity space distribution, showing well-developed hole, bump, and loss cone features, is shown in a contour plot format in Figure 3. This distribution was generated from two spin periods of data beginning at 1014:30.2 UT on day 81292. The appearance of the hole and bump features in this type of distribution function plot has been described previously by Omidi et al. (1984). Both of these features are contained within the "forbidden region," which is encircled roughly by the f = 1.20 s^3 km^{-6} contour. This contour marks the inner boundary of the region of closely spaced contours that map out the accelerated plasma sheet electron distribution. The bump is a rather weak enhancement extending along the $V_\perp$ axes, while the hole is the deep depression located about 4/10 of the way up the $+V_\parallel$ axis. The outer boundary of the electron hole along the $+V_\parallel$ axis provides a measure of an inferred potential drop above the satellite, while the upward ion beam energy can be used to estimate the potential drop below it. Although, as found by Reiff et al. (1986), such an estimate may be low by 30% to 50%. Throughout the core of the inverted-V structure, where the hole and bump distributions are observed, the inferred potential drop below the satellite was significantly larger than that above it. Typical values of 700 V above and 2500 V below the satellite are used in the model simulation described in the next section.

Model Simulation

To study the formation of the trapped particle population we developed a computer simulation model. The simulation is one dimensional, using single dimensional magnetic and electric field vectors, and particles are traced using the guiding center approximation. Both the field-aligned particle position and velocity are traced using a Hamming predictor/corrector algorithm. The perpendicular velocity is computed at each step from the conservation of energy (H = T + V) and magnetic moment ($T_\perp/B$).

The magnetic field used was functionally identical to the earth's dipolar field. The simulation was bounded at 1000 and 50,0000 km altitude. At the lower bound, particles which exited the simulation were considered lost to the atmosphere, while at the upper bound particles which exited the simulation were considered to have escaped the potential. It is assumed that the initial distribution populating the field line at the time of onset of the potential drop has the same velocity space distribution at all points.

Two basic quantities can be input into the simulation, defining a run. The first is the form of the electrostatic potential which is to be applied to the field line, and the second is the position at which the distribution function is to be followed. With these items defined, particles are injected at the point of the observation in 49 logarithmically-spaced energy steps between 5 and 4000 eV and in pitch angle from 0° to 180° in 1° steps. The field line is divided into two hundred 250-km segments. Each particle, in traveling along the field line, has its velocity and number of time steps taken recorded at the beginning of each segment. This approach gives a full time history of each particle along the field line, which can be used to reconstruct the distribution at the point of observation at any given time beginning with the time of injection. Because the particle motion is fully reversible under the conditions set up in the simulation (the phase space path for a particle to travel from a to b is identical to that in traveling from b to a), particles need only be followed until they either mirror or are lost to the system.

The simulation model was run using several different potential profiles. The initial runs used an exponential potential profile of the form

$$\phi = Ae^{-\alpha(r - r_s)},$$

where r is the location on the field line and r_s is satellite location. The constants A and α are computed by fitting the exponential to an estimated potential at the satellite and an estimated potential at an altitude below the satellite. The potential at the satellite is estimated as

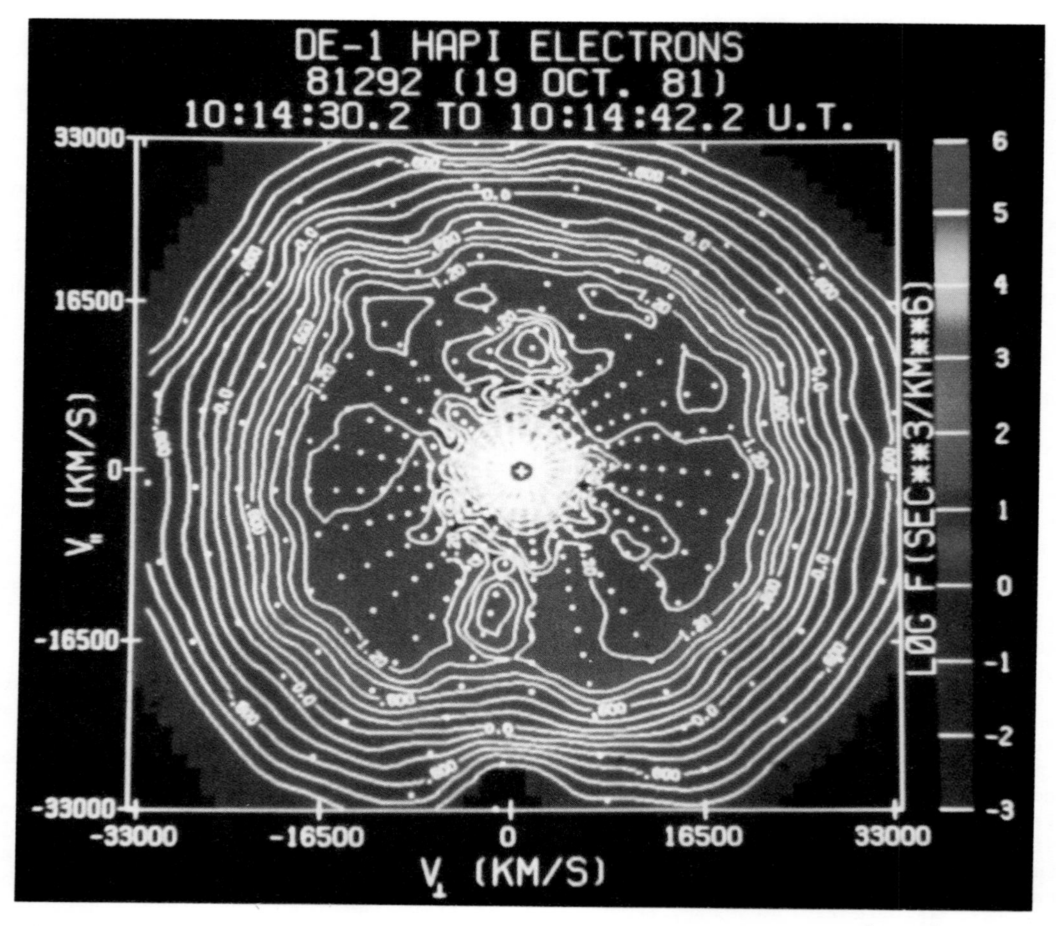

Fig. 3. Contour plot of the electron distribution function [$\log_{10}$ f(s^3 km^{-6})] for a two-spin data period near the middle of the time period plotted in Figure 1.

the voltage defined by the upper boundary of the electron hole, which is the lower edge of the plasma sheet distribution. The potential at the lower edge of the potential region is the sum of the potential above the satellite plus that below it, as defined by the upward ion beam energy.

The reason for initially choosing a potential distribution that varies exponentially with geocentric radial distance is the fact that the hole and bump distributions are commonly observed over a wide range of altitudes with the potential drop beneath the satellite always being much larger than that above it. An exponentially varying potential, or a similar function, seemed to be most consistent with this observation. However, even though the trapped region was populated in the simulations which used the exponentially varying potentials, the resulting equilibrium electron distribution functions at the satellite altitude did not resemble the observed distributions. For example, rather than having peaks at the largest pitch angles (i.e., the bump distribution), there were local maxima near the edges of the loss cone. For details of these simulations the reader is referred to Gurgiolo and Burch (1987).

Another possible potential structure is one in which two rather localized potential drops are widely separated in altitude, with the smaller of the two located well above the typical satellite altitude and the larger one well below it. In fact, it was found that this type of potential structure can produce equilibrium electron distributions that are very similar to the observed ones. An example of the results of a simulation run using such a separated potential structure is shown in Figure 4. For this particular case the satellite geocentric distance was chosen to be 20,620 km; a linear potential of 700 V was imposed over a 1000-km altitude range centered at 46,870 km geocentric; and a potential difference of 2500 V, also over 1000 km in altitude, was centered at 13,870 km. Figure 4 contains four contour plots of electron velocity space distribution, in units of $\log_{10}$ f(s^3 km^{-6}), versus $v_\parallel$ and $v_\perp$. The contour plot in the upper left-

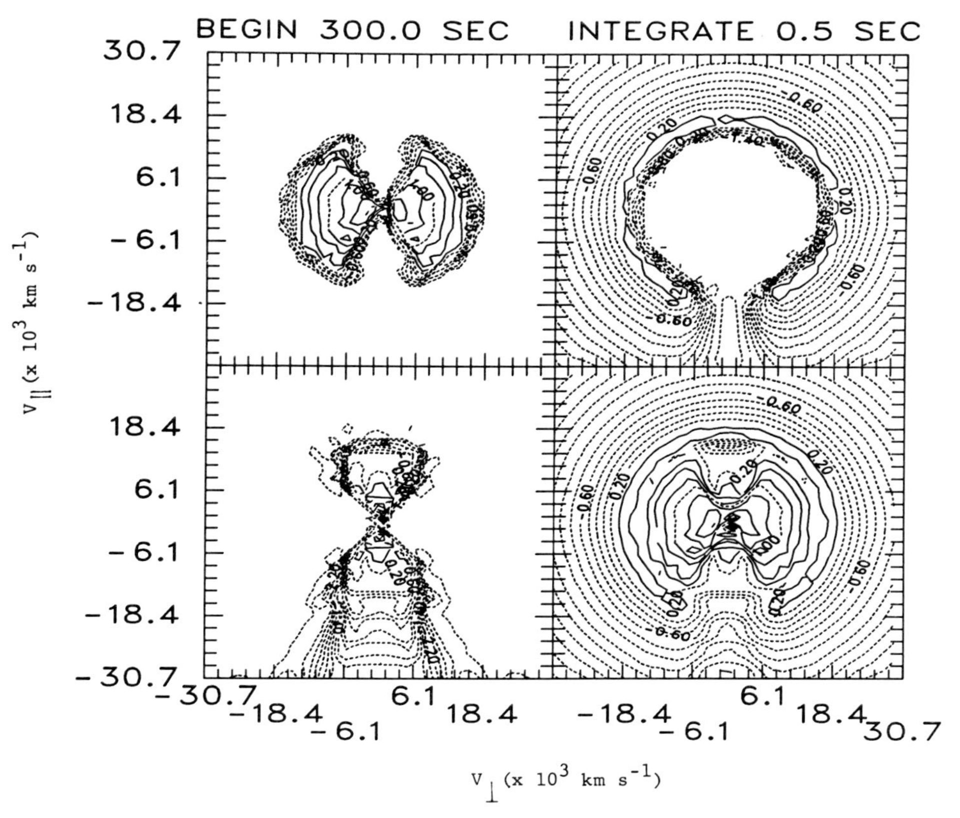

Fig. 4. Electron contour plots showing simulation results for a plasma sheet electron distribution on which are imposed two potential drop regions as described in the text. One potential drop is located well above the satellite altitude and the other well below it. In the four panels are plotted the distribution of trapped electrons (upper left), accelerated plasma sheet electrons (upper right), backscattered and secondary electrons (lower left), and the composite distribution of all three populations.

hand panel displays the trapped electron distribution. The distribution in the upper right-hand panel is the newly injected plasma sheet population, which has been accelerated through the potential drop above the satellite. The lower left-hand panel shows the secondary and backscattered primary electron distribution (Evans, 1974). Finally, the electron distribution in the lower right-hand panel is a composite of the first three distributions (trapped, plasma sheet, and backscattered). This final distribution is the one that should be observed at equilibrium at the satellite altitude.

Summary and Conclusions

Comparison of the electron distribution functions in Figure 3 and in the lower right-hand panel of Figure 4 shows very good qualitative agreement. The hole and loss cone distributions are clearly present, and the trapped region is populated with a bump-type distribution.

The good agreement between Figures 3 and 4, coupled with the fact that only very poor agreement was found for any of a number of distributed potential structures with parallel electric fields at the satellite altitude, argues strongly for the separated potential model. Furthermore, we have found in numerous simulation runs that the good agreement with the measured distributions is obtained over a wide range of altitudes that are chosen for the two potential differences, as long as one is well above the satellite and the other well below it.

Based on the experimental and simulation results of this study one can conclude that the trapping of plasma sheet electrons by two potential drop regions, at high and low altitudes, is a candidate source for the hole, bump, and loss cone types of electron distribution that are often observed at mid-altitudes over the auroral oval. More work remains to be done on turbulent cross-field diffusion processes that may lead to other observed features of electron distributions which are sometimes observed in the vicinity of auroral potential structures.

Acknowledgment. This research was supported by NASA contract NAS5-28711.

References

Croley, D. R., Jr., P. F. Mizera, and J. F. Fennell, Signature of a parallel electric field in ion and electron distributions in velocity space, J. Geophys. Res., 83, 2701, 1978.

Evans, D. S., Precipitating electron fluxes formed by a magnetic field aligned potential difference, J. Geophys. Res., 79, 2853, 1974.

Gurgiolo, C., and J. L. Burch, Simulation of electron distributions within auroral acceleration regions, J. Geophys. Res., in press, 1987.

Knight, S., Parallel electric fields, Planet. Space Sci., 21, 741, 1973.

Lin, C. S., J. L. Burch, J. D. Winningham, J. D. Menietti, and R. A. Hoffman, DE-1 observations of counterstreaming electrons at high altitudes, Geophys. Res. Lett., 9, 925, 1982.

Lin, C. S., J. L. Burch, C. Gurgiolo, and C. S. Wu, DE-1 observations of hole electron distribution functions and the cyclotron maser resonance, Ann. Geophys., 86, 33, 1986.

Omidi, N., C. S. Wu, and D. A. Gurnett, Generation of auroral kilometric and z-mode radiation by the cyclotron maser mechanism, J. Geophys. Res., 89, 883, 1984.

Reiff, P. H., H. L. Collin, E. G. Shelley, J. L. Burch, and J. D. Winningham, Heating of upflowing ionospheric ions on auroral field lines, in Ion Acceleration in the Magnetosphere and Ionosphere, Geophys. Monogr. Ser., vol. 38, edited by T. Chang, p. 83, AGU, Washington, D.C., 1986.

Whipple, E. C., Jr., The signature of parallel electric fields in a collisionless plasma, J. Geophys. Res., 82, 1525, 1977.

INTERNAL SHEAR LAYERS IN AURORAL DYNAMICS

W. Lotko

Dartmouth College, Hanover, New Hampshire 03755

C. G. Schultz*

Laboratory of Plasma Studies, Cornell University, Ithaca, New York 14853

Abstract. An important part of the magnetosphere-ionosphere interaction takes place in narrow layers where the vorticity is locally enhanced and where the internal dynamics are largely influenced by effects of viscosity, field-aligned potential drops, and ionospheric friction. Within these layers the ionospheric impression of magnetospheric electric fields and field-aligned currents is selectively filtered and dissipated and, depending on the magnetic field mapping, can evolve anisotropically. The processes by which energy is transported in the layer also depend sensitively on the wave number spectrum of the electric fields that exist within it. These effects are described in the context of a two-dimensional, time-dependent model which is characterized by two intrinsic parameters: a flux tube anisotropy factor and an effective Hartmann number that measures the ratio of (ionospheric) resistive friction to (magnetospheric) viscous friction. Illustrative numerical calculations indicate that turbulent magnetospheric flows may organize into either relatively isotropic eddies or striated shear layers depending on the values of these parameters. The possibility of using the model (or improved versions of it) to interpret observations of two-dimensional mesoscale turbulence at auroral latitudes is discussed briefly.

1. Introduction

In attempting to simulate the interaction between the magnetosphere and ionosphere, a certain practical issue arises, which concerns the feasibility of modeling a variety of physical processes that may occur at vastly different scale lengths, but which are not necessarily unrelated. Even if the effects of microprocesses are lumped into effective transport coefficients, the range of physically relevant scale lengths may span many orders of magnitude--from a few kilometers in the ionosphere to 100 R_E or more in the outer magnetosphere. When confronted with a disparity in scale of this magnitude, one soon realizes that it is not possible to model the magnetosphere-ionosphere interaction in a fully comprehensive manner that includes both large- and small-scale processes and their interrelationship. This dilemma has led to two different, although complementary, approaches to the problem of magnetosphere-ionosphere coupling: the global modeling approach and what might be characterized as the mesoscale modeling approach.

Global features of the magnetosphere are generally described by the equations of one-fluid magnetohydrodynamics, including some type of viscous and resistive transport, as well as some rudimentary features of magnetosphere-ionosphere coupling. Global models have now been implemented on large-scale computers and are beginning to provide detailed information on the large-scale morphology of magnetospheric fields and currents and on the dynamics of long-time-scale magnetospheric disturbances such as magnetic storms. Their principal limitations, in addition to those implied by the MHD equations themselves, are due to the memory capacity and computational speed of modern computers. It is not possible at the present time to solve the three-dimensional MHD equations on sufficiently fine grids or, what is equivalent, at large mechanical and magnetic Reynolds numbers. Consequently, narrow layers such as the magnetopause and its adjoining boundary layer, the tail current sheet, the plasma sheet boundary layers, and various internal shear layers that may form, for example, in response to local convection enhancements, are not adequately resolved in global models.

The aim of mesoscale models is to describe with sufficient spatial and/or temporal resolution the physical processes that occur within these narrow layers, one ultimate goal being to provide a lumped description of their behavior (in the form

*Now at CRPP-EPFL, Lausanne, Switzerland.

of effective overall transport coefficients, for example) which can then be incorporated into global models in the form of boundary conditions. This is accomplished by modeling a subregion of the magnetosphere or coupled magnetosphere-ionosphere system, usually in two dimensions. The equations governing the mesoscale layer of course depend on the particular region being modeled.

This paper is concerned with mesoscale features of the magnetosphere-ionosphere interaction, specifically in regions where vorticity becomes concentrated in relatively narrow layers, and where the dynamics of the interaction are largely influenced by the presence of these layers. Intense field-aligned currents, broadband turbulence, and auroral precipitation are generally associated with velocity shear layers, and for this reason, they are thought to be of central importance in magnetospheric and ionospheric dynamics.

The objectives of the paper are to examine some of the underlying physical processes governing the dynamics of magnetosphere-ionosphere coupling in internal shear layers and to illustrate how numerical simulations might be used to facilitate the interpretation of turbulent fields associated with the shear layers. A brief overview of some of the theoretical issues and their observational basis is given in the following section. This overview is not intended to be comprehensive--it includes a limited discussion of previous work that bears on the two-dimensional boundary layer model recently proposed by Lotko et al. (1987). The basic equations of this model are described in section 3. Some consequences of the model, including modal filtering, scale-dependent and anisotropic dissipation, and energy transport, are discussed in section 4. Illustrative numerical results are presented in section 5, including numerical simulations of two-dimensional turbulence. Of particular interest are cases where organized internal shear layers develop from an initially random state. The paper concludes with a brief discussion of the basic parameters that characterise non-steady magnetospheric flows and how knowledge of these parameters may help organize complex observations of mesoscale turbulence in magnetosphere-ionosphere coupling.

2. Theoretical and Observational Background

One of the first theoretical models of a mesoscale shear layer extending between the magnetosphere and ionosphere was proposed independently by Chiu and Cornwall (1980) and Lyons (1980). The model was originally developed to describe the steady state structure of inverted-V precipitation regions and to evaluate the influence of parallel electric fields on the structure of such regions. The effect of the parallel field is manifested in the model in a resistive scale length for the inverted-V precipitation region, which depends on the ratio of the ionospheric Pedersen conductance to the magnetic field-aligned conductance. The significance of this characteristic scale length can be understood in terms of a low pass filter, which operates on wave number rather than frequency components of the electric field spectrum. Perpendicular magnetospheric electric fields with scale lengths exceeding the resistive scale length (small wave numbers) are transmitted, essentially unattenuated, to the ionosphere. The smaller scale length perpendicular electric fields are filtered out and effectively dissipated before reaching the ionosphere because they induce parallel electric fields and associated inverted-V precipitation.

Although the existence of the inverted-V precipitation region is the primary experimental evidence for the Chiu-Cornwall-Lyons (CCL) model and associated filtering effect, indirect evidence has also been found in fluctuations in the horizontal components of electric and magnetic fields observed at auroral latitudes. Using data from Dynamics Explorer (DE) 1 and 2 at times of approximate magnetic conjunction, Weimer et al. (1985) have shown that low-frequency electric field fluctuations occurring at scales larger than about 10 km (referenced to 1 R_E) map between the two spacecraft approximately in accordance with the CCL model for assumed wave number cut-offs in the range of observed inverted-V scale lengths. On the basis of HILAT data, Vickrey et al. (1986) have also suggested that the difference in intensity of 3- to 80-km scale ionospheric velocity fluctuations between the winter and summer hemispheres and the lack thereof in the observed magnetic fluctuations (and, therefore, presumably in the field-aligned currents) can be attributed to scale-size-dependent potential drops along magnetic field lines. Since the ionospheric conductivity varies with season, they conclude that the velocity fluctuations depend on the ionospheric conductivity whereas the magnetic fluctuations are independent of it. This result can be shown to be consistent, if not in quantitative agreement, with the CCL model if the resistive scale length exceeds a reasonable value of 80 km in both winter and summer hemispheres and if the magnetospheric sources of these fluctuations are relatively independent of season. In view of these results, it appears that the CCL model may apply not only to coherent auroral forms, as initially proposed, but perhaps more generally to intermediate and large turbulence observed at auroral latitudes.

The source of the turbulence of fluctuations is not indicated by these observations or, indeed, by the CCL model. Kintner (1976) and Kelley and Kintner (1978) have reported a correlation between non-steady horizontal shear flows at altitudes of a few thousand kilometers and an essentially two-dimensional turbulence similar to that reported in the above studies. Such flows are a natural consequence of magnetospheric convection and are essentially the high-latitude, ionospheric projection of convection reversals or vorticity in the equatorial magnetosphere (Vasyliunas, 1970; Sonnerup, 1980). As demonstrated by recent computer simulations of magnetosphere-ionosphere coupling (Miura and Sato, 1980; Lysak and Dum, 1983; Lysak, 1985, 1986; Kan and Sun, 1985, Watanabe et al., 1986), they are also associated with a variety of auroral phenomena including auroral arcs, Alfvén waves, and propagating ionospheric disturbances. These numerical studies have focused primarily on transient responses to imposed magnetospheric or ionospheric convection patterns, however, and do not specifically address the problem of turbulent convection, either in the ionosphere or in the equatorial magnetosphere.

Recently, Lotko et al. (1987) developed a simple two-dimensional model for the horizontal flow dynamics in internal boundary layer regions of the equatorial magnetosphere. The model incorporates the main features of the electrostatic models described by Chiu and Cornwall (1980) and Lyons (1980), namely, some simple ionospheric physics and a linear current-voltage relation (Lyons et al., 1979) to characterize the auroral acceleration region. It differs from the CCL model (and various electrodynamic models) in the treatment of the transverse equatorial flow (or electric field), which is described self-consistently by the incompressible momentum equation including viscosity and a static magnetic field. In particular, the magnetospheric convection pattern is not fixed externally but is determined through internal coupling to the ionosphere and by the boundary conditions at the edges of the two-dimensional domain in the equatorial plane. The relation between the field-aligned current and equatorial potential resulting from the CCL model is used to determine the field-aligned current in the equatorial momentum equation. Because of this feature, the model should apply to the observations reported by Weimer et al. (1985) and Vickrey et al. (1986) which, as indicated above, appear to support the basic hypotheses of the CCL model. Since the equatorial flow is also allowed to evolve with time, the model or an improved version of it can be used to study turbulent processes in magnetosphere-ionosphere coupling. It applies mainly to fluctuations occurring on time scales larger than the Alfvén travel time between the ionosphere and magnetosphere because transit time effects between the two regions are neglected.

3. Model Equations

In the model described by Lotko et al. (1987), convective flow in a low-latitude region of the equatorial magnetosphere is treated as a two-dimensional velocity field with components in the $\hat{\mathbf{x}}_e$ and $\hat{\mathbf{y}}_e$ directions. (The subscript e is used to refer to equatorial coordinates and fields.) The equatorial dc magnetic field $\mathbf{B}_e$ is in the $\hat{\mathbf{z}}$ direction and for simplicity is taken to be uniform within the equatorial region. It is assumed that the dc field is sufficiently strong so internally generated magnetic fields may be neglected. A generalization of the model to include weakly nonlinear magnetic effects is described in the paper by Song and Lysak (1987).

The equations describing the equatorial flow are the incompressible MHD momentum equation, the frozen field condition, and the equation for current continuity:

$$\rho(\partial_t + \mathbf{v}_e \cdot \nabla_e - \nu\nabla_e^2)\mathbf{v}_e = -\nabla_e P + \mathbf{j}_e \times \mathbf{B}_e, \quad (1)$$

$$\mathbf{v}_e = \mathbf{E}_e \times \hat{\mathbf{z}}/B_e, \quad (2)$$

$$j_{\parallel e} = H\nabla_e \cdot \mathbf{j}_{\perp e}. \quad (3)$$

∇_e is a two-dimensional differential operator with components in the $\hat{\mathbf{x}}_e$ and $\hat{\mathbf{y}}_e$ directions. P is a scalar pressure, ρ is the mass density, ν is a scalar viscosity, $\mathbf{v}_e$ is the two-dimensional velocity field, and $j_{\parallel e}$ and $j_{\perp e}$ are the currents parallel and perpendicular to the dc magnetic field, respectively. The dependent variables may depend on x_e, y_e, and t. The effective height of the low-latitude region is 2H. In (3), the partial derivative ∂/∂_z has been replaced by 1/H, so $j_{\parallel e}$ should be interpreted as an average field-aligned current. Furthermore, the winter and summer hemispheres are assumed to be symmetric. For asymmetric conditions, an additional term would be included on the right-hand side of (3), which describes the exchange of field-aligned current between the two hemispheres.

The incompressibility condition implies that the characteristic flow speed is much smaller than the sound speed. A further simplification results if the transit time of shear Alfvén waves between the ionosphere and the equatorial magnetosphere is small compared to the time scale for variations in the flow. Inductive effects can therefore be neglected, and $\mathbf{E}_e$ can be defined in terms of a scalar potential: $\mathbf{E}_e = -\nabla_e \phi_e$. This implies that the magnetospheric response is capacitive; i.e., space charge is stored (or generated) in the magnetosphere.

The ionosphere is treated as a two-dimensional substrate whose plane is perpendicular to the dc magnetic field. The equations describing the ionosphere are Ohm's law and current continuity:

$$\mathbf{I}_{\perp i} = \Sigma_P(\mathbf{E}_i + \mathbf{v}_n \times \mathbf{B}_i) + \Sigma_H \hat{\mathbf{z}} \times (\mathbf{E}_i + \mathbf{v}_n \times \mathbf{B}_i), \quad (4)$$

$$j_{\parallel i} = -\nabla_i \cdot \mathbf{I}_{\perp i}. \quad (5)$$

The subscript i is used to refer to ionospheric variables. $\mathbf{I}_{\perp i}$ is the height-integrated perpendicular current in the substrate, and Σ_P and Σ_H are height-integrated Pedersen and Hall conductivities, respectively. $\mathbf{v}_n$ is the neutral wind velocity. The differential operator ∇_i has components in the $\hat{\mathbf{x}}_i$ and $\hat{\mathbf{y}}_i$ directions. The divergence of the magnetic flux tube implies that $\nabla_i \neq \nabla_e$. The ionospheric electric field is also related to an electric potential through the relation $\mathbf{E}_i = -\nabla_i \phi_i$.

Equations (4) and (5) are simplified considerably when the neutral wind velocity and conductivities are constant. It then follows that

$$j_{\|i} = \Sigma_P \nabla_i^2 \phi_i . \tag{6}$$

The remainder of the paper is restricted to this case. As a consequence, the auroral conductivity enhancements that occur in regions of strong upward field-aligned current are not included. This effect is probably more important in the nightside and winter ionosphere than in the sunlit and summer ionosphere since, in the former cases, the decreased ionization state of the ionosphere makes auroral enhancements proportionately more dramatic.

The ionosphere and equatorial magnetosphere are coupled as a consquence of the field-aligned current. When the coupling is perfect, $\phi_i = \phi_e$ on any given magnetic field line. The field lines are no longer equipotentials, however, when parallel electric fields arise. Following Lyons (1980) and Chiu and Cornwall (1980), field-aligned potential drops are included by assuming the potential difference between the equatorial region and the ionosphere depends on the field-aligned current at the ionosphere through the linear relation

$$j_{\|i} = K(\phi_i - \phi_e) + j_{\|0}. \tag{7}$$

The parameter K is a constant conductance density which, in regions where the field-aligned current is directed out of the ionosphere, lies in the range 0.1 to 1 (μA m^{-2})/kV (Fridman and Lemaire, 1980). It turns out that the offset current $j_{\|0}$ has no effect on the structure or dynamics of the fields, so it can be disregarded. It is not yet known whether relation (7) applies in regions where the field-aligned current is directed into the ionosphere but if it does, one might expect K to be larger than the above value in such regions by a factor of 10^2 to 10^3. This increase is required to limit the magnitude of field-aligned potential drops that would tend to produce kilovolt electron beams moving up the magnetic field line; such electrons are not observed.

A relationship between $j_{\|e}$ and $j_{\|i}$ can be inferred by noting that the plasma is both collisionless and strongly magnetized (very low β) in the intermediate altitude region between the ionosphere and the low-altitude boundary of the magnetospheric region. The amount of field-aligned current diverted into perpendicular current in this region is therefore minimal in comparison to what is diverted in the collisional ionosphere or the higher (though not necessarily high) β equatorial magnetosphere. Neglecting any diversion in the intermediate altitude region, conservation of magnetic flux and field-aligned current implies

$$j_{\|i} = \frac{B_i}{B_e} j_{\|e}. \tag{8}$$

To close equations (1)-(8), a magnetic field mapping is required to relate the ionospheric (x_i, y_i) coordinates to the equatorial (x_e, y_e) coordinates. For simplicity, we will use the orange segment mapping, which preserves the orthogonality between the x and y coordinates in both regions and determines the stretching between magnetic longitudinal meridians in going from the ionosphere to the equatorial plane. The $\hat{\mathbf{x}}_i$ direction is defined to be locally tangent to a contour of constant magnetic latitude in the ionosphere. Orange segment mapping then implies

$$\frac{dx_i}{dx_e} = \frac{\cos \Lambda}{L} , \tag{9}$$

where Λ is the latitude of the foot of the field line in the ionosphere and L is the radial distance, measured in earth radii R_E, to the intersection of a field line with the equatorial plane. The relation between dy_i and dy_e follows from (9) and conservation of magnetic flux as

$$\frac{dy_i}{dy_e} = \frac{B_e}{B_i} \frac{dx_e}{dx_i} . \tag{10}$$

Finally, derivatives in the two regions are related by

$$\partial_{x_e} = \frac{dx_i}{dx_e} \partial_{x_i} \quad \text{and} \quad \partial_{y_e} = \frac{dy_i}{dy_e} \partial_{y_i} . \tag{11}$$

It should be emphasized that orange segment mapping may not be very reliable in the vicinity of the magnetopause or in the geomagnetic tail. In these regions the field lines become highly stretched. For example, empirical magnetic field models imply that ionospheric meridians of magnetic latitude and local time map to non-orthogonal curves in the equatorial plane, which are nearly parallel in the vicinity of the magnetopause.

Equations (1)-(3) and (6)-(11), together with boundary and initial conditions, are sufficient to determine the state of the system. Various subsets of these equations have been previously investigated. The CCL model is contained in equations (6) and (7). The steady state, nonviscous version of equations (1)-(5) and (8), i.e., excluding the effects of field-aligned potential drops but including the effects of ionospheric conductivity variations, is the basis for the magnetosphere-ionosphere circulation model described by Vasyliunas (1970) and implemented, for example, by Harel et al. (1981). The effects of viscosity on the coupled system have been previously described by Sonnerup (1980) in the context of a one-dimensional, steady state model; one-dimensional equilibrium and time-dependent solutions to these equations have also been described by Lotko et al. (1987). Hasegawa and

Sato (1980) have discussed the effects of time-varying vorticity in the equations.

In closing this section, it is noted that the incompressibility condition ($\nabla_e \cdot \mathbf{v}_e = 0$) implies that the pressure term in (1) has no effect on the system dynamics, except possibly at the boundaries. Taking the curl of equation (1) yields an equation for the scalar vorticity field ω defined as

$$\omega = (\nabla_e \times \mathbf{v}_e) \cdot \hat{\mathbf{z}} = \nabla_e^2 \phi_e / B_e , \qquad (12)$$

which is independent of P. Notice that ω is essentially the local charge density. The enslaved pressure distribution is determined by the Poisson equation resulting from the divergence of (1). The vorticity and pressure equations are

$$\rho(\partial_t + \mathbf{v}_e \cdot \nabla_e - \nu\nabla_e^2)\omega = -B_e j_{\parallel e}/H \qquad (13)$$

and

$$\nabla_e^2 P = \nabla_e \cdot (\mathbf{j}_e \times \mathbf{B}_e - \rho \mathbf{v}_e \cdot \nabla_e \mathbf{v}_e). \qquad (14)$$

For the spectral calculations described in the following sections, it is more convenient to work with equations (13) and (14) than with equation (1).

4. Filtering, Anisotropy, and Energy Transport

4.1 Modal Filtering

As discussed earlier, the CCL model may be regarded as a low pass filter for the transmission of magnetospheric electric fields. The easiest way to understand the effect is to introduce a Fourier representation of equations (6) and (7). Following Weimer et al. (1985), we find after mapping the high-altitude potential to the ionospheric reference altitude and transforming the ionospheric spatial variables

$$\phi_i(\mathbf{k}) = \frac{1}{1 + k^2\lambda_i^2} \phi_e(\mathbf{k}), \qquad (15)$$

$$\Delta\phi(\mathbf{k}) = -\frac{k^2\lambda_i^2}{1 + k^2\lambda_i^2} \phi_e(\mathbf{k}) , \qquad (16)$$

where $\Delta\phi = j_{\parallel i}/K$ is the field-aligned potential drop and

$$\lambda_i = (\Sigma_P/K)^{1/2} \qquad (17)$$

is the resistive scale length introduced by CCL. As to order of magnitude, $\lambda_i \sim 100$ km in inverted-V precipitation regions but may be smaller than this by a factor of 10 or more in large-scale regions of downward field-aligned current.

Relation (15) shows that small-scale ($k\lambda_i >> 1$) magnetospheric structure is filtered out of the ionospheric field while larger-scale ($k\lambda_i << 1$) electric fields are fully impressed on the ionosphere. Alternatively, relation (16) shows that the largest field-aligned currents and potential drops are generted at the smallest scale lengths present in the magnetospheric spectrum. The magnitude of the field-aligned potential drop is nearly equal to the magnitude of the potential difference across equatorial field lines when the (mapped) scale length of the magnetospheric field is much less than λ_i. The implication is that large-scale magnetospheric structure is manifested in the ionosphere in the ionospheric electric field, whereas small-scale magnetospheric structure is manifested in the electron precipitation pattern, at least in regions of upward field-aligned current. In such regions, the filter roll-off is roughly 100 km; in regions of downward field-aligned current, the break point may be as small as a few kilometers. Weimer et al. (1985) have shown examples of magnetic conjugate electric fields at both high and low altitudes in large-scale regions where the average field-aligned current is upward in two cases and downward in another. In the former cases, they infer a filter roll-off of order 100 km; in the latter case, they find no evidence for a filtering effect at scales larger than 10 km. Since their measurement technique was unable to resolve structure less than 10 km, their results appear to be consistent with the CCL model.

4.2 Anisotropic Decay

Equations (15) and (16) are incomplete because the equatorial potential appearing on the right-hand side is not determined self-consistently. When the full set of equations given in section 2 is expressed in a Fourier representation and the field-aligned conductance parameter K in relation (7) is taken to be constant, the following equation is obtained for the Fourier coefficients of the equatorial vorticity distribution:

$$\frac{d}{dt}\omega_{\mathbf{k}}(t) = \sum_{\mathbf{p}} \frac{\hat{\mathbf{z}} \cdot (\mathbf{k} \times \mathbf{p})}{p^2} \omega_{\mathbf{p}}(t)\omega_{\mathbf{k-p}}(t) - \Gamma_{\mathbf{k}}\omega_{\mathbf{k}}(t). \qquad (18)$$

The (linear) decay rate $\Gamma_{\mathbf{k}}$ of the kth Fourier mode is given by

$$\Gamma_{\mathbf{k}} = \Gamma(k,\theta) = \gamma\left[\frac{k^2\lambda_e^2}{M^2} + \frac{A(\theta)}{1 + k^2\lambda_e^2 A(\theta)}\right] , \qquad (19)$$

TABLE 1. Illustrative Parameters

Parameters	Value
Basic Parameters	
Height-integrated Pedersen conductivity	$\Sigma_P = 6$ mho
Field line conductance density	$K = 10^{-9}$ mho m^{-2}
Ionospheric latitude	$\Lambda = 73°$
Equatorial distance†	$L = 12$
Ionospheric magnetic field†	$B_i = 6 \times 10^{-5}$ t
Equatorial magnetic field†	$B_e = 2 \times 10^{-8}$ t
Equatorial height	$H = 10\ R_E$
Kinematic viscosity	$\nu = 10^8\ m^2\ s^{-1}$
Magnetospheric density	$n_0 = 10^7$ protons m^{-3}
Derived Parameters	
Longitudinal stretching factor	$dx_e/dx_i = 42$
Latitudinal stretching factor	$dy_e/dy_i = 71$
Flux tube anisotropy factor	$\delta = 0.35$
Ohmic coupling length (at ionosphere)	$\lambda_i = 80$ km
Ohmic coupling length (at equator)	$\lambda_e = 5700$ km
Ohmic damping rate	$\gamma = 3.8 \times 10^{-3}\ s^{-1}$
Effective Hartmann number	$M = 35$

†Calculated for a dipolar magnetic field.

with $k^2 = k_x^2 + k_y^2$, $\theta = \arctan(k_y/k_x)$, and

$$A(\theta) = \delta \cos^2 \theta + \sin^2 \theta. \tag{20}$$

Other parameters in (19) are:

$$\gamma = \frac{\Sigma_P}{H\rho}\frac{B_e^3}{B_i}\frac{dy_e^2}{dy_i^2} \qquad \lambda_e = \lambda_i \frac{dy_e}{dy_i}$$

$$M^2 = \frac{\gamma\lambda_e^2}{\nu} \qquad \delta = \left(\frac{dx_e}{dx_i}\frac{dy_i}{dy_e}\right)^2 \tag{21}$$

Note that k refers here to the wave number in the equatorial region, whereas in (15) and (16) it refers to the wave number in the ionospheric substrate. The relation between the two is $dx_e k_{xe} = dx_i k_{xi}$ and $dy_e k_{ye} = dy_i k_{yi}$.

In (21), γ is a characteristic damping rate associated with ionospheric drag on the convecting magnetic field lines, λ_e is the resistive scale length in (17) stretched by the magnetospheric (radial) length element, M is an effective Hartmann number, and δ is a flux tube anisotropy factor. The Hartmann number is a measure of the relative dissipation associated with resistive drag, which is seated in the ionosphere, and viscous damping at the scale length λ_e (cf. Cowling, 1976). The ionospheric drag or magnetic friction is dominant when $M \gg 1$. This situation is expected in low-latitude regions of the inner magnetosphere where the field-aligned current is directed out of the ionosphere (see Table 1 for some numerical estimates); the value of M is uncertain in the outer magnetosphere, particularly near the magnetopause or in the geomagnetic tail where the magnetic field lines become highly stretched. In regions where the field-aligned current is directed into the ionosphere, M may be of order 1 or less because of the larger value of the field-aligned conductance in such regions.

Figure 1 shows how the modal decay time, $\tau_d = 1/\Gamma_{\mathbf{k}}$, varies with k and θ for a Hartmann number of M = 35 (the value given in Table 1). The lower curve is a plot of the normalized damping time, $\gamma/\Gamma(k, \frac{\pi}{2})$, which is independent of the flux tube anisotropy factor δ. The middle curve is a plot of $\gamma/\Gamma(k,0)$ for an anisotropy factor of $\delta = 0.35$ (dipolar mapping). For comparison, a more anisotropic case for $\gamma/\Gamma(k,0)$ is shown for $\delta = 0.028$ (upper curve). It is apparent that the modal damping is scale-dependent and can be highly anisotropic at large scales ($k\lambda_e < 1$). The damping at large scales is attributed to ionospheric friction; the anisotropy is due to the magnetic field mapping. The relatively long decay time at intermediate scales is a direct consequence of the filtering effect discussed in the previous section. At intermediate scale lengths, magnetospheric flow energy is converted into fast particle energy by the field-aligned potential drop. For a given angle θ, the maximum in the decay time occurs at the wave number

$$k_*(\theta)\lambda_e = [M - 1/A(\theta)]^{1/2} \tag{22}$$

when $M > 1/A(\theta)$. If $M < 1/A(\theta)$, the maximum occurs at k = 0 as for the upper curve in Figure

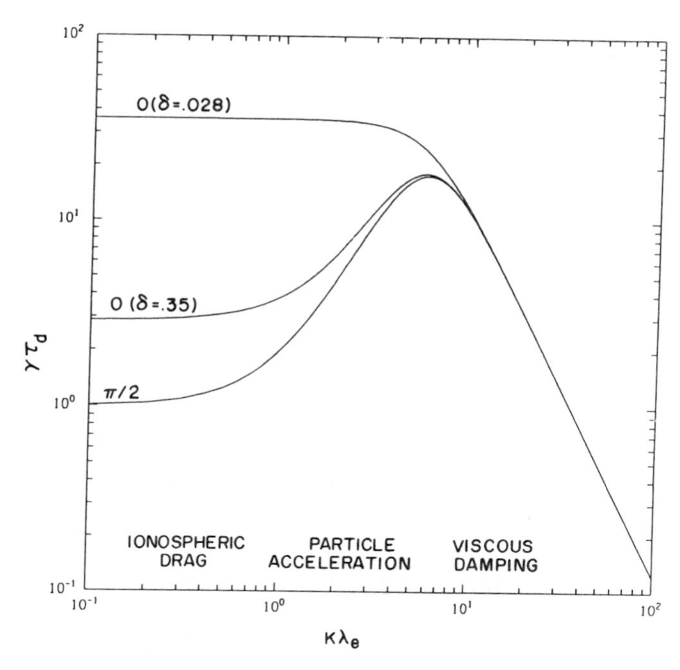

Fig. 1. Normalized damping time $\gamma\tau_d$ versus $k\lambda_e$ for wave vectors lying in the longitudinal ($\theta = 0$) and latitudinal ($\theta = \frac{\pi}{2}$) directions in the ionosphere for $M = 35$. $\delta = 0.028$ (0.35) for the upper (lower) $\theta = 0$ curves. Dissipation due to particle acceleration dominates (for latitudinal perturbations) at wave numbers $1 < k\lambda_e < 6$. Ionospheric drag and viscous damping are dominant to the left and right of this interval.

1. At the smallest scales, the local viscosity--which is assumed to act isotropically in the equatorial region--dominates the flow dynamics. The three wave number regimes for $\theta = \frac{\pi}{2}$, where the dissipation is attributed primarily to ionospheric drag (finite Σ_P), particle acceleration (finite K), or viscous damping (finite ν), are also indicated in the figure. (See also the following discussion on energy transport.)

The curves in Figure 1 illustrate only two of six distinct regimes in the M-δ parameter space defined by relation (22). The expected variations in M have already been discussed. A small value of δ (i.e., smaller than the dipolar value) might be expected in flux tubes passing through the nightside plasma sheet because of the tailward stretching of magnetic field lines in the local $\mathbf{y}_e$ direction. Given the range of possible values of M and δ and the implied variation in the dissipation properties and flow dynamics, it would be useful to have a reference model of the average values of these parameters as functions of magnetic latitude and local time.

4.3 Energy Transport

Equations (1)-(11) possess an energy integral of the form (cf. the Appendix)

$$\frac{d}{dt}E_{kin} = S_{visc} - S_{flux} - S_{elec} - P_{wind} - P_{visc} - P_{\parallel} - P_{ohm}, \tag{23}$$

which describes energy transport within a subregion of the coupled magnetosphere-ionosphere-thermosphere system. The various terms appearing in (23) are defined as

$$E_{kin} = \int \frac{1}{2}\rho v^2 \, dV_e,$$

$$S_{visc} = \rho\nu \oint (\mathbf{v} \times \boldsymbol{\omega}) \cdot d\mathbf{S}_e,$$

$$S_{flux} = \rho \oint (\frac{1}{2} v^2 + T)\mathbf{v} \cdot d\mathbf{S}_e,$$

$$S_{elec} = \oint \phi_e \mathbf{j}_{\perp e} \cdot d\mathbf{S}_e + \oint \phi_i \mathbf{j}_{\perp i} \cdot d\mathbf{S}_i,$$

$$P_{wind} = \int \mathbf{v}_n \cdot (\mathbf{j}_{\perp i} \times \mathbf{B}_i) dV_i,$$

$$P_{visc} = \rho\nu \int \omega^2 \, dV_e,$$

$$P_{\parallel} = \frac{1}{Kh} \int j_{\parallel i}^2 dV_i,$$

$$P_{ohm} = \frac{\Sigma_P h}{\Sigma_P^2 + \Sigma_H^2} \int j_{\perp i}^2 dV_i.$$

Here, h is the height of the ionosphere, dV_e and dV_i are volume elements in the magnetospheric and ionospheric regions, respectively, and $d\mathbf{S}_e$ and $d\mathbf{S}_i$ are corresponding surface element vectors in the outward normal direction ($\hat{\mathbf{x}}_{e,i}$ or $\hat{\mathbf{y}}_{e,i}$).

The first four terms on the right-hand side of (23) represent external processes that either supply energy to the system or absorb internal energy. These include friction at the magnetospheric boundaries (S_{visc}), thermal and kinetic energy flux into the magnetospheric boundaries (S_{flux}), an external generator or load which may maintain a voltage or induce a (perpendicular) current into either the ionospheric or magnetospheric region or both (S_{elec}), and friction with the neutral wind (P_{wind}). Notice that the net energy transport at the boundaries is zero when the system is periodic, as in equation (18). The last three terms are all negative definite and represent internal dissipative processes. These include energy transformed into heat by friction in the magnetospheric region (P_{visc}), Ohmic heating in the ionosphere (P_{ohm}), and the production of fast particles by the field-aligned potential drop ($P_{\parallel}$).

5. Turbulent Flows

In this section, some numerical solutions to equation (18) are described. These solutions are

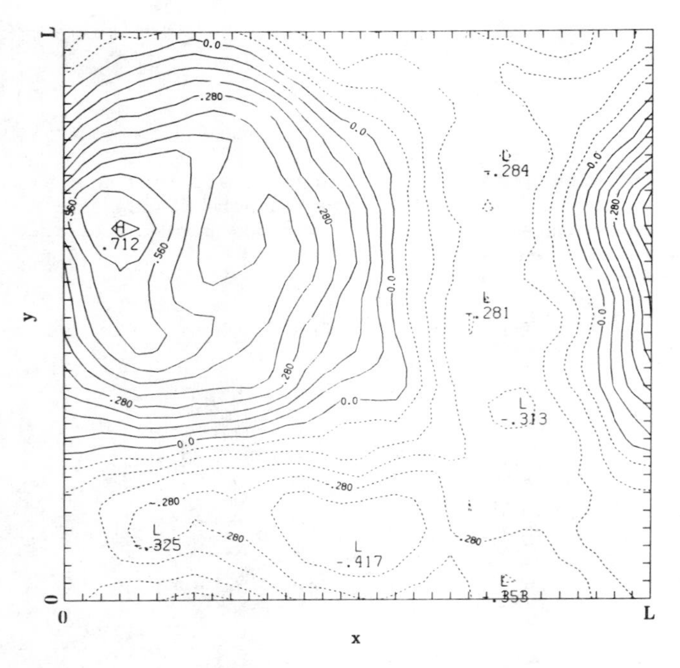

Fig. 2. The randomly selected initial state (t = 0) for the two-dimensional numerical calculations. Streamlines or contours of constant electric potential are shown in the equatorial plane of the magnetosphere. The circulation direction is clockwise along solid lines and counterclockwise along dotted lines. The high and low values and some selected contour values are indicated. The box size is $L = 4\pi\lambda_e$. The x axis maps to a line in the ionosphere that is locally tangent to a contour of constant magnetic latitude; the mapped y axis is tangent to an ionospheric meridian of constant magnetic local time or longitude.

periodic and, therefore, represent purely decaying states; i.e., all of the boundary terms in the energy transport equation are zero. Since the system is not driven by external agents, its evolution is determined solely by the initial state and by intrinsic properties such as the Hartmann number (M) and the flux tube anisotropy factor (δ). The behavior described here may be relevant to magnetospheric turbulence or shear layers that form in response to local convection enhancements, for example, during magnetic substorms. If the enhanced flow is not maintained as the magnetosphere relaxes to a quiet state, then the energy of the perturbed state must be dissipated by internal processes. We are interested in the evolution and organization of the relaxing system.

For simplicity, we do not include the dependence of the field-aligned conductance parameter K on the field-aligned current; K is taken to be constant regardless of the direction (and magnitude) of the field-aligned current. As a consequence, the results described here should be regarded as illustrative rather than applicable in detail to the magnetosphere.

We now consider the evolution of two-dimensional flows. Equation (18) is solved in a box of length $L = 4\pi\lambda_e$ per side. A pseudo-spectral method with dealiasing (Orszag, 1971) is used to time-advance 64 x 64 Fourier coefficients of the (equatorial) electric potential. The numerical algorithm is a modified version of the code ASTER (Schultz, 1986) developed to simulate two-dimensional Navier-Stokes turbulence. The modification consists of replacing the viscous damping coefficient in the Navier-Stokes equations with the modal damping coefficient given by equation (19). The time-advancing algorithm is a variable time step Adams-Moulton predictor-corrector scheme.

The initial state is specified in terms of the (one-dimensional) energy spectrum defined as

$$\varepsilon_k = 2\pi k |v_k|^2. \tag{24}$$

The initial spectrum was chosen to be isotropic and followed the power law $\varepsilon_k = 0.1\ k^{-2}$ (dimensionless units: $kL/2\pi$ is the mode number; $2\pi v/\gamma L$ is a dimensionless velocity). The initial phases of the Fourier coefficients were chosen randomly. The largest coordinate space velocity resulting from the initial spectrum was approximately 1.7 $\gamma\lambda_e$. For the parameters in Table 1, this corresponds to a modest amplitude of about 37 km s^{-1} in the equatorial region. The initial flow pattern in the equatorial magnetosphere is shown in Figure 2. The magnetic field is perpendicular to the plane of the figure. The Hartmann number was fixed at M = 35 for the two examples described below.

Figure 3 shows the flow state at time $\gamma t = 10$ in a weakly anisotropic flux tube ($\delta = 0.35$). Most of the initial energy in the large-scale field has been absorbed by the ionosphere at this time, and what remains are a few intermediate scale isotropic eddies. These eddies are the two-dimensional analogues of the one-dimensional localized oscillations described by Lotko et al. (1987). The eddy size is approximately $\frac{3}{32}L$ (about 1-2 λ_e), which corresponds roughly to the wavelength of the spectral peak in Figure 4. The maximum velocity at this time is about 20% of the initial amplitude.

The three curves shown in Figure 4 represent the energy spectrum (corresponding to Figure 3) averaged over angles within 30° of $k = k_x$ (dotted line), 30° of $k = k_y$ (dashed line), and over all angles (solid line). The anisotropy at large scales is consistent with the damping times in Figure 1 for the $\theta = \pi/2$ case ($k = k_y$) and the δ = 0.35 curve for $\theta = 0$ corresponding to $k = k_x$. The relatively broad spectral peak occurs at $k\lambda_e \approx 4$ which is about 2/3 of the value in Figure 1. The difference is probably due to mode couplings arising from the nonlinear term in equation (18). At large k, the spectrum is isotropic and follows the power law $\varepsilon_k \sim k^{-3}$. This power law is expected for the inertial range of a two-dimensional

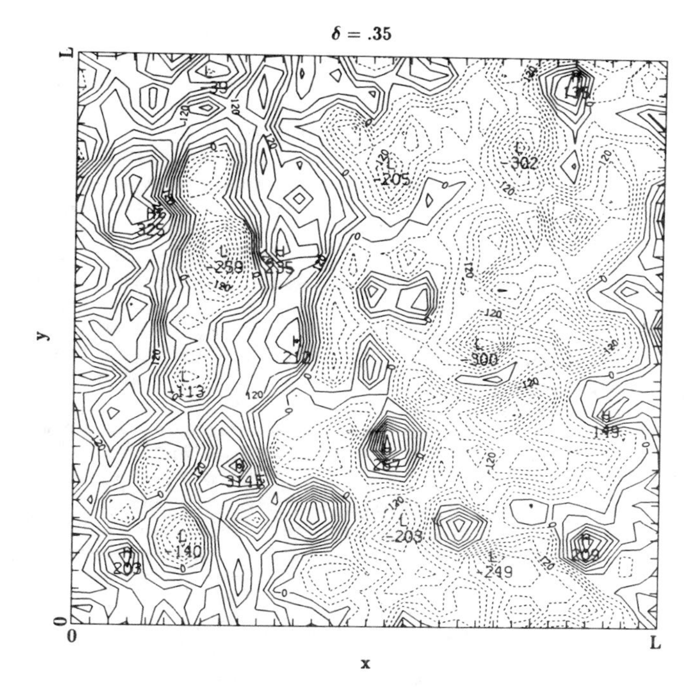

Fig. 3. Streamline distribution after time γt = 10 for a weakly anisotropic flux tube (δ = 0.35) (same format as Figure 2). The Hartmann number is M = 35. High and low values should be multiplied by 10^{-4}.

enstrophy cascade (Kraichnan and Montgomery, 1980). Two-dimensional turbulence theory also predicts a $k^{-5/3}$ inertial range at small wave numbers, corresponding to an inverse energy cascade. If such an inverse cascade is occurring, there is no evidence in Figure 4 for the $k^{-5/3}$ inertial range. This is not surprising since the ionosphere is a strong absorber of energy at large scales.

The flow dynamics in a more strongly anisotropic flux tube (δ = 0.028) were also investigated. The flow state for this case is shown in Figure 5 at time γt = 10. Although the initial state was the same as in the previous example, there is a striking difference in the evolution of the system. The initial energy is again dissipated but at a slower rate. The velocity amplitude at this time has been reduced to only 43% of the initial amplitude. The flow is also more ordered than in the previous case. Its structure is dominated by the (k_x,k_y) = (1,0) mode. Recall that the **x** and **y** directions map onto ionospheric contours of constant magnetic latitude and longitude, respectively, for orange segment mapping. The striations are therefore north-south aligned in the ionosphere. (Note that this is not the sort of alignment one would expect for east-west aligned auroral arcs.) Imbedded within the larger-scale striations in Figure 5 are a number of smaller-scale, somewhat elongated eddies. The eddy size in the **x** direction is comparable to that in the previous example.

The energy spectrum for the strongly anisotropic case is shown in Figure 6. The anisotropy at large scales is again consistent with the damping times depicted in Figure 1. Essentially all of the large-scale modal energy within 30° of $k = k_y$ is absorbed by the ionosphere at this time. The isotropic k^{-3} inertial range at large k is apparent, and the amplitude of the energy spectrum above $kL/2\pi$ = 10 is very nearly the same as in the previous example. At small k, the dotted and solid curves appear to follow a $k^{-5/3}$ power law which may be consistent with an inverse energy cascade. If so, this is somewhat surprising since one would not necessarily expect the theory of isotropic two-dimensional turbulence to apply when the spectrum is so anisotropic. As in the previous example, the spectral peak in the dashed curve occurs at a value of k that is somewhat less than the value in Figure 1.

The dominance of the (k_x,k_y) = (1,0) mode in Figure 5 is a direct consequence of relation (22). For δ = 0.028 and M = 35, k_* is imaginary for θ = 0, so the spectral peak in the $k = k_x$ spectrum should occur at the smallest k_x mode in the system. This is the (1,0) mode. For example, if δ = 0.032, then relation (22) implies that the dominant energy containing mode will be (4,0) when M = 35. (Note: $L = 4\pi\lambda_e$ implies $kL/2\pi = 2k\lambda_e$.) The striations would be more closely spaced in this case. On the other hand, if δ^{-1} < M <1, the dominant mode becomes (0,1). In this case, striations would occur as in Figure 5, but their orientation would be orthogonal to those in Figure 5.

6. Concluding Remarks

Clearly, we have only begun to probe the various kinds of turbulent behavior that arise in the

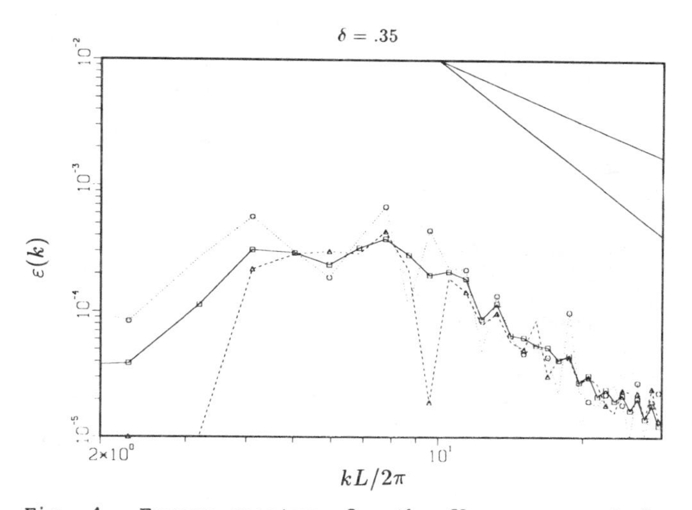

Fig. 4. Energy spectrum for the flow represented in Figure 3. Dotted line: averaged over values within 30° of θ = 0° or $k = k_x$; dashed line: averaged over values of 30° of $\theta = \frac{\pi}{2}$ or $k = k_y$; solid line: averaged over all angles. The straight slanted lines in the upper right corner refer to $k^{-5/3}$ and k^{-3} power laws.

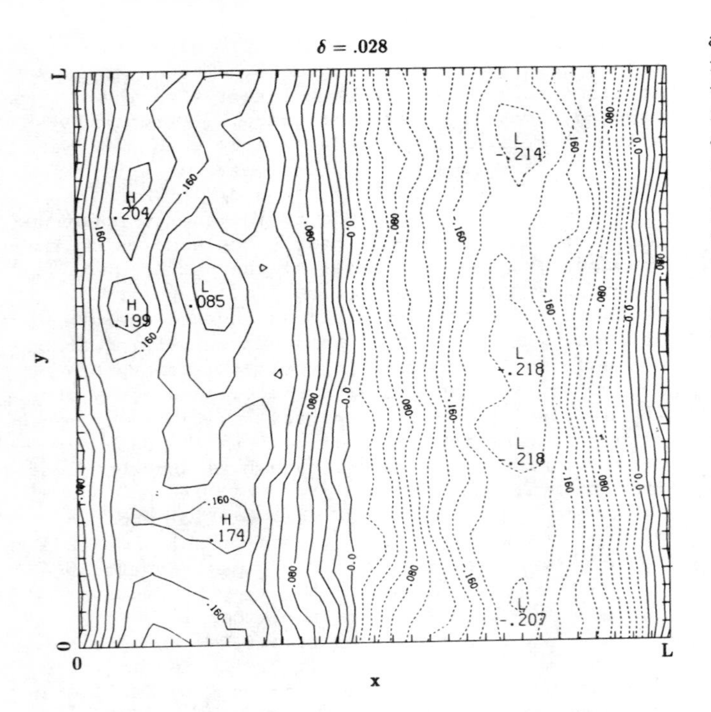

Fig. 5. Streamline distribution after a time $\gamma t = 10$ for a strongly anisotropic flux tube ($\delta = 0.028$. The Hartmann number is $M = 35$ (same format as Figure 2).

interaction between the magnetosphere and ionosphere. The nature of the turbulence can vary significantly with magnetic latitude and local time, magnetospheric activity, and season and may influence the transport of mass, momentum, and energy within the magnetosphere in ways that are not well characterized by large-scale or average models of magnetospheric circulation. In order to reach a fundamental understanding of magnetosphere-ionosphere coupling, it is therefore necessary to develop models that include not only electrodynamic coupling between the two regions but also nonlocal turbulent interactions. The simple model described in this paper represents a first step in this direction. It illustrates how the effects of viscosity, field-aligned potential drops, and ionospheric friction, as well as some rudimentary features of magnetic field mapping, can influence the turbulent dynamics of the coupled system. In its present configuration, the model applies mainly to regions of the magnetosphere where the magnetic field may be considered as known (see also section 3). Like large-scale convection models (e.g., Harel et al., 1981), it is of limited use in describing boundary layer processes occurring in the outer regions of the magnetosphere where local currents often produce large deformations of the magnetic field and where the transit time of Alfvén waves between the ionosphere and magnetosphere may not be negligible.

Despite their inherent limitations, models such as the one described here may be useful in organizing compelx satellite data of non-steady magnetospheric flows and their manifestations in the ionosphere. For example, in addition to boundary and initial conditions, the state of the system represented by equation (18) is characterized by two intrinsic parameters: an effective Hartmann number M and a magnetic anisotropy parameter δ. The Hartmann number determines whether the flow is viscous-dominated ($M \ll 1$) or resistive-dominated ($M \gg 1$). The magnetic anisotropy parameter is a measure of the relative stretching of the east-west and north-south ionospheric flux tube dimensions in the equatorial plane. The illustrative calculations presented in section 5 show that the flow dynamics and associated spectral properties can vary dramatically with these parameters, forming, in some cases, isotropic eddies and, in others, organized striations. Given a magnetic field mapping and the variation of the Hartmann number (at least in an average sense), one could use this model to determine statistical properties of mesoscale length fluctuations in the ionospheric electric field and field-aligned current at various magnetic latitudes and local times. This capability would add a new dimension to our understanding of the interaction between the magnetosphere and ionosphere.

All of the numerical studies presented here assume periodic boundary conditions and therefore apply to absolutely decaying flows. An obvious extension is to examine driven or externally maintained shear flows, as occur in the low-latitude boundary layer (Eastmann et al., 1976; 1985; Schopke et al., 1981) or near the inner edge of the nightside plasma sheet (Wolf and Harel, 1980; Olsen et al., 1986). For a given energy input at the shear layer boundaries, it should be possible on the basis of the energy transport equation, e.g., relation (23), to estimate the relative energy dissipation associated with ionospheric

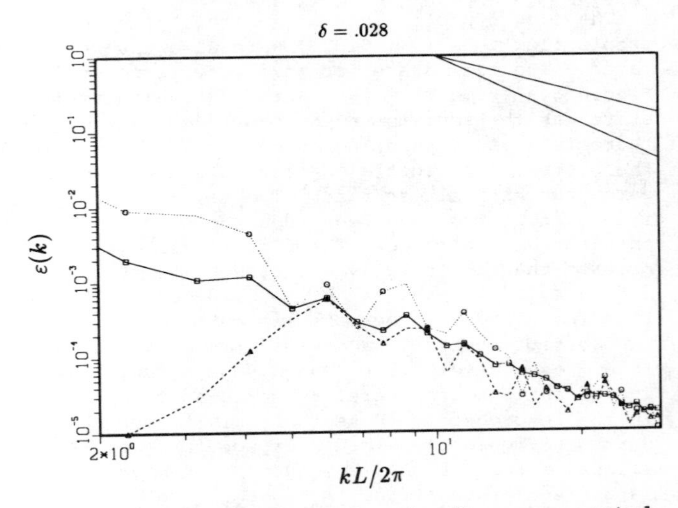

Fig. 6. Energy spectrum for the flow represented in Figure 5 (same format as Figure 4).

Ohmic heating, the production of fast particles in regions of field-aligned potential drop, and viscous damping in the equatorial magnetosphere. Of course, an accurate treatment of the low-latitude boundary layer must ultimately include the effects of internally-generated magnetic fluctuations and mass transport across the magnetopause--effects that make it difficult to isolate the magnetosphere-ionosphere interaction from the magnetosphere-solar wind interaction and that may cause the boundary layer to straddle closed and open magnetic field lines. In this regard, we are still at a somewhat rudimentary level in modeling--and understanding--the dynamics of the interaction between the magnetosphere and ionosphere.

Appendix: Energy Transport Equation

The incompressible momentum equation in the equatorial region is

$$\rho(\partial_t + \mathbf{v} \cdot \nabla_e - \nu\nabla_e^2)\mathbf{v} = -\nabla_e P + \mathbf{j_e} \times \mathbf{B_e}. \quad \text{(A1)}$$

The dot product of $\mathbf{v}$ with equation (A1) is formed. Using vector calculus identities, the resulting equation can be expressed as

$$\partial_t(\frac{1}{2}\rho v^2) + \nabla_e \cdot (\frac{1}{2}\rho\mathbf{v}v^2 + \rho\mathbf{v}T) \quad \text{(A2)}$$

$$= \rho\nu\nabla_e \cdot (\mathbf{v} \times \boldsymbol{\omega}) - \rho\nu\Omega^2 + \mathbf{j_e} \cdot \mathbf{E_e},$$

where $T = P/\rho$, $\boldsymbol{\omega} = \nabla_e \times \mathbf{v}$, and $\mathbf{E_e} = -\mathbf{v} \times \mathbf{B_e}$. Integrating (A2) over the volume of the equatorial region yields

$$\frac{d\varepsilon}{dt} = S_{visc} - S_{pres} - P_{visc} + \int \mathbf{j_e} \cdot \mathbf{E_e} dV_e, \quad \text{(A3)}$$

where

$$\varepsilon = \int \frac{1}{2}\rho v^2 dV_e ,$$

$$S_{flux} = \rho \oint (\frac{1}{2} v^2 + T)\mathbf{v} \cdot d\mathbf{S_e} ,$$

$$S_{visc} = \rho\nu \oint (\mathbf{v} \times \boldsymbol{\omega}) \cdot d\mathbf{S_e},$$

$$P_{visc} = \rho\nu \int \omega^2 dV_e .$$

Using $\mathbf{E_e} = -\nabla_e\phi_e$ and integrating the last term on the right-hand side of (A3) by parts yields

$$\int \mathbf{j_e} \cdot \mathbf{E_e} dV_e = \int\phi_e(\nabla_e \cdot \mathbf{j_e})dV_e - \oint \phi_e\mathbf{j_e} \cdot d\mathbf{S_e}. \quad \text{(A4)}$$

Using the relations between ionospheric and equatorial variables

$$j_{\parallel e} = H\nabla_e \cdot \mathbf{j_e} = (B_e/B_i)j_{\parallel i},$$

$$\phi_e = \phi_i - \frac{1}{K} j_{\parallel i},$$

$$dV_e = \frac{B_i}{B_e}\frac{H}{h} dV_i,$$

where H is the height of the equatorial region and h is the height of the ionosphere, the volume integral on the right-hand side of (A4) becomes

$$\int \phi_e(\nabla_e \cdot \mathbf{j_e})dV_e = \frac{1}{h}\int\phi_i j_{\parallel i} dV_i - \frac{1}{Kh}\int j_{\parallel i}^2 dV_i. \quad \text{(A5)}$$

Note that the offset current $j_{\parallel 0}$ in relation (7) has been neglected. Using $\mathbf{E_i} = -\nabla_i\phi_i$ and $j_{\parallel i} = -h\nabla_i \cdot \mathbf{j_i}$, the first volume integral on the right-hand side of (A5) becomes

$$\frac{1}{h}\int \phi_i j_{\parallel i} dV_i = -\int \mathbf{j_i} \cdot \mathbf{E_i} dV_i - \oint \phi_i\mathbf{j_i} \cdot d\mathbf{S_i}. \quad \text{(A6)}$$

Using Ohm's law, the volume integral on the right-hand side of (A6) can be divided into a neutral wind contribution and a purely Ohmic contribution given by

$$P_{wind} = \int \mathbf{v_n} \cdot (\mathbf{j_i} \times \mathbf{B_i})\, dV_i ,$$

$$P_{ohm} = \frac{\sigma_P}{\sigma_P^2 + \sigma_H^2} \int j_i^2\, dV_i .$$

With the additional definitions,

$$P_\parallel = \frac{1}{Kh}\int j_{\parallel i}^2 dV_i,$$

$$S_{elec} = \oint \phi_e\mathbf{j_e} \cdot d\mathbf{S_e} + \oint \phi_i\mathbf{j_i} \cdot d\mathbf{S_i},$$

the energy transport equation for the coupled magnetosphere-ionosphere system can be written as

$$\frac{d\varepsilon}{dt} = S_{visc} - S_{flux} - S_{elec} \quad \text{(A8)}$$

$$- P_{visc} - P_\parallel - P_{ohm} - P_{wind}.$$

Acknowledgments. A number of ideas discussed in this paper were brought into focus during summary discussions at the First Huntsville Workshop on Magnetosphere/Ionosphere Plasma Models. Tom

Moore and Hunter Waite should be congratulated for their efforts in organizing the workshop and stimulating some of the critical discussions that took place. Bengt Sonnerup and Bob Lysak contributed significantly to the conceptual development of this paper. The numerical code ASTER was developed at Cornell University in collaboration with R. N. Sudan. The research at Dartmouth was made possible by National Science Foundation, Atmospheric Sciences Division grant numbers ATM-8445010 and ATM-8619019.

References

Chiu, Y. T., and J. M. Cornwall, Electrostatic model of a quiet auroral arc, J. Geophys. Res., 85, 543, 1980.

Cowling, T. G., Magnetohydrodynamics, Adam Hilger Ltd., Bristol, England, p. 11, 1976.

Eastman, T. E., E. W. Hones, Jr., S. J. Bame, and J. R. Asbridge, The magnetospheric boundary layer: Site of plasma, momentum, and energy transfer from the magnetosheath into the magnetosphere, Geophys. Res. Lett., 3, 685, 1976.

Eastman, T. E., B. Popielawska, and L. A. Frank, Three-dimensional plasma observations near the outer magnetospheric boundary, J. Geophys. Res., 90, 9519, 1985.

Fridman, M., and J. Lemaire, Relationship between auroral electron fluxes and field-aligned electric potential differences, J. Geophys. Res., 85, 664, 1980.

Harel, M., R. A. Wolf, P. H. Reiff, R. W. Spiro, W. J. Burke, F. J. Rich, and M. Smiddy, Quantitative simulation of a magnetospheric substorm 1. Model logic and overview, J. Geophys. Res., 86, 2217, 1981.

Hasegawa, A., and T. Sato, Generation of field-aligned current during substorm, in Dynamics of the Magnetosphere, edited by S.-I. Akasofu, p. 529, D. Reidel Publ. Co., Boston, Massachusetts, 1980.

Kan, J. R., and W. Sun, Simulation of the westward traveling surge and Pi 2 pulsations during substorms, J. Geophys. Res., 90, 10,911, 1985.

Kelley, M. C., and P. M. Kintner, Evidence for two-dimensional inertial turbulence in a cosmic-scale low-β plasma, Astrophys. J., 220, 339, 1978.

Kintner, P. M., Observations of velocity shear driven plasma turbulence, J. Geophys. Res., 81, 5114, 1976.

Kraichnan, R. H., and D. Montgomery, Two-dimensional turbulence, Rep. Prog. Phys., 43, 547, 1980.

Lotko, W., B.U.Ö. Sonnerup, and R. L. Lysak, Nonsteady boundary layer flow including ionospheric drag and parallel electric fields, J. Geophys. Res., in press, 1987.

Lyons, L. R., Generation of large-scale regions of auroral currents, electric potentials, and precipitation by the divergence of the convection electric field, J. Geophys. Res., 85, 17, 1980.

Lyons, L. R., D. S. Evans, and R. Lundin, An observed relation between magnetic field-aligned electric fields and downward electron energy fluxes in the vicinity of auroral forms, J. Geophys. Res., 84, 457, 1979.

Lysak, R. L., Auroral electrodynamics with current and voltage generators, J. Geophys. Res., 90, 4178, 1985.

Lysak, R. L., Coupling of the dynamic ionosphere to auroral flux tubes, J. Geophys. Res., 91, 7047, 1986.

Lysak, R. L., and C. T. Dum, Dynamics of magnetosphere-ionosphere coupling including turbulent transport, J. Geophys. Res., 88, 365, 1983.

Miura, A., and T. Sato, Numerical simulation of global formation of auroral arcs, J. Geophys. Res., 85, 73, 1980.

Olsen, R. C., T. L. Aggson, and B. G. Ledley, Observations of electric fields near the plasmapause at midnight, J. Geophys. Res., 91, 12,017, 1986.

Orszag, S. A., Studies in Appl. Math, vol. L, 293, 1971.

Sckopke, N., G. Paschmann, G. Haerendel, B.U.O. Sonnerup, S. J. Bame, T. G. Forbes, E. W. Hones, Jr., and C. T. Russell, Structure of the low latitude boundary layer, J. Geophys. Res., 86, 2099, 1981.

Schultz, C. G., A users guide for the code ASTER: 2-D turbulence using DIA techniques, LPS Report No. 360, Laboratory of Plasma Studies, Cornell University, Ithaca, New York, 1986.

Song, Y., and R. L. Lysak, Turbulent generation of auroral currents and fields--a spectral simulation of two-dimensional MHD turbulence, this volume, 1987.

Sonnerup, B.U.Ö., Theory of the low latitude boundary layer, J. Geophys. Res., 85, 2017, 1980.

Vasyliunas, V. M., Mathematical models of magnetospheric convection and its coupling to the ionosphere, in Particles and Fields in the Magnetosphere, edited by B. M. McCormac, p. 60, D. Reidel Publ. Co., Dordrecht, Holland, 1970.

Vickrey, J. F., R. C. Livingston, N. B. Walker, T. A. Potemra, R. A. Heelis, M. C. Kelley, and F. J. Rich, On the current-voltage relationship of the magnetospheric generator at intermediate spatial scales, Geophys. Res. Lett., 13, 495, 1986.

Watanabe, K., M. Ashour-Abdalla, and T. Sato, A numerical model of magnetosphere-ionosphere coupling: Preliminary results, J. Geophys. Res., 91, 6973, 1986.

Weimer, D. R., C. K. Goertz, D. A. Gurnett, N. C. Maynard, and J. L. Burch, Auroral zone electric fields from DE 1 and 2 at magnetic conjunctions, J. Geophys. Res., 90, 7479, 1985.

Wolf, R. A., and M. Harel, Dynamics of the magnetospheric plasma, in Dynamics of the Magnetosphere, edited by S.-I. Akasofu, p. 143, D. Reidel Publ. Co., Boston, Massachusetts, 1980.

MULTIMOMENT FLUID SIMULATIONS OF TRANSPORT PROCESSES IN THE AURORAL ZONES

P. J. Palmadesso

Naval Research Laboratory, Washington, D.C. 20375

S. B. Ganguli and H. G. Mitchell, Jr.

Science Applications International Corporation, McLean, Virginia 22102

Abstract. We have performed multimoment, time-dependent, and steady state multifluid plasma simulations to study the equilibria and dynamics of the auroral field line plasma in the presence of upward and downward field-aligned currents and anomalous transport processes. This work is intended to provide experience in the use of generalized fluid models developed for use in the study of large-scale magnetospheric-ionospheric dynamics. The results of the auroral field line studies and practical issues arising from the use of the multimoment models are reviewed and discussed.

1. Introduction

The development of global scale models of the interaction of the magnetosphere with the solar wind and the ionosphere has proceeded rapidly over the last few years. These models use ideal magnetohydrodynamic (MHD) equations which assume near-Maxwellian distribution functions and neglect temperature anisotropies, heat flows, and kinetic effects. In the collisionless magnetosphere it is difficult to justify these approximations, yet, in spite of this, it is generally necessary to use fluid theory to construct numerical models of such large-scale dynamic phenomena. Even though kinetic effects and non-Maxwellian distribution functions can play an important role in the dynamics, at least in some regions of the magnetosphere, a macroscale dynamic kinetic model is clearly not feasible. Thus, while the present generation of large-scale simulation models represent many of the magnetospheric processes quite well, more general fluid models that can handle larger deviations from the Maxwellian and which do a better job of modeling the interaction of large- and small-scale phenomena will eventually be needed to deal with non-ideal phenomena that influence macroscopic transport.

We have been conducting research focussed on a particular aspect of the ionosphere-magnetosphere (I-M) coupling problem, i.e., the transport of plasma and energy along auroral field lines, in the presence of field-aligned currents and some electrostatic ion-cyclotron-related collective effects. In addition to providing insights into the physics of auroral transport processes, this work has led to an increased understanding of several issues relating to the potential benefits and the practical difficulties associated with the use of the generalized fluid theory approach (Schunk, 1977; Schunk and Watkins, 1981, 1982; Ganguli et al., 1985a, 1987) with anomalous transport coefficients in large-scale dynamic modeling. The purpose of this paper is to discuss these method-related issues in the context of the I-M coupling application.

We shall, in the next section, review very briefly the mathematical basis of the transport equations and display the equations we are currently using. Section 3 contains a qualitative discussion of the results of our I-M coupling research with emphasis on those aspects which serve to illustrate the advantages of using this approach, and the synergism between higher moment and multispecies information and the anomalous transport terms. Some of the practical difficulties associated with use of the multispecies/multimoment approach and methods for dealing with these are discussed in section 4. Finally, we summarize and suggest some directions for future research in this area in section 5.

2. Discussion of the Approach

2.1 Mathematical Basis of the Transport Equations

Generalized transport equations for use in aeronomy and space physics were developed by

Schunk and co-workers using moment sequence methods based on the work of Grad (1949, 1958) and Mintzer (1965). A thorough review of the mathematical basis for these equations has been provided by Schunk (1977); we offer a short review here as background for the subsequent discussions.

In general, a finite system of equations obtained by taking moments of the Boltzmann equation does not constitute a closed set since the equation governing the velocity moment of order m contains the velocity moment of order m + 1. It is therefore necessary to adopt a closure approximation. This is accomplished by representing the distribution function by an explicit analytic expression in the form of the product of a Maxwellian or other base function and a truncated series of orthogonal polynomials. This analytic function contains a fixed number, N, of free parameters, chosen so that the parameters are also moments. The additional equations needed to supplement the truncated sequence of moment equations and close the system are obtained by substituting the analytic distribution function into the defining equations for the moments and integrating. For example, in Grad's eight-moment approximation, the distribution function for each species is assumed to have the form

$$f(\mathbf{r},\mathbf{v},t) = f_0(\mathbf{r},\mathbf{v},t)$$

$$\left[1 - \frac{m_s}{kT_s p_s}\left(1 - \frac{m_s(\mathbf{v}-\mathbf{v}_s)^2}{5kT_s}\right)\mathbf{q}_s \cdot (\mathbf{v}-\mathbf{v}_s)\right] , \qquad (1)$$

where $p_s = nkT_s$ and $f_0(\mathbf{r},\mathbf{v},t) = n_s[m_s/(2\pi kT_s)]^{3/2} \cdot \exp[-m_s(\mathbf{v}-\mathbf{v}_s)^2/(2kT_s)]$, and the eight moments are:

$$n_s = \int f\, d^3v \qquad \text{density} \qquad (2)$$

$$\mathbf{v}_s = \langle \mathbf{v} \rangle \quad \text{bulk velocity (three components)} \qquad (3)$$

$$T_s = 1/3(m_s/k)\langle(\mathbf{v}-\mathbf{v}_s)^2\rangle \qquad \text{temperature} \qquad (4)$$

$$\mathbf{q}_s = 1/2\, n_s m_s \langle(\mathbf{v}-\mathbf{v}_s)^2(\mathbf{v}-\mathbf{v}_s)\rangle$$

heat flow vector (three components) (5)

where

$$\langle M \rangle \equiv \left(\int f\, M\, d^3v\right)/\left(\int f\, d^3v\right) . \qquad (6)$$

One obtains eight equations for n_s, $\mathbf{v}_s$, T_s, and $\mathbf{q}_s$ by taking moments of the Boltzmann equation in the usual way. The moment equations for the components of $\mathbf{q}_s$ contain components of the higher order moments

$$\mathbf{Q}_s = n_s m_s \langle(\mathbf{v}-\mathbf{v}_s)(\mathbf{v}-\mathbf{v}_s)(\mathbf{v}-\mathbf{v}_s)\rangle$$

and

$$\boldsymbol{\mu}_s = 1/2\, n_s m_s \langle(\mathbf{v}-\mathbf{v}_s)^2(\mathbf{v}-\mathbf{v}_s)(\mathbf{v}-\mathbf{v}_s)\rangle .$$

The system is closed by expressing these in terms of the lower order moments by use of (1), (6), and the equations which define $\mathbf{Q}_s$ and $\boldsymbol{\mu}_s$. Grad's thirteen-moment approximation is based on an expression similar to (1), with an additional term in the polynomial factor involving the stress tensor. The symmetric, traceless stress tensor $(\mathbf{P}_s - p_s\mathbf{I})$, where $\mathbf{P}_s = n_s m_s\langle(\mathbf{v}-\mathbf{v}_s)(\mathbf{v}-\mathbf{v}_s)\rangle$ and $\mathbf{I}$ is the unit matrix, adds five additional moment variables to the system. A ten-moment approximation is obtained by adding the stress tensor term to the polynomial factor in (1) and dropping the heat flow term. When only n_s, T_s, and $\mathbf{v}_s$ are retained, one has the familiar five-moment approximation.

The base function f_0 can be replaced by a more general function (Mintzer, 1965), such as a bi-Maxwellian (St.-Maurice and Schunk, 1976; Barakat and Schunk, 1982). The bi-Maxwellian analogue of the thirteen-moment approximation has sixteen moments and forms the basis for the equations in section 2.2. The essential differences, for our purposes, between the thirteen- and sixteen-moment equations are (1) that the anisotropy is built into the base function and therefore need not be small, and (2) that there are separate heat flow vectors for parallel and perpendicular heat transport. In any case, the base function acts as a weight function for the orthogonal polynomials, which must be determined via an appropriate orthogonalization procedure.

One has freedom of choice in selection of the base function, and, in addition, some flexibility is available in the determination of which polynomial terms to retain, as illustrated by the 5, 8, 10, 13, and higher level approximations associated with the Maxwellian f_0, each of which emphasizes a different set of physical processes. The approximations work when the physical situation being modeled produces distribution functions not too different from the base function and when the distortions about the base function are well represented by the terms retained in the polynomial expansion, so that the expansion may be considered converged.

In the next section we explicitly write down one version of the transport equations used in our auroral modeling research to illustrate their structure.

2.2 Reduced Sixteen-Moment Transport Equations

The discussion up to this point applies in a completely general geometry. In practice, when one applies this method to macroscopic transport in a well-magnetized plasma, it is convenient to assume gyrotropic symmetry, which reduces the actual number of moment equations one needs to deal with. In addition, we have considered electrostatic models of the auroral flux tube plasma which allow variation in only one spatial dimension (parallel to the magnetic field). The magnitude of **B** varies along the field line, but curvature effects, which are not expected to be important for the relatively cold plasma we have dealt with so far, are not modeled. The sixteen-moment formalism is thus reduced to six moments: number density, fluid velocity, parallel and perpendicular temperatures, and two heat flows (parallel energy flowing along **B** and perpendicular energy flowing along **B**).

The resulting transport equations, for the special case of a plasma of mobile electrons and protons and immobile oxygen ions, are as follows:

continuity

$$\frac{\partial n}{\partial t} + \frac{\partial nv}{\partial r} + \dot{A}nv = \frac{\delta n}{\delta t}, \qquad (7)$$

momentum

$$\frac{\partial v}{\partial t} + v\frac{\partial v}{\partial r} + \frac{k}{mn}\frac{\partial nT_{\parallel}}{\partial r} + \frac{GM}{r^2} - \frac{e\mathbf{E}}{m} + \frac{\dot{A}k}{m}(T_{\parallel} - T_{\perp}) = \frac{\delta v}{\delta t}, \qquad (8)$$

parallel energy

$$k\frac{\partial T_{\parallel}}{\partial t} + kv\frac{\partial T_{\parallel}}{\partial r} + \frac{2}{n}\frac{\partial nh_{\parallel}}{\partial r} + 2\dot{A}(h_{\parallel} - h_{\perp}) + 2kT_{\parallel}\frac{\partial v}{\partial r} = k\frac{\delta T_{\parallel}}{\delta t}, \qquad (9)$$

perpendicular energy

$$k\frac{\partial T_{\perp}}{\partial t} + kv\frac{\partial T_{\perp}}{\partial r} + \frac{1}{n}\frac{\partial nh_{\perp}}{\partial r} + \dot{A}(2h_{\perp} + kT_{\perp}v) = k\frac{\delta T_{\perp}}{\delta t}, \qquad (10)$$

heat flow per particle for parallel energy

$$\frac{\partial h_{\parallel}}{\partial t} + v\frac{\partial h_{\parallel}}{\partial r} + 3h_{\parallel}\frac{\partial v}{\partial r} + \frac{3}{2}\frac{k^2T_{\parallel}}{m}\frac{\partial T_{\parallel}}{\partial r} = \frac{\delta h_{\parallel}}{\delta t}, \qquad (11)$$

heat flow per particle for perpendicular energy

$$\frac{\partial h_{\perp}}{\partial t} + v\frac{\partial h_{\perp}}{\partial r} + h_{\perp}\frac{\partial v}{\partial r} + \frac{k^2T_{\parallel}}{m}\frac{\partial T_{\perp}}{\partial r} + \dot{A}\left(vh_{\perp} + \frac{T_{\perp}k^2}{m}(T_{\parallel} - T_{\perp})\right) = \frac{\delta h_{\perp}}{\delta t}. \qquad (12)$$

In writing these equations we have dropped the species subscript for the sake of simplicity and used the symbols $h_{\parallel}$ and $h_{\perp}$ to represent the flow along the field line of parallel and perpendicular thermal energy per particle, defined by $h_{\parallel} = 1/2\, m_s\langle(v_{\parallel}-v_s)^3\rangle$ and $h_{\perp} = 1/2\, m_s\langle v_{\perp}^2(v_{\parallel}-v_s)\rangle$. The scale of this model is large compared to the electron Debye length, so the transport equation (7) for electron number density may be replaced by an expression for charge quasi-neutrality. In a three component plasma, this is $n_e = n_p + n_o$.

In the above equations n,v are number density and species velocity, A is the cross-sectional area of a flux tube, **B** is the earth's magnetic field, **E** is the electric field parallel to the field line, GM/r^2 is the gravitational acceleration, m is species mass, $T_{\parallel}$ and $T_{\perp}$ are temperatures parallel and perpendicular to the field line, k is Boltzmann's constant, and the subscripts e, p, and o denote electrons, protons, and oxygen ions.

For a given moment F of the distribution function, $\delta F/\delta t$ represents the change in F due to the effects of collisions and anomalous transport effects associated with plasma turbulence. The collision terms used in this model are Burgers' (1969) collision terms for the case of Coulomb collisions with corrections for finite species velocity differences.

We assume that the total flux tube current I remains constant along the tube. Thus, when the oxygen velocity vanishes, $v_e = (n_p v_p - I/eA)/n_e$. Using these facts and the quasi-neutrality condition, the electric field **E** parallel to the field line is calculated from the electron momentum equation.

$$E = \frac{m_e}{en_eA}\frac{\partial}{\partial r}(n_pv_p^2A - n_ev_e^2A) - \frac{k}{e}\left[\frac{\partial T_{e\parallel}}{\partial r} + \frac{T_{e\parallel}}{n_e}\frac{\partial n_e}{\partial r} + \frac{(T_{e\parallel} - T_{e\perp})}{A}\frac{\partial A}{\partial r}\right] - \frac{n_om_eGM}{n_eer^2} + \frac{m_e}{e}\left[\frac{\delta v_e}{\delta t} - \frac{n_p}{n_e}\frac{\delta v_p}{\delta t}\right] \qquad (13)$$

2.3 Anomalous Transport Coefficients

We consider here the simplest model that illustrates essential features of the interaction between microscopic and macroscopic processes and the role played by the higher moments in this interaction. We have developed an anomalous transport model incorporating one current-driven ion process and one current-driven electron process: the electrostatic ion-cyclotron (EIC) instability, which is well known to produce strong ion heating and is suspected to play a role in ion conic formation for ions, and an EIC-related anomalous resistivity process for electrons. The EIC instability converts the kinetic energy of electron current flow into turbulent fluctuation energy, most of which resides with the ions, and ultimately appears as ion transverse thermal energy and transverse energetic ion tails (Ashour-Abdalla and Okuda, 1983). The anomalous resistivity process heats electrons when a critical field-aligned current threshold value is exceeded in the presence of the EIC or any other instability producing strong ion density fluctuations. The anomalous resistivity mechanism has been described in a series of papers by Rowland et al. (Rowland and Palmadesso, 1987, 1983; Rowland et al., 1981a,b).

The effective collision frequency for the EIC instability is assumed to have a simple form

$$\nu^*_{eiH} = \alpha_H \Omega_i \left(\frac{n_i}{n_e}\right) \left(\frac{V_D}{V_{cH}} - 1\right)^2 \tag{14}$$

if $V_D > V_{cH}$ and zero otherwise. Here Ω_i is the ion-cyclotron frequency for the ion species carrying the EIC wave, and α_H is an adjustable parameter which is chosen for numerical stability and to express qualitatively the effects of the anomalous collision processes.

The critical velocity for exciting the EIC instability (V_{cH}) was given by Lee (1972). For a plasma with a single ion species, and in the parameter regime of most interest here,

$$\frac{V_{cH}}{V_{ther_{\parallel i}}} \sim \left[1 + 5\left(\frac{T_{i\perp}}{T_{e\parallel}}\right)\right] \left[\ln\left\{17\left(\frac{T_{i\parallel}}{T_{i\perp}}\right)\left(\frac{T_{e\parallel}}{T_{i\parallel}}\right)^{\frac{3}{2}}\right\}\right]^{\frac{1}{2}}. \tag{15}$$

The perpendicular ion heating rate due to the EIC instability is given in terms of the effective collision frequency by

$$k \frac{\delta T_{i\perp}}{\delta t} = \nu^*_{ieH} m_i V_D^2 . \tag{16}$$

The anomalous collision frequency for the resistivity is given by

$$\nu^*_{eiR} = \alpha_R \Omega_e \left(\frac{n_i}{n_e}\right) \left(\frac{V_D}{V_{cR}} - 1\right)^2 \tag{17}$$

if $V_D >$ both V_{cH} and V_{cR}, and zero otherwise, where α_R is an adjustable parameter. In principle, the magnitudes of the parameters α_H and α_R appropriate to a given situation should be determined approximately, as functions of macroscopic variables accessible within the model, by analyzing the results of particle simulations or other independent studies of the instabilities involved. The values we have chosen in the simulations conducted to date reflect a compromise between the very large realistic rates, which make stable numerical computation difficult, and small but less realistic values. The factors limiting the magnitudes of these coefficients are discussed in the section 4.

The critical velocity for anomalous resistivity (V_{cR}) is assumed to scale as follows (based on the work of Haber et al., 1978):

$$V_{cR} = \frac{1}{4} \left(\frac{\Omega_e}{\omega_e}\right) V_{ther_{\parallel e}} , \tag{18}$$

where ω_e is the electron plasma frequency, Ω_e is the electron cyclotron frequency, and $V_{ther_{\parallel e}}$ is the electron parallel thermal velocity.

The electron heating is taken to be isotropic and the heating rate due to anomalous resistivity is also expressed in terms of the effective collision frequency given in equation (17)

$$k \frac{\delta T_e}{\delta t} = \nu^*_{eiR} m_e V_D^2 . \tag{19}$$

3. Some Research Results

We present in this section a brief qualitative review of those aspects of the results of our I-M coupling research which serve to illustrate the need for and the advantages of the generalized fluid approach with anomalous transport coefficients, and the synergism which exists between multimoment information, multispecies information, and the anomalous transport terms. For a more complete and quantitative presentation of the results themselves we refer the reader to works by Mitchell and Palmadesso (1983, 1984), Ganguli et al. (1985a,b), Ganguli (1986), and Ganguli and Palmadesso (1987).

3.1 Upward Current Simulations

Mitchell and Palmadesso (1983, 1984) used dynamic numerical models based on the gyrotropically reduced thirteen- and sixteen-moment systems of equations to study the plasma along an auroral field line extending from 800 km to 10 R_E. The plasma consisted of the electrons and hydrogen and oxygen ions. Simulations were performed for the case of a current-free polar wind and the case in which an upward field-aligned current was applied along the field line. The upward current simulations provided the experience on which the discussions of multimoment dynamics issues in section 4 are based, established time scales for auroral macroscopic transport processes, and exhibited several interesting dynamic effects. We shall discuss two examples of the latter.

The first effect was noticed in simulations of the dynamic response of the flux tube plasma to the sudden application of an upward field aligned current. Electrons experiencing compressive heating as they moved downward along the converging magnetic field lines were shown to produce a transient localized hot spot and an accompanying

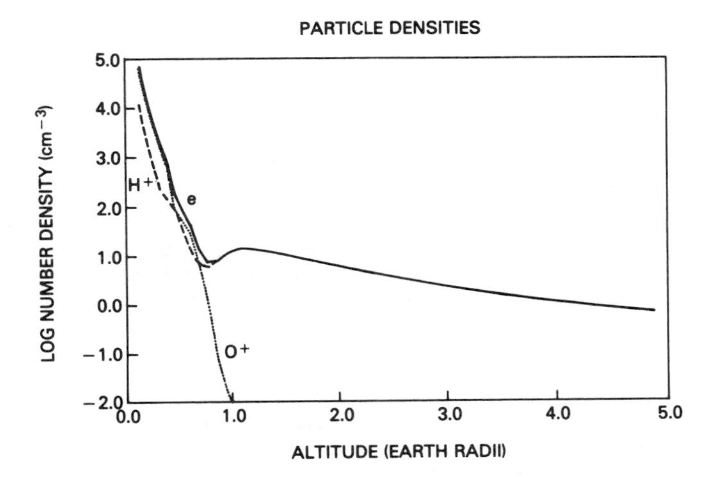

Fig. 1. Particle densities at 6 min after the sudden onset of a 1 μA m^{-2} upward current.

density depletion just above the crossover point for O^+-H^+ dominance (Figure 1). This density depletion is a result of the locally increased electron pressure and ambi-polar electric field. It appeared shortly after the application of current and was observed to be quite large (order of magnitude reduction in density) in some cases. The effect has important consequences when instabilities are included in the model, because the electron-ion drift velocity V_D associated with the field-aligned current increases dramatically within a deep density cavity (Figure 2), and the onset of current-driven instability within the cavity can lead to further heating and thus provide a positive feedback.

While the density depletion was a persistent feature in our upward current runs, its magnitude was observed to depend sensitively on the heat flow model being used. We found much deeper density cavities with the thirteen-moment approximation than with the sixteen-moment approximation. This is understandable because heat flow away from the hot spot opposes the effect and the electron temperature change experienced by a downward moving fluid element is very anisotropic. Total temperature increases because perpendicular energy is gained via first-adiabatic invariant conservation while the self-consistent parallel electric field, which maintains downward electron flow against the mirror forces prevents the loss of parallel energy which would otherwise occur. The sixteen-moment approximation permits freer flow of transverse thermal energy in such a highly anisotropic plasma than the thirteen-moment system, in which the flow of transverse and parallel thermal energy is implicitly coupled. Use of the thirteen-moment description in this case exaggerates the effect, and clearly, use of a fluid model without heat flow would be even worse. In this case the need to retain an adequate heat flow model arises simply because electrons are treated as a separate species and we have field-aligned current flow into a strong magnetic field gradient, a common feature in magnetospheric dynamics.

The upward current simulations exhibited another potentially important effect in the form of a mechanism for preferential field-aligned acceleration of minority ion species. The effect arises because anomalous friction due to microscopic wave-particle interactions is highly species dependent, while the parallel electric field supported by the friction affects all species in the same way. For details, see Mitchell and Palmadesso (1984).

3.2 Return Current Simulations

Ganguli et al. (1985a, 1986) conducted steady state and dynamic simulations of an auroral flux tube plasma extending along the geomagnetic field lines from 1500 km to 10 R_E and supporting a field-aligned return current carried by cold upflowing ionospheric electrons. Electrons and hydrogen ions were treated as dynamic species in these simulations, while oxygen ions were assumed to form a static background population at a constant temperature. These simulations have provided insights into the physics of electron and ion temperature anisotropies and the relationship of these to plasma instabilities on auroral field lines. In particular, they have shown that strong anisotropies arise naturally in association with convective flow or strong heat flow, and that the threshold for current-driven EIC instability is considerably lower than the value estimated by Kindel and Kennel (1971), as a result of the large ion temperature anisotropy associated with the ion outflow [see equation (15)]. They have also yielded useful information relevant to the relationship between current and polar wind bulk ion velocity and the electro-

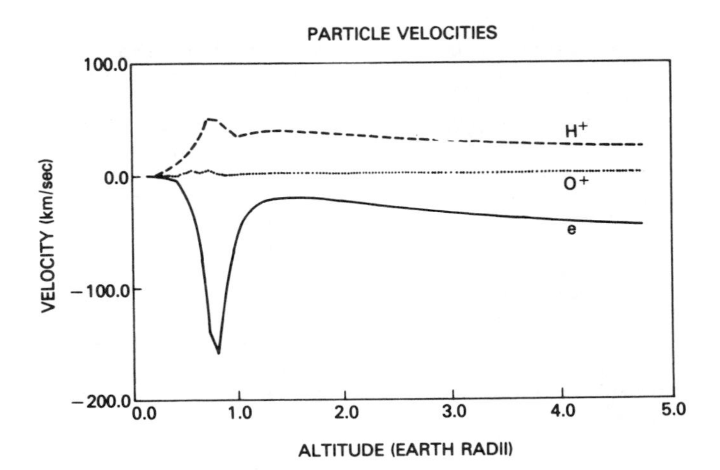

Fig. 2. Particle velocities for the case corresponding to Figure 1.

dynamics of flux tubes in the return current region. We refer the reader to the companion paper in this monograph by Ganguli and Palmadesso (1987) for a more quantitative discussion of some of these issues.

3.3 Arguments for the Use of Generalized Fluid Models

These and other results of our simulations suggest that the multimoment multispecies approach is a promising vehicle for modeling the macroscopic effects of plasma microprocesses via anomalous transport coefficients, within the framework of a fluid model. A key requirement for the success of this enterprise is the ability to represent the characteristic instability parameters, such as critical thresholds for onset of turbulence, the magnitudes of the effective collision frequencies, etc., as functions of the macroscopic variables available within the fluid formalism. The instabilities feed off free energy in the system; hence excitation thresholds are invariably sensitive to free-energy bearing velocity distribution function distortions which can only be described, within the context of a fluid model, by the retention of additional moments. The EIC instability considered here is a good example of this. It is clear from the discussions above that thermal anisotropies are a prominent feature of the auroral return current system even in the absence of instability and that the threshold current required to excite the EIC instability would have appeared to be considerably higher if we had neglected these and attempted to represent V_{cH} as a function of a single isotropic ion temperature. The inclusion of temperature anisotropy in a fluid formulation offers another benefit as well: mirror forces, which clearly play an important role in I-M coupling, are suppressed when the species temperature is forced to be isotropic but are properly modeled as a bulk force acting on fluid elements in a fluid model with anisotropic temperature.

Once excited, the instabilities themselves alter the velocity distribution functions by generating or modifying anisotropies, heat flow asymmetries, or more complex distortions. This also argues for the retention of appropriate higher moment information. It would make little sense, for example, to add terms describing an anomalous heat source such as the EIC instability to the fluid equations for a plasma species without also allowing heat flow.

The gyroscopically reduced sixteen-moment formulation discussed here, which adds thermal anisotropy and separate parallel and perpendicular heat flow to the ideal fluid treatment, appears to offer a reasonable mix of generality and tractability.

4. Practical Issues

Having discussed the advantages of using the generalized fluid approach with anomalous transport coefficients, we now turn our attention to some of the difficulties one encounters with present versions of this method. We emphasize at the outset that while some of these issues are troublesome, none seems at this point to be insurmountable.

4.1 Steady State Solutions

One obtains a set of equations suitable for direct determination of steady state solutions of the transport equations by discarding the time derivative terms and then algebraically eliminating all derivatives in a given equation except one. The result is a system of coupled ordinary differential equations with distance along the field line as the independent variable. These equations are stiff in the sense discussed by Gear (1969) and Shampine and Gear (1979). We have obtained our steady state return current region solutions by using the general procedure employed by Watkins (1981a,b). The practical consequence of stiffness in this case is that the stability of the computation depends sensitively on the direction of integration; moreover the equations for some of the variables must be integrated from top to bottom, while the rest must be integrated in the opposite direction. For example, in the steady state return current region problem, the density and fluid velocity equations and the supersonic H^+ temperature and heat flow equations are integrated upward while the electron temperature and heat flow equations must be integrated from top to bottom. Therefore, since the equations are coupled, it is not possible to obtain a solution in a single integration pass. Instead, one must first provide an initial guess for each moment profile, based on observations or a converged solution from a previous case. The equations are then solved iteratively: the density, velocity, and H^+ thermal equations are integrated upward using the initial guess or previous iteration profiles for the electron temperatures; then the electron thermal equations are integrated downward using previous iterate profiles for the other quantities. This process is repeated until a set of profiles that simultaneously satisfies all the moment equations in steady state is obtained.

A second practical difficulty encountered in the solution of the steady state equations is the sensitivity of the solutions to boundary conditions, particularly boundary conditions on the heat flow moments. For example, we start the electron temperature and heat flow integrations with the parallel and perpendicular electron temperatures appropriate to the upper boundary and initial guesses for the boundary values on the parallel and perpendicular heat flows. The converged solutions after iteration, as described above, will not, in general, satisfy the lower boundary conditions on electron temperature ($T_{e\perp} = T_{e\parallel}$ = 1000 K in the return current simulations), and one must then repeat the entire process with new values for the heat flow moments at the upper boundary. Similarly, one iterates the H^+ heat flow at the lower boundary in order to satisfy a lower boundary condition on the H^+ temperature gradient. Thus, one performs a double iteration to obtain converged solutions satisfying all the equations and both upper and lower boundary conditions. In practice, extremely small changes in the heat flows specified at the boundaries make large changes in the solutions, and a fair amount of experimentation is usually necessary before a converged solution which satisfies all requirements is obtained.

The third area of difficulty encountered with the steady state equations arises from the fact that these equations are singular at sonic points (Mach number = 1). If a sonic transition occurs within the simulation region, one must generate solutions above and below this point and match these across the singularity by means of an appropriate jump condition. Given the complexity of obtaining converged steady state solutions in the absence of sonic points, we have opted to avoid this additional complexity by placing our lower boundary above the sonic transition for the polar wind so that H^+ is always supersonic and all other species are always subsonic.

The sonic point problem can impose limitations on the magnitudes of the effective collision frequencies associated with the anomalous transport coefficients: instabilities which heat the plasma can increase the local H^+ sound speed and thus push the Mach number toward the sonic point. Limitations on the magnitude of heat flows tolerable within the framework of the thirteen- and sixteen-moment approximations can also lead to limitations on the strength of the anomalous transport coefficients and on the magnitude of the electron temperature at the upper boundary, which drives heat flow down the flux tube. The heat flow limitation arises in part because the base functions (Maxwellian or bi-Maxwellian) underlying the thirteen- and sixteen-moment approximations do not support heat flow (see section 2). Thus, heat flow is described in these approximations via the truncated orthogonal polynomial series factor, which converges, generally speaking, when the actual distribution function is not very different from the base function. It might be possible to relax this limitation somewhat by using a base function with a heat flow asymmetry built into it, but we have not yet attempted to do this.

The issues discussed above force us to use a steady state solution procedure which is somewhat awkward but nevertheless practicable, at least in one dimension. Fortunately, almost all of these problems disappear when the time derivative terms are retained and the equations are solved dynamically: iterative solution is no longer necessary; the physical two point boundary conditions are always satisfied, and the equations are no longer singular at sonic points.

4.2 Time-Dependent Solutions

Unfortunately, the heat flow limitation does not disappear when one retains the time derivative terms; in fact the heat flow issue takes on some new dimensions in the context of a dynamical problem described via the thirteen- or sixteen-moment approximations. In general, the process of truncating a moment sequence results in a loss of information relative to a Boltzmann level description, and this may result in an alteration of the dynamical properties of a system. Analysis of the manner in which this happens in the case of heat transport leads to some insights into the general case, and so we will discuss this issue in some detail.

Suppose we wish to model a two-species, collisionless plasma in a constant magnetic field (we neglect gravity, etc., for simplicity). The moment approximation we choose is the sixteen-moment approximation, reduced via the gyrotropic assumption. As discussed in section 2, the sixteen-moment approximation is based on the assumption that the species distribution function can be written in the form $f_s = f_{so}(1 + \phi_s)$, where f_{so} is a bi-Maxwellian and ϕ_s is a truncated orthogonal polynomial expansion. In terms of the moments, f_{so} and ϕ_s have the following form

$$f_{so}(v_\parallel, v_\perp^2, r) = (2\pi)^{-3/2} n_s\, v_{s\parallel}^{-1}\, v_{s\perp}^{-2}\, e^{-\frac{1}{2}(c_{s\parallel}^2 + c_{s\perp}^2)}$$

and

$$\phi_s(v_\parallel, v_\perp^2, r) = H_{s\parallel} c_{s\parallel}\left(\frac{1}{3} c_{s\parallel}^2 - 1\right) + H_{s\perp} c_{s\parallel}(c_{s\perp}^2 - 1),$$

where m_s is the mass of species s, r is the distance parallel to the magnetic field, and

$$v_{s\parallel}^2 = kT_{s\parallel}/m_s \qquad v_{s\perp}^2 = kT_{s\perp}/m_s$$

$$c_{s\parallel} = (v_\parallel - v_s)/v_{s\parallel} \qquad c_{s\perp} = v_\perp/v_{s\perp}$$

$$H_{s\parallel} = h_{s\parallel}/(m_s v_{s\parallel}^3) \qquad H_{s\perp} = h_{s\perp}/(m_s v_{s\parallel} v_{s\perp}^2)$$

With the distribution function in this form, the moment equations for species s become, in this simplified system

$$\frac{\partial n_s}{\partial t} + \frac{\partial n_s v_s}{\partial r} = 0$$

$$\frac{\partial v_s}{\partial t} + v_s \frac{\partial v_s}{\partial r} + \frac{k}{n_s m_s} \frac{\partial n_s T_{s\parallel}}{\partial r} - \frac{e_s}{m_s} E_\parallel = 0$$

$$\frac{\partial T_{s\parallel}}{\partial t} + v_s \frac{\partial T_{s\parallel}}{\partial r} + 2T_{s\parallel} \frac{\partial v_s}{\partial r} + \frac{2}{kn_s} \frac{\partial n_s h_{s\parallel}}{\partial r} = 0$$

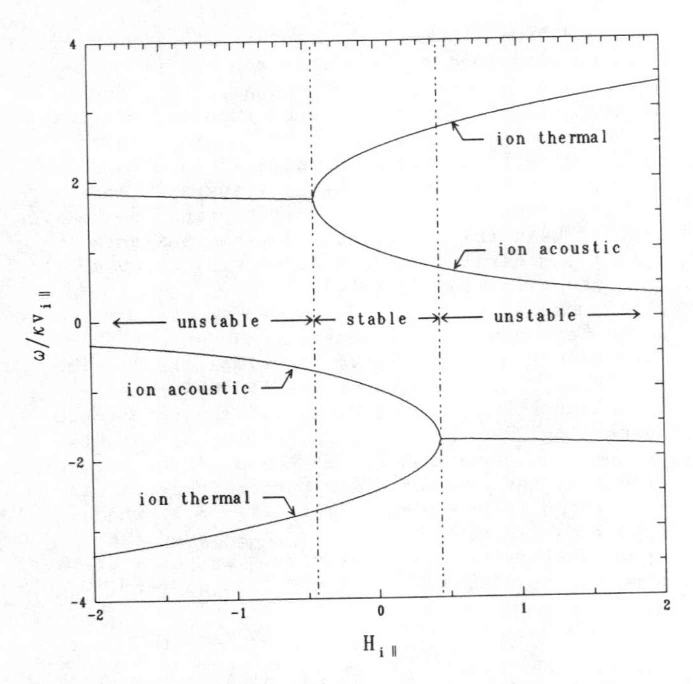

Fig. 3. Dispersion diagram for ion acoustic/ thermal wave interactions.

$$\frac{\partial T_{s\perp}}{\partial t} + v_s \frac{\partial T_{s\perp}}{\partial r} + \frac{1}{kn_s} \frac{\partial n_s h_{s\perp}}{\partial r} = 0$$

$$\frac{\partial h_{s\parallel}}{\partial t} + v_s \frac{\partial h_{s\parallel}}{\partial r} + 3h_{s\parallel} \frac{\partial v_s}{\partial r} + \frac{3k^2 T_{s\parallel}}{2m_s} \frac{\partial T_{s\parallel}}{\partial r} = 0$$

$$\frac{\partial h_{s\perp}}{\partial t} + v_s \frac{\partial h_{s\perp}}{\partial r} + h_{s\perp} \frac{\partial v_s}{\partial r} + \frac{k^2 T_{s\parallel}}{m_s} \frac{\partial T_{s\perp}}{\partial r} = 0 ,$$

where e_s is the charge of species s, $E_\parallel$ is the electric field parallel to B, and the plasma is assumed to be electrostatic.

In an electrostatic plasma consisting of electrons and one ion species, $E_\parallel$ is determined by the two conditions of quasi-neutrality and uniform current parallel to the magnetic field. We will assume that the parallel current is constant in time. Therefore, the electron density n_e and velocity v_e become dependent variables in the electrostatic system and we are left with ten independent quantities. This implies that a characteristic analysis of this system should yield ten waves. By assuming that all of the independent quantities have the form $M(r,t) = M + \delta M \exp i(\kappa r - \omega t)$, where the background quantities M are constant along the magnetic field, it is possible to derive a dispersion relation for the waves in this system. This dispersion relation has the form

$$0=[z_e^2-1][z_i^2-1][(z_e^2-3)(z_i^4-6z_i^2+8H_{i\parallel}z_i+3)$$
$$+ T(z_i^2-3)(z_e^4-6z_e^2+8H_{e\parallel}z_e+3)] ,$$

where $z_s = (v_s - \omega/\kappa)/v_{s\parallel}$ and $T = T_{e\parallel}/T_{i\parallel}$. The first two bracketed quantities represent perpendicular electron and ion thermal waves, i.e., waves in the perpendicular temperatures and heat flows which have phase velocities of $v_s \pm v_{s\parallel}$. The perpendicular thermal waves represent four of the ten waves in this system.

The final bracketed quantity may be simplified somewhat by assuming that the electron thermal velocity $v_{e\parallel}$ is much greater than the ion thermal velocity $v_{i\parallel}$. This assumption implies that either z_i is very large, so that

$$0 = (z_e^2 - 3) z_i^4 ,$$

or z_e is very small, so that

$$0 = -3(z_i^4 - 6z_i^2 + 8H_{i\parallel}z_i + 3) + 3T(z_i^2 - 3) .$$

The first case represents a parallel electron thermal wave, a wave in the parallel electron temperature, and heat flow at a phase velocity of $v_e \pm v_{e\parallel}\sqrt{3}$. The dispersion relation for the final four waves is contained in the second case. If the ion heat flow is assumed to vanish in the equilibrium about which we are perturbing (i.e., the term containing $H_{i\parallel} = 0$), the dispersion relation has the solution

$$\frac{\omega}{k} = v_i \pm v_{i\parallel}\left(3 + \frac{T}{2} \pm [6 + \frac{T^2}{4}]^{1/2}\right)^{1/2} .$$

This expression can be seen to describe two waves whose frequency ω is real for any value of T. If T is taken to be much larger than 1, the expression simplifies considerably and we can see that we have a parallel ion thermal wave, with phase velocity $v_i \pm v_{i\parallel}\sqrt{3}$ and an ion acoustic wave, with phase velocity $v_i \pm (T_{e\parallel}/m_i)^{1/2}$.

A problem arises when the zero-order parallel ion heat flow does not vanish. The plot of phase velocity versus $H_{i\parallel}$ in Figure 3 for T = 1 shows that the ion acoustic and parallel ion thermal waves couple when the magnitude of $H_{i\parallel}$ is above about 0.44. As a result, there is an unstable wave in this system for $H_{i\parallel}$ above this value. For T > 1, the instability threshold rises due to the increase in the ion acoustic velocity with increasing $T_{e\parallel}$, but there is a threshold at all values of T. For T = 1 and $H_{i\parallel} = 1$, the growth rate of the unstable wave is about $0.9\ \kappa v_{i\parallel}$.

Very often, when running dynamic fluid plasma simulations which retain heat flow terms as dynamic quantities, we have encountered unstable behavior when a heat flow has a magnitude on the order of the thermal energy of the plasma multiplied by the parallel thermal velocity, i.e., $H_{s\parallel}$ on the order of 1.

One explanation for this instability could be that this range of values for $H_{i\parallel}$ is exactly the range for which the distribution correction function ϕ_s is on the order of 1, so that the distribution function approximation may be in the process of breaking down. If this is the root of the problem, then an improved approximation with a heat flow asymmetry built into f_{s0} might well solve it, as noted above. However, another interpretation is possible.

It turns out that there is a direct way of gaining some insight into the nature of the breakdown of the moment approximation for large heat flows. Suppose that the model is applied to the following case:

$$n_e,\ n_i,\ T_{e\parallel},\ T_{i\parallel},\ T_{e\perp} \text{ are constant}$$

$$v_e,\ v_i,\ h_{e\parallel},\ h_{i\parallel},\ h_{e\perp},\ E_\parallel = 0$$

$$h_{i\perp}(r,t) = -k\delta T_{i\perp} v_{i\parallel} \cos(\kappa r)\sin(\kappa v_{i\parallel} t)$$

$$T_{i\perp}(r,t) = T_{i\perp o} + \delta T_{i\perp} \sin(\kappa r)\cos(\kappa v_{i\parallel} t)$$

The last two relationships above satisfy the pair of transport equations below, and all the other moment equations are trivial in this situation:

$$\frac{\partial T_{i\perp}}{\partial t} + \frac{1}{kn_i}\frac{\partial h_{i\perp}}{\partial r} = 0$$

$$\frac{\partial h_{i\perp}}{\partial t} + \frac{k^2 T_{i\perp}}{m_i}\frac{\partial h_{i\perp}}{\partial r} = 0$$

This case corresponds to a superposition of two perpendicular ion thermal waves with phase velocities $\pm v_{i\parallel}$ to create a perpendicular ion thermal standing wave. A unique characteristic of this particular situation is that the kinetic equations which describe the evolution of a system initially set up in this way are analytically soluble. Since the parallel electric field is always exactly zero, all of the particles have ballistic trajectories so that

$$f_i(v_\parallel, v_\perp^2, r, t) = f_{io}(v_\parallel, v_\perp^2, r - v_\parallel t).$$

The exact values of the quantities $T_{i\perp}$ and $h_{i\perp}$ can be calculated as functions of time

$$T_{i\perp}(r,t) = T_{i\perp o} + \delta T_{i\perp} \sin(\kappa r)\exp\left(-(\kappa v_{i\parallel} t)^2/2\right)$$

$$h_{i\perp}(r,t) = -(\kappa v_{i\parallel} t)\delta T_{i\perp} v_{i\parallel} \cos(\kappa r) \cdot \exp\left(-(\kappa v_{i\parallel} t)^2/2\right) .$$

The perpendicular ion temperature perturbations for both the moment and the kinetic representations can now be compared. The two expressions agree for t much less than $\tau = 1/\kappa v_{i\parallel}$, but are radically different for $t \geq \tau$. It is clear that the perpendicular ion thermal wave is an artifact of the moment approximation, since a kinetic analysis shows that the wave is damped on a time scale roughly equal to the time it takes an ion traveling at the parallel ion thermal velocity to cross one wavelength.

The reason for the wave damping is clear from the kinetic analysis. The initial spatial perturbation in $T_{i\perp}$ rapidly becomes a velocity space perturbation on the distribution function at any point in space due to phase mixing of particles from different spatial regions of the field line. Increasingly fine structure (scale size $1/\kappa t$) is generated in velocity space, and the distribution function becomes highly filamented until some weak velocity space diffusion process smooths away the fine structure and leaves the system in its asymptotic state, which is a uniform plasma at the average temperature.

The moment approximation does not allow this effect because the increasingly fine structure in velocity space would require increasingly higher moments to model accurately. In effect, the phase mixing process described by the kinetic theory is a process in which a spatial perturbation in one moment damps as the perturbation cascades to higher moments and is finally diffused away. Since a moment approximation artificially truncates the moment sequence, the cascade is halted at the highest moments, creating spurious waves in those moments. These waves may, in turn, interact with real waves, such as ion acoustic waves, and create spurious instabilities.

If this is the correct explanation for the thermal wave instability problem, then building heat flow into the base function might not eliminate it. Moreover, adding a few new higher order moments would simply generate a few new spurious wave modes and a lot a mathematical complexity, so this is not the most promising approach for dealing with the situation.

A pragmatic but somewhat unsatisfying way of dealing with this problem involves artificially limiting the values of the heat flows so that $H_{i\parallel}$ is always less than some value, say 0.5, on the assumption that a realistic distribution function would be restricted to this range by the effect of phase-space instabilities on the distribution function. A more promising approach is based on the argument that the kinetic theory indicates that the phase mixing process, which is the natural thermal damping mechanism in the kinetic plasma, ultimately acts like diffusion, even in a collisionless plasma. If this effect could be included in a moment approximation, it might be possible to obtain reasonable results for the dynamics of the lower order moments without modeling higher and higher moments. One possible fix to the moment equations would be to recognize that the quantity $\kappa v_{i\parallel}$ is an effective diffusion rate for thermal waves and to add a diffusion term to heat flow equations, e.g., a term such as $-(v_{i\parallel}/\kappa)\partial^2 h/\partial r^2$. This term produces the same effective thermal diffusion as in the kinetic analysis, but it has the disadvantage of having explicit κ dependence. In some of the simulations that we have performed, we have added such a term and used the simulation cell size as a measure of κ^{-1}. This adds the diffusion correction only for waves with cell-size wavelengths without altering the dynamics of longer wavelength perturbations. The result has been to greatly reduce the difficulties we have had with thermal wave instabilities. A better method might be to dynamically calculate the appropriate κ so as to track the thermal diffusion more accurately.

The obvious drawback to simply adding a term to the transport equations in an ad hoc manner is that the mathematical link between the transport equations and the analytic distribution function used to close the moment equations is broken. If the effective diffusion idea is to be made truly useful, a way needs to be found to self-consistently incorporate the concept into the generalized fluid formalism, so that the mathematical foundations of the transport equations are not threatened.

4.3 General Comments

The generalized fluid approach is clearly more complex than ideal fluid theory, and solution of the multimoment transport equations requires more CPU time and memory. While these factors may argue against use of this method for multidimensional large-scale dynamics studies at present, our feeling is that this is a transient situation. As computers evolve and gain speed and memory, these issues will fade in importance.

5. Summary Comments

We have discussed the merits and drawbacks of multimoment fluid models with anomalous transport coefficients as tools for the study of classical and anomalous transport phenomena and the interaction of large- and small-scale phenomena in space plasmas. The arguments on both sides of this question are substantial, in our opinion, but the decisive factor may be the apparent lack of easier alternatives to this approach to going beyond present methods of simulating large-scale dynamic phenomena.

The most serious of the outstanding problems seems to be the heat flow problem discussed in section 4. This issue sets severe limits on the magnitudes of the temperature gradients and anomalous heating rates that one can deal with. Clearly, additional research is needed in this area.

There are a large number of other microprocesses which could profitably be incorporated into models of this type. A wide variety of kinetic (i.e., nonfluid) processes have been studied in order to explain wave, particle, and electromagnetic field observations on auroral field lines. Instabilities driven by field-aligned current, ion-ion streaming, and heat flow, among others, have all been inferred to be significant (Kindel and Kennel, 1971; Dusenbery and Lyons, 1985). In addition to plasma microinstabilities, which are essentially kinetic, there are a variety of laminar processes which can also play a role in auroral transport, but which are not easily modeled within the context of simple ideal fluid approximations. Mirror forces, which, as noted in section 4, are modeled by the velocity moment equations only when the species temperature is allowed to be anisotropic, can support field-aligned potential drops (Chiu et al., 1983, and references therein). The sharp differences in the temperatures and other characteristics of ionospheric and magnetospheric electrons may also contribute to the formation of field-aligned potential drops. In principle, the major macroscopic effects of most, if not all, of these processes could be modeled within the framework of a multimoment, multispecies simulation code with anomalous transport coefficients, but each process would need to be studied separately and incorporated in such a way that the essential physics is preserved.

In conclusion, the generalized fluid method with anomalous transport coefficients offers both promise and problems and is a fertile area for future research.

Acknowledgments. This work was supported by the Office of Naval Research and the National Aeronautics and Space Administration.

References

Ashour-Abdalla, M., and H. Okuda, Transverse acceleration of ions on auroral field lines, in Energetic Ion Composition in the Earth's Magnetosphere, edited by R. G. Johnson, p. 43, Terra Scientific Publ. Co., Tokyo, 1983.

Barakat, A. R., and R. W. Schunk, Transport equations for multicomponent anisotropic space plasmas: A review, Plasma Phys., 24, 389, 1982.

Burgers, J. M., Flow Equations for Composite Gases, Academic Press, New York, 1969.
Chiu, Y. T., M. Schulz, J. F. Fennel, and A. M. Kishi, Mirror instability and the origin of morningside auroral structure, J. Geophys. Res., 88, 4041, 1983.
Dusenbery, P. B., and L. R. Lyons, The generation of electrostatic noise in the plasma sheet boundary layer, J. Geophys. Res., 90, 10,935, 1985.
Ganguli, S. B., Plasma transport in the auroral return current region, Ph.D. Dissertation, Boston College, 1986.
Ganguli S. B., H. G. Mitchell, Jr., and P. J. Palmadesso, Behavior of ionized plasma in the high latitude topside ionosphere, NRL Memo. Rept. 5623, August 1985a; also Planet. Space Sci., 35, 703 (1987).
Ganguli, S. B., H. G. Mitchell, Jr., and P. J. Palmadesso, Dynamics of the large-scale return currents on auroral field lines, NRL Memo. Rept. 5673, November 1985b.
Ganguli, S. B., and P. J. Palmadesso, Simulations of classical and anomalous transport processes in the auroral return current region, this volume, 1987.
Gear, C. W., Numerical Initial Value Problems in Ordinary Differential Equations, Prentice-Hall, Inc., Englewood Cliffs, New Jersey, 1969.
Grad, H., On the kinetic theory of rarefied gases, Comm. Pure Appl. Math., 2, 331, 1949.
Grad, H., Principles of the Kinetic Theory of Gases, Handbook of Physics, vol. XII, p. 205, 1958.
Haber, I., J. D. Huba, P. J. Palmadesso, and K. Papadopoulos, Slope reversal of a monotonically decreasing electron tail in a strong magnetic field, Phys. Fluids, 21, 1013, 1978.
Kindel, J. M., and C. F. Kennel, Topside Current Instabilities, J. Geophys. Res., 76, 3055, 1971.
Lee, K. F., Ion cyclotron instability in current-carrying plasmas with anisotropic temperatures, J. Plasma Phys., 8, 379, 1972.
Mintzer, D., Generalized orthogonal polynomial solutions of the Boltzmann equation, Phys. Fluids, 8, 1076, 1965.
Mitchell, H. G., and P. J. Palmadesso, A dynamic model for the auroral field line plasma in the presence of field aligned current, J. Geophys. Res., 88, 2131, 1983.
Mitchell, H. G., and P. J. Palmadesso, O^+ acceleration due to resistive momentum transfer in the auroral field line plasma, J. Geophys. Res., 89, 7576, 1984.
Rowland, H. L., and P. J. Palmadesso, Anomalous resistivity due to low frequency turbulence, J. Geophys. Res., 88, 7997, 1983.
Rowland, H. L., and P. J. Palmadesso, Spiky parallel dc electric fields in the aurora, J. Geophys. Res., 92, 299, 1987.
Rowland, H. L., P. J. Palmadesso, and K. Papadopoulos, One-dimensional direct current resistivity due to strong turbulence, Phys. Fluids, 24, 833, 1981a.
Rowland, H. L., P. J. Palmadesso, and K. Papadopoulos, Anomalous resistivity on auroral field lines, Geophys. Res. Lett., 8, 1257, 1981b.
Schunk, R. W., Mathematical structure of transport equations for multispecies flows, Rev. Geophys. Space Phys., 15, 429, 1977.
Schunk, R. W., and D. S. Watkins, Electron temperature anisotropy in the polar wind, J. Geophys. Res., 86, 91, 1981.
Schunk, R. W., and D. S. Watkins, Proton temperature anisotropy in the polar wind, J. Geophys. Res., 87, 171, 1982.
Shampine, L. F., and C. W. Gear, A user's view of solving stiff ordinary differential equations, SIAM Rev., 21, 1, 1979.
St.-Maurice, J.-P., and R. W. Schunk, Use of generalized orthogonal polynomial solutions of Boltzmann's equation in certain aeronomy problems: Auroral ion velocity distributions, J. Geophys. Res., 81, 2145, 1976.
Watkins, D. S., Efficient initialization of stiff systems with one unknown initial condition, SIAM J. Numer. Anal., 18, 794, 1981a.
Watkins, D. S., Determining initial values for stiff systems of ordinary differential equations, SIAM J. Numer. Anal., 18, 13, 1981b.

AURORAL ZONE ION COMPOSITION

W. K. Peterson

Lockheed Palo Alto Research Laboratory, Palo Alto, California 94304

Abstract. Ion composition measurements provide valuable information about the sources, energization, transport, and loss mechanisms of space plasmas. In the past 10 years the initial series of plasma composition measurements from the S3-3, GEOS, ISEE, SCATHA, PROGNOZ, AUREOL-3, DE, and AMPTE satellites have provided insights which have significantly improved our understanding and models of the magnetosphere and ionosphere. In particular, we now know that the auroral and polar ionosphere are important sources of the magnetospheric plasma. We have learned that the ionospheric source of magnetospheric plasma varies systematically in its mass and energy composition with respect to local time, season of the year, and the solar cycle, as well as with respect to magnetic activity. There is also some evidence that variations in the ionospheric source composition can lead to the initiation of plasma sheet instabilities. Today we realize that auroral field lines are the sites of many different types of acceleration and energization processes, some of which are known to be mass dependent. Detailed measurements of ion distribution functions have been useful in exploring the importance of various ion energization and thermalization processes. Some recent observational results obtained from satellite-borne ion mass spectrometers that are relevant to magnetosphere/ionosphere plasma models will be discussed.

Introduction

Understanding the physical processes involved in creating the spectacular visual displays called the aurora has been a goal for several centuries. Since the start of the systematic exploration of the near earth plasma environment using rocket and satellite probes, we have made rapid progress toward this goal. The in situ measurements of ion mass, energy, and angle distributions on auroral field lines have provided important insights that have lead us to our present understanding of the ionosphere and magnetosphere and the models we use to characterize them. Some recent observations of auroral ion composition that are relevant to magnetospheric and ionospheric models will be discussed.

The paper is organized as follows: The discussion begins with a description of the auroral zone and its role as the region of exchange of warm and hot plasmas between the magnetosphere and ionosphere. The ion mass, energy, and angle distributions observed on a typical mid-altitude auroral zone crossing are then presented and discussed, followed by a discussion on the central role of oxygen ions as a tracer of the processes involved in coupling of ionospheric and magnetospheric plasmas. Rather than review the extensive statistical studies of the upflowing ionospheric ions on auroral field lines, which are discussed in detail elsewhere (Ghielmetti et al., 1978; Gorney et al., 1981; Yau et al., 1985a, 1987), discussions will be given on some recent observations that have been made using the very high resolution ion mass spectrometer data that have improved our understanding of magnetosphere/ionosphere interactions. The thermal (i.e., less than ~10 eV) ion plasma observed on auroral field lines is not emphasized here since it is discussed elsewhere (Yau and Lockwood, 1987; Schunk, 1987). It is also important to note that the Swedish VIKING satellite was successfully operated in 1986. The first series of papers from the VIKING high resolution plasma instruments were not available at the time this paper was written.

Plasma Exchange in the Auroral Zone

Plasma is continuously exchanged at all latitudes between the ionosphere and magnetosphere. In fact the often sharp gradients in the energy and mass composition of the plasma exchanged between the ionosphere and magnetosphere as a function of latitude are used to identify distinct regions of the magnetosphere. The auroral zone can be thought of as the region where plasma is exchanged between the plasma sheet and ionosphere. This exchange of plasma on auroral field lines is, of course, directly associated with the visible aurora. The relationship between the

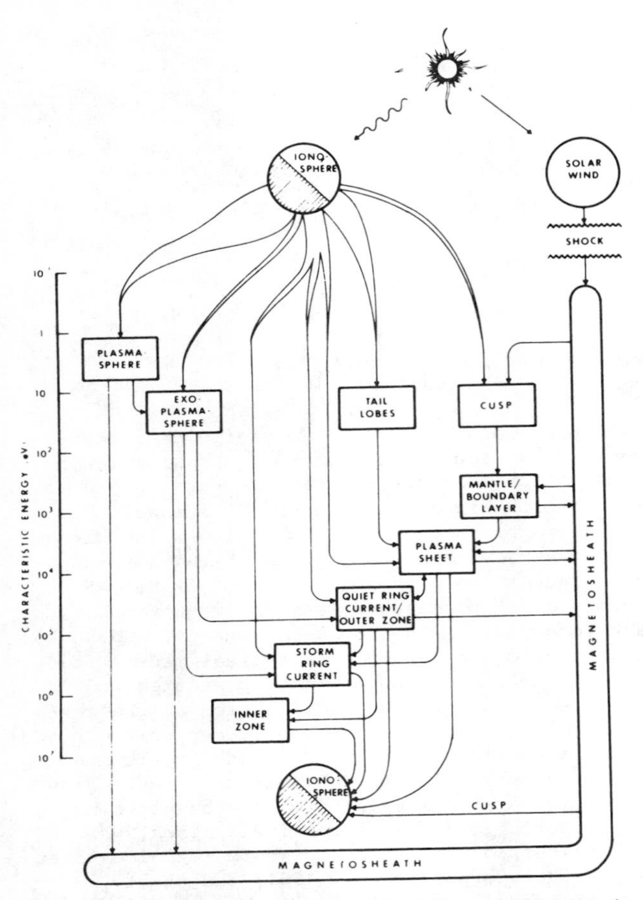

Fig. 1. Schematic relationship of magnetospheric particle populations from Young (1986). See text.

aurora, auroral zone particle populations, and magnetospheric plasma regions has been recently discussed by Feldstein and Galperin (1985) and Young (1983).

Figure 1 (reproduced from Young, 1986) schematically shows the relationship of magnetospheric particle populations to their source regions in the ionosphere and solar wind. The ionosphere and magnetosheath (solar wind) are shown as both sources (top) and sinks (bottom). The relative latitudes of ionospheric sources and sinks as well as the characteristic energies of the various particle populations are also indicated on this chart. At low magnetic latitudes there is the plasmaspheric (and exoplasmaspheric) exchange of thermal and warm plasmas of primarily ionospheric origin. Very high magnetic latitudes are characterized by the magnetospheric cusp, the escape of polar ionospheric plasma, and by the entry of solar wind (electron) plasma in the form of polar rain. The auroral zone is between the plasmasphere and magnetotail lobes (polar cap) and is characterized by heated and/or accelerated ionospheric plasma flowing up magnetic field lines, intermixed with trapped and precipitating energetic plasma sheet plasma.

Mass composition measurements in the auroral zone are used to sort out the complex physical processes that occur when cool ionospheric plasma and energetic plasma sheet plasma mix on auroral field lines. Because the energy and angular distributions of ions in the ionospheric and solar wind source regions are reasonably well known, and the result of plasma processes operating in the auroral zone is in general to modify the energy and angular distributions of the various plasma constituents, ion mass composition measurements can be a very powerful tool. Unfortunately the major species, hydrogen, is the dominant constituent of the solar wind (magnetosheath) and, above ~1000 km, is the dominant ionospheric ion. H^+ energy and angular distributions provide clues that can, at times, be used to identify its source. For example the relatively cool (~100 eV) H^+ beams moving up auroral field lines with streaming energies of ~1 keV could not have come from the magnetosheath or plasma sheet, while the very intense streams of H^+ with narrow energy distributions with energies of ~1 keV seen flowing downward in the cusp region (see, for example, Peterson, 1984) are definitely from the magnetosheath (solar wind).

Other singly charged ions encountered in the magnetosphere are generally of ionospheric origin; multiply charged ions are generally of solar wind origin. However, there are important qualifications to this simplified source identification rule. Young et al. (1977) identified an equatorial suprathermal plasma of ionospheric constituents that contains significant densities of low-energy, doubly charged ions, He^{++} and O^{++}. Chappell et al. (1982) also reported detectable fluxes of low-energy O^{++} and N^{++} flowing up auroral field lines.

The mass composition of both the solar wind and ionosphere is variable. The solar wind charge state distribution is one that is characteristic of variable solar coronal temperatures (see, for example, Bame et al., 1983). Helium is the most common minor constituent (typically a few percent of hydrogen) and oxygen the next most abundant (typically 10^{-4} as abundant as hydrogen). Other species are typically less common. Except for unusual impulsive events, solar wind helium is doubly charged. Ionospheric plasma composition is more variable than that of the solar wind; it depends on altitude, solar lighting (local time), and other energy inputs such as particle precipitation in the auroral region. Typical ionospheric constituents are H^+, He^+ (singly ionized), O^+, N^+, and NO^+.

As indicated in Figure 1, very little solar wind plasma flows directly from the magnetosheath onto auroral zone magnetic field lines. The trapped and precipitating energetic plasma sheet ions encountered in the auroral zone have resided for variable times in the plasma sheet after their entry from either the ionosphere or magnetosheath. In the plasma sheet, the energy, angle, and charge state distributions of the plasma are modified on various time scales, but a discussion of the processes acting in the plasma sheet is beyond the scope of this paper. (See, for example, the recent reviews by Cornwall, 1986, and Young, 1986.)

It is essential in constructing models of the magnetosphere to have an idea of the relative strengths of the ionospheric and solar wind contributions to magnetospheric plasmas. Shelley (1985, 1986) and Chappell et al. (1987) have recently addressed the question of the relative importance of what Shelley has called the "geogenic" and "heliogenic" plasma sources. Chappell et al. (1987) argued that the ionospheric source is fully adequate to explain observed plasma densities in all regions of the magnetosphere. In a recent presentation Shelley et al. (1986) extended the calculations presented in Shelley (1986) and calculated the ionospheric or geogenic fraction of the plasma sheet as a function of magnetic and solar activity. At high levels of magnetic and solar activity Shelley et al. (1986) found the plasma sheet to be dominated by ions of ionospheric origin and that most of these ions are transported along auroral field lines to the plasma sheet. Clearly, the composition of the magnetosphere is not static and models should reflect this.

Energetic Ion Plasma Typically Observed in the Mid-Altitude Auroral Zone

In addition to identifying the origin of the plasma, magnetospheric models must also concern themselves with the details of plasma transport. Ion composition measurements obtained on auroral field lines have provided several key insights into our current understanding of plasma transport in the magnetosphere. Figure 2 presents a typical example of the composition of the energetic ion plasma encountered in the evening auroral zone at a distance of about 3 earth radii (geocentric) by Dynamics Explorer 1 (DE 1). The data in Figure 2 were acquired on October 25, 1981, by the energetic ion composition spectrometer (EICS) (Shelley et al., 1981). The universal time (UT) and orbital information are displayed at the bottom. Figure 2 presents angle-time and energy-time spectrograms which are only briefly described here; a complete description of the spectrogram formats is given in Peterson and Shelley (1984). The top four angle-time spectrogram panels display hydrogen (first and third panels) and oxygen (second and fourth panels) number fluxes encoded in units of cm^{-2} s^{-1} sr^{-1} using the gray scale bar on the right. The number fluxes displayed in the top four panels are integrated over two different energy ranges (1 to 17 keV in the top two panels and 10 eV to 1 keV in the third and fourth panels) and are displayed as a function of instrumental look direction with respect to the satellite motion. In this spin-phase angle coordinate system, ions flowing up magnetic field lines from the ionosphere appear near 90°; ions flowing down magnetic field lines toward the ionosphere appear near 270°. The bottom two energy-time spectrogram panels display hydrogen and oxygen differential number fluxes for ions flowing upward with pitch angles that lie within 15° of the magnetic field direction. These differential fluxes are also encoded using the gray scale on the right, but the units are in cm^{-2} s^{-1} sr^{-1} keV^{-1}.

In Figure 2 the energy and angular distribution of the ion constituents of the plasma are seen to vary dramatically from the polar cap region before ~1700 UT to the plasmasphere ~1730 UT. In fact the particle distributions observed on DE 1 from ~1700 to ~1708 UT indicate that the satellite passed above the region of primary auroral acceleration. During this period relatively intense upward beams of kiloelectron volt ions were observed. (Particle beams are characterized by peaked energy distributions and narrow angular distributions centered on the magnetic field direction.) Electron data from the DE high altitude plasma instrument for this time presented by Persoon et al. (1987) show an energetic, quasi-isotropic electron distribution, similar to that found in the plasma sheet, rather than an electron inverted-V which is seen on auroral field lines below the primary auroral acceleration region. The rising and falling of the energy of the upward flowing hydrogen and oxygen beams from ~1700 to ~1708 UT (bottom two panels) reflects variations of the electrostatic potential seen by ions below the satellite. Ion beams accelerated upward and electron beams accelerated downward are in fact the basic, remote plasma signatures of the primary auroral acceleration region (see, for example, Sharp et al., 1983).

Not all of the ion distributions on auroral field lines shown in Figure 2 can be characterized as ion beams, however. Near 1708 and 1700 UT, the low-energy angular distributions in Figure 2 (center two panels) show the angular distribution peaking on either side of the magnetic field direction. These distributions have been called ion conic distributions and are indicative of ion accelerations process(es) acting perpendicular to the local magnetic field (see, for example, Sharp et al., 1977). Conic-type ion distributions are frequently found on the "edges" of ion inverted-V events, such as those shown in Figure 2, but are also observed in isolation, such as near 1650 UT in the data presented in Figure 2.

Auroral and sub-auroral field lines are also populated by energetic kiloelectron volt quasi-isotropic ions. The top two panels in Figure 2 show such distributions. After ~1712 UT, depletions of ions with pitch angles nearly aligned with the magnetic field direction are observed. (In the spin-phase angle coordinate system used in Figure 2, 90° corresponds to upflowing ions

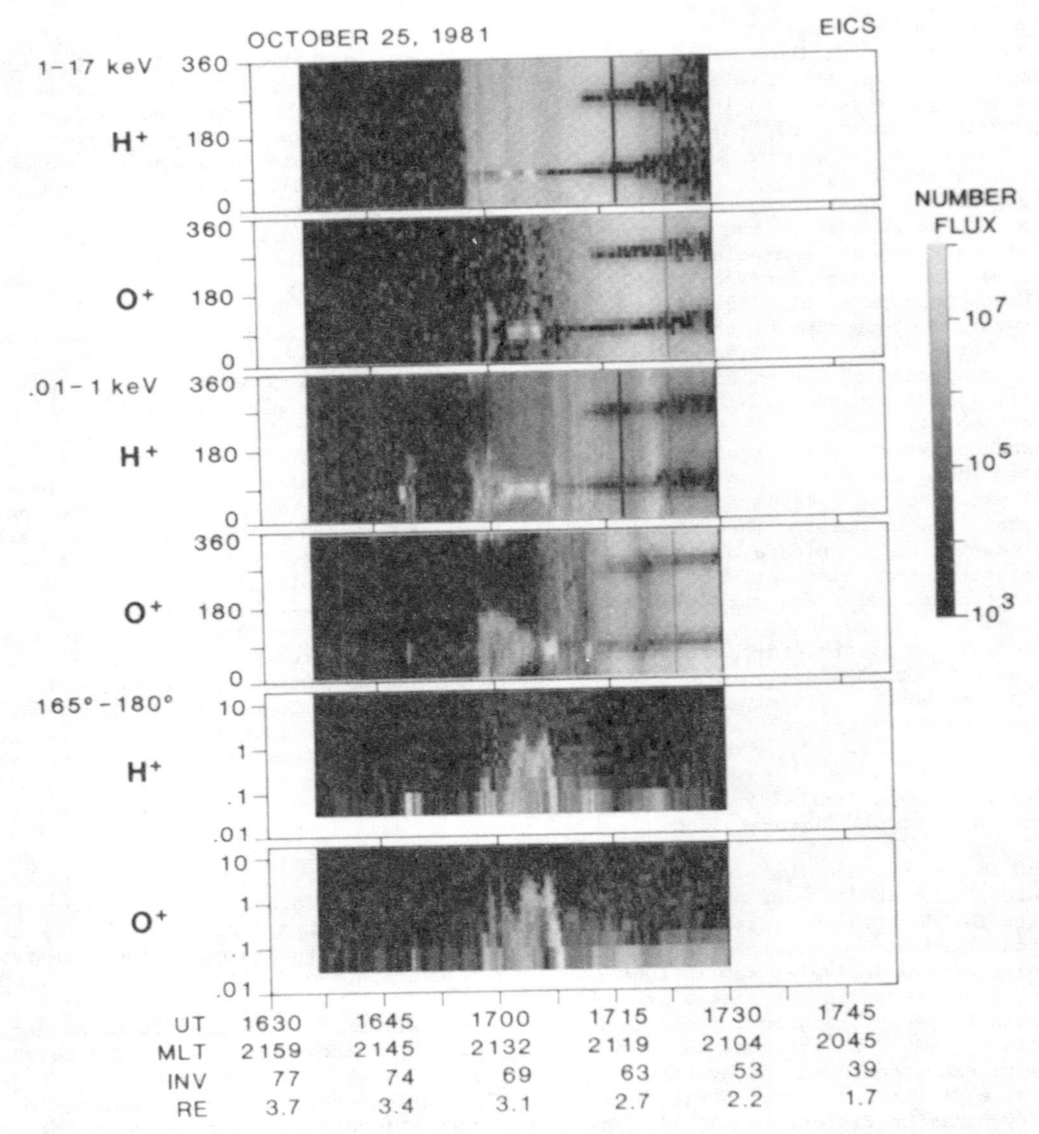

Fig. 2. Energetic ion composition data observed on mid-altitude auroral magnetic field lines from the DE 1 satellite on October 25, 1981. The universal time (UT) and orbital parameters, magnetic local time (MLT), invariant latitude (INVL), and geocentric distance in earth radii (R_E) are displayed given at 15-min intervals at the bottom. Angle-time spectrograms for two energy ranges are presented in the top four panels and energy-time spectrograms for hydrogen and oxygen ions flowing upward with pitch angles within 15° of the magnetic field are displayed in the bottom two panels. The spectrogram format is further described in the text.

with a pitch angle of 180°, and a 270° spin-phase angle corresponds to downflowing ions with pitch angles near 0°.) These depletions are the well-known particle "loss cones" caused by scattering and loss of ions in the ionosphere. Particles with pitch angles away from the loss cones mirror above the ionosphere and so are trapped. The upward accelerated ion beams seen from ~1700 to ~1708 UT appear in the downward looking loss cone, indicating their ionospheric origin.

The upward looking loss cone (i.e., the one at 270° spin-phase angle in Figure 2) is the result of ions being lost in the magnetically conjugate hemisphere, while the downward looking loss cone monitors particle losses as they mirror in the ionosphere below the satellite. From ~1700 to ~1712 UT, the upward looking loss cone is clearly filled. Our current understanding is that the ions are scattered into pitch angles within the loss cone by plasma waves as they follow magnetic

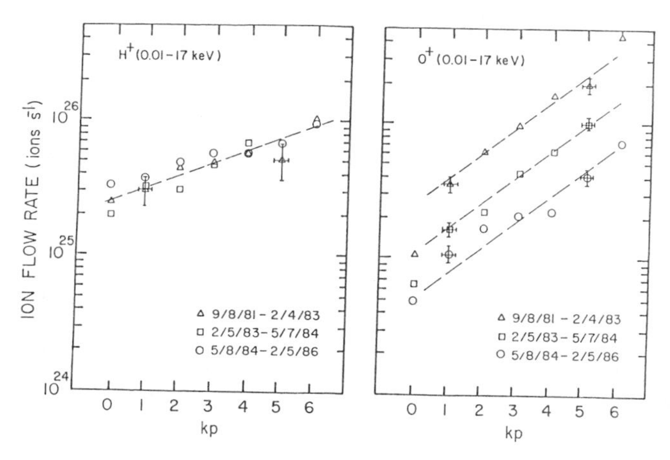

Fig. 3. Average ion outflow from the earth's auroral zones and polar caps observed from the DE 1 satellite for three time periods during the declining phase of the current solar cycle. Hydrogen data on the left and oxygen data on the right are presented as a function of the magnetic activity index Kp.

field lines between hemispheres, but as discussed below we really do not fully understand the processes involved.

The Excess Oxygen Problem

The intense fluxes of oxygen flowing up magnetic field lines out of the ionosphere first observed 10 years ago were unexpected. Subsequent investigation of the morphology of upflowing ions established that there were at least two broad classes of ion acceleration mechanisms operating on auroral field lines: one class accelerates ions primarily parallel to the local magnetic field and another class imparts energy to ions perpendicular to the magnetic field. The region of primary auroral acceleration parallel to the magnetic field was found to be at an altitude of 5000 to 10,000 km, while evidence for acceleration of ions perpendicular to the magnetic field line has been found at altitudes ranging from 400 km to deep in the magnetotail. These results have been discussed in detail by Sharp et al. (1983) and Johnson (1983). The beams of oxygen observed streaming parallel to the magnetic field above about 5000 km have on average comparable oxygen and hydrogen fluxes. However, ionospheric models predict that the hydrogen is the dominant thermal ion with little or no thermal oxygen present in the region of parallel acceleration above 5000 km. To account for the high-altitude parallel acceleration of large numbers of oxygen ions, a low-altitude (<1000 km) process is required to transport sufficient oxygen to the region of parallel acceleration. Experimental and theoretical work has now firmly established the existence of such low-altitude processes, although no single mechanism has been identified as the source of all or even a major part of the low-altitude preheating (see, for example, Whalen et al., 1978; Moore, 1980; Hultqvist, 1983; Moore et al., 1986; Chang et al., 1986).

One of the consequences of an ionospheric source for magnetospheric oxygen is a variation of oxygen ion outflow during the solar cycle. Young et al. (1982) demonstrated a long-term increase in oxygen ions near geosynchronous altitude with no associated increase in hydrogen ions during the rising phase of the current solar cycle. They suggested that these results are expected from the known increase in oxygen scale height in the ionosphere with solar activity if the auroral ionosphere is the source of magnetospheric oxygen. Yau et al. (1985a,b, 1987) investigated the long-term variation in outflowing hydrogen and oxygen ions from auroral and polar regions using data obtained since 1981 by the DE satellite and confirmed this hypothesis.

Figure 3 presents the observed hydrogen and oxygen ion outflow rates as a function of the magnetic activity index, Kp, obtained from three consecutive periods during the declining phase of the current solar cycle from DE. These data were compiled by Yau et al. (1987) and the method used is described there. The total ion outflow rate for both hydrogen and oxygen is seen to increase with the magnetic activity index, Kp. The total oxygen ion outflow rate retained approximately the same magnetic activity dependence, but the level observed in the 1984 to 1986 period had decreased by at least a factor of 5 from the near solar maximum level observed from 1981 to 1983. Young et al. (1982) noted that the daily solar radio flux index, $F_{10.7}$, is a good indicator of the total solar ultraviolet flux and is therefore the appropriate index with which to monitor solar cycle variations. Yau et al. (1987) extended their earlier report (Yau et al., 1985b) to include data obtained through the current solar minimum in early 1986 and have presented the data shown here in Figure 3 as a function of the solar radio flux index, $F_{10.7}$. They have shown that from minimum to maximum values of the solar radio flux index, at a given level of magnetic activity, the total oxygen outflow increases by a factor of ~5 while the hydrogen outflow rate decreases by less than a factor of 2. The result is that there is a significant variation in the average oxygen-to-hydrogen ratio of upflowing ions over the solar cycle.

In addition to the variation in mass composition of ionospheric ion outflow with solar cycle discussed above there is experimental evidence that suggests that the altitude of the primary auroral acceleration region rises with increasing solar activity and that microphysical processes on auroral field lines are modulated by the ionospheric response to variations in the input of solar energy. Ghielmetti et al. (1984) found that the frequency of occurrence of upward streaming hydrogen and oxygen ions with energies

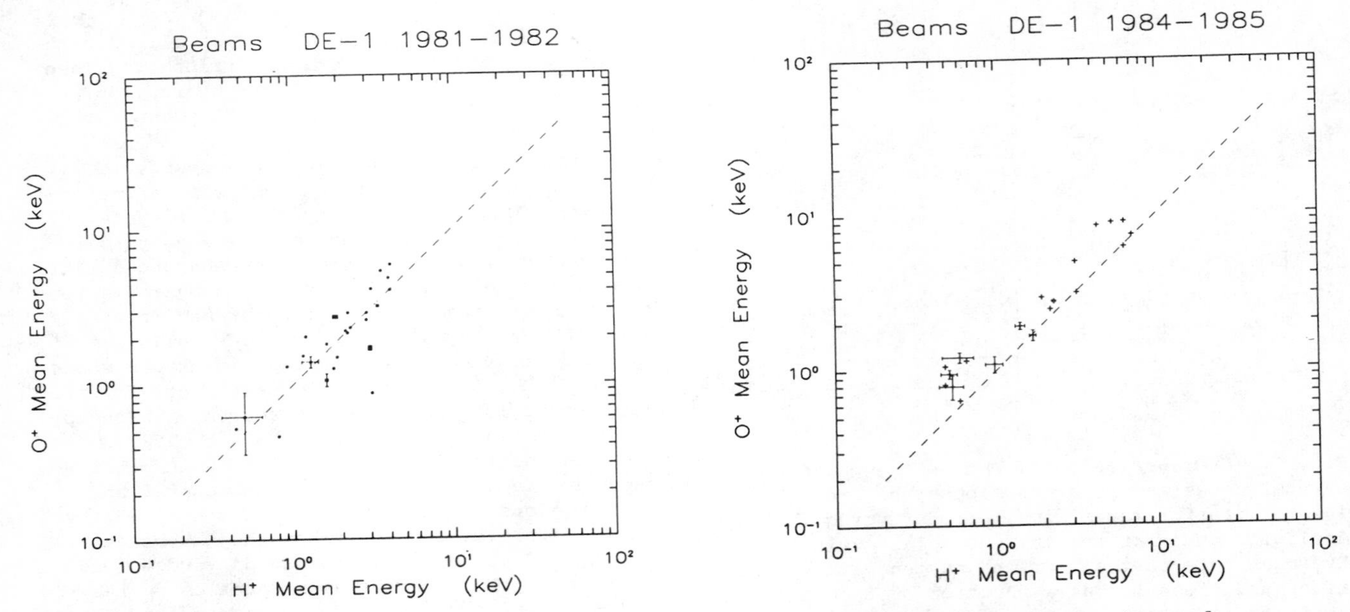

Fig. 4. Average energy of selected upflowing hydrogen and oxygen ion beams observed on the DE 1 satellite in two intervals. 1981-1982 was a period of high solar activity and 1984-1985 was a period of low solar activity. The average energy of oxygen beams is seen to be higher than that of the hydrogen beams during the 1984-1985 interval (from Collin et al., 1987).

greater than 500 eV and at altitudes less than 8000 km by the S3-3 satellite decreased from 1976 to 1979. Ghielmetti et al. (1984) also noted systematic changes in the composition, energy, and pitch angle characteristics of these upward streaming ions over this period of rising solar activity. During the same general time period (1979) of decreased S3-3 observations of upward flowing ions, field-aligned hydrogen and oxygen structures were commonly observed near geosynchronous orbit by the SCATHA satellite implying continuing injection from the ionosphere. The fluxes and energies of the ions detected by the SCATHA instrument were generally sufficient to have been detectable at the S3-3 orbit. Ghielmetti et al. (1984) concluded that these observations were consistent with the displacement of the principal auroral acceleration region to altitudes above 8000 km near the peak of the solar cycle.

There are some surprising consequences that follow from the variation in the number and energy of oxygen ions leaving the ionosphere. Measurements from the S3-3 satellite in 1976 and 1977 showed that oxygen beams observed above 5000 km had, on average, 1.7 times the energy of simultaneously observed hydrogen beams (Collin et al., 1981). Observations at higher altitude by Lundin et al. (1982) made in 1978 showed that both oxygen and the singly charged helium ion beams had more energy than accompanying hydrogen beams. These results contrast with the results of Reiff et al. (1987) who studied several beam events and reported that oxygen and hydrogen beams had nearly the same energy. The Reiff et al. observations were made in 1981, near solar maximum, from DE 1. Reiff et al. (1987) also showed that, in their beam events, there was a transfer of ion beam "streaming" energy into increased random thermal energy in the streaming frame of reference. Collin et al. (1987) examined the relative energies of many simultaneously observed oxygen and hydrogen beams observed by S3-3 at solar minimum and DE 1 at both solar maximum (1981-1982) and in the declining phase of the current solar cycle (1984-1985). Figure 4, reproduced from their paper, shows that the relative energies of the hydrogen and oxygen components of upflowing ion beams for the two periods are different. It shows that near solar maximum the oxygen and hydrogen ion beam components had comparable mean energies. Near the 1976-1977 solar minimum, Collin et al. (1986a) found that oxygen beams had more energy than could have been acquired from the potential drop determined from the simultaneously observed electron loss cone widening. No simultaneous electron loss cone measurements are available during the 1984-1985 period, but oxygen beams were found to have more energy than could have been acquired from the potential drop determined from the hydrogen beam energy.

Collin et al. (1987) suggested that the observations summarized in Figure 4 are consistent with the ion beams being partially thermalized through the two-stream instability between hydro-

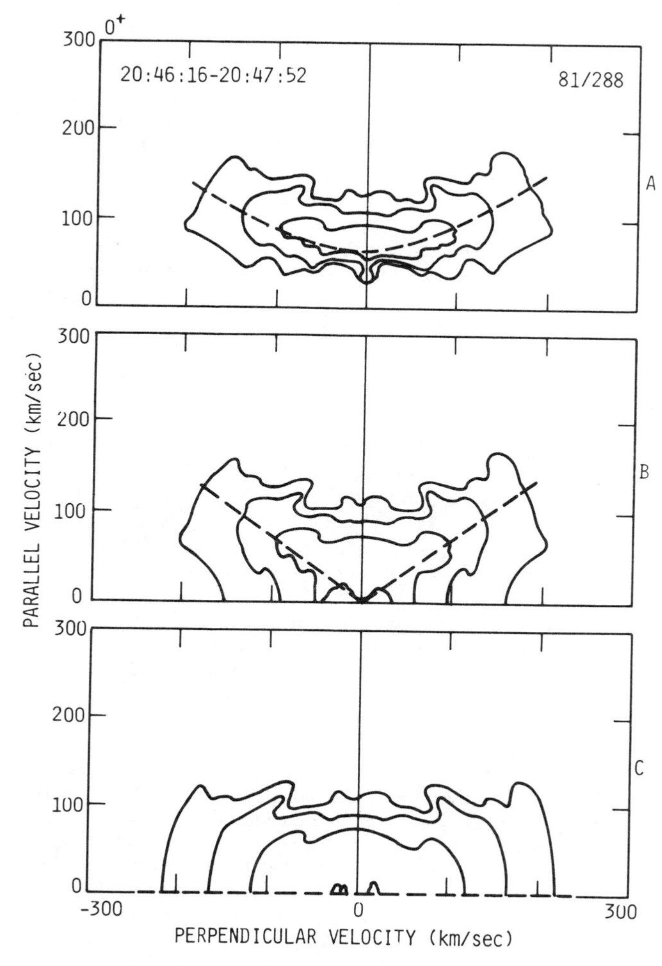

Fig. 5. Velocity space density contours reproduced from Klumpar et al. (1984). See text.

gen and oxygen with the effect being modulated by the beam composition. The two-stream instability arises because of the velocity difference between hydrogen and oxygen beams which, initially, acquire the same energy when they pass through a magnetic field-aligned potential drop (Kaufmann et al., 1986; Bergmann and Lotko, 1986). Collin et al. (1986b) examined the multi-species dispersion relation and showed how it is possible for oxygen to acquire extra energy from hydrogen beams with higher velocities when oxygen is a minor species.

Transverse Acceleration and Ion Conics

The early results from the S3-3 satellite (Sharp et al., 1977) clearly established that energization of ions perpendicular to the local magnetic field is an important process in the coupling of ionospheric and magnetospheric plasmas. Since this initial recognition, considerable work on both the macro- and microphysical aspects of perpendicular acceleration processes has been done. Klumpar (1986) compiled a comprehensive bibliography on this subject. It is beyond the scope of this report to discuss all aspects of transverse acceleration; instead discussions will concentrate on recent results that suggest that the expected ion energy-angle "signature" of transverse acceleration may not be unique and that more than one microphysical process is important for imparting transverse energy to significant numbers of ions on auroral field lines.

The starting point for this discussion is to note the extreme difficulty experimentalists have had in finding regions where transverse ion energization is the dominant process. The expected "signature" of such regions is the simultaneous occurrence of intense low-frequency plasma waves and energetic ions with their energy primarily perpendicular to the local magnetic field, a so-called 90° conic distribution. Kintner and Gorney (1984) were able to find only one such example in the entire S3-3 data set. Whalen et al. (1978), Yau et al. (1983), and Kintner et al. (1986) have been remarkably successful in observing the so-called 90° conics at ionospheric altitudes from rockets. However, mapping ion pitch angle distributions from ion conic distributions measured at satellite altitudes indicates that almost all such conical distributions have a source region at altitudes well above the ionosphere. This point is illustrated in Figure 5 from Klumpar et al. (1984). The top panel of Figure 5 presents the observations made at an altitude of ~22,000 km and shows contours of the logarithm of observed velocity space density. Considering, for the moment, the pitch angles of only the most energetic particles, Klumpar et al. showed that conservation of the first adiabatic invariant in the earth's magnetic field indicates that the energetic ions would have had 90° pitch angles where the strength of the magnetic field was ~1.49 times larger, or at ~18,200 km.

More importantly, because Klumpar et al. (1984) had observations of ion velocity distributions to low energies (i.e., ~50 eV), they were able to show, as indicated in the second two panels of Figure 5, that the observed ion distribution is inconsistent with simple adiabatic transport of ions from a source at 18,200 km to the observation altitude of ~22,000 km. The interpretation implied in Figure 5 is that a parallel acceleration of 310 eV and adiabatic transport, a two-step or bimodal process, is responsible for producing the observed ion distribution. There have been three other suggestions of processes that could produce ion distributions such as those shown in Figure 5. Chang et al. (1986) and Temerin and Roth (1986) showed how such distributions can be caused by a relatively weak transverse acceleration mechanism acting over a range of altitudes. Retterer et al. (1987) extended the work of Chang et al. and were able to reproduce the type of ion distribution discussed by Klumpar et al. (1984). It should also be noted that Horwitz (1986) has shown that some of the features of the observed ion distribution in Figure 5 can be explained by a velocity filter mechanism (i.e., simply particle kinematics).

The point of the above discussion is to show that our understanding of the microphysical processes that are responsible for the exchange of ionospheric and magnetospheric plasmas, in spite of the large number of papers devoted to the subject, is still incomplete and evolving. Lennartsson (1983) provided a short introduction to and catalog of the types of acceleration mechanism that have been considered to be operating in the auroral region. It is important to realize that most of the mechanisms cataloged by Lennartsson probably do occur on auroral field lines and it is difficult to distinguish among them using state-of-the-art in situ plasma diagnostics.

Ion Velocity Dispersion Events

In addition to ion "beams" and "conics," there is a third type of ion energy-pitch angle distribution observed on auroral field lines that has been less intensively studied. As noted below, some of these distributions are unstable and can lead to the emission of plasma waves. Observable velocity dispersion effects can be seen after the introduction in a localized region and/or in a short period (compared to ion travel times) of a plasma population with a rather broad energy and/or pitch angle distribution. The difference in travel times caused by the different spiral path distances traveled by ions with different pitch angles to a remote location leads to unique energy-pitch angle, or energy-latitude distributions at locations well removed from the source.

The first intensively studied ion velocity dispersion events were observed in the low-latitude magnetospheric cusp (Shelley et al., 1976; Reiff et al., 1977). In the cusp there is localized entry of a broad energy and pitch angle spectrum of magnetosheath ions and a well defined anti-sunward drift (independent of energy) of ions as they move down to the ionosphere. The spread in ion velocities parallel to cusp field lines coupled with anti-sunward drift, independent of parallel velocity, results in the characteristic energy-latitude signature identified by Shelley et al. (1976) and Reiff et al. (1977). At higher altitudes in the cusp Burch et al. (1982) showed that a characteristic "V" in ion energy-pitch angle distributions is explained by the same kinematic effects. Gorney (1983) noted that these downstreaming cusp ion distributions have a characteristic ring or shell shape in velocity space that could lead to plasma wave instabilities. Subsequently Roth and Hudson (1985) showed there is enough energy in these distributions to support significant ion heating.

A second velocity dispersion effect has also been found in the cusp region. The Marshall group has recently found in their DE 1 retarding ion mass spectrometer (RIMS) data an interesting and unexpectedly intense source of low-energy ion outflow from a limited latitudinal region in the dayside cusp/cleft which they have called the "upwelling ion region" (see, for example, Lockwood et al., 1985a). This spatially restricted low-energy ion source and the normal anti-sunward high-latitude plasma convection pattern result in effects which have been called a "fountain in a wind" (Lockwood et al., 1985a) where the upwelling ion energies are small compared to the gravitational potential energy at high altitudes, or a "geomagnetic ion mass spectrometer" (Lockwood et al., 1985b) where the energy of the upflowing low-energy ions is slightly higher. The further acceleration of some of these ions on auroral and cusp field lines by a mechanism such as that recently proposed by Cladis (1986) suggests that this upwelling ion region could be an important source of low-energy (~10-50 eV) plasma sheet ions (Chappell et al., 1987).

Ion velocity dispersion signatures not associated with the dayside cusp or cleft region have also been reported on auroral magnetic field lines. Figure 6 is from a presentation made by Klumpar et al. (1983) and shows in the middle panel an oxygen energy-pitch angle velocity dispersion feature. The data presented in Figure 6 were obtained by the EICS instrument on DE 1 and are in the form of mass resolved energy-time spectrograms. Data for about 7 min on July 14, 1982, when DE 1 was on mid-altitude ($R/R_E \sim 4$) auroral field lines ($L \sim 5.6$) near 1600 magnetic local time are shown. The top two panels display mass-resolved hydrogen and oxygen ion count rate presented at the highest time resolution encoded with the gray bar shown on the right. The EICS instrumental response function is such that the counting rate is approximately proportional to number flux. The bottom panel, labeled ED, is from an ion detector sensitive to all ion species. The EICS mode used during the interval presented in Figure 6 obtained complete pitch angle coverage for oxygen and hydrogen ions over the energy range (10 eV to 17 keV) every 24 s. The data in Figure 6 are displayed in pitch angle order for each measurement cycle instead of in strict time order. The pitch angle characteristic of each energy spectrum is displayed as a function of time in the bottom panel. Klumpar et al. (1983) analyzed the oxygen energy-time-pitch angle signature in Figure 6 and have found that it is consistent with an impulsive isotropic equatorial source with a broad energy and pitch angle spectrum which occurred for a short (compared to the ion bounce time) period at 23:26:25. Note that the low-energy trace (i.e., the slow ions) have come directly from the equator, while the higher energy dispersion signature starting near 23:34 comes from ions which have traveled through the opposite hemisphere, mirrored, and

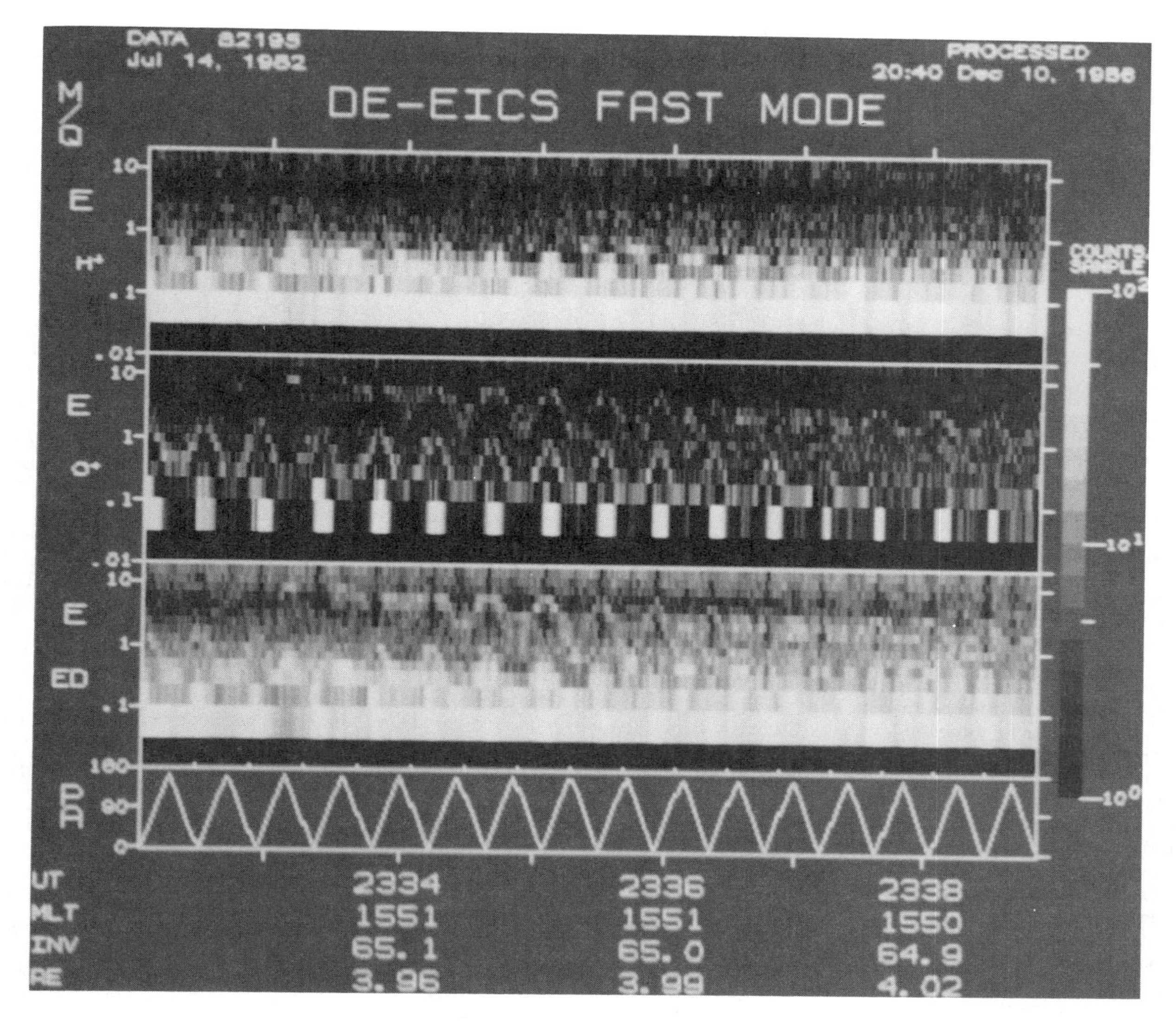

Fig. 6. Energy-time spectrograms of hydrogen (top panel) and oxygen (middle panel) illustrating an oxygen ion dispersion event. See text.

are coming back down magnetic field lines. Quinn and McIlwain (1979) reported the not infrequent occurrence of ion energy pitch angle dispersion signatures consistent with those shown in Figure 6 at geosynchronous altitudes. They analyzed the ion dispersion signatures for a number of events and found two classes of dispersion signature, one consistent with an equatorial source and the other with an ionospheric source for the initial population responsible for what they have called "bouncing ion clusters." In a subsequent paper Quinn and Southwood (1982) proposed a parallel ion acceleration mechanism for the equatorial ion source.

Quinn and McIlwain (1979) noted that a very large fraction of the "bouncing ion clusters" they observed survived multiple reflections at low altitudes with very little reduction in intensity. At the time of their observations this was consistent with the lack of detection of significant numbers of downflowing ion events (i.e., ion beams precipitating into the ionosphere) by instrumentation on the S3-3 satellite (Ghielmetti et al., 1979). Since that time there have been reports from the DE 2 and AUREOL-3 satellites of significant numbers of downflowing ion events (Winningham et al., 1984; Frahm et al., 1986; Bosqued et al., 1986).

Figure 7, reproduced from Bosqued et al. (1986), illustrates the observational details and interpretation of a downward flowing ion event observed on AUREOL-3 on June 27, 1982. The observations are of a precipitating ion distribution with a maximum intensity at an energy that

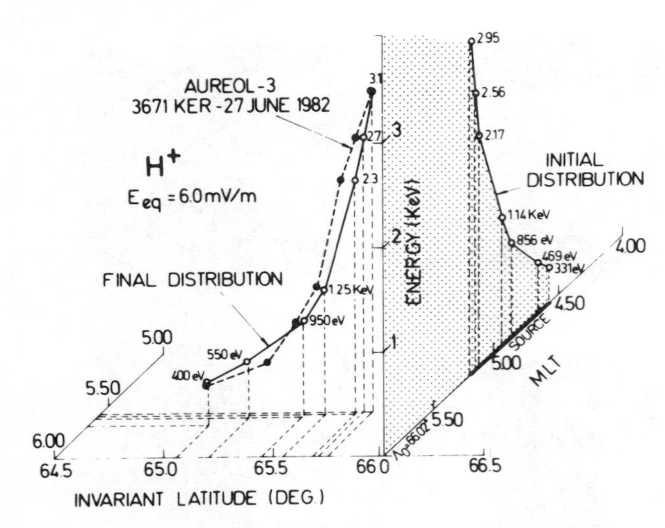

Fig. 7. Schematic representation of the observed energy-latitude dispersion (left) of an ion band observed from the AUREOL-3 satellite on June 27, 1982. The inferred source distribution is shown in the right-hand panel (from Bosqued et al., 1986).

systematically decreases with invariant latitude as the satellite moves across auroral field lines. The curve labeled "final distribution" in Figure 7 shows the observed peak energy as a function of invariant latitude. The interpretation is that the final energy-latitude distribution is the result of the transport of ions accelerated from a range of local times at more or less the same invariant latitude in the magnetically conjugate hemisphere, including the effects of ion drift in a uniform dawn-dusk electric field. The curve labeled "initial distribution" in Figure 7 shows the ion energies as a function of magnetic local time inferred using this interpretation.

Bosqued et al. (1986) had a data base consisting of 600 auroral zone crossings in 1982 and 1983. They found ion energy-latitude dispersion features similar to the one shown in Figure 7 in 68 auroral crossings (~11%). Frahm et al. (1986) analyzed 227 low-altitude auroral crossings of the DE 2 satellite from the fall of 1981 and found 40 such signatures (~18%). This contrasts with the results of Ghielmetti et al. (1979) who found only 20 downward flowing ion events in 570 auroral crossings of the S3-3 satellite between 1976 and 1979 (~4%). The relatively low number of downward flowing ion events seen by the S3-3 satellite might be explained by the large differences in sensitivity and energy threshold between the S3-3 instrument and the AUREOL-3 and DE 2 instruments. It is interesting to note however, that the DE 2 observations were obtained closer to the recent solar maximum than either the S3-3 or AUREOL-3. The ATS 6 results of Quinn and McIlwain (1979) were also obtained during relatively low levels of solar activity. It is, of course, not possible to demonstrate a solar cycle variation in the occurrence frequency of downward flowing ion events from the data discussed here, but since Yau et al. (1985a,b, 1987) have demonstrated a solar cycle dependence in upflowing ion events, a solar cycle dependence to the observation of downward flowing ion events would not be surprising.

The above discussion shows a large gap in our understanding of the physical processes involved in the coupling of ionospheric and magnetospheric plasmas. Ionospheric ions stream out of the ionosphere as collimated beams or, in the absence of other interactions, become collimated beams near the equator as the strength of the magnetic field decreases because of the conservation of the first adiabatic invariant. At some time in their transport through the plasma sheet, these directed ion streams are converted into the hot quasi-isotropic plasma sheet population. We do not understand, at this time, the processes occurring on auroral field lines in the plasma sheet well enough to be able to explain why some streams of ions retain their identity and perform multiple bounces, others precipitate into the ionosphere, and still others are pitch-angle scattered to become the quasi-isotropic plasma sheet population. Perhaps the key to understanding these complicated relationships will come from the detailed studies of counterstreaming ion populations observed on mid-altitude auroral field lines (see, for example, Horita et al., 1987; Sagawa et al., 1987) and near the equator (see, for example, Quinn and Johnson, 1985).

Concluding Remarks

In situ plasma diagnostics and particular energetic ion composition measurements on auroral field lines in the past 10 years have greatly improved our understanding of the magnetosphere and the ionosphere and their interactions, but we are still far from a complete picture. Current models of the magnetosphere and ionosphere are starting to reflect the importance and dynamic nature of the ionospheric contribution to magnetospheric plasmas. Our present level of understanding is illustrated by considering the recent suggestion by Baker et al. (1982) that variations in the ionospheric source composition can lead to the initiation of plasma sheet instabilities. We have observations that lead to such suggestions that large-scale physical processes are important in coupling ionospheric and magnetospheric plasmas, but we cannot confirm or rule out these suggestions because of the very global nature of the processes and the very limited, in space and time, simultaneous observational data available to test them. On a more fundamental level, the recent discovery of the frequent occurrence of cross polar cap emission features, the so-called "theta aurora," during extended periods of northward interplanetary magnetic field (see, for example, Frank et al., 1986; Peterson and Shelley, 1984) has raised some basic questions about our understanding of the topology of the magnetosphere.

On auroral field lines we also have a very incomplete understanding of the relative importance of the many possible mechanisms that could be involved in the transport of atomic oxygen from the ionosphere to the magnetosphere and why some ion populations pass through the equatorial magnetosphere to precipitate in the magnetically conjugate hemisphere and some populations become trapped.

Until we can answer questions such as these, there will have to be doubts about the generality of any empirical models we use to describe the magnetosphere/ionosphere system. We are making progress, however. The Dynamics Explorer project has illustrated the importance and difficulty of looking at the complete range of in situ parameters. The continued analysis of the high resolution Dynamics Explorer data, the very recent successful operation of the Swedish VIKING satellite with its complement of high resolution plasma instruments, and multi-satellite collaboration such as the recently initiated PROMIS campaign (Hones, 1985) should continue to provide new insights into both the micro- and macrophysics involved in the interaction of magnetospheric and ionospheric plasmas on auroral field lines.

Acknowledgments. Thanks go to D. Klumpar, A. Ghielmetti, H. Collin, J. Drake, and A. Yau for their helpful comments. This work has been supported by NASA under contracts NAS5-28710 and NAS8-36323.

References

Baker, D. N., E. W. Hones, Jr., D. T. Young, and J. Birn, The possible role of ionospheric oxygen in the initiation and development of plasma sheet instabilities, Geophys. Res. Lett., 9, 1337, 1982.

Bame, S. J., W. C. Feldman, J. T. Gosling, D. T. Young, and R. D. Zwickl, What magnetospheric workers should know about solar wind composition, in Energetic Ion Composition in the Earth's Magnetosphere, edited by R. G. Johnson, p. 73, Terra Scientific Publ. Co., Tokyo, 1983.

Bergmann, R., and W. Lotko, Transition to unstable flow in parallel electric fields, J. Geophys. Res., 91, 7033, 1986.

Bosqued, J. M., J. A. Sauvaud, K. Delcourt, and R. A. Kovrazhkin, Precipitation of suprathermal ionospheric ions accelerated in the conjugate hemisphere, J. Geophys. Res., 91, 7006, 1986.

Burch, J. L., P. H. Reiff, R. A. Heelis, J. D. Winningham, W. B. Hanson, C. Gurgiolo, J. D. Menietti, R. A. Hoffman, and J. N. Barfield, Plasma injection and transport in the mid-altitude polar cusp, Geophys. Res. Lett., 9, 921, 1982.

Chang, T., G. B. Crew, N. Hershkowitz, J. R. Jasperse, J. M. Retterer, and J. D. Winningham, Transverse acceleration of oxygen ions by electromagnetic cyclotron resonance with broad band left-hand polarized waves, Geophys. Res. Lett., 13, 636, 1986.

Chappell, R. C., C. R. Olsen, J. L. Green, J.F.E. Johnson, and J. H. Waite, Jr., The discovery of nitrogen ions in the earth's magnetosphere, Geophys. Res. Lett., 9, 937, 1982.

Chappell, C. R., T. E. Moore, and J. H. Waite, Jr., The ionosphere as a fully adequate source of plasma for the earth's magnetosphere, J. Geophys. Res., 92, 5896, 1987.

Cladis, J. B., Parallel acceleration and transport of ions from polar ionosphere to plasma sheet, Geophys. Res. Lett., 13, 893, 1986.

Collin, H. L., R. D. Sharp, E. G. Shelley, and R. G. Johnson, Some general characteristics of upflowing ion beams and their relationship to auroral electrons, J. Geophys. Res., 86, 6820, 1981.

Collin, H. L., E. G. Shelley, A. G. Ghielmetti, and R. D. Sharp, Observations of transverse and parallel acceleration of terrestrial ions at high latitudes, in Ion Acceleration in the Magnetosphere and Ionosphere, Geophys. Monogr. Ser., vol. 38, edited by T. Chang, p. 67, AGU, Washington, D.C., 1986a.

Collin, H. L., E. G. Shelley, and R. Bergmann, The heating of upflowing ion beams by an H^+/O^+ two-stream instability (abstract), Eos Trans. AGU, 67, 1164, 1986b.

Collin, H. L., W. K. Peterson, and E. G. Shelley, Solar cycle variation of some mass dependent characteristics of upflowing beams of terrestrial ions, J. Geophys. Res., 92, 4757, 1987.

Cornwall, J. M., Magnetospheric ion acceleration processes, in Ion Acceleration in the Magnetosphere and Ionosphere, Geophys. Monogr. Ser., vol. 38, edited by T. Chang, p. 3, AGU, Washington, D.C., 1986.

Feldstein, Y. I, and Yu. I. Galperin, The auroral luminosity structure in the high-latitude upper atmosphere: Its dynamics and relationships to the large-scale structure of the earth's magnetosphere, Rev. Geophys., 23, 217, 1985.

Frahm, R. A., P. H. Reiff, J. D. Winningham, and J. L. Burch, Banded ion morphology: Main and recovery storm phases, in Ion Acceleration in the Magnetosphere and Ionosphere, Geophys. Monogr. Ser., vol. 38, edited by T. Chang, p. 98, AGU, Washington, D.C., 1986.

Frank, L. A., J. D. Craven, D. A. Gurnett, S. D. Shawhan, D. R. Weimer, J. L. Burch, J. D. Winningham, C. R. Chappell, J. H. Waite, R. A. Heelis, N. C. Maynard, M. Sugiura, W. K. Peterson, and E. G. Shelley, The theta aurora, J. Geophys. Res., 91, 3177, 1986.

Ghielmetti, A. G., R. G. Johnson, R. D. Sharp, and E. G. Shelley, The latitudinal, diurnal, and altitudinal distributions of upward flowing energetic ions of ionospheric origin, Geophys. Res. Lett., 5, 59, 1978.

Ghielmetti, A. G., R. D. Sharp, E. G. Shelley, and R. G. Johnson, Downward flowing ions and evidence for injection of ionospheric ions into the plasma sheet, J. Geophys. Res., 84, 5781, 1979.

Ghielmetti, A. G., J. M. Quinn, E. G. Shelley, R. D. Sharp, and R. G. Johnson, Decreased auroral acceleration below ~2 R_E altitude--A solar cycle effect? (abstract), Eos Trans. AGU, 65, 1061, 1984.

Gorney, D. J., An alternative interpretation of ion ring distributions observed by the S3-3 satellite, Geophys. Res. Lett., 10, 417, 1983.

Gorney, D. J., A. Clark, D. Croley, J. Fennell, J. Luhmann, and P. Mizera, The distribution of ion beams and conics below 8000 km, J. Geophys. Res., 86, 83, 1981.

Hones, E. W., Jr., Project PROMIS to coordinate satellites (abstract), Eos Trans. AGU, 66, 1369, 1985.

Horita, R. E., E. Ungstrup, E. G. Shelley, R. R. Anderson, and R. J. Fitzenreiter, Counter-streaming ion (CSI) events in the magnetosphere, J. Geophys. Res., in press, 1987.

Horwitz, J. L., Velocity filter mechanism for ion bowl distributions (bimodal conics), J. Geophys. Res., 91, 4513, 1986.

Hultqvist, B., On the origin of the hot ions in the disturbed dayside magnetosphere, Planet. Space Sci., 31, 173, 1983.

Johnson, R. G., The hot ion composition, energy, and pitch angle characteristics above the auroral zone ionosphere, in High-Latitude Space Plasma Physics, edited by B. Hultqvist and T. Hagfors, p. 271, Plenum Publ. Co., New York, 1983.

Kaufmann, R. L., G. R. Ludlow, H. L. Collin, W. K. Peterson, and J. L. Burch, Interaction of up-going auroral H^+ and O^+ beams, J. Geophys. Res., 91, 1080, 1986.

Kintner, P. M., and D. J. Gorney, A search for the plasma processes associated with perpendicular ion heating, J. Geophys. Res., 89, 937, 1984.

Kintner, P. M., J. LaBelle, W. Scales, A. W. Yau, and B. A. Whalen, Observations of plasma waves within regions of perpendicular ion acceleration, Geophys. Res. Lett., 13, 1113, 1986.

Klumpar, D. M., A digest and comprehensive bibliography on transverse auroral ion acceleration, in Ion Acceleration in the Magnetosphere and Ionosphere, Geophys. Monogr. Ser., vol. 38, edited by T. Chang, p. 389, AGU, Washington, D.C., 1986.

Klumpar, D. M., W. K. Peterson, E. G. Shelley, and J. M. Quinn, Localized magnetospheric ion injection outside the cusp (abstract), Eos Trans. AGU, 64, 297, 1983.

Klumpar, D. M., W. K. Peterson, and E. G. Shelley, Direct evidence for two-stage (bimodal) acceleration of ionospheric ions, J. Geophys. Res., 89, 10,779, 1984.

Lennartsson, W., Ion acceleration mechanisms in the auroral regions, general principles, in Energetic Ion Composition in the Earth's Magnetosphere, edited by R. G. Johnson, p. 23, Terra Scientific Publ. Co., Tokyo, 1983.

Lockwood, M., J. H. Waite, Jr., T. E. Moore, J.F.E. Johnson, and C. R. Chappell, A new source of suprathermal O^+ ions near the dayside polar cap boundary, J. Geophys. Res., 90, 4099, 1985a.

Lockwood, M., T. E. Moore, J. H. Waite, Jr., C. R. Chappell, J. L. Horwitz, and R. A. Heelis, The geomagnetic mass spectrometer--Mass and energy dispersions of ionospheric ion flows into the magnetosphere, Nature, 316, 612, 1985b.

Lundin, R., B. Hultqvist, E. Dubinin, A. Zackarov, and N. Pissarenko, Observations of outflowing ion beams on auroral field lines at altitudes of many earth radii, Planet. Space Sci., 30, 715, 1982.

Moore, T. E., Modulation of terrestrial ion escape flux composition (by low-altitude acceleration and charge exchange chemistry), J. Geophys. Res., 85, 2021, 1980.

Moore, T. E., C. J. Pollock, R. L. Arnoldy, and P. M. Kintner, Preferential O^+ heating in the topside ionosphere, Geophys. Res. Lett., 13, 901, 1986.

Persoon, A. M., D. A. Gurnett, W. K. Peterson, J. H. Waite, Jr., J. L. Burch, and J. L. Green, Electron density depletions in the nightside auroral zone, J. Geophys. Res., in press, 1987.

Peterson, W. K., Ion injection and acceleration in the polar cusp, in The Polar Cusp, edited by J. A. Holtet and A. Egeland, p. 67, Reidel Publ. Co., Dordrecht, Holland, 1984.

Peterson, W. K., and E. G. Shelley, Origin of the plasma in a cross-polar cap auroral feature (theta aurora), J. Geophys. Res., 89, 6729, 1984.

Quinn, J. M., and C. E. McIlwain, Bouncing ion clusters in the earth's magnetosphere, J. Geophys. Res., 84, 7365, 1979.

Quinn, J. M., and D. J. Southwood, Observations of parallel ion energization in the equatorial region, J. Geophys. Res., 87, 10,536, 1982.

Quinn, J. M., and R. G. Johnson, Observation of ionospheric source cone enhancements at the substorm injection boundary, J. Geophys. Res., 90, 4211, 1985.

Reiff, P. H., T. W. Hill, and J. L. Burch, Solar wind plasma injection at the dayside magnetospheric cusp, J. Geophys. Res., 82, 479, 1977.

Reiff, P. H., H. L. Collin, J. D. Craven, J. L. Burch, J. D. Winningham, E. G. Shelley, L. A. Frank, and M. A. Friedman, Determination of auroral electrostatic potentials using high- and low-altitude particle distributions, J. Geophys. Res., in press, 1987.

Retterer, J. M., T. Chang, G. B. Crew, J. R. Jasperse, and J. D. Winningham, Monte Carlo modeling of large-scale ion-conic generation, this volume, 1987.

Roth, I., and M. K. Hudson, Lower hybrid heating of ionospheric ions due to ion ring distributions in the cusp, J. Geophys. Res., 90, 4191, 1985.

Sagawa, E., A. W. Yau, B. A. Whalen, and W. K. Peterson, Pitch-angle distributions of low-energy ions in the near-earth magnetosphere, J. Geophys. Res., in press, 1987.

Schunk, R. W., The polar wind, this volume, 1987.

Sharp, R. D., R. G. Johnson, and E. G. Shelley, Observations of an ionospheric acceleration mechanism producing energetic (keV) ions primarily normal to the geomagnetic field direction, J. Geophys. Res., 82, 3324, 1977.

Sharp, R. D., A. G. Ghielmetti, R. G. Johnson, and E. G. Shelley, Hot plasma composition results from the S3-3 spacecraft, in Energetic Ion Composition in the Earth's Magnetosphere, edited by R. G. Johnson, p. 167, Terra Scientific Publ. Co., Tokyo, 1983.

Shelley, E. G., Circulation of energetic ions of terrestrial origin in the magnetosphere, Adv. Space Res., 5, 401, 1985.

Shelley, E. G., Magnetospheric energetic ions from the earth's ionosphere, Adv. Space Res., 6(3), 121, 1986.

Shelley, E. G., R. D. Sharp, and R. G. Johnson, He^{++} and H^{+} flux measurements in the dayside cusp: Estimates of convection electric field, J. Geophys. Res., 81, 2363, 1976.

Shelley, E. G., D. A. Simpson, T. C. Sanders, E. Hertzberg, H. Balsiger, and A. Ghielmetti, The energetic ion mass spectrometer (EICS) for the Dynamics Explorer-A, Space Sci. Instrum., 5, 443, 1981.

Shelley, E. G., H. L. Collin, J. F. Drake, W. Lennartsson, and A. W. Yau, Origin of plasma sheet ions: Substorm and solar cycle dependence (abstract), Eos Trans. AGU, 67, 1133, 1986.

Temerin, M., and I. Roth, Ion heating by waves with frequencies below the ion gyrofrequency, Geophys. Res. Lett., 13, 1109, 1986.

Whalen, B. A., W. Bernstein, and P. W. Daly, Low altitude acceleration of ionospheric ions, Geophys. Res. Lett., 5, 55, 1978.

Winningham, J. D., J. L. Burch, and R. A. Frahm, Bands of ions and angular V's: A conjugate manifestation of ionospheric acceleration, J. Geophys. Res., 89, 1749, 1984.

Yau, A. W., and M. Lockwood, Vertical ion flow in the polar ionosphere, this volume, 1987.

Yau, A. W., B. A. Whalen, A. G. McNamera, P. J. Kellogg, and W. Bernstein, Particle and wave observations of low-altitude ionospheric ion acceleration events, J. Geophys. Res., 88, 341 1983.

Yau, A. W., P. H. Beckwith, W. K. Peterson, and E. G. Shelley, Long-term (solar cycle) and seasonal variations of upflowing ionospheric ion events at DE-1 altitudes, J. Geophys. Res., 90, 6395, 1985a.

Yau, A. W., E. G. Shelley, W. K. Peterson, and L. Lenchyshyn, Energetic auroral and polar ion outflow at DE 1 altitudes: Magnitude, composition, magnetic activity dependence, and long-term variations, J. Geophys. Res., 90, 8417, 1985b.

Yau, A. W., W. K. Peterson, and E. G. Shelley, Quantitative parametrization of energetic ionospheric ion outflow, this volume, 1987.

Young, D. T., Near-equatorial magnetospheric particles from ~1 eV to ~1 MeV, Rev. Geophys. Space Phys., 21, 402, 1983.

Young, D. T., Experimental aspects of ion acceleration in the earth's magnetosphere, in Ion Acceleration in the Magnetosphere and Ionosphere, Geophys. Monogr. Ser., vol. 38, edited by T. Chang, p. 17, AGU, Washington, D.C., 1986.

Young, D. T., J. Geiss, H. Balsiger, P. Eberhardt, A. Ghielmetti, and H. Rosenbauer, Discovery of He^{++} and O^{++} ions of terrestrial origin in the outer magnetosphere, Geophys. Res. Lett., 4, 561, 1977.

Young, D. T., H. Balsiger, and J. Geiss, Correlations of magnetospheric ion composition with geomagnetic and solar activity, J. Geophys. Res., 87, 9077, 1982.

KINETIC TREATMENT OF OXYGEN CONIC FORMATION IN THE CENTRAL PLASMA SHEET BY BROADBAND WAVES

G. B. Crew and Tom Chang

Center for Theoretical Geoplasma Physics
Center for Space Research, Massachusetts Institute of Technology
Cambridge, Massachusetts 02139

Abstract. The acceleration of ions by broadband waves has recently been proposed as an explanation for the simultaneous observation of low-frequency electromagnetic turbulence and oxygen-dominated ion conics within the central plasma sheet. Here we develop a kinetic description for the evolution of a minority (oxygen) ionic species within a hydrogenic plasma. In consideration of some of the properties of the broadband wave spectrum in this plasma, we find that the oxygen heating is described by a quasilinear, velocity space diffusion operator for the oxygen ions. With the aid of suitable approximations, we obtain a simplified kinetic equation for the gyrotropic distribution of ions including the effects of perpendicular velocity space diffusion and the mirror force due to the geomagnetic field. In many cases the spectrum of resonant turbulence is well described as a power law, and one finds that the results are rather independent of the initial conditions imposed on these equations. This is due to the existence of a self-similar solution of the kinetic equation in which the ion distribution relaxes toward a fixed shape in velocity space, with the velocity scaling determined by the altitude. We present a numerical technique for constructing the conic distribution function.

Introduction

Conics have been observed in various parts of the earth's ionosphere-magnetosphere system for a number of years and continue to remain a topic of some theoretical interest. While a number of mechanisms have been proposed to explain the observations [including wave-particle interactions with electrostatic ion-cyclotron waves (Lysak et al., 1980; Ashour-Abdalla and Okuda, 1984 and references therein) and lower hybrid waves (Chang and Coppi, 1981; Crew and Chang, 1985; Retterer et al., 1986)], it is becoming clear that no single mechanism is responsible for all of the conics that have been observed. In particular, the observation of conics in the central plasma sheet (CPS) portion of the earth's auroral zone by the plasma instrument aboard Dynamics Explorer 1 (DE 1) (Winningham and Burch, 1984) was something of a puzzle in that none of the signatures of other acceleration mechanisms were present. On the other hand, there was intense power observed at low frequencies, and Chang et al. (1986) were able to make use of this to construct a new theory for the formation of ion conics.

The treatment of Chang et al. (1986) made use of a heuristic model of the electromagnetic cyclotron resonance heating mechanism which took into account some of the specifics of the event. In particular, the conics were observed to be fairly broad in pitch angle, typically around 115° to 140°. They were moderately energetic, with energies in the tens to hundreds of electron volts. Also, the conics were determined to be predominantly oxygen (D. M. Klumpar, W. K. Peterson, and E. G. Shelley, private communication, 1985) and embedded in a predominantly hydrogenic nonconic background plasma. The low-frequency turbulence observed with the plasma wave instrument (D. A. Gurnett and M. Mellott, private communication, 1985) on DE 1 was similar to the turbulence that has been frequently observed in the auroral zone. Gurnett et al. (1984) determined that some of this turbulence was electromagnetic and was probably the signature of Alfven waves mediating between the magnetosphere and ionosphere. Chang et al. (1986) suggested that at each altitude, some fraction of the total power at the local oxygen cyclotron frequency could be in a left-hand polarized mode. There would then be an efficient transfer of power to the oxygen ions and an explanation for the energization necessary to form conics.

Subsequent work (Retterer et al., 1987a,b) demonstrated the efficacy of the process using a realistic kinetic description with a solution via a Monte Carlo procedure (Retterer et al., 1983). This work made use of the observation that the power spectrum varied with altitude in roughly a power law fashion. In this paper we shall pursue analytic results which follow from this same assumption. However, the results have a broader applicability and, in particular, may allow quantitative comparisons with the data, as well as offer insights which facilitate an understanding of the conic formation process. We shall see in the next section how these basic ideas lead to a kinetic formulation for the conic formation process. Given the power law nature of the turbulence, further progress can be made by scaling the conic properties with altitude. Finally we discuss a numerical scheme that allows one to obtain the shape of the conic.

Kinetic Equation

A kinetic description for the conic formation process rests heavily on the responsible physical processes. In this case, the conic arises as a compromise between the separate demands of the magnetic mirror force due to the inhomogeneous geomagnetic field and the heating due to the wave-particle interactions. Accordingly, we neglect here all secondary effects such as the behavior of the system with time and spatial variations transverse to the magentic field. The result is a kinetic equation for the conic ion distribution function $f(l,v_\parallel,v_\perp)$ of the form

$$v_\parallel\frac{\partial f}{\partial l} - \frac{v_\perp^2}{2B}\frac{dB}{dl}\left(\frac{\partial f}{\partial v_\parallel} - \frac{v_\parallel}{v_\perp}\frac{\partial f}{\partial v_\perp}\right) = \frac{1}{v_\perp}\frac{\partial}{\partial v_\perp}\left(v_\perp D_\perp\frac{\partial f}{\partial v_\perp}\right). \quad (1)$$

Here the surviving spatial variable l is the arc length along the geomagnetic field **B**, and the velocity variable has been decomposed into components $v_\parallel$ and $v_\perp$, parallel and perpendicular to **B**, respectively. The distribution f has been averaged over the third component of velocity, the gyrophase, since we are interested in time scales longer than the gyroperiod. The first term in equation (1) is the surviving convective term due to the flow of ions up the magnetic field. The second term expresses the action of the mirror force on the distribution. If there is a gradient in the magnetic field (and there is since **B** decreases with altitude roughly as l^{-3}), then in the absence of other effects, an ion will attempt to conserve its energy and magnetic moment. For upward motion, this results in a conversion of perpendicular energy to parallel energy.

The right-hand side of equation (1) contains the additional physics of the wave-particle interactions. The general case of this interaction for electromagnetic waves has been worked out in some detail (Kennel and Engelmann, 1966); however, for our present purposes, we do not require such complexity. For long wavelengths, a much simpler calculation (Sagdeev and Galeev, 1969) results in a diffusion term of the form appearing in equation (1), where

$$D_\perp = (2\pi q/m)^2 \int_0^\infty d\omega \left|E_L\right|^2(\omega)\,\delta\left(\omega - k_\parallel v_\parallel - \Omega(l)\right) \quad (2)$$

expresses the diffusion coefficient for ions of charge q and mass m in terms of the spectral power in left-hand polarized waves, $|E_L|^2$. The Doppler shift is small enough to allow us to write

$$D_\perp \approx (2\pi q/m)^2 \left|E_L\right|^2(\Omega(l)) , \quad (3)$$

where $\Omega(l)$ is the local gyrofrequency. The last factor in this equation is to be interpreted as a frequency average about a small band at the oxygen cyclotron frequency. Thus, while in the observational data there may be a slight dip near the local cyclotron frequency, one should recognize that this dip is quite narrow and the Doppler shift is sufficient to bring the ions into resonance with waves on the "shoulders" of the dip.

With this formulation in hand, one can proceed to solve equation (1) using the Monte Carlo technique (Retterer et al., 1987a,b). One of the interesting results of that work is that while the energy increases with increasing altitude, the distribution appears to relax onto a definite "shape" in velocity space. A related behavior was observed in the heuristic theory of Chang et al. (1986).

A second observation made for both of these approaches is that the initial conditions assumed at the base of the flux tube do not seem to be important for the final form of the conic. In the heuristic treatment, one could impose at an arbitrary initial altitude a variety of pitch angles or energies without affecting the results at sufficiently high altitude. Similarly, the choice of initial distribution or altitude does not seem to matter in the Monte Carlo runs.

We can put this behavior into mathematical terms by observing that equation (1) has a similarity transformation whenever the magnetic field and turbulence are expressible as power laws. To be explicit, we note that

$$\Omega(l) = \frac{q}{mc} B(l) \sim l^{-3} , \quad (4)$$

and take

$$\left|E_L\right|^2(\omega) \propto \omega^{-\alpha} , \quad (5)$$

with the result that

$$D_\perp = D_o(l/l_o)^{3\alpha} , \quad (6)$$

where $D_o = D_\perp(l_o)$ and l_o is a convenient reference point. It is then possible to show that at large altitudes, one asymptotically obtains

$$f(l,v_\parallel,v_\perp) \sim F[x(v_\perp,l),y(v_\parallel,l)](l/l_o)^{-(4\sigma+3)}, \quad (7)$$

where the variables x and y scale the velocity dependence with altitude as

$$x(v_\perp,l) \equiv (v_\perp/v_o)(l/l_o)^{-\sigma} \quad (8a)$$

$$y(v_\parallel,l) \equiv (v_\parallel/v_o)(l/l_o)^{-\sigma} \quad (8b)$$

with the parameters $\sigma \equiv \alpha + 1/3$ and $v_o \equiv (D_o l_o)^{1/3}$. Other scalings are possible, but this is the only one that conserves particles in their flow up along the magnetic field. Not contained in the function F(x,y) and implicit in the absence of a third independent variable is the relaxation with altitude of the distribution onto this asymptotic form.

Conic Shape

To obtain this new distribution F(x,y), we start with the partial differential equation for F which may be obtained from the kinetic equation (1) via the change of variables implied by equations (7) and (8). The result is

$$\frac{1}{x}\frac{\partial}{\partial x}\left(x\frac{\partial F}{\partial x}\right) + \frac{1}{x}\frac{\partial}{\partial x}\left[\left(\frac{3}{2}+\sigma\right)x^2 yF\right] + \frac{1}{x}\frac{\partial}{\partial y}\left[\left(\sigma y^2 - \frac{3}{2}x^2\right)xF\right] = 0 \ . \quad (9)$$

In the interest of brevity, we note that this function F(x,y) depends only on σ, and pass over its other properties.

This equation may be cast into a somewhat more comprehensible form with the introduction of a "density" $N(x,y) \equiv xF(x,y)$ and a prescribed "flow" $\mathbf{u}(x,y)$ with components in x and y directions

$$u_x \equiv \frac{1}{x} - \left(\frac{3}{2}\right)xy, \quad u_y \equiv \frac{3}{2}x^2 - \sigma y^2 \ . \quad (10)$$

These definitions cast equation (9) into the form

$$\frac{\partial N}{\partial t} + \nabla \cdot (\mathbf{u}N) = \frac{\partial^2 N}{\partial x^2} \quad (11)$$

in which we have suggestively introduced the term $\partial N/\partial t$. Since N is independent of this dummy "time" t, this term is identically zero in the original equation for F. The inclusion of this term makes equation (11) describe the time evolution of the density of conserved particles subject to diffusion in a spatial variable x and flow in a prescribed field **u**. Since the time variable t is artificial, only stationary solutions are of interest. In particular, one might imagine an evolution toward a stationary state as $t \to \infty$. (In this sense, t plays the role of l which was abandoned when we transferred our attention from f to F.)

To understand equation (11), we begin with the flow field $\mathbf{u}(x,y)$ and note that it converges on a fixed point where $\mathbf{u} = 0$. This occurs at the point

$$x = \left(\frac{2\sigma}{3}\right)^{1/6}\left(\frac{3}{2}+\sigma\right)^{-1/3}, \qquad y = \left(\frac{2\sigma}{3}\right)^{-1/3}\left(\frac{3}{2}+\sigma\right)^{-1/3} \ . \quad (12)$$

One may gain an appreciation for the left-hand side of equation (11) by plotting streamlines for representative values of σ. This is done in Figure 1 for $\sigma = 2.53$, which corresponds to the value of α used in Chang et al. (1986).

The convergence that these flow patterns exhibit is then balanced by the diffusive action of the right-hand side of equation (11). Thus, as the density tends to pile up at the fixed point, the action of this term is to spread out the pile into something of a mound. Although the diffusion acts only in the x direction, the flow pattern produces a spread, albeit smaller, in the y direction. A contour plot for the function N(x,y) is also included in Figure 1. Division by x then skews the result to obtain the distribution F(x,y). One can determine that $N(x \to 0,y)$ goes linearly to zero, with the result that F maximizes at x = 0. Figure 2 displays the resulting distribution F(x,y).

We note that this distribution has all of the qualitative features of the observed conic. One feature of special note is the absence of ions at small $v_\parallel$; i.e., we find $F(x,y \to 0) \to 0$ rapidly. Conics of this character have been observed previously, but the appearance of this feature was interpreted as a signature of a field-aligned potential (Klumpar et al., 1984). Here we find an alternative explanation which does not require the presence of such a potential. This is in fact a general feature of distributions generated via wave-particle interactions where the diffusion coefficient is nonvanishing at small velocities $[D_\perp(v_\perp \to 0) \neq 0]$, as was noted recently by Temerin (1986) without a heating mechanism.

The numerical technique used to calculate the density N rests on the fact that equation (9) may be viewed as a Fokker-Planck equation. To this

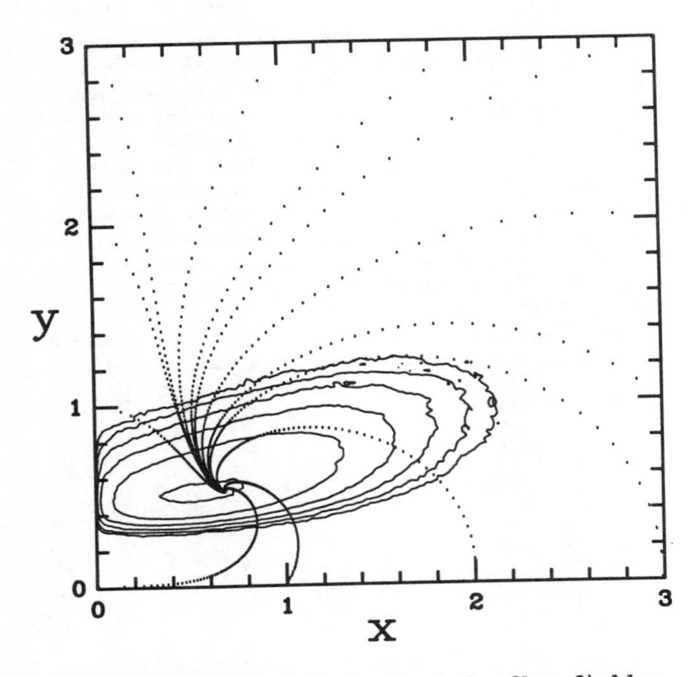

Fig. 1. Streamlines (dotted) of the flow field **u** defined by equation (10), superimposed on a contour plot (logarithmically spaced contours) of the density N(x,y) determined by equation (11). The flow converges from the periphery to the fixed point given by equation (12), where N(x,y) peaks.

equation corresponds a Langevin equation--namely, an evolution equation for a system of variables governed by a deterministic term and a noise term. For our case we may express this equation in a time-discretized form:

$$x_{n+1} = x_n + \Delta x, \; \Delta x = \Delta t u_x + \sqrt{2\Delta t}\, \xi \, , \quad (13a)$$

$$y_{n+1} = y_n + \Delta y, \; \Delta y = \Delta t u_y \, . \quad (13b)$$

Here the time step $\Delta t << 1$, and ξ is a random variable with Gaussian statistics. Clearly the terms involving **u** express the deterministic part of the evolution and ξ expresses the noise. The steady state distribution we seek may then be constructed as

$$N(x,y) = \lim_{T\to\infty} \frac{1}{T} \sum_{n=1}^{T} \delta(x - x_n)\delta(y - y_n) \, . \quad (14)$$

In practice, numerical results are obtained by averaging equation (14) over a grid in the x-y plane.

Summary

We have presented some analytic results available in the theory of ion conic formation due to cyclotron resonance with left-hand polarized waves. The results presented here are developed for the case where the altitudinal dependence of the turbulent wave spectrum is given by a power law. We have seen that this is rather useful from a pedagogic standpoint in that it allows one to analyze the conic formation process to a point where it is possible to understand the velocity space structure that the conic ultimately assumes. However, the utility of the theory developed here is not simply a pedagogic exercise.

In fact, results derived from this restricted set of assumptions have relevance to the observed conic populations for the following reason. Since the energization of the conic increases with altitude and since the form of the spectrum is such that the greatest wave power is resonant with the ions at the highest altitudes, it is only in the final stages of the conic's evolution to the satellite that the bulk of the heating takes place. This restricts the range of altitudes over which one requires the turbulence to be reasonably well approximated as a power law.

Thus, we have constructed a theory which makes specific predictions about the conics which should be observed in the presence of low-frequency electromagnetic turbulence. These predictions may be compared and contrasted to the observational data in order to develop an appreciation for what is transpiring along the flux

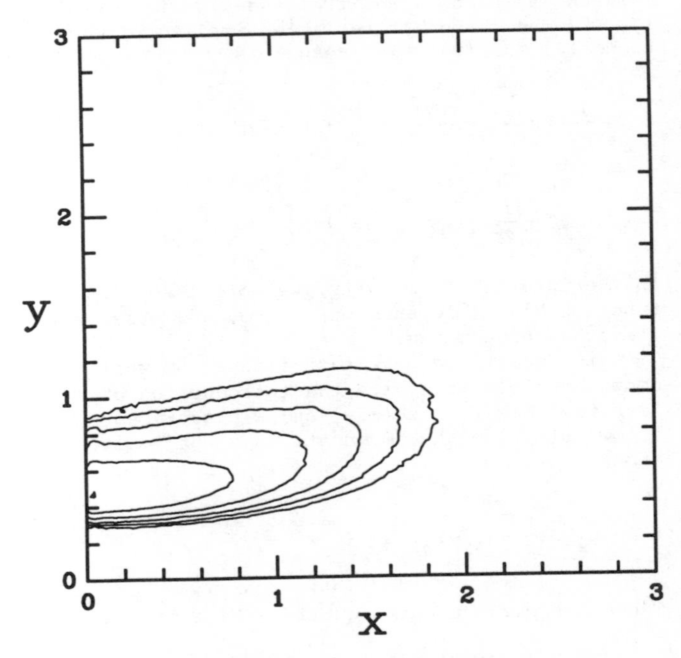

Fig. 2. Contour plot (logarithmically spaced contours) of the scaled distribution F(x,y) obtained from N(x,y) shown in Figure 1. The horizontal and vertical axes are directly proportional to the perpendicular and parallel velocities, respectively.

tube. Indeed, as the conics are energized by the wave turbulence, so do they also bear useful information concerning the interactions of the ionosphere with the magnetosphere.

Acknowledgments. This work is one aspect of a larger problem which has been tackled in collaboration with several other colleagues. We specifically wish to acknowledge fruitful discussions with John Retterer, Jack Jasperse, Noah Hershkowitz, and Dave Winningham. This work has been partially supported by AFGL contracts F19628-83-C-0060 and F19628-86-K-0005, and partially sponsored by the Air Force Office of Scientific Research under contract F49620-86-C-0128. The United States Government is authorized to reproduce and distribute reprints for governmental purposes notwithstanding any copyright notation hereon.

References

Ashour-Abdalla, M., and H. Okuda, Turbulent heating of heavy ions on auroral field lines, J. Geophys. Res., 89, 2235, 1984.

Chang, T., and B. Coppi, Lower hydrid acceleration and ion evolution in the suprauroral region, Geophys. Res. Lett., 8, 1253, 1981.

Chang, T., G. B. Crew, N. Hershkowitz, J. R. Jasperse, J. M. Retterer, and J. D. Winningham, Transverse acceleration of oxygen ions by electromagnetic ion cyclotron resonance with broad band left-hand polarized waves, Geophys. Res. Lett., 13, 636, 1986.

Crew, G. B., and T. Chang, Asymptotic theory of ion conic distributions, Phys. Fluids, 28, 2382, 1985.

Gurnett, D. A., R. L. Huff, J D. Menietti, J. L. Burch, J. D. Winningham, and S. D. Shawhan, Correlated low-frequency electric and magnetic noise along the auroral field lines, J. Geophys. Res., 89, 8971, 1984.

Kennel, C. F., and F. Engelmann, Velocity space diffusion from weak plasma turbulence in a magnetic field, Phys. Fluids, 9, 2377, 1966.

Klumpar, D. M., W. K. Peterson, and E. G. Shelley, Direct evidence for two-stage (bimodal) acceleration of ionospheric ions, J. Geophys. Res., 89, 10,779, 1984.

Lysak, R. L., M. K. Hudson, and M. Temerin, Ion heating by strong electrostatic ion cyclotron turbulence, J. Geophys. Res., 85, 678, 1980.

Retterer, J. M., T. Chang, and J. R. Jasperse, Ion acceleration in the suprauroral region: A Monte Carlo model, Geophys. Res. Lett., 10, 583, 1983.

Retterer, J. M., T. Chang, and J. R. Jasperse, Ion acceleration by lower hybrid waves in the suprauroral region, J. Geophys. Res., 91, 1609, 1986.

Retterer, J. M., T. Chang, G. B. Crew, J. R. Jasperse, and J. D. Winningham, Monte Carlo modeling of ionospheric oxygen accelerations by cyclotron resonance with broadband electromagnetic turbulence, Phys.Rev.Lett., 59,148,1987a.

Retterer, J. M., T. Chang, G. B. Crew, J. R. Jasperse, and J. D. Winningham, Monte Carlo modeling of large-scale ion-conic generation, this volume, 1987b.

Sagdeev, R. Z., and A. A. Galeev, in Nonlinear Plasma Theory, pp. 54-55, W. A. Benajamin, Inc., New York, 1969.

Temerin, M., Evidence for a large bulk ion conic heating region, Geophys.Res.Lett., 13,1059,1986.

Winningham, J. D., and J. Burch, Observation of large scale ion conic generation with DE 1, in Physics of Space Plasmas, edited by J. Belcher, H. Bridge, T. Chang, B. Coppi, and J. R. Jasperse, p. 137, Scientific Publishers, Inc., Cambridge, Massachusetts, 1984.

ELECTRIC FIELDS AND PARTICLE PRECIPITATION DURING THE SUBSTORM OF JANUARY 18, 1984

O. de la Beaujardière,[1] D. S. Evans,[2] Y. Kamide,[3] and R. Lepping[4]

Abstract. The purpose of this paper is to examine the transition between quiet and active conditions in the early afternoon high-latitude ionosphere. How the electric field and particle precipitation respond to a slow monotonic decrease in the interplanetary magnetic field B_z component is investigated. Observations from the Sondrestrom incoherent-scatter radar and the NOAA 7 satellite are presented. It is shown that the electric field intensifies very quickly (~15 min) after the change in B_z. Around 1430 MLT, the latitude of the convection reversal moves poleward during the time when B_z is decreasing, but still positive. During this time, the precipitation from the central plasma sheet electrons also moves poleward, although at a slower rate. On two consecutive passes, there is no central plasma sheet electron precipitation. These observations are contrary to what is predicted by steady state calculations and empirical models. They may be explained in terms of the temporal and spatial variations of the adiabatic motion of ~1-keV electrons, under the influence of a convection electric field that is abruptly increased. Other observations concern the relative position of the plasma sheet boundary layer and the convection reversal. The plasma convection reversal appears to move from the equatorward to the poleward then back to the equatorward edge of the boundary layer precipitation. This relative motion is also interpreted in terms of a transient response. The speculation is that the convection responds quickly to changes in the solar wind convection, whereas the electrons are slower to adapt to the new configuration of the convection.

[1]Geoscience and Engineering Center, SRI International, Menlo Park, California 94025.
[2]Space Environment Laboratory, NOAA, Department of Commerce, Boulder, Colorado 80303.
[3]Kyoto Snagyo University, Kamigamo, Kita-ku, Kyoto, Japan.
[4]Laboraory for Extraterrestrial Physics, NASA Goddard Space Flight Center, Greenbelt, Maryland 20771.

Introduction

In the high-latitude ionosphere, the convection and the particle precipitation are interconnected. Their dependence on one another has been investigated both from a theoretical modeling standpoint (Fontaine et al., 1985) and through deriving empirical relationships (Foster et al., 1986). However, because these studies deal either with steady state situations or with average properties, time-dependent and transient features are often masked or ignored. As an example of the importance of treating time-dependent situations, Gorney and Evans (1987) showed that when the magnetospheric convection is increased by a step function, the invariant latitude of precipitation from the central plasma sheet (CPS) in the early afternoon sector behaves in a way opposite to that predicted by steady state calculations and empirical models. They showed that at this local time the equatorward boundary of the auroral oval responds to the change by initially moving to higher latitudes instead of moving to lower latitudes as would be expected from increased convection. Their study illustrated the importance of doing case studies which are essential to understanding the transient response of the magnetosphere/ionosphere system to changes in input conditions.

The purpose of this paper is to study substorm effects observed in the early afternoon, high-latitude sector. In particular, we are interested in the transition from quiet to active conditions. We present a detailed study of the early afternoon behavior of the electric field and of auroral electron precipitation during such a transition. Particle precipitation from both the central plasma sheet and the plasma sheet boundary layer is considered. We present results from a substorm that occurred on January 18, 1984. This event was selected because the substorm occurred after a prolonged quiet period and was associated with a very slow and monotonic change in the B_z component of the interplanetary magnetic field (IMF) from northward to southward. The data used in this study were collected as part of the GISMOS campaign (Global Ionospheric Simul-

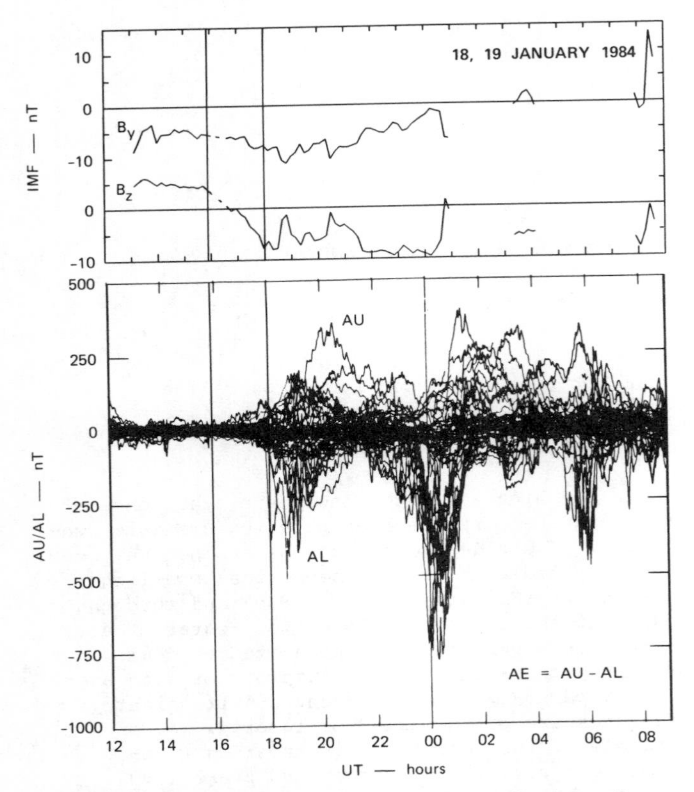

Fig. 1. IMF from IMP-8 and AU/AL variations for January 18-19, 1984.

taneous Measurements Of Substorms) that was designed to acquire a global and comprehensive set of ground- and space-based observations. The measurements reported here are limited to those from the Sondrestrom incoherent radar in Greenland, high-latitude magnetometers, and the NOAA 7 satellite. The IMF parameters were from the IMP-8 and ISEE 2 satellites.

Observations of the January 18, 1984, Substorm

The interplanetary magnetic field B_y and B_z components from IMP-8 and the stacked-H components of about 60 high-latitude magnetograms are shown in Figure 1. The B_y (dawn-dusk) component was negative and almost constant throughout the period of interest. At ~1600 UT, the IMF B_z component, which had been positive and fairly constant for more than 4 hours, started to decrease at an almost constant rate, going from about +5 to -8 nT in 2 hours. (During the data gap near 1600 UT, ISEE 2 measurements show that B_z decreased smoothly during this interval.)

Figure 1, lower panel, shows that the AE index was very small for several hours; it increased slightly at ~1630 UT, and at 1800 UT it increased sharply to 600 nT. This increase marks the onset of the substorm expansive phase, and the growth phase presumably took place between 1630 UT and 1800 UT. The geomagnetic conditions remained active to moderately active during the following 24 hours. In this paper, we limit ourselves to this first substorm, which is of interest because it was initiated on a quiet background, and because it occurred as a result of a slow, gradual turning of the B_z component.

The electric field between Λ = 69° and 80° was measured by the Sondrestrom radar. The radar and the operating modes are described in Wickwar et al. (1984). Figure 2 shows the north-south (N-S) electric field component at a latitude of Λ = 75°, close to the radar. The dashed curve represents the average electric field value expected for this orientation of the IMF B_y component. It is derived from the empirical model of de la Beaujardière and Wickwar (1986). This figure shows that the electric field intensified at ~1615 UT, very soon after B_z started to decrease. Thus, even though B_z was still positive, the electric field intensified appreciably within minutes of the decrease in B_z. (The experiment cycle time was 13 min.) The rapidity of the change in the convection pattern is consistent with previous observations (de la Beaujardière et al., 1983, 1986; Clauer and Banks, 1986).

At the 75° latitude, the convection was first sunward (west) and then anti-sunward (east). The convection reversal passed rapidly across this latitude at 1815 UT.

The plasma velocity at several latitudes is plotted on Figure 3. Before the substorm onset at 1800 UT and while the B_z component is still positive, the latitude of the velocity reversal changes in an unexpected way. It moves poleward from 1600 UT to 1710 UT. Then, close to the time when B_z becomes negative, this reversal moves equatorward, as would be expected from an increase in the rate of erosion of magnetic flux from the dayside magnetosphere (e.g., McPherron,

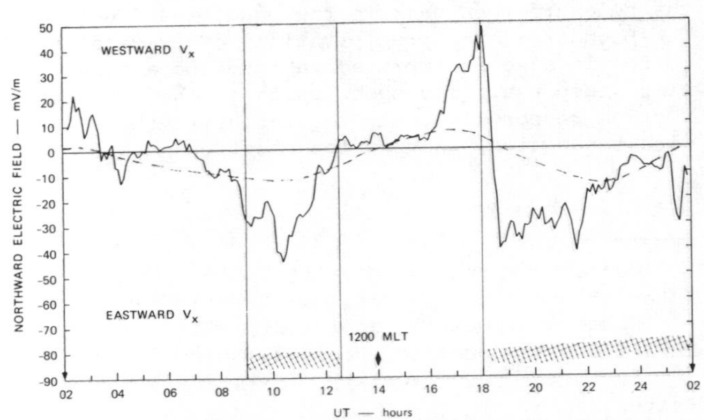

Fig. 2. Northward electric field measured by the Sondrestrom radar (solid line) and average electric field for B_y negative (dashed line). Magnetic noon is at 1400 UT (diamond) and the hatched areas represent times of substorm activity.

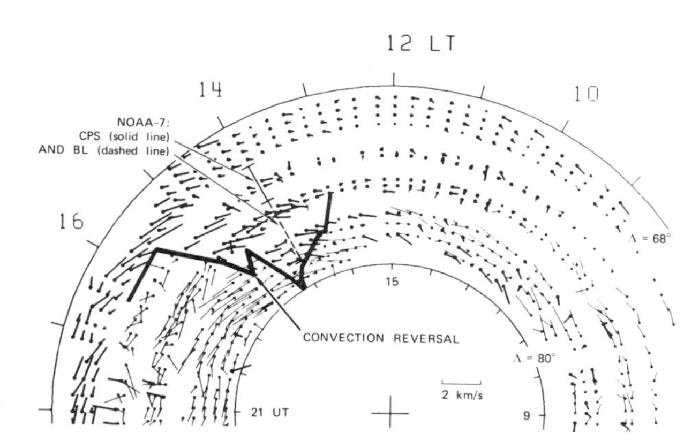

Fig. 3. Ion drifts measured by the Sondrestrom radar and NOAA 7 oval boundaries. Magnetic midnight is at 1400 UT. The heavy and light lines are from westward and eastward convection, respectively.

1979; Holzer and Reid, 1975; Holzer and Slavin, 1978).

This local time sector is sampled by the NOAA 7 spacecraft, which is sunsynchronous in a 1430-0230 local time orbit (Hill et al., 1985; Senior et al., 1987). The 0.3- to 20-keV particle data for one pass are shown in Figure 4. Two distinct electron populations are recognized. In the diffuse aurora region marked "A," precipitation is fairly smooth and has characteristic energies at or above the kiloelectron volt level. These are CPS electrons (Winningham et al., 1975). Poleward of the diffuse precipitation there is a patchy precipitation, marked "B," whose characteristic energy is lower. These electrons originate from the region called boundary plasma sheet (BPS) in the region close to midnight (Winningham et al., 1975) or, more generally, plasma sheet boundary layer (BL) (Eastman and Hones, 1979). In the ionosphere, the boundary between these electron populations has been called the transition boundary (Hardy et al., 1981; Meng and Makita, 1986).

The latitudes of the equatorward boundary of CPS precipitation, the transition between the CPS and BPS, and the poleward boundary of precipitation from the boundary layer plasma are all determined from the NOAA 7 particle observations for several hours around the time of the substorm. The north hemisphere boundaries are plotted versus UT in Figure 5, as viewed from above the pole. The lower part of the plot is for the early afternoon crossings. Because the total time covered by this plot is fairly short and only data from one spacecraft are shown, no attempt has been made to normalize these latitudes at a fixed MLT location, taking into consideration the offset of the poles and of the auroral zone center, as done in Meng et al. (1977).

Figure 5 shows that in the early afternoon sector both the transition boundary and the equatorward CPS electron boundary move poleward during the time when a poleward motion was also observed in the convection reversal. The motion, however, is more modest than that of the reversal: ~1.3° and 0.5° for the equatorward and poleward edge of the diffuse aurora, respectively, between the passes at 1510 UT and 1655 UT.

The most notable feature in Figure 5 is that during two consecutive passes (1835 UT and 2016 UT), there was no CPS electron precipitation in the early afternoon. However, substantial proton precipitation was present equatorward of the BL. The CPS precipitation reappeared in the 2200 UT pass.

Observation Summary

Observations were made of the transition from quiet to moderately active conditions, while the IMF B_z slowly decreased and became negative. These observations can be summarized as follows:

1. The convection intensifies very soon after the IMF B_z starts to decrease from an initial value of ~5 nT.

2. The latitude of the convection reversal moves poleward during the time when the IMF is decreasing, but still positive. (It moved by

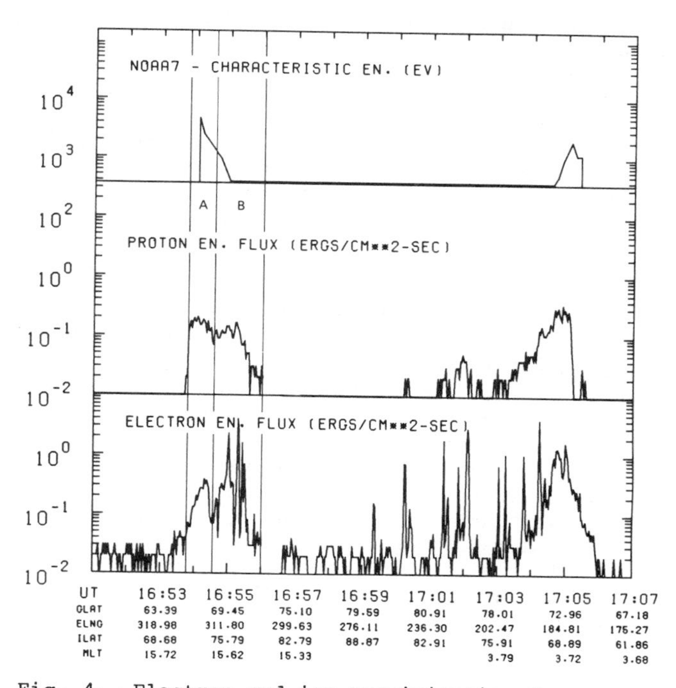

Fig. 4. Electron and ion precipitation for a northern hemisphere NOAA 7 pass. The lower and middle panels are the electron and proton energy fluxes. The upper panel is the electron characteristic energy. The vertical lines indicate the limits of CPS and BL electron precipitation (regions A and B, respectively).

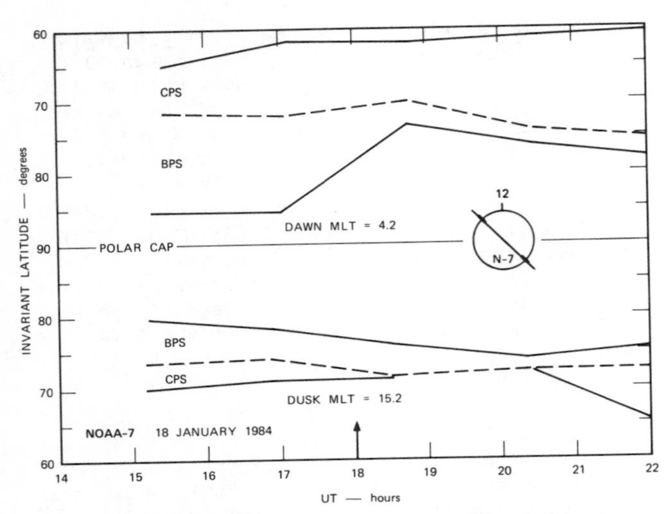

Fig. 5. Time variation of the oval boundaries measured by NOAA 7.

4.5° from 1605 UT to 1715 UT, which corresponds to a rate of 3.9 deg/hr.)

3. The CPS equatorward boundary and the transition boundary also moved poleward at this time. The rate of motion was slower, ~1.4 deg/hr for the equatorward boundary, if we take into consideration the fact that the electric field intensification occurred around 1605 UT.

4. There was no CPS precipitation for two consecutive passes after the substorm expansive phase onset.

5. The latitude of the convection reversal moved equatorward when IMF became negative. (It moved by 8.5° from 1715 UT to 1835 UT, which corresponds to a rate of 6.4 deg/hr.)

Discussion

Data from other MLTs around the auroral oval are available from other satellites, such as NOAA 8 and DMSP and other auroral zone incoherent-scatter radars. Space does not permit presenting those data here. However, it should be noted that the poleward motion and CPS disappearance (items 2, 3, and 4 above) were only observed in this early afternoon local time sector. The other points above (items 1 and 5) were also observed at the other local times.

Our observation of a poleward motion of the convection reversal, during what appears to be a substorm growth phase, is not an isolated case at this local time. Another similar case was seen on other Sondrestrom data. Also, observations by Sandholt et al. (1986) are probably related. In their Figure 4, precipitation is seen to move rapidly north before moving south. A similar case was reported by Gorney and Evans (1987).

Many of the results shown in this paper may be interpreted as the transition from quiet to active conditions. In a recent paper, Gorney and Evans (1987) examined the shape of the equatorward boundary of CPS precipitation as a function of magnetic local time in terms of the adiabatic motion of ~1-keV electrons under the influence of the geomagnetic field and an externally applied convection electric field. They extended this analysis to simulating the change in the adiabatic motion of these particles and, presumably, the shape of the equatorward boundary of plasma sheet precipitation, introduced by a step function increase in magnetospheric convection. Their time-dependent simulation showed that in a narrow local time sector around 1300, the equatorward precipitation boundary (identified from the orbits of plasma sheet electrons) initially moved poleward for several hours after the electric field increased, and, only afterwards, moved equatorward. A new steady state boundary was reached only after about 24 hours. The calculated rate of poleward migration was about 1 deg/hr in the simulation, which is close to the rate found in the observations reported here. The poleward excursion of this precipitation boundary could carry it to the point where its location would map to the magnetopause. Under these circumstances, no precipitation would be expected in the early afternoon sector because the plasma sheet particles that had been available would have escaped the magnetosphere when their orbits intersected the magnetopause. This approach would account for both the poleward motion of the equatorward boundary of precipitation and the disappearance of central plasma sheet precipitation that was observed by NOAA 7 in the afternoon sector during the initial phases of this substorm.

Therefore, it appears that part of the observations in this paper can be explained by the model of Gorney and Evans (1987). However, a more precise simulation using the actual conditions in effect during our observations needs to be done. In particular, the trajectories of the ions also need to be simulated to see if the model can explain the presence of ion precipitation in the narrow region depleted by CPS electrons. Furthermore, the complex motion of the convection reversal latitude is not understood at this point (the model assumes that it remains constant).

The relative position of the convection reversal and the BL precipitation is another aspect of these observations that can be interpreted as a transient feature rather than a steady state configuration. The NOAA 7 pass illustrated in Figure 4 came very close to Sondrestrom at 1655 UT, i.e., almost an hour after the electric field intensified. The two regions of electron precipitation are shown as A and B in Figure 4, and their limits are indicated on the plot of ionospheric drift (Figure 3). Both regions of hard and soft precipitation are entirely within the region of sunward convection. The convection reversal from sunward to anti-sunward marks almost exactly the poleward edge of electron precipitation. This situation is not what is most generally observed or predicted. For example, Holzer et al. (1986) placed the BL electrons entirely on open magnetic field lines; Bythrow et al. (1981) and Heikkila (1986) placed the convection reversal within the BL. However, in the systematic work of Heelis et al. (1980), the poleward edge of the BL was observed at the convection reversal on two occasions.

It may be reasonable to assume that the fact that the poleward edge of the boundary layer is colocated with the convection reversal is a transient feature. In the Sondrestrom magnetic meridian, electron density profiles were measured by elevation scans. The antenna was cycled through a series of nine fixed positions followed by an elevation scan in the magnetic meridian. The elevation scan measurements indicate that the convection reversal was probably close to the equatorward edge of the BL precipitation at ~1600 UT, i.e., at the start of the convection intensification, as well as at 2000 UT, i.e., at the end of the southward excursion of the convection reversal. The convection reversal moved to the poleward edge of the BL precipitation at the time when the reversal was moving poleward; then, the convection reversal migrated back to the equatorward edge of the the BL precipitation. Therefore, the fact that two boundaries are colocated at 1655 UT, is a transient feature, not a situation encountered in steady state conditions. It is as if the convection boundary were moving too fast and the precipitation could not catch up with it. In other words, the changes in solar wind conditions impose changes in the convection topology that occur very rapidly. The electrons, however, need much longer to react, as indicated, for example, by the work of Gorney and Evans (1987).

Acknowledgments. The authors express their thanks to D. Gorney for helpful discussions, and to S. Gussenhoven, C. Senior, J. Foster, R. Elphic, and E. Smith who provided data from DMSP, EISCAT, Millstone Hill, and ISEE 2, respectively. This work was supported by NSF cooperative agreement ATM85-16436.

References

Bythrow, P. F., R. A. Heelis, W. B. Hansen, and R. A. Power, Observational evidence for a boundary layer source of dayside region 1 field-aligned currents, J. Geophys. Res., 86, 5577, 1981.

Clauer, C. R., and P. M. Banks, Relationship of the interplanetary electric field to the high latitude ionospheric electric field and currents: Observation and model simulation, J. Geophys. Res., 91, 6959, 1986.

de la Beaujardière, O., and V. B. Wickwar, IMF control of plasma drift, ion temperature and neutral wind, in Proceedings of the Third Finnish-American Auroral Workshop, edited by E. Turunen and E. Kataja, The Finnish Academy of Science and Letters, Finland, 1986.

de la Beaujardière, O., J. Holt, and E. Nielsen, Early MITHRAS results: The electric field response to substorms, Radio Sci., 18, 981, 1983.

de la Beaujardière, O., V. B. Wickwar, and J. H. King, Sondrestrom radar observations of the effect of the IMF B_y component on polar cap convection, in Solar Wind-Magnetosphere Coupling, edited by Y. Kamide and J. A. Slavin, p. 495, Terra Scientific Publ. Co., Tokyo, 1986.

Eastman, T., and E. W. Hones, Jr., Characteristics of the magnetospheric boundary layer and magnetopause layer as observed by Imp 6, J. Geophys. Res., 84, 2019, 1979.

Fontaine, D., M. Blanc, L. Reinhart, and R. Glowinski, Numerical simulations of the magnetospheric convection including the effects of electron precipitation, J. Geophys. Res., 90, 8343, 1985.

Foster, J. C., J. M. Holt, R. G. Musgrove, and D. S. Evans, Ionospheric convection associated with discrete levels of particle precipitation, Geophys. Res. Lett., 13, 656, 1986.

Gorney, D. J., and D. S. Evans, The low-latitude auroral boundary: Steady state and time-dependent representations, J. Geophys. Res., in press, 1987.

Hardy, D. A., W. J. Burke, M. S. Gussenhoven, N. Heinemann, and E. Holeman, DMSP/F2 electron observations of equatorward auroral boundaries and their relationship to the solar wind velocity and the north-south component of the interplanetary magnetic field, J. Geophys. Res., 86, 9961, 1981.

Heelis, R. A., J. D. Winningham, W. B. Hanson, and J. L. Burch, The relationships between high-latitude convection reversals and the energetic particle morphology observed by Atmosphere Explorer, J. Geophys. Res., 85, 3315, 1980.

Heikkila, W. J., Transport of plasma across the magnetopause, in Solar Wind-Magnetosphere Coupling, edited by Y. Kamide and J. A. Slavin, p. 337, Terra Scientific Publ. Co., Tokyo, 1986.

Hill, V. J., D. S. Evans, and H. H. Sauer, TIROS/NOAA satellites space environment monitor archive tape documentation, NOAA Tech. Memo., ERL SEL-71, 1985.

Holzer, T. E., and G. C. Reid, The response of the dayside magnetosphere-ionosphere system to time-varying field line reconnection at the magnetopause. 1. Theoretical model, J. Geophys. Res., 80, 2041, 1975.

Holzer, R. E., and J. A. Slavin, Magnetic flux transfer associated with expansions and contractions of the dayside magnetosphere, J. Geophys. Res., 83, 3831, 1978.

Holzer, R. E., R. L. McPherron, and D. A. Hardy, A quantitative empirical model of the magnetospheric flux transfer process, J. Geophys. Res., 91, 3287, 1986.

McPherron, R. L., Magnetospheric substorms, Rev. Geophys. Space Phys., 17, 657, 1979.

Meng, C.-I., and K. Makita, Dynamic variations of the polar cap, in Solar Wind-Magnetosphere Coupling, edited by Y. Kamide and J. A. Slavin, p. 605, Terra Scientific Publ. Co., Tokyo, 1986.

Meng, C.-I., R. H. Holzworth, and S.-I. Akasofu, auroral circle-delineating the poleward boundary of the quiet auroral belt, J. Geophys. Res., 82, 164, 1977.

Sandholt, P. E., A. Egeland, and B. Lybekk, On the spatial relationship between auroral emissions and magnetic signatures of plasma convection in the midday polar cusp and cap ionospheres during negative and positive IMF B_z: A case study, J. Geophys. Res., 91, 12,108, 1986.

Senior, C., J. R. Sharber, O. de la Beaujardière, R. A. Heelis, D. S. Evans, J. D. Winningham, M. Sugiura, and W. R. Hoegy, E- and F-region study of the evening sector auroral oval: A Chatanika/Dynamics Explorer-2/NOAA-6 comparison, J. Geophys. Res., 92, 2477, 1987.

Wickwar, V. B., J. D. Kelly, O. de la Beaujardière, C. A. Leger, F. Steenstrup, and C. H. Dawson, Sondrestrom overview, Geophys. Res. Lett., 11, 883, 1984.

Winningham, J. D., F. Yosuhara, S.-I. Akasofu, and W. J. Heikkila, The latitudinal morphology of 10-eV to 10-keV electron fluxes during magnetically quiet and disturbed times in the 2100-0300 MLT sector, J. Geophys. Res., 80, 3148, 1975.

CLASSICAL AND ANOMALOUS TRANSPORT PROCESSES IN THE AURORAL RETURN CURRENT REGION

S. B. Ganguli

Science Applications International Corporation, McLean, Virginia 22102

P. J. Palmadesso

Naval Research Laboratory, Washington, D.C. 20375

Abstract. The classical and anomalous transport properties of a multifluid plasma consisting of H^+, O^+, and electron populations in the presence of auroral field-aligned return currents are investigated using a multimoment fluid model with anomalous transport coefficients. This approach offers the possibility of simulating large-scale dynamic phenomena without neglecting the important macroscopic consequences of microscopic processes such as anomalous resistivity, turbulent heating, etc. The macroscopic effects of the electrostatic ion-cyclotron (EIC) instability (perpendicular ion heating) and of an EIC-related anomalous resistivity mechanism which heats the electrons are included in the present version of the model. The responses of the outflowing ionospheric plasma to the application of current and instabilities are exhibited. Downward electron heat flow competes with upward convection and adiabatic effects to determine the direction of the electron temperature anisotropy. Resistive electron heating lowers the critical drift velocity for marginal EIC stability and leads to enhanced ion heating.

Introduction

We have developed a multimoment fluid model with anomalous transport coefficients which offers the possibility of simulating large-scale dynamic phenomena without neglecting the important macroscopic consequences of microscopic processes such as turbulent heating, anomalous resistivity, etc. In addition to providing improved understanding of plasma transport phenomena in the return current region, this work is also intended to contribute to the development of improved fluid models for use in the study of large-scale magnetospheric-ionospheric dynamics.

The effects of field-aligned upward currents on the plasma outflow were investigated in recent years by Mitchell and Palmadesso (1983). The physics of plasma transport in the return current regions, when compared to the upward current regions, is expected to exhibit both similarities and differences, and these have not been thoroughly explored. We find, for example, that the cold upflowing ionospheric electrons excite the EIC instability with surprisingly low currents and may also generate anomalous resistivity.

A brief description of the model is given in the following section. Discussions of the mathematical basis of the model, the equations used to study the classical and anomalous transport processes, the approximations used in deriving the equations, and the advantages and disadvantages of using this approach are given by Palmadesso et al. (1987). In the third section we explain the simulation results followed by a summary in the last section.

The Model

The model (Ganguli et al., 1985, 1987) simulates the steady state behavior of a fully ionized plasma (H^+, O^+, and the electrons) along the geomagnetic field lines in the high-latitude topside ionosphere (extending from 1500 km to 10 R_E). The theoretical formulation is based on the sixteen-moment system of transport equations of Barakat and Schunk (1982). The gyrotropic approximation reduces the 16 moments to 6 and the resulting equations are solved in one dimension for the H^+ ions and electrons, while the O^+ ions form a static background population at a constant temperature. In order to simulate ion heating due to EIC instability and anomalous resistivity, anomalous friction terms are introduced into the model and the transport equations are modified to maintain momentum and energy conservation with these processes.

Simulations with Ion Heating and Anomalous Resistivity

The steady state model and the treatment of boundary conditions for this simulation correspond to the zero current plasma outflow case of Ganguli et al. (1987).

Previous studies (Ganguli and Palmadesso, 1987) have shown that the application of return current to the outflowing ionospheric plasma in the steady state, even in the absence of instabilities, significantly modifies the equilibrium plasma profiles. We shall briefly summarize the conclusions of these earlier simulations of return current flow without anomalous transport so that the additional effects of the instabilities discussed below can be viewed in the context of the classical effects. The increase in return current means faster electron upflow which implies faster expansion of the electron gas and cooler electron temperature. The reduced electron temperature leads to a reduction in the ambi-polar electric field and a concomitant decrease in the hydrogen ion flow velocity. Since the total hydrogen ion flux is determined by processes which are insensitive to return current effects, the decrease in hydrogen ion outflow velocity implies an increase in hydrogen ion density and, via the quasi-neutrality condition, the electron density also. The H^+ velocity decrease also leads to an increase in the H^+ temperature relative to its zero current value. The electron temperature anisotropy reverses from $T_\perp > T_\parallel$ to $T_\parallel > T_\perp$ with an increase in the magnitude of the return current.

In this section we investigate the effects of ion heating and anomalous resistivity on the transport properties of the multifluid plasma in the auroral return current region. These processes are incorporated into the transport equations through the addition of terms containing "effective collision frequencies," as discussed in Palmadesso et al. (1987). In principle, the magnitudes of these parameters appropriate to a given situation can be determined approximately, as functions of macroscopic variables accessible within the model, by analyzing the results of particle simulations or other independent studies of the instabilities involved. The range of values we have considered here, $\alpha_H = \alpha_R = 0.001$-$0.005$ [see equations (14) and (17) in Palmadesso et al., (1987)], reflects a compromise between the very large realistic rates, which make stable numerical computation difficult, and small but less realistic values. For a more detailed discussion of the factors affecting the choice of values for the α parameters, see the companion paper by Palmadesso et al. (1987). Since the realistic values tend to drive the system toward marginal stability, the results should, at least qualitatively, be insensitive to the exact values chosen as long as these are not too small. It should be noted, however, that these values of α do not saturate the ion temperature profiles.

In the figures profiles generated with the EIC instability and anomalous resistivity active are shown in comparison with their previous values with current only. We shall use the expression "current only" frequently in this section. By current only we mean the plasma flow with effects due to currents, but with the effective collision frequencies for ion heating and anomalous resistivity set to zero.

An increase in current increases the relative drift velocity between the electrons and ions and lowers the altitude at which the EIC instability is excited. The theoretical calculations of Lee (1972) showed that the critical velocity for exciting the EIC instability decreases as the ratio $T_{p\perp}/T_{e\parallel}$ decreases. Kindel and Kennel (1971) concluded that the EIC waves are unstable to cold electron fluxes of the order of 10^9-10^{10} cm^{-2} s^{-1}. This implies that the magnitude of the critical current for exciting the EIC instability in the auroral ionosphere is 1.6 $\mu A\ m^{-2}$ or higher. However, our calculations show that the EIC instability is excited at a much lower current. The discrepancy is primarily due to the fact that in making their threshold estimates Kindel and Kennel (1971) assumed isotropic and equal ion and electron temperatures ($T_{p\perp}/T_{e\parallel} = 1$). In fact, the self consistent transverse ion temperature associated with the outflowing plasma decreases rapidly; the ions are reheated only as a result of the onset of instability. Thus, $T_{p\perp}/T_{e\parallel} \ll 1$ prior to the onset of instability and the instability threshold is reduced accordingly. We find that for a return current of 0.65 $\mu A\ m^{-2}$ the EIC instability turns on at an altitude of 1.16 R_E.

We will discuss the case of return current 1.5 $\mu A\ m^{-2}$ in some detail. For this case we have used $\alpha_H = \alpha_R = 0.001$. The EIC instability turns on at 0.4 R_E for this current. The sharp increase in the steady state H^+ ion perpendicular temperature due to ion heating is shown in Figure 1a. The resistivity turns on at 1.6 R_E. According to our preimposed conditions, the resistivity turns on only after the EIC instability is excited. The electron temperature profiles are shown in Figure 1b. The effects of collisions at low altitudes, thermal conduction, convective and adiabatic processes, and anomalous heating combine to determine the direction of the electron temperature anisotropy.

The critical drift velocity V_{cH} increases as the instability increases the ion temperature until $V_{cH}(T_{p\perp}/T_{e\parallel}) \simeq V_D$, whereupon the instability is almost shut off, as discussed by Palmadesso et al. (1974). Since V_{cH} is a decreasing function of $T_{e\parallel}$, the onset of anomalous resistivity and the consequent heating of electrons is expected to result in an increase in the rate of ion heating and in the equilibrium ion temperature; a detailed study of this effect will be the subject of a future paper.

The electric field is due to the ambi-polar field, E_A, and the resistive electric field E_R

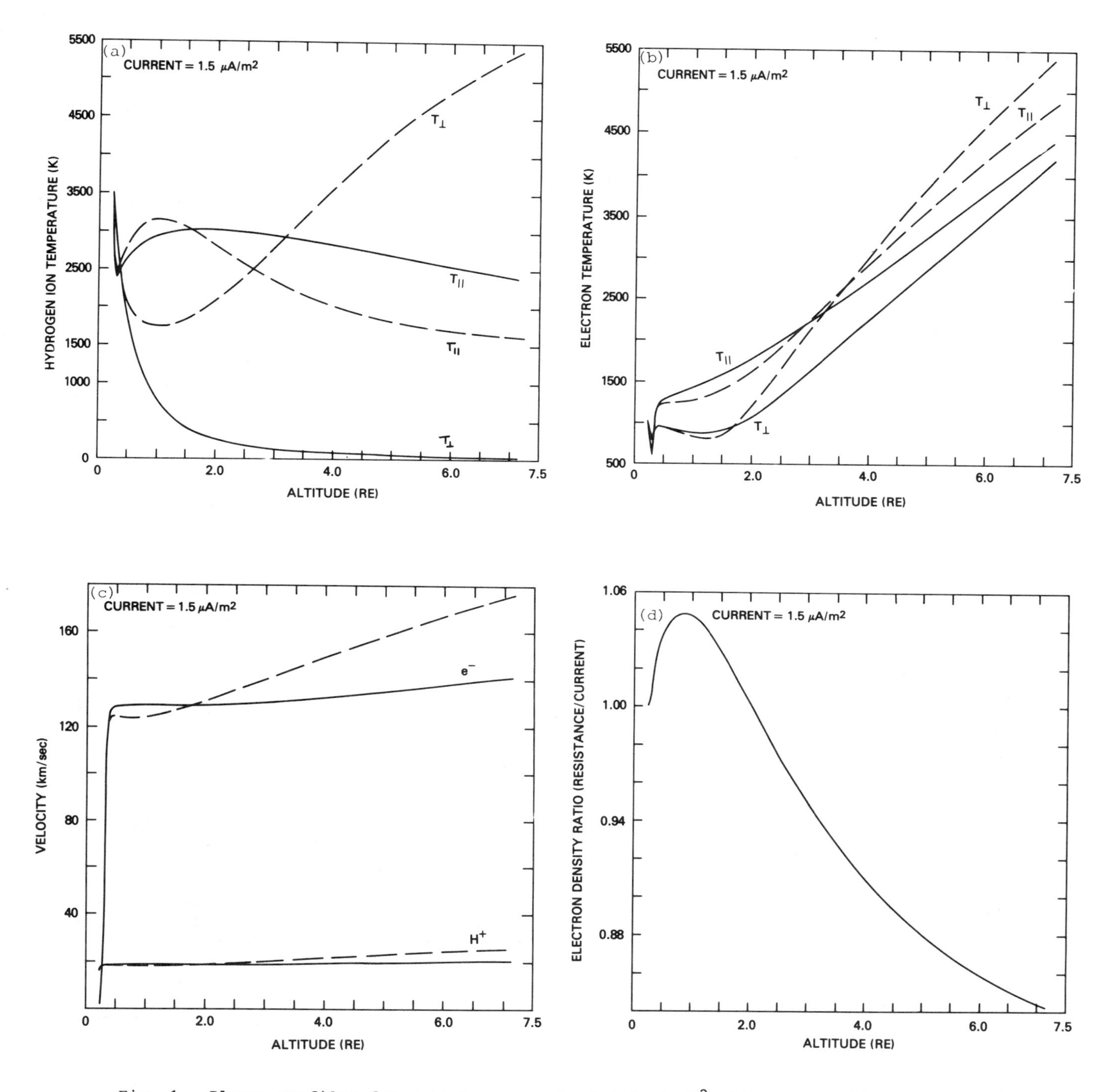

Fig. 1. Plasma profiles for a return current of 1.5 μA m^{-2} without the effects of ion heating and anomalous resistivity (solid curves) are compared with the corresponding profiles for the same current including these effects (dashed curves). (a) H^+ ion temperature parallel and perpendicular to the geomagnetic field. (b) Electron temperature parallel and perpendicular to the geomagnetic field. (c) H^+ ion and electron velocity profiles. (d) Electron density ratio (with ion heating and resistivity/current only).

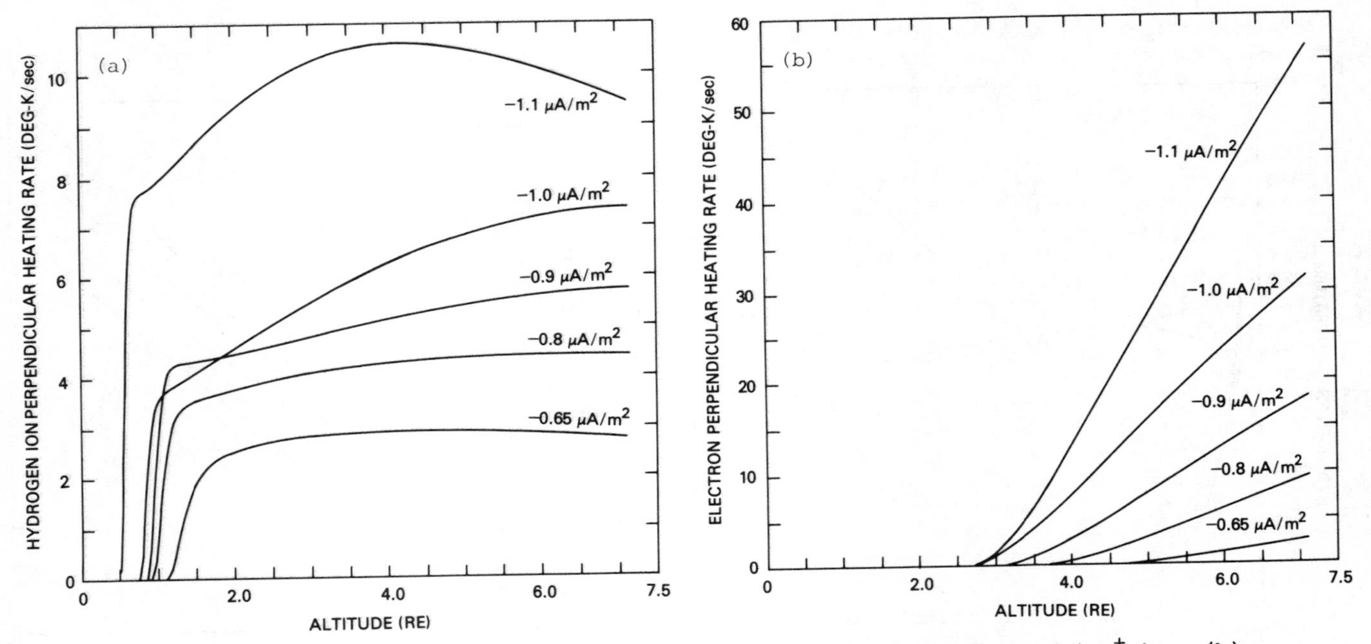

Fig. 2. Perpendicular heating rate versus altitude for all currents: (a) H^+ ion, (b) Electron.

(IR drop) which is unimportant at low altitudes. Potential profiles are shown in Figure 3. The ion heating tends to decrease the ambi-polar electric field $E_A \simeq -\Delta\Phi_A/\Delta L$ by increasing the scale height ΔL; whereas an increase in electron temperature would tend to increase E_A by increasing the ambi-polar potential $\Delta\Phi_A$. The ion heating effect dominates at low altitude and the field decreases as a result. The decrease in E_A decreases the hydrogen ion velocity at low altitudes (Figure 1c). The H^+ ion density increases as required for H^+ flux conservation. This increases the electron density to maintain charge neutrality. The total current remains constant, so the increased low-altitude density implies reduced electron velocity (Figure 1c). The ratio of the electron density with the instabilities relative to the current only case is shown in Figure 1d.

At high altitudes the electric field is mainly resistive and directed downward. The average force on an ion due to the resistive electric field is balanced by the anomalous friction terms, however, so that there is no net ion deceleration (i.e., decrease in bulk flow velocity) associated with the resistivity. The increase in the high-altitude H^+ ion velocities relative to their previous values with current only is due to effect of mirror forces acting on the transversely heated ions and, to a lesser extent, the weak high-altitude ambi-polar electric field.

The H^+ ion EIC heating rates are shown in Figure 2a for several different return current levels. The electron transverse instability-induced heating rates are shown in Figure 2b.

The potential drop along the tube for all currents is plotted in Figure 3. With increase in current and altitude, the contribution of resistive terms to the electric field increases. Burch et al. (1983) concluded from their analysis of Dynamics Explorer 1 (DE 1) high altitude plasma instrument (HAPI) data that the cold ionospheric electrons which carry the region 1 Birkeland return currents on the morning side are accelerated upward by potential drops of a few tens of electron volts at altitudes above 1 R_E. These potential drops may be thermal contact potentials produced by the interaction of hot magnetospheric and cold ionospheric plasmas in the presence of mirror forces, or by other effects not yet contained in our model. However, the resistive electric field in our model is in the right direction to accelerate electrons upward and may contribute to the observed acceleration. As noted earlier, the strength of the anomalous resistivity used in the model is limited by numerical stability considerations and the actual values are likely to be larger. The resistive electric field is balanced by anomalous friction for the average electron, but tail electrons will run away and be accelerated by the field.

We can calculate the total or lumped-sum resistance of a flux tube of 1 square meter cross section at the lower boundary from Ohm's law by dividing the total potential drop along the field line by the current. Based on such a calculation, a plot of resistivity versus current is presented in Figure 4. The resistance changes approximately linearly with current. Note that the flux tube is a "negative resistor" at lower return current levels: the ambi-polar electric field, which is a "thermoelectric battery" wherein electron thermal energy is converted to electric potential energy, dominates at low current. The relationship between return current and resistance may play a role in determining the distribution of return current in space in the direction transverse to the magnetic field (see, for example, Burke et al., 1983).

Summary

We have described a multimoment fluid model with anomalous transport coefficients designed to study classical and anomalous transport phenomena in the auroral return current region and to provide a test bed for the development of more general large-scale fluid models that are able to do a better job of modeling the interaction of large- and small-scale phenomena.

We have also investigated the effects of the EIC instability and an EIC-related anomalous resistivity mechanism on return current region plasma transport.

Our model at present incorporates only a very limited subset of the possible types of plasma turbulence and nonturbulent kinetic effects which could be important in ionosphere-magnetosphere coupling. In addition, we remind the reader that

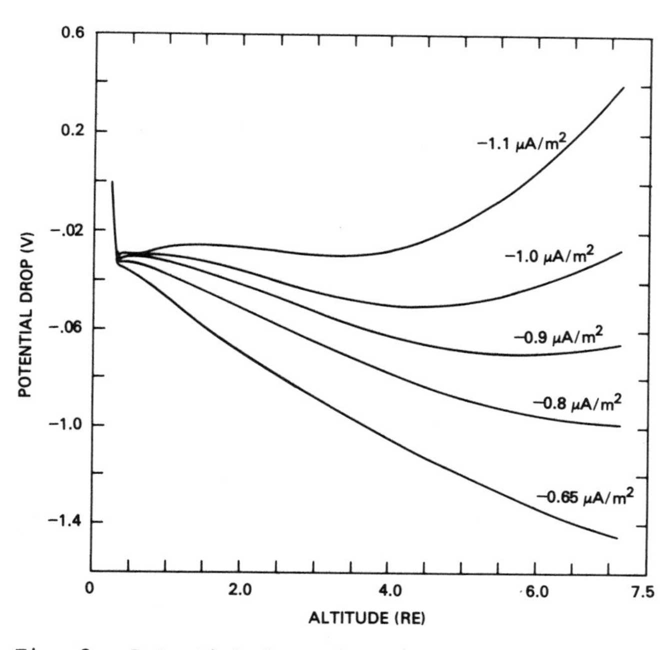

Fig. 3. Potential drop along the tube for all currents.

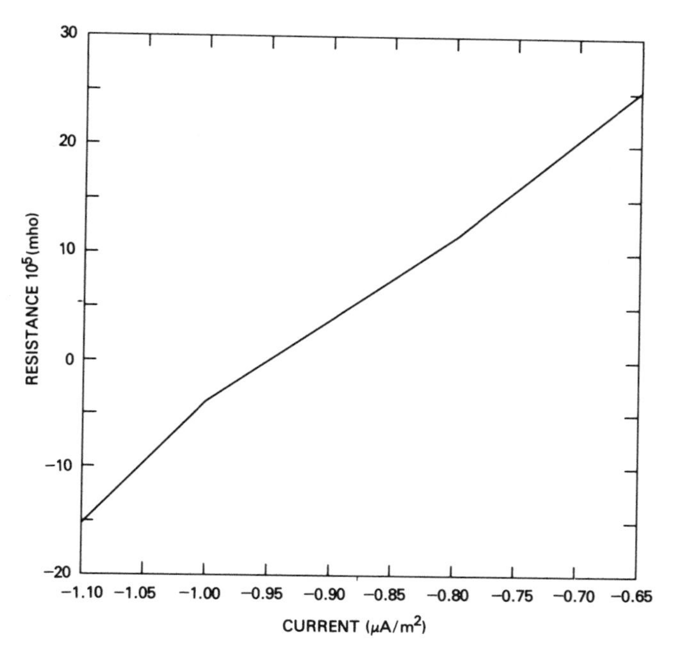

Fig. 4. Flux tube resistivity versus current.

the effective collision frequencies presently in the model are smaller than the realistic physical values; hence one should consider the qualitative features of these results without placing great weight on absolute magnitudes. With these caveats, the return current region model results appear to be physically reasonable and qualitatively consistent with behavior expected on the basis of the available relevant observational data.

Acknowledgments. This work is supported by the Office of Naval Research and the National Aeronautics and Space Administration.

References

Barakat, A. R., and R. W. Schunk, Transport equations for multicomponent anisotropic space plasmas: A review, Plasma Phys., 24, 389, 1982.

Burch, J., P. H. Reiff, and M. Sugiura, Upward electron beams measured by DE-1: A primary source of dayside region-1 Birkeland currents, Geophys. Res. Lett., 8, 753, 1983.

Burke, W. J., M. Silevitch, and D. A. Hardy, Observations of small-scale auroral vortices by the S3-2 Satellite, J. Geophys. Res., 88, 3127, 1983.

Ganguli, S. B., and P. J. Palmadesso, Plasma transport in the auroral return current region, J. Geophys. Res., 92, 8673, 1987.

Ganguli S. B., H. G. Mitchell, Jr., and P. J. Palmadesso, Behavior of ionized plasma in the high latitude topside ionosphere, NRL Memo. Rept. 5623, August 1985; also Planet. Space Sci., 35, 703 (1987).

Kindel, J. M., and C. F. Kennel, Topside current instabilities, J. Geophys. Res., 76, 3055, 1971.

Lee, K. F., Ion cyclotron instability in current-carrying plasmas with anisotropic temperatures, J. Plasma Phys., 8, 379, 1972.

Mitchell, H. G., and P. J. Palmadesso, A dynamic model for the auroral field line plasma in the presence of field-aligned current, J. Geophys. Res., 88, 2131, 1983.

Palmadesso, P. J., T. P. Coffey, S. L. Ossakow, and K. Papadopoulos, Topside ionosphere ion heating due to electrostatic ion cyclotron turbulence, Geophys. Res. Lett., 1, 105, 1974.

Palmadesso, P. J., S. B. Ganguli, and H. G. Mitchell, Jr., Multimoment fluid simulations of transport processes in the auroral zones, this volume, 1987.

DYNAMIC EVOLUTION OF LOW-ENERGY IONS IN THE TERRESTRIAL MAGNETOSPHERE

B. L. Giles,[1] C. R. Chappell,[1] J. H. Waite, Jr.,[1] T. E. Moore,[1] and J. L. Horwitz[2]

Abstract. Results of a statistical study of low-energy (0-50 eV) field-aligned ion pitch angle distributions for H^+, He^+, and O^+ observed by the Dynamics Explorer retarding ion mass spectrometer instrument are presented. Ion distributions are characterized as uni- and bi-directional field-aligned and bi-directional conic distributions. The spatial relationships of these distributions can be interpreted as dynamic evolution of low-energy ion flow in which injected uni-directional field-aligned streams originating in the nightside auroral zone evolve first into bi-directional flows and then into bi-directional conical flows under the influence of convection in the mirror magnetic field configuration. Characteristic convection times are evaluated with a pitch angle diffusion model in which field-aligned flows evolve into conical distributions through charge exchange loss of particles to the atmosphere during particle mirroring periods.

Introduction

Since the discovery of energetic O^+ ions in the magnetosphere by Shelley et al. (1972), ion composition measurements have revealed that a significant amount of plasma in the earth's magnetosphere is of ionospheric origin (see, for example, Chappell et al., 1987 and the review by Horwitz, 1987, and Yau and Lockwood, 1987). The importance of this ionospheric source is confirmed by the presence of energized ionospheric ions in all magnetospheric regions outside the inner plasmasphere (Ghielmetti et al., 1978; Johnson, 1979; Balsiger et al., 1980; Gorney et al., 1981; Lundin et al., 1982; Young et al., 1982; Collin et al., 1984; Eastman et al., 1984; Yau et al., 1985; Lennartsson and Shelley, 1986). Knowledge of the pitch angle distribution variations of these ions can provide insight into the large-scale plasma transport processes and acceleration mechanisms carrying ionospheric ions into the magnetosphere regions.

Several studies of angular distributions of low-energy ions have been made in the past few years. Particularly important for comparison is a survey of ISEE 1 low-energy ion pitch angle distributions at mid-latitudes by Nagai et al. (1983) introducing the possible mechanism of a nightside uni-directional ion outflow which evolves into bi-directional field-aligned and bi-directional conic distributions during sunward convection due to mirroring and loss cone formation. A recent study by Sagawa et al. (1987) examined various types of ion pitch angle distributions at energies below 1 keV using Dynamics Explorer 1 (DE 1) energetic ion composition spectrometer (EICS) data. These results are similar to those of Nagai et al. (1983). In addition, a study of DE retarding ion mass spectrometer (RIMS) O^+ field-aligned distributions by Lockwood et al. (1985a) has identified a dominant dayside source of O^+ from the polar cleft ionosphere. Studies of equatorially trapped ion distributions (Olsen, 1981; Olsen et al., 1987) found trapped plasma populations confined to within a few degrees latitude of the equator. Observations at synchronous orbit have shown that low-energy ions in this region exhibit mainly field-aligned distributions (see, for example, Horwitz and Chappell, 1979; Comfort and Horwitz, 1981; Fennell et al., 1981; Olsen, 1982).

This report will present a statistical survey of DE 1 RIMS low-energy ion field-aligned pitch angle distribution observations that suggests global mirroring and convection processes operative on H^+, He^+, and O^+ ions in the terrestrial magnetosphere. The DE 1 RIMS instrument has angular resolution and mass discrimination (1-32 amu) capabilities well suited to this type of study. The results complement and extend the aforementioned studies due to these instrument capabilities and the nearly complete orbital coverage in local time and latitude out to 4.6 R_E.

[1]Space Science Laboratory, NASA Marshall Space Flight Center, Huntsville, Alabama 35812.
[2]Department of Physics, The University of Alabama in Huntsville, Huntsville, Alabama 35099.

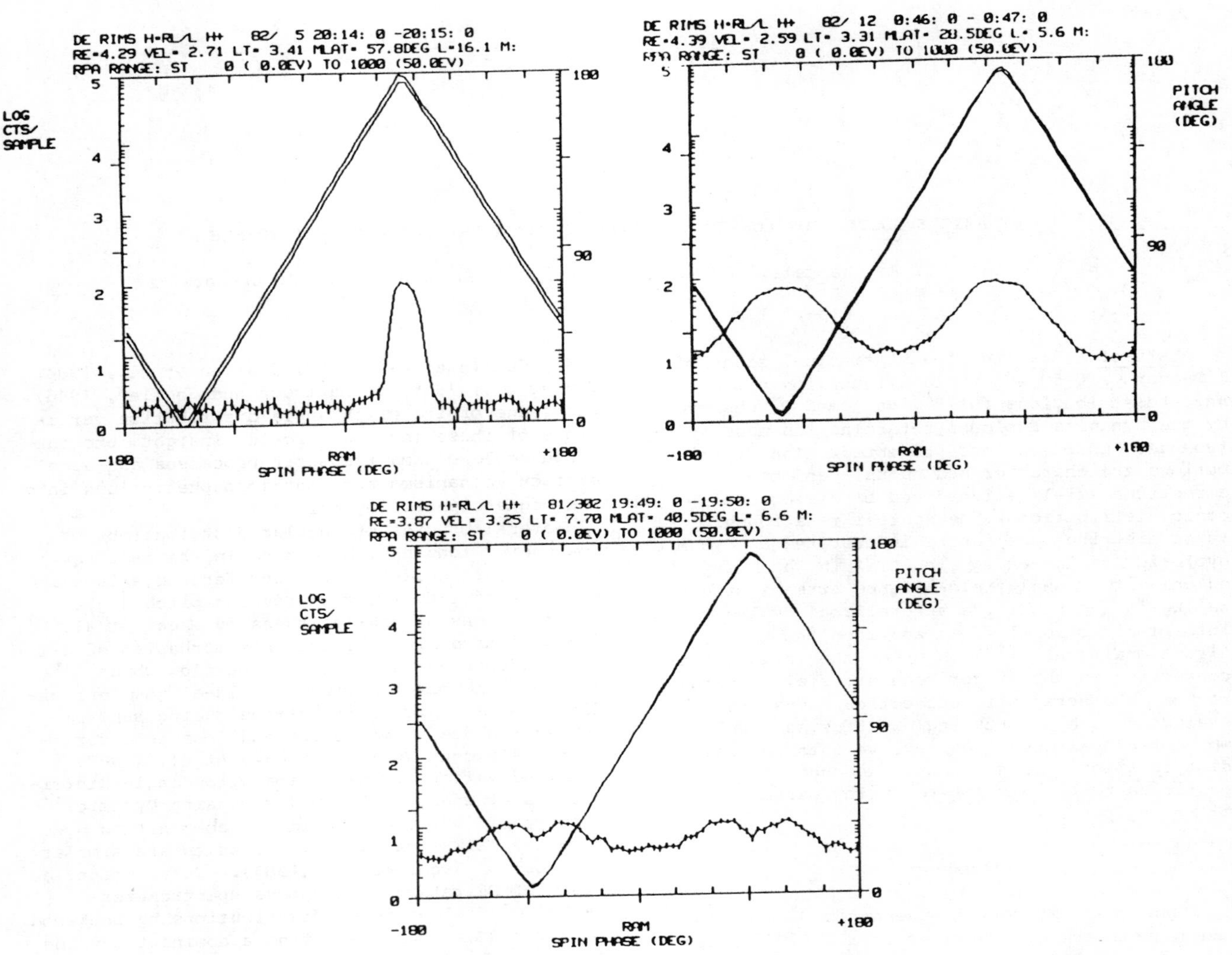

Fig. 1. Typical examples of low-energy (<50 eV) ion pitch angle distributions observed by DE 1 RIMS. Total accumulated ion count rates, corresponding to the vertical scale on the left-hand side, are shown as a function of spin phase angle. The sinusoidal-type curves give the pitch angle with respect to the geomagnetic field and correspond to the vertical scale on the right-hand side.

Data Analysis

The data set used in this study was derived from the RIMS instrument onboard the DE 1 satellite. A full description of the RIMS instrument is given by Chappell et al. (1981). Flows of H^+, He^+, and O^+ were classified according to their pitch angle distributions as determined from the radial sensor head of the instrument as it rotated through a full range of pitch angles in each 6-s satellite spin period. An example of such a pitch angle scan is shown in Figure 1.

The characteristic pitch angle distribution for each ion sample is assigned in a two-step automated process. First, a software algorithm scans the accumulated count rates averaged over a 1-min period to identify local maxima and minima as a function of spin phase angle. An intermediate data base is created storing centroid and width information for each peak encountered along with position of parallel and anti-parallel magnetic field line direction, relative minima between peaks, and peak skew calculations. Next, a software algorithm scans this intermediate data base for each 1-min period and assigns a pitch angle distribution type based on location of the maxima and minima relative to the magnetic field direction and the spacecraft ram direction. All available data from the period October 19, 1981, to January 1, 1983, were used in the study re-

TABLE 1. Numbers of 1-min pitch angle distribution samples for H^+, He^+, and O^+

Latitude	LT	Hydrogen		Helium		Oxygen	
		$1.0<R_E<3.0$	$3.0\leq R_E\leq 4.6$	$1.0<R_E<3.0$	$3.0\leq R_E\leq 4.6$	$1.0<R_E<3.0$	$3.0\leq R_E\leq 4.6$
ILAT > 45	0-6	2209	18140	1860	14793	1363	9889
	6-12	3310	16746	2991	9814	2289	5647
	12-18	6069	7581	5303	11969	3571	6678
	18-24	6825	10163	5611	7476	4347	5192
ILAT < 45	0-6	2483	16814	2202	15527	1629	12000
	6-12	4221	11855	3991	7546	2747	5014
	12-18	2157	9603	1936	7178	1386	3910
	18-24	5046	21505	4795	19601	3011	16888

gardless of RIMS operational mode, including those data when the aperture planes surrounding the detectors were set for 0, 2, 4, or 8 V negative potential relative to spacecraft. In the case of spacecraft potentials in excess of the magnitude of the aperture bias, it is possible that RIMS fails to detect the lowest-energy ions.

The elliptical orbit of DE, with apogee at 24,875 km altitude and perigee at 675 km and an orbital precession rate of 0.3° per day, results in an uneven coverage of local time and latitude during the seasons of the year. Total samples obtained for H^+, He^+, and O^+ are 144,727, 122,593, and 85,561 samples, respectively. Table 1 lists the number of H^+, He^+, and O^+ 1-min pitch angle distribution samples in four local time quadrants, two latitude levels, and two geocentric distance levels. Note that 48% of the data set represents Kp < 3o, 48% represents 3o < Kp < 6o, and 4% represents Kp > 6o, so that sampling of the higher activity levels is relatively sparse.

Classification of Pitch Angle Distributions

This study concentrates on the probability of occurrence for three basic types of pitch angle distributions: the uni-directional field-aligned distribution, the bi-directional field-aligned distribution, and the bi-directional conic distribution. Figure 1 shows accumulated counts versus spin phase angle plots of data recorded by the RIMS instrument. These data are given as examples of the three types of field-aligned flow observed. The uni-directional field-aligned distribution is characterized by a single flux peak in one direction with respect to the magnetic field and low flux in the other field-aligned direction. In some cases the peak flux is slightly shifted from the field direction. The bi-directional field-aligned distributions are characterized by flux peaks in both the parallel and anti-parallel field-aligned directions. The bi-directional conic distributions are similar to bi-directional field-aligned distributions in that there are flux peaks in the parallel and anti-parallel field-aligned directions, but there are also relative minima at the pitch angle minimum and maximum.

Observations

L Versus MLT Variations

Figures 2a-c show the occurrence probability variations with local time and L value for H^+ uni-directional field-aligned, bi-directional field-aligned, and bi-directional conic distributions. The data include all magnetic latitudes and Kp values. Also sampled, but not shown, are the distributions for He^+ and O^+. The uni-directional flows are dominant at L values greater than 8 at all local times except the evening auroral zone. There is a marked absence of uni-directional flows at the lower L shells in all local times and generally low occurrences in the dusk to midnight sector. The H^+ bi-directional cases are dominant at local times from midnight to noon. Near midnight, the distributions are found primarily at L shells between 4 and 6 and progressively extend to higher L shells on the dayside. These bi-directional distributions have higher probability than the uni-directional cases by as much as 20%. H^+ bi-directional conic flows are confined to the morning sector between 6 and 11 MLT and L shells between 4 and 8 and exhibit significantly lower probability of occurrence than the other distributions.

The uni-directional flows for He^+ are generally confined to L shells greater than 6 on the dayside and occurrences are generally low on the nightside for all L shell values. The He^+ bi-directional field-aligned distributions follow the pattern of H^+ but have lower probability by as much as 20%. The probability for He^+ bi-directional conic distributions is less than 10% for all local times and L shells.

The uni-directional distributions for O^+ are generally confined to L shells greater than 4 from the midnight region to noon. The O^+ bi-

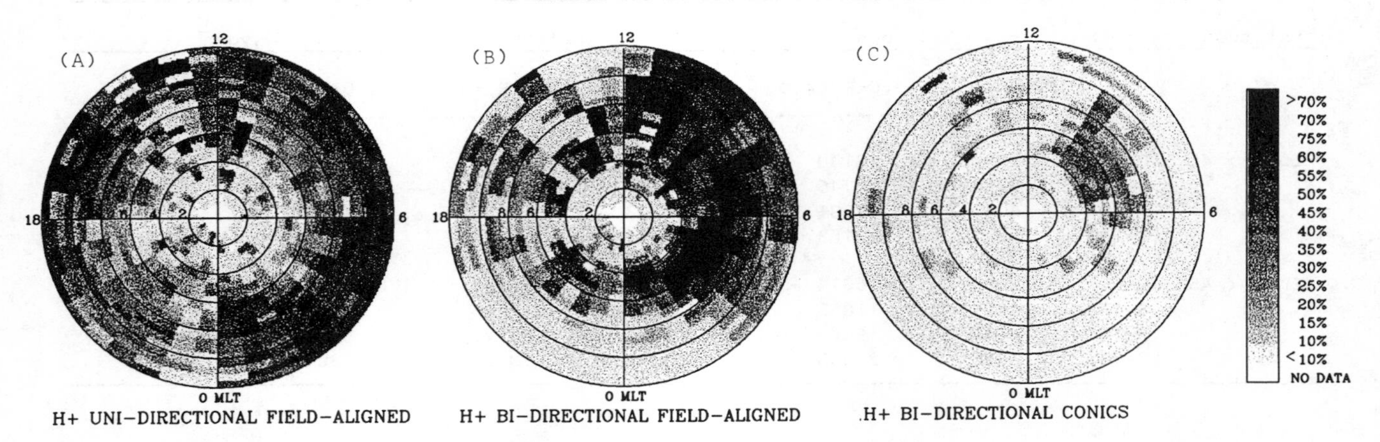

Fig. 2. Occurrence probabilities for H^+ low-energy (<50 eV) ion pitch angle distributions plotted in dipole L value versus magnetic local time diagrams. The L value is represented by the radius, increasing from L = 2 to L = 12 outward from the center of the figure.

directional field-aligned distributions occur at L shells between 4 and 6 for all local times and have significantly lower probability as compared to H^+ and He^+. The probability for O^+ bi-directional conic distributions is less than 1% for all local times and L shells.

ILAT Versus MLT Variations

Figure 3 shows the occurrence probability variations with local time and invariant latitude for H^+, He^+, and O^+ uni-directional and bi-directional field-aligned distributions. The data include all altitudes and Kp values. The uni-directional field-aligned distributions show substantial probability on the dayside sector for all ions. H^+ and O^+ also have a contribution on the nightside confined between 55° and 70° invariant latitude. He^+ has comparatively low probability on the nightside. The distributions extend to lower invariant latitudes as a function of mass. The H^+ flows are generally confined above 70° invariant latitude. He^+ is generally confined within 65° and O^+ within 60° invariant latitude. The occurrence probabilities for H^+, He^+, and O^+ bi-directional field-aligned flows form a narrow ring in local time, widening with mass.

Discussion

The uni-directional field-aligned distributions correspond to pitch angle distributions that have not undergone magnetic mirroring and are thought to consist of ions supplied from the auroral zone ionosphere and the dayside polar cusp. The data analysis showed that all the uni-directional field-aligned distributions were traveling up and not down the magnetic field lines. These distributions were shown to be dominant at high latitudes for all ion species. Traveling upward from the polar ionosphere regions, these ions flow along high-latitude magnetic field lines closed in a highly elongated fashion such that bouncing occurs over long time periods and may be disrupted by nonadiabatic processes. Those uni-directional flows originating at very high L shells in the midnight sector may be lost to the plasma sheet region or they may become bi-directional during convection to the dayside. Those originating at somewhat lower L shells have shorter bounce times and evolve into bi-directional field-aligned distributions during sunward convection. The heavier O^+ ions show appreciably larger occurrences at lower L shells on the nightside due to longer residence times along field lines after injection. The longer residence time results from the slower O^+ velocities compared to the lighter ions at a given energy.

In view of the 1-min accumulation time per sample, the bi-directional field-aligned flows probably correspond to pitch angle distributions that have undergone one or more bounce periods. These bouncing distributions are seen at mid-latitudes and lower L shells corresponding to the region of closed field lines nearer to the earth. The higher occurrence probabilities for bi-directional flow as compared to the uni-directional flow for the lighter ions may be explained by the significantly longer ion drift time at these lower L shells than the corresponding bounce time as the ions move sunward from the nightside source region. Typical bounce times for 10-eV H^+ ions injected at midnight are 24.5 min for a dipole L shell value of 4.1 and

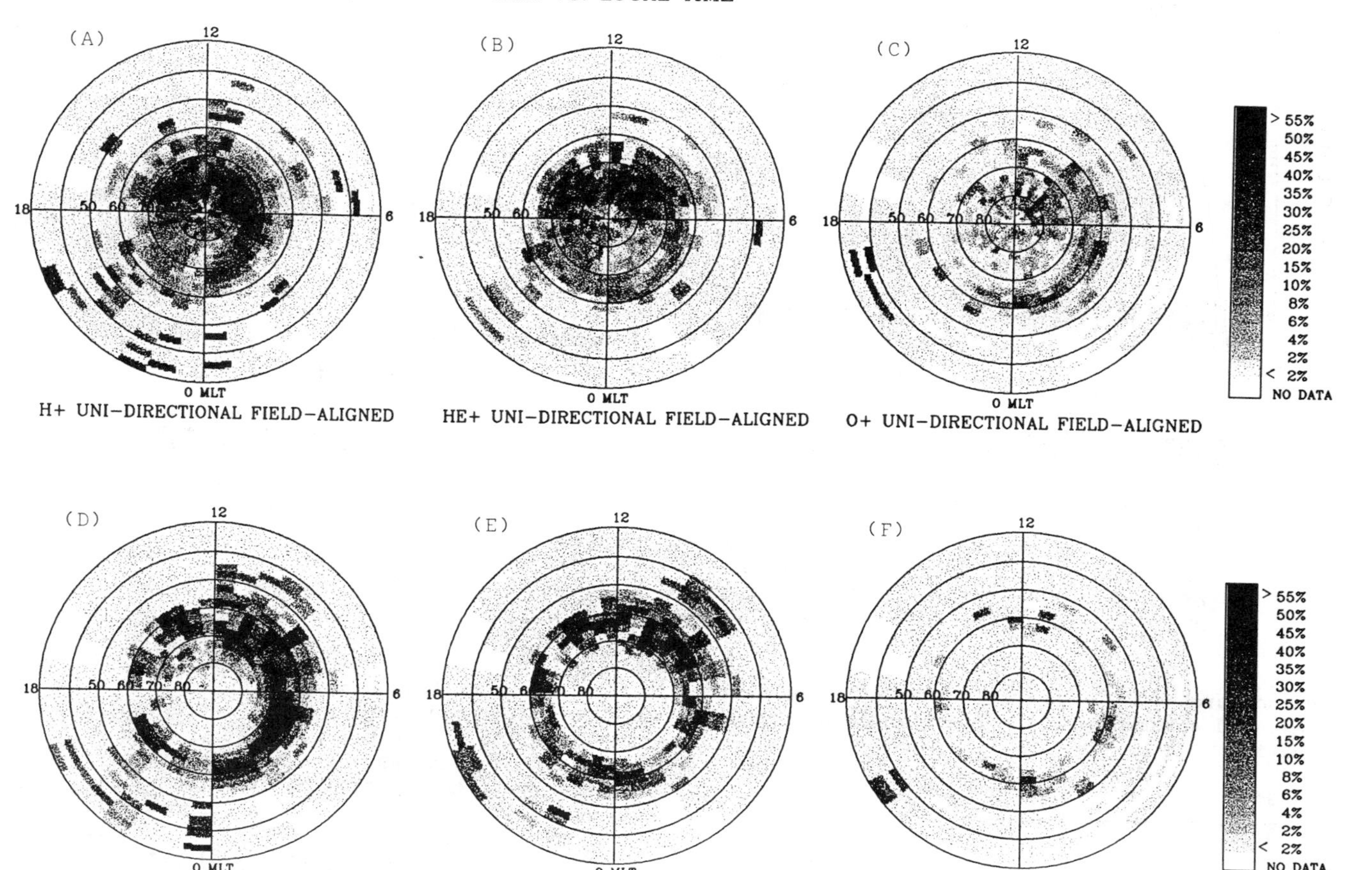

Fig. 3. Occurrence probabilities for H^+, He^+, and O^+ low-energy (<50 eV) ion pitch angle distributions plotted in invariant latitude versus magnetic local time diagrams. The invariant latitude is represented by the radius decreasing from 90 to 30 outward from the center of the figure.

51.0 min for a dipole L shell value of 8.4 (D. Delcourt, private communication, 1987).

The H^+ bi-directional conic distributions in the dayside trough region may be evidence of loss cones developing from bi-directional field-aligned distributions during sunward convection. After several bounce periods, the bi-directional field-aligned distributions can experience the loss cone formation through charge exchange in the dayside upper atmosphere.

Examination of invariant latitude versus local time occurrence probability plots indicates similar ionospheric source regions for H^+, He^+, and O^+. A nightside auroral ionosphere source is evident for all ion species. This nightside injection of ions provides those uni-directional field-aligned distributions which by the suggestion of Nagai (1985) can evolve into bi-directional field-aligned and then bi-directional conic distributions during sunward convection or may contribute to the plasma sheet region. In addition to this nightside source, all ion species are seen to have significant occurrences in the dayside polar cusp and are thought to be upwelling from the lower altitudes. The dayside source of O^+ ions may be compared to the work of Waite et al. (1985) and Lockwood et al. (1985b). Waite et al. (1985), using electic field measurements from DE 2, mapped nightside polar cap O^+ observations to an ionospheric dayside polar cap source. Lockwood et al. (1985a) presented a statistical survey of this new dayside source region for O^+ that should be viewed as a subset of the present study due to Lockwood's more restrictive selection criteria.

As a crude test of the hypothesis that the bi-directional conic distributions in the dayside trough region are the result of pitch angle dif-

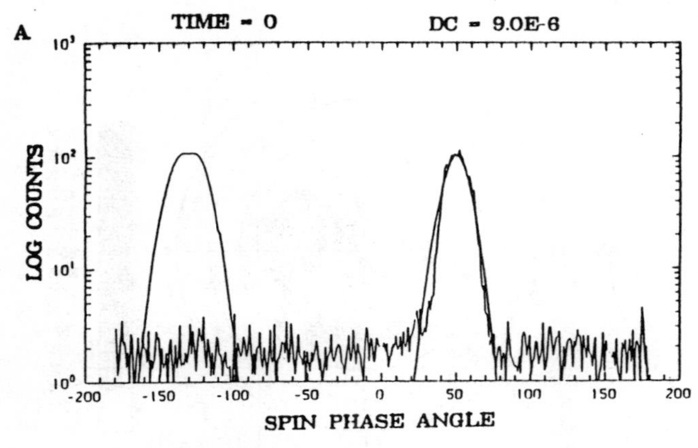

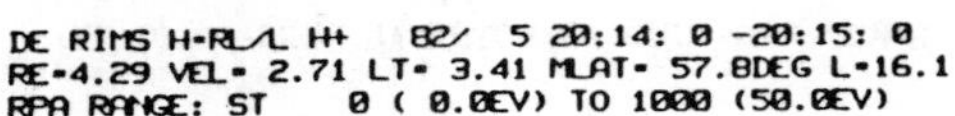

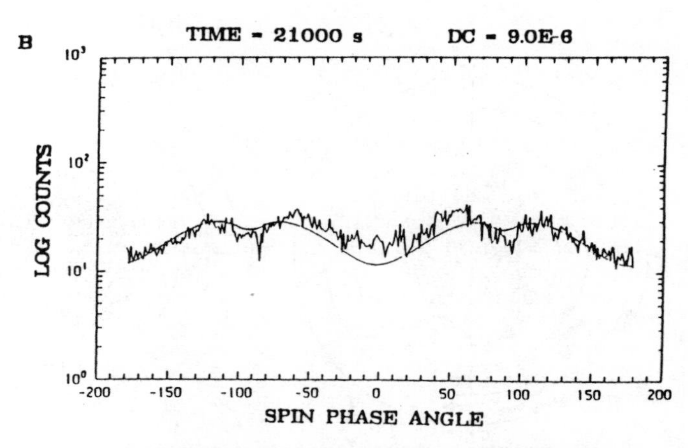

Fig. 4. Count versus spin phase angle plots illustrating the evolution of nightside uni-directional field-aligned streams into bi-directional conical flows by loss cone formation by the dayside upper atmosphere. In panel a, the initial pitch angle diffusion curve is shown superimposed over RIMS data observed at the assumed location of injection on the nightside. In panel b, the final pitch angle diffusion curve is shown superimposed over bi-directional conic RIMS data observed on the dayside.

fusion through an atmospheric loss cone, we have evaluated solutions to the pitch angle diffusion equation

$$\frac{\partial f_o}{\partial t} = \frac{1}{s(\alpha_o)\sin\alpha_o\cos\alpha_o}\frac{\partial}{\partial\alpha_o}\left[s(\alpha_o)\sin\alpha_o\cos\alpha_o\langle D_{\alpha\alpha}\rangle\frac{\partial f_o}{\partial\alpha_o}\right] - \frac{f_o}{\tau_{atm}},$$

where f_o is the particle distribution function, α_o is the particle pitch angle at the equator, D_α^o is the diffusion coefficient, $s(\alpha)$ is the variation of particle bounce period with pitch angle, and T_{atm} is the time scale for losses to the atmosphere. The equation is time averaged over a particle's bounce trajectory between mirror points (Lyons and Williams, 1984).

The initial distribution function is taken to be a bi-Maxwellian distribution with $T_\parallel = 10$ eV and $T_\perp = 1$ eV ions. The integral flux of this distribution over the 0-50 eV RIMS energy range matches to H^+ RIMS data for a uni-directional field-aligned distribution located at 2.4 hours magnetic local time. The parallel temperature enhancement presumably reflects the interpenetration of counterstreaming beams and not a parallel heating process at low altitudes. In Figure 4a, the initial distribution is shown superimposed over the RIMS data--the transition from a uni-directional field-aligned distribution to bi-directional field-aligned distribution is assumed to occur during the bounce period immediately following injection. Adopting an approximately corotating frame consistent with Ondoh and Aikyo (1986) and a diffusion coefficient of 9.0×10^{-6}, the diffusion equation is solved numerically for the evolution of pitch angle distribution with time. The results show that after 5.8 hours of sunward drift the initial bi-directional field-aligned distribution can evolve, through pitch angle diffusion, into a bi-directional conic distribution. In Figure 4b, the final pitch angle diffusion curve is shown superimposed over RIMS data for a bi-directional conic distribution located at 9.4 hours magnetic local time.

Conclusions

The low-energy (<50 eV) ion pitch angle distributions observed by the DE 1 RIMS experiment have been studied statistically at all local times and latitudes in the near-earth magnetosphere. The uni-directional field-aligned distributions are observed at high latitudes for all ion species. The bi-directional field-aligned distributions are observed at the mid-latitudes, at lower L shells than the uni-directional distributions. Bi-directional conic distributions are strictly confined to the dayside trough region.

These observations suggest the following scenario. Ions originating at very high L shells in the midnight sector are either lost to the plasma sheet region or become bi-directional during sunward convection. Those originating at somewhat lower L shells evolve into bi-directional field-aligned distributions during sunward convection and then into bi-directional conic distributions by pitch angle diffusion after several bounce periods with loss cone formation by the dayside upper atmosphere.

References

Balsiger, H., P. Eberhardt, J. Geiss, and D. T. Young, Magnetic storm injection of 0.9- to 16-keV/e solar and terrestrial ions into the high-altitude magnetosphere, J. Geophys. Res., 85, 1645, 1980.

Chappell, C. R., S. A. Fields, C. R. Baugher, J. H. Hoffman, W. B. Hanson, W. W. Wright, H. D. Hammack, G. R. Carignan, and A. F. Nagy, The retarding ion mass spectrometer on Dynamics Explorer-A, Space Sci. Instrum., 5, 477, 1981.

Chappell, C. R., T. E. Moore, and J. H. Waite, Jr., The ionosphere as a fully adequate source of plasma for the earth's magnetosphere, J. Geophys. Res., 92, 5896, 1987.

Collin, H. L., R. D. Sharp, and E. G. Shelley, The magnitude and composition of the outflow of energetic ions from the ionosphere, J. Geophys. Res., 89, 2185, 1984.

Comfort, R. H., and J. L. Horwitz, Low energy ion pitch angle distributions observed on the dayside at geosynchronous orbit, J. Geophys. Res., 86, 1621, 1981.

Eastman, T. E., L. A. Frank, W. K. Peterson, and W. Lennartsson, The plasma sheet boundary layer, J. Geophys. Res., 89, 1553, 1984.

Fennell, J. F., D. R. Croley, Jr., and S. M. Kaye, Low-energy ion pitch angle distributions in the outer magnetosphere: Ion zipper distributions, J. Geophys. Res., 86, 3375, 1981.

Ghielmetti, A. G., R. G. Johnson, R. D. Sharp, and E. G. Shelley, The latitudinal, diurnal, and altitudinal distributions of upward flowing energetic ions of ionospheric origin, Geophys. Res. Lett., 5, 59, 1978.

Gorney, D. J., A. Clarke, D. Croley, J. Fennell, J. Luhmann, and P. Mizera, The distribution of ion beams and conics below 8000 km, J. Geophys. Res., 86, 83, 1981.

Horwitz, J. L., The kinetic approach in magnetospheric plasma transport modeling, this volume, 1987.

Horwitz, J. L., and C. R. Chappell, Observations of warm plasma in the dayside plasma trough at geosynchronous orbit, J. Geophys. Res., 84, 7075, 1979.

Johnson, R. G., Energetic ion composition in the earth's magnetosphere, Rev. Geophys. Space Phys., 17, 696, 1979.

Lennartsson, W., and E. G. Shelley, Survey of 0.1- to 16-keV/e plasma sheet ion composition, J. Geophys. Res., 91, 3061, 1986.

Lockwood, M., J. H. Waite, Jr., T. E. Moore, J.F.E. Johhnson, and C. R. Chappell, A new source of suprathermal O^+ ions near the dayside polar cap boundary, J. Geophys. Res., 90, 4099, 1985a.

Lockwood, M., M. O. Chandler, J. L. Horwitz, J. H. Waite, Jr., T. E. Moore, and C. R. Chappell, The cleft ion fountain, J. Geophys. Res., 90, 9736, 1985b.

Lundin, R., B. Hultgvist, E. Dubinin, A. Zackarov, and N. Pissavenko, Observations of outflowing ion beams on auroral field lines at altitudes of many earth radii, Planet. Space Sci., 30, 715, 1982.

Lyons, L. R., and D. J. Williams (editors), Quantitative Aspects of Magnetospheric Physics, D. Reidel Publ. Co., Dordrecht, Holland, 1984.

Nagai, T., J.F.E. Johnson, and C. R. Chappell, Low-energy (<100 eV) ion pitch angle distributions in the magnetospere by ISEE 1, J. Geophys. Res., 88, 6944, 1983.

Olsen, R. C., Equatorially trapped plasma populations, J. Geophys. Res., 86, 11,235, 1981.

Olsen, R. C., Field-aligned ion streams in the earth's midnight region, J. Geophys. Res., 87, 2301, 1982.

Olsen, R. C., S. D. Shawhan, D. L. Gallagher, J. L. Green, C. R. Chappell, R. R. Anderson, Plasma observations at the earth's magnetic equator, J. Geophys. Res., 92, 2385, 1987.

Ondoh, T., and K. Aikyo, Equatorial drift paths of plasma particles in the Mead-Fairfield magnetosperic model, J. Radio Res. Lab., 33, 1, 1986.

Sagawa, E., A. W. Yau, B. A. Whallen, and W. K. Peterson, Pitch angle distributions of low energy ions in the near-earth magnetosphere, J. Geophys. Res., in press, 1987.

Shelley, E. G., R. G. Johnson, and R. D. Sharp, Satellite observations of energetic heavy ions during a geomagnetic storm, J. Geophys. Res., 77, 6104, 1972.

Waite, J. H., Jr., T. Nagai, J.F.E. Johnson, C. R. Chappell, J. L. Burch, T. L. Killeen, P. B. Hays, G. R. Carignan, W. K. Peterson, and E. G. Shelley, Escape of suprathermal O^+ ions in the polar cap, J. Geophys. Res., 90, 1619, 1985.

Yau, A. W., and M. Lockwood, Vertical ion flow in the polar ionosphere, this volume, 1987.

Yau, A. W., E. G. Shelley, W. K. Peterson, and L. Lenchyshyn, Energetic auroral and polar ion outflow at DE 1 altitudes: Magnitude, composition, magnetic activity dependence, and long-term variations, J. Geophys. Res., 90, 8417, 1985.

Young, D. T., H. Balsiger, and J. Geiss, Correlations of magnetospheric ion composition with geomagnetic and solar acitivity, J. Geophys. Res., 87, 9077, 1982.

MONTE CARLO MODELING OF LARGE-SCALE ION-CONIC GENERATION

J. M. Retterer,[1] T. Chang,[2] G. B. Crew,[2] J. R. Jasperse,[3] and J. D. Winningham[4]

Abstract. Cyclotron resonance with observed electric field fluctuations is demonstrated to be responsible for production of the oxygen ion conics that are observed by the Dynamics Explorer 1 satellite in the central plasma sheet region of the earth's auroral zone. The ion velocity distribution is described by a quasi-linear diffusion equation which we solve using the Monte Carlo technique. The acceleration produced by the observed wave spectrum agrees well with the ion observations, in both form and magnitude. To our knowledge, this represents the first successful comparison of an observed conic with any theoretical model.

Introduction

Wave-particle interaction with low-frequency auroral plasma turbulence has been commonly invoked (Lysak et al., 1980; Ashour-Abdalla and Okuda, 1984; Chang and Coppi, 1981; Crew and Chang, 1985; Retterer et al., 1986) to explain the transverse acceleration of observed (Mizera et al., 1981; Mozer et al., 1980), intense fluxes of energetic ions flowing out of the ionosphere in the auroral zone, which are known as ion conics because of the form of the ion distribution in velocity space. Until recently (Kintner et al., 1986), however, attempts to verify the theories by correlating ion flux and plasma turbulence data have been unsuccessful (Kintner and Gorney, 1984). Because of the small spatial scale of energetic conics and the regions in which they form, it has proven difficult to observe simultaneously both an ion conic and the waves which are responsible for its generation.

[1]Space Data Analysis Laboratory, Boston College, Chestnut Hill, Massachusetts 02167.
[2]Center for Space Research, Massachusetts Institute of Technology, Cambridge, Massachusetts 02139.
[3]Air Force Geophysics Laboratory, Bedford, Massachusetts 01731.
[4]Southwest Research Institute, San Antonio, Texas 78284.

By examining a special class of conic events, namely the oxygen-dominated, less energetic conics produced in the broad central plasma sheet (CPS) region of the auroral zone, we are able to report the first successful description of an observed ion conic by a theoretical model of ion acceleration through wave-particle interaction (Retterer et al., 1987a). The theoretical ion velocity distribution is calculated using a Monte Carlo technique, allowing us to compare not only the overall magnitude of the acceleration produced by the observed turbulence, but also to compare the form of the ion velocity distribution produced by the wave-particle interaction combined with the effect of the static, but inhomogeneous, geomagnetic field.

Such comparisons can be realized only where it is possible to measure the plasma turbulence and ion fluxes simultaneously. Intense, broadband, low-frequency, electric and magnetic field noise has been commonly observed at low altitudes over the earth's auroral zone by nearly all the satellites that have flown in this region (Gurnett et al., 1984). Particle measurements (Winningham and Burch, 1984) performed onboard the Dynamics Explorer 1 (DE 1) satellite have revealed the existence of a population of oxygen-dominated ion conics extending in latitude throughout the equatorward portion of the auroral zone (which maps out to the CPS in the earth's magnetotail) during times of magnetic storm activity. These energetic particle fluxes are coincident with intense, low-frequency auroral zone turbulence (D. Gurnett and M. Mellott, personal communications, 1986), and it has been suggested (Chang et al., 1986) that wave-particle interaction with this turbulence is responsible for the transverse acceleration of the ions to form the observed conics. Chang et al. (1986) found that if the turbulence contained a modest fraction of left-hand polarized waves, then its observed amplitude could easily account for the energies of the measured conics, making this event an ideal one for our detailed scrutiny.

The Transverse Acceleration Mechanism

In the absence of wave-particle interaction, the evolution of the ion velocity distribution, f, would be described by the Liouville equation (Roederer, 1970); the ions move adiabatically, conserving their magnetic moment and energy in the static geomagnetic and electric fields. In the presence of the observed broadband turbulence, the effect on the ions of their interaction with the turbulence is adequately described by a diffusion equation

$$\frac{df}{dt} = \frac{1}{v_\perp} \frac{\partial}{\partial v_\perp} \left(v_\perp D_\perp \frac{\partial f}{\partial v_\perp} \right) , \tag{1}$$

where $D_\perp$ is the quasi-linear velocity diffusion rate perpendicular to the geomagnetic field (Kennel and Englemann, 1966). We have used df/dt to symbolize the convective time derivative terms of the Liouville equation, which includes the effect of the magnetic mirror force. The diffusion coefficient $D_\perp$ depends on the electric field spectrum of the turbulence, but the generation and maintenance of the turbulent steady state is a complicated, nonlinear, nonlocal problem that is beyond the scope of this paper. In the present instance, we proceed using the observed wave spectrum to estimate the diffusion coefficient.

Before presenting the formula for $D_\perp$, we simplify it and express it in terms of the quantities measured by the instruments onboard the satellite. More details of the derivation are presented by Retterer et al. (1987b). For interaction between electromagnetic plasma modes and ions, we can take advantage of the approximation that wavelengths are sufficiently long so that the components of wave vector and ion velocity perpendicular to the magnetic field, $k_\perp$ and $v_\perp$, satisfy $k_\perp v_\perp \ll \Omega$, where Ω is the ion gyrofrequency and similarly that the parallel components satisfy $k_\parallel v_\parallel \ll \Omega$. This implies that the dominant contribution to $D_\perp$ comes from the spectral density of the electric field in the left-hand polarization with frequencies ω near Ω. The satellite measures no wave vector information for the electric field spectral density, so the result must be expressed in terms of the spectral density integrated over wave vector space, leaving only its dependence on frequency. Polarization information from the satellite observations is also limited, so let us take the spectral density of left-hand polarized waves, $|E_L|^2(\omega)$, to be some fraction η of $|E_x|^2(\omega)$, the observed spectral density of one orthogonal component of the electric field in the plane perpendicular to the geomagnetic field: $|E_L|^2(\omega) = \eta|E_x|^2(\omega)$. Thus, we have

$$D_\perp \approx \frac{\eta q^2}{4m^2} |E_x|^2(\omega = \Omega) , \tag{2}$$

where q and m are the ion charge and mass, respectively, and we have assumed that $|E_x|^2(\omega)$ is smooth enough near Ω that the small Doppler shift $k_\parallel v_\parallel$ can be neglected in evaluating $D_\perp$; to first approximation the diffusion coefficient is independent of velocity. The local heating rate, $2mD_\perp$, calculated using equation (2) agrees with the result obtained earlier (Chang et al., 1986) by a heuristic argument.

Over the relevant range of frequencies, the observed electric field spectral density is well approximated by a power law: $|E_x|^2(\omega) = |E_o|^2(\omega_o/\omega)^\alpha$, where $|E_o|^2$ is the spectral denstiy at the frequency ω_o. Because the spectral density is smaller at the gyrofrequencies of lighter-mass ion species, we expect that heavy species, such as oxygen, will be perferentially accelerated by this mechanism (Chang et al., 1986). The survey of DE 1 observations of the low-frequency auroral zone turbulence (Gurnett et al., 1984) indicates that $|E_o|^2$ is either roughly constant or grows with altitude; for simplicity let us assume that both $|E_o|^2$ and η are constant with altitude. Although the velocity diffusion rate is independent of velocity, it does depend on position through the variation of the ion gyrofrequency along the geomagnetic field line. Evaluating the spectral density at the ion gyrofrequency and taking the gyrofrequency to fall with the cube of the geocentric distance, $\Omega(s) \propto s^{-3}$, we find that the velocity diffusion rate increases with altitude, $D_\perp \propto s^{3\alpha}$, in the relevant region of the geomagnetic field. It is this power law form for $D_\perp$ which produces the interesting asymptotic properties of the solution of (1) for the ion velocity distribution discussed below.

The Monte Carlo Model

We will solve the kinetic equation for the ion velocity distribution using a Monte Carlo model which was developed to investigate problems of this kind (Retterer et al., 1983). Because the number of accelerated ions is small (oxygen is a minority constituent of the plasma), we treat them as test particles in externally imposed fields rather than calculating the fields self-consistently. (In addition, a self-consistent kinetic simulation of the microphysics could not practically describe the ion evolution over the distances of hundreds or thousands of kilometers that are relevant for the ion conic phenomenon.) From an initial distribution in velocity and space, the calculation of the evolution of the distribution proceeds by following the trajectories of a large number of ions with time. Between the velocity perturbations caused by interaction with the waves, it is assumed that the ions travel in the static geomagnetic field with constant energy and magnetic moment. The wave-particle interactions are taken into account by perturbing the ion velocities with random increments Δv such that $\langle \Delta v \Delta v \rangle = 2D\Delta t$ where the time step is Δt, and D is the velocity diffusion tensor.

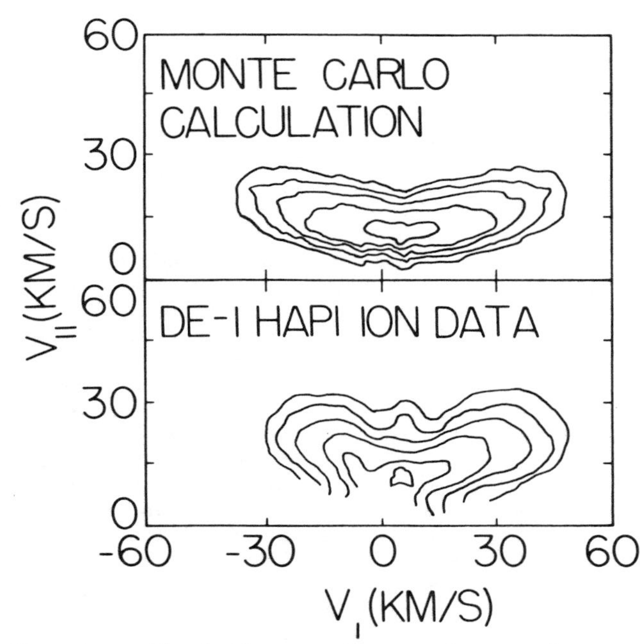

Fig. 1. The bottom panel presents a contour diagram of the observed ion-conic distribution function, measured by the HAPI instrument on DE 1, while the top panel presents our theoretical ion velocity distribution, plotted in the same way as the observed conic distribution.

To study the event observed at 23:46 UT on day 318, 1981 by DE 1, we follow the ion trajectories along a portion of a CPS geomagnetic field line at an invariant latitude of 60°, extending up to the geocentric altitude of the satellite, roughly 2.0 R_E. Because of the form of the velocity diffusion rate, most of the ion acceleration observed will have occurred near the altitude of the observation point, and the results of the calculation are insensitive to the initial conditions chosen for the oxygen ions; we started the ions at s = 1.2 R_E, thermally distributed with a temperature of 0.2 eV. The velocity diffusion rate was evaluated using the simultaneously measured amplitude of the electric field spectral density. The fitted power law parameters are $|E_o|^2 = 2.2 \times 10^{-8}$ $(V/m)^2$/Hz at $\omega_o/2\pi$ = 45 Hz, with α = 2.2. Retterer et al. (1987a,b) used the actual frequency spectrum rather than this fitted power law, but the results differ only in minor detail. Mean particle calculations (Chang et al., 1986) indicate that η = 1/8 will produce approximately the level of observed heating; this is the value of η used in the Monte Carlo calculations. As ions pass the observation point at 2.0 R_E, statistics on their velocities are accumulated to calculate the ion velocity distribution at this point. Several velocity moments are also calculated from the particle velocities to compare with the mean particle calculations of Chang et al. (1986).

The ion velocity distribution, calculated as described above, is presented in the top panel of Figure 1, in the same format as the satellite data to be discussed later. Positive $v_\parallel$ denotes ions that are traveling up the geomagnetic field line; a transverse velocity of 5 km s^{-1} was added to place the theoretical distribution in the satellite reference frame. The velocity distribution is plotted using contours of constant phase-space density in the $v_\parallel$-$v_\perp$ plane, uniformly spaced with an increment of 0.4 in the logarithm of phase-space density between contours. Without a simultaneous measurement of the ion density in the source region, the absolute normalization of the Monte Carlo velocity distribution is arbitrary, and has been chosen to match the level of the observed distribution. We do find that normalization of the calculated density to the observed density (~10 cm^{-3}) (D. Klumpar, W. K. Peterson, and E. G. Shelley, personal communications, 1986), yields a density of ~10^3 cm^{-3} when mapped back to the source; this is typical of the oxygen ion density at the source altitude. In Figure 1, we see that the theoretical velocity distribution indeed has the characteristic conic form, with the energetic ion flux peaking at pitch angles between 110° and 140°. The form of the velocity distribution is determined by the competition of the transverse acceleration and magnetic mirroring process. For the case of a power law heating rate considered here, the velocity distribution exhibits similarity scaling: one may obtain f at another altitude by scaling the velocity coordinates in magnitude by the power α + 1/3 of the ratio of the geocentric altitudes. This fact is being pursued in an effort to obtain a simpler, asymptotic solution to the kinetic equation (Crew and Chang, 1987).

Despite the absence of a parallel electric field in this calculation, noticeable parallel acceleration can be observed in Figure 1. All transverse acceleration schemes must cause some parallel acceleration, because the effect of the mirror force is to convert the perpendicular energy gained into parallel energy. The fact that the velocity distribution contours in this case resemble the hyperbolic curves that are a signature of the effect of a parallel electric field is a consequence of the simultaneous action of the magnetic mirror force and a heating process which is effective at all perpendicular velocities (Temerin, 1986). Because of this alternative means of parallel acceleration, evidence for parallel electric field acceleration of ion conics (Klumpar et al., 1984) must be interpreted carefully, at least for conics produced by this mechanism. Flatter spectra (i.e., smaller α) produce a wider spread in $v_\parallel$, while steeper spectra produce flatter velocity distributions because transverse heating is more localized in the latter case.

The mean particle theory (Chang et al., 1986) predicts the values of the parallel and perpendicular energies as a function of geocentric

TABLE 1. Comparison of results of Monte Carlo (MC) calculations with the simple analytic estimates for several wave spectra

α	$2mD_\perp$ (eV/s)	$W_\perp/W_\parallel$ MC	$W_\perp/W_\parallel$ Theory	W(eV) MC	W(eV) Theory
1.0	0.58	1.06	0.89	109.0	107.0
1.5	0.58	1.29	1.22	91.6	91.3
2.2	0.29	1.69	1.69	48.1	49.4
2.2	0.58	1.70	1.69	77.2	78.3
2.2	1.16	1.66	1.69	119.0	123.0
5.0	0.58	3.78	3.56	56.3	56.8

distance by including the mean heating rate per ion in a set of equations which describes the motion of an ion guiding center in the geomagnetic field. Solutions of these equations (Chang et al., 1986) show that the ratio of perpendicular to parallel energies $W_\perp/W_\parallel$ rapidly approaches the constant value $(6\alpha + 2)/9$. This behavior is consistent with the similarity scaling of the velocity distributions discussed above. In this limit, the equations for the ion energies can be easily integrated, providing results that are insensitive to the choice of initial conditions. The result for the total ion energy, $W = W_\parallel + W_\perp$, is

$$W(s) = (3\alpha + 11/2)^{1/3} m[sD_\perp(s)/(3\alpha + 1)]^{2/3}. \quad (3)$$

Table 1 presents a summary of several Monte Carlo runs, one set of runs made with different heating rates $2mD_\perp$ and another set with different power law exponents for the spectral density. The table gives the mean ion energy and the ratio of mean $W_\perp$ and $W_\parallel$ at $s = 2.0\ R_E$ and compares these results to the results of the asymptotic mean particle formulas given above. We see that the predictive power of the mean particle calculations is quite good; the statistical error inherent in the Monte Carlo results and the asymptotic approximation in the mean particle formulas easily account for the small differences between the results in the table.

Comparison with Observations

We now compare the predicted ion velocity distribution to the particle observations. The bottom panel of Figure 1 contains a contour plot of phase-space density, from the particle observations made with the high altitude plasma instrument (HAPI) instrument onboard DE 1. The velocity scales are drawn assuming that the ions are singly charged oxygen, as determined by the DE 1 energetic ion composition spectrometer (EICS) (D. Klumpar, W. K. Peterson, and E. G. Shelley, personal communications, 1986). In comparing the theory and the observations, we should note the following points. The contour plot of the observed velocity distribution is distorted for $v_\parallel < 8$ km s^{-1}, where no ion flux was detected but artificial flux counts were introduced in the contour plot. Along with the main ion-conic component of the velocity distribution, there is a superimposed, unresolved (cold), field-aligned component from some other source. There is also a slight asymmetry between the two lobes of the observed conic, which is probably due to spatial variations along the satellite trajectory which are beyond the scope of this model. With these considerations, we find that the main ion-conic components of the observed ion velocity distribution are well represented by the theoretical calculation, both in form and magnitude. This conclusion has been reinforced by the study of other observed conics (Retterer et al., 1987b) in which equally good or better fits were obtained. Taking advantage of simultaneously observed wave spectrum and ion flux data, we have obtained the first (to our knowledge) successful comparison of an observed conic and a theoretical conic model.

Discussion

Although the intense, low-frequency electric and magnetic field noise has been observed over the auroral zones for many years, the nature and origin of the turbulence are still not thoroughly understood, and are the subjects of ongoing research. During the event we modeled, there were no obvious local sources of free energy for the turbulence, and field-aligned currents measured by the magnetometer and the plasma instrument on DE 1 were weak and varied irregularly without correlation with variations in the ion conics (M. Sugiura, personal communication, 1986). The event did occur during a moderate magnetic substorm (similar events have been observed during other magnetic substorms), and the conics were observed equatorward of intense, high-energy electron precipitation, accompanied by ion beams, in the boundary plasma sheet (BPS). The polarization data for the low-frequency auroral zone turbulence in the DE 1 survey appear to be consistent with a nonlocal source mechanism for the turbulence (Gurnett et al., 1984) in which disturbances generated at higher altitudes propagate to low altitude as Alfvén waves (Goertz and Boswell, 1979; Lysak and Dum, 1983). Such waves could accelerate heavy ions as the waves approached gyroresonance after being reflected from the ionosphere, but the propagation of left-hand polarized waves at frequencies near the ion gyrofrequencies, with the resulting possibility of mode conversion phenomena (Swanson, 1985), remains largely unexplored in the suprauroral region. The uncertainty in the origin of the low-frequency turbulence does not alter the conclusion, based on the success of our modeling work, that this turbulence can explain the observed oxygen CPS conics.

Acknowledgments. We wish to thank I. Roth and M. Temerin for pointing out that the effects of transverse acceleration combined with adiabatic folding can sometimes mimic those of a parallel electric field. We also thank the Dynamics Explorer experimentalists, D. Klumpar, W. Peterson, E. Shelley, D. Gurnett, M. Mellott, and M. Sugiura, for providing the data that made this project possible. This research is partially supported by AFOSR contract F49620-86-C-0128, AFGL contracts FY7121-84-0-0006, F19628-83-C-0060, and F19628-86-K-0005, and NASA contract NAS5-28712.

References

Ashour-Abdalla, M., and H. Okuda, Turbulent heating of heavy ions on auroral field lines, J. Geophys. Res., 89, 2235, 1984.

Chang, T., and B. Coppi, Lower hybrid acceleration and ion evolution in the suprauroral region, Geophys. Res. Lett., 8, 1253, 1981.

Chang, T., G. B. Crew, N. Hershkowitz, J. R. Jasperse, J. M. Retterer, and J. D. Winningham, Transverse acceleration of oxygen ions by electromagnetic ion cyclotron resonance with broadband left-hand polarized waves, Geophys. Res. Lett., 13, 636, 1986.

Crew, G. B., and T. Chang, Kinetic treatment of oxygen conic formation in the central plasma sheet by broadband waves, this volume, 1987.

Crew, G. B., and T. Chang, Asymptotic theory of ion conic distributions, Phys. Fluids, 28, 2382, 1985.

Goertz, C. K., and R. W. Boswell, Magnetosphere-ionosphere coupling, J. Geophys. Res., 84, 7239, 1979.

Gurnett, D. A., R. L. Huff, J. D. Menietti, J. L. Burch, J. D. Winningham, and S. D. Shawhan, Correlated low-frequency electric and magnetic noise along auroral field lines, J. Geophys. Res., 89, 8971, 1984.

Kennel, C. F., and F. Engelmann, Velocity space diffusion from plasma turbulence in a magnetic field, Phys. Fluids, 9, 2377, 1966.

Kintner, P. M., and D. J. Gorney, A search for the plasma processes associated with perpendicular ion heating, J. Geophys. Res., 89, 937, 1984.

Kintner, P. M., J. Labelle, W. Scales, A. W. Yau, and B. A. Whalen, Observations of plasma waves within regions of perpendicular ion acceleration, Geophys. Res. Lett., 13, 1113, 1986.

Klumpar, D. M., W. K. Peterson, and E. G. Shelley, Direct evidence for two-stage (bimodal) acceleration of ionospheric ions, J. Geophys. Res., 89, 10,779, 1984.

Lysak, R. L., M. K. Hudson, and M. Temerin, Ion heating by strong electrostatic ion cyclotron turbulence, J. Geophys. Res., 85, 678, 1980.

Lysak, R. L., and C. T. Dum, Dynamics of magnetosphere-ionosphere coupling including turbulent transport, J. Geophys. Res., 88, 365, 1983.

Mizera, P. F., J. F. Fennell, D. R. Croley, A. L. Vampola, F. S. Mozer, R. B. Torbert, M. Temerin, R. Lysak, M. Hudson, C. A. Cattell, R. J. Johnson, R. D. Sharp, A. Ghielmetti, and P. M. Kintner, The aurora inferred from S3-3 particles and fields, J. Geophys. Res., 86, 2329, 1981.

Mozer, F. S., C. A. Cattell, R. L. Lysak, M. K. Hudson, M. Temerin, and R. B. Torbert, Satellite measurements and theories of low altitude auroral particle acceleration mechanisms, Space Sci. Rev., 27, 155, 1980.

Retterer, J. M., T. Chang, and J. R. Jasperse, Ion acceleration in the suprauroral region: A Monte Carlo model, Geophys. Res. Lett., 10, 583, 1983.

Retterer, J. M., T. Chang, and J. R. Jasperse, Ion acceleration by lower hybrid waves in the suprauroral region, J. Geophys. Res., 91, 1609, 1986.

Retterer, J. M., T. Chang, G. B. Crew, J. R. Jasperse, and J. D. Winningham, Monte Carlo modeling of oxygen ion conic acceleration by cyclotron resonance with broadband electromagnetic turbulence, Phys. Rev. Lett., 59, 148, 1987a.

Retterer, J. M., T. Chang, G. B. Crew, J. R. Jasperse, and J. D. Winningham, Monte Carlo modeling of oxygen ion conic acceleration by cyclotron resonance with broadband electromagnetic turbulence, in Physics of Space Plasmas (1985-87), SPI Conference Proceedings and Reprint Series, vol. 6, edited by T. Chang, J. Belcher, J. R. Jasperse, and G. Crew, Scientific Publishers, Inc., Cambridge, Massachusetts, 1987b.

Roederer, J. G., Dynamics of Geomagnetically Trapped Radiation, p. 89, Springer-Verlag, Berlin, 1980.

Swanson, D. G., Radio frequency heating in the ion cyclotron range of frequencies, Phys. Fluids, 28, 2645, 1985.

Temerin, M., Evidence for a large bulk ion conic heating region, Geophys. Res. Lett., 13, 1059, 1986.

Winningham, J. D., and J. Burch, Observations of large-scale ion conic generation with DE 1, in Physics of Space Plasmas (1982-84), SPI Conference Proceedings and Reprint Series, edited by J. Belcher, H. Bridge, T. Chang, B. Coppi, and J. R. Jasperse, p. 137, Scientific Publishers, Inc., Cambridge, Massachusetts, 1984.

POLAR CUSP ELECTRODYNAMICS - A CASE STUDY

P. E. Sandholt

Institute of Physics, University of Oslo, P.O. Box 1048 Blindern, N-0316 Oslo 3, Norway

Abstract. Combined satellite and ground-based observations provided the basis for investigating the electrodynamics associated with auroral structures within the polar cusp. A ~45-km-wide structure of enhanced electron precipitation with peak energy flux ~5 erg cm^{-2} s^{-1} (5 x 10^{-3} W m^{-2}) and average energy ~0.1 keV was observed to be associated with a strong, red-dominated auroral arc (I 630.0 nm ≃ 8 kR) with typical spectral ratio I 630.0 nm/I 557.7 nm ≃ 5. Within this arc the northward electric field component reached a peak value of ~175 mV m^{-1} A pair of field-aligned currents was observed, with downward current at the arc equatorward boundary and upward current further north. From correlated variations in electric and magnetic field components, the height-integrated Pedersen conductivity and joule heat dissipation rate within the auroral form could be estimated. The joule heat dissipation in the center of the structure was found to be ~5 times the particle energy input rate.

Introduction

A systematic classification of auroral arcs in different time sectors, according to their E-field and current characteristics, has been carried out by Marklund (1984). His two main categories were Birkeland current arcs and polarization arcs, based on the relative contributions to the horizontal E-field across the arc from the two current continuity mechanisms, Birkeland currents and polarization electric fields. His study was based on radar and rocket electric field observations.

Vickrey et al. (1986) investigated the auroral electrodynamics of summer and winter hemisphere structures with scale sizes between 3 and 80 km, using data from the drift meter and magnetometer on HILAT. In the summer cases they found that in general the $\Sigma_p \nabla \cdot \mathbf{E}$ term dominated the $\nabla\Sigma$ terms in the current continuity equation, due to high and relatively homogeneous conductivity. In the winter hemisphere more structured electric fields compensated the lower conductivity. Maynard et al. (1982) (cf. also references therein) reported small-scale E-field variations with amplitudes ~100 mV m^{-1} and associated particle precipitation in the polar cusp region, possibly related to localized merging injection of plasma in the cusp.

In this study we combine satellite and ground-based observations of structures in the polar cusp ionosphere. Photometric observations from the ground provide information on emission characteristics and dynamics, with the possibility of separating time and space variations in the particle flux (cf. Sandholt et al., 1986). The satellite observations (cf. The Johns Hopkins APL Technical Digest, 1984) included particle precipitation, magnetic field variations associated with Birkeland currents, as well as ion drift providing information on electric field components. We will focus on the electrodynamics of intermediate scale structures (~50 km latitudinal extent) in the polar cusp. From the satellite data we are also able to estimate the Poynting flux vector above the ionosphere and the joule heat dissipation rate in the ionosphere.

The cusp auroral structure analyzed here was found to be a typical Birkeland current arc with the joule heat dissipation rate in the center being 5 times higher than the energy input rate represented by the electron precipitation flux.

Observations

Figure 1 shows one passage of the HILAT satellite above the cusp aurora, slightly to the west of Svalbard, close to the magnetic mid-day (1107 MLT) meridian, on December 10, 1983. Also shown is the scanning direction of a four-channel meridian scanning photometer system at Longyearbyen (LYR), Svalbard (geographic coordinates: 78.2°N, 15.7°E; geomagnetic coordinates: 74.4°, 130.9°). By this technique the midday aurora is observed within the range 70°-80° geomagnetic latitude. Local magnetic noon at the optical recording site corresponds to ~0830 UT. An overview of the auroral situation above Svalbard and Alaska (simultaneous evening aurora recorded at 65° geomagnetic latitude) during a 3-hour period

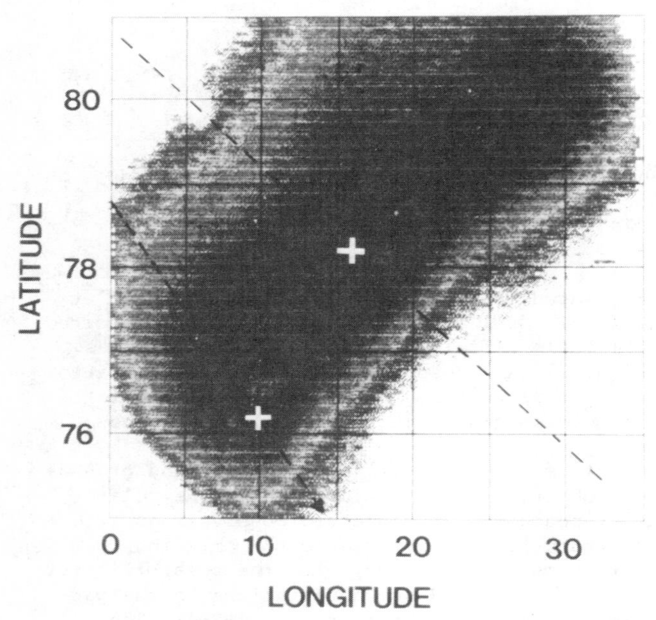

Fig. 1. Cusp aurora (red oxygen line at 630.0 nm) above Svalbard at 0804:45 UT on December 10, 1983, photographed by an image-intensified, all-sky camera in Longyearbyen (white cross in the center). The light intensities (digitized from photographic film) have been projected down to a flat earth, with the geographic coordinates marked in the figure. The scanning direction of the Longyearbyen photometer system (dashed line through the zenith) and the HILAT trajectory (dashed line in lower left corner) are shown. The white cross along the HILAT satellite track marks the foot of field line coordinates of the satellite at 0804:45 UT. The decreasing intensity toward magnetic west (lower left corner) is due to reduced camera sensitivity at the boundary of the field-of-view.

including the satellite pass is shown in Figure 1a of Sandholt et al. (1986). A large and stable southward component in the external magnetic field, within the magnetosheath, was observed between ~0730 and ~0855 UT. A steady southward motion of day and nightside auroral arcs occurred between ~0750 and ~0820 UT. A major substorm onset with the associated poleward expansion occurred on the nightside at 0820 UT. A more short-lived, poleward moving structure is observed in the Svalbard record near the time of the nightside breakup. This dayside activity included distinct signatures in the local magnetic fields at Ny Alesund (NYA) and Hornsund (HSD). For further details on the optical phenomenon and the physical interpretation, cf. Sandholt et al. (1986).

Figure 2 shows more details of the photometer recordings during the time of the satellite pass through the cusp region between 0804 and 0805 UT. We notice certain time variations of the aurora within the 3-min period covered in the figure. Between 0804 and 0805 UT two optical peaks are observed, one on each side of the zenith. A typical spectral ratio I 630.0 nm/I 557.7 nm for those peaks close to the zenith is ~4.5. This ratio is in good agreement with calculations by Stamnes et al. (1985) using Maxwellian spectra with average energy ~0.1 keV. The luminosity shows a rather sharp equatorward boundary, in contrast to the more gradual decrease toward the north.

Figure 3 shows HILAT measurements of east-west components of the ion drift and magnetic field, as well as electron precipitation (average energy, energy flux, number flux). Positive drift velocity component V_y (westward E x B drift) corresponds to the northward electric field, with 1 km s^{-1} equivalent to ~50 mV m^{-1}.

From the magnetometer trace B_y in the second panel of Figure 3, local values of current density can be derived, using the infinite current sheet approximation. Positive gradient in B_y (along the satellite trajectory) means downward current flow.

From the properties of the downward (zenith detector) electron flux, the cusp region is defined by the two vertical lines in the figure. An alternative definition of the cusp poleward boundary is near 76° MLAT (sub-satellite coordinate) which is associated with the decreasing energy flux toward north. The energy flux in the cusp, following the cusp definition in Figure 3,

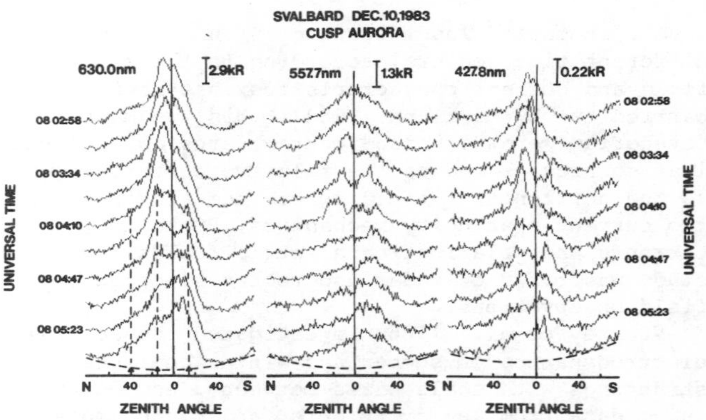

Fig. 2. North-south meridian photometer scans of auroral emissions at 630.0 nm (OI), 557.7 nm (OI), and 427.8 nm (N_2^+) observed from LYR, Svalbard during a 3-min period including the HILAT satellite pass between 0804 and 0805 UT (cf. Figure 1). Arrows and dashed vertical lines along the horizontal axis in the left panel mark the zenith angle locations of emission maxima at the time of the satellite pass.

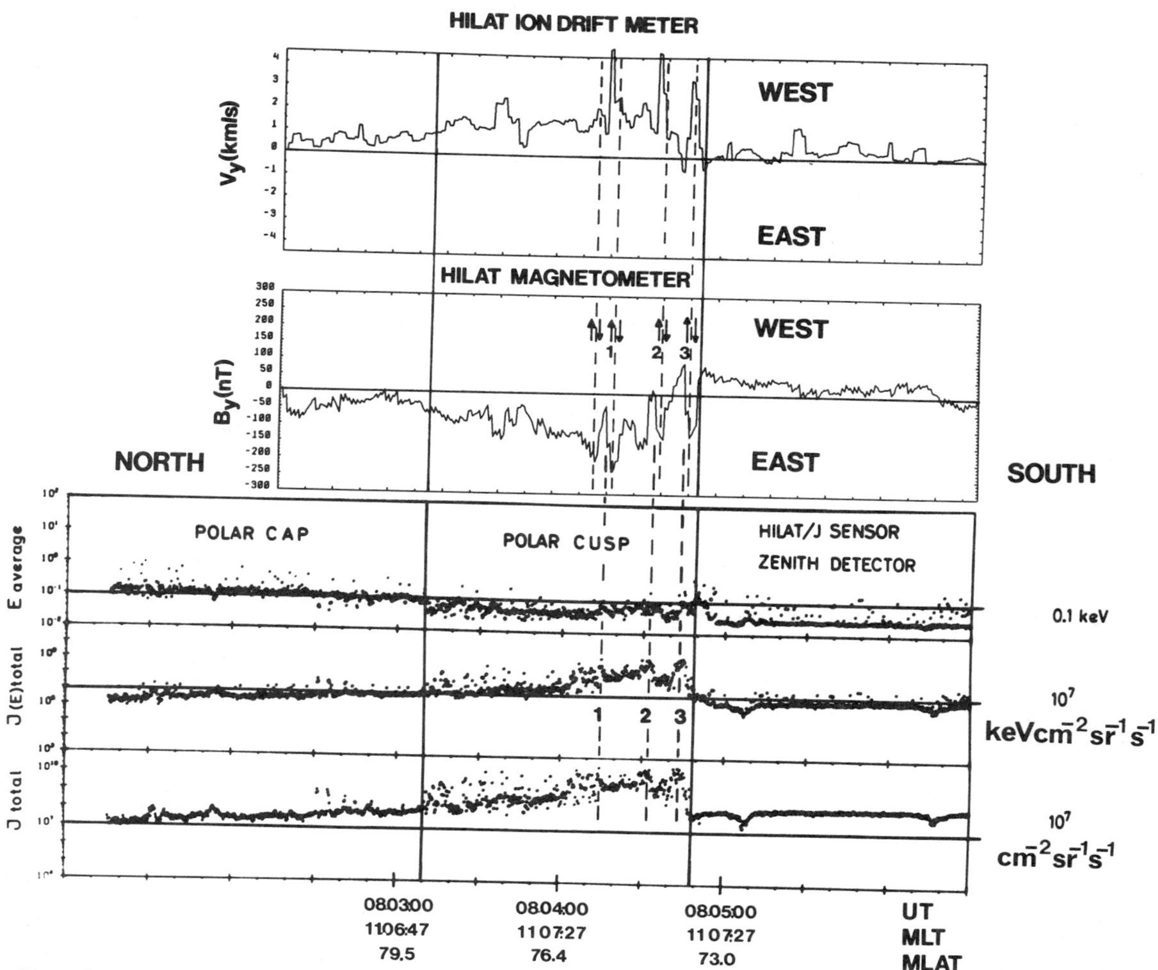

Fig. 3. Upper panel: ion drift transverse to the satellite trajectory (positive values corresponding to westward drift). Second panel: magnetic field east-west component deflections (positive westward). Arrows indicate the directions of Birkeland currents. Lower panels: electron precipitation data (average energy, energy flux, and number flux). Numbers 1, 2, and 3 mark the latitudinal locations of three main structures within the cusp (cf. Figures 2-5). MLAT is the sub-satellite coordinate.

is between 10^7 and 10^9 keV cm^{-2} sr^{-1} s^{-1} (or 5 x 10^{-2} and 5 erg cm^{-2} s^{-1}) with average energies generally ≈0.1 keV. Three of the most prominent precipitation structures are marked by labels 1 to 3. One-second averaged electron energy spectra show distinct shoulders in these regions of flux enhancements, indicating particle acceleration.

In the discussion section we will concentrate on feature 3, located close to the cusp equatorward boundary. The total width of that structure is ~45 km. All three flux enhancements are associated with strong intensifications of the westward drift velocity and consequently enhanced northward electric field, reaching peak values between 175 and 250 mV m^{-1}. Pairs of field-aligned currents in the same regions are observed, with peak density values ~20 μA m^{-2}, in a region with a more large-scale background downward current of ~1 μA m^{-2}.

In the northern part of the cusp the average B_y trend indicates a less intense (average value ~0.2 μA m^{-2}) upward current. Figure 4 shows electric field and magnetic field components along and transverse to the satellite trajectory, respectively. Positive E_x corresponds to northward E-field, while B_y is positive toward east.

Figure 5 illustrates the geometry of the observations from the ground and above the ionosphere (cf. Figure 1). We have focused on three different auroral structures observed at different zenith angles (cf. Figure 2), with the

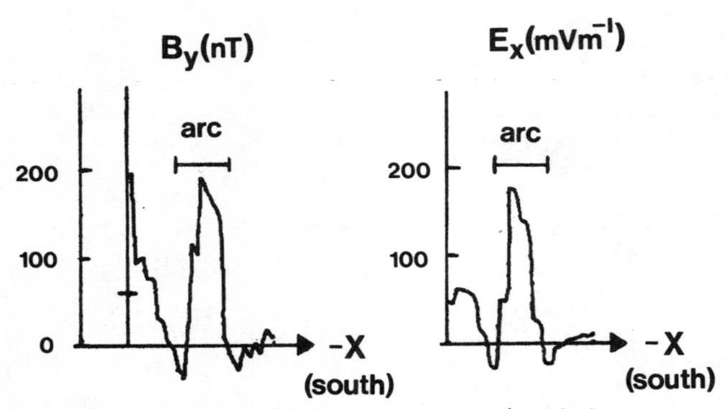

Fig. 4. Magnetic field east-west (positive eastward) and electric field north-south (positive northward) components through structure 3 in Figure 3.

corresponding precipitation structures shown in Figure 3. Representative values of different parameters observed in feature 3 are given to the right.

Discussion

We are going to focus on the electrodynamics of structure 3, located at the equatorward boundary of the cusp, showing enhanced particle energy flux, optical emission intensities, electric field, and field-aligned current density (cf. Figures 3-5).

Figure 4 shows a remarkably good correlation between the B_y and E_x profiles across the arc. The two curves are almost exactly overlapping within the arc.

An analytical expression for the relationship between these parameters follows from the general current continuity equation. In a cartesian coordinate system with the x, y, and z axes pointing toward north, east, and downward, respectively, and assuming $\delta/\delta y = 0$, we obtain (cf. Sugiura et al., 1984)

$$\mu_o^{-1} \frac{\delta \Delta B_y}{\delta x} = \Sigma_p \frac{\delta E_x}{\delta x} - E_y \frac{\delta \Sigma_H}{\delta x} . \tag{1}$$

Equation (1) can be integrated to give

$$\Delta B_y = \mu_o(\Sigma_p E_x - \Sigma_H E_y) + \text{const.} \tag{2}$$

From the assumptions $\delta/\delta y \simeq 0$ and $\nabla \times E \simeq 0$, it follows that $E_y \simeq$ const. A good correlation between ΔB_y and E_x then indicates that the gradients in the Hall and Pedersen conductivities at the equatorward boundary of the arc are small. Thus, we have:

$$\Delta B_y \simeq \mu_o \Sigma_p E_x + \text{const} , \tag{3}$$

where $\Sigma_p \simeq$ const. With B in nT, E in mV m^{-1}, and Σ_p in mhos, equation (3) reads (cf. Smiddy et al., 1980)

$$\Delta B_y(\text{nT}) = 1.25\ \Sigma_p(\text{mho})\ E_x(\text{mV m}^{-1}) + \text{const.} \tag{4}$$

Equation (4) can only be applied in cases when effects of neutral winds can be ignored. From Figure 4 we obtain $\Sigma_p \simeq 0.8$ mho for the case studied here. This value may be checked against the empirical formula obtained by Spiro et al. (1982)

$$\Sigma_p(\text{mhos}) \simeq \frac{20\ E_o}{(4 + E_o^{2})} J_E^{1/2} \tag{5}$$

with the average energy E_o in keV and the energy flux J_E in erg cm^{-2} s^{-1}. Inserting typical values for our case, $E_o \simeq 0.1$ keV, $J_E \simeq 5$ erg cm^{-2} s^{-1} gives $\Sigma_p \simeq 1$ mho.

From the satellite measurements we make the following estimate of the Poynting flux vector

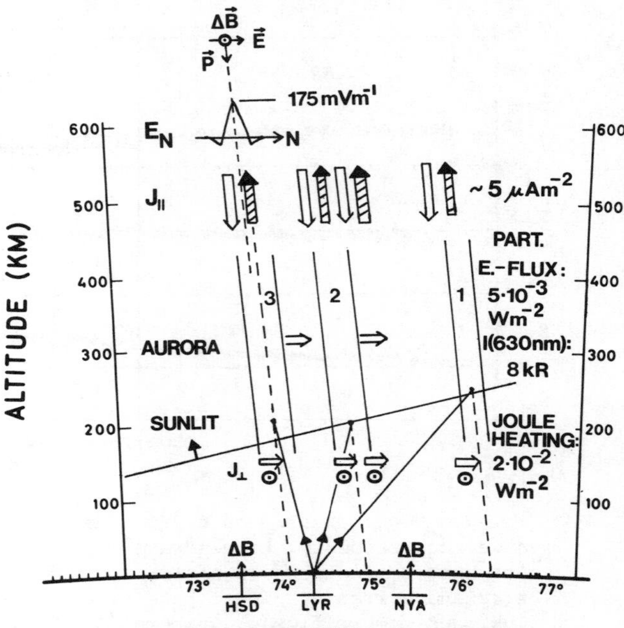

Fig. 5. Illustration of three structures within the cusp region as observed from the ground (scanning photometers) and from the HILAT satellite at 800 km altitude. Columns mark the locations of enhanced auroral emissions (cf. Figure 2) associated with electron precipitation maxima (cf. Figure 3). Also shown are the north-south electric field profile in structure 3 and pairs of Birkeland currents inferred from the east-west magnetic deflections in Figure 3. Representative values of northward electric field, Birkeland current density, electron energy flux, auroral emission intensity, and joule dissipation rate in structure 3 are shown on the right side. Directions of magnetic deflection, electric field, and Poynting vector above the ionosphere, in the center of strucure 3, are shown in the upper left corner.

associated with the northward E-field and the eastward magnetic deflection in the center of structure 3:

$$\mathbf{P} = \mu_o^{-1}\ \mathbf{E} \times \Delta\mathbf{B}$$
$$P_z \simeq \mu_o^{-1}\ E_x \cdot \Delta B_y. \qquad (6)$$

Using equation (3) we obtain

$$P_z \simeq \Sigma_p E_x^2.$$

Thus, within structure 3 we have a downward Poynting flux with the amplitude approximately equal to the ionospheric joule heat dissipation rate, which is of ~2 x 10^{-2} W m^{-2}. This is 5 times the value of the maximum particle energy flux in the same region of space.

Current continuity at the arc equatorward boundary gives the following expression for the northward electric field within the arc (cf. Marklund, 1984)

$$E_x^A = \frac{\Sigma_p^E}{\Sigma_p^A} \cdot E_x^E + \frac{\Sigma_H^A - \Sigma_H^E}{\Sigma_p^A} \cdot E_y^E + \frac{J_\parallel}{\Sigma_p^A}, \qquad (7)$$

$$\simeq E_x^E + \frac{J_\parallel}{\Sigma_p^A},$$

where the values inside and outside the arc are marked by superscripts A and E, respectively. In our case $J_\parallel (A\ m^{-1}) = j_\parallel (A\ m^{-2}) \cdot \Delta x \simeq 0.15$ $(A\ m^{-1})$ having used the average value of $j_\parallel$ = 5 $\mu A\ m^{-2}$ and the latitudinal width of the downward flowing current filament, Δx = 30 km. (The corresponding values for the upward current farther north are $j_\parallel$ = 10 $\mu A\ m^{-2}$, Δx = 15 km and $J_\parallel$ = 0.15 A m^{-1}. In both regions $j_\parallel \simeq \Sigma_p \cdot |\delta E_x/\delta x|$, with $\Sigma_p \simeq 0.8$ mho.) Thus, it follows that the last term in equation (7) is ~185 mV m^{-1}. Since the measured peak value of E_x is ~175 mV m^{-1}, the Birkeland current term in equation (7) is the most important one. According to the terminology introduced by Marklund (1984), the cusp structure studied here is a Birkeland current arc.

Based on the observed good correlation between ΔB_y and E_x we have already concluded that conductivity gradients and associated polarization electric fields play a minor role in the electrodynamics of this arc. This is so because of the soft particle precipitation in the mid-day cusp compared to other local time sectors of the auroral oval where conductivity gradients and polarization effects are much more important (e.g., Doyle et al., 1986).

Acknowledgment. It is a great pleasure to thank S. Basu, D. A. Hardy (principal investigator), and F. Rich of the Air Force Geophysics Laboratory (AFGL), Boston; P. F. Bythrow and T. A. Potemra of APL, Johns Hopkins University, Maryland; and R. Heelis, of the University of Texas at Dallas for providing particle, magnetic field, and ion drift data from the HILAT satellite. The all-sky camera operated at Svalbard was provided by H. C. Carlson (AFGL). Thanks are also due to Alv Egeland and Björn Lybekk, University of Oslo, for respectively organizing the optical observation program on Svalbard and taking the optical data/digitizing the all-sky picture used in this study. Part of this investigation was done when the author visited APL, Johns Hopkins University, Maryland, sponsored by NATO grant RG85/0521.

References

Doyle, M. A., W. J. Burke, D. A. Hardy, P. F. Bythrow, F. J. Rich, and T. A. Potemra, A simple model of auroral electrodynamics compared with HILAT measurements, J. Geophys. Res., 91, 6979, 1986.

The Johns Hopkins APL Technical Digest, 5, 1984.

Marklund, G., Auroral arc classification scheme based on the observed arc-associated electric field pattern, Planet. Space Sci., 32, 193, 1984.

Maynard, N. C., J. P. Heppner, and A. Egeland, Intense, variable electric fields at ionospheric altitudes in the high latitude regions as observed by DE-2, Geophys. Res. Lett., 9, 981, 1982.

Sandholt, P. E., C. S. Deehr, A. Egeland, B. Lybekk, R. Viereck, and G. J. Romick, Signatures in the dayside aurora of plasma transfer from the magnetosheath, J. Geophys. Res., 91, 10,063, 1986.

Smiddy, M., W. J. Burke, M. C. Kelley, N. A. Saflekos, M. S. Gussenhoven, D. A. Hardy, and F. J. Rich, Effects of high-latitude conductivity on observed convection electric fields and Birkeland currents, J. Geophys. Res., 85, 6811, 1980.

Spiro, R. W., P. H. Reiff, and L. J. Maher, Jr., Precipitating electron energy flux and auroral zone conductances: An empirical model, J. Geophys. Res., 87, 8215, 1982.

Stamnes, K., M. H. Rees, B. A. Emery, and R. G. Roble, Modeling of cusp auroras: The relative impact of solar EUV radiation and soft electron precipitation, in The Polar Cusp, NATO Adv. Study Inst. Ser. C, vol. 145, edited by J. A. Holtet and A. Egeland, p. 137, D. Reidel Publ. Co., Hingham, Massachusetts, 1985.

Sugiura, M., T. Iyemori, R. A. Hoffman, N. C. Maynard, J. L. Burch, and J. D. Winningham, Relationships between field-aligned currents, electric fields, and particle precipitation as observed by Dynamics Explorer-2 in Magnetospheric Currents, Geophys. Monogr. Ser., vol. 28, edited by T. A. Potemra, p. 96, AGU, Washington, D.C., 1984.

Vickrey, J. E., R. C. Livingston, N. B. Walker, T. A. Potemra, R. A. Heelis, M. C. Kelley, and F. J. Rich, On the current-voltage relationship of the magnetospheric generator at intermediate spatial scales, Geophys. Res. Lett., 13, 495, 1986.

TURBULENT GENERATION OF AURORAL CURRENTS AND FIELDS - A SPECTRAL SIMULATION OF TWO-DIMENSIONAL MHD TURBULENCE

Y. Song and R. L. Lysak

School of Physics and Astronomy, University of Minnesota, Minneapolis, Minnesota 55455

Abstract. The nonlinear evolution of large-scale turbulent boundary layer flow with magnetosphere-ionosphere coupling is investigated by a two-dimensional, forced, time-dependent MHD model of the disturbed flux tube. It is suggested that the nonlinear effect, especially the current nonlinear effect, plays an important role in breaking large-scale vortices and currents into medium and small ones. Spectral simulation results show that the large-scale turbulent magnetospheric vortices can be connected with highly structured auroral forms by an energy cascade process. The results are subject to a wavelength-dependent damping rate, which is caused by field-aligned anomalous resistivity and Pedersen conductivity, and fluctuating driving terms (due to, for example, the Kelvin-Helmholtz instability). Our results indicate that the evolution of a disturbed flux tube is determined by the nonlinear effect, the scale length dependence of the damping rate, and the structure of the driving terms.

Introduction

The auroral magnetosphere is a turbulent region with multiple time and length scales. Turbulent plasma sheet flow can be organized into vortices with scale lengths of a few earth radii (Hones et al., 1978, 1981; Saunders et al., 1981). The region 1 field-aligned current would result from the equatorial twin convection pattern (Sato and Iijima, 1979). Kintner (1976) indicated that the scale size of observed velocity shears within the large convection pattern from Hawkeye 1 implies magnetic Reynolds numbers of 10^3 to 10^4. Plasma flow data analysis from IMP-6 showed that slow plasma flows, which are caused by driving mechanisms other than reconnection, are independent of direction (Hayakawa and Nishida, 1982), consistent with randomly directed turbulent flow. Auroral spirals, curls, and folds of the order of 1-1000 km is a common part of active auroral display (Hallinan and Davis, 1970; Davis and Hallinan, 1976). High resolution electric field and magnetic field data obtained by S3-3 (Mozer et al., 1977; Mozer et al., 1980) and Dynamics Explorer 1 and 2 (DE 1 and 2) (Sugiura, 1984; Sugiura et al., 1982; Weimer et al., 1985; Weimer et al., 1987) showed auroral zone electric field and current structures at many different scale lengths extending from 0.1 to ~1000 km (at the field line base). The mechanism of the generation of highly structured auroral currents and fields is still an unanswered question.

An experimental demonstration showed that an instability in current and charge sheets may play a part in the formation of auroral folds and curls (Webster and Hallinan, 1972; Buneman et al., 1966). A dual cascade of magnetospheric electric field power spectral density from high-altitude balloons and the Hawkeye 1 satellite agrees with the power spectrum from two-dimensional Navier-Stokes fluid turbulence. It is suggested that auroral beam instabilities may inject energy at scales of 3 ~ 30 km. The energy inverse cascade produces a system with energy at large wavelength compared to the auroral widths (~2000 km) (Kelley and Kintner, 1978). Swift (1977) has argued that large-scale, field-aligned currents are presumably connected with the magnetospheric circulation and an inverse cascade process, in which turbulent energy at scales from the gyroradius of earthward streaming ions to the mesosize scale may cause highly structured aurora. Miura and Pritchett (1982) suggested that the MHD Kelvin-Helmholtz instability could serve as a dynamo process driving small-scale, field-aligned current in the presence of sheared plasma flow in the magnetosphere.

Many of these theories and simulations of current generation do not depend on the condition of the whole auroral flux tube. As is known, the discrete auroral arcs are associated with upward Birkeland current sheets caused by equatorial twin convection cells. The current is carried by means of a shear Alfvén wave, which connects ionosphere and magnetosphere and closes by Pédersen and Hall currents. Highly structured auroral currents and fields (including folds, curls, and spirals) depend on dynamics of the whole disturbed auroral flux tube: turbulent dynamo energy input, field-aligned conductivity,

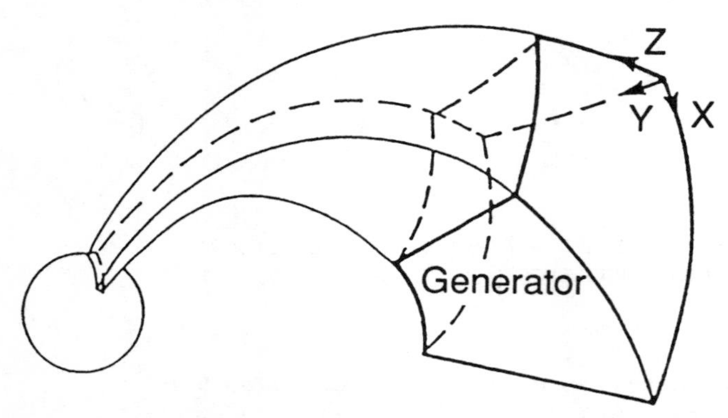

Fig. 1. Sketch of the model geometry. The generator region simulated is in the x,y plane. x and y directions are the longitudinal and the latitudinal direction, respectively.

and the ionosphere and generator region conductivity (Lysak, 1985, 1986).

Some MHD models include coupling between magnetosphere and ionosphere (Goldstein and Schindler, 1978; Goertz and Boswell, 1979; Lysak and Dum, 1983; Watanabe and Ashour-Abdalla, 1986), especially parallel electric fields produced by microscopic turbulence in an auroral flux tube (Lysak and Dum, 1983). The magentospheric convection pattern is generally imposed as a boundary condition, although Lysak (1985) treated a more general generator boundary condition. More recently, a two-dimensional model for time-dependent coupling between magnetospheric shear layers and the ionosphere, in which the magnetospheric convection pattern is allowed to evolve through self-consistent interaction with the ionosphere, was presented (Lotko et al., 1987; Lotko, 1986). Their results successfully showed that nonlocal dissipation is a possible mechanism for producing longitudinally elongated auroral forms. The dipole field mapping effect leads to an anisotropy in the decay rate of longitudinal and latitudinal disturbances, and consideration of field-aligned conductivity leads to a minimum in the decay rate. The perturbation of the magnetic field is neglected; e.g., the magnetic nonlinear term is not included in their model. However, results from two-dimensional MHD turbulence studies indicate that the magnetic nonlinear term plays an important role in the formation of current filaments and small-scale vorticities, causing an enhanced cascade and more rapid decay (Kraichnan and Montgomery, 1980; Orszag and Tang, 1979; Matthaeus and Montgomery, 1981).

The purpose of this simulation is to investigate the nonlinear evolution of non-steady boundary layer flow to highly structured auroral currents and fields. As the first step, a two-dimensional, forced, time-dependent MHD model has been used to study the dynamics of whole disturbed flux tube. Results show nonlinear evolution of large-scale currents to small-scale currents. The wavelength-dependent damping rate given by Lotko et al. (1987) is observed.

Simulation Model

We adopt a two-dimensional simplified Strauss equation to describe the dynamics of turbulent boundary layer flow. It is known that energy transfer in wave number space in the direction parallel to background magnetic field is almost inhibited (Shebalin et al., 1983). The Strauss equations (Strauss, 1976; Montgomery, 1982) are a reduced set of three-dimensional MHD equations between two and three dimensinos for anisotropic MHD plasmas, where the dynamics in the direction perpendicular to dc magnetic field is dominated by the convective nonlinearity, and the dynamics in the direction parallel to dc magnetic field is dominated by Alfven waves. The Strauss equations are:

$$\frac{\partial \omega}{\partial t} + (\mathbf{u}_\perp \cdot \nabla_\perp)\omega = (\mathbf{b}_\perp \cdot \nabla_\perp)j + B_o \frac{\partial j}{\partial z} + \frac{\nabla^2 \omega}{R} \quad (1)$$

and

$$\frac{\partial a}{\partial t} + (\mathbf{u}_\perp \cdot \nabla_\perp)a = B_o \frac{\partial \psi}{\partial z} + \frac{\nabla^2 a}{R_M}, \quad (2)$$

where $j = -\nabla^2 a$ is the electric current density and $\omega = -\nabla^2\psi$ is the vorticity (both are only in the z direction). Standard dimensionless units are used in the simulation (Montgomery, 1982). $\mathbf{B}_o = B_o\hat{z}$ is the dc magnetic field, and R and R_M are the mechanical and the magnetic Reynolds numbers, respectively.

Figure 1 shows a sketch of a whole disturbed flux tube (Lysak, 1985). The generator region to be simulated is simplified in the x,y plane. The variation of current within the generator in the z direction is assumed to be linear. A set of nonlocal equations including Ohm's law, current continuity in the ionosphere, the frozen-in condition in the generator region, and the linear relationship between field-aligned current and the potential drop in the acceleration region is used instead of the second Strauss equation.

$$\frac{\partial \omega}{\partial t} + (\mathbf{u}_\perp \cdot \nabla_\perp)\omega = \frac{(\mathbf{b}_\perp \cdot \nabla_\perp)}{c\rho} j + \frac{B_e j}{c\rho H} + \nu\nabla^2\omega + F \quad (3)$$

$$\omega = \frac{c}{B_e} \nabla_e^2 \phi_e \quad (4)$$

$$j = -(\nabla_i^2 \phi_i)\Sigma_p \quad (5)$$

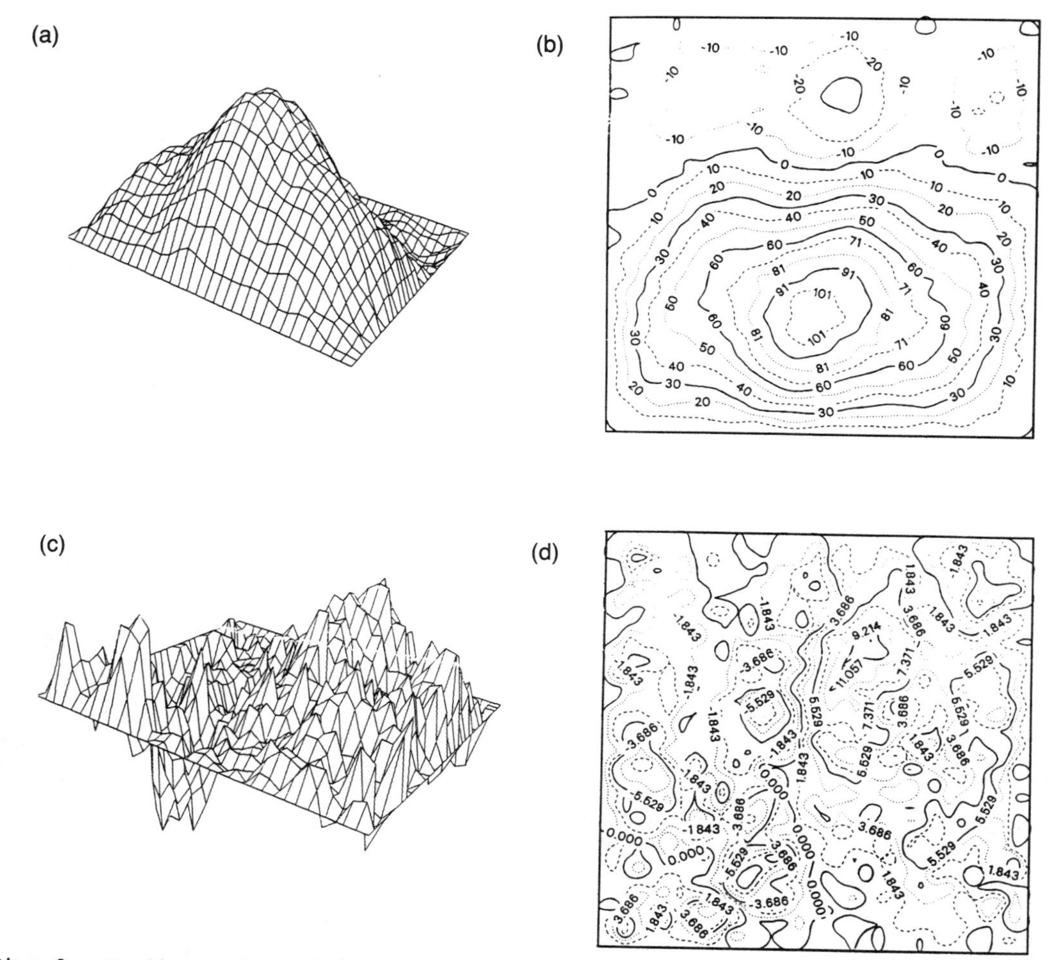

Fig. 2. Nonlinear decay current evolution. (a) Current contour plot at t = 0. (b) Current surface plot at t = 0. (c) Current contour plot at t = 72. (d) Current surface plot at t = 72. Note small-scale currents appear at t = 72.

$$j = \alpha(\phi_i - \phi_e) \tag{6}$$

where i and e represent the physical quantities in the ionosphere and the equator (generator region), respectively. The magnetic field perturbation $\mathbf{b}_\perp(x,y,t)$ and velocity field $\mathbf{u}_\perp(x,y,t)$ lie in the x,y plane and are assumed to be independent of z. They satisfy the free slip boundary condition in a square box. The equatorial electric field $\mathbf{E}_e = -\nabla\phi_e$, B_{oe} is the background magnetic field, and H is the effective height of the boundary layer (Lotko et al., 1987). The height-integrated Pedersen conductivity and the field-aligned conductivity are Σ_p and α, respectively. Incompressibility is used in our model. Constant density and light speed are ρ and c. F is a fluctuating forcing term. Initially, most of the vorticity lies in the lowest k mode plus low level noise in the high k modes. Numerical simulation of equations (3)-(6) was accomplished by means of the spectral method (Gottlieb and Orszag, 1977). The free slip boundary condition is executed by an expansion of the Fourier sine-sine series. The box size is π x 5000 km, and runs with 64 x 64 and 32 x 32 modes were performed. Results using the free slip boundary in latitude and periodic boundary condition in longitude will be presented in the next paper (Song and Lysak, 1986).

For time advancing, we use the second-order Adams-Bashforth method on the nonlinear terms and the Crank-Nicolson method on the linear terms. Fast Fourier Transform (FFT) was used to communicate between configuration and Fourier space. Aliasing errors are removed by the Patterson and Orszag (1971) method. Our runs were performed on the Cray 2 of the University of Minnesota Supercomputer Institute.

Figure 2 shows the evolution of current on the boundary layer in a nonlinear run. Figures 2a

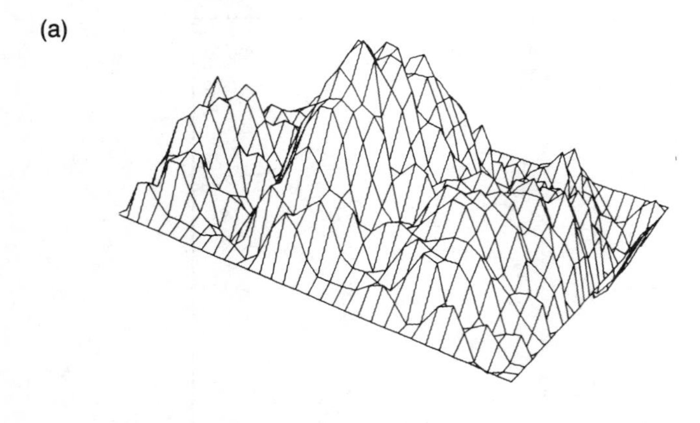

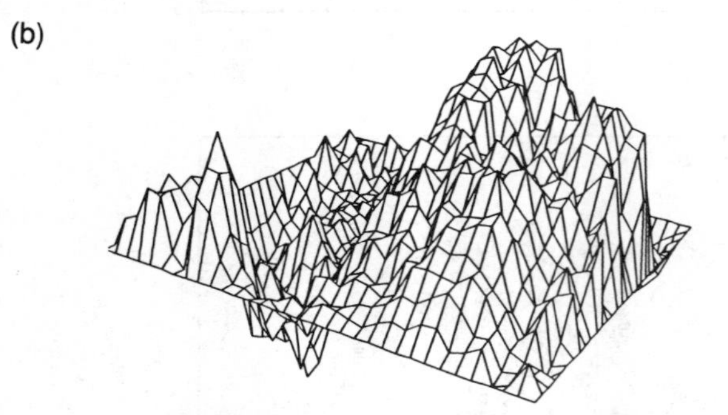

Fig. 3. (a) Current surface plot for a linear decay run at t ~ 30. (b) Current surface plot for a nonlinear decay run at t ~ 30. The initial condition for both runs is the same as in Figure 2a. Note current surface plot profile did not change for the linear run. The current fine structure in the linear case is caused by the k-dependent decay rate.

and 2b show the initial condition of current for these runs. Current filaments are formed by the nonlinear effect and the k-dependent decay rate. Comparison of the current and stream function pattern in the nonlinear run (Figures 2c, 2b, and 3b, and 4c to 4d) and linear run (Figures 3a and 4e to 4f) shows that no significant structural changes were observed in linear runs with the only effect being damping of the large- and small-scale modes. However, smaller-scale currents are observed at later times in the nonlinear case (Figures 2c, 2b, and 3b). Figure 5 (see also Lotko et al., 1987) shows that the nonlocal diffusion terms including field-aligned conductivity and ionospheric Pedersen conductivity can cause a maximum in the decay time. The solid line is for L = 15 R_E and the dashed line is for L = 10 R_E, where L is the equatorial distance of the field line. The maximum decay time occurs at scales of 70 ~ 160 km at the field line base. This corresponds to the scale of the strucure in Figure 3a. Figure 4 presents surface plots for the velocity stream function in nonlinear and linear cases. The large vortices rapidly break into medium and small loops in the nonlinear case. Again, we note that the integrated current does not vary much from t = 30 (Figure 3) to t = 72 (Figure 2). This indicates that the nonlinear effect decreases as the currents decay. The external fluctuating forcing function F has a Fourier transform f(k,t). A forcing term is chosen (Hossain et al., 1983; Fyfe et al., 1977a,b) in the low k range with the low level noise in the higher k modes, corresponding to a large-scale boundary dynamo. Results show that small-scale fluctuating forcing terms, such as those produced by the Kelvin-Helmholtz instability, can also cause small-scale currents (Figure 6). Figure 7 shows the energy evolution with time in a forcing run. The energy increases much more slowly in the nonlinear run than in the linear run, since the nonlinear effect can move energy in k space to the dissipative range by the energy cascade process.

Summary and Discussion

The interaction of the magnetosphere with the solar wind results in an extremely turbulent cir-

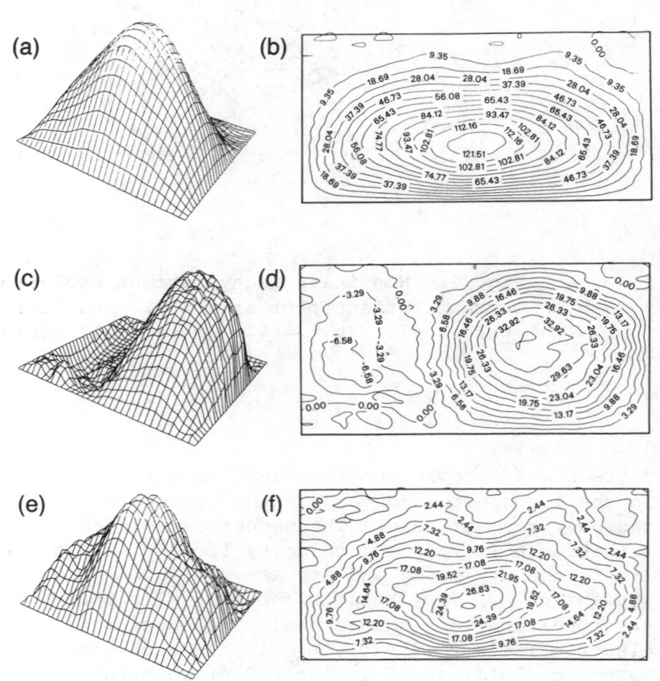

Fig. 4. Velocity stream function evolution. (a) Initial condition surface plot at t = 0. (b) Initial condition contour plot at t = 0. (c) Nonlinear decay surface plot at t ~ 30. (d) Nonlinear decay contour plot at t ~ 30. (e) Linear decay surface plot at t ~ 30. (f) Linear decay contour plot at t ~ 30. Note the nonlinear effect can break the large vortices and cause them to move.

culation which causes field-aligned currents. Nonlinear effects (especially the current nonlinearity) can effectively break large-scale vortices into small eddies. Parallel electric fields and drag due to the finite ionosphere conductivity cause a filter effect, which causes a k-dependent decay rate. Propagation of high k driving terms can also produce small-scale structure.

A simplified model for describing a disturbed flux tube which couples large-scale convection in the boundary layer to the ionosphere is presented. Our interest here lies in the nonlinear turbulent evolution of large-scale vortices in the presence of coupling between magnetosphere and ionosphere.

The considerable similarity between the results of this simulation and two-dimensional MHD turbulence simulations has been noted. It comes from the similarity of the equations used. The effect of the background magnetic field in our simplified equations introduces a wavelength-

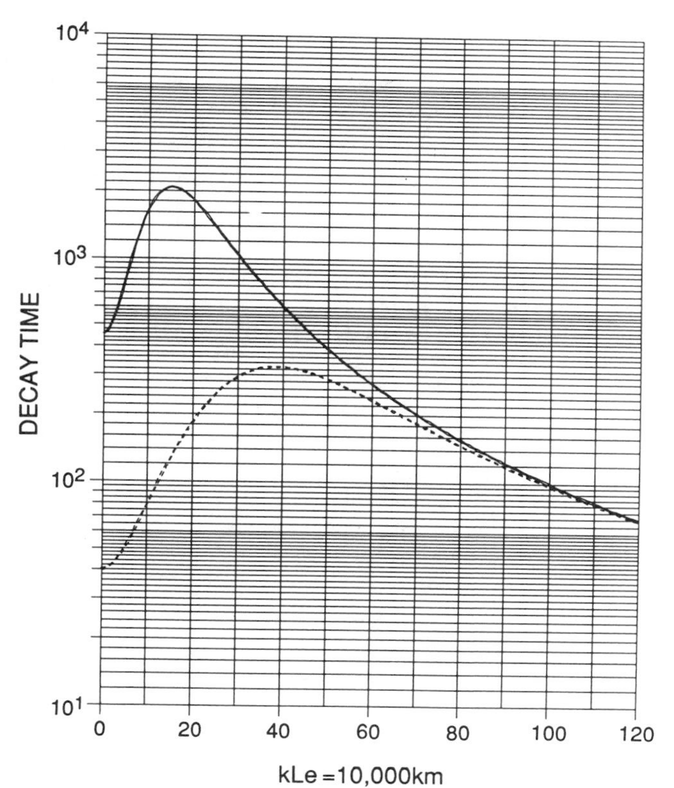

Fig. 5. Decay time versus wave number, where $\Sigma_p = 9 \times 10^{12}$, $\alpha = 0.9$ (cgs units). The solid line is for $L = 15\ R_E$; the dashed line is for $L = 10\ R_E$. The maximum decay time is in the scale $kL_e = 17 \sim 38$, which corresponds to 70 to ~160 km at the field line base. L is the distance between the generator region and the earth.

(a)

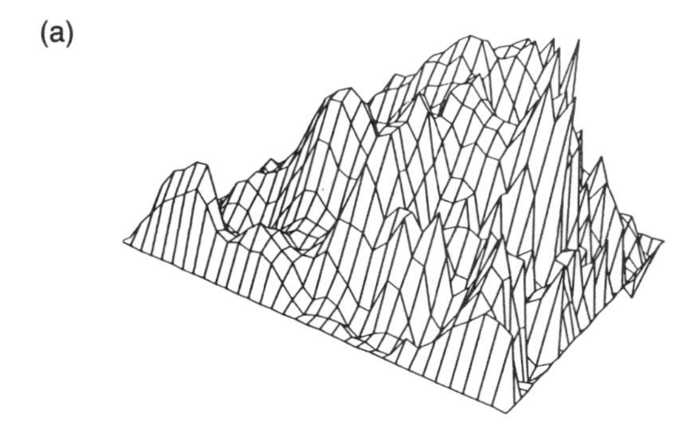

(b)

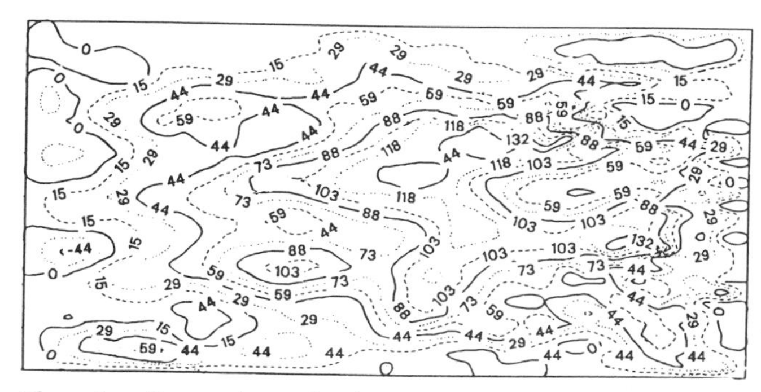

Fig. 6. Current evolution of a forcing run. (a) Current surface plot at t = 108. (b) Current contour plot at t = 108.

dependent decay term. We observe the formation of current filaments and small scale vortices, as in two-dimensional MHD turbulence (Orszag and Tang, 1979; Matthaeus and Montgomery, 1981; Matthaeus, 1982; Hasegawa, 1985; Pouquet, 1978).

We have attempted to explain how large-scale boundary layer nonsteady vortices cause highly structured auroral arcs. (1) Magnetic field lines are wrapped into ropes by the vorticity in the velocity field from the turbulent dynamo effect and field-aligned currents are generated by $\mathbf{j} = \nabla \times \mathbf{B}$. The wrapping effect (current dynamo effect) of different size tubes is subject to a wavelength-dependent effective damping rate, which is determined by field-aligned and Pedersen conductivities. (2) Two flux tubes carrying currents of the same sign attract each other by the $\mathbf{j} \times \mathbf{B}$ force. More precisely, this is the volume force acting on the currents in the flux tube which concentrates currents in the direction perpendicular to the maximum ∇b. This attractive force pushes magnetic field lines closer to each other. The local current density is increased by the increased magnetic field. This process is repeated, forming strong current filaments. (3) Vorticity around current filaments increases by the $(\mathbf{b} \cdot \nabla)\mathbf{j}$ term.

This is an oversimplified explanation. First of all, the turbulent energy input has different

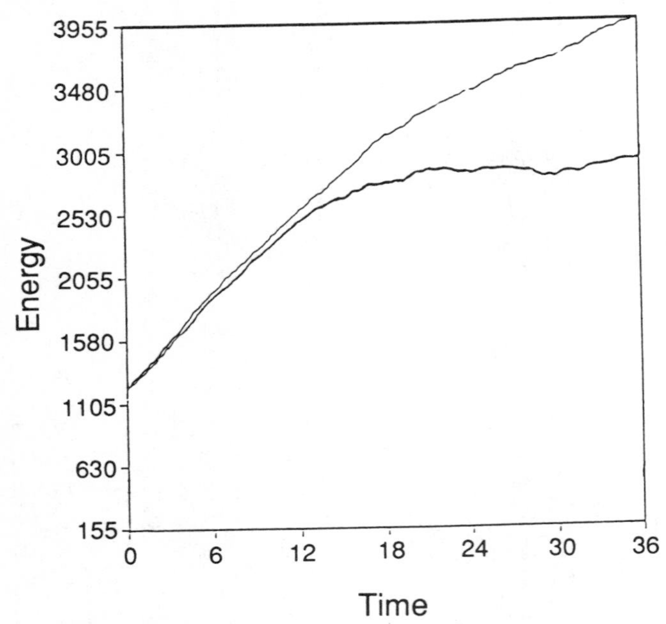

Fig. 7. Energy versus time for a forcing run. The thin line is in the linear case; the thick line is in the nonlinear case.

strengths at different scale lengths (such as the dynamo effect from Kelvin-Helmholtz instability). Second, the conductivity may be wavelength dependent. Consideration of the nonlinear term, which gives rise to the random turbulent nature of the cascade process, shows that this is a complicated picture. However, the appearance of small-scale currents and vortices in the model indicates that the essential physics of the situation has been included.

This model has certain limitations. (1) It is an action-at-a-distance boundary model. It is a fair approximation for these long time scale vortices. On shorter spatial scales than can be resolved by the ISEE spacecraft, however, it is likely that Alfvén transit time effects may be important (Lotko et al., 1987). Alfvén wave effects in the turbulent boundary layer model will be included in future work. Many new physics phenomena, such as flux tube bending, are not yet included in the model. (2) Our preliminary model gives only a qualitative description of nonlinear evolution of large-scale vortices. It is important to remember that cascade arguments and Kolmogoroff exponents are valid only for homogeneous, isotropic, reflection-invariant turbulence (Moffatt, 1978). The analogy between the electrostatic guiding center plasma and two-dimensional Navier-Stokes turbulence does not extend to dissipative effects (Kraichnan and Montgomery, 1980). It becomes more complicated to determine an inertial range, since the decay rate is wavelength dependent. (3) Compressibility is not yet included in our model. (4) Field-aligned and Pedersen conductivities have been set to constants. An improved model with k-dependent conductivity and multiple scale equations will be presented later.

The existence of the small-scale, field-aligned auroral zone current filaments should be accompanied with twisted magnetic flux tubes (Elphic et al., 1986). If we consider the model geometry used in this simulation (Figure 1) as half of a large closed magnetic tube (Cowley, 1982; Figure 1), our simulation results show that the evolution of the current and vortices in a disturbed magnetic flux tube is subjected to the nonlinear effect, scale length dependence of the physical quantities and the structure of the driving terms.

Acknowledgments. This work has benefitted from discussions with Bill Lotko. The work was supported in part by NSF grants ATM-8451168 and ATM-8508949. Computing time was provided by the University of Minnesota Supercomputer Institute.

References

Buneman, D., R. H. Levy, and L. M. Linson, Stability of cross-field electron beams, J. Appl. Phys., 37, 3202, 1966.

Cowley, S.W.H., The cause of convection in the earth's magnetosphere: A review of development during the IMS, Rev. Geophys. Space Phys., 20, 531, 1982.

Davis, T. N., and T. J. Hallinan, Auroral spirals 1. Observations, J. Geophys. Res., 81, 3953, 1976.

Elphic, R. C., C. A. Cattell, K. Takahashi, S. J. Bame, and C. T. Russell, ISEE-1 and 2 observations of magnetic flux ropes in the magnetotail: FTE's in the plasma sheet, Geophys. Res. Lett., 13, 648, 1986.

Fyfe, D., G. Joyce, and D. Montgomery, Magnetic dynamo action in two-dimensional turbulent magnetohydrodynamics, J. Plasma Phys., 17, 317, 1977a.

Fyfe, D., D. Montgomery, and G. Joyce, Dissipative, forced turbulence in two-dimensinal magnetohydrodynamics, J. Plasma Phys., 17, 369, 1977b.

Goertz, C. K., and R. W. Boswell, Magnetospheric-ionosphere coupling, J. Geophys. Res., 84, 7329, 1979.

Goldstein, H., and K. Schindler, On the role of the ionosphere in substorms: Generation of field-aligned currents, J. Geophys. Res., 83, 2574, 1978.

Gottlieb, D., and S. A. Orszag, Numerical Analysis of Spectral Methods: Theory and Application, SIAM, Philadelphia, CBMS-NSF Monogr., vol. 26, 1977.

Hallinan, T. J., and T. N. Davis, Small-scale auroral arc distortions, Planet. Space Sci., 18, 1735, 1970.

Hasegawa, A., Self-organization processes in continuous media, Advances in Physics, 34, 1, 1985.

Hayakawa, H., and A. Nishida, Statistical characteristics of plasma flow in the magnetotail, J. Geophys. Res., 87, 277, 1982.

Hones, E. W., Jr., G. Paschmann, S. J. Bame, J. R. Asbridge, N. Sckopke, and K. Schindler, Vortices in magnetospheric plasma flow, Geophys. Res. Lett., 5, 1059, 1978.

Hones, E. W., Jr., J. Birn, S. J. Bame, J. R. Asbridge, G. Paschmann, N. Sckopke, and G. Haerendel, Further determination of the characteristics of magnetospheric plasma vortices with ISEE 1 and 2, J. Geophys. Res., 86, 814, 1981.

Hossain, M., W. H. Mattheaus, and D. Montgomery, Long-time states of inverse cascades in the presence of maximum length scale, J. Plasma Phys., 30, 479, 1983.

Kelley, M. C., and P. M. Kintner, Evidence for two-dimensional inertial turbulence in a cosmic-scale low-β plasma, Ap. J., 220, 339, 1978.

Kintner, P. M., Jr., Observations of velocity shear driven plasma turbulence, J. Geophys. Res., 81, 5114, 1976.

Kraichnan, R. H., and D. Montgomery, Two-dimensional turbulence, Rep. Prog. Phys., 43, 547, 1980.

Lotko, W., Mesoscale turbulence in magnetosphere-ionosphere coupling (abstract), Eos, 67, 1179, 1986.

Lotko, W., B.U.Ö. Sonnerup, and R. L. Lysak, Nonsteady boundary layer flow including ionospheric drag and parallel electric fields, J. Geophys. Res., in press, 1987.

Lysak, R. L., and C. T. Dum, Dynamics of magnetosphere-ionosphere coupling including turbulent transport, J. Geophys. Res., 88, 365, 1983.

Lysak, R. L., Auroral electrodynamics with current and voltage generators, J. Geophys. Res., 90, 4178, 1985.

Lysak, R. L., Coupling of the dynamic ionosphere to auroral flux tubes, J. Geophys. Res., 91, 7047, 1986.

Matthaeus, W. H., and D. Montgomery, Nonlinear evolution of the sheet pinch, J. Plasma Phys., 25, 11, 1981.

Matthaeus, W. H., Reconnection in two dimensions: Localization of vorticity and current near magnetic x-points, Geophys. Res. Lett., 9, 660, 1982.

Miura, A., and P. L. Pritchett, Nonlocal stability analysis of the MHD Kelvin-Helmholtz instability in a compressible plasma, J. Geophys. Res., 87, 7431, 1982.

Moffatt, H. K., Magnetic Field Generation in Electrically Conducting Fluids, Cambridge University Press, Cambridge, Massachusetts, 1978.

Montgomery, D., Major disruptions, inverse cascades, and the Strauss equations, Physica Scripta, 12, 83, 1982.

Mozer, F. S., C. W. Carlson, M. K. Hudson, R. B. Torbert, B. Parady, J. Yatteau, and M. C. Kelley, Observations of paired electrostatic shocks in the polar magnetosphere, Phys. Rev. Lett., 38, 292, 1977.

Mozer, F. S., C. W. Carlson, M. K. Hudson, R. L. Lysak, M. Temerin, and R. B. Torbert, Satellite measurements and theories of auroral particle acceleration, Space Sci. Rev., 27, 155, 1980.

Orszag, S. A., and C.-M. Tang, Small-scale structure of two-dimensional magnetohydrodynamic turbulence, J. Fluid Mech., 90, 129, 1979.

Patterson, G. S., Jr., and S. A. Orszag, Spectral calculations of isotropic turbulence, Phys. Fluids, 14, 2538, 1971.

Pouquet, A., On two-dimensional magnetohydrodynamic turbulence, J. Fluid Mech., 88, 1, 1978.

Sato, T., and T. Iijima, Primary source of large-scale Birkeland currents, Space Sci. Rev., 24, 347, 1979.

Saunders, M. A., D. J. Southwood, E. W. Hones, Jr., and C. T. Russell, A hydromagnetic vortex seen by ISEE 1 and 2, J. Atm. Terr. Phys., 43, 927, 1981.

Shebalin, J. V., W. H. Matthaeus, and D. Montgomery, Anisotropy in MHD turbulence due to a magnetic field, J. Plasma Phys., 29, 525, 1983.

Strauss, H. R., Nonlinear, three-dimensional magnetohydrodynamics of noncircular tokamaks, Phys. Fluids, 19, 134, 1976.

Song, Y., and R. L. Lysak, A two-dimensional spectral simulation of the auroral dynamics (abstract), Eos, 67, 1170, 1986.

Sugiura, M., N. C. Maynard, W. H. Farthing, J. P. Happner, B. G. Ledley, and L. J. Cahill, Jr., Initial results on the correlation between the magnetic and electric fields observed from the DE-2 satellite in the field-aligned current regions, Geophys. Res. Lett., 9, 985, 1982.

Sugiura, M., A fundamental magnetosphere-ionosphere coupling mode involving field-aligned current as deduced from DE-2 observations, Geophys. Res. Lett., 11, 877, 1984.

Swift, D. W., Turbulent generation of electrostatic fields in the magnetosphere, J. Geophys. Res., 82, 5143, 1977.

Watanabe, K., and M. Ashour-Abdalla, A numerical model of magnetosphere-ionosphere coupling: Preliminary results, J. Geophys. Res., 91, 6973, 1986.

Webster, H. F., and T. J. Hallinan, Instabilities in charge sheets and current sheets and their possible occurrence in the aurora, Radio Sci., 8, 475, 1972.

Weimer, D. R., C. K. Goertz, D. A. Gurnett, N. C. Maynard, and J. L. Burch, Auroral zone electric fields from DE 1 and 2 at magnetic conjunctions, J. Geophys. Res., 90, 7479, 1985.

Weimer, D. R., D. A. Gurnett, C. K. Goertz, J. D. Menietti, J. L. Burch, and M. Sugiura, The current-voltage relationship in auroral current sheets, J. Geophys. Res., 92, 187, 1987.

ALFVÉN ION-CYCLOTRON HEATING OF IONOSPHERIC O^+ IONS

R. M. Winglee*

Institute of Geophysics and Planetary Physics, University of California, Los Angeles
Los Angeles, California 90024

R. D. Sydora

Department of Physics, University of California, Los Angeles, Los Angeles, California 90024

M. Ashour-Abdalla

Institute of Geophysics and Planetary Physics, University of California, Los Angeles
Los Angeles, California 90024

Abstract. Transversely heated ionospheric ions, in particular O^+ ions, are often observed flowing upward along auroral field lines. Currents observed in association with the tranversely heated ions can drive shear Alfvén waves and electrostatic ion-cyclotron waves unstable which can, in turn, be resonantly absorbed by the ions to produce the heating. Particle simulations are used to examine self-consistently the excitation of these waves and the associated heating. It is shown that the growth of the electrostatic ion-cyclotron waves quickly becomes suppressed as the ions become heated and the dominant wave fields are those of the shear Alfvén wave. The resultant transverse ion heating is larger and faster than that produced by solely electrostatic ion-cyclotron wave heating. Due to trapping of ions by the shear Alfvén wave, the temperature of the O^+ ions remains comparable to that of the H^+ ions.

Introduction

Energetic ions from the ionosphere, including H^+ and O^+, are often observed flowing upward along auroral field lines (Shelley et al., 1976; Ghielmetti et al., 1978; Klumpar et al., 1984; Yau et al., 1984; Collin and Johnson, 1985). These energetic ions often have velocity distributions which are confined to pitch angles nearly perpendicular to the magnetic field. Such distributions are called conics and suggest that there is some mechanism present which preferentially heats ionospheric ions perpendicular to the magnetic field.

Observations indicate that the perpendicular heating is often associated with currents (Heelis and Winningham, 1984; Kintner and Gorney, 1984; Moore et al., 1985). The heating is also sometimes associated with intense electrostatic and electromagnetic waves near the hydrogen and oxygen cyclotron frequencies (Bering, 1984; Kintner and Gorney, 1984; Gurnett et al., 1984).

Theories for the perpendicular heating include electrostatic ion-cyclotron waves (Ashour-Abdalla and Okuda, 1984 and references therein) and by lower hybrid waves (Chang and Coppi, 1981; Singh and Schunk, 1984). Recently, Chang et al. (1986) and Winglee et al. (1987) pointed out that Alfvén ion-cyclotron waves can also play an important role in the perpendicular heating. Chang et al. (1986) examined the heating produced by a given wave spectrum while Winglee et al. (1987) used linear theory to examine the excitation of Alfvén waves by currents and their interaction with the background ions.

In this paper, particle simulations are used to quantitatively examine the self-consistent excitation of shear Alfvén waves (SAWs) and electrostatic ion-cyclotron waves (EICWs) by currents and the perpendicular ion heating produced by these waves.

Simulation Model

The simulations use a one-dimensional (three velocity components) magnetostatic particle simulation model. Full ion dynamics is used with

---

*Present address: Department of Astrophysical, Planetary and Atmospheric Sciences, University of Colorado, Boulder, Colorado 80309.

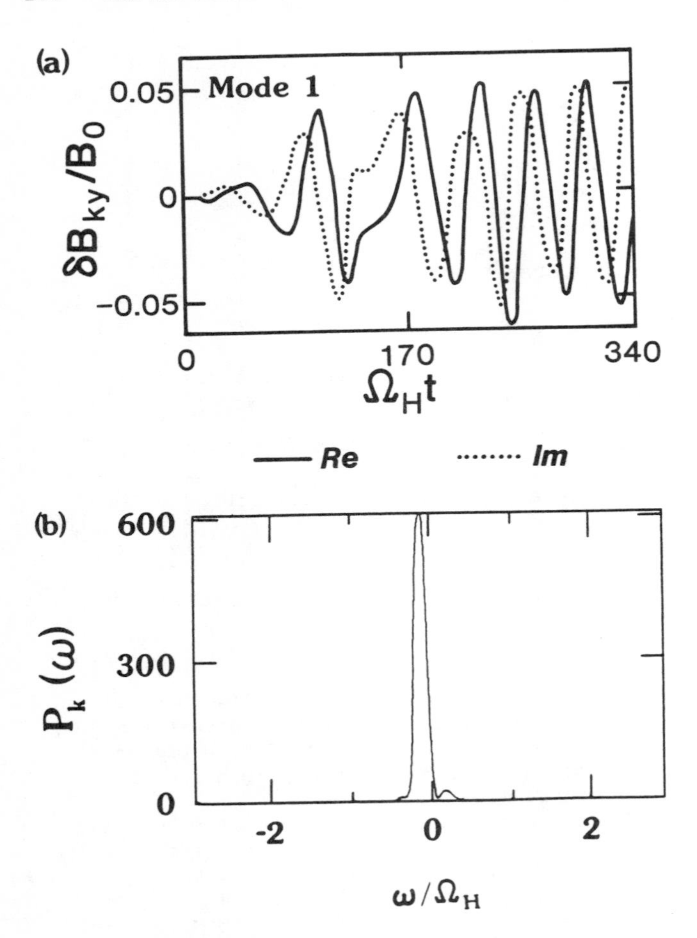

Fig. 1. Time history of the transverse wave magnetic field for mode 1 and the corresponding power spectrum. This mode corresponds to the shear Alfvén wave and is the most intense.

realistic mass ratios, i.e., $m_H/m_e = 1837$ and $m_O/m_H = 16$. Full dynamics for the electron motion parallel to the magnetic field is also used but the electron motion transverse to the magnetic field is approximated by their guiding centers (cf. Ashour-Abdalla and Okuda, 1984). This approximation is valid for wavelengths $k_\perp \rho_e << 1$ and frequencies $\omega << \Omega_e$ and eliminates unwanted high-frequency oscillations associated with the electron gyromotion. The magnetostatic approximation (i.e., neglecting the transverse part of the displacement current) eliminates high-frequency oscillations associated with the light branch $\omega \approx kc$. This approximation does not affect the properties of the SAW or EICW.

The electrons are assumed to have a drifting Maxwellian distribution with a drift velocity along the magnetic field v_d assumed to be equal to their thermal velocity v_{Te}. In order to model the continual flow of current through the source region, uniform recycling of the electron distribution is utilized; i.e., at each time step the velocity of a few electrons is rechosen from the initial distribution (Okuda, 1984). The entire distribution is recycled in a time $\Omega_H t = 10$. The recycling on average maintains the electron distribution and thereby the electron current through the system.

Both H^+ and O^+ are assumed to be present and of equal density, i.e., $n_H = n_O = 0.5\ n_e$. The initial distributions of the ions are assumed to be Maxwellian with $T_H = T_O = T_e$. The oxygen Alfvén speed v_{AO} is taken to be $0.825\ v_{Te}$.

The coordinate system for the simulations is as follows. The wave vector **k** is in the X-direction and the magnetic field is in the X-Z plane inclined at an angle $\theta = 87.2°$ to **k** where the growth rates of the SAW and EICW are near a maximum. The component of velocity parallel to **B** is denoted by $v_\parallel$ and the component perpendicular to **B** in the X-Z plane by v'_x. The simulations use a 256 grid with a total length equivalent to 60 ρ_H (where ρ_H is the hydrogen Larmor radius) and with a total of 320 particles per cell. The wavenumber for the modes in the system is then $k\rho_H = 0.1$ m or $k\rho_O = 0.4$ m, where m is the mode number. The time step is $\Omega_H \Delta t = 0.1$.

Simulation Results

The parameters described in the previous section are the same as in the simulation study of the perpendicular heating of ionospheric ions by current-driven EICWs by Ashour-Abdalla and Okuda (1984). However, as discussed by Winglee et al. (1987), the electron current can drive not only EICWs unstable but also SAWs unstable. In order to identify the dominant wave mode, the time history of the Fourier modes was examined. The temporal evolution of the electrostatic potential and perturbed magnetic field for the most unstable mode, $k_\perp \rho_O = 0.4$ (m = 1), is shown in Figure 1. This mode coresonds to the shear Alfvén wave with a frequency slightly below Ω_O. The mode reaches saturation at approximately $\Omega_H t = 180$ with the maximum of $|e\Phi_k|/T_e \approx 2.5$ and $|\delta B_{yk}|/B_O \approx 0.06$.

Mode 2 which corresponds to an oxygen EICW with $\omega > \Omega_O$ is the next most intense mode. The maximum values of $|e\Phi_k|$ and $|\delta B_{yk}|$ are about one-fifth of those of the SAW. There are also hydrogen EICWs present at higher mode numbers with peak amplitudes being less than one-tenth of that of mode 1.

Thus, the dominant mode is the SAW. EICWs are present but at much smaller amplitudes. This is in contrast to the pure electrostatic case (Ashour-Abdalla and Okuda, 1984) where hydrogen EICWs dominate initially. The oxygen waves only dominate after the hydrogen waves saturated (by reaching marginal stability through the development of temperature anisotropy in the hydrogen ions).

Because the excited waves have frequencies near the cyclotron frequencies of the H^+ and O^+ ions, the waves can heat the ions by gyro-

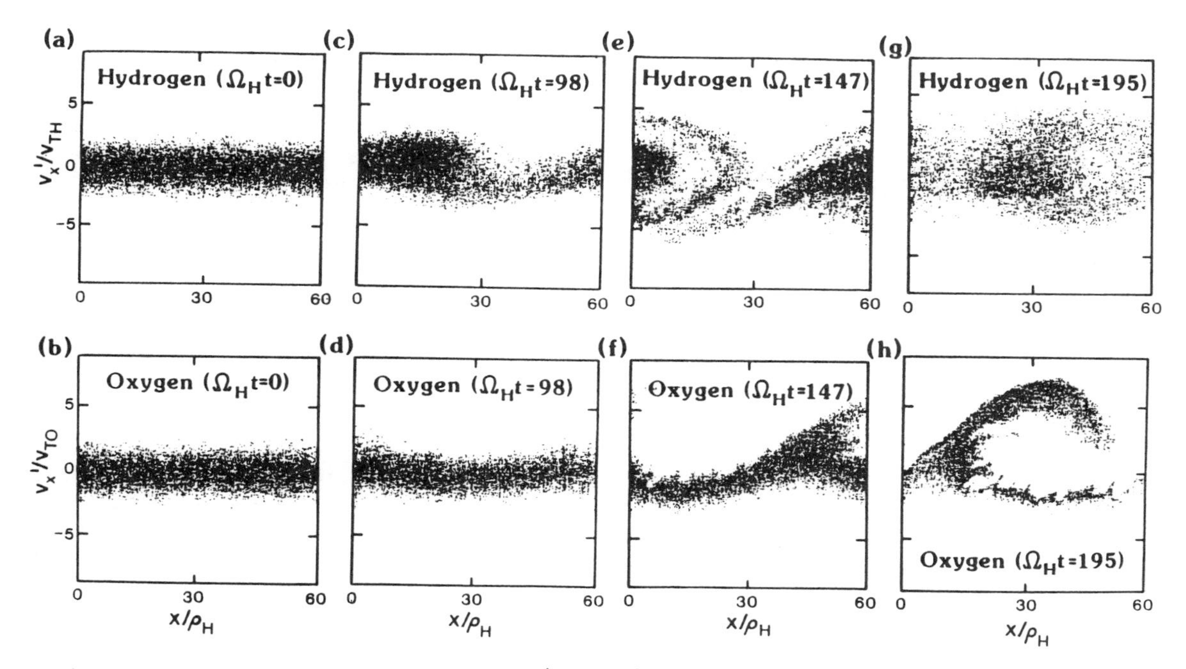

Fig. 2. The v_x'-x phase space of the H^+ and O^+ ions at $\Omega_H t$ equal to 0, 98, 147, and 195. When the amplitude of the shear Alfvén becomes sufficiently large, trapping of the ions occurs, producing large increases in v_x' ($\Omega_H t$ = 147, 195).

resonant interactions. In order to show the heating produced, Figure 2 shows the v_x' -x phase space of the H^+ and O^+ ions for several times during the simulation. The initial phase space of the ions is shown in Figures 2a and 2b. It is seen in Figures 2c and 2d that the H^+ ions first become heated. This heating is produced by the low-amplitude hydrogen EICWs. It occurs first because of the relatively light mass of the H^+ ions.

However, when the amplitude of the SAW becomes large, strong heating of the H^+ ions as well as the O^+ ions occurs. This is seen in Figures 2e to 2h where both the H^+ and O^+ ions are accelerated to perpendicular velocities nearly 5 times their initial thermal velocity. This acceleration is produced by phase bunching as evidenced by the coherent phase space structures. This phase bunching has the same wavelength as the SAWs and coincides with the saturation of the SAWs (Figure 1). In other words, the electron current generates large-amplitude SAWs which are able to trap O^+ and H^+ and produce strong perpendicular ion acceleration. At later times (Figure 2h) the trapped H^+ ions become thermalized due to the presence of the low-amplitude hydrogen EICWs, while the O^+ ions still remain phase bunched. The reason that growth of the SAW dominates over the EICW is that as the perpendicular temperature of the ions increases, growth of the EICW is suppressed unlike that of the SAW (Winglee et al., 1987).

Trapping and phase bunching are microscopic

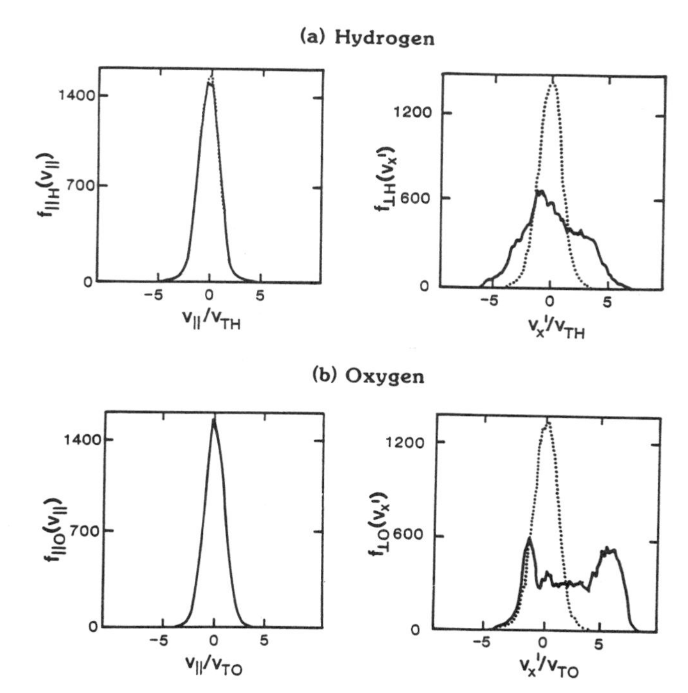

Fig. 3. The perpendicular and parallel distributions of the H^+ ions (a) and the O^+ ions (b) at $\Omega_H t$ = 195. The dotted curves indicate the initial distributions. Bulk perpendicular heating is evident in both species but there is little parallel heating.

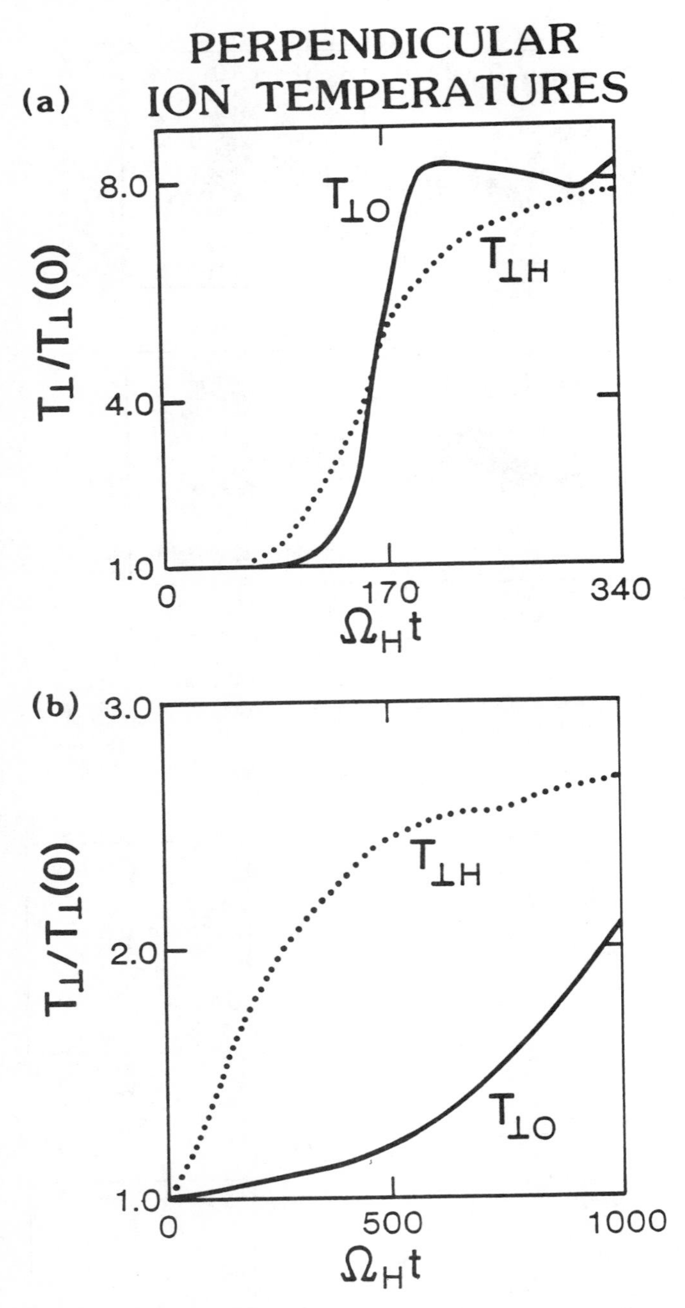

Fig. 4. Time histories for the perpendicular ion temperatures normalized to their initial temperature. The histories for the present case are shown in (a); those for a purely electrostatic case (i.e., where the shear Alfvén wave is absent) are shown in (b). The heating is larger and occurs in a shorter time when the Alfvén wave is present.

processes. Actual observations of particle distributions taken by satellites are averaged over length and time scales much larger than those for trapping. As a measure of the macroscopic properties of the ions that might be observed, Figure 3 shows the perpendicular and parallel distributions of the H^+ and O^+ ions at $\Omega_H t = 195$ taken over the entire simulation system. The dotted distributions are the initial values. It is seen that both ion species undergo significant bulk heating perpendicular to the magnetic field, with the energy reached by both ion species being comparable. This equipartition is due to the fact that a single mode dominates and produces trapping of both species. Due to the thermalization by electrostatic hydrogen-cyclotron waves the distribution of the H^+ ions tends to be isotropic in the perpendicular direction while the O^+ ions, which remain phase bunched, have a net drift in v_x'. The perpendicular ion heating is not accompanied by any appreciable parallel heating as seen in Figure 3.

Figure 4a shows the time histories of the perpendicular temperature of both ion species normalized to the initial value. For comparison, the time histories for the purely electrostatic case (i.e., when the SAW is absent) are shown in Figure 4b. The electrostatic results are essentially the same as in Ashour-Abdalla and Okuda (1984). It is seen that when the SAW is present both ion species are strongly heated in the perpendicular direction with $T_\perp/T_\perp(0)$ increasing by a factor of about 8 in $\Omega_H t \simeq 200$. The heating rate is much slower when the SAW is absent with $T_\perp/T_\perp(0)$ for the oxygen increasing by only a factor of about 2 in $\Omega_H t = 1000$. Thus, the SAW is clearly able to produce enhanced perpendicular heating of both O^+ and H^+ ions.

The above simulation results are limited in that only one SAW mode is present in the system. In order to investigate the effects of a broader spectrum of SAWs, simulations were also performed at $\theta = 89.2°$ with all other parameters the same. In this case there are four SAW modes in the system. Initially the high $\mathbf{k}$ modes dominate but as the perpendicular temperature of the ions increases, the dominant mode shifts to lower $\mathbf{k}$ with mode 2 being the dominant mode at the end of the run. The heating in this case remains qualitatively the same as the above results with the bulk perpendicular heating of the ions occurring much faster than in the purely electrostatic case. The absolute magnitude of the heating is smaller for the larger θ because of the reduction in the growth rate as θ approaches 90°.

Conclusion

Particle simulations have been used to investigate self-consistently the excitation of shear Alfvén and electrostatic ion-cyclotron waves by an electron current and the ion heating produced by the waves. It has been shown that both waves can be excited but growth of the SAW dominates. These waves reach large amplitudes with $\delta B/B \approx 0.06$.

The excited waves are able to produce strong perpendicular heating of both H^+ and O^+ ions by wave trapping. In the example presented the perpendicular temperature of both ions increased by a factor of 8 in a time $\Omega_H t \approx 200$. This heating rate is substantially enhanced over that for heating solely by current-driven EICWs. This is because growth of the EICW becomes suppressed as the ion perpendicular temperature increases, whereas the SAW is not subject to this suppression.

Because both H^+ and O^+ ions are trapped by the same wave fields, the perpendicular temperature of both ion species tends to be comparable. This is consistent with the observations of Collin and Johnson (1985).

Acknowledgments. This research was supported by National Science Foundation grant ATM-8503434, Air Force contract F19628-85-k-0027, and NASA Solar Terrestrial Theory Program grant NAGW-78. The simulations were performed on the CRAY X-MP at the San Diego Supercomputer Center which is funded by NSF.

References

Ashour-Abdalla, M., and H. Okuda, Turbulent heating of heavy ions on auroral field lines, J. Geophys. Res., 89, 2236, 1984.

Bering, E. A., The plasma wave environment of an auroral arc: Electrostatic ion cyclotron waves in the diffuse aurora, J. Geophys. Res., 89, 1635, 1984.

Chang, T. T., and B. Coppi, Lower hybrid acceleration and ion evolution in the supra-auroral region, Geophys. Res. Lett., 8, 1253, 1981.

Chang, T. T., G. B. Crew, N. Hershkowitz, J. R. Jasperse, J. M. Rettener, and J. D. Winningham, Transverse acceleration of oxygen ions by electromagnetic ion cyclotron resonance with broad left-hand polarized waves, Geophys. Res. Lett., 13, 636, 1986.

Collin, H. L., and R. G. Johnson, Some mass dependent featuers of energetic ion conics over the auroral regions, J. Geophys. Res., 90, 9911, 1985.

Ghielmetti, A. G., R. G. Johnson, R. D. Sharp, and E. G. Shelley, The latitudinal diurnal and altitudinal distributions of upward flowing energetic ions of ionospheric origin, Geophys. Res. Lett., 5, 59, 1978.

Gurnett, D. A., R. L. Huff, J. D. Menietti, J. L. Burch, J. D. Winningham, and S. D. Shawhan, Correlated low-frequency electric and magnetic noise along the auroral field lines, J. Geophys. Res., 89, 8971, 1984.

Heelis, R. A., and J. D. Winningham, Particle acceleration parallel and perpendicular to the magnetic field, J. Geophys. Res., 89, 3893, 1984.

Kintner, P. M., and D. J. Gorney, A search for the plasma processes associated with perpendicular ion heating, J. Geophys. Res., 89, 937, 1984.

Klumpar, D. M., W. K. Peterson, and E. G. Shelley, Direct evidence for two-state (bimodal) acceleration of ionospheric ions, J. Geophys. Res., 89, 10,779, 1984.

Moore, T. E., C. R. Chappell, M. Lockwood, and J. H. Waite, Jr., Superthermal ion signatures of auroral acceleration processes, J. Geophys. Res., 90, 1611, 1985.

Shelley, E. G., R. D. Sharp, and R. G. Johnson, Satellite observations of an ionospheric acceleration mechanism, J. Geophys. Lett., 3, 654, 1976.

Singh, N., and R. W. Schunk, Energization of ions in the auroral plasma by broadband waves: Generation of ion conics, J. Geophys. Res., 89, 5538, 1984.

Winglee, R. M., M. Ashour-Abdalla, and R. D. Sydora, Heating of ionospheric O^+ ions by shear Alfvén waves, J. Geophys. Res., 92, 5911, 1987.

Yau, A. W., B. A. Whalen, W. K. Peterson, and E. G. Shelley, Distribution of upflowing ionospheric ions in the high-altitude polar cap and auroral ionosphere, J. Geophys. Res., 89, 5507, 1984.

QUANTITATIVE PARAMETRIZATION OF ENERGETIC IONOSPHERIC ION OUTFLOW

A. W. Yau

Herzberg Institute of Astrophysics, National Research Council Canada,
Ottawa, Ontario, K1A 0R6 Canada

W. K. Peterson and E. G. Shelley

Lockheed Palo Alto Research Laboratory, Palo Alto, California 94304

Abstract. The magnitude and composition of energetic upflowing ionospheric ions are key parameters in a merged magnetospheric-ionospheric model. Their magnetic and solar activity dependences have been parametrized in terms of the magnetic Kp and AE indices and the solar radio flux index $F_{10.7}$ using data from the Lockheed energetic ion composition spectrometer on Dynamics Explorer 1 between September 1981 and May 1986. The data period extends from near the maximum to the minimum of solar cycle 21 when $F_{10.7}$ varied between ~70 and ~300. For a given solar activity level, the ion outflow rate of H^+ and O^+ in the 0.01-17 keV range was found to increase exponentially with the magnetic Kp index, by a factor of 4 and 20, respectively, from Kp = 0 to 6. The exponential dependence prevails at all solar activity levels. Empirically, $F_{O^+} \propto \exp(0.50\ Kp)$, $F_{H^+} \propto \exp(0.23\ Kp)$. In addition, it approximately follows a power law dependence with the AE index. Empirically, $F_{O^+} \propto AE^{0.8}$ and $F_{H^+} \propto AE^{0.4}$ for AE above 100 nT. From solar minimum to near maximum ($F_{10.7}$ from ~70 to ~250), the O^+ rate for a given magnetic activity level (Kp, Dst, or AE) increases by a factor of ~5 while the H^+ rate decreases by a factor of ~2, resulting in an order of magnitude increase in the O^+/H^+ upflowing ionospheric ion composition ratio.

Introduction

The magnetosphere and the ionosphere were previously treated as two essentially separate entities consisting of plasmas of distinctly different origins and characteristics, and the charged particles in the terrestrial environment were accordingly classified as magnetospheric and ionospheric. In contrast, energetic ion composition measurements from satellites in the last decade have brought forth the realization that (a) ionospheric ions (ions of ionospheric origin) constitute a substantial fraction of the magnetospheric plasma population under certain geomagnetic and solar conditions, and (b) the magnetosphere and the ionosphere are in fact two intimately coupled parts of the terrestrial environment. The realization points to the merger of current magnetospheric and ionospheric models as the necessary next step in our quantitative understanding of the near earth space.

At Dynamics Explorer 1 (DE 1) altitudes, a variety of ion outflow populations from the polar ionosphere have been observed, including the supersonic polar wind (Nagai et al., 1984); the upwelling ions, which are ions of different species transversely heated to 10-20 eV in the dayside cusp and then upwelling and convecting anti-sunward across the polar cap, forming a cleft ion fountain under the combined influence of gravitation and convection electric fields (Lockwood et al., 1985; Moore et al., 1986); polar cap O^+ ions, which appear to originate from inside the polar cap distinct from the cleft ion fountain (Shelley et al., 1982; Waite et al., 1985; Lockwood et al., 1986); and auroral upflowing ion beams and conics, in energies up to several kiloelectron volts (Klumpar et al., 1984; Yau et al., 1984; Collin et al., 1987) and down to a few electron volts (near the gravitational escape energy). The few electron volt auroral upflowing ionospheric ions (UFIs) are expected to follow parabolic trajectories unless they are further energized at higher altitudes (see, for example, Cladis, 1986), resulting in an auroral ion fountain in analogy to the cleft ion fountain.

The morphology and occurrence characteristics of energetic UFIs have been the subject of a number of statistical studies using data on S3-3 below 8000 km (Ghielmetti et al., 1978; Gorney et al., 1981; Collin et al., 1981, 1984; Ghielmetti et al., 1987) and data from the energetic ion composition spectrometer (EICS) on DE 1 up to

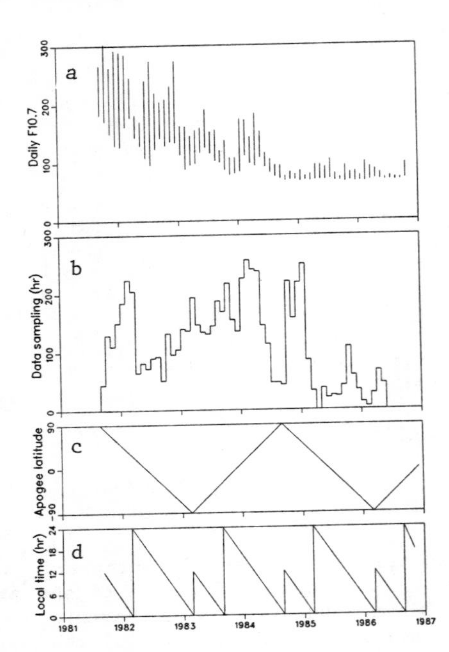

Fig. 1. (a) Monthly minimum and maximum of daily solar radio flux at 10.7 cm, $F_{10.7}$ (10^{-22} Wm^{-2} Hz^{-1}) at 1 AU. (b) Monthly EICS data coverage in present study (auroral data only). (c) Geographic latitude of DE 1 apogee. (d) Local time of DE 1 orbital plane. As a result of the 18-month apsidal cycle and the 12-month local time drift period, apogee latitude and local time visited in a particular season in the first apsidal cycle are revisited in the same season in the subsequent cycles.

23,300 km (Yau et al., 1984, 1985a,b; Collin et al., 1987). The morphology of upwelling ion events up to ~50 eV was also studied using data from the retarding ion mass spectrometer (RIMS) on DE 1 (Lockwood et al., 1985).

The Yau et al. (1985a,b) studies, in particular, were based on data from a 3-year period between 1981 and 1984 and were focussed on the occurrence morphology and outflow of energetic (0.01-17 keV) UFIs near the solar maximum and in the early declining phase of the solar cycle. These studies revealed systematic changes in the UFI occurrence morphology and mass composition with geomagnetic activity, as well as significant long-term variations which were correlated with variations in the solar activity. The present study is a follow-on to these studies; its aim is to quantify and empirically parametrize these variations and to explore the physical consequences of the parametric dependences.

UFIs from the high-latitude (auroral and polar cap) ionosphere represent the primary source of energetic ionospheric ions in the magnetosphere. Therefore, their magnitude, composition, and energy distributions are key parameters in a quantitative magnetospheric-ionospheric model. The present study represents a first step toward the synthesis of quantitative observational results into analytic inputs for such a model.

Data Analysis

DE 1 was launched on August 3, 1981, into a highly elliptic orbit. The DE 1 orbit has an apogee of 4.6 R_E geocentric and an inclination of 90°. Its orbital plane has a local time drift period of 12 months and a line-of-apsides drift period of 18 months. Consequently, the latitude and local time of the apogee for a given season in the first apsidal cycle are revisited in the same season in subsequent cycles. The revisits make it possible to readily delineate long-term variations in the observed data from seasonal and other variations.

Data in the present data base have been acquired from the EICS on DE 1 between September 15, 1981, and May 31, 1986, from near the maximum to the minimum of solar cycle 21. On DE 1, EICS typically measures ions in the ~0.01-17 keV/q range and operates in a number of specialized modes (see Shelley et al., 1981 for details). Data in this study were acquired when the instrument was operating in its fast and drum modes in which the instrument steps through 15 energy steps in the 0.01-17 keV/q range and completes a full instrument (mass-energy-angular) scan in 24 s (fast modes) or 96 s (drum modes).

The data base in the present study consists of 96-s averages of integral ion fluxes of H^+ and O^+ in three energy bands (0.01-1, 1-4, and 4-17 keV) and nine pitch angle bins. At auroral and polar cap latitudes, data coverage was fairly complete at high altitude (above 16,000 km, within ~1 R_E of DE 1 apogee) for all magnetic local times. The data base includes over 10^4 hours of data. Figure 1 shows the monthly data sampling distribution (a) and the solar radio flux index $F_{10.7}$ (monthly minimum and maximum of daily values) (b) from 1981 to 1986 inclusive, together with the apogee latitude (c) and local time (d). Over the data period, $F_{10.7}$ varied from a high of >300 in October 1981 to a low of <70 in 1986. Near the solar maximum and in the early declining phase, it exhibited large variations within a given month. The solar radio flux is correlated with the solar EUV flux and is therefore also an indicator of solar EUV activities. Naturally, the month-to-month data coverage is not uniform, peaking in periods when the apogee was at auroral latitudes, and typically ranges from ~50 to 200 hours of data per month.

The total ion outflow rate of UFIs between 10 eV and 17 keV was determined as a function of geomagnetic and solar activity conditions using the data base and following the computational procedure in Yau et al. (1985b). Briefly, data samples acquired within a particular data period and during specific magnetic (one of Kp, Dst, and AE) and solar activity (daily $F_{10.7}$) conditions were binned into 2°-invariant latitude and 3-hour

MLT bins. For each 96-s data sample, the net ion flux J was computed from the measured pitch angle distributions of H^+ and O^+ ion fluxes at 0.01-17 keV, and then normalized to a reference altitude (1000 km),

$$J = 2\pi \int_{\beta_{1c}}^{\pi} I(m,\beta)\cos(\pi - \beta)\sin\beta d\beta \qquad (1)$$

$$J_r = JB_r/B , \qquad (2)$$

where $I(m,\beta)$ is the integrated ion flux, β is the pitch angle, m is the ion mass, β_{1c} is the loss cone angle, J_r is the normalized net ion flux, and B and B_r are the magnetic fields at the observation and reference altitudes, respectively. In equation (1), an energy-independent loss cone angle was assumed. Also, where the pitch angle sampling did not extend to 180°, approximate corrections were made to take into account ion fluxes in unsampled pitch angle regions. The normalized flux J_r was averaged over all data samples in a given bin. The bin-averaged normalized flux $\bar{J}_r$ was integrated over all invariant latitudes (above 56°) and MLT in both hemispheres to obtain the total ion outflow rate F as a function of magnetic and solar activity indices.

$$F = (\pi/6)\Delta\tau R_r^2 \sum_{MLT,\Lambda} \bar{J}_r(MLT,\Lambda)$$

$$\{[- (R_r R_E)\cos^2\Lambda_2]^{1/2} \qquad (3)$$

$$- [1 - (R_r/R_E)\cos^2\Lambda_1]^{1/2} \} ,$$

where $\Delta\tau$ is the range of MLT bin in hours, Λ_1 and Λ_2 are the limits of the invariant latitude bin Λ, R_E is the earth radius, and R_r is the geocentric distance at the reference altitude.

Insofar as UFI activities are concerned, the relative merit between the magnetic Kp, Dst, and AE indices as a primary magnetic activity indicator is by no means established. Each index has its inherent limitations (see, for example, Rostoker, 1972). To the extent that energetic auroral UFI events are associated with auroral substorms, the auroral electrojet (AE) index, being a measure of the auroral electrojet and therefore of the auroral substorm, is theoretically the most appropriate indicator. However, the index is not as readily available as the Kp and Dst indices. Also, it tends to underestimate the strengths of auroral electrojets associated with small, highly localized substorms (that are situated between the sparsely spaced auroral observatories) and those associated with substorms well equatorward of the observatories.

On the other hand, the Kp index contains contributions from both the auroral electrojet and the ring current, while the Dst index is primarily an indicator of ring current strength (level of magnetic storm). In the earlier studies, the Kp and Dst indices were used as the magnetic activity indicator in the absence of available AE data. Strong correlation was found between O^+ UFI occurrence and intensity with the Kp index. However, the correlation in itself does not necessarily mean that the index is the primary magnetic activity indicator insofar as UFI activity is concerned. In the present work, the three indices were used where available on a purely empirical basis (Kp and Dst for the full data period and AE up to December 1983), and no attempt was made to establish their a priori theoretical relevance to UFI.

The solar radio flux index $F_{10.7}$ was used as an indicator of solar activity since it is known to correlate with the solar EUV flux, and the latter modulates the upper atmospheric and ionospheric densities and temperatures. Theoretically (Moore, 1984), the change in atmospheric scale heights during the solar cycle results in a corresponding change in the UFI composition. Qualitatively, Yau et al. (1985a) found that near solar maximum, O^+ UFIs occur more frequently and are statistically more intense. Also, the conic-to-beam ratio in UFIs was higher, consistent with the S3-3 results (Ghielmetti et al., 1987). In contrast, the H^+ UFI morphology did not display similar long-term variations.

In Yau et al. (1985b), the H^+ and O^+ ion outflow rates were computed as a function of the magnetic Kp index, independently for each of the first two apsidal cycles. It was found that for O^+, the rates in the two periods had similar Kp dependences, but the rate in the first period (near solar maximum) was consistently a factor of 2 larger than that of the second period (in the early declining phase of the solar cycle) for all Kp. In contrast, the H^+ rates in the two periods were equal within statistical errors. The decrease in the O^+ rate in the second period was correlated with the corresponding decrease in the solar EUV flux.

In the present study, the ion outflow rates were computed for the third apsidal cycle (June 1984 to January 1986) near the solar minimum. A factor of ~4-5 decrease in the O^+ rate relative to the first apsidal cycle was found, while the H^+ rate remained near its earlier value. The consistent, quantitative decrease in the O^+ rate with decreasing solar activity ($F_{10.7}$) and the consistent lack of corresponding decrease in the H^+ rate point to a quantitative dependence of the ion outflow composition on $F_{10.7}$ and on the magnetic activity indices.

In the following, ion outflow rates are computed as a function of $F_{10.7}$ for selected ranges of Kp, Dst, or AE, and conversely as a function of Kp, Dst, or AE, for selected ranges of $F_{10.7}$. Note that despite the large size and

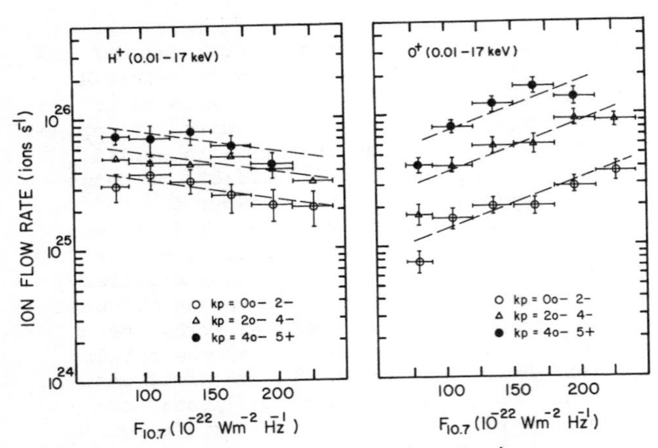

Fig. 2. Ion outflow rate of H^+ and O^+ ions at 0.01-17 keV, as a function of the solar radio flux index $F_{10.7}$, for different ranges of Kp values.

the generally uniform invariant latitude and magnetic local time coverage of the full data base, data gaps in invariant latitude MLT bins exist for some of the $F_{10.7}$-Kp, $F_{10.7}$-Dst, and $F_{10.7}$-AE parameter pairs. The computation was limited to parameter pairs where all invariant latitude bins between 66° and 86° invariant were sampled in each of the four 6-hour MLT sectors.

Ion Flow Dependences

In Figure 2, the ion outflow rate of energetic (0.01-17 keV) H^+ and O^+ was computed as a function of $F_{10.7}$, for three different Kp ranges. In O^+, there is a clearly statistically significant, factor of ~5 increase in the outflow rate from near solar minimum ($F_{10.7} \approx 70$) to near solar maxium ($F_{10.7} \approx 250$). For all three Kp ranges, the dependence of the rate on $F_{10.7}$ is very similar. In contrast, the H^+ rate displays a statistically marginal decrease with $F_{10.7}$. The choice of displaying the flow rate in logarithmic scale versus $F_{10.7}$ in linear scale is somewhat arbitrary in that (a) there is not a known theoretical functional relationship between the two quantities, and (b) other $F_{10.7}$-related quantities (such as the exospheric temperature) may be just as appropriate as the independent parameter. However, a simple exponential dependence appears to be adequate for empirical representation of the data. Explicitly,

$$F_{O^+} \propto \exp(+1.0 \times 10^{-2} F_{10.7}) \tag{4}$$

$$F_{H^+} \propto \exp(-2.7 \times 10^{-3} F_{10.7}) . \tag{5}$$

In Figure 3, the H^+ and O^+ outflow rates were computed as a function of Kp for three $F_{10.7}$ ranges. For clarity, the error bars in this and the succeeding figures have been omitted. The O^+ rate increases exponentially witht Kp by a factor of 20 from Kp = 0 to 6. The rate exceeds 3 x 10^{26} ions s^{-1} at times of high magnetic (Kp = 6) and solar ($F_{10.7} > 150$) activity. As may be anticipated from Figure 2, the rate at low solar activity ($F_{10.7}$ between 70 and 100) is about a factor of 4 smaller than that at higher activity ($F_{10.7}$ between 150 and 250). For all three $F_{10.7}$ ranges, the dependence of the rate on Kp is similar. The similarity justifies the separated treatment of the magnetic (Kp) and solar activity ($F_{10.7}$) variations in the ion outflow rate. In comparison, the H^+ rate increases with Kp more moderately, by a factor of 4, from Kp = 0 to 6. Also, the rates for the three $F_{10.7}$ ranges are similar.

Since the Kp index is a quasi-logarithmic measure of magnetic disturbance, the exponential increase of the H^+ and O^+ outflow rates suggests a quasi-linear dependence of UFI outflow with magnetic activity (as measured by magnetic field perturbation). Empirically, the dependence may be represented as

$$F_{O^+} \propto \exp(0.50 \text{ Kp}) \tag{6}$$

$$F_{H^+} \propto \exp(0.23 \text{ Kp}) . \tag{7}$$

In Figure 4, the H^+ and O^+ outflow rates were computed as a function of the Dst index for $F_{10.7}$ in the range of 70-100 and 100-150, respectively. Again, the rates for the two $F_{10.7}$ ranges are equal within statistical errors for H^+ but significantly different for O^+. Given the large scatters in the data, part of which were attributed to nonuniform data sampling, the O^+ rates for the two $F_{10.7}$ ranges exhibit qualitatively

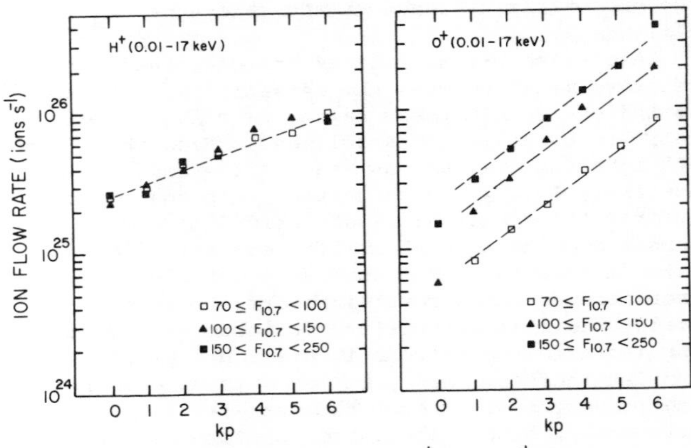

Fig. 3. Ion outflow rate of H^+ and O^+ ions at 0.01-17 keV, as a function of the magnetic Kp index, for different ranges of $F_{10.7}$ values.

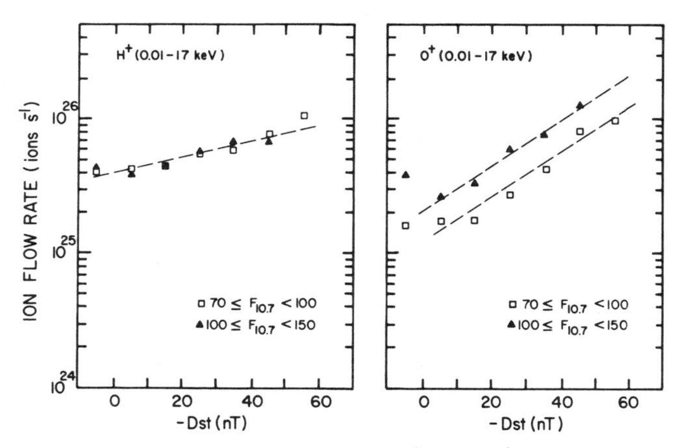

Fig. 4. Ion outflow rate of H^+ and O^+ ions at 0.01-17 keV, as a function of the magnetic Dst index, for different ranges of $F_{10.7}$ values.

similar Dst dependences. The similarity further supports the separation of the magnetic Kp and the solar $F_{10.7}$ dependences in Figures 2 and 3. As noted earlier, the Dst index is primarily a measure of the ring current strength. A quantitative relationship between UFI activity and Dst is not expected theoretically, and the reason for the apparent, exponential correlation between the rates and the Dst index under negative Dst conditions (dashed lines in figure) is not clear. Note that the exponential trend does not appear to extend to positive Dst conditions.

In Figure 5, the outflow rates were computed as a function of the AE index for $F_{10.7}$ below and above 150, respectively, using data samples acquired in the period between September 1981 and December 1983 for which AE index data were available. For all AE values, the rate for high solar activity ($F_{10.7}$) is consistently higher for H^+ and marginally lower for H^+ than the corresponding low $F_{10.7}$ value. In both H^+ and O^+, the rate increases with AE and, except for low AE values, approximately follows a power law dependence. Over the range of AE between 10 and 10^3 nT, the O^+ rate varies between 2×10^{25} and 2×10^{26} ions s^{-1} and the H^+ rate between 2×10^{25} and 7×10^{25} ions s^{-1}. Empirically,

$$F \propto AE^{\alpha} , \qquad (8)$$

where $\alpha \approx 0.8$ for O^+ and ≈ 0.4 for H^+. Since the AE index is an approximate, linear measure of the auroral electrojet strength, and UFI events are associated with auroral substorms, a simple, quasi-linear statistical correlation between the UFI outflow rate and the AE index is perhaps not surprising. From the point of view of energy input-to-output ratio (the free energy flux associated with a substorm available for ionospheric ion energization to the energy flux associated with the outflowing ions), a power law index of unity would be expected if the input energy flux was the only controlling factor of UFI production and the overall efficiency of the ion energization process was independent of the input source strength. Empirically, a power law index slightly less than unity simply reflects that such is probably not the case, and that other factors play a significant role in moderating the total throughput. These include, for example, possible changes in the relative importance of parallel and perpendicular energization, changes in the acceleration altitude distribution, and corresponding changes in the amount of ambient ions available for energization, for changing levels of substorm activity. The factor of 2 difference between the H^+ and O^+ power law indices is likely to be related to these factors.

The apparent dependence of the outflow rates at low AE values must be interpreted with caution. As explained earlier, the AE index may underestimate the actual auroral substorm activity, particularly in the case of small, highly localized substorms that are situated between two sparsely spaced auroral observatories, and of very large substorms that are highly equatorward of the observatories. Thus, whereas a large AE value indicates the presence of high substorm activity, a small AE value does not necessarily imply the absence of activity. Also, the transit time for a 10-eV (low-energy limit of the observed ions) O^+ originating as a transversely accelerated ion at 1000 km altitude to reach the DE 1 apogee (23,300 km) is of the order of 1 hour. Thus, UFIs injected at low altitudes during the late recovery phase of a substorm may be present at DE 1 altitude for up to the order of 1 hour following the substorm, when the hourly AE index may have returned to near zero values. For these two reasons, non-negligible ion outflow may exist at times of near zero AE values, despite the fact that ionospheric

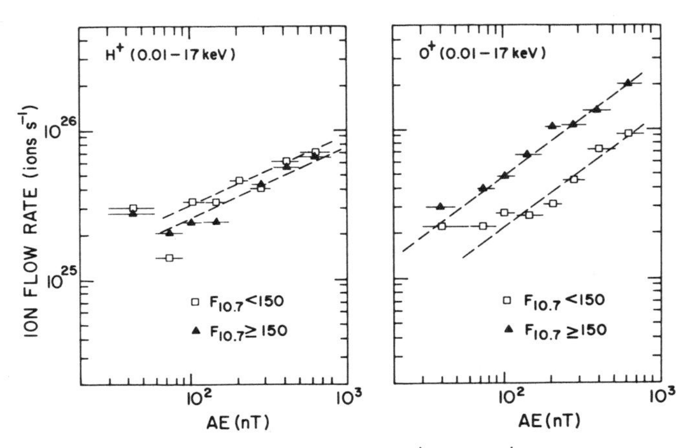

Fig. 5. Ion outflow rate of H^+ and O^+ ions at 0.01-17 keV, as a function of the auroral electrojet index AE, for different ranges of $F_{10.7}$ values.

ion energization may not be occurring at times of no substorm activity.

Conclusions and Discussions

Observed UFI data from the DE 1 EICS, acquired from near the maximum to the minimum of solar cycle 21 ($F_{10.7}$ from <70 to >300), were used to determine the UFI outflow rate as a function of geomagnetic and solar activity conditions and to empirically parametrize its solar ($F_{10.7}$) and magnetic (Kp and AE) dependences. In both H^+ and O^+, ion outflow rate was calculated for ions in the 0.01-17 keV range using the computational procedure described in Yau et al. (1985b). The ion outflow rate increases exponentially with Kp by a factor of 20 for O^+ and 4 for H^+ from Kp = 0 to 6. The exponential dependence prevails at all solar EUV flux levels. Similar exponential increase is found with the Dst index. Also, the rate increases with the hourly AE index. The increase is proportional to AE^α where $\alpha \approx 0.8$ and 0.4 for O^+ and H^+, respectively. For a given magnetic activity condition, the O^+ rate increases by a factor of ~5 in the $F_{10.7}$ range of 70-250. In contrast, the H^+ rate displays a statistically marginal factor of 2 decrease with increasing $F_{10.7}$. The very different magnetic and solar activity dependences between the H^+ and O^+ rates result in an order of magnitude increase in the O^+/H^+ UFI composition ratio from solar minimum to maximum for a given magnetic activity level, as well as from magnetically quiet to active time for a given solar EUV activity level. The variations may be empirically parametrized as follows:

$$F_{O^+}(Kp, F_{10.7}) = 1.0 \times 10^{25} \exp[+1.0 \times 10^{-2} (F_{10.7} - 100)] \exp(0.50\ Kp) \quad (9)$$

$$F_{H^+}(Kp, F_{10.7}) = 2.5 \times 10^{25} \exp[-2.7 \times 10^{-3} (F_{10.7} - 100)] \exp(0.23\ Kp) \quad (10)$$

$$F_{O^+}(AE, F_{10.7}) = 4.2 \times 10^{25} \exp[+1.0 \times 10^{-2} (F_{10.7} - 100)] (AE/100)^{0.8} \quad (11)$$

$$F_{H^+}(AE, F_{10.7}) = 4.3 \times 10^{25} \exp[-2.7 \times 10^{-3} (F_{10.7} - 100)] (AE/100)^{0.4} \quad (12)$$

in the ranges of Kp = 0-6, $F_{10.7}$ = 70-250, and AE $\geq$ 100.

Young et al. (1982) found similar $F_{10.7}$ dependence of 0.9-16 keV equatorially trapped O^+ density at geosynchronous altitudes. They found an exponential coefficient between 1.0×10^{-2} (all Kp data) and 1.8×10^{-2} (Kp $\leq$ 2 data) for O^+ and a coefficient between ~0 (all Kp data) and 3.2×10^{-3} (Kp $\leq$ 2 data).

The magnitude and composition of energetic UFI are key parameters in a coupled magnetospheric-ionospheric model. Equations (9)-(12) represent important input terms to such a model. The equations do not include UFIs below 10 eV or above 17 keV. The contribution of >17-keV ions is believed to be small since the outflow in the 0.01-17 keV range is dominated by <1-keV ions. On the other hand, the contributions from ions below 10 eV, particularly the cleft ion fountain, are probably non-negligible. Recently, Chappell et al. (1987) inferred the relative contribution between ions below and above 10 eV to the cleft ion fountain, using typical upwelling ion energy spectra taken near solar maximum. They found a ratio of about 4 for O^+ and about 0.25 for H^+. Furthermore, they made a rough estimate of the contribution of the <10-eV cleft ion fountain to the total ion outflow by equating the observed outflow of UFI above 10 eV (by EICS) near the dayside cusp to the >10-eV component of the cleft ion fountain. A quantitative determination of the contribution of upwelling ions below 10 eV to the total ion outflow (as well as their fate at high altitude) will be very useful for the model.

The empirically separate dependences of UFI outflow rate on magnetic and solar activities in equations (9)-(12) simply reflect the fact that overall UFI activity--and ultimately the coupling between the ionosphere and the magnetosphere--are intimately modulated by the neutral atmosphere.

Finally, the quantitative AE and $F_{10.7}$ dependences of UFIs above may be related to their magnetospheric ion composition counterparts. On ISEE 1, Lennartsson and Shelley (1986) found order of magnitude changes in energetic O^+ ion densities in the plasma sheet with AE, and lesser but nevertheless significant changes with moderate change in $F_{10.7}$. Quantitative comparison of the dependences between the present UFI and the ISEE 1 plasma sheet results provides a definitive means with which to assess the terrestrial contribution to the hot magnetospheric plasma population.

Acknowledgments. This research was supported in part by NASA contract NAS5-28710.

References

Chappell, C. R., T. E. Moore, and J. H. Waite, Jr., The ionosphere as a fully adequate source of plasma for the earth's magnetosphere, J. Geophys. Res., 92, 5896, 1987.
Cladis, J. B., Parallel acceleration and trans-

port of ions from polar ionosphere to plasma sheet, Geophys. Res. Lett., 13, 893, 1986.

Collin, H. L., R. D. Sharp, E. G. Shelley, and R. G. Johnson, Some general characteristics of upflowing ion beams over the auroral zone and their relationship to auroral electrons, J. Geophys. Res., 86, 6820, 1981.

Collin, H. L., R. D. Sharp, and E. G. Shelley, The magnitude and composition of the outflow of energetic ions from the ionosphere, J. Geophys. Res., 89, 2185, 1984.

Collin, H. L., W. K. Peterson, and E. G. Shelley, Solar cycle variation of some mass dependent characteristics of unflowing beams of terrestrial ions, J. Geophys. Res., in press, 1987.

Ghielmetti, A. G., R. G. Johnson, R. D. Sharp, and E. G. Shelley, The latitudinal, diurnal, and altitudinal distributions of upward flowing energetic ions of ionospheric origin, Geophys. Res. Lett., 5, 59, 1978.

Ghielmetti, A. G., E. G. Shelley, and D. M. Klumpar, Correlation between number flux and energy of upward flowing ion beams, Physica Scripta, in press, 1987.

Gorney, D. J., A. Clarke, D. Croley, J. F. Fennell, J. Luhmann, and P. Mizera, The distribution of ion beams and conics below 8000 km, J. Geophys. Res., 86, 83, 1981.

Klumpar, D. M., W. K. Peterson, and E. G. Shelley, Direct evidence for two-stage (bimodal) acceleration of ionospheric ions, J. Geophys. Res., 89, 10799, 1984.

Lennartsson, W., and E. G. Shelley, Survey of 0.1-16 keV/e plasmasheet ion composition, J. Geophys. Res., 91, 3061, 1986.

Lockwood, M., M. O. Chandler, J. L. Horwitz, J. H. Waite, Jr., T. E. Moore, and C. R. Chappell, The cleft ion fountain, J. Geophys. Res., 90, 9736, 1985.

Lockwood, M., A. P. van Eyken, B.J.I. Bromage, J. H. Waite, Jr., T. E. Moore, and J. R. Doupnik, Low-energy ion outflows from the ionosphere during a major cap expansion - Evidence for equatorward motion of inverted-V structures, Adv. Space Res., 6, 93, 1986.

Moore, T. E., Superthermal ionospheric outflows, Rev. Geophys. Space Phys., 22, 264, 1984.

Moore, T. E., M. Lockwood, M. O. Chandler, J. H. Waite, Jr., C. R. Chappell, A. Persoon, and M. Sugiura, Upwelling O^+ ion source characteristics, J. Geophys. Res., 91, 7019, 1986.

Nagai, T., J. H. Waite, Jr., J. L. Green, and C. R. Chappell, First measurements of supersonic polar wind in the polar magnetosphere, Geophys. Res. Lett., 11, 669, 1984.

Rostoker, G., Geomagnetic indices, Rev. Geophys. Space Phys., 10, 935, 1972.

Shelley, E. G., D. A. Simpson, T. C. Sanders, E. Hertzberg, H. Balsiger, and A. Ghielmetti, The energetic ion composition spectrometer (EICS) for the Dynamics Explorer-A, Space Sci. Instrum., 5, 443, 1981.

Shelley, E. G., W. K. Peterson, A. G. Ghielmetti, and J. Geiss, The polar ionosphere as a source of energetic magnetospheric plasma, Geophys. Res. Lett., 9, 941, 1982.

Waite, J. H. Jr., T. Nagai, J.F.E. Johnson, C. R. Chappell, J. L. Burch, T. L. Killeen, P. B. Hayes, G. R. Carignan, W. K. Peterson, and E. G. Shelley, Escape of suprathermal O^+ ions in the polar cap, J. Geophys. Res., 90, 1619, 1985.

Yau, A. W., B. A. Whalen, W. K. Peterson, and E. G. Shelley, Distribution of upflowing ionospheric ions in the high-altitude polar cap and auroral ionosphere, J. Geophys. Res., 89, 5507, 1984.

Yau, A. W., P. H. Beckwith, W. K. Peterson, and E. G. Shelley, Long-term (solar-cycle) and seasonal variations of upflowing ionospheric ion events at DE-1 altitudes, J. Geophys. Res., 90, 6395, 1985a.

Yau, A. W., E. G. Shelley, W. K. Peterson, and L. Lenchyshyn, Energetic auroral and polar ion outflow at DE-1 altitudes: Magnitude, composition, magnetic activity dependence and long-term variations, J. Geophys. Res., 90, 8417, 1985b.

Young, D. T., H. Balsiger, and J. Geiss, Correlations of magnetospheric ion composition with geomagnetic and solar activity, J. Geophys. Res., 87, 9077, 1982.

THE POLAR WIND

R. W. Schunk

Center for Atmospheric and Space Sciences, Utah State University, Logan, Utah 84322

Abstract. The classical polar wind is an ambi-polar outflow of thermal plasma from the terrestrial ionosphere at high latitudes. As the plasma escapes along diverging geomagnetic flux tubes, it undergoes four major transitions, including a transition from chemical to diffusion dominance, a transition from subsonic to supersonic flow, a transition from collision-dominated to collisionless regimes, and a transition from a heavy to a light ion. A further complication arises because of horizontal convection of the flux tubes because of magnetospheric electric fields. The characteristic features of this flow are reviewed, with emphasis on recent theoretical predictions. The review covers the following topics: (1) H^+ and He^+ outflow characteristics both with and without convection electric field effects, (2) the collisionless polar wind characteristics including the formation of temperature anisotropies, (3) the problems with the transition from collisional to collisionless regimes, (4) the stability of the polar wind velocity distribution functions, (5) the creation of suprathermal O^+ and H^+ ions in the polar wind via elevated thermal electron temperatures and hot magnetospheric electrons, and (6) the time-dependent polar wind behavior in response to localized density perturbations and localized ion heat sources. The recent experimental evidence which supports some of the theoretical predictions is also briefly mentioned.

Introduction

In the early 1960's, it was recognized that the interaction of the solar wind with the earth's dipole magnetic field acts to significantly modify the magnetic field configuration in a vast region close to the earth (Axford and Hines, 1961; Dungey, 1961). The dynamo action associated with the flow of solar wind plasma across magnetic field lines generates an intense current system which acts to compress the earth's magnetic field on the sunward side and stretch it into a long comet-like tail on the anti-sunward side. The magnetic field lines which form the tail originate in the earth's polar regions, and since the pressure in the ionosphere is much greater than that in the distant tail, it was suggested that a continual escape of thermal plasma (H^+ and He^+) should occur along these open field lines (Dessler and Michel, 1966). Nishida (1966) also proposed an escape of thermal plasma in combination with horizontal plasma convection to explain the experimental discovery of the plasmapause by whistler measurements.

These early suggestions of light ion outflow were based on the well-known theory of thermal evaporation, which had been successfully applied to the escape of neutral gases from planetary atmospheres. Via thermal evaporation, the light ions would escape the topside ionosphere with velocities close to their thermal velocities and then flow along magnetic field lines to the magnetospheric tail. However, it was subsequently argued that the outflow should be supersonic and it was termed the polar wind in analogy to the solar wind (Axford, 1968). A hydrodynamic model was then used that emphasized the supersonic nature of the flow, thereby elucidating its basic characteristics (Banks and Holzer, 1969; Marubashi, 1970).

It is now well known that the classical polar wind is an ambi-polar outflow of thermal plasma from the high-latitude ionosphere and that it undergoes four major transitions, including a transition from chemical to diffusion dominance, a transition from subsonic to supersonic flow, a transition from collision dominated to collisionless regimes, and a transition from a heavy to a light ion. Also, recent measurements have indicated that the polar wind contains suprathermal components of both light and heavy ions (Lockwood et al., 1985; Yau et al., 1985; Moore et al., 1986). Because of the complicated nature of the flow, numerous mathematical models have been constructed over the years, including hydrodynamic (Banks and Holzer, 1969; Raitt et al., 1975), hydromagnetic (Holzer et al., 1971], generalized transport (Schunk and Watkins, 1981, 1982), kinetic (Lemaire and Scherer, 1973), and semikinetic (Barakat and Schunk, 1983, 1984). These models have been used primarily to study

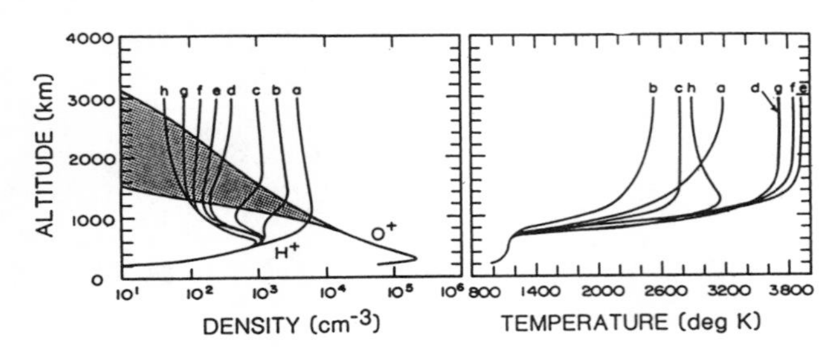

Fig. 1. Theoretical H^+ density and temperature profiles as a function of altitude for different H^+ escape velocities at 3000 km. The H^+ velocities at 3000 km are: (a) 0.06, (b) 0.34, (c) 0.75, (d) 2.0, (e) 3.0, (f) 5.0, (g) 10.0, and (h) 20.0 km s^{-1}. The shaded region shows the range of O^+ densities (Raitt et al., 1975).

the steady state characteristics of the flow, with emphasis on elucidating physical processes.

The classical polar wind occurs in the polar cap along horizontally convecting flux tubes and at latitudes equatorward of the auroral zone along depleted corotating flux tubes. Since the classical polar wind models are based on charge neutrality and charge conservation with no field-aligned current (ambi-polar flow), the models cannot be applied to auroral field lines. Also, the presence of waves and instabilities on auroral field lines acts to further invalidate polar wind models in this region. Consequently, this review will not be concerned with auroral processes of any kind, but instead will concentrate on the classical polar wind.

Hydrodynamic Solutions

In the initial studies of the polar wind, the hydrodynamic continuity and momentum equations for H^+ and O^+ ions were solved by assuming an isothermal ionosphere and a gravitationally bound O^+ population (Banks and Holzer, 1969; Marubashi, 1970). Subsequently, the polar wind energy balance and the effects of magnetospheric convection electric fields were taken into account (Raitt et al., 1975, 1977, 1978).

H^+ Outflow

Typical polar wind density and temperature profiles are shown in Figure 1 for the case when convection electric field effects are neglected. The left panel shows the effect on the H^+ and O^+ densities of different H^+ escape velocities at 3000 km. Curve a represents near diffusive equilibrium, with H^+ becoming the dominant ion at 900 km (the O^+ density in this case follows the lower curve of the shaded region). As the upper boundary velocity is increased, the H^+ density is progressively reduced, with a peak in the H^+ density profile appearing near 600-700 km altitude. Curves b-e correspond to subsonic outflow, while for curves g-h the flow is supersonic. Curve f covers the transonic region with the Mach number equal to 1.17 at 1400 km and reducing to 0.89 at 3000 km. For curve h, which is for a flow velocity of 20 km s^{-1} at 3000 km, the H^+ escape flux is 8.5 x 10^7 cm^{-2} s^{-1}.

The H^+ temperature is strongly affected by the H^+ flow, as shown in the right panel of Figure 1. The behavior of the H^+ temperature is also fairly complicated. As the escape velocity is increased, the H^+ temperature at high altitudes first decreases, then increases, and then decreases again. This behavior is related to the relative contributions made to the H^+ thermal balance by convection, advection, thermal conduction, frictional heating, and collisional cooling, and the complete details are given by (Raitt et al., 1975). However, we note that the general trend of increasing H^+ temperatures at high altitudes as the H^+ escape velocity increases in the subsonic regime (curves b-e) is due primarily to enhanced frictional heating as H^+ moves through a gravitationally bound O^+ population with an increasing speed. The decrease in the H^+ temperature with increasing H^+ escape velocity in the supersonic regime (curves f-h) is due both to a decrease in frictional heating as the plasma becomes collisionless and a change in the shape of the velocity profile, which acts to increase the importance of convective cooling.

The effect of a convection electric field is to heat the plasma through frictional interaction between the ions and neutrals, which acts to increase the plasma pressure differential between the ionosphere and magnetosphere and to enhance the outward flux of H^+ ions. In practice, this effect is opposed by a higher topside O^+ density, which reduces the H^+ diffusion coefficient, and by the depletion of O^+ in the F_2 region by chemical effects associated with the ion heating (Schunk et al., 1975), which reduce the production rate of H^+ ions.

Raitt et al. (1977) studied the effect that convection electric fields have on the polar wind, but they separately studied the electric field heating and O^+ depletion effects. Figure 2 shows the effect of electric field heating on the

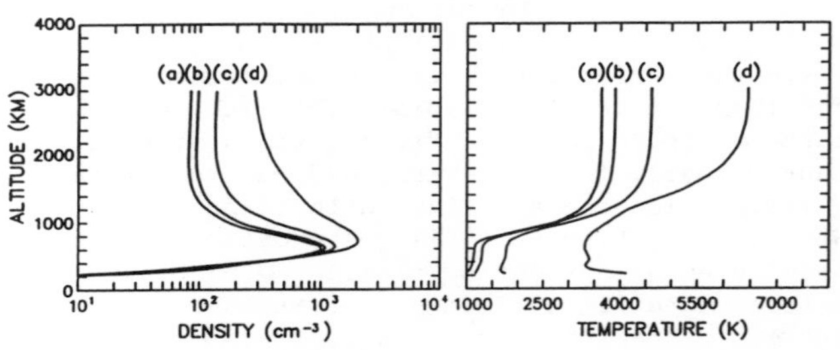

Fig. 2. Theoretical H^+ density and temperature profiles as a function of altitude for a 10 km s^{-1} H^+ outflow velocity at 3000 km and for different convection electric fields: (a) 0, (b) 25, (c) 50, and (d) 100 mV m^{-1} (Raitt et al., 1977).

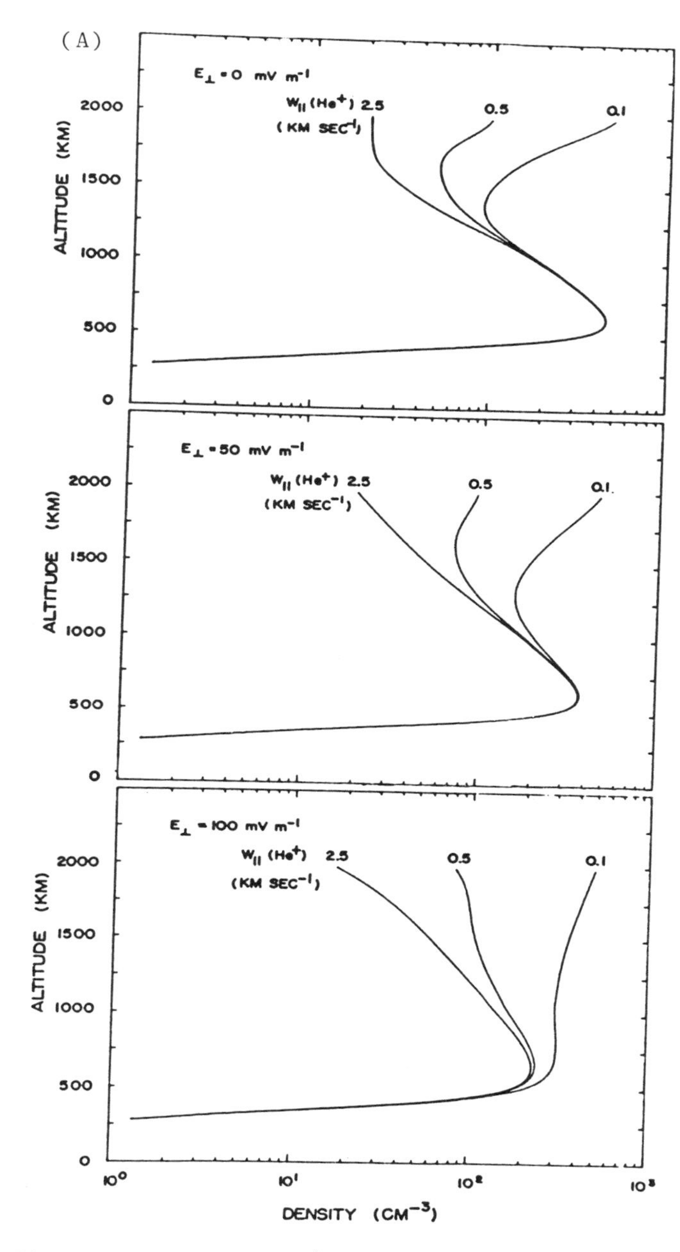

Fig. 3a. Theoretical He^+ density profiles as a function of altitude for He^+ upper boundary outflow velocities of 0.1, 0.5, and 2.5 km s^{-1}. The upper, middle, and lower panels show the profiles for convection electric fields of 0, 50, and 100 mV m^{-1}, respectively (Raitt et al., 1978).

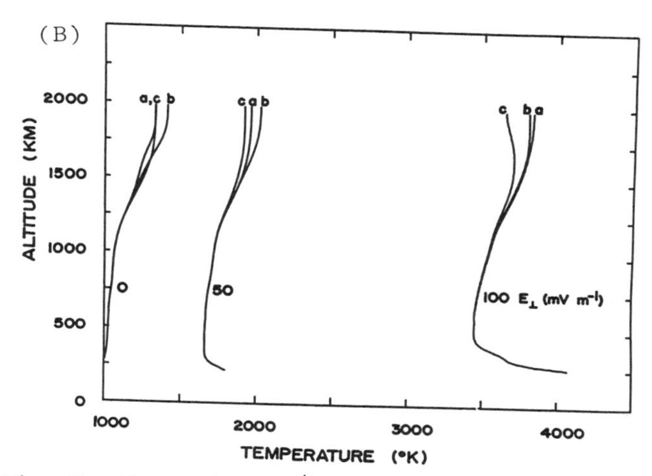

Fig. 3b. Theoretical He^+ temperature profiles as a function of altitude for He^+ upper boundary velocities of: (a) 0.1, (b) 0.5, and (c) 2.5 km s^{-1}. Each family of curves corresponds to a different convection electric field, as indicated by the labels (Raitt et al., 1978).

H^+ density and temperature profiles for an H^+ outflow velocity of 10 km s^{-1} at 3000 km, an F_2 peak O^+ density of 2.1 x 10^5 cm^{-3}, and for convection electric fields of 0, 25, 50, and 100 mV m^{-1}. As the convection electric field increases, the general trend is for an increased H^+ density, a decreased H^+ drift velocity, and an enhanced H^+ temperature. The electric field also acts to increase the O^+ temperature and, hence, topside O^+ density (not shown in Figure 2). The increased O^+ density, in turn, acts to slow the H^+ outflow and thereby increase the H^+ density. As far as the H^+ temperature is concerned, the electric field provides a direct heat source below 600 km because of the frictional interaction between H^+ ions and the neutral atmosphere. Above 600-800 km, the H^+ temperature increases rapidly with altitude for all four electric field cases, and this results from the additional H^+-O^+ frictional heating that occurs as H^+ flows up and out of the topside ionosphere. At high altitudes, the H^+ temperature profiles are isothermal because of the dominance of thermal conduction.

He^+ Outflow

Raitt et al. (1978) have made extensive calculations of the characteristics of the He^+ outflow from the high-latitude topside ionosphere. Steady state solutions to the hydrodynamic continuity, momentum, and energy equations were obtained self-consistently, yielding He^+ density, drift velocity, and temperature profiles over the altitude range from 200 to 2000 km. To cover the wide variation expected at high latitudes, several parameters were varied, including the He^+ and H^+ outflow velocities, the F region peak electron density, the convection electric field, and the He and N_2 neutral density profiles.

Figures 3a and 3b show corresponding He^+ density and temperature profiles for convection electric fields of 0, 50, and 100 mV m^{-1} and for upper boundary He^+ outflow velocities of 0.1, 0.5, and 2.5 km s-1. For these calculations the O^+ density at the F_2 peak was 2.1 x 10^5 cm^{-3}, and the H^+ outflow velocity at 3000 km was 10 km s^{-1}.

The general characteristics of the He^+ density profiles are similar to those of H^+. There is a region below about 600 km where equilibrium between production and loss dominates; whereas at higher altitudes diffusion is more important, resulting in a peak He^+ density in the vicinity of 600 km. The higher He^+ outflow cases qualitatively resemble the flux-limited profiles computed by Banks and Holzer (1969). For the 0 and 50 mV m^{-1} perpendicular electric fields, this peak is unaffected by the upper boundary He^+ outflow velocity. When the perpendicular electric field is 100 mV m^{-1} there is a slight reduction in the He^+ peak density as the He^+ upper boundary outflow velocity increases from 0.1 to 2.5 km s^{-1}.

The reason for this behavior is a consequence of the increasing ion temperature as the perpendicular electric field increases. The higher ion temperature causes a decreased Mach number for a given outflow at the upper boundary. The result is that the profiles become nearer to diffusive equilibrium profiles above the peak, this effect being particularly noticeable in the most extreme case of $E_\perp$ = 100 mV m^{-1} and 0.1 km s^{-1} He^+ outflow velocity (Figure 3a, bottom panel). For this case, the increased peak density and much higher upper boundary density result from the He^+ outward flux being below the saturated flux of 3.8×10^6 cm^{-2} s^{-1} at 3.2×10^6 cm^{-2} s^{-1}. This again is due to the high ion temperature and the inhibiting effect on the He^+ diffusion coefficient of the higher H^+ and O^+ densities for $E_\perp$ = 100 mV m^{-1}.

As far as the He^+ temperature profiles are concerned, the zero electric field case shows that below about 1300 km the boundary He^+ outflow velocity has very little effect on the He^+ temperature. Above 1300 km a more rapid increase in velocity occurs, which results in flow terms such as advection, convection, thermal conduction, and joule heating by collisions with O^+ being more important in the energy balance. These terms do not, however, dominate the energy balance, and coupling to O^+ remains an important term up to the top boundary. This prevents a large variation of He^+ temperature with He^+ outflow velocity in contrast to the marked effect for H^+ ions. The main reason for the small effect of frictional heating is that the He^+ velocity shows a rapid increase only above 1300 km, while for H^+ the rapid increase in outflow velocity begins at about 800 km. At 1300 km the O^+ density is typically a factor of 5 lower than at 800 km and the joule heating is consequently reduced. It should be noted, however, that the behavior of the joule heating agrees with that observed for H^+ in that the subsonic-transonic cases of 0.1 and 0.5 km s^{-1} outflow velocity show an increase in upper boundary temperature, while the supersonic outflow case of 2.5 km s^{-1} shows a reduction. This can be attributed to the same reason as that for H^+ temperatures, a reduction in the velocity-dependent collision frequency as the Mach number exceeds unity.

The groupings of the families of He^+ temperature profiles for different He^+ outflow velocities for each of the three perpendicular electric fields show that this latter parameter has the strongest influence on the He^+ temperature. The increase in frictional heating of He^+ ions near the lower boundary by collisions with the neutral gas as $E_\perp$ increases can be seen. This frictional heating rapidly declines with altitude as the neutral gas density decreases until the strong coupling to the O^+ ions takes over. Like the zero perpendicular electric field case, the effect of the He^+ outflow is minimal until about 1300 km when the flow, conduction, and He^+-O^+ frictional heating terms become significant.

Collisionless Polar Wind Characteristics

The polar wind results that have been discussed in the previous section are valid at the altitudes where the flow is collision dominated. As a rough guide, the flow is effectively collision dominated when $U_i/H_i \nu_i \ll 1$, where U_i is the ion field-aligned drift velocity, H_i is the ion density scale height, and ν_i is the appropriate ion collision frequency. For H^+, this condition generally begins to break down at 1000 km and is clearly violated at 2000 km (Raitt et al., 1975). When the plasma is not collision dominated, the H^+ pressure distribution becomes anisotropic and the H^+ heat flow vector is not simply related to the gradient in the H^+ temperature.

The collisionless characteristics of the polar wind can be described by kinetic, semikinetic, hydromagnetic, and generalized transport models. The hydromagnetic and generalized transport equations are obtained by taking velocity moments of the Boltzmann equation in an effort to derive conservation equations for the physically significant moments of the distribution function, such as density, drift velocity, temperature, stress tensor, and heat flow vector. The hydromagnetic equations correspond to the collisionless moment equations. The generalized transport equations are similar to the hydromagnetic equations, except that collisional terms are retained; therefore, these equations provide a continuous transition from the collision-dominated regimes to the collisionless regimes. The kinetic models, on the other hand, are obtained by directly integrating the collisionless Boltzmann equation; consequently, they satisfy the full hierarchy of moment equations deduced from the collisionless Boltzmann equation. Semikinetic models correspond to a hybrid approach whereby some species are described by transport equations and others by kinetic equations.

With regard to the polar wind, the hydromagnetic, generalized transport, kinetic, and semikinetic models produce density and drift velocity profiles that are similar to those obtained from the hydrodynamic equations for the case of supersonic flow. However, the ion temperature distributions are different, with the collisionless

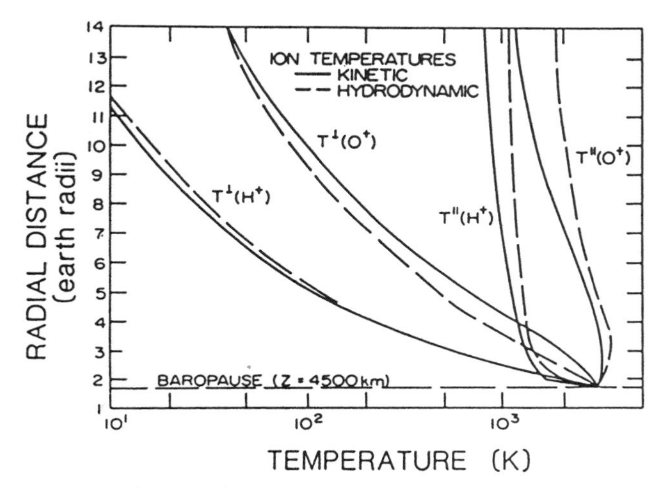

Fig. 4. O^+ and H^+ temperatures parallel and perpendicular to the geomagnetic field obtained from kinetic and hydromagnetic models of the collisionless, supersonic polar wind (Holzer et al., 1971).

models yielding large temperature anisotropies at high altitudes. Typical results are shown in Figure 4, where the H^+ and O^+ temperatures parallel and perpendicular to the geomagnetic field are plotted as a function of altitude for collisionless, supersonic H^+ outflow. The ion temperature distributions were calculated with both kinetic and hydromagnetic models and the results are similar. The parallel ion temperatures are essentially constant with altitude at high altitudes, while the perpendicular ion temperatures decrease monotonically with altitude. The net result is a parallel-to-perpendicular temperature anisotropy that grows with altitude, reaching nearly a factor of 50 for H^+ ions at a distance of 10 earth radii.

Transition From Collisional to Collisionless Regimes

The hydrodynamic formulation is valid in the collision-dominated regime and the kinetic and hydromagnetic formulations are valid in the collisionless regime. There have been several attempts to bridge the gap between the formulations and obtain polar wind solutions that are valid at all altitudes. Figure 5 shows the hybrid approach developed by Lemaire (1972) which involves matching hydrodynamic and kinetic solutions at the exobase. Shown in this figure are two H^+ velocity profiles from a kinetic model with imposed ion and electron temperatures of 2500 K (full solid curve) and 3000 K (dotted curve). Also shown in the figure are several hydrodynamic solutions for the low-altitude region (barosphere). For each solution, the exospheric level (indicated by a vertical bar) is determined from the equality of the density scale height and the collision mean free path. From the family of hydrodynamic solutions, the one that insures continuity of density and flux with the exospheric kinetic solution is selected as the appropriate hydrodynamic solution.

The above hybrid approach was an important first step toward obtaining polar wind solutions that are continuous from the collision-dominated regime at low altitudes to the collisionless regime at high altitudes. However, in this example, the ion and electron temperatures were imposed so that solutions were required only for the density and drift velocity moments of the particle distribution function. In theory, there are an infinite number of moments that must be matched across the exobase in order to get a truely continuous description of the polar wind. At the very least, one should try to obtain continuous solutions for the parallel and perpendicular temperatures and the heat flow in addition to the density and flux.

Generalized transport equations were used by Schunk and Watkins (1981, 1982) to obtain polar wind solutions that are continuous through the transition from collision-dominated regimes to

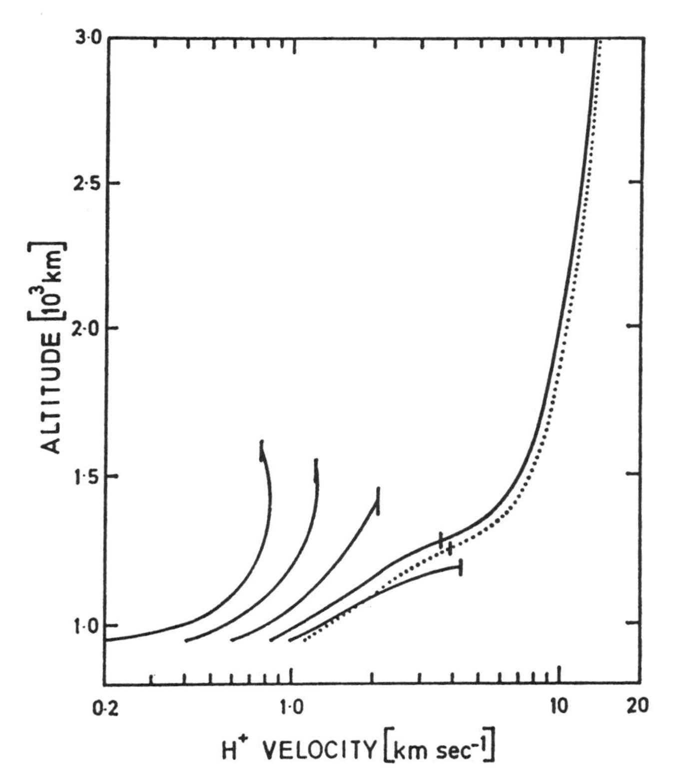

Fig. 5. H^+ velocity profiles from a kinetic model with ion and electron temperatures of 2500 K (full solid curve) and 3000 K (dotted curve). Also shown are the family of hydrodynamic solutions calculated for different escape fluxes. The one that insures continuity of density and flow at the exobase (vertical bar) with the kinetic solution is selected as the appropriate hydrodynamic solution (Lemaire, 1972).

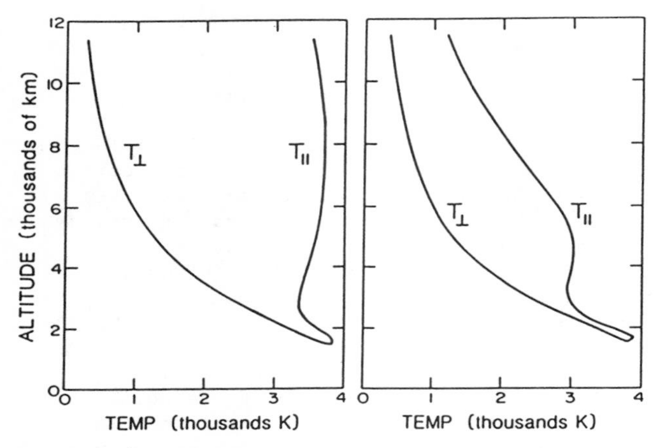

Fig. 6. H^+ temperatures parallel and perpendicular to the geomagnetic field versus altitude for supersonic flow and for both cold (left panel) and hot (right panel) electron temperature distributions (Schunk and Watkins, 1982).

collisionless regimes. Not only are the density and flow velocity continuous through the exobase, but so are the parallel and perpendicular temperatures and the heat flow. The dramatic new result to emerge from these studies was that the character of the solution to the generalized transport equations was different for subsonic and supersonic H^+ outflows. Figure 6 shows the H^+ temperatures parallel and perpendicular to the geomagnetic field for supersonic H^+ outflow and for both cold (left panel) and hot (right panel) electron temperature distributions. For both cases, there is an appreciable H^+ temperature anisotropy with $T_{p\parallel} > T_{p\perp}$ at all altitudes above 1500 km. The perpendicular H^+ temperature displays a rapid decrease with altitude at low altitudes and then tends to go constant at high altitudes. The variation of $T_{p\parallel}$ with altitude, on the other hand, is more complicated. At low altitudes, $T_{p\parallel}$ displays a decrease with altitude that is similar to the $T_{p\perp}$ decrease. Above about 3000 km, $T_{p\parallel}$ exhibits little variation with altitude all the way to the top boundary for the low electron temperature case. However, for the high electron temperature case, $T_{p\parallel}$ is roughly constant with altitude between 3000 and 5000 km and then decreases with altitude above this altitude range.

The behavior of the parallel and perpendicular H^+ temperatures shown in the left panel of Figure 6 for the cold T_e distribution is in good qualitative agreement with that obtained from the kinetic and hydromagnetic models (Figure 4). The main qualitative difference is that the generalized transport equations produce an H^+ temperature anisotropy that tends to go constant at high altitudes, whereas the anisotropy obtained from the kinetic and hydromagnetic models does not display this tendency. Also, solutions to the generalized transport equations for supersonic H^+ outflow can be obtained only with an upward H^+ heat flow from the lower ionosphere.

For subsonic H^+ outflow, it is possible to obtain solutions to the generalized transport equations with a downward H^+ heat flow, which implies that a high-altitude heat source exists above the altitude range of interest. For subsonic H^+ outflow with a downward heat flow, the variation of the parallel and perpendicular H^+ temperatures is different from that found for supersonic outflow. For subsonic outflow, the anisotropy is opposite to that found for supersonic outflow, with $T_{p\parallel} > T_{p\perp}$ below about 3500 km and $T_{p\perp} > T_{p\parallel}$ above this altitude. Also, in contrast to what was found for supersonic outflow, for this case the parallel and perpendicular H^+ temperatures increase with altitude above 3500 km because of the high-altitude heat source.

Stability of the Polar Wind

At altitudes above about 2000 km, the polar wind becomes supersonic and collisionless, and the H^+ velocity distribution becomes non-Maxwellian. The evolution of the H^+ distribution function with altitude is shown in Figure 7 for a collisionless supersonic H^+ outflow. At the starting altitude of 1.8 R_E (earth radii), the H^+ velocity distribution is basically an upward drifting Maxwellian with a drift velocity of about 10 km s^{-1} except for the absence of negative (downward) velocities. As the plasma expands in the diverging geomagnetic field, the H^+ distribution becomes non-Maxwellian. The non-Maxwellian features include a temperature anisotropy, with the parallel H^+ temperature greater than the perpendicular temperature, and an asymmetry, with an elongated tail in the upward direction. These distortions from a Maxwellian increase with altitude, and at 10 R_E the parallel-to-perpendicular temperature ratio is about 50 and the tail is sufficiently long to move the drift velocity point off the peak of the distribution function.

The highly non-Maxwellian H^+ velocity distributions described above are potentially unstable. If the polar wind were to become unstable, this could significantly affect ionosphere-magnetosphere coupling. For example, the resulting wave-particle interactions might act to energize heavy ions in the polar wind, and this could increase their escape flux within the limitations decided by the circumstances at lower altitudes. Therefore, Barakat and Schunk (1987) studied the stability of these non-Maxwellian distributions with regard to the excitation of electrostatic waves. Despite their highly non-Maxwellian character, the distributions were found to be remarkably stable for a wide range of conditions. This indicates that the various macroscopic formulations of the classical polar wind are valid.

Suprathermal Ions in the Polar Wind

There are several mechanisms that can lead to the presence of suprathermal ions in the polar wind. First, the suprathermal ions can be created when the flux tubes of plasma convect into

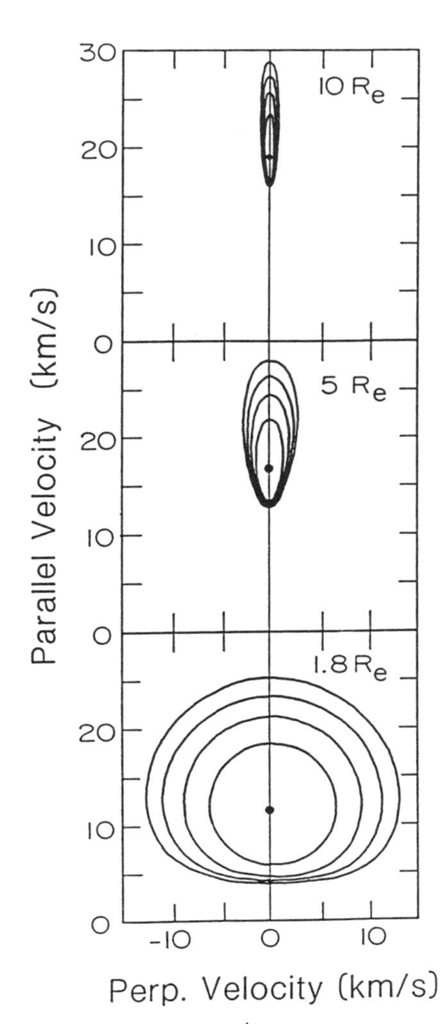

Fig. 7. Contours of the H^+ distribution function at three altitudes. The altitudes are 1.8 R_E (bottom), 5 R_E (middle), and 10 R_E (top), where R_E is the earth's radius. The contour levels decrease by a factor of e starting from the maximum, which is shown by the dot (Barakat and Schunk, 1987).

the dayside cusp and nocturnal auroral oval because of a variety of auroral acceleration processes. The subsequent convection of the plasma flux tubes into the polar cap coupled with the rapid outflow of the heated ions will act to produce suprathermal ion components passing through the classical polar wind. Because of a page limitation, this topic will not be discussed further in this review. However, there are two energization mechanisms that have been proposed that operate in the polar cap and these are discussed below.

High Thermal Electron Temperatures

Although the early polar wind models predicted a negligible O^+ escape flux, satellite measurements clearly indicate that the magnetosphere contains a significant population of both energetic and suprathermal O^+ ions. Motivated by these measurements, a systematic parameter study was conducted by Barakat and Schunk (1983) using a semikinetic model to describe the steady state, collisionless polar wind. This study indicated that for high electron temperatures O^+ is not gravitationally bound and significant escape fluxes ($\sim 10^7$ cm^{-2} s^{-1}) of suprathermal ($\sim$2 eV) O^+ ions can occur.

Figure 8 shows representative ion density and Mach number profiles for supersonic H^+ outflow and for electron temperatures of 3000 and 10,000 K. The case of T_e = 3000 K is essentially the same case considered by Holzer et al. (1971) except for a constant multiplier of the ion densities at the lower boundary (4500 km). For this case, the O^+ density decreases rapidly with altitude, the O^+ escape flux is negligibly small, and the O^+ Mach number is less than unity over most of the altitude range. For T_e = 10,000 K, on the other hand, the O^+ density profile tends to follow the H^+ profile, and significantly elevated O^+ densities occur at high altitudes. The O^+ escape flux also increases dramatically, from 1.8 x 10^2 cm^{-2} s^{-1} for T_e = 3000 K to 0.44 x 10^7 cm^{-2} s^{-1} for T_e = 10,000 K. In addition, the O^+ Mach number, drift velocity, and energy increase markedly when T_e is increased from 3000 to 10,000 K. The O^+ outflow is supersonic at most altitudes and the Mach number approaches 20 at the upper boundary.

Hot Magnetospheric Electrons

Precipitating hot electrons of magnetospheric origin are a common feature of the polar cap. Specifically, three hot electron populations have been identified, including the polar rain, polar showers, and polar squall (Winningham and Gurgiolo, 1982). Recently, the effect of such hot electron populations on the polar wind has been studied by Barakat and Schunk (1984) using a semikinetic model to describe the collisionless regime at high altitudes. Estimates of hot electron parameters based on characteristic energy

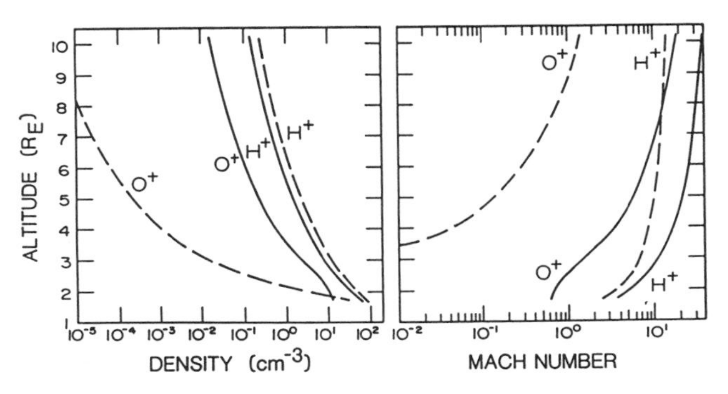

Fig. 8. Ion density (left panel) and Mach number (right panel) profiles versus altitude for supersonic H^+ outflow and for T_e = 3000 K (dashed curves) and T_e = 10,000 K (solid curves) (Barakat and Schunk, 1983).

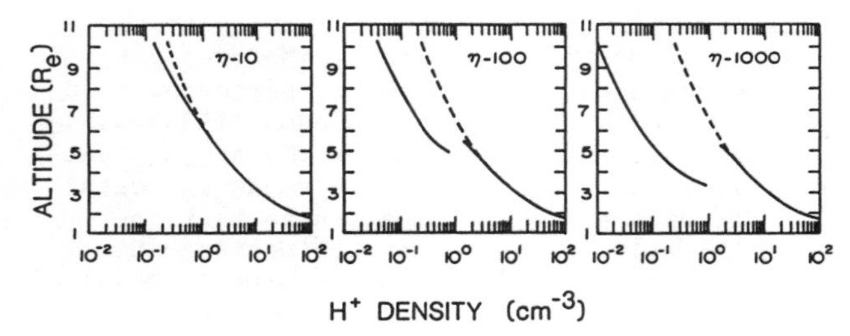

Fig. 9. H^+ density profiles versus altitude for supersonic H^+ outflow, for a hot/total electron density ratio of 1% at 4500 km, and for hot/cold electron temperature ratios of $\eta = T_h/T_c = 10$ (left panel), 100 (middle panel), and 1000 (right panel). The dashed curves correspond to the solutions obtained without including the effect of hot electrons (Barakat and Schunk, 1984).

and flux measurements indicate that the hot/cold electron temperature ratio varies from 10 to 10^4 and that the percentage of hot electrons varies from 0.1% to 10% at 4500 km.

For ratios at the lower ends of these ranges, the polar wind solutions with hot electrons are similar to those obtained previously without hot electrons. However, for higher hot electron temperatures and a greater percentage of hot electrons, there is a discontinuity in the kinetic solution, which indicates the presence of a sharp transition. This is shown in Figure 9, where H^+ density profiles are shown for the case of supersonic H^+ outflow, for a hot/total electron density ratio of 1% at 4500 km, and for hot/cold electron temperature ratios of 10, 100, and 1000. The transition corresponds to a contact surface between the hot and cold electrons. Along this surface, an outwardly directed, parallel electric field exists which reflects cold ionospheric electrons and prevents their escape. The double layer electric field also acts to increase the supersonic H^+ outflow velocity and escape energy. The H^+ energy gain may be as large as 1 to 2 keV, depending on the ionospheric and magnetospheric conditions. With regard to O^+, the hot electrons act to reduce the potential barrier, thereby allowing more O^+ ions to escape. A significant enhancement in the O^+ escape flux

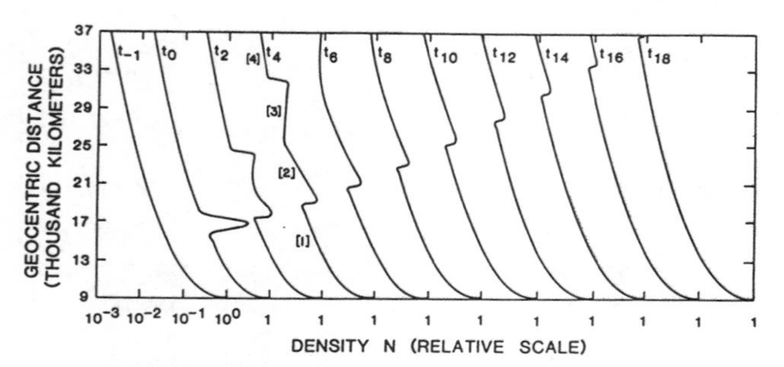

Fig. 10. Temporal evolution of a localized density enhancement created at $t = t_0$ in the collisionless polar wind (Singh and Schunk, 1985).

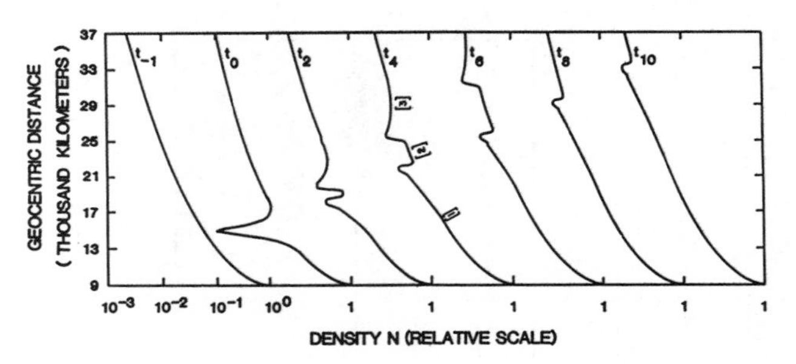

Fig. 11. Temporal evolution of a localized density depletion created at $t = t_0$ in the collisionless polar wind (Singh and Schunk, 1985).

can occur depending on the hot electron density and temperature.

Time-Dependent Polar Wind

All of the polar wind studies discussed up to this point were for steady state conditions; therefore, the energization that occurs during the initial plasma expansion was not modeled. Recently, however, the temporal evolution of density perturbations in the supersonic, collisionless polar wind was modeled with the aid of time-dependent hydrodynamic equations (Singh and Schunk, 1985). Extended density depletions as well as localized density bumps and holes were considered.

Localized Density Bump

When the polar wind is perturbed by a localized density bump or hole, the perturbation propagates upward in the direction of the polar wind, but it undergoes a considerable modification, evolving into forward-reverse shock pairs. Figure 10 shows the temporal evolution of a density bump. The density profile at t_{-1} is a typical, steady state, supersonic polar wind profile. At t_0 the polar wind is perturbed by a density bump that peaks at a geocentric distance of 17,000 km and is approximately Gaussian with a width of about 500 km. The density bump is initially given a zero drift velocity along the flux tube to simulate the release of plasma from a spacecraft moving horizontally through the polar ionosphere. At later times (t_2, t_4), distinct forward and reverse shocks can be clearly seen. The forward shock moves with a velocity of about 36 km s^{-1} until it crosses the upper boundary, while the reverse shock undergoes a continual acceleration, crossing the upper boundary with a speed of about 14.5 km s^{-1}. After the reverse shock crosses the upper boundary, the polar wind returns to a steady state situation, which typically takes only tens of minutes.

Localized Density Hole

Figure 11 shows the temporal evolution of a density hole created at $t = t_0$ in a manner analogous to the density bump shown in Figure 10. As

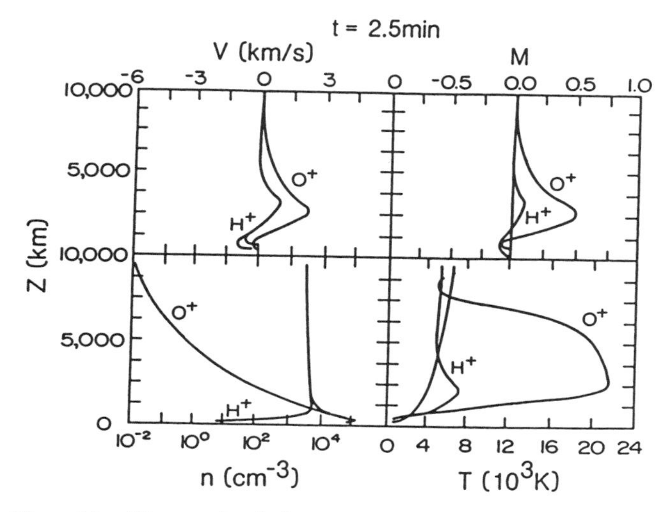

Fig. 12. Flow velocities, Mach numbers, number densities, and temperature distributions in an ion heated ionosphere 2.5 min after ion heating was initiated (Gombosi et al., 1985).

the perturbation propagates upward, a considerable filling of the hole occurs at its center, dividing the hole into two segments. The leading hole propagates upward faster than the trailing hole, and the density profile between the two holes turns out to be identical to that obtained in the steady state, as shown at t_{-1}. The two holes appear as rarefaction waves, which transmit the information that a density hole was created in the polar wind to increasing distances. As was found for the density bump, the creation of a density hole in the supersonic polar wind leads to the formation of forward and reverse shocks. The reverse shock joins the trailing rarefaction wave (hole) in region [1] to the steady state profile in region [2], while the forward shock joins the latter to the leading rarefaction wave (hole).

Localized Ion Heat Source

The effect of localized ion heating on the polar ionosphere has been studied by Gombosi et al. (1985) with the aid of a time-dependent hydrodynamic model. The ionosphere was initially assumed to be in a state of diffusive equilibrium and then a localized ion heat source was applied. The distributed ion heating had a Gaussian shape with the peak at 2000 km and a half width of 250 km. The ion heating rate was increased up to 0.025 erg cm^{-3} s^{-1} over a time period of 150 s. The absorbed heat was divided between H^+ and O^+ according to their mass densities. The topside pressure was kept at a very low value to simulate open geomagnetic field lines. Figure 12 shows the effect of the ion heating on the ionosphere 2.5 min after the heating was initiated. At this time, the O^+ temperature already exceeds 20,000 K between 2000 and 5000 km, which causes a significant (but subsonic) upward O^+ flow. Consequently, intense ion heating applied for brief periods of time can produce subsonic upward O^+ flows, but supersonic O^+ outflows are obtained only if the ion heating is applied for 1 hour or longer. In this latter time period, the flux tube of plasma can convect a significant distance across the polar cap because of magnetospheric electric fields; therefore, it is highly unlikely that a naturally occurring, localized ion heat source can produce supersonic O^+ outflows of the polar wind type.

Conclusions

In recent years, theoretical models of the polar wind have produced a number of new predictions with regard to its characteristics and temporal evolution. As in the past, most of the theoretical predictions are beyond experimental verification because of the lack of detailed measurements of polar wind parameters. However, some important characteristics of the polar wind have recently been confirmed. First, the supersonic nature of the polar wind has been confirmed by individual direct measurements made by instruments on the Dynamics Explorer (DE) satellites (Nagai et al., 1984; Gurgiolo and Burch, 1985). Also, DE 1 satellite data were used to show indirectly that the polar wind is supersonic in the polar cap most of the time (Persoon et al., 1983). Altitude profiles of electron density were constructed from measurements made on many traversals of the polar cap, and the results were compared to both the subsonic and the supersonic solutions calculated by Schunk and Watkins (1982). The decrease of the measured electron density with altitude was in good agreement with that predicted for supersonic flow, and the predicted ion densities for supersonic flow were within 25% of the median densities measured by DE 1.

More recently, DE 1 satellite data obtained at an altitude of 4500 km by Biddle et al. (1985) have directly verified the predictions regarding ion heat flux asymmetries (Schunk and Watkins, 1982). On open field lines, the light ions (H^+ and He^+) were observed to be supersonic, and the shape and direction of the asymmetry of the distribution functions were consistent with the presence of upward heat fluxes. On closed field lines, on the other hand, the outward flows were subsonic and the heat fluxes were downward (Schunk and Watkins, 1982). In addition, measurements of photoelectron spectra by Winningham and Gurgiolo (1982) suggest the presence of a spatially and/or temporally varying large-scale, outwardly directed parallel electric field over the polar cap. This suggestion is consistent with the prediction of a contact surface (double layer electric field) between the hot magnetospheric and the cold ionospheric electron populations (Barakat and Schunk, 1984).

Acknowledgments. This research was supported by NASA grant NAGW-77 and AFOSR contract F49620-86-C-0109 to Utah State University.

References

Axford, W. I., The polar wind and the terrestrial helium budget, J. Geophys. Res., 73, 6855, 1968.

Axford, W. I., and C. O. Hines, A unifying theory of high-latitude geophysical phenomena and geomagnetic storms, Can. J. Phys., 39, 1433, 1961.

Banks, P. M., and T. E. Holzer, Features of plasma transport in the upper atmosphere, J. Geophys. Res., 74, 6304, 1969.

Barakat, A. R., and R. W. Schunk, O^+ ions in the polar wind, J. Geophys. Res., 88, 7887, 1983.

Barakat, A. R., and R. W. Schunk, Effect of hot electrons on the polar wind, J. Geophys. Res., 89, 9771, 1984.

Barakat, A. R., and R. W. Schunk, Stability of the polar wind, J. Geophys. Res., in press, 1987.

Biddle, A. P., T. E. Moore, and C. R. Chappell, Evidence for ion heat flux in the light ion polar wind, J. Geophys. Res., 90, 8552, 1985.

Dessler, A. J., and F. C. Michel, Plasma in the geomagnetic tail, J. Geophys. Res., 71, 1421, 1966.

Dungey, J. W., Interplanetary magnetic field and the auroral zones, Phys. Rev. Lett., 6, 47, 1961.

Gombosi, T. I., T. E. Cravens, and A. F. Nagy, A time-dependent theoretical model of the polar wind: Preliminary results, Geophys. Res. Lett., 12, 167, 1985.

Gurgiolo, C., and J. L. Burch, Composition of the polar wind - not just H^+ and He^+, Geophys. Res. Lett., 12, 69, 1985.

Holzer, T. E., J. A. Fedder, and P. M. Banks, A comparison of kinetic and hydrodynamic models of an expanding ion-exosphere, J. Geophys. Res., 76, 2453, 1971.

Lemaire, J., O^+, H^+, and He^+ ion distributions in a new polar wind model, J. Atmos. Terr. Phys., 34, 1647, 1972.

Lemaire, J., and M. Scherer, Kinetic models of the solar and polar winds, Rev. Geophys. Space Phys., 11, 427, 1973.

Lockwood, M., J. H. Waite, T. E. Moore, J.F.E. Johnson, and C. R. Chappell, A new source of suprathermal O^+ ions near the dayside polar cap boundary, J. Geophys. Res., 90, 4099, 1985.

Marubashi, K., Escape of the polar-ionospheric plasma into the magnetospheric tail, Rep. Ionosph. Space Res. Japan, 24, 322, 1970.

Moore, T. E., M. Lockwood, M. O. Chandler, J. H. Waite, Jr., C. R. Chappell, A. Persoon, and M. Suguira, Upwelling O^+ ion source characteristics, J. Geophys. Res., 91, 7019, 1986.

Nagai, T., J. H. Waite, Jr., J. L. Green, C. R. Chappell, R. C. Olsen, and R. H. Comfort, First measurements of supersonic polar wind in the polar magnetosphere, Geophys. Res. Lett., 11, 669, 1984.

Nishida, A., Formation of a plasmapause or magnetospheric plasma knee by combined action of magnetospheric convection and plasma escape from the tail, J. Geophys. Res., 71, 5669, 1966.

Persoon, A. M., D. A. Gurnett, and S. D. Shawhan, Polar cap electron densities from DE 1 plasma wave observations, J. Geophys. Res., 88, 10, 123, 1983.

Raitt, W. J., R. W. Schunk, and P. M. Banks, A comparison of the temperature and density structure in high and low speed thermal proton flows, Planet. Space Sci., 23, 1103, 1975.

Raitt, W. J., R. W. Schunk, and P. M. Banks, The influence of convection electric fields on thermal proton outflow from the ionosphere, Planet. Space Sci., 25, 291, 1977.

Raitt, W. J., R. W. Schunk, and P. M. Banks, Helium ion outflow from the terrestrial ionosphere, Planet. Space Sci., 26, 255, 1978.

Schunk, R. W., and D. S. Watkins, Electron temperature anisotropy in the polar wind, J. Geophys. Res., 86, 91, 1981.

Schunk, R. W., and D. S. Watkins, Proton temperature anisotropy in the polar wind, J. Geophys. Res., 87, 171, 1982.

Schunk, R. W., W. J. Raitt and P. M. Banks, Effect of electric fields on the daytime high-latitude E and F regions, J. Geophys. Res., 80, 3121, 1975.

Singh, N., and R. W. Schunk, Temporal evolution of density perturbations in the polar wind, J. Geophys. Res., 90, 6487, 1985.

Winningham, J. D., and C. Gurgiolo, DE-2 photoelectron measurements consistent with a large-scale parallel electric field over the polar cap, Geophys. Res. Lett., 9, 977, 1982.

Yau, A. W., E. G. Shelley, W. K. Peterson, and L. Lenchyshyn, Energetic auroral and polar ion outflow at DE 1 altitudes: Magnitude, composition, magnetic activity dependence, and long-term variations, J. Geophys. Res., 90, 8417, 1985.

VERTICAL ION FLOW IN THE POLAR IONOSPHERE

A. W. Yau

Herzberg Institute of Astrophysics, National Research Council Canada, Ottawa, Canada K1A 0R6

M. Lockwood

Rutherford Appleton Laboratory, Chilton, Didcot OX11 0QX UK

Abstract. Outflowing ions from the polar ionosphere fall into two categories: the classical polar wind and the suprathermal ion flows. The flows in both these categories vary a great deal with altitude. The classical polar wind is supersonic at high altitude: at ~3 R_E geocentric, the observed polar wind is H^+ dominated and has a Mach number of 2.5-5.1. At 400-600 km, thermal and suprathermal upward O^+ ion fluxes frequently occur at the poleward edge of the nightside auroral oval during magnetically active times. Above 500 km, ions are accelerated transverse to the local geomagnetic field. At 1400 km, transversely accelerated ions are frequently observed in winter nights but rarely appear in the summer. In the dayside cleft above ~2000 km, ions of all species are transversely heated and upwell with significant number and heat fluxes, forming a cleft ion fountain as they convect across the polar cap. Upwelling ions are observed most (least) frequently in the summer (winter). At yet higher altitudes, energetic (>10 eV to several kiloelectron volts) upflowing H^+ and O^+ ions are frequently observed, their active time occurrence frequency being as high as 0.7 at auroral latitudes and 0.3 in the polar cap. Their composition, intensity, and angular characteristics vary quantitatively with solar activity, being O^+ dominant and more intense near solar maximum. Their resulting ion outflow is dominated by ions below 1 keV and reaches 3.5×10^{26} O^+ and 7×10^{25} H^+ ions s^{-1} at magnetically active times ($Kp \geq 5$) near solar maximum. In comparison, the estimated polar wind ion outflow at times of moderate solar activity is 7×10^{25} H^+ and 4×10^{24} He^+ ions s^{-1}. The estimated <10-eV cleft ion fountain flow is 3.8×10^{25} O^+ and 8.6×10^{23} H^+ ions s^{-1} near solar maximum.

Introduction

The terms epithermal, superthermal, suprathermal, heated, nonthermal, and energetic have been used in the literature in different contexts to describe the various ion flow processes in the polar ionosphere. In this paper, all ion flow populations in the polar ionosphere with the exception of the classical polar wind are referred to as suprathermal, the rationale being that the polar wind is the consequence of hydrodynamic expansion of plasma at typical ionospheric temperatures; whereas the others result from ion acceleration or heating to above thermal energy involving an external free energy source.

The first direct evidence of outflowing suprathermal heavy ions from the polar ionosphere was the observation of kiloelectron volt upflowing O^+ ions, both field-aligned (ion beams; Shelley et al., 1976a) and non-field-aligned (ion conics; Sharp et al., 1977), above 5000 km altitude on S3-3. Subsequent observations have revealed a variety of ion outflow populations in different parts of the polar ionosphere. These include transversely accelerated ions down to 500 km: these are ions accelerated transversely to the geomagnetic field and then injected upward as a result of the grad B mirror force (Klumpar, 1979; Whalen et al., 1978); large flows of thermal O^+ ions at 500-600 km near the poleward edge of the nightside auroral oval (Lockwood and Titheridge, 1981); hundred electron volt ion conics and beams above 2000 km (Gorney et al., 1981); upwelling ion events at 2000-5000 km in the dayside auroral zone: these are ions of different mass species transversely heated and carrying upward heat flux (Moore et al., 1986); upflowing H^+ and O^+ ions in the 0.01-1 keV range inside the high-altitude polar cap (Shelley et al., 1982); and downward flowing ion bands: these appear as narrow, decreasing energy bands in energy-latitude spectrograms at low altitudes (Winningham et al., 1984). A coherent picture of suprathermal ion flows is starting to emerge. We review observations from a number of satellites that are key to understanding the interrelationship and relative importance of the different ion flow populations. Note that the earlier observations were limited to the higher ion energies. As these observations were extended to lower energies, a number

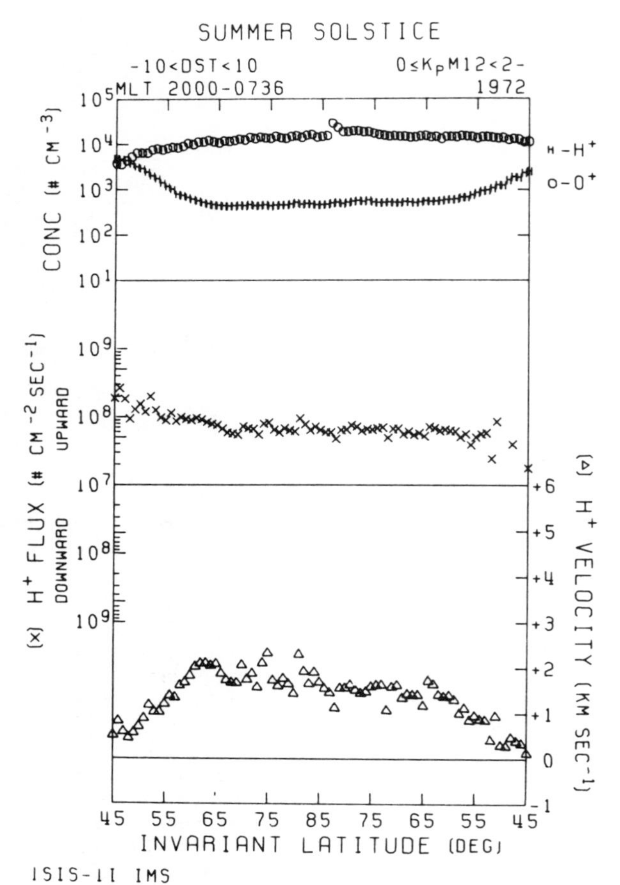

Fig. 1. ISIS 2 polar wind observation at 1400 km for summer conditions in 1972 (Hoffman and Dodson, 1980).

of important, lower-energy ion acceleration processes in the topside ionosphere (~500-2000 km) gradually became apparent. Also, the respective satellite data sets have been acquired in different time periods of the last two solar cycles. Significant long-term variations in ion flow characteristics exist; hence the time period of a data set may be an important factor.

The Polar Wind

Numerous steady state theoretical models for the polar wind have been developed; see the excellent review by Raitt and Schunk (1983). Recently, time-dependent polar wind models have been presented by Gombosi et al. (1985).

The polar wind ion flux is limited by the production rate of the outflowing ions and their Coulomb collisions with the stationary (heavier) ions. The dominant source of the outflowing H^+ is the accidentally resonant charge-exchange reaction between O^+ and H. The source of the He^+ is the photoionization of neutral He. Theoretically, the polar wind is primarily H^+, with only a few percent He^+; O^+ represents a negligible component, except at very high (>10^4 K) electron temperature (Barakat and Schunk, 1984). The limiting H^+ flux decreases with increasing exospheric temperature and is ~3 x 10^8 cm^{-2} s^{-1} near solar minimum and ~1 x 10^8 cm^{-2} s^{-1} near solar maximum (see, for example, Raitt and Schunk, 1983). The limiting He^+ flux is dependent primarily on the neutral atmospheric He and N_2, which affect the photoionization rate of He and the charge-exchange loss rate of He^+, respectively. It varies from a low of 1 x 10^5 cm^{-2} s^{-1} for extreme solar minimum, summer, and magnetically active conditions to a high of 2 x 10^7 cm^{-2} s^{-1} for solar maximum, winter, and quiet conditions. The winter-to-summer ratio is ~20-30. The solar maximum-to-minimum ratio is ~2.

The polar wind was observed from in situ ion measurements, first on Explorer 31 in the 500-3000 km altitude range and later on ISIS 2 at 1400 km (Hoffman and Dodson, 1980). Figure 1 shows the measured H^+ and O^+ ion densities and outward fluxes on ISIS 2 for summer conditions in 1972 (at the declining phase of the solar cycle). The latitudinal variations of the H^+ ion density and flow velocity were anti-correlated; hence the outward flux remained constant, indicative of limiting flux conditions. The averaged H^+ flux above 60° invariant was ~7 x 10^7 cm^{-2} s^{-1}. The flow was subsonic. The H^+ velocity peaked at 2.5 km s^{-1}. In the winter, the H^+ flux peaked at its summer value at 75°-85° invariant but dipped to 3 x 10^7 cm^{-2} s^{-1} at the pole. The velocity peaked at 4 km s^{-1}. The peak winter He^+ flux was ~2 x 10^7 cm^{-2} s^{-1} on the dawnside and ~1 x 10^7 cm^{-2} s^{-1} on the duskside. The averaged summer solstice value was 1.5 x 10^6 cm^{-2} s^{-1}. The winter-to-summer ratio of the He^+ flux was ~10. Note that the corresponding ratio for the neutral atmospheric He density is 20 to 30.

At high altitude, where a spacecraft is typically charged to a few volts positive, the polar wind ions are repelled by the spacecraft potential. Nagai et al. (1984) used a negative bias aperture in their retarding ion mass spectrometer (RIMS) on DE 1 to overcome the positive spacecraft potential and successfully detected both H^+ and He^+ polar wind ions down to ~0 eV in the nightside polar cap (65°-81° invariant) near 2 R_E altitude, during a large substorm in the northern winter. The observed H^+ flux was fairly stable. In contrast, the He^+ flux was quite variable. The He^+/H^+ ratio varied between $\leq$0.01 and unity. The polar wind was supersonic: the observed data were consistent with a flow velocity of 16-25 km s^{-1}, a Mach number of 2.5-5.1, a temperature of 0.12-0.2 eV, and an equivalent escape flux of 2.6 x 10^8 cm^{-2} s^{-1} (at 1000 km). No O^+ measurements in the electron volt energy range were made, and the question remains as to the possible presence of O^+ polar wind. Data in the 10-100 eV range suggest the presence of higher-energy O^+ flow, which Gurgiolo and Burch (1982) interpreted as polar wind ions perpendicularly heated at lower altitude (8000-12000 km). However, Green and Waite (1985) argued that the flow was not a component of the classical polar wind, but was rather a distribution with 10-eV perpendicular and 1-eV parallel temperatures originating from the dayside cusp.

It is not possible to quantitatively compare the few DE 1 observations available at high altitude with the ISIS 2 measurements at low altitude. The former were limited to the light ion species and magnetically active, winter conditions near solar maximum. The latter were made during magnetically quiet times in the declining phase of the solar cycle.

Suprathermal Ion Flows

Topside Ionospheric Flow (Above 400 km)

From ionospheric plasma scale height profiles observed by Alouette I in 1962-1968 (within 3 years of solar minimum), Lockwood and Titheridge (1981) obtained evidence of suprathermal O^+ ion fluxes in the lower topside ionosphere (~500-600 km) at the poleward edge of the nightside auroral oval. They found that the normalized mean dayside flux (inferred from the plasma scale height profiles) was ~1×10^8 cm^{-2} s^{-1} for Kp < 2 and slightly higher at active times, which is comparable with the light ion escape flux at higher altitude, which it supports via charge exchange. In the nightside, the flux was upward and peaked at the poleward edge of the statistical auroral oval. For Kp $\geq$ 2, the observed peak flux often exceeded 6×10^8 cm^{-2} s^{-1}; i.e., it was sufficiently large to support not only the thermal light ion flow but also suprathermal O^+ outflows of the order of $\leq 5 \times 10^8$ cm^{-2} s^{-1} originating at (slightly) higher altitude. The latter were associated with the auroral oval in that their local time variation was characteristic of the oval at both quiet and active times. Moreover, they were relatively rare in the summer and, where found, were clustered around the dayside cleft; during the winter, they covered the nightside between 1800 and 0900 LT (Lockwood, 1982). Their occurrence frequency was as high as 0.5 at 70°-80° invariant for winter nights. Events with fluxes exceeding 2.5×10^9 cm^{-2} s^{-1} were found.

Low-Altitude Transverse Ionospheric Acceleration (~500-1400 km)

On ISIS 2 at 1400 km altitude, where transversely accelerated ions (TAIs) were first identified, Klumpar (1979) found an almost total absence of TAIs during the northern hemispheric summer but common occurrence in the winter. Only 8 events were observed between day 120 and 250 in 1971, compared with >500 events between day 250 in 1971 and day 120 in 1972. Figure 2 shows the occurrence probability of TAIs in the winter months from November 1971 through February 1972.

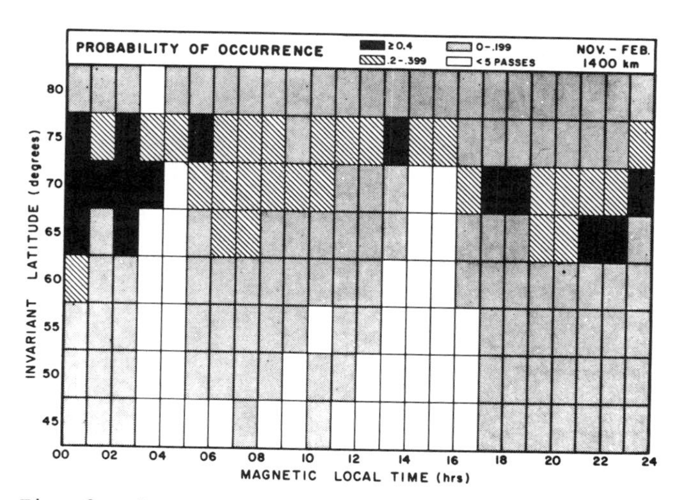

Fig. 2. Occurrence frequency of transversely accelerated ions at 1400 km on ISIS 2 in November 1971-February 1972 (Klumpar, 1979).

The occurrence probability was based on the number of TAI events divided by the number of satellite passes through an MLT-invariant latitude bin, where an event was defined as an observation of TAIs in at least one but possibly several spins in a satellite pass through the bin. The probability was between 0.3 and 0.6 in the nightside (1800-0600 MLT) between ~65° and 75° invariant, and it was lower in the dayside.

Klumpar (1979) also surveyed the data on ISIS 1 between February and October 1969 (near solar maximum). Klumpar found that TAI events were observed only above 2750 km in the summer months (between days 120 and 200). Moreover, the vast majority of the observed events came from the dayside cleft.

Aside from the ISIS 1 and 2 data, observations of low-altitude TAIs so far have been limited to those of a few sounding rockets. Both the ISIS and rocket observations reveal that TAIs typically have power law-like intensity spectrum, i.e., number intensity $j(E) \propto E^{-\alpha}$, where $\alpha \approx 2$; the ion velocity distributions display a high-energy tail. Also, they are often very intense and highly collimated. From a rocket flight in the expansive phase of a pre-midnight auroral substorm, Whalen et al. (1978) observed TAIs up to 500 eV which peaked at 90° pitch angle near 400 km altitude. The peak pitch angle increased systematically with altitude in a manner that was consistent with the adiabatic expansion of ions injected at 90° in the 400-500 km altitude range. The measured angular width of the TAIs was of the order of the instrument resolution (10°) and indicates a source altitude width of ~100 km. Similarly narrow angular widths were observed on ISIS and other rocket flights (Yau et al., 1983; Kintner et al., 1986). On ISIS 1 and 2, the integrated TAI flux above 10 eV often exceeded 10^7 cm^{-2} s^{-1}.

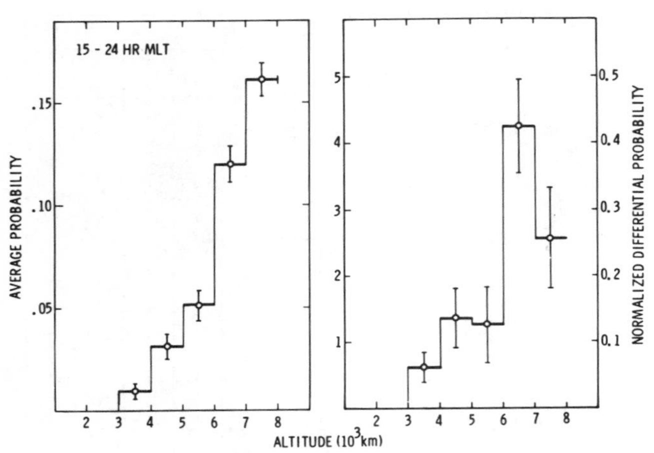

Fig. 3. Left: Averaged occurrence frequency of upflowing ionospheric ions above 0.5 keV, below 8000 km on S3-3, in the peak latitude region in the 1500-2400 MLT sector. Right: Normalized differential frequency obtained by subtracting the frequencies of adjacent altitude bins (Sharp et al., 1983).

Mid-Altitude (Below ~8000 km) Energetic Ion Beams and Conics

The occurrence and morphological characteristics of upflowing ion (UFI) beams and conics (above 100 eV) below 8000 km were the subject of a number of studies based on the ion composition spectrometer and electrostatic analyzer data on S3-3. Ghielmetti et al. (1978) studied the occurrence frequency of UFI occurrence using the ion composition data above 0.5 keV. Their study did not distinguish between ion beams and conics; however, it was weighted in favor of beams because of its high-energy threshold. The left panel of Figure 3 shows the altitude distribution of energetic (>0.5 keV) UFIs in 1000-km altitude bins, averaged over the peak latitude region in the 1500-2400 MLT sector (Ghielmetti et al., 1978; Sharp et al., 1983). In the absence of strong angular diffusion, each upflowing beam should be observable at any altitude above its point of origin, and the differential of this curve (right panel of Figure 3) should characterise the location of the source region. The figure shows that ion acceleration to energies above 0.5 keV occurred primarily at altitudes greater than 4000 km.

Gorney et al. (1981) surveyed several hundred orbits of the S3-3 electrostatic analyzer data between July 1976 and December 1977 and determined the occurrence frequencies of ion beams and conics in three energy ranges (below 0.4 keV, 0.4-2 keV, and above 2 keV). They found that the conics and beams displayed very different local time distributions. The distribution for conics showed a broad daytime maximum at quiet times and became relatively uniform at active times. The active time frequency was comparable to the quiet time dayside peak value. The beams occurred primarily in the dusk and midnight sectors, and they were a factor of 2 more frequent at active times than at quiet times in the dusk sector. The majority of conics were below 0.4 keV, whereas the ion beams were more energetic. During quiet times, conics were observed uniformly in altitude above 1000 km. Also, perpendicular ion conics were observed primarily in the 1500-2000 km altitude region. In contrast, beams occurred primarily above 5000 km, with increasing frequency with altitude, consistent with the earlier results of Ghielmetti et al. (1978). Collin et al. (1981) surveyed the data in the evening local time sector above 6000 km altitude and found that the O^+ beams were on the average more energetic than the H^+ beams by a factor of 1.7. Also, they had significantly wider pitch angle distributions (24° medium half width) than H^+ (15° half width). The H^+ were generally more intense, with fluxes up to 10^9 cm^{-2} s^{-1} sr^{-1}; the O^+ flux was up to 5 x 10^7 cm^{-2} s^{-1} sr^{-1}.

Sharp et al. (1983) found that the H^+ and O^+ UFI occurrence frequencies had almost identical dependences on Kp, increasing threefold from Kp = 0 to 5. The increase in peak energy with Kp was also very similar. There was, however, a marked difference in their outflow intensity dependence on magnetic activity. Collin et al. (1984) found that the total H^+ and O^+ ion outflow in the 0.5-16 keV range during quiet times from both hemispheres was ~1.3 x 10^{25} s^{-1} with a O^+/H^+ ratio of 0.25. During the storm times, the total was ~7.2 x 10^{25} s^{-1}, with a larger O^+/H^+ ratio of 1.4.

Mid-Altitude Superthermal Upwelling Ions and Transversely Accelerated Core Ions

In the superthermal energy range (10-100 eV) not sampled by the S3-3 spectrometers, the DE 1 RIMS measurements revealed differences in the energization characteristics between the dayside and the nightside auroral upflowing ions. In the nightside, the observed energization was in the perpendicular direction. The transverse heating affected the bulk of the ion distribution, and perpendicular temperatures of the order of 10 eV were produced. In some of the events, essentially all of the thermal ions participated in the transverse acceleration, leaving no cold rammed plasma, and resulting in a toroidal distribution with a minimum at zero speed in the plasma frame. Lockwood et al. (1985c) termed these transversely accelerated core ions (TACIs). Often, the transverse acceleration region extends over a broad latitudinal region; Moore et al. (1985) analyzed in detail an event extending from L = 7.3 to 13. Also, the rate of folding of the evolving conical distribution decreases with increasing mass per unit charge, reflecting the larger gravitational effects on the heavier ions in counteracting the magnetic mirror force. In contrast, the dayside events display the effects of both parallel and perpendicular accelerations. When observed at 2000-5000 km, the core of O^+ distribution exhibits transverse heating to ~10 eV and carries significant upward number flux (>10^8 cm^{-2} s^{-1}) and heat flux (>3 x 10^{-4} erg cm^{-2} s^{-1}); all observed species (H^+, He^+, O^+, O^{++}, and N^+) are warmed. Hence, they are termed upwelling ion events (Lockwood et al., 1985c). The distinction between upwelling ions in the dayside and transversely accelerated ions in the nightside is not sharp--some overlap exists--and is a matter of degree to which transverse acceleration affects the thermal core of the ion distribution. The distinction was not relevant in the ISIS and S3-3 observations since only the high-energy tails of the distributions were sampled and the mass composition was not resolved.

Upwelling ions are the most persistent suprathermal ion outflow feature in the cleft region. Moore et al. (1986) presented a detailed study of an upwelling ion event which is summarized in Figure 4. The event carried large (~10^9 cm^{-2} s^{-1}) H^+ and O^+ ion fluxes out of the ionosphere (panel d). The classical polar wind fluxes of H^+ and He^+ can be seen equatorward of the event (after the UT labeled 3) but not poleward of it (before the UT labeled 1). The sharp equatorward edge of the event was closely associated with that of field-aligned current regions (panel c). The edge was also associated with a strong eastward convection channel (Moore et al., 1986). The density of O^+ observed by RIMS was virtually equal to the total plasma density, indicating that the plasma was dominated by upwelling O^+ ions.

Using RIMS data below 3 R_E geocentric between October 1981 and October 1983, Lockwood et al. (1985c) surveyed the occurrence of upwelling ions, TACIs, and field-aligned O^+ outflows in the polar cap which, as will be explained below, originate as upwelling ions convecting from the cusp anti-sunward into the high-altitude polar cap. The occurrence frequency of field-aligned flow, f_a, has a maximum of 0.3 and that of TACIs, f_c, 0.2. Note that the latter is considerably smaller than the frequency of low-altitude TAIs seen by ISIS 2 at 1400 km and is likely the result of ions being accelerated to beyond the energy range of RIMS (~60 eV). The upwelling O^+ ions were observed exclusively in the morning sector of the auroral oval and the lower latitudes of the polar cap. Integrated over the altitude range below 3 R_E, the occurrence frequency of upwelling ions, f_u, was up to 0.6; the integrated frequency over invariant latitudes in the morning sector was very close to unity; i.e., upwelling ions were almost always present but their location varied. The sum of f_a and f_u indicates that upward field-aligned O^+ flow was a regular feature on the dayside, with peak frequencies between 75° and 80° invariant. This is very similar to the occurrence of thermal upwellings in the topside ionosphere (Lockwood, 1982;

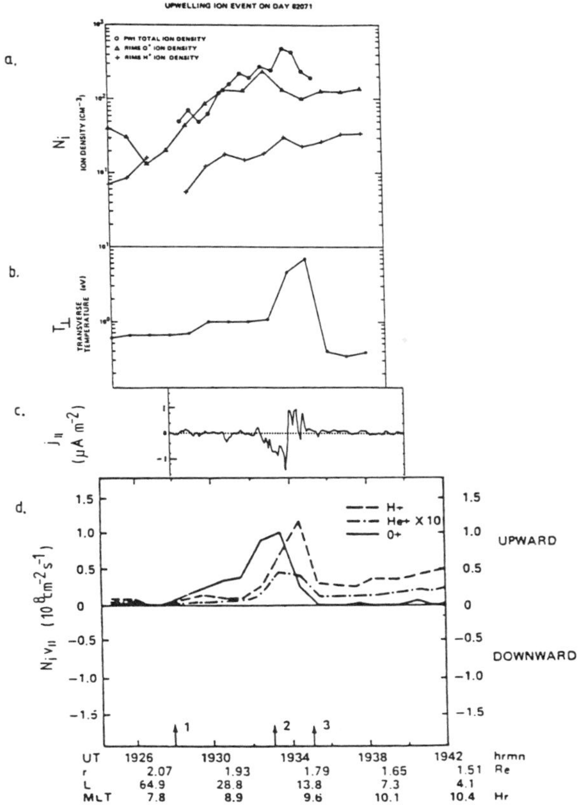

Fig. 4. DE 1 RIMS observation of upwelling ion event on March 12, 1982. (a) H^+, O^+, and total ion densities; (b) transverse O^+ ion temperature; (c) inferred field-aligned current, $j_\parallel$; (d) field-aligned fluxes of O^+, H^+, and He^+, $N_i V_\parallel$ (Lockwood, 1986).

Tsunoda et al., 1987). The sum was season-dependent: it was highest in the summer, smaller at equinox, and zero in winter for low Kp conditions. The low-latitude edge of the upwelling ions was found to be closely associated with that of field-aligned currents, as inferred from the east-west component of the magnetic field perturbations measured in situ (Lockwood et al., 1985c).

Geomagnetic Mass Spectrometer Effect

From the narrow nature of the upwelling ion source region, Lockwood et al. (1985b) concluded that ions should be spatially dispersed across the polar cap according to their time of flight. Waite et al. (1985) observed energy dispersion of O^+ ions within the polar cap extending continuously from several kiloelectron volts observed near the cleft to several electron volts observed in the nightside polar cap. The mass dispersion of upgoing ionospheric ions of the same energy

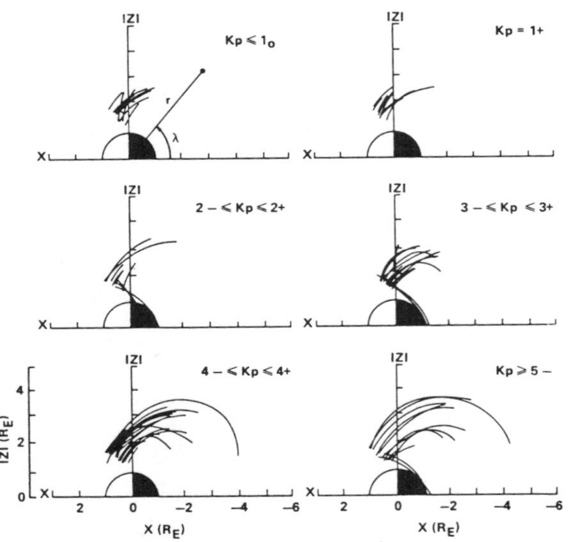

Fig. 5. Locations of polar cap O^+ ions observed on DE 1 contiguous with upwelling ion events, for various Kp ranges (Lockwood et al., 1985a).

was noted by Moore et al. (1985), who termed the time-of-flight effect the "geomagnetic mass spectrometer." It is very similar in nature to ion dispersion in the cusp (Shelley et al., 1976b) and in the plasma mantle (Rosenbauer et al., 1975). A first-order quantitative behavior of the effect was described by Horwitz (1984) using a two-dimensional kinetic trajectory model. Lockwood et al. (1985b) illustrated the effect on ions observed by RIMS in the high-altitude polar cap by tracing the ion trajectories earthward from DE 1 using the convection velocity observed on DE 2, and the trajectory model of Horwitz (1984) and showed that all had originated from a narrow source in the topside ionosphere where DE 2 indeed observed strong thermal plasma flow ($>10^9$ cm^{-2} s^{-1}). These outfluxes have been confirmed by statistical studies using data on HILAT (Tsunoda et al., 1987).

The analogy of the geomagnetic ion mass spectrometer does not allow for the significant effect of gravity on heavy low-energy ionospheric ions. The lowest-energy upwelling ions are below the escape energy and are expected to fall in the polar cap. Lockwood et al. (1985a) observed O^+, N^+, O^{++}, and even He^+ flowing downward at low altitudes in the polar cap; they introduced the fountain in a wind analogy and termed the outflow the "cleft ion fountain."

High-Altitude Polar Cap Ions

Initial DE 1 observations in the high-altitude polar cap revealed the frequent outflow of 10-100 eV O^+ and H^+ (O^+/H^+ ~10) (Shelley et al., 1982; Chappell et al., 1982). Probably a substantial fraction, but not all, of the upflowing ions observed at the high-altitude polar cap originate from the cleft ion fountain. The role of convection in filling the high-altitude polar cap with the lower-energy O^+ from the cleft ion fountain is demonstrated by Figure 5 which shows segments of DE 1 orbits in which RIMS observed O^+ ions in the polar cap contiguous with upwelling ion events. The data were binned into six Kp ranges. For low Kp, the O^+ was found only on the dayside. However, for Kp > 4, the O^+ extended through the cap, consistent with the variation of occurrence probability of O^+ in the polar cap with Kp found by Waite et al. (1985). Lockwood (1986) found that for southward IMF (B_z < 0), the occurrence probability of O^+ outflow above 80° invariant was as high as unity in some MLT sectors at times of high Kp. There was a strong dawn-dusk asymmetry in the O^+ flows, dependent on the sense of B_y. For positive B_y, the occurrence probability was at or near unity on the dawnside only. For negative B_y, it was lowest near dawn and the flows filled a larger portion of the polar cap. For northward IMF, the same B_y asymmetry was found; the occurrence probability of O^+ flow was lower for a larger part of the polar cap, presumably due to a region of sunward convection. No clear B_y effect on the location of the upwelling ions was found. Hence, the effect on the polar cap O^+ flow morphology must be due to the differences between the convection patterns.

There is evidence that some of the UFIs observed at the high-altitude polar cap, particularly the higher-energy ones and those associated with polar cap arcs, originate from within the polar cap. For example, the events presented in Shelley et al. (1982) probably mapped down to the nightside polar cap at ionospheric height, assuming realistic convection electric fields. Recently, Lockwood et al. (1986) observed a burst of upflowing O^+ ions in the 0200 MLT sector within the polar cap during a major polar cap expansion event. These ions could not have originated from the cleft ion fountain, since they were not observed throughout the polar cap and did not display any energy dispersion. This points to an ionospheric source within the polar cap. The relative contributions between the cleft ion fountain and within the polar cap to the superthermal upflowing ions observed in the high-altitude polar cap are uncertain.

Perpendicular and Parallel Auroral Acceleration Signatures

In the auroral zone, Moore et al. (1985) described representative signatures of suprathermal auroral ions (~10-60 eV) and noted the occurrence of ion conics sandwiching an upgoing ion beam in the nightside auroral oval. On RIMS spin phase angle-time spectrograms, the continuous folding of the conics into the beam gives an X-shaped form. The occurrence frequency of these X-events is not clear. However, large-scale ion conics of suprathermal energy (10-100 eV) are commonplace, particularly at magnetically active times.

Klumpar et al. (1984) reported frequent observations of hybrid conical distributions of ions that showed clear evidence of ions injected from lower altitude and having experienced both transverse and parallel acceleration. These ion distributions are field-aligned and upgoing at the lowest energy and gradually become conical with increasing energy. On high resolution DE 1 energetic ion composition spectrometer (EICS) energy versus time spectrograms, these distributions appear as truncated Y-shaped structures with the apex of the Y centered on the magnetic field direction. Klumpar et al. (1984) showed EICS data for a pass at an altitude of ~21,000 km at ~0810 MLT and ~74° invariant latitude in which the Y-shaped signature was evident in the oxygen data. The O^+ ion flux exhibited a sharp low-energy cutoff below 350 eV, and the field alignment of the ions at the lowest energy gradually gave way to a progressively wider pitch angle distribution with increasing energy. O^+ ions of energy in excess of 5 keV were observed in the extremes of the conic distribution. The observed distribution was consistent with a two-stage acceleration process wherein the ions were transversely accelerated within a restricted region some distance below the satellite, producing a broad energy distribution in the perpendicular direction, and were subsequently pushed upward by the mirror force into a region of parallel potential, thereby gaining a fixed amount of additional energy. This is illustrated in Figure 6, which shows (a) the measured phase space density distribution of the O^+ ions; (b) the pre-parallel acceleration distribution by transforming the measured distributions backward through a parallel potential of 310 V; and (c) the source distribution by adiabatic transformation (of the pre-parallel acceleration distribution) down the field line from the satellite altitude of 21,400 km to the mirror point at 18,200 km. The source distribution (c) is consistent with expected ion distributions inside a TAI source region.

The unique physical processes of transverse ion acceleration have led to many proposed acceleration mechanisms, including those by electrostatic ion-cyclotron waves (Lysak et al., 1980), lower hybrid waves (Chang and Coppi, 1981), narrow oblique potential jumps (Yang and Kan, 1983), and broadband electromagnetic ion-cyclotron waves (Chang et al., 1986). However, thus far, no conclusive direct experimental evidence exists to favor one mechanism over another. Kintner and Gorney (1984) searched through the S3-3 plasma wave data within TAI regions and found one event in which ion-cyclotron waves were present; however, the wave amplitude was somewhat smaller than expected. Recently, Kintner et al. (1986) reported large-amplitude VLF waves inside regions of two TAI events at 500-600 km, propagating near and above the lower hybrid frequency and well correlated with the perpendicular ion accelera-

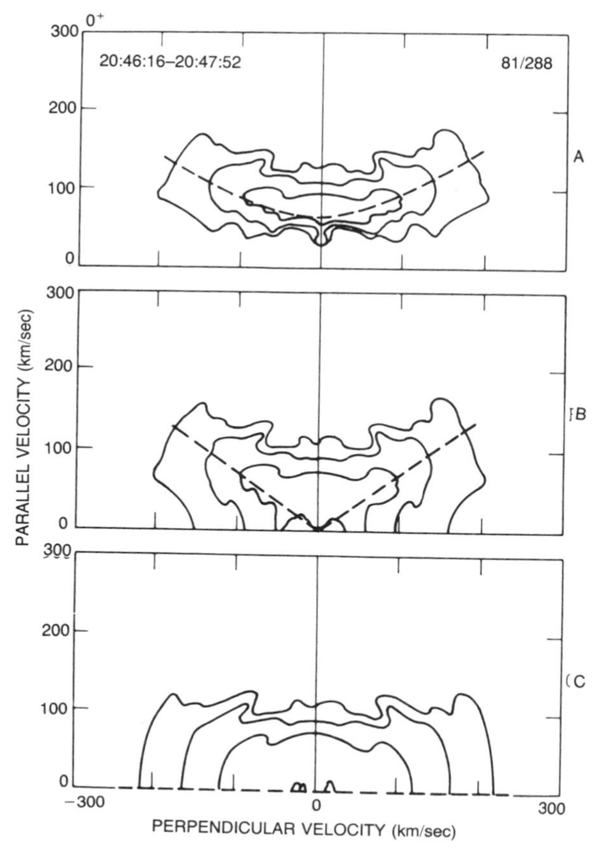

Fig. 6. DE 1 EICS observation of two-stage ion acceleration. O^+ phase space density contours (at half-decade intervals below the peak value). (a) Measured distribution. (b) Transformed distribution back through a parallel potential of 310 V. (c) Adiabatic transformation of (b) from 21,400 km (altitude) to 18,200 km (mirror point) (Klumpar et al., 1984).

tion. The data suggested an acceleration mechanism employing lower hybrid waves, but were not conclusive. On ISIS 2, TAIs have been associated with both VLF waves and with regions of depleted ion density and large field-aligned currents (i.e., regions unstable to electrostatic ion-cyclotron wave growth). James (1976) and Klumpar (1985) found one-to-one as well as statistical correlations between TAI and VLF saucers. The latter were observed down to 500 km in the winter but were observed only above 3000 km in the summer. In other words, they displayed seasonal variations essentially identical to those of TAIs at ISIS altitudes.

Polar Ionospheric Ion Outflow

Yau et al. (1984, 1985a) surveyed the occurrence of upflowing ions above 10 eV using the DE

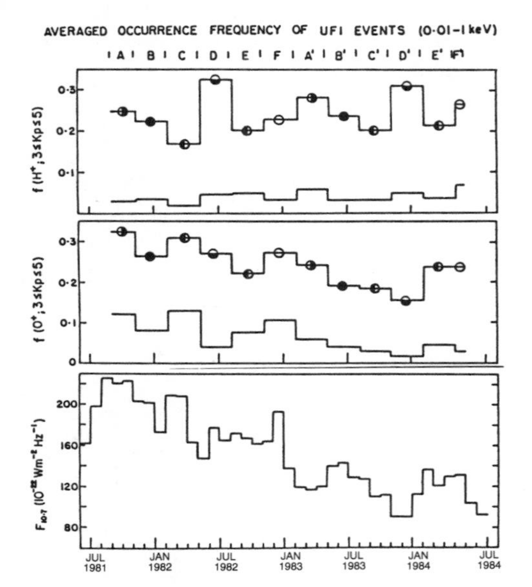

Fig. 7. Averaged occurrence frequency of 0.01-1 keV UFI during active times (Kp between 3 and 5) from September 1981 to May 1984. (a) H^+ frequency. (b) O^+ frequency. (c) Monthly mean solar radio flux at 10.7 cm. In (a) and (b), top histogram: events with $>10^6$ cm^{-2} s^{-1} sr^{-1} fluxes; bottom histogram: events with $>10^7$ fluxes. Different circle types denote different seasons of apogee data sampling (Yau et al., 1985a).

1 EICS data above 8000 km, acquired between September 1981 and May 1984. In their study, the pitch angle distributions of H^+ and O^+ ion fluxes over 0.01-1, 1-4, and 4-17 keV were used, and UFIs were identified in terms of each of the six pitch angle distributions independently. At both quiet and active times, low-energy (<1 keV) UFIs occurred about 5 times more frequently than kiloelectron volt ones. For both H^+ and O^+, the overall occurrence frequency increased gradually with altitude and did not vary significantly with magnetic activity. At low Kp, the H^+ and O^+ frequencies were comparable, being ~0.3 at low altitude and ~0.4 at high altitude; the O^+ frequency at high Kp was slightly higher. However, the O^+ intensity distribution displayed a marked magnetic activity dependence that was absent in H^+. The occurrence frequency of very intense O^+ events below 1 keV (those exceeding 10^7 cm^{-2} s^{-1} sr^{-1} in flux) was a factor of 3 higher at high Kp. A similar but less dramatic increase was apparent in the higher-energy channels. Such was not the case for H^+ UFIs, where the intensity distribution remained fairly unchanged with magnetic activity. The observed UFIs at high altitudes were predominantly field-aligned. Above 22,000 km, 60% of H^+ and 70% of O^+ UFIs were ion beams (and <20° conics). The occurrence of both H^+ and O^+ UFIs appeared to be associated with the statistical auroral oval in that their frequencies peaked at lower latitude at active times and they had a broader latitudinal extent and a more equatorward peak in the nightside than in the dayside (see Figures 11 and 12 in Yau et al., 1984). The oval was energy dependent, the low-energy UFI oval being more poleward than the energetic UFI oval. Also, the energetic O^+ UFI oval was equatorward of the H^+ oval. At polar cap latitudes, UFIs were confined predominantly to low energy (below 1 keV), unlike their auroral counterparts; also, O^+ were more frequent than H^+. At auroral latitudes, the occurrence frequencies of H^+ and O^+ peaked at 0.5 during quiet times and 0.6-0.7 during active times.

The data period from September 1981 to May 1984 coincided with the declining phase of the solar cycle. The observed occurrence frequency, intensity, and angular characteristics of O^+ UFIs exhibited marked variations which correlated with variations in the solar radio flux. In contrast, the H^+ UFI morphology did not display any observable long-term variations. Figure 7 shows the seasonally averaged occurrence frequencies of H^+ and O^+ UFIs below 1 keV at moderately active times (3- $\leq$ Kp $\leq$ 5+) as a function of time in the data period. In panels a and b, the top histograms are occurrence frequencies of events with $>10^6$ $(cm^2\ s\ sr)^{-1}$ fluxes; the bottom ones show the portion of events with $>10^7$ $(cm^2\ s\ sr)^{-1}$ fluxes; the different circle types denote different seasons of apogee data sampling. Panel c shows the monthly mean 10.7 cm solar radio flux at 1 AU. The solar radio flux index $F_{10.7}$ was used as an indicator of solar activity since it is known to correlate with the solar EUV flux, and the latter modulates the upper atmospheric and ionospheric densities and temperatures. The monthly mean of $F_{10.7}$ decreased from a high of 222 in September 1981 to a low of 93 in November 1983. The O^+ UFI frequency displayed a continual trend of overall decrease. In contrast, no long-term trend of decrease was apparent in H^+ UFIs. The long-term trend of decrease in the O^+ frequency was correlated with the solar radio flux, with the O^+/H^+ ratio peaking near solar maximum. The short-term variation in the O^+ frequency also appeared to track the corresponding variation in the solar flux.

In both H^+ and O^+, 0.01-1 keV ions constituted over 90% of the total ion outflow in the 0.01-17 keV ion energy range (energy range of EICS). Overall, the outflow was much higher at active times than at quiet times. Figure 8 displays the 0.01-1 keV H^+ and O^+ ion outflow rates for the September 1981-January 1983 and February 1983-May 1984 periods as a function of Kp and Dst. The outflow rates were determined by integrating the averaged upward ion flux over invariant latitude and magnetic local time, weighted by the surface areas at a reference altitude (1000 km) in both hemispheres. At all Kp values, the H^+ outflow rates in the two periods were equal, within statistical error. In contrast, the O^+ outflow rate for a given Kp condition was higher in the earlier period by about a factor of 2. However, the Kp dependences in the two periods were similar. Likewise, the O^+ outflow rates in the two periods had similar dependence with Dst, even though they differed by about a factor of 2 in magnitude. Both the O^+ and H^+ outflow rates increased exponentially with Kp, as did the O^+/H^+ composition ratio.

Recently, Yau et al. (1987) extended their analysis to data acquired at the solar minimum (in 1986) and found the exponential dependences of both the H^+ and O^+ outflow rates to prevail at all solar activity levels (from September 1981 to May 1986 when $F_{10.7}$ varied between ~70 and ~300). For a given solar activity level, the H^+ and O^+ outflow rates in the 0.01-17 keV range were found to increase exponentially by a factor of 4 and 20, respectively, from Kp = 0 to 6. Empirically, they found $F(O^+) \propto \exp(0.50\ Kp)$ and $F(H^+) \propto \exp(0.23\ Kp)$. Furthermore, from solar minimum to near maximum, the O^+ rate for a given Kp level increased by a factor of ~5, while the H^+ rate decreased by a factor of ~2, resulting in an order of magnitude increase in the O^+/H^+ composition ratio.

The ion outflow values in Figure 8 exclude ions below 10 eV. From typical upwelling ion distributions measured near the solar maximum, Chappell et al. (1987) estimated the relative contributions of ion fluxes below and above 10 eV to the cleft ion fountain to be 4:1 for O^+ and 1:4 for H^+. Also, they estimated the >10-eV component of the cleft ion fountain from the upward ion flux data in Yau et al. (1985b), from which they inferred the active time <10-eV cleft ion fountain to be 3.8×10^{25} O^+ and 8.6×10^{23} H^+ ions s^{-1}.

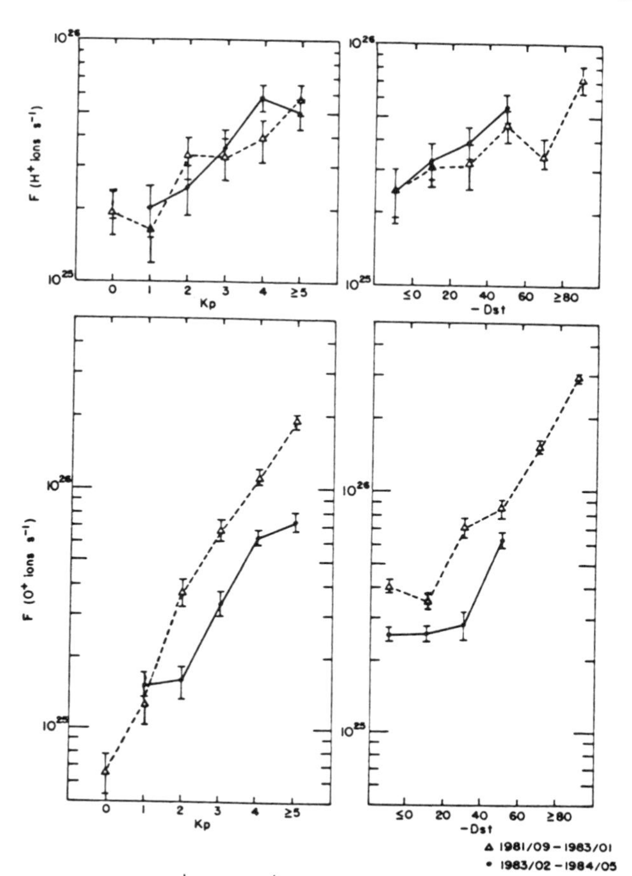

Fig. 8. DE 1 H^+ and O^+ outflow rates 0.01-1 keV for September 1981 to January 1983 (triangles) and February 1983 to May 1984 (dots) (Yau et al., 1985b).

Summary and Discussions

Outflowing ions from the polar ionosphere fall into two categories: the classical polar wind and the suprathermal ion flows. The polar wind is supersonic at high altitude: at ~3 R_E geocentric, the observed polar wind is H^+ dominated with a flux of 2.6×10^8 cm^{-2} s^{-1} and a Mach number between 2.5 and 5.1. At the poleward edge of the nightside auroral oval in the lower topside ionosphere, upward O^+ ion fluxes exceeding the limiting light polar wind ion flux and indicative of suprathermal O^+ ion outflow frequently occur at magnetically active times. Above 500 km, ions are accelerated transverse to the local geomagnetic field to tens and hundreds of electron volts. The acceleration has been associated with both VLF waves and regions of large field-aligned currents. At 1400 km, TAIs are frequently observed in winter nights but rarely appear in the summer. In the dayside cleft above ~2000 km, ions of all species are transversely heated to ~10 eV and upwell with significant number flux (>10^8 cm^{-2} s^{-1}) and heat flux (>3×10^4 erg cm^{-2} s^{-1}), forming a cleft ion fountain as they convect across the polar cap. Their low-latitude edge is closely associated with that of the field-aligned current region. These upwelling ions are observed most (least) frequently in the summer (winter). At high altitude, intense, energetic (>10 eV) H^+ and O^+ ion beams and conics are common, the active time occurrence frequency being as high as 0.7 at auroral latitudes and 0.3 in the polar cap. Their composition, intensity, and angular characteristics display marked variations with solar activity, being O^+ dominant, more intense, and richer in conics near solar maximum. The outflowing flux is dominated by ions below 1 keV. The outflow rate increases exponentially with the magnetic Kp and Dst indices, the O^+ and H^+ rates by a factor of 20 and 4, respectively, from Kp = 0 to 6. Near solar maximum, the active time rate reaches 3.5×10^{26} O^+ and 7×10^{25} H^+ ions s^{-1}. In comparison, the light polar wind ion outflow, from ISIS 2 observations in moderate solar conditions, averaged over the four seasons and extrapolated to all local times and invariant latitudes down to 60°, is 6.5 x 10^{25} H^+. Thus, the polar wind represents an important source of magnetospheric H^+ plasma. The <10-eV cleft ion fountain is estimated to be ~3.8 x 10^{25} O^+ and 8.6 x 10^{23} H^+ ions s^{-1}.

Ion flows of ~10-50 eV in the polar cap generally map back to a cleft ion fountain source, but much of the higher-energy O^+ flows are too energetic for this to be the case. This suggests that either a source exists within the polar cap, perhaps driven by the polar rain or polar cap arc electrons, or that the cleft ions undergo secondary acceleration at high altitude within the polar cap. Cladis (1986) showed that ions in the cusp regions may be accelerated to kiloelectron volt energies by the component of the curvature drift of the ion along the convection electric field. Unfortunately, an accurate estimate for the fraction of upwelling cleft ions at low altitude that becomes energized at source or at higher altitude is not yet available. Our present knowledge of the relative contributions from the various source regions to the total polar ionospheric outflow is uncertain in the following respects. (1) Above 10 eV, the estimates for the contribution of the cleft ion fountain to ion outflow at high altitude were derived from typical upwelling ion energy spectra at low altitude, assuming no magnetic activity dependence, no long-term variation, and no secondary ion acceleration at higher altitude. (2) Only limited observations are available for polar wind at high altitudes. The polar wind is theoretically expected to be relatively uniform in local time and latitude down to the plasmapause, but observational evidence of this uniformity does not yet exist; neither does the observational evidence of theoretically expected solar activity and magnetic activity dependences. (3) Without observation down to 0 eV, it is not possible to determine the flux of low-energy O^+ fed across the polar cap from the cleft ion fountain into the nightside auroral oval.

There is no question that the outflowing auroral and polar cap ions constitute the primary source of energetic magnetospheric plasma of terrestrial origin. Recently, Shelley (1986) reviewed energetic ions of ionospheric origin throughout the magnetosphere and on its boundaries and concluded that O^+ is the dominant identifiable ionospheric component in most regions, probably exceeding the ionospheric H^+ component except during periods of low geomagnetic and solar activities. Only in the dayside magnetopause boundary layer does He^+ appear to be a significant contributor. The O^+ observed in the magnetotail boundary layer and the mantle is believed to be injected directly from the dayside cusp ionosphere. The magnetotail lobe O^+ and H^+ streams are supplied by the energetic (>10 eV) polar cap ion outflow; the two populations display striking similarities in intensity, mass composition, and magnetic activity dependence (Sharp et al., 1981; Yau et al., 1984). The ultimate source for at least a fraction of the energetic polar cap ions is the cleft ion fountain. The terrestrial component of the plasma sheet is adequately maintained by the energetic upflowing auroral ions. The energetic auroral ion outflow rate can reach 3 x 10^{26} ions s^{-1}; it is comparable to the plasma sheet loss rate and provides an estimated 50% of the plasma sheet ions during periods of high magnetic activity. Recently, Chappell et al. (1987) raised the possibility that if the polar wind and the cleft ion fountain ions below 10 eV are included in the ionospheric input to the magnetosphere, the observed magnetospheric densities can be attained under all geomagnetic conditions without any contribution from the solar wind plasma. A detailed analysis of this possibility requires detailed modeling of the quantitative behavior of thermal-energy ionospheric ions in the magnetosphere.

The observed solar cycle variation and magnitude of energetic O^+ outflow at high altitude underscores the importance of ion acceleration at the O^+-dominated low altitudes. The neutral atmosphere has the tendency to prevent appreciable O^+ outflow as the neutral atomic hydrogen atoms charge-exchange with the O^+ ions and convert them to H^+ ions in the vicinity of the H-O crossover altitude (altitude of equal O and H density). However, modest acceleration below the crossover altitude will permit the O^+ to escape without charge exchange, by reducing their effective path length through the neutral atmosphere. Moore (1980) presented model calculations showing that for O^+, transverse acceleration to ~10 eV or parallel acceleration to a few electron volts below the crossover altitude is adequate. The crossover altitude is an increasing function of the neutral atmospheric temperature. Moore (1984) concluded that (1) for a given acceleration altitude, an increase in the crossover altitude increases the O^+ content (fraction of O^+) in the escaping flux, and therefore (2) one would expect greatly enhanced O^+ outflow due to transverse acceleration in the 1000-km altitude region near the solar maximum, when the solar EUV flux is greatest, the neutral atmosphere is hottest, and the crossover altitude is highest. Note that the presence of molecular ions in the cleft ion fountain (Craven et al., 1985) indicates that acceleration can indeed occur at very low altitudes.

The upward transport of polar ionospheric ions is necessarily accompanied be electrons of comparable (and in the long term, equal) rate, except possibly in some regions of upward field-aligned current. Both upflowing electron beams and conics have been observed, sometimes in association with TAIs and conics (Klumpar and Heikkila, 1982; Menietti and Burch, 1985), the latter in both the polar cap and the nightside auroral oval. The possible importance of upgoing electron beams and conics in the overall vertical polar plasma transport warrants further investigation.

The body of observations from the DE and earlier satellites underscores the importance of ion acceleration at the O^+-dominated low altitudes.

As well, it establishes semi-quantitatively the sources of energetic ions of terrestrial origin in different parts of the magnetosphere. A comprehensive, quantitative understanding of the geomagnetic and solar activity dependence of polar ion flow must await routine ion composition measurements down to polar wind ion energy (electron volt) at high altitude and at superthermal energy (few to tens of electron volts) in the low-altitude acceleration region, as well as detailed modeling of ion transport and charge exchange in the topside ionosphere.

References

Barakat, A. R., and R. W. Schunk, Effect of hot electrons in the polar wind, J. Geophys. Res., 89, 9771, 1984.

Chang, T., and B. Coppi, Lower hybrid acceleration and ion evolution in the suprauroral region, Geophys. Res. Lett., 8, 1253, 1981.

Chang, T., G. B. Crew, N. Hershkowitz, J. R. Jasperse, J. M. Retterer, and J. D. Winningham, Transverse acceleration of oxygen ions by electromagnetic ion cyclotron resonance with broad band left-hand polarized waves, Geophys. Res. Lett., 13, 636, 1986.

Chappell, C. R., J. L. Green, J.F.E. Johnson, and J. H. Waite, Jr., Pitch angle variations in magnetospheric thermal plasma--Initial observations from Dynamics Explorer-1, Geophys. Res. Lett., 9, 933, 1982.

Chappell, C. R., T. E. Moore, and J. H. Waite, Jr., The ionosphere as a fully adequate source of plasma for the earth's magnetosphere, J. Geophys. Res., 92, 5896, 1987.

Cladis, J. B., Parallel acceleration and transport of ions from polar ionosphere to plasma sheet, Geophys. Res. Lett., 13, 893, 1986.

Collin, H. L., R. D. Sharp, E. G. Shelley, and R. G. Johnson, Some general characteristics of upflowing ion beams over the auroral zone and their relationship to auroral electrons, J. Geophys. Res., 86, 6820, 1981.

Collin, H. L., R. D. Sharp, and E. G. Shelley, The magnitude and composition of the outflow of energetic ions from the ionosphere, J. Geophys. Res., 89, 2185, 1984.

Craven, P. D., R. C. Olsen, C. R. Chappell, and L. Kakani, Observations of molecular ions in the earth's magnetosphere, J. Geophys. Res., 90, 7599, 1985.

Ghielmetti, A. G., R. G. Johnson, R. D. Sharp, and E. G. Shelley, The latitudinal, diurnal, and altitudinal distributions of upward flowing energetic ions of ionospheric origin, Geophys. Res. Lett., 5, 59, 1978.

Gombosi, T., T. E. Cravens, and A. F. Nagy, A time-dependent theoretical model of the polar wind: Preliminary results, Geophys. Res. Lett., 12, 167, 1985.

Gorney, D. J., A. Clarke, D. Croley, J. F. Fennell, J. Luhmann, and P. Mizera, The distribution of ion beams and conics below 8000 km, J. Geophys. Res., 86, 83, 1981.

Green, J. L., and J. H. Waite, Jr., On the origin of polar ion streams, Geophys. Res. Lett., 12, 149, 1985.

Gurgiolo, C., and J. L. Burch, DE 1 observations of the polar wind--A heated and unheated component, Geophys. Res. Lett., 9, 945, 1982.

Hoffman, J. H., and W. H. Dodson, Light ion concentrations and fluxes in the polar regions during magnetically quiet times, J. Geophys. Res., 85, 626, 1980.

Horwitz, J. L., Features of ion trajectories in the polar magnetosphere, Geophys. Res. Lett., 11, 1111, 1984.

James, H. G., VLF saucers, J. Geophys. Res., 81, 501, 1976.

Kintner, P. M., and D. J. Gorney, A search for the plasma processes associated with perpendicular ion heating, J. Geophys. Res., 89, 937, 1984.

Kintner, P. M., J. LaBelle, W. Scales, A. W. Yau, and B. A. Whalen, Observations of plasma waves within regions of perpendicular ion acceleration, Geophys. Res. Lett., 13, 1113, 1986.

Klumpar, D. M., Transversely accelerated ions: An ionospheric source of hot magnetospheric ions, J. Geophys. Res., 84, 4229, 1979.

Klumpar, D. M., Characteristics of the high-latitude ionospheric particle sources: Transversely accelerated ions (TAI) at 1400 km, Adv. Space Res., 5, 145, 1985.

Klumpar, D. M., and W. J. Heikkila, Electrons in the ionospheric source cone: Evidence for runaway electrons as carriers of downward Birkeland currents, Geophys. Res. Lett., 9, 873, 1982.

Klumpar, D. M., W. K. Peterson, and E. G. Shelley, Direct evidence for two-stage (bimodal) acceleration of ionospheric ions, J. Geophys. Res., 89, 10779, 1984.

Lockwood, M., Thermal ion flows in the topside auroral ionosphere and the effects of low-altitude, transverse acceleration, Planet. Space Sci., 30, 595, 1982.

Lockwood, M., Low-energy ion flows into the magnetosphere, Adv. Space Res., 6, 63, 1986.

Lockwood, M., and J. E. Titheridge, Ionospheric origin of magnetospheric O^+ ions, Geophys. Res. Lett., 8, 381, 1981.

Lockwood, M., M. O. Chandler, J. L. Horwitz, J. H. Waite, Jr., T. E. Moore, and C. R. Chappell, The cleft ion fountain, J. Geophys. Res., 90, 9736, 1985a.

Lockwood, M., T. E. Moore, J. H. Waite, Jr., C. R. Chappell, J. L. Horwitz, and R. A. Heelis, The geomagnetic mass spectrometer--Mass and energy dispersions of ionospheric ion flows into the magnetosphere, Nature, 316, 612, 1985b.

Lockwood, M., J. H. Waite, Jr., T. E. Moore, J. F. E. Johnson, and C. R. Chappell, A new source of suprathermal O^+ ions near the dayside polar cap boundary, J. Geophys. Res., 90, 4099, 1985c.

Lockwood, M., A. P. van Eyken, B.J.I. Bromage, J. H. Waite, Jr., T. E. Moore, and J. R. Doupnik,

Low-energy ion outflows from the ionosphere during a major cap expansion - Evidence for equatorward motion of inverted-V structures, Adv. in Space Res., 6, 93, 1986.

Lysak, R. L., M. K. Hudson, and M. Temerin, Ion heating by strong electrostatic ion cyclotron turbulence, J. Geophys. Res., 85, 678, 1980.

Menietti, J. D., and J. L. Burch, "Electron conic" signatures observed in the nightside auroral zone and over the polar cap, J. Geophys. Res., 90, 5345, 1985.

Moore, T. E., Modulation of terrestrial escape flux composition (by low-altitude acceleration and charge exchange chemistry), J. Geophys. Res., 85, 2011, 1980.

Moore, T. E., Superthermal ionospheric outflows, Rev. Geophys. Space Phys., 22, 264, 1984.

Moore, T. E., C. R. Chappell, M. Lockwood, and J. H. Waite, Jr., Superthermal ion signatures of auroral acceleration processes, J. Geophys. Res., 90, 1611, 1985.

Moore, T. E., M. Lockwood, M. O. Chandler, J. H. Waite, Jr., C. R. Chappell, A. Persoon, and M. Sugiura, Upwelling O^+ ion source characteristics, J. Geophys. Res., 91, 7019, 1986.

Nagai, T., J. H. Waite, Jr., J. L. Green, and C. R. Chappell, First measurements of supersonic polar wind in the polar magnetosphere, Geophys. Res. Lett., 11, 669, 1984.

Raitt, W. J., and R. W. Schunk, Composition and characteristics of the polar wind, in Energetic Ion Composition in the Earth's Magnetosphere, edited by R. G. Johnson, pp. 99-141, Terra Scientific Publ. Co., Tokyo, 1983.

Rosenbauer, H., H. Grunwaldt, M. D. Montgomery, G. Paschmann, and N. Sckopke, HEOS-2 plasma observations in the distant polar magnetosphere: The plasma mantle, J. Geophys. Res., 80, 2723, 1975.

Sharp, R. D., R. G. Johnson, and E. G. Shelley, Observations of an ionospheric acceleration mechanism producing energetic (keV) ions primarily normal to the geomagnetic field direction, J. Geophys. Res., 82, 3324, 1977.

Sharp, R. D., D. L. Carr, W. K. Peterson, and E. G. Shelley, Ion streams in the magnetotail, J. Geophys. Res., 86, 4639, 1981.

Sharp, R. D., A. G. Ghielmetti, R. G. Johnson, and E. G. Shelley, Hot plasma composition results from the S3-3 spacecraft, in Energetic Ion Composition in the Earth's Magnetosphere, edited by R. G. Johnson, pp. 167-193, Terra Scientific Publ. Co., Tokyo, 1983.

Shelley, E. G., Magnetospheric energetic ions from the earth's ionosphere, Adv. Space Res., 6, 121, 1986.

Tsunoda, R., R. C. Livingston, J. F. Vickrey, C. L. Rino, F. J. Rich, and P. F. Bythrow, Dayside observations of thermal ion upwelling at 800 km altitude: An ionospheric signature of the cleft ion fountain, J. Geophys. Res., in press, 1987.

Waite, J. H. Jr., T. Nagai, J.F.E. Johnson, C. R. Chappell, J. L. Burch, T. L. Killeen, P. B. Hays, G. R. Carignan, W. K. Peterson, and E. G. Shelley, Escape of suprathermal O^+ ions in the polar cap, J. Geophys. Res., 90, 1619, 1985.

Whalen, B. A., W. Bernstein, and P. W. Daly, Low altitude acceleration of ionospheric ions, Geophys. Res. Lett., 5, 55, 1978.

Winningham, J. D., J. L. Burch, and R. A. Frahm, Bands of ions and angular V's: A conjugate manifestation of ionospheric ion acceleration, J. Geophys. Res., 89, 1749, 1984.

Yang, W. H., and J. R. Kan, Generation of conic ions by auroral electric fields, J. Geophys. Res., 88, 465, 1983.

Yau, A. W., B. A. Whalen, A. G. McNamara, P. J. Kellogg, and W. Bernstein, Particle and wave observations of low-altitude ionospheric ion acceleration events, J. Geophys. Res., 88, 341, 1983.

Yau, A. W., B. A. Whalen, W. K. Peterson, and E. G. Shelley, Distribution of upflowing ionospheric ions in the high-altitude polar cap and auroral ionosphere, J. Geophys. Res., 89, 5507, 1984.

Yau, A. W., P. H. Beckwith, W. K. Peterson, and E. G. Shelley, Long-term (solar-cycle) and seasonal variations of upflowing ionospheric ion events at DE 1 altitudes, J. Geophys. Res., 90, 6395, 1985a.

Yau, A. W., E. G. Shelley, W. K. Peterson, and L. Lenchyshyn, Energetic auroral and polar ion outflow at DE 1 altitudes: Magnitude, composition, magnetic activity dependence and long-term variations, J. Geophys. Res., 90, 8417, 1985b.

Yau, A. W., W. K. Peterson, and E. G. Shelley, Quantitative parametrization of energetic ionospheric ion outflow, this volume, 1987.

Shelley, E. G., R. D. Sharp, and R. G. Johnson, Satellite observations of an ionospheric acceleration mechanism, Geophys. Res. Lett., 3, 654, 1976a.

Shelley, E. G., R. D. Sharp, and R. G. Johnson, He^+ and H^+ flux measurements in the dayside cusp: Estimates of convection electric field, J. Geophys. Res., 81, 2363, 1976b.

Shelley, E. G., W. K. Peterson, A. G. Ghielmetti, and J. Geiss, The polar ionosphere as a source of energetic magnetospheric plasma, Geophys. Res. Lett., 9, 941, 1982.

O^+ AND H^+ ESCAPE FLUXES FROM THE POLAR REGIONS

A. R. Barakat,[1] R. W. Schunk,[1] T. E. Moore,[2] and J. H. Waite, Jr.[2]

Abstract. The coupled continuity and momentum equations for H^+, O^+, and electrons were solved for the terrestrial ionosphere in order to determine the limiting ion escape fluxes at high latitudes. The effects of solar cycle, season, geomagnetic activity, and the altitude of the acceleration region on the ion escape fluxes were studied for average conditions. The main conclusions of the study are as follows: (1) As solar activity increases, the general trend is for an increase in the limiting O^+ escape flux and a decrease in the limiting H^+ escape flux. (2) In winter, the limiting escape fluxes of both O^+ and H^+ are larger than those in summer, particularly for low geomagnetic activity. (3) The O^+ content of the ion outflow increases with increasing demand imposed on the ionosphere by a high-altitude acceleration process, increasing solar activity, increasing geomagnetic activity, increasing solar elevation from winter to summer, and a lowering of the altitude of the acceleration region. The general trends obtained for average conditions appear to mimic the qualitative behavior determined from statistically averaged data for comparable absolute escape flux magnitudes.

Introduction

Up until the 1980's, the theories of ionospheric outflow were based on the assumption that the heavy ions, including O^+, were gravitationally bound, while the lighter ions (H^+ and He^+) flowed freely outward in response to their thermal energy and the thermal energy of the electrons, as communicated by the ambi-polar electric field. However, recent observations (Waite et al., 1985; Lockwood et al., 1985; Yau et al., 1985a,b; Moore et al., 1986) and theories (Barakat and Schunk, 1983, 1984; Singh and Schunk, 1985; Gombosi et al., 1985; Ashour-Abdalla and Okuda, 1984) have clearly shown that energetic processes come into play in the topside ionosphere and provide sufficient energy to the O^+ ions so that they also escape rather freely to the magnetotail.

The outflow of ionospheric plasma is expected to be influenced by the collisional and chemical processes which occur both in the F region and in the topside ionosphere below the region where substantial acceleration of ions occurs. Moore (1980, 1984) has advanced rudimentary arguments to this effect. The purpose of this paper is to define the constraints which are imposed upon ionospheric outflows by these low-altitude effects. Therefore, we adopt a theoretical framework which properly describes the ambipolar, collisional, subsonic, chemically active flow of the ionospheric plasma through the neutral atmosphere to higher altitudes. All of the effects of various nonthermal energetic plasma processes, including ion acceleration processes, are lumped into imposed flux conditions assumed to be effected at the upper boundary of our computational space. Therefore, the upper boundary of our computational space is to be identified as the lower boundary of the ion acceleration region.

H^+ and O^+ Escape Fluxes

The theoretical formulation describing the ion escape fluxes, including transport equations, photochemical reactions, and collision frequencies, is given by Barakat et al. (1987) and is not repeated here. However, in order to solve for the ion escape fluxes, some input parameters were needed. We used the mass spectrometer and incoherent scatter (MSIS) atmospheric model to compute the neutral gas densities and temperatures (Hedin et al., 1977), and the parameters needed for this model ($F_{10.7}$, Ap, day number) are given in Table 1 for the different solar cycle, seasonal, and geomagnetic activity conditions that we considered. Our model also required the O^+ ionization frequency, β (see Table 1).

Ion Flux Variation with Season, Solar Cycle, and Geomagnetic Activity

Figure 1 shows possible O^+ and H^+ escape fluxes for two different values of the upper

[1]Center for Atmospheric & Space Sciences, Utah State University, Logan, Utah 84322.
[2]NASA Marshall Space Flight Center, Huntsville, Alabama 35812.

TABLE 1. Range of Parameters Used

Condition	Parameters
Solar maximum	$F_{10.7} = 150 \times 10^{-22}$ Wm^{-2} Hz^{-1} $\beta = 4.5 \times 10^{-7}$ s^{-1}
Solar minimum	$F_{10.7} = 60 \times 10^{-22}$ Wm^{-2} Hz^{-1} $\beta = 1.7 \times 10^{-7}$ s^{-1}
Summer	day = 183
Winter	day = 1
High geomagnetic activity	Ap = 70
Low geomagnetic activity	Ap = 0

boundary altitude (Z_T = 600 and 1000 km) and for solar minimum, summer, and high geomagnetic activity. Z_T can be thought of as the altitude above which ion energization takes place. The borders shown in the figure separate two regions in the F_{O+}-F_{H+} plane. The points within these borders correspond to physically permissible results, and the points on the borders correspond to computer runs where F_{O+} and/or F_{H+} attain their limiting flux values. In particular, the upper border corresponds to limiting H^+ outflows, while the lower border corresponds to limiting O^+ outflows.

For an upper boundary at 600 km, the limiting H^+ escape flux is about 3×10^7 cm^{-2} s^{-1} if there is no O^+ outflow. As O^+ begins to escape, the limiting H^+ escape flux decreases because of the decrease in the O^+ density; hence the H^+ production rate via the charge exchange reaction $O^+ + H \rightleftharpoons H^+ + O$. In contrast to the H^+ result, the limiting O^+ escape flux does not depend on H^+ for Z_T = 600 km because at and below this altitude H^+ is a minor ion and has little effect on O^+. In this case, the limiting O^+ flux is ~3×10^8 cm^{-2} s^{-1}.

For Z_T = 1000 km, which corresponds to an altitude above the O/H crossover altitude, the limiting O^+ escape flux is not changed significantly, but there is a factor of 1.7 increase in $(F_{H+})_L$ for the case of no O^+ outflow. This increase in $(F_{H+})_L$ results from the fact that between 600 and 1000 km the $O^+ + H \rightleftharpoons H^+ + O$ reaction is still important when O^+ is close to a diffusive equilibrium distribution.

In order to show the solar cycle effect, we present in Figure 2 the possible O^+ and H^+ escape fluxes for solar maximum, summer, and high geomagnetic activity (MX-S-H), which is the same as the previous case (MN-S-H) except for the solar activity level. The increase in solar activity has a twofold effect on the ionospheric behavior. First, the O^+ photoionization frequency (β) increases due to the enhancement of the EUV radiation during the solar maximum period. Second, the neutral atmosphere changes (e.g., exospheric temperature (T_∞) increases, O/H crossover altitude increases, [O] increases, and [H] decreases). The increases in β and [O] result in an O^+ density increase (Schunk and Raitt, 1980); consequently $(F_{O+})_L$ increases. The results we obtained tend to indicate that, holding everything else constant, $(F_{O+})_L \propto \beta$, but we only studied this variation for one geophysical situation (see Figure 2 in Barakat et al., 1987). Although the increase in the O^+ density tends to increase $(F_{H+})_L$, the decrease in [H] tends to reduce it. The latter effect is stronger; therefore, a net decrease in $(F_{H+})_L$ results.

Contrary to the previous case (MN-S-H), for solar maximum we considered values of Z_T up to 1500 km in order to include the O/H crossover altitude at 1230 km. As Z_T increases, more H^+ can be generated; hence $(F_{H+})_L$ increases. Also, since [H]/[O] is greatly increased (compared to the previous case), O^+ dominates H^+ to the extent that as we try to deplete O^+, less H^+ is generated. This effect is strong enough to keep O^+ as the major ion. Hence, the lower border, which

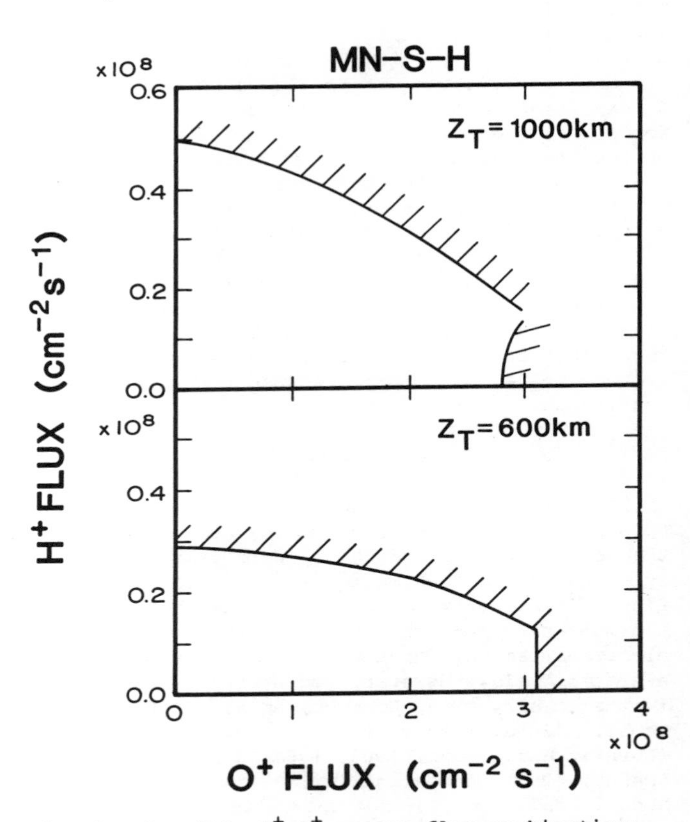

Fig. 1. Possible H^+-O^+ escape flux combinations for solar minimum, summer, and high geomagnetic activity conditions. In each panel, the top boundary shows the limiting H^+ escape flux for a given O^+ escape flux. The lower boundary shows the limiting O^+ escape flux for a given H^+ escape flux. The possible ion escape flux combinations correspond to points within these boundaries.

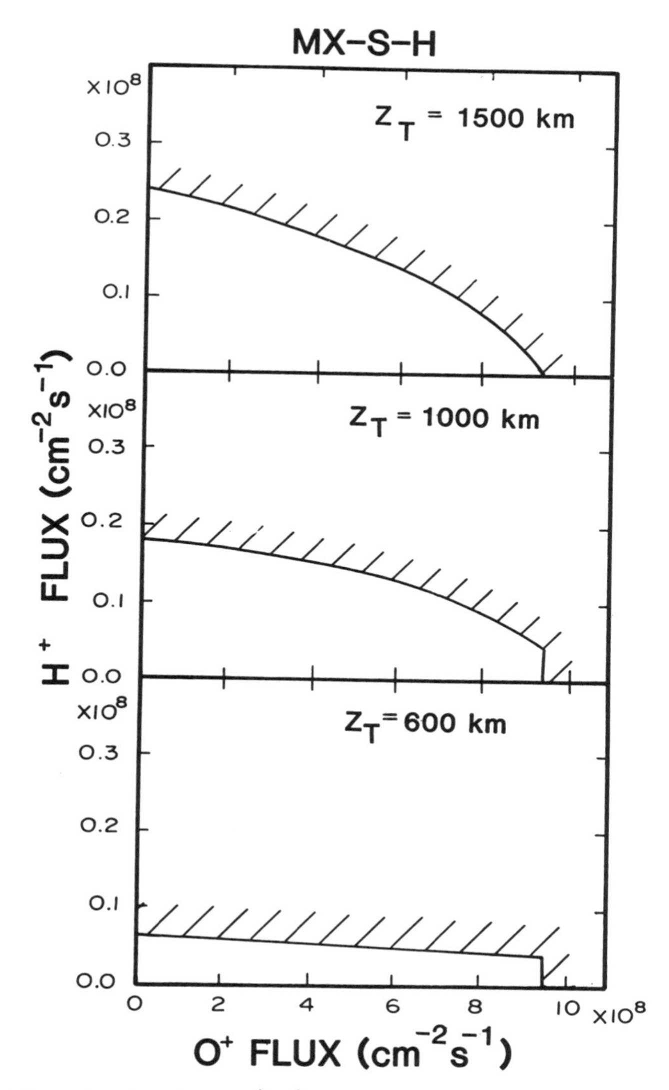

Fig. 2. Possible H^+-O^+ escape flux combinations for solar maximum, summer, and high geomagnetic activity conditions. The format is the same as that for Figure 1.

corresponds to the situation when H^+ becomes the major ion, moves to the right and eventually disappears.

Next, we present in Figure 3 the results for solar maximum, winter, and low geomagnetic activity (MX-W-L). The bottom panel shows the results for Z_T = 600 km. The comparison of these results with those in Figure 2 (bottom) shows the combined effect of season and geomagnetic activity. In comparison with the previous case (MX-S-H), we notice that the H^+ limiting escape flux increases by a factor of ~20 and the O^+ limiting escape flux increases by a factor of ~3. The decrease of $[O_2]$ and $[N_2]$ for MX-W-L causes the O^+ loss rate to decrease. This increases the O^+ density and, hence, its escaping flux. The increase of both the O^+ density and [H] explains the large increase in F_{H+}.

These increases could be overestimates. In fact, this model is valid only in sunlit regions, which eliminates most of the polar cap region in winter. To address this problem properly, we need a full time-dependent model that accounts for the **E** x **B** drift of the F region and the solar zenith angle changes. If we could have taken account of these effects, smaller increases in $(F_{O+})_L$ and $(F_{H+})_L$ would have occurred. The limiting value of the O^+ escape flux might have even decreased (see Schunk and Raitt, 1980).

Escape Flux Composition

In order to show the effects of solar cycle, season, geomagnetic activity, and the altitude of the acceleration mechanism (Z_T) on the escape flux composition, we will present the results of the cases in Figures 1-3 in a different format. We will be mainly concerned with the ion escape along the open field lines above the polar cap. In this region, the H^+ escape flux usually saturates. Therefore, we consider only runs for which F_{H+} attains its limiting value. These runs

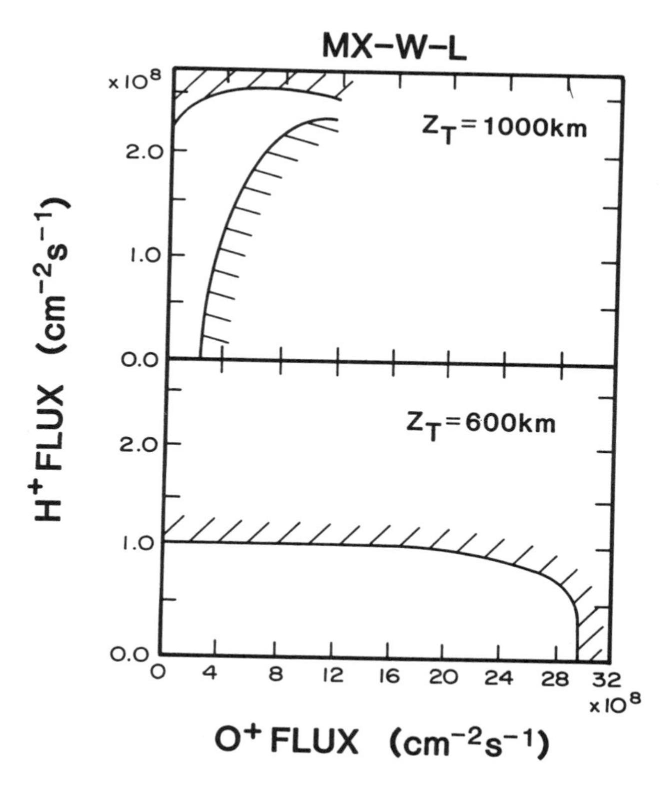

Fig. 3. Possible H^+-O^+ escape flux combinations for solar maximum, winter, and low geomagnetic activity. The format is the same as that for Figure 1.

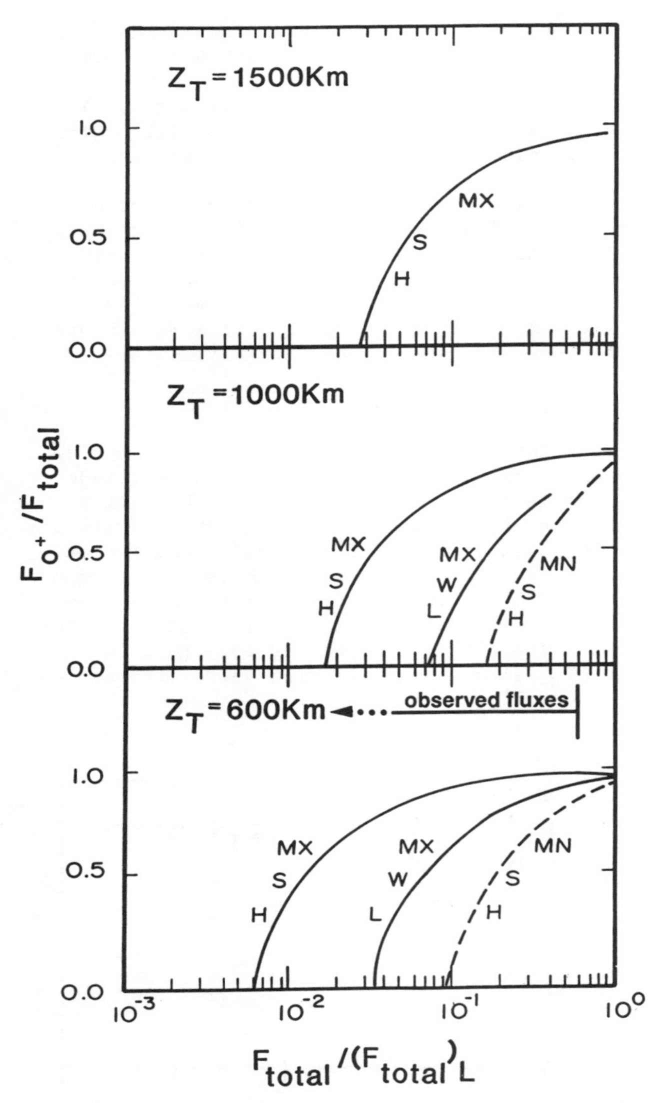

Fig. 4. O^+ fraction of the total ion outflow at three altitudes as a function of the demand. Here, demand is defined to be (total ion flux)/(limiting total ion flux). Also shown are the flux values deduced by Lockwood and Titheridge (1982).

correspond to the upper border of the permissible region shown in Figures 1-3.

Figure 4 shows the O^+ fraction of the total ion outflow $[F_{O+}/(F_{O+} + F_{H+})]$ versus demand for flux for different geophysical parameters (solar cycle, season, and geomagnetic activity) and for different altitudes of ion acceleration (Z_T). The demand for the ion flux is a parameter that is used to represent the role of the magnetosphere in deciding the ion escape flux (e.g., via ion energization). Here, demand is represented by the ratio of the total flux to its limiting value $(F_{O+} + F_{H+})/(F_{O+} + F_{H+})_L$. The figure shows that the curves corresponding to solar maximum are higher than those corresponding to solar minimum; i.e., at solar maximum the ion flux composition is changed in favor of O^+. This is due to the increase in the O^+ density (Schunk and Raitt, 1980) coupled with the decrease in [H] and, hence, the production of H^+ via the charge exchange reaction. We notice also that the ratio of O^+ flux is larger for summer and high magnetic activity than it is for winter and low magnetic activity. As the demand for flux increases, the fraction of O^+ flux increases because the already saturated H^+ escape flux does not change; hence any increase in the total flux must occur via an increase of F_{O+}. Finally, the O^+ to total flux ratio increases as the ions are energized at lower altitudes. This is expected because the O^+ relative abundance is higher at low altitudes; consequently the resulting ion escape flux should include more O^+.

Conclusions

We solved the hydrodynamic transport equations for H^+ and O^+ including the important dynamic, collisional, and chemical effects that operate in the F region ionosphere below regions of ion acceleration. From our study, we found the following: (1) An important parameter controlling the amount of O^+ in plasma outflows is the total ion flux demand imposed on the ionosphere by the higher altitude acceleration region. (2) The O^+ content is further modulated by the temperature of the exosphere and the resultant composition in the topside ionosphere. Higher exospheric temperatures are produced by increased solar activity, by high solar elevation in summer, and by increased geomagnetic activity, leading to a greater O^+ content at constant total ion outflux. (3) The O^+ content is also modulated by the location of the lower boundary of the ion acceleration region relative to the crossover altitude where O and H have equal densities. (4) As solar activity increases, the limiting escape flux for O^+ increases, but the limiting escape flux for H^+ decreases. (5) In summer, the limiting escape fluxes for both H^+ and O^+ are smaller than those in winter in sunlit regions.

In general, the results we obtained are in agreement with the recent measurements of Yau et al. (1985a,b). In particular, the increase of the O^+ outflow with solar activity is consistent with the observations. Also, the model predictions of an increasing O^+ content of the total ion escape flux with increasing solar activity and magnetic activity are consistent with the observations. The major difference between the observations and our results is the variation of the H^+ flux, which we would expect to decline with solar activity and summer and high magnetic activity; whereas the observations show a small but significant increase with geomagnetic activity and no significant change with solar activity. We ascribe this discrepancy to the observational low-energy threshold at 10 eV which makes the instrument sensitive to the more energetic ions.

Acknowledgments. This research was supported by NASA grant NAGW-77 and NSF grant ATM-8417880 to Utah State University.

References

Ashour-Abdalla, M., and H. Okuda, Turbulent heating of heavy ions on auroral field lines, J. Geophys. Res., 89, 2235, 1984.

Barakat, A. R., and R. W. Schunk, O^+ ions in the polar wind, J. Geophys. Res., 88, 7887, 1983.

Barakat, A. R., and R. W. Schunk, Effect of hot electrons on the polar wind, J. Geophys. Res., 89, 9771, 1984.

Barakat, A. R., R. W. Schunk, T. E. Moore, and J. H. Waite, Jr., Ion escape fluxes from the terrestrial high latitude ionosphere, J. Geophys. Res., in press, 1987.

Gombosi, T. I., T. E. Cravens, and A. F. Nagy, A time-dependent theoretical model of the polar wind: Preliminary results, Geophys. Res. Lett., 12, 167, 1985.

Hedin, A. E., J. E. Salah, J. V. Evans, C. A. Reber, G. P. Newton, N. W. Spencer, D. C. Kayser, D. Alcaydé, P. Bauer, L. Cogger, and J. P. McClure, A global thermospheric model based on mass spectrometer and incoherent scatter data, MSIS-1, N_2 density and temperature, J. Geophys. Res., 82, 2139, 1977.

Lockwood, M., and J. E. Titheridge, Departures from diffusive equilibrium in the topside ionosphere, J. Atmos. Terr. Phys., 44, 425,1982.

Lockwood, M., J. H. Waite, T. E. Moore, J.F.E. Johnson, and C. R. Chappell, A new source of suprathermal O^+ ions near the dayside polar cap boundary, J. Geophys. Res., 90, 4099, 1985.

Moore, T. E., Modulation of terrestrial escape flux composition (by low-altitude acceleration and charge exchange chemistry), J. Geophys. Res., 85, 2011, 1980.

Moore, T. E., Superthermal ionospheric outflows, Rev. Geophys. Space Phys., 22, 264, 1984.

Moore, T. E., M. Lockwood, M. O. Chandler, J. H. Waite, Jr., C. R. Chappell, A. Persoon, and M. Suguira, Upwelling O^+ ion source characteristics, J. Geophys. Res., 91, 7019, 1986.

Schunk, R. W., and W. J. Raitt, Atomic nitrogen and oxygen ions in the daytime high-latitude F region, J. Geophys. Res., 85, 1255, 1980.

Singh, N., and R. W. Schunk, Temporal evolution of density perturbations in the polar wind, J. Geophys. Res., 90, 6487, 1985.

Waite, J. H., Jr., T. Nagai, J.F.E. Johnson, C. R. Chappell, J. L. Burch, T. L. Killeen, P. B. Hays, G. R. Carignan, W. K. Peterson, and E. G. Shelley, Escape of suprathermal O^+ ions in the polar cap, J. Geophys. Res., 90, 1619, 1985.

Yau, A. W., P. H. Beckwith, W. K. Peterson, and E. G. Shelley, Long-term (solar cycle) and seasonal variations of upflowing ionospheric ion events at DE 1 altitudes, J. Geophys. Res., 90, 6395, 1985a.

Yau, A. W., E. G. Shelley, W. K. Peterson, and L. Lenchyshyn, Energetic auroral and polar ion outflow at DE 1 altitudes: Magnitude, composition, magnetic activity dependence, and long-term variations, J. Geophys. Res., 90, 8417, 1985b.

OPEN FLUX MERGING IN AN EXPANDING POLAR CAP MODEL

J. J. Moses

Space Science Laboratory, The Aerospace Corporation, P.O. Box 92957, Los Angeles, California 90009

G. L. Siscoe

Department of Atmospheric Science, UCLA, Los Angeles, California 90024

Abstract. We present calculated ionospheric convection patterns that result from dayside magnetic merging when open and closed flux tubes are cut. In the model a boundary representing the boundary between open and closed field lines expands as newly open flux enters through a gap on the dayside. The gap which represents the dayside merging region mapped to the ionosphere is extended into the polar cap to represent open field line merging. It has been suggested that it is necessary to extend the ionospheric image of the merging line into the polar cap to account for sunward flows observed there. We show how the length and the merging potential associated with the extended merging line affects the resulting flow patterns. We compare the results to relevant published data and find that an extended merging line indeed induces sunward polar cap flow, but the observed sunward flow is better explained by a day-night ionospheric conductivity gradient.

Introduction

Taking the polar cap to be the region containing open magnetic flux, Siscoe and Huang (1985) showed that the usual symmetrical two-celled convection pattern in the high-latitude ionosphere results from pure dayside merging when the polar cap expands to accommodate the newly opened magnetic flux. In their model, the polar cap boundary was a circle with a gap on its dayside through which newly merged flux entered. The gap represented the ionospheric image of the merging line located on the dayside magnetopause.

To accommodate the reported effects of IMF B_y on high-latitude convection, Moses et al. (1987) modified the above model by deforming the polar cap boundary into a partial spiral, closed on the dayside by a merging gap running north and south. The flux thus entered the polar cap deflected eastward or westward, as desired, however, the initial deflection decayed so quickly with distance into the polar cap that they could not reproduce published patterns showing the B_y effect. But by adding the natural decrease in the ionospheric conductivity from day to night across the polar cap, they found a distortion in the convection pattern that reproduced well the published convection flow data. The model-produced patterns have an antisymmetry not commonly expected for the negative B_y case.

The Moses et al. (1978) study has therefore led to the possibility that away from the merging gap, the convection pattern attributed to the B_y effect exists only for the positive B_y case and is caused by the day-night gradient in ionospheric conductivity rather than by the IMF B_y. This possibility will be confirmed, rejected, or modified by thoroughly comparing computed flow patterns against unpublished ionospheric flow and electric field data taken when B_y was negative.

Our purpose here is to determine whether the putative B_y effect can be imposed on the expanding polar cap model by extending the merging gap into the polar cap, as suggested by Reiff and Burch (1985). According to Reiff and Burch (1985), a closed circulation within the polar cap, and therefore on open field lines, called a lobe cell, is driven by merging on open field lines tailward of the polar cusp. In their model, merging on open field lines occurs by a tailward extension of the dayside merging line; thus open and closed field line merging occurs together. In our model, the extended merging line (EML) on the magnetopause itself extends the merging gap into the polar cap. The EML accounts naturally for one well-established, B_y -related phenomenon: the Svalgaard-Mansurov effect, which is otherwise absent in the present expanding polar cap model. The Svalgaard-Mansurov effect is localized to the dayside portion of the polar cap. The question is whether the EML pattern, which can account for the Svalgaard-Mansurov

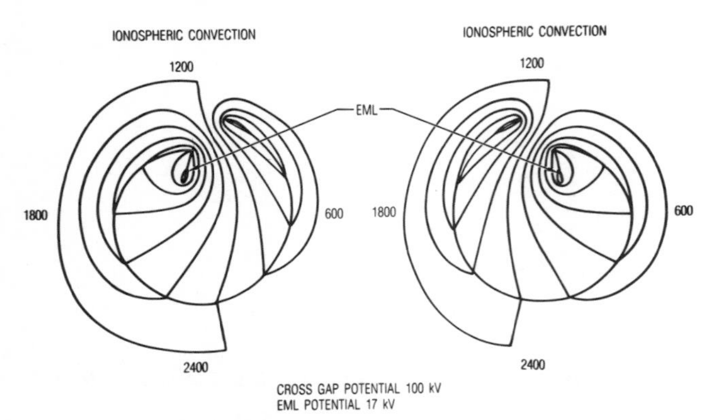

Fig. 1. Ionospheric convection pattern for $B_y > 0$ (left) and $B_y < 0$ (right). The polar cap boundary and extended merging lines are drawn in. The effect of the extended merging line is to exclude flow from the region around it. A circular B_y-dependent flow pattern results around the extended merging line. The potential of the extended merging line is 17 kV and the potential across the gap is 100 kV.

effect, extends far enough to explain the convection pattern distortions attributed to the B_y effect.

Procedure

When the IMF B_z component is southward, the polar cap is observed to expand. Like Siscoe and Huang (1985) and Moses et al. (1987), we assume that the boundary between open and closed field lines expands uniformly. Uniform expansion requires a constant electric field component tangent to the boundary. Our boundary is a spiral rather than a circle, a geometry which allows flux to enter the boundary gap at an angle consistent with the B_y-dependent polar cap flows. We assign potential values across the gap and distribute the potential linearly around the boundary. We impose a potential value across the gap proportional to the dayside merging rate and distribute the potential linearly around the boundary for uniform expansion. For a given potential, the flow velocity into the polar cap is fixed by the gap's width. We solve the current continuity equation in a plane for the above boundary conditions. Our model shows flow patterns for a steady merging rate.

To model the effect of Reiff and Burch's EML, we draw the boundary into the polar cap (Figure 1). This effectively lengthens the merging line (represented by the gap) into the open field region. A potential is arbitrarily assigned to the EML. A rough estimate of the EML's potential can be deduced by looking at published ionospheric potential and current patterns (Heelis et al., 1983; Friis-Christensen et al., 1985; Reiff and Luhmann, 1986). The EML length in the calculations here is comparable to that pictured in Burch et al.'s (1985) model.

Results

Figure 1 shows model results for $B_y \lessgtr 0$ with the EML added. Conductivity is uniform in these calculations. Potential contours encircle the EML and create an island where entering flow avoids the gap. The potential imposed across the gap is 100 kV. The EML tip potential is 17 kV, a typical value picked from potential contours shown in Friis-Christensen et al. (1985) for $B_y \lessgtr 0$ and $B_z < 0$. The effect of the EML appears localized to the dayside ionosphere.

We compare these solutions and DE 2 ion flow data from Burch et al. (1985) in Figures 2 and 3. Figure 2 shows the $B_y > 0$ case and includes the boundary and extended merging line in our results. In both cases inside the morning convection reversal, strong flow moves toward the

Fig. 2. A cut from our model pictured in Figure 1 for $B_y > 0$ and DE 2 ion flow data from Burch et al. (1985) when IMF B_y was strongly positive for several hours.

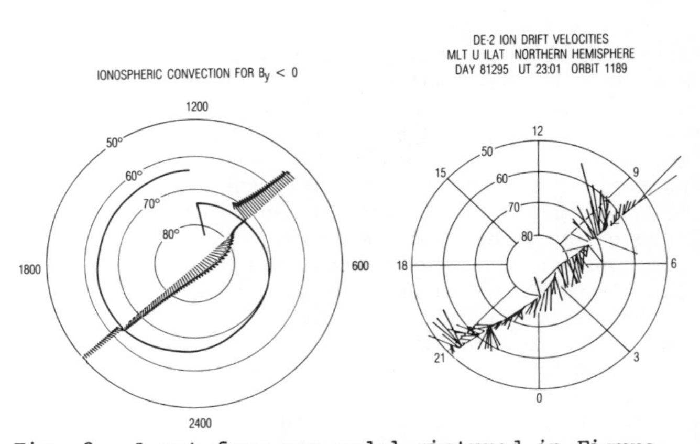

Fig. 3. A cut from our model pictured in Figure 1 for $B_y < 0$ and DE 2 ion flow data from Burch et al. (1985) when IMF B_y was negative for several hours.

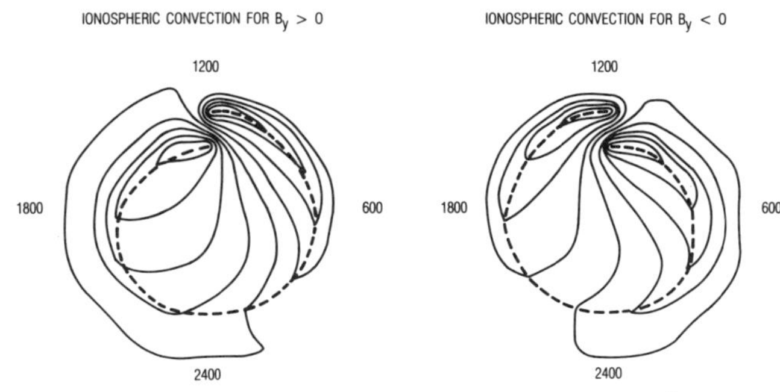

Fig. 4. Ionospheric flow patterns for $B_y \lessgtr 0$ when there is a day-night ionospheric conductivity gradient and no extended merging line. The terminator stretches from 1800 to 600 hours. The conductivity gradient concentrates the electric field towards dawn and causes flow to curve back toward evening.

nightside, and in the evening polar cap the flow rotates toward the evening boundary. However, the data show an evening sunward flow component missing in our results.

Figure 3 shows the $B_y < 0$ comparison. Inside the evening convection reversal, the flow in the data and our model goes into the nightside. In our model, the 2100- to 900-hour cut passes close to the EML and picks up the flow rotation associated with it. The data show no rotation in the morning.

For further comparison, Figure 4 includes a day-night ionospheric conductivity gradient and shows the flow pattern for $B_y \gtrless 0$ and $B_y < 0$ when there is no EML. The terminator stretches from 1800 to 600 hours. The method for adding a day-night ionospheric conductivity gradient to the model is also given by Moses et al. (1987). The conditions are equinox. The conductivity falls off slower in front of the terminator and faster behind it across the polar cap. Conductivity gradient parameters are determined from electron densities published by Singh et al. (1985). The conductivity gradient concentrates the electric field toward dawn and causes flow to curve toward the evening. This pattern produces an anti-symmetry not reported in data (Heppner, 1972; Burch et al., 1985; Heelis et al., 1983) or predicted by theory (Crooker, 1979; Reiff and Burch, 1985) for the $B_y < 0$ case.

Figure 5 contains the same data as Figure 2 and a cut through the $B_y > 0$ solution as in Figure 4. The agreement between the data and the model is good. Inside the morning convection reversal, strong flow goes toward the nightside boundary. Both cases show weaker flow and a sunward component to the flow in the evening. The data are for a day near equinox.

In Figure 6 the data again come from a day near equinox. The $B_y < 0$ solution (Figure 4) here is compared to the same data as in Figure 3. Inside the morning convection reversal, strong flow heads toward the nightside in both cases; then the flow weakens and turns toward the evening boundary.

Discussion

We chose an EML length corresponding to the length presented in Burch et al.'s (1985) model. Actually the EML may be shorter. If one maps a merging region tailward of the cusp on the magnetopause to the ionosphere, one finds a very small area in the polar cap. Conversely, a long merging line within the polar cap would connect to supersonic magnetosheath flow out of communication with the ionosphere. Shortening the EML would further localize its flow.

We presented results for an average EML potential of 17 kV. Friis-Christensen (1983) showed potential and ionospheric current contours derived from ground magnetometer data for disturbed conditions and $B_y \gtrless 0$. These contours show that EML potentials ~40 kV may be present during disturbed times. A greater EML potential than we presented would produce a larger effect on polar

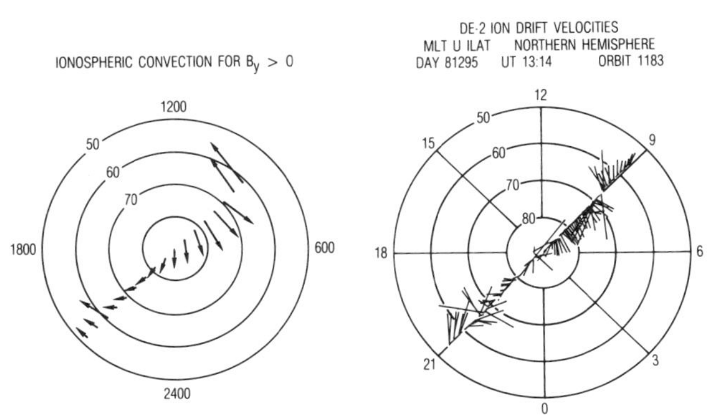

Fig. 5. A cut from our model with the day-night ionospheric conductivity gradient and the same data as in Figure 2 for $B_y > 0$.

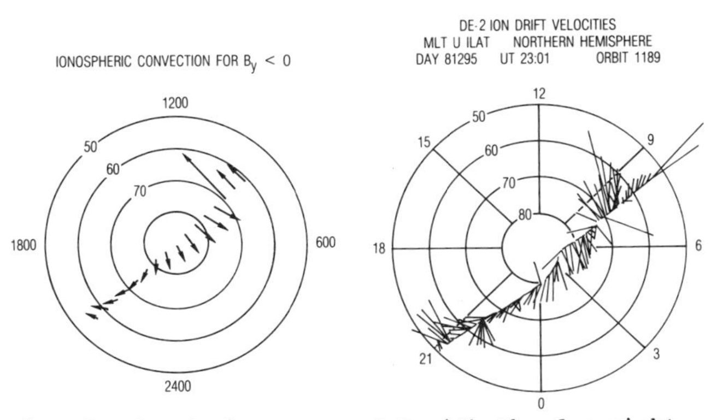

Fig. 6. A cut from our model with the day-night ionospheric conductivity gradient and the same data as in Figure 3 for $B_y < 0$.

cap convection. A smaller potential would reduce the effect.

As shown in the Figure 3 comparison, adding the EML to the polar cap produces sunward flow inside the polar cap on the dawnside. This sunward flow is not present in the data shown here for negative B_y. We conclude that in those cases either the EML is very short with a small potential or nonexistent. Also, particle signatures within the polar cap may provide clues to the EML location which can deviate from those pictured by Burch et al. (1985) (P. H. Reiff, personal communication, 1987). Figure 6 from Moses et al. (1987) shows a comparison between the model with a day-night ionospheric conductivity gradient and the same data as in Figure 3. The solution with the conductivity gradient and no EML seems to model well the morning quadrant in the data. No rotation toward morning occurs in either pattern, only flow toward the nightside.

The data for $B_y > 0$ and the model agree well in the morning quadrant in Figure 2. However, in the evening a sunward flow component occurs in the data but not in the model. In Figure 5, comparing the data and the model with the added day-night conductivity gradient shows flow with the sunward component in the evening for both the model and the data. In this case, the conductivity gradient adds curvature to the flow, a shape which fits the data better. There are no published data that show flow vectors in the afternoon quadrant of the polar cap where the EML is located for $B_y > 0$. For this case the polar cap flow is also sunward in our model.

Conclusion

We have incorporated an extended merging line into the expanding polar cap model to produce a circular B_y-dependent flow pattern. For an EML length taken from Burch et al. (1985) and an average potential determined from Friis-Christensen et al. (1985), sunward flow is obtained on the dayside polar cap near the boundary for $B_y \lessgtr 0$. This flow pattern is absent from the $B_y < 0$ data presented. Instead, we are forced to conclude that the EML is very short and has a small potential or does not exist. In both sets of flow vectors studied, the inclusion of a day-night ionospheric conductivity gradient effectively models polar cap flows. The possibility that a day-night gradient can account for sunward flow observed in the polar cap will be confirmed, rejected, or modified by comparing computed flow patterns with unpublished ionospheric flow and electric field data.

References

Burch, J. L., P. H. Reiff, J. D. Menietti, R. A. Heelis, W. B. Hanson, S. D. Shawhan, E. G. Shelley, M. Sugiura, D. R. Weimer, and J. D. Winningham, IMF B_y-dependent plasma flow and Birkeland currents in the dayside magnetosphere, 1, Dynamics Explorer observations, J. Geophys. Res., 90, 1577, 1985.

Crooker, N. U., Dayside merging and cusp geometry, J. Geophys. Res., 84, 951, 1979.

Friis-Christensen, E., Polar cap current systems, in Magnetospheric Currents, Geophys. Monogr. Ser., vol. 28, edited by T. A. Potemra, pp. 86-95, AGU, Washington, D.C., 1983.

Friis-Christensen, E., Y. Kamide, A. D. Richmond, and S. Matsushita, Interplantetary magnetic field control of high-latitude electric fields and currents determined from Greenland magnetometer data, J. Geophys. Res., 90, 1325, 1985.

Heelis, R. A., J. C. Foster, O. de la Beaujardière, and J. Holt, Multistation measurements of high latitude ionospheric convection, J. Geophys. Res., 88, 10,111, 1983.

Heelis, R. A., The effects of interplanetary magnetic field orientation on dayside high-latitude ionospheric convection, J. Geophys. Res., 89, 2873, 1984.

Heppner, J. P., Polar cap electric field distributions related to the interplantetary magnetic field direction, J. Geophys. Res., 77, 4877, 1972.

Moses, J. J., G. L. Siscoe, N. U. Crooker, and D. J. Gorney, IMF B_y and day-night conductivity effects in the expanding polar cap convection model, J. Geophys. Res., 92, 1193, 1987.

Reiff, P. H., and J. L. Burch, IMF B_y-dependent plasma flow and Birkeland currents in the dayside magnetosphere, 2, A global model for northward and southward IMF, J. Geophys. Res., 90, 1595, 1985.

Reiff, P. H., and J. G. Luhmann, Solar wind control of the polar-cap voltage, in Solar Wind-Magnetosphere Coupling, edited by Y. Kamide and J. Slavin, pp. 453-476, Terra Scientific Publ. Co., Tokyo, Japan, 1986.

Singh, M., P. Rodriguez, and E. P. Szuszczewicz, Spectral classification of medium-scale high latitude F region plasma density irregularities, J. Geophys. Res., 90, 6525, 1985.

Siscoe, G. L., and T. S. Huang, Polar cap inflation and deflation, J. Geophys. Res., 90, 543, 1985.

LARGE-SCALE INSTABILITIES AND DYNAMICS OF THE MAGNETOTAIL PLASMA SHEET

J. Birn

Los Alamos National Laboratory, Los Alamos, New Mexico 87545

K. Schindler

Ruhr-Universität Bochum, 4630 Bochum 1, FRG

Abstract. We review the stability properties of the magnetotail current sheet against large-scale modes in the framework of ideal MHD, resistive MHD, and collisionless Vlasov theory. It appears that the small deviations from a plane sheet pinch (in particular, a magnetic field component normal to the sheet) are important to explain the transition of the tail from a quiet stable state to an unstable dynamic state. It is found that the tail is essentially stable in ideal MHD, but unstable in resistive MHD, while both stable and unstable configurations are found within collisionless theory. The results favor an interpretation where the onset of magnetotail dynamics leading to a sudden thinning of the plasma sheet and the ejection of a plasmoid is caused by the onset of a collisionless instability that either directly leads to the growth of a collisionless tearing mode or, via microscopic turbulence, leads to the growth of a resistive mode. The actual onset conditions are not fully explored yet by rigorous methods. The onset may be triggered by local conditions as well as by boundary conditions at the ionosphere or at the magnetopause (resulting from solar wind conditions).

1. Introduction

The geotail plasma sheet is certainly the best-studied example of a current sheet in space plasmas. Since current sheets, their formation, and breakup are usually considered as being eminently important for a variety of dynamic eruptive processes in space as well as in laboratory plasmas, it is worthwhile to study the stability properties of this best-known example in considerable detail. While the general existence of the tail current sheet is a manifestation of its overall stability, its occasional unstable behavior is probably most clearly demonstrated by the sudden and drastic thinning that occurs close to the onset of a substorm expansive phase and by the later thickening associated with the onset of substorm recovery (Hones et al., 1984b). The statistic evidence, as well as individual current sheet observations, clearly shows that the observed effects near substorm onset are due to thinning to probably less than one-tenth of the average sheet thickness during quiet times (e.g., Fairfield et al., 1981; McPherron and Manka, 1985). The time scale of this process is apparently not directly related to an external change. The thinning and later thickening is observed at distances from around 15 R_E, covered by the VELA and ISEE satellites, up to at least 35 R_E downtail as covered by the IMP satellites. At much larger distances of around 200 R_E as explored by ISEE 3, the consequences of the near-earth thinning seem to be the passage of a thick plasmoid, which is considered as a severed part of the near-earth plasma sheet (Hones et al., 1984a), roughly 30 min after substorm onset. These observations seem to clearly suggest the existence of a large-scale instability involving a major portion of the tail current sheet and possibly related to changes in magnetic topology which would allow for the severance of a part of the plasma sheet that was originally magnetically connected with the earth. In this paper we will review our present knowledge of the stability properties of the tail current sheet concentrating on large-scale modes. After a short presentation of stability tools in section 2, we will start in section 3 with instabilities of a one-dimensional sheet pinch, because the tail current sheet resembles such a pinch to a high degree. We will see, however, that in particular the transition from stability to instability and the dynamic changes in the nonlinear evolution of the tail cannot be understood without taking the more realistic two-dimensional (or even three-dimensional) geometry into account. This will be discussed in section 4. While these two sections basically follow a classical stability analysis disregarding boundary effects, section 5 will be concerned with modifications primarily associated with boundaries but also with details of the plasma population that can possibly lead to triggering or enhancement of the growth of the instabilities. We would like to stress that we do not intend to give a complete review of all relevant papers but rather will illuminate typical results by selected examples of the literature.

2. Stability Tools

Linear stability properties are usually studied by variational approaches or by a classical perturbation method studying small perturbations of an equilibrium or quasi-equilibrium. By quasi-equilibrium we mean configurations that may change slowly (as compared to the growth of the instabilities of interest) due to either external forces or an internal process such as diffusion. This concept is particularly useful within resistive magnetohydrodynamics where the requirement of an exact equilibrium

$$\nabla \times \mathbf{E} = \nabla \times (\eta \mathbf{j}) = 0$$

is often too restrictive (e.g., Barston, 1972). The nonlinear evolution is most often studied by a numerical integration of the dynamic equations. We do not want to go into detail about linear perturbation theory and numerical integration. Both methods are straightforward in principle, although they may become very tricky in details. We would like, however, to make a few comments about the existing variational methods because they seem to shed some light on an underlying free energy concept and on different stabilizing effects due to certain restrictions on particle or fluid motion. Variational principles have been derived within the framework of ideal MHD (Bernstein et al., 1958), resistive MHD (Tasso, 1975), and Vlasov theory (Laval et al., 1966; Schindler, 1966; Schindler et al., 1973). With the exception of the ideal MHD criterion, these principles are restricted to a spatial dependence on two coordinates only, both for the equilibrium and the perturbations. We will, therefore, restrict our discussion to this case keeping in mind that the derived instability thresholds may be sufficient for instability only. The interesting feature of these variational principles is that they can all be written in the form

$$\delta^2 W = \delta^2 F + \delta^2 Q > 0 \text{ necessary for stability,} \quad (1)$$

where $\delta^2 F$ is common for all cases while $\delta^2 Q$ is different for the different models, but generally positive (Schindler et al., 1983). The term $\delta^2 F$ is the second variation of a functional F given by

$$F = \int \left[\left(\frac{\nabla A}{2\mu_o}\right)^2 - P(A) \right] d^2 r \ , \quad (2)$$

where A is the flux function (y-component of the vector potential) of the two-dimensional magnetic field given by $\mathbf{B} = \nabla A(x,z) \times \mathbf{y}$, and P is the plasma pressure using a constraint that keeps the function P(A) fixed (for more details, see also Schindler and Birn, 1987). One finds

$$\delta^2 F = \frac{1}{2\mu_o} \int \left[|\nabla A_1|^2 - \mu_o \frac{dJ_o}{dA_o} |A_1|^2 \right] dxdz \ , \quad (3)$$

where A_1 is the first-order perturbation of A and J_o the zero-order current density connected with the zero-order flux function, A_o, and the pressure function through the equilibrium condition

$$J_o = -\frac{1}{\mu_o} \nabla^2 A_o = \frac{dP(A_o)}{dA_o} \ . \quad (4)$$

Equation (4) can be derived from the first variation of (2). For the explicit form of the additional positive terms $\delta^2 Q$, we refer to Schindler et al. (1983). A dynamic model that keeps the pressure fixed on perturbed field lines (as identified by the value of A) is the most unstable one, while other constraints add stabilizing terms. In the case of ideal MHD this most unstable case corresponds to isobaric variations where the pressure stays constant in a moving fluid element.

3. Stability of a Plane (One-Dimensional) Current Sheet

The stability of a plane sheet pinch within ideal MHD is well known (a proof is given, for example, by Schindler et al., 1983; unfortunately we do not know the first published proof). It is the major reason why the magnetotail plasma/current sheet keeps existing to distances well beyond 200 R_E. Nonideal effects generally destabilize the current sheet. The linear instability of the plane sheet pinch has been first shown within resistive MHD by Furth et al. (1963), within collisionless Vlasov theory by Furth (1962), and later through variational approaches by Laval et al. (1966) and Schindler (1966). The most relevant unstable modes, called tearing modes, are symmetric with respect to the symmetry plane of the sheet pinch. Because nonideal effects are important only in a thin layer around the symmetry plane (where $\mathbf{B} = 0$ or at least $\mathbf{k} \cdot \mathbf{B} = 0$, where $\mathbf{k}$ is the wave vector parallel to the sheet), the modes are essentially identical in resistive MHD and collisionless theory. They produce localized thinning and the growth of magnetic islands with fast flow from the thinning regions into the island regions. This is illustrated in Figure 1 which shows results from a resistive MHD simulation by Biskamp (1982). The thinning regions are characterized by their x-type magnetic field topology (for $t \gtrsim 400$; or by the flat sheet-like structure at $t \approx 200$).

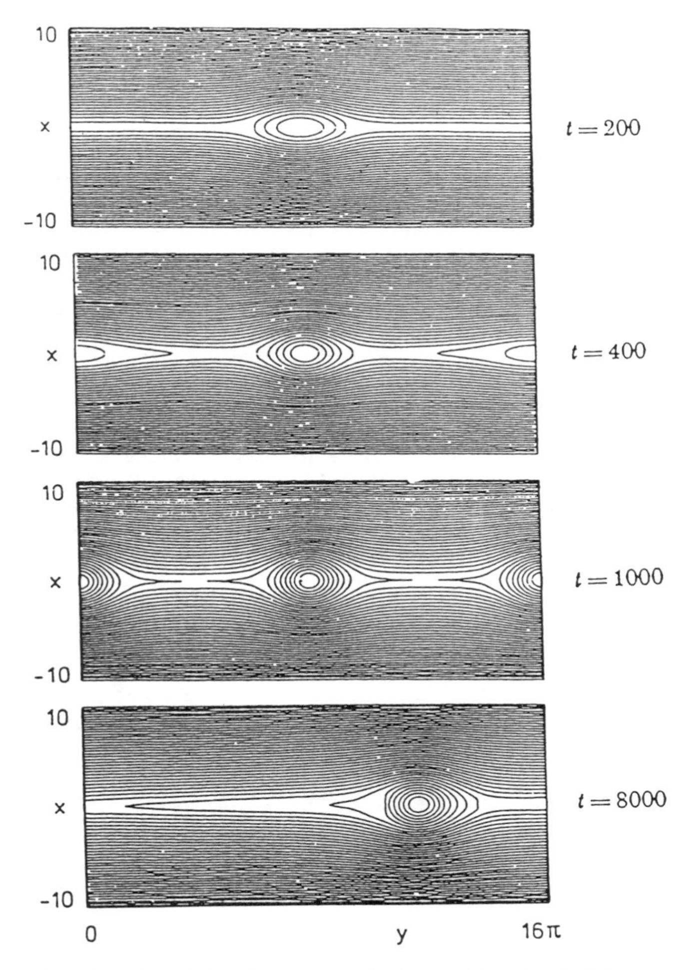

Fig. 1. Results from a resistive MHD simulation by Biskamp (1982) showing the time evolution of the magnetic field configuration.

At this point we may also stress two important stability limits valid for both the collisional (i.e., resistive) and the collisionless regime. Tearing modes are stable for $kL > 1$, where L is the characteristic scale of the current sheet [defined, for example, by a variation of the main magnetic field component $B_x \propto \tanh(z/L)$], and wall stabilization becomes important if a solid conducting wall is present parallel to the sheet at a distance close to L. These conditions impose certain minimum requirements on the size of numerical simulation boxes if unstable tearing modes are to be treated.

The instability of the collisionless sheet pinch was used by Coppi et al. (1966) to explain auroral activity. The problem with this interpretation, however, is that it does not explain the transition from stability to instability because the one-dimensional sheet pinch is always unstable within collisionless Vlasov theory. One, therefore, has to look for ways to stabilize these effects and possibly overcome them. One line of thought that involves the more refined geometry of the sheet, including a small normal magnetic field component, will be discussed in the following section. A possible alternative is to look for the nonlinear evolution of the tearing instabilities and saturation mechanisms and ways to possibly overcome them.

The nonlinear growth of resistive tearing modes has been studied with analytical and numerical methods. While shorter wavelength modes may saturate at a relatively low level (White et al., 1977) or exhibit a reduced (algebraic instead of exponential) growth (Rutherford, 1973), larger wavelengths are found to grow to finite size with magnetic island widths comparable to the sheet thickness (e.g., Biskamp, 1982, see Figure 1; Steinolfson and Van Hoven, 1984).

Nonlinear results within the collisionless regime seem less certain. Numerical simulations suffer from the fact that electron contributions are either neglected or are treated within unrealistic parameter regimes characterized by high electron/ion mass ratios, a limited sheet thickness of a few electron Debye length, and integration times of hundreds of electron plasma oscillation times only.

Using analytic estimates, Drake and Lee (1977) predicted the transition of collisionless tearing break modes into a semicollisional regime characterized by electron diffusion, where the growth becomes algebraic similar to the resistive results of Rutherford (1973). Their approach, however, also excludes the very long wavelengths. A more speculative approach by Galeev et al. (1978) predicts the transition into an explosive growth. A similar result was found from numerical simulations by Terasawa (1981) considering, however, only one particle component, while Katanuma and Kamimura (1980) have found only an algebraic growth.

From the linear and nonlinear stability results for a plane one-dimensional current sheet, it is not obvious how the transition from a quiet stable plasma sheet to a dynamic unstable state can be accomplished. However, promising models can be based on the results discussed in the following section, which include the effects of the actual geometry of the geotail.

4. Stability of the (Two- or Three-Dimensional) Magnetotail Current Sheet

Although the magnetotail plasma sheet resembles a plane current sheet to a good degree of approximation, the slight differences are apparently quite important for its stability properties and dynamic behavior. These properties are demonstrated by Figure 2 which shows magnetic field lines of a self-consistent equilibrium model of the quiet tail similar to those obtained by Birn (1979). The most important difference from a plane current sheet, as far as stability is concerned, is the presence of a magnetic field

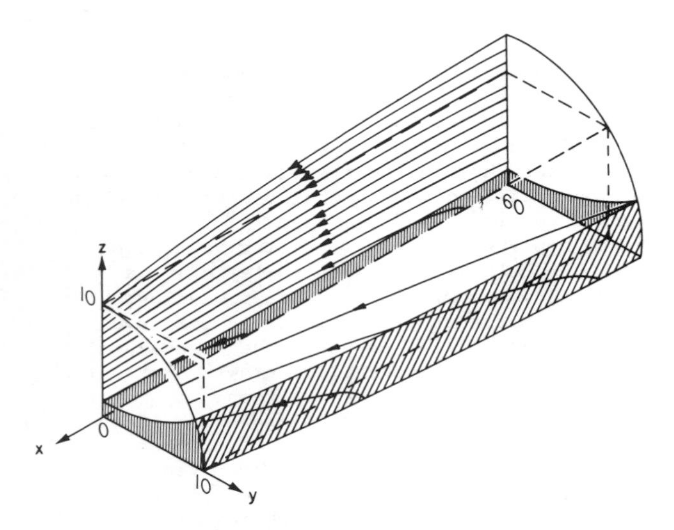

Fig. 2. Self-consistent magnetotail equilibrium showing magnetic field lines in the northern dusk sector of the tail and the width of the plasma/current sheet (hatched region). Length values are normalized by the scale length L_z equal to the characteristic half-thickness of the plasma/current sheet at the near earth (left) side at local midnight (y = 0).

component B_z perpendicular to the sheet, which is positive in the center of the sheet and becomes negative in the lobes producing closed field lines within the plasma sheet and flaring of the lobes. While the presence of $B_z \neq 0$ at the center of the plasma sheet has an important stabilizing effect, as we will discuss later, the change of the sign of B_z near the plasma sheet/lobe boundary is in fact a feature necessary to allow for instability. It is the consequence of a general criterion that a (two-dimensional) plasma configuration is stable, if any Cartesian magnetic field component can be found which does not change sign within a closed area (Schindler, 1970; Birn et al., 1975). The presence of $B_z \neq 0$ is closely related to variations with x along the tail axis, since the $j_y \times B_z$ force is typically balanced by a pressure gradient $\partial p/\partial x$.

Other features that involve a y dependence across the tail are a flaring of field lines in that direction also and the plasma sheet thickening toward the tail flanks, which is also related to an increase of B_z with $|y|$.

No rigorous ideal MHD stability analysis has been done for such a two- or three-dimensional tail configuration. The numerical MHD simulations of Birn (1980) and Birn and Hones (1981), however, indicate stability, as in the one-dimensional case.

A stability analysis within resistive MHD, similar to that of Furth et al. (1963), has been performed by Janicke (1980) for a two-dimensional (i.e., x and z dependent) quasi-equilibrium. In the wavelength regime $L_z \ll \lambda \ll L_x$, where L_z and L_x denote the characteristic length scales for variations of the equilibrium with z and x, respectively, he found no change of the tearing growth rates although the eigenmodes became affected by the presence of B_z. The nonlinear evolution in the resistive regime was studied via numerical simulation by Birn (1980) in the two-dimensional case and by Birn and Hones (1981) in the three-dimensional case. The transition from stability to instability was accomplished by turning on a resistive term representing a sudden occurrence or increase of (anomalous) resistivity. The growth of tearing instabilities developed out of an initial slow diffusion. These simulations produced features such as plasma sheet thinning in the near tail and the formation and tailward ejection of a plasmoid, whose size was limited in the y direction, consistent with observations and empirical substorm models (e.g., Hones, 1979). Results from Birn and Hones (1981) are shown in Figures 3 and 4. It appears that

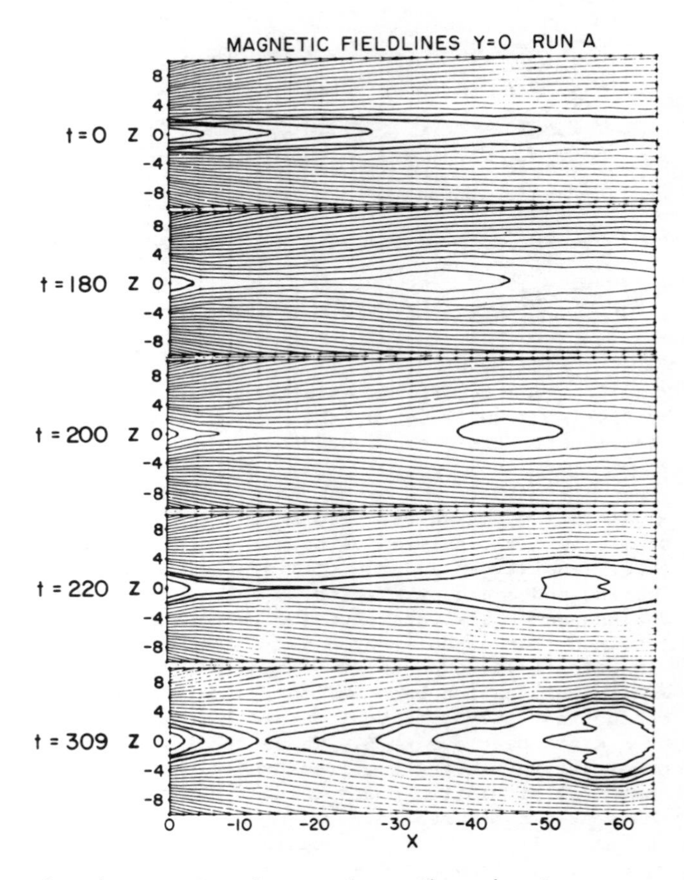

Fig. 3. Results from a three-dimensional resistive MHD simulation of magnetotail dynamics by Birn and Hones (1981) showing magnetic field lines in the midnight meridian plane. Length values are normalized as in Figure 2.

the pressure gradients in the x direction, though generally small, play an important role in the generation of the fast tailward flow associated with the ejection of the plasmoid.

While a resistive model may be applicable to the unstable evolution, a discussion of the stability transition requires collisionless theory. Schindler (1974) first considered the role of the normal component in a stability analysis of two-dimensional equilibria. He found a transition from stability to instability when ions became non-adiabatic in the center of the plasma sheet with a stability criterion involving the product $B_z L_z^{5/2}$. Destabilization thus occurs not only by a reduction of B_z but also by a compression of the plasma sheet (i.e., a reduction of L_z). Schindler's analysis, however, neglected the stabilizing effect from electron contribution to the term $\delta^2 Q$ in equation (1). A more detailed, yet still not fully rigorous, treatment by Galeev and Zelenyi (1976) confirmed Schindler's results but also gave a more restricted unstable regime which required a scale length L_z below a few

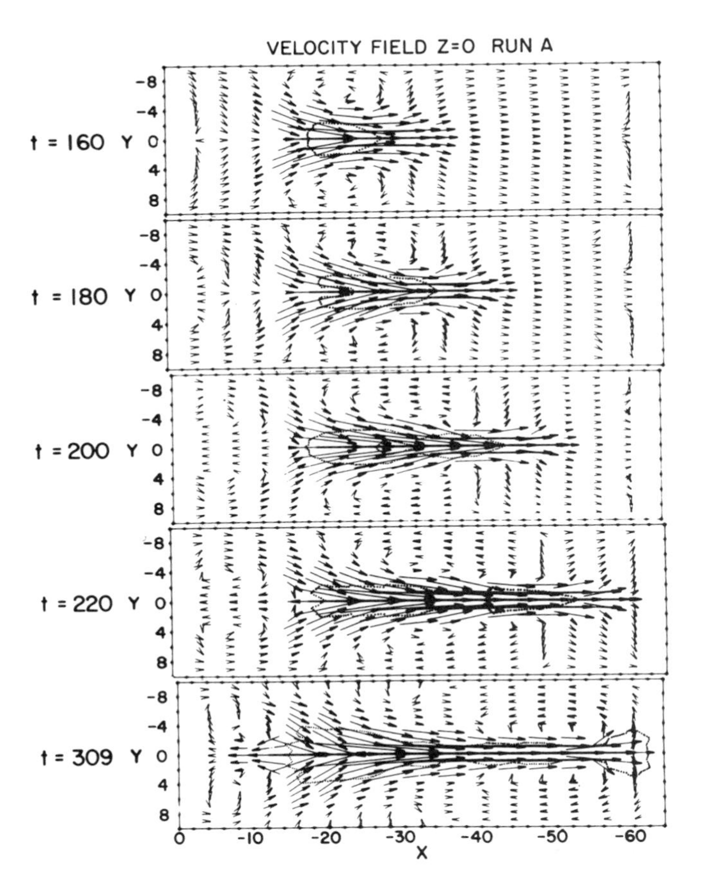

Fig. 4. Velocity vectors and magnetic neutral lines (dotted lines) in the equatorial plane in the three-dimensional computer simulation of Birn and Hones (1981).

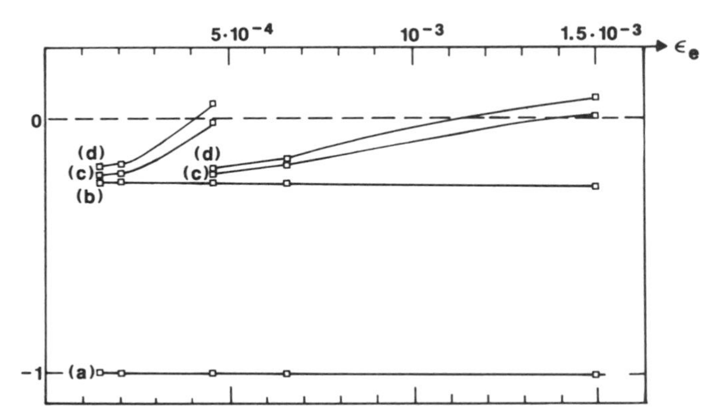

Fig. 5. Results from a numerical analysis of tail stability using Vlasov theory by Goldstein and Schindler (1982) showing different contributions to the energy integral in equation (1) as a function of the normalized electron Larmor radius ε_e. (a) The term $\delta^2 F$, normalized to 1; (b) the sum of (a) and the ion contribution to $\delta^2 Q$; (c) the sum of (b) and the electron contribution; and (d) the total sum $\delta^2 W$ including also an electrostatic potential term.

times the ion Larmor radius. A rigorous WKB approach in the wavelength regime $L_z \ll \lambda \ll L_x$ predicted general stability (Lembège and Pellat, 1982). Goldstein and Schindler (1982), however, showed by numerical evaluation of the full variational integral including all stabilizing terms that unstable solutions exist in the regime where $\lambda \approx L_x$, i.e., for large wavelengths of the order of tens of earth radii. Figure 5, taken from that paper, shows that a stability transition exists even in the regime where electrons are adiabatic (left parts of curves c and d). The actual stability properties result from several contributions to the energy principle (1). The term $\delta^2 F$ is negative for any reasonable plasma sheet configuration which is a necessary requirement for an instability to exist. Stabilizing effects that appear in different contributions to $\delta^2 Q$ are due to restrictions of the accessible phase space for each particle species and to electrostatic potentials resulting from charge separation effects. The configuration is unstable if the minimum of $\delta^2 W = \delta^2 F + \delta^2 Q$ becomes negative. The finite element representation of trial functions in the variational integral was somewhat coarse such that the transition parameters may not be very accurate. A more refined analysis, however, can only expand the unstable region.

The results from numerical particle simulations starting from two-dimensional equilibrium (Hamilton and Eastwood, 1982; Swift, 1983) are not yet conclusive. They suffer not only from the limitations in space and time, mentioned in the previous section, but also from the fact that initial configurations did not include the flaring of the lobes which is necessary for instability of a closed system.

5. Refinements

The results discussed in the previous two sections were based essentially on the classical (linear or nonlinear) stability concept where it is assumed that perturbations of the equilibrium (or quasi-equilibrium) vanish at the boundaries of a given box. Since the magnetotail is obviously not enclosed by a solid box and it is impossible to treat the infinite problem, it is important to look for refinements which result from widening the boundary conditions. Variational principles typically are no longer valid for generalized boundary conditions. We must, therefore, rely largely on numerical results based on some ad hoc assumptions and on more speculative arguments.

Two ways are suggested in which the ionosphere might influence the onset and/or the growth rate of a collisionless tearing mode. (1) The stabilizing space charges in the magnetotail could be discharged by field-aligned currents closing through the ionosphere (Goldstein and Schindler, 1978; Swift, 1986). This would imply a control of ionospheric resistivity over the onset of a tail tearing mode and a close relation of the occurrence of sheets of field-aligned currents with this onset. (2) The presence of heavy ions (e.g., oxygen of ionospheric origin) can lower the threshold for the onset of an ion tearing mode (Baker et al., 1982). Since O^+ ions are pulled out from the ionosphere preferably in the pre-midnight sector (Sharp et al., 1981), this can be related to an east-west asymmetry of the occurrence of substorms. A self-consistent treatment of ionospheric effects on the dynamics of the magnetotail within a realistic model has not been done. The problems that one encounters are illustrated by the fact that there is not even a satisfactory equilibrium model that covers both the inner magnetosphere and the tail.

The most important role of the solar wind boundary conditions can be ascribed to a growth phase of magnetospheric substorms (McPherron, 1970). The gradual deformation of the tail (e.g., stretching and compression of the plasma sheet resulting from addition of magnetic flux from the front side magnetosphere to the tail) either directly forces the tail plasma sheet beyond a certain stability threshold (as modeled within resistive MHD by Sato and Hayashi, 1979, or Ugai, 1980) or brings it into a metastable state (Schindler and Birn, 1987) which might be subject to unstable ion tearing modes but is stabilized by electron effects. A destabilization and the onset of the ion tearing mode could then result from ionospheric discharging or an occurrence of fluctuations scattering the electrons anywhere along the field lines (Coroniti, 1980). After the onset of a tearing instability the influence of driving fields at the boundary is probably less important. Computer results within resistive MHD show that the major effect is to modify the tearing growth by changing the relevant scales such as plasma sheet width and a typical Alfvén wave travel time (Birn and Schindler, 1986). Continuous driving, however, may also be responsible for repetitive ejection of plasmoids (Lee et al., 1985).

In addition to boundary effects, one might consider modifications due to local properties of the plasma population. Such properties might, of course, also be related to boundary effects (as for instance, the role of heavy ions discussed above). We mentioned already that the presence of fluctuations, scattering the pitch angles of electrons, can considerably destabilize (Coroniti, 1980). Recently, Büchner and Zelenyi (1986) analyzed the effect of stochastic particle motion and concluded a substantial increase of growth rates and of the unstable wavelength regime. There are several ways how such fluctuations could be generated. An obvious way is the onset of microinstabilities not necessarily in the center of the plasma sheet. An alternative would be the resonant absorption of waves generated, e.g., near the magnetopause (Smith et al., 1986).

Anisotropies of the ion distribution can lead to an increased growth rate of the collisionless tearing mode (Chen et al., 1986; Ambrosiano et al., 1986). The presence of plasma flow (beams) can also influence stability and growth rates. Fast beams near the plasma sheet/lobe interface can generate Kelvin-Helmholtz instability, i.e., ideal MHD surface modes. To produce an effect on the body of the plasma sheet these modes have to penetrate to the center of the plasma sheet and interact with the modes from the opposite surface. Such an interaction seems more likely to lead to an enhanced anti-symmetric flapping mode than to a sausage-type mode although, to our knowledge, a rigorous analysis has not been done. There is, however, a destabilizing effect of the plasma flow on resistive tearing modes (e.g., Paris and Sy, 1983) that also seems to exist in the collisionless limit (G. S. Lakhina and K. Schindler, private communications, 1986).

6. Summary and Discussion

The most dramatic dynamic behavior of the geomagnetic tail is the sudden thinning in the tail region from about 15 R_E to more than 30 R_E which is apparently related to effects in the distant tail identified as plasmoid ejections and also a cause of auroral activity. This dynamic behavior is most likely the result of a large-scale instability that may or may not be triggered externally but proceeds on an intrinsic time scale. Since the tail current sheet is stable within ideal MHD, the instability must involve nonideal effects, e.g., resistivity or particle inertia effects. Its appearance should be that of a tearing mode which involves the formation of magnetic islands (in a more realistic two- or three-dimensional geometry, however, first at an advanced stage of the evolution).

Our present limited knowledge of the stability of current sheets allows the following suggestions for the transition from stability to large-scale instability, which are discussed in several places in the literature. In all cases we must assume that during the quiet stage the collisionless tearing instability is stabilized by the presence of a sufficiently large normal magnetic field component B_z or has saturated at a sufficiently low level not affecting the average large-scale geometry. It is conceivable that the former holds in the near tail up to the distance of the suggested quasi-steady distant neutral line, while the latter is the case beyond that distance. A destabilization might occur through one of the following mechanisms which may be neither exclusive nor complete. (a) A gradual change of magnetotail conditions (e.g., stretching or compression as suggested to be characteristic of the substorm growth phase; McPherron, 1970) leads to the onset of a microinstability which generates turbulence and thereby an effective resistivity that allows a large-scale resistive tearing mode to grow as simulated, e.g., by Birn (1980) and Birn and Hones (1981). The problem with this concept is that an appropriate microinstability has not yet been identified. Current-driven instabilities typically require too narrow a current sheet; the lower hybrid drift instability (Huba et al., 1977) can be exited for thicker sheets, however, near the plasma sheet boundary. It does not penetrate into the center where the resistivity is needed unless the plasma sheet becomes sufficiently thin (of the order of a typical ion Larmor radius of a few hundred kilometers). (b) A gradual stretching of the tail reduces B_z until the threshold for a collisionless tearing instability is exceeded, which grows to finite amplitude. Although such a transition is shown to exist in principle (Goldstein and Schindler, 1982), it is not yet proven rigorously that the parameters at the transition point are realistic nor that the growth after the transition is fast enough. (c) A gradual externally driven process as in (a) and (b) brings the tail current sheet into a stage that is unstable to an ion tearing mode but stabilized by electron effects as discussed in section 4. There are several possibilities for a destabilization of these electron effects. (1) The stabilizing electrostatic potentials could be discharged at the ionosphere (Goldstein and Schindler, 1978; Swift, 1986). This process would closely associate the onset of the unstable evolution with the buildup of sheets of field-aligned currents. (2) A loading of field lines with heavy ions of ionospheric origin could make the onset criterion for an ion tearing mode less stringent (Baker et al., 1982). (3) A sudden onset of fluctuations leading to pitch angle scattering of electrons would also destabilize (Coroniti, 1980; Büchner and Zelenyi, 1986). In contrast to the resistive model (a), this scattering need not happen near the center of the plasma sheet but could occur anywhere along a field line, in particular also near the boundary of the plasma sheet which seems to be more easily subject to microinstabilities due to strong gradients and beaming of particles. The fluctuations may also result from the resonant absorption of an externally generated wave field (Smith et al., 1986) instead of from a local microinstability.

These possibilities show that the instability although relying on nonideal effects in the center of the plasma sheet might actually be triggered not only by conditions in the center but as well by conditions near the plasma sheet/lobe boundary, at the ionosphere, or in the external (magnetopause, magnetosheath) region, i.e., by the solar wind.

So far, we have discussed the transition from stability to instability essentially as a temporal problem, relevant for substorm onset. There is a spatial aspect, however, as well. As ISEE 3 observations have shown (Slavin et al., 1985), the average magnetotail configuration makes a transition from a flaring tail with $B_z > 0$ to a nonflaring one with $B_z \approx 0$, at a distance of about 100 R_E. The nearly one-dimensional average current sheet configuration beyond about 100 R_E should be unstable to collisionless tearing modes, according to results discussed in section 3. The fact that the distant tail at about 200 R_E does not continuously undergo major disruptions indicates that the tearing modes must saturate at a level that does not affect the overall structure. The occurrence of small-scale islands, however, seems quite consistent with the observations of roughly equal amounts of positive and negative B_z.

This discussion, again, shows the important role of B_z, the tail flaring, and the associated variations along the tail for large-scale dynamic effects.

Acknowledgments. This work was supported by the U.S. Department of Energy through the Office for Basic Energy Sciences and by the Deutsche Forschungsgemeinschaft through the Sonderforschungsbereich 162.

References

Ambrosiano, J., L. C. Lee, and Z. F. Fu, Simulation of the collisionless tearing instability in an anisotropic neutral sheet, J. Geophys. Res., 91, 113, 1986.

Baker, D. B., E. W. Hones, Jr., D. T. Young, and J. Birn, The possible role of ionospheric oxygen in the initiation and development of plasma sheet instabilities, Geophys. Res. Lett., 9, 1337, 1982.

Barston, E. M., Stability of non-equilibria of

the resistive sheet pinch, Commun. Pure Appl. Math. , 25, 63, 1972.

Bernstein, I. B., E. A. Friemann, M. D. Konskal, and R. M. Kulsrud, Proc. Roy. Soc. London, A244, 17, 1958.

Birn, J., Self-consistent magnetotail theory: General solutions for the quiet tail with vanishing field-aligned currents, J. Geophys. Res., 84, 5143, 1979.

Birn, J., Computer studies of the dynamic evolution of the geomagnetic tail, J. Geophys. Res., 85, 1214, 1980.

Birn, J., and E. W. Hones, Jr., Three-dimensional computer modeling of dynamic reconnection in the geomagnetic tail, J. Geophys. Res., 86, 6802, 1981.

Birn, J., and K. Schindler, On the influence of an external electric field on magnetotail reconnection, J. Geophys. Res., 91, 8817, 1986.

Birn, J., R. Sommer, and K. Schindler, Open and closed magnetospheric tail configurations and their stability, Astrophys. Space Sci., 35, 389, 1975.

Biskamp, D., Dynamics of a resistive sheet pinch, Z. Naturforsch., 37a, 840, 1982.

Büchner, J., and L. M. Zelenyi, A stochastic mechanism for the triggering of explosive reconnection, Zentralinstitut für Astrophysik Preprint PRE-ZIAP 86-05, Potsdam Babelsberg, 1986.

Chen, J., P. J. Palmadesso, and Y. C. Lee, Magnetic reconnection in a non-Maxwellian neutral sheet, NRL Memorandum Report 5787, 1986.

Coppi, B., G. Laval, and R. Pellat, Dynamics of the geomagnetic tail, Phys. Rev. Lett., 16, 1207, 1966.

Coroniti, F. V., On the tearing modes in quasi-neutral sheets, J. Geophys. Res., 85, 6719, 1980.

Drake, J. F., and Y. C. Lee, Nonlinear evolution of collisionless and semicollisional tearing modes, Phys. Rev. Lett., 39, 453, 1977.

Fairfield, D. M., E. W. Hones, Jr., and C.-I. Meng, Multiple crossings of a very thin plasma sheet in the earth's magnetotail, J. Geophys. Res., 86, 11,189, 1981.

Furth, H. P., The "mirror instability" for finite particle gyro-radius, Nucl. Fusion Suppl., 1, 169, 1962.

Furth, H. P., J. Killeen, and M. N. Rosenbluth, Finite-resistivity instabilities of a sheet pinch, Phys. Fluids, 6, 459, 1963.

Galeev, A. A., and L. M. Zelenyi, Tearing instability in plasma configurations, Sov. Phys. JETP, 43, 1113, 1976.

Galeev, A. A., F. V. Coroniti, and M. Ashour-Abdalla, Explosive tearing mode reconnection in the magnetospheric tail, Geophys. Res. Lett., 5, 707, 1978.

Goldstein, H., and K. Schindler, On the role of the ionosphere in substorms: Generation of field-aligned currents, J. Geophys. Res., 83, 2574, 1978.

Goldstein, H., and K. Schindler, Large-scale collision-free instability of two-dimensional plasma sheets, Phys. Rev. Lett., 48, 1468, 1982.

Hamilton, J.E.M., and J. W. Eastwood, The effect of a normal magnetic field component on current sheet stability, Planet. Space Sci., 30, 293, 1982.

Hones, E. W., Jr., Plasma flow in the magnetotail and its implications for substorm theories, in Dynamics of the Magnetosphere, edited by S.-I. Akasofu, p. 545, D. Reidel Publ. Co., Dordrecht, Holland, 1979.

Hones, E. W., Jr., D. N. Baker, S. J. Bame, W. C. Feldman, J. T. Gosling, D. J. McComas, R. D. Zwickl, J. Slavin, E. J. Smith, and B. T. Tsurutani, Structure of the magnetotail at 220 R_E and its response to geomagnetic activity, Geophys. Res. Lett., 11, 5, 1984a.

Hones, E. W., Jr., T. Pytte, and H. I. West, Jr., Associations of geomagnetic activity with plasma sheet thinning and expansion: A statistical study, J. Geophys. Res., 89, 5471, 1984b.

Huba, J. D., N. T. Gladd, and K. Papadapolous, The lower-hybrid-drift instability as a source of anomalous resistivity for magnetic field line reconnection, Geophys. Res. Lett., 4, 125, 1977.

Janicke, L., Resistive tearing mode in weakly two-dimensional neutral sheets, Phys. Fluids, 23, 1843, 1980.

Katanuma, I., and T. Kamimura, Simulation studies of the collisionless tearing instabilities, Phys. Fluids, 23, 2500, 1980.

Laval, G., R. Pellat, and M. Vuillemin, Instabilities electromagnetiques des plasmas sans collisions, in Plasma Physics and Controlled Nuclear Fusion Research vol. II, p. 259, Int. Atomic Energy Agency, Vienna, 1966.

Lee, L. C., Z. F. Fu, and S.-I. Akasofu, A simulation study of forced reconnection processes and magnetospheric storms and substorms, J. Geophys. Res., 90, 10,896, 1985.

Lembège, B., and R. Pellat, Stability of a thick two-dimensional quasineutral sheet, Phys. Fluids, 25, 1995, 1982.

McPherron, R. L., Growth phase of magnetospheric substorms, J. Geophys. Res., 75, 5592, 1970.

McPherron, R. L., and R. H. Manka, Dynamics of the 1054 UT March 22, 1979, substorm event: CDAW 6, J. Geophys. Res., 90, 1175, 1985.

Paris, R. B., and W.N.-C. Sy, Influence of equilibrium shear flow along the magnetic field on the resistive tearing instability, Phys. Fluids, 26, 2966, 1983.

Rutherford, P. H., Nonlinear growth of the tearing mode, Phys. Fluids, 16, 1903, 1973.

Sato, T., and T. Hayashi, Externally driven magnetic reconnection and a powerful magnetic energy converter, Phys. Fluids, 22, 1189, 1979.

Schindler, K., A variational principle for one-dimensional plasmas, in Proceedings of the Seventh International Conference on Phenomena in Ionized Gases, vol. II, p. 736, Beograd, Yugoslavia, 1966.

Schindler, K., Tearing instabilities in the magnetosphere, in Intercorrelated Satellite Observations Related to Solar Events, edited by V. Manno and D. E. Page, p. 309, D. Reidel Publ. Co, Dordrecht, Holland, 1970.

Schindler, K., A theory of the substorm mechanism, J. Geophys. Res., 79, 2803, 1974.

Schindler, K., and J. Birn, Magnetotail theory, Space Sci. Rev., 44, 307, 1987.

Schindler, K., D. Pfirsch, and H. Wobig, Stability of two-dimensional collision-free plasmas, Plasma Phys., 15, 1165, 1973.

Schindler, K., J. Birn, and L. Janicke, Stability of two-dimensional pre-flare structures, Solar Phys., 87, 102, 1983.

Sharp, R. D., D. L. Carr, W. K. Peterson, and E. G. Shelley, Ion streams in the magnetotail, J. Geophys. Res., 86, 4639, 1981.

Slavin, J. A., E. J. Smith, D. G. Sibeck, D. N. Baker, R. D. Zwickl, and S.-I. Akasofu, An ISEE 3 study of average and substorm conditions in the distant magnetotail, J. Geophys. Res., 90, 10,875, 1985.

Smith, R. W., C. K. Goertz, and W. Grossmann, Thermal catastrophy in the plasma sheet boundary layer, Geophys. Res. Lett., 13, 1380, 1986.

Steinolfson, R. S., and G. Van Hoven, Nonlinear evolution of the resistive tearing mode, Phys. Fluids, 27, 1207, 1984.

Swift, D. W., A two-dimensional simulation of the interaction of the plasma sheet with the lobes of the earth's magnetotail, J. Geophys. Res., 88, 125, 1983.

Swift, D. W., The effect of the ionosphere on the growth of tearing mode instabilities in the magnetotail, J. Geophys. Res., 91, 4256, 1986.

Tasso, H., Energy principle for two-dimensional resistive instabilities, Plasma Phys., 17, 1131, 1975.

Terasawa, T., Numerical study of explosive tearing mode instability in one-component plasmas, J. Geophys. Res., 86, 9007, 1981.

Ugai, M., Spontaneously developing magnetic reconnections in a current-sheet system under different sets of boundary conditions, Phys. Fluids, 25, 1027, 1980.

White, R. B., D. Monticello, M. N. Rosenbluth, and B. V. Waddell, Saturation of the tearing mode, Phys. Fluids, 20, 800, 1977.

DYNAMICS OF THE NEAR-EARTH MAGNETOTAIL - RECENT OBSERVATIONS

L. A. Frank

Department of Physics and Astronomy, The University of Iowa, Iowa City, Iowa 52242

Abstract. Recent observations that are relevant to models of the magnetotail and its coupling to the ionosphere are summarized. In particular observations in the near-earth plasma sheet and magnetotail lobes are reviewed. The two principal current sheets in the plasma sheet are located at the neutral sheet and the lobe-plasma sheet interface. The concept of plasmoid formation in the near-earth plasma sheet is critically examined in the context of new measurements of auroral oval motions and of plasma convection near the center plane of the plasma sheet. Global images of earth's polar cap are employed, together with simultaneous determination of the solar wind parameters, to demonstrate the storage and release of magnetotail energy prior to and following the onset of a magnetic substorm. During sustained periods of northward-directed interplanetary magnetic fields, auroral activity is most significant at latitudes poleward of a relatively quiescent auroral oval. The striking configuration of auroral luminosities known as a theta aurora can appear during these periods. It seems unlikely that the sometimes sought ground state of little or no ionospheric coupling with magnetotail plasmas and convection occurs with periods of northward interplanetary field. Instead the ionosphere-magnetotail system is actively coupled in the polar ionosphere.

Introduction

A review of the general character of the magnetotail plasmas, e.g., plasma domains, plasma flows, ion and electron temperatures and densities, etc., is given by Frank (1985). The purpose of this discussion is not to extensively review again the general observational knowledge of the magnetotail, but to focus on some recent findings and current problems in understanding the dynamical behavior of the magnetotail and its coupling with the ionosphere. What are the properties of the primary current systems in the plasma sheet? Are plasma flows in the center plane of the plasma sheet consistent with those expected for plasmoid formation and transport? Is the expected poleward leap of the aurora observed after the plasmoid is released from the near-earth plasma sheet into the distant magnetotail? Does the boundary layer of the plasma sheet map into discrete arcs at the poleward edge of the auroral oval? What is the nature of the coupling of the magnetotail with the polar ionosphere during those lesser studied periods of northward-directed interplanetary fields? Recent observations give us further insight into these important issues.

Currents in the Plasma Sheet

There are two major currents in the plasma sheet at geocentric radial distances ~10 to 100 R_E (earth radii). Specifically the definition of major current is taken as any current that can be detected directly with plasma instrumentation or by means of a major topological feature of the magnetic fields. These two zones of currents are identified in Figure 1. The most obvious current must be located at the reversal of the magnetic fields in the magnetotail, i.e., the current sheet associated with the magnetotail neutral sheet. Direct detection of the electron and ion motions that give rise to this current proves to be difficult. Traversals of the neutral sheet are not accompanied by easily detectable fluctuations in plasma densities, electron or ion temperatures, or motions of the electron and ion plasmas. Only one example of direct detection of plasma drifts responsible for the neutral sheet current is available in the literature (Frank et al., 1984). The corresponding magnetospheric condition is quiescence that allows a sufficiently long sampling of the plasmas near the neutral sheet during a period of minimum motions associated with flapping or twisting of the magnetotail. Simultaneous observations of the angular distributions of positive ions and electrons in a plane nearly parallel to the ecliptic plane and at a position near the neutral sheet are shown in Figures 2 and 3, respectively. The remarkable feature is the difference in directions of the maxima in ion and electron intensities. Thus,

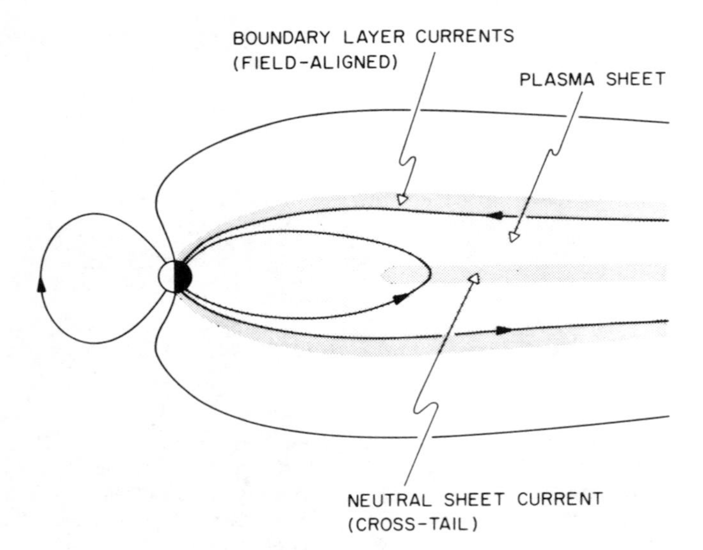

Fig. 1. Two regions of large current densities in the near-earth magnetotail.

the directions of fluid motions for the ion and electron plasmas significantly differ, i.e., in the direction of solar ecliptic longitude $\phi \simeq 40°$ (generally duskward) for the ion plasma and along $\phi \simeq 240°$ (generally dawnward) for the electron plasma. These motions give rise to a current density $j \simeq 2.4 \times 10^{-8}$ A m^{-2} directed perpendicular to the local magnetic field. Because the

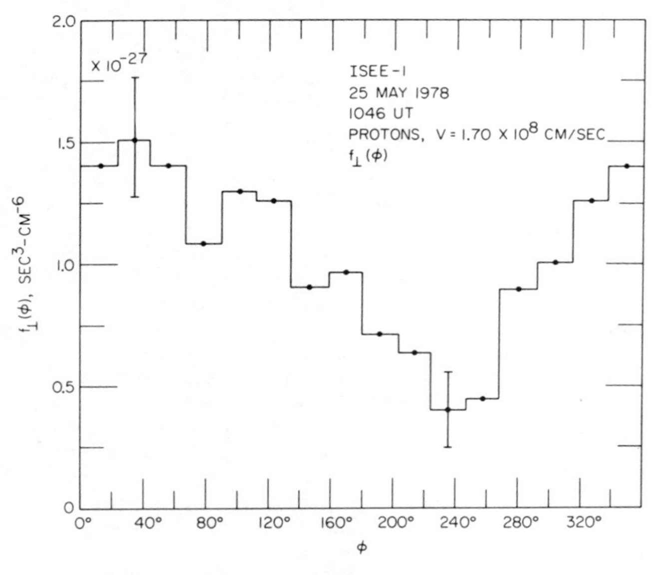

Fig. 2. The angular variation of the proton velocity distribution, f(**v**) in a plane perpendicular to **B** and for a proton speed of 1.7×10^8 cm s^{-1}. ISEE 1 is located near the neutral sheet (after Frank et al., 1984).

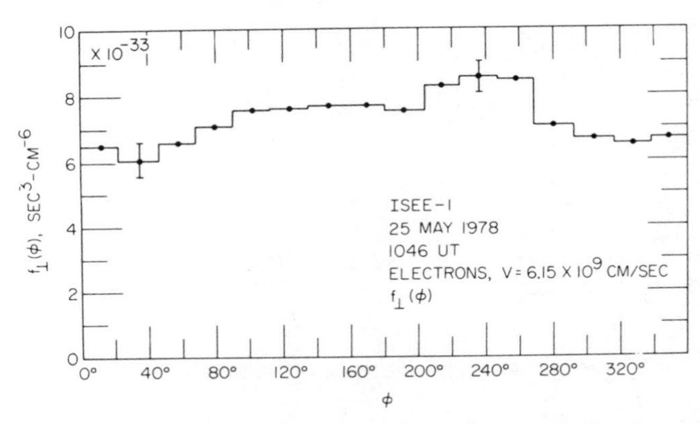

Fig. 3. Continuation of Figure 2 for the electron velocity distribution. The electron speed for this slice of the velocity distribution is 6.2×10^9 cm s^{-1} (after Frank et al., 1984).

motions of two plasmas are not parallel, **E** × **B** drifts are not solely responsible for the angular distributions shown in Figures 2 and 3. Gradient drifts in an inhomogeneous magnetic field are not observable with particle detectors and do not contribute to the current density (cf. Spitzer, 1952). Plasma drifts due to pressure gradients, i.e., a diamagnetic current, are eliminated by Frank et al. (1984) on the basis of the measured gradients in plasma pressures. Thus, the ion plasma drift, if not also that of the electron plasma, must be due to nonadiabatic particle motion in the magnetic and electric fields near the neutral sheet.

Recently McComas et al. (1986) reported the current densities in the cross-tail current sheet from an analysis of magnetic field measurements with the two International Sun-Earth Explorer (ISEE) spacecraft. Current densities for one of the three examples are reproduced in Figure 4.

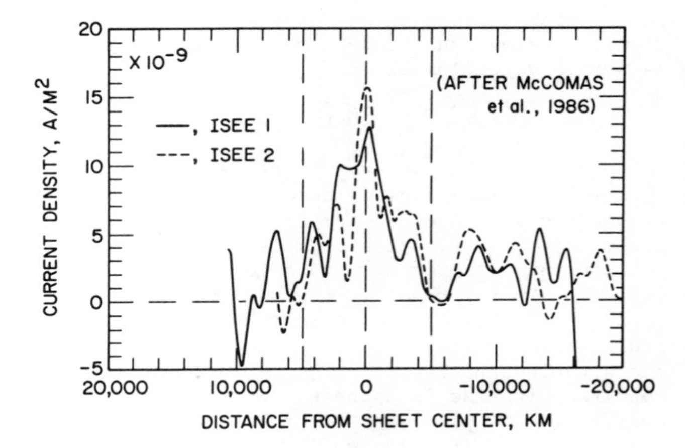

Fig. 4. Current densities as a function of distance from the center of the neutral sheet as determined with the ISEE 1 and 2 magnetometers.

The procedure that is employed to derive these current densities requires a strong disturbance in the solar wind to produce rapid motion of the current sheet past the two spacecraft. The observations displayed in Figure 4 were taken at -22.0, -1.4, and 4.0 R_E in geocentric solar magnetospheric (GSM) coordinates at about 0200 UT on April 5, 1979. Peak current densities for the three crossings range from ~0.4 to 4.5 x 10^{-8} A m^{-2} and bracket the value noted above for the direct plasma measurements. These magnetic field measurements provide a further parameter for the cross-tail current sheet: its thickness. The thickness of this current sheet varies for the three examples and is ~10,000 to 20,000 km.

At the present date, now more than 20 years past the discovery of the cross-tail current sheet (Ness, 1965), our direct knowledge of the motions of ion and electron plasmas in the current sheet is meager. The direct evidence that is available suggests that a model of this current sheet based upon single-particle nonadiabatic motion in weak magnetic fields and cross-tail electric fields is consistent with the observations (Lyons and Speiser, 1982).

Direct observations of the plasma motions that provide the currents in the boundary layer of the plasma sheet (cf. Figure 1) are considerably more detailed than those outlined above for the cross-tail current. These boundary layer currents are magnetically field-aligned and the drift speeds of the electron plasma, the main contributor to the current densities, are sufficiently large that the angular anisotropies are easily measured. Typical current densities are in the range of ~10^{-8} A m^{-2}, and multiple, parallel sheets of currents flowing toward and away from the ionosphere are observed (Frank et al., 1981b). The magnetic signatures of these field-aligned currents, i.e., fluctuations in the y-component of the magnetic field, are easily detected and interpreted in terms of current sheets (Elphic et al., 1985). These currents are associated with the low-altitude Region 1 current system (cf. Potemra, 1979).

The plasma sheet boundary layer is also characterized by beams of (1) ions from the more distant magnetotail that flow toward the ionosphere at speeds comparable to or greater than the ion characteristic thermal speeds (cf. DeCoster and Frank, 1979) and (2) ionospheric ions that are transported tailward (cf. Eastman et al., 1984). The specific mechanism for generating the high-speed ion beams directed toward the ionosphere is not identified. Possible mechanisms suggested in the literature are (1) injection of an ion plasma from a reconnection site in the more distant magnetotail (Birn et al., 1981), (2) field-aligned electric fields (DeCoster and Frank, 1979), and (3) cross-tail current sheet acceleration (Lyons and Speiser, 1982). Delineation of the ion acceleration mechanism is important in establishing the topological and dynamical relationship between these field-aligned currents in the boundary layer and the cross-tail current sheet. Because the boundary layer maps into the system of discrete arcs at the poleward boundary of the auroral oval (see next section), this relationship is particularly relevant to magnetospheric models that are intended to give physical insight into the flow of energy from the magnetotail into the auroral ionosphere.

Mapping the Plasma Sheet Boundary Layer into the Auroral Oval

The availability of global imaging of the auroral oval and simultaneous, in situ observations of the near-earth plasma sheet allow an identification of auroral features with the plasma sheet boundary layer. A realistic model of the distorted geomagnetic field is necessary in order to determine the magnetic footprint for the in situ measurements as mapped into the auroral ionosphere. A specific example of this mapping of the boundary layer of the plasma sheet into the auroral ionosphere is summarized here (J. D. Craven et al., unpublished manuscript, 1987). The position of the plasma sheet boundary layer is located by the decrease of energetic electron intensities as observed during ~1750-1800 UT on February 9, 1983, with the University of California/Berkeley detectors onboard ISEE 2. These measurements are shown in the center panel of Figure 5. The spacecraft position in GSM coordinates is given at the top of this figure. The top panel of Figure 5 displays the positions of the poleward arc and the equatorward boundary of the auroral oval as determined simultaneously with the global auroral images from Dynamics Explorer 1 (DE 1). The geomagnetic latitude and local time corresponding to the ISEE 2 position for these auroral observations are found by using the magnetic field model developed by Tsyganenko and Usmanov (1982). The latitudes of this spacecraft footprint for values of the Bartels planetary magnetic index K_p = 0 and 3^+ are given in the top panel of Figure 5 for direct comparison with the observed auroral zone boundaries. In order to avoid the difficulties introduced by rapid fluctuations of the auroral configuration, these observations are chosen specifically for a period of relatively stable auroral luminosities. The constancy of the latitudes for the poleward arc and the equatorward boundary is evidence for the stability of the auroral configuration. It is clearly demonstrated in the top panel of Figure 5 that the plasma sheet boundary layer is mapped into the poleward arc of the auroral oval. The usefulness of global auroral images for investigation of the ionosphere-magnetosphere interaction is demonstrated by the image shown in Figure 6. This image is one of a set of global auroral images that is employed to provide the auroral boundaries displayed in Figure 5. The image is taken at ~1745 UT with the DE 1 spacecraft as ISEE 2 traverses the boundary layer. The ultraviolet wavelength range for this image

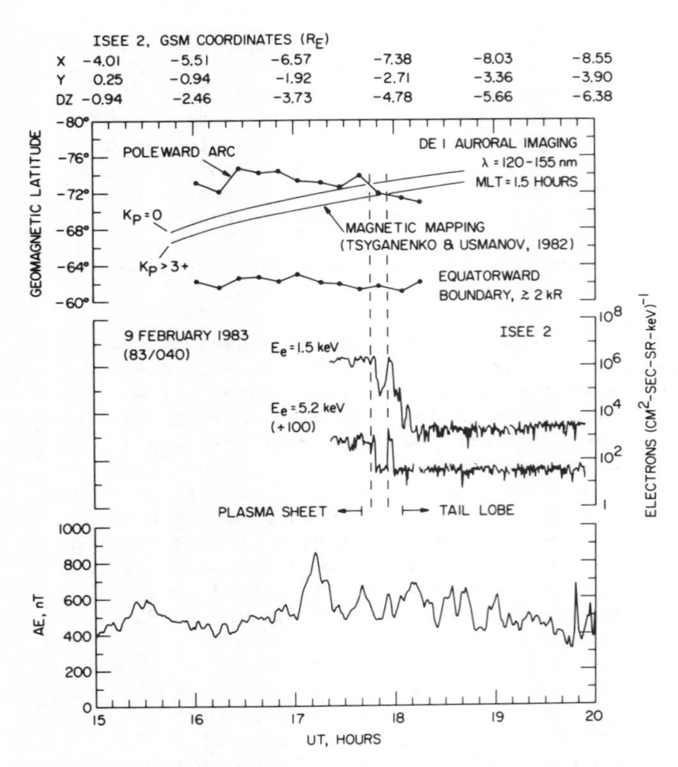

Fig. 5. Energetic electron intensities for the ISEE 2 crossing of the plasma sheet boundary layer on February 9, 1983 (center panel). Instantaneous position of ISEE 2 relative to the boundaries of the auroral oval as determined from auroral images obtained with DE 1 (upper panel).

is wide and includes Lyman α, atomic oxygen emissions at 130.4 and 135.6 nm, and N_2 emissions from the LBH bands. The auroral oval and the sunlit atmosphere are viewed primarily at 130.4 nm, whereas the glow above earth's limb is due to resonantly scattered solar Lyman α in the exosphere. For this image of the southern hemisphere, local dawn is to the left, and evening to the right. The footprint of the ISEE 2 spacecraft as it crossed the plasma sheet boundary layer is also shown in Figure 6. This position lies on the poleward discrete arc that extends over a broad range of local times in the evening sector. The equatorward zone of luminosities is mapped into the near-earth plasma sheet. Thus the single-point, in situ observations of plasmas with ISEE 2 are located with respect to the reference frame of the auroral oval.

Elphic and coworkers (1987) reported the observational tracing of the field-aligned current intercepted in this plasma sheet boundary layer with the high-altitude spacecraft ISEE and at lower altitudes with DE 1. For this example of current tracing the boundary layer is mapped into the poleward discrete arc of the auroral oval.

The global auroral images allow the extension of the in situ measurements into a two-dimensional map of the instantaneous spatial extent and inhomogeneities of field-aligned currents in the boundary layer of the plasma sheet. These luminosity maps must also take into account the low-altitude auroral acceleration mechanism.

Plasmoids and Boundary Layers

Currently there are two greatly dissimilar models for the behavior of the near-earth plasma sheet during magnetic substorms, i.e., the plasmoid model and the boundary layer model. In order to discuss the relevance of recent observations to these two models it seems necessary to briefly outline the features of these models. One of the most notable characteristics of the magnetotail plasmas is the presence of ion beams with bulk speeds comparable to or greater than the ion thermal speeds (Hones et al., 1974; Frank et al., 1976). Because these early observations of the ion flows were acquired at a single point in the plasma sheet and its immediate environs, the presence of these flows could be interpreted as either (1) the bulk flow for the entire, or a major part of the plasma sheet or (2) the passage of a spatial structure, e.g., a boundary layer, over the position of the spacecraft. The occasionally observed correlation of earthward ion flows with northward-directed magnetic fields and of tailward ion flows with southward magnetic fields during magnetic substorms, together with interpretation (1) above, provides the basis for the concept of plasmoid formation in the near-earth plasma sheet (Hones, 1979). The early foundations for the application of reconnection to the magnetotail during substorms are reviewed by Russell and McPherron (1973) and by McPherron (1979).

A series of sketches depicting several major features of plasmoid formation in the plasma sheet during magnetic substorms is given in Figure 7. Reconnection at an X-line in the magnetic field topology occurs across the plasma sheet, a section of the plasma sheet is pinched off to form a plasmoid, and the plasmoid subsequently convects tailward. Recovery of the plasma sheet proceeds by rapid, massive transport of plasma from the magnetotail into the near-earth plasma sheet. A plasma sheet boundary layer, such as that discussed here in previous sections, does not play a significant role in the plasmoid model. A more recent description of the plasmoid model is given by Hones (1984).

The origins of the boundary layer model for magnetotail substorms are not as clearly defined. Frank and coworkers (1976) suggested that boundary layers adjacent to the plasma sheet may be important regions with respect to plasma acceleration and transport [see also Frank (1976); Lui et al. (1977)]. This concept that the ion beams observed in the magnetotail are confined to a boundary layer adjacent to the plasma sheet remained relatively dormant until dual-spacecraft

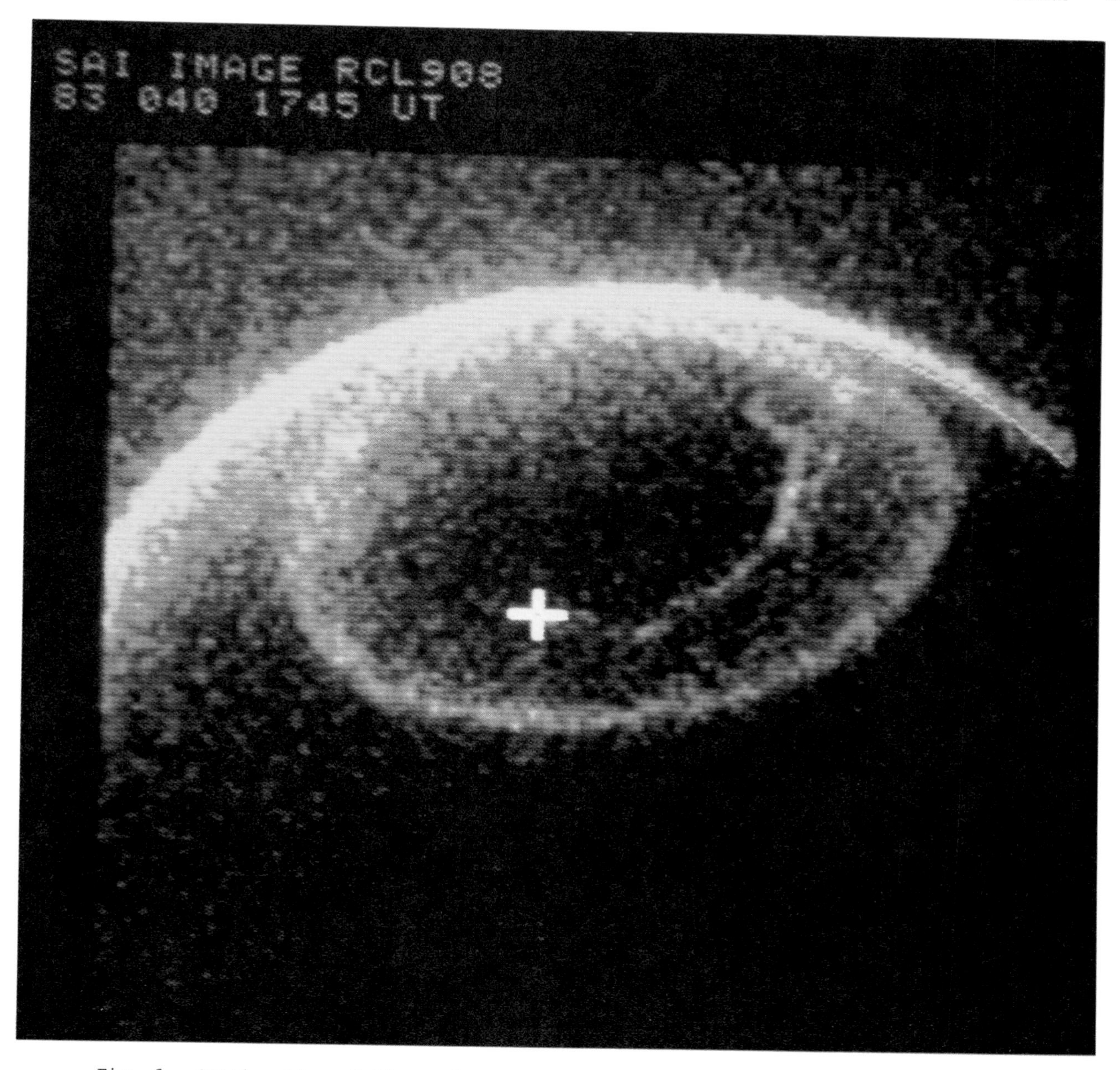

Fig. 6. Continuation of Figure 5. Position of ISEE 2 during its traversal of the plasma sheet boundary layer as mapped into the auroral oval with a model of the distorted geomagnetic field.

observations of the plasma sheet became available. With such measurements DeCoster and Frank (1979) showed that the ion beams are a durable feature of the plasma sheet, i.e., occurring also in the absence of magnetic substorms, and that these ion beams are confined to a boundary layer positioned at the plasma sheet-magnetotail lobe interface. Subsequently, the boundary layer model for the magnetotail slowly emerged with a series of extensive analyses of observations (cf. Eastman and Frank, 1984; Eastman et al., 1984, 1985). The most definitive description of the boundary layer model for magnetic substorms is given by Eastman and colleagues (1987) and Rostoker and Eastman (1987). The behavior of the plasma sheet during substorms, as proposed with the boundary layer model, is depicted in the sequence of sketches in Figure 8. Prior to substorm onset, the observing spacecraft S is located within the plasma sheet. Ion beams and

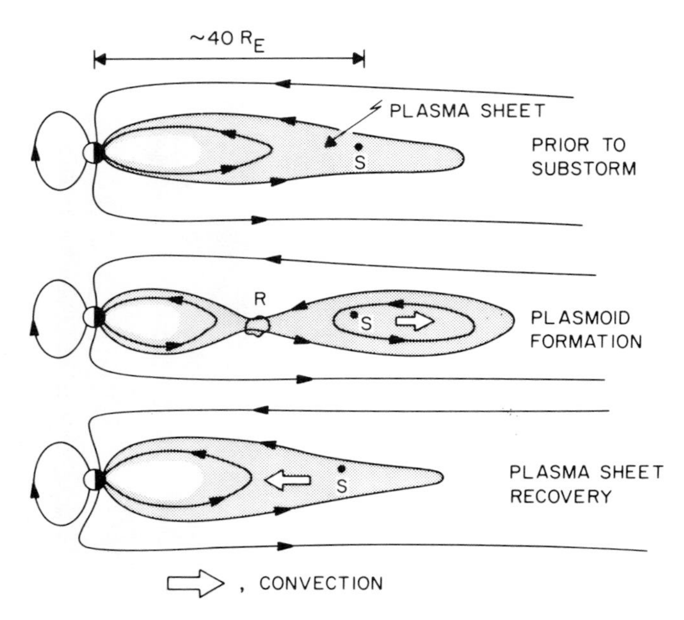

Fig. 7. Plasmoid model for magnetotail substorms (after Frank, 1985).

field-aligned currents exist at the plasma sheet-magnetotail lobe interface. With substorm onset the plasma sheet thins and the boundary layer, with beams of ions from the magnetotail and out-flowing ionospheric ion beams, passes over the spacecraft position. During recovery the plasma sheet slowly fills and expands and the boundary layer once again passes over the spacecraft position. Strongly flowing ions at speeds $\gtrsim$100 km s^{-1} and intense field-aligned currents are found within the boundary layer during the substorm sequence. Bulk motions of plasma within the plasma sheet are characterized by considerably lesser speeds.

The region F indicates the position for plasma acceleration in the plasma boundary layers adjacent to the plasma sheet (Frank, 1976). The mechanism could be reconnection in magnetic fields at locations outside of the plasma sheet. The plasma acceleration region(s) is presumed to be magnetically linked to the discrete arcs at the poleward boundary of the auroral oval (Frank, 1976). Magnetospheric substorms arise from an intensification of plasma acceleration in these boundary layers adjacent to the plasma sheet, and not in hot, quasi-isotropic plasmas of the plasma sheet. The plasmas of the plasma sheet are the debris from the acceleration mechanism in the boundary layers (Frank et al., 1976).

These brief descriptions of the substorm behavior of the near-earth plasma sheet as anticipated from the two models are sufficient for our purposes of the following comparisons with two recent series of observations. It is important that the reader recognize that our arguments pertain only to plasmoid formation within the near-earth plasma sheet. The approximate average downstream location for the boundary of the plasma sheet is at ~100 R_E as determined by indirect sounding with energetic particles (Williams, 1981) and direct surveys of magnetic fields (Slavin et al., 1985). At distances downstream in the magnetotail beyond the plasma sheet, e.g., at and beyond region F of Figure 8, substantial evidences exist in favor of reconnection phenomena such as plasmoid formation (cf. Baker et al., 1987). It is not necessary that the origins of plasmoids, or closed magnetic loop structures, be the near-earth plasma sheet. Frank and coworkers (1976) proposed that the sites of reconnection, or fireballs, are located in relatively confined regions within the entrained magnetosheath plasmas along the flanks of the plasma sheet (see also Frank, 1976). Because these entrained magnetosheath plasmas, or low-latitude boundary layers, exhibit both open and closed topologies of the geomagnetic field (Mitchell et al., 1987), closed loop structures, or plasmoids, can be created in these boundary layers at the edges of the plasma sheet during or in the absence of substorms. These boundary layer plasmoids are subsequently transported into the distant magnetotail.

Plasma Flow in the Central Plasma Sheet

If plasmoids are formed in the near-earth plasma sheet during magnetic substorms then high-speed streaming of plasmas is expected near the center plane of the plasma sheet (see Figure 7). Thus substantial increases in the bulk flow velocities of positive ions are expected with

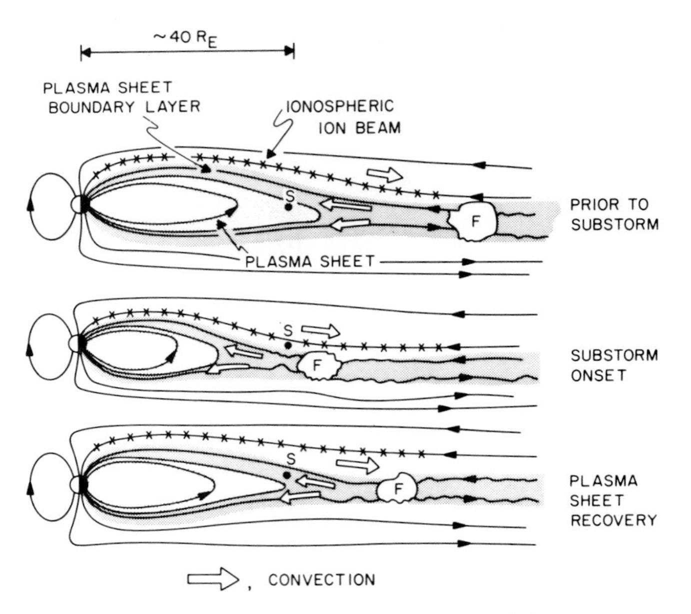

Fig. 8. Boundary layer, or fireball (F) model for magnetotail substorms (after Frank, 1985).

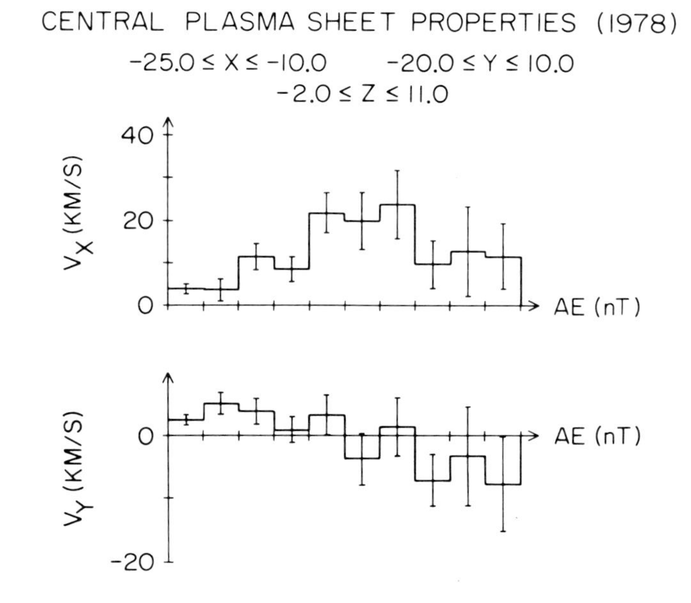

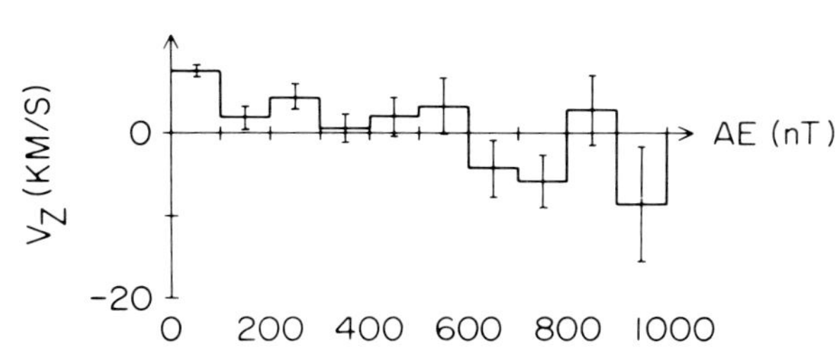

Fig. 9. The three components of ion bulk velocity in the plasma sheet as functions of the AE index. The spatial range is indicated at the top of the graph. Error bars give the standard deviation for each averaged bin (after Huang and Frank, 1986).

increasing magnetic activity. Huang and Frank (1986) reported a statistical study of the bulk velocities of ion plasmas in the central plasma sheet. These measurements were acquired during 1978 with a plasma analyzer onboard ISEE 1. Approximately 3000 individual samples within the central plasma sheet were available for this time period. Observations within the plasma sheet boundary layers and the streams of ionospheric ions in the magnetotail lobes were excluded from this survey. However all measurements within 1.5 R_E of the mid-plane of the magnetotail [taken here to be at Z(GSM) = 0] were retained in the statistical analysis. The averaged values of the three components of the ion bulk velocities, in the above coordinate system, are shown in Figure 9. These averaged values are given as a function of the magnetic AE index for the auroral electrojet. The magnitudes of the velocity components are small, $\lesssim 20$ km s^{-1}, relative to the ion thermal speeds, ~1000 km s^{-1}. These small bulk speeds of the ion plasmas are observed over the entire range of AE values of 0 to 1000 nT (nanotesla). There are 393 observations of plasma flow during periods of substantial magnetic activity with AE = 500 to 1000 nT. No increases in the magnitudes of the bulk speeds of the ion plasmas are evident for these AE values in the observations shown in Figure 9. However, a tendency for flows directed toward the center plane ($-V_z$) and toward local dawn ($-V_y$) is suggested by these measurements. The magnitudes of the V_x component of flow are expected to be in the range of about a few hundreds of kilometers per second to 1000 km s^{-1} during formation and transport of a near-earth plasmoid (see Figure 7). If only a few percent of the 393 samples of the V_x component of plasma flows are 500 km s^{-1}, then the presence of these infrequent plasmoid-associated flows would be readily discernible in the measurements summarized in Figure 9. The near-earth plasmoid model appears to be incompatible with the results of this statistical study.

The plasma sheet boundary layer model assumes that the high-speed plasma streams are confined to the plasma sheet-magnetotail lobe interface. Bulk flows within the central plasma sheet are relatively slow and these plasmas may be the debris from the boundary layer processes (cf. Frank et al., 1976). Apparent injection times into the ring current, for example, at geosynchronous orbital altitudes can still be short because the earthward motion of the plasma sheet is ~1 R_E in 300 s at a speed $V_x \simeq 20$ km s^{-1} (see upper panel, Figure 9).

However, one parameter for the ion plasmas in the central plasma sheet does vary substantially with increasing magnetic activity as gauged with the AE index. This parameter is the average ion temperature as shown in Figure 10 (Huang and Frank, 1986). The characteristic thermal energy increases from ~4 keV for AE = 0-100 nT to ~10 keV for AE = 800-1000 nT. On the other hand, the bulk flow speed and the number density, ~0.3 cm^{-3}, show no strong correlation with the AE index. With the exception of the ion temperatures, the ion velocity distributions in the plasma sheet appear to be generally unresponsive to magnetic activity.

Search for the Poleward Leap of the Aurora

A poleward leap of the poleward boundary of the auroral oval during magnetic substorms is one of the anticipated signatures of the near-earth plasmoid model (Hones, 1985, 1986). The reader is referred to the bottom two panels of Figure 7. The near-earth plasmoid forms and is launched tailward during the beginning of the expansion phase of a substorm (second panel). During the recovery phase of the substorm the plasma sheet rapidly refills (third panel). With this inrush of plasma into the plasma sheet, the poleward boundary of the auroral oval similarly responds, i.e., leaps to poleward latitudes. This poleward

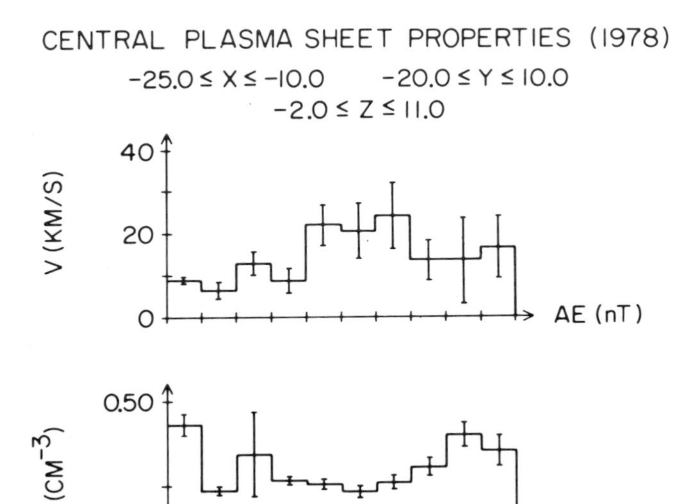

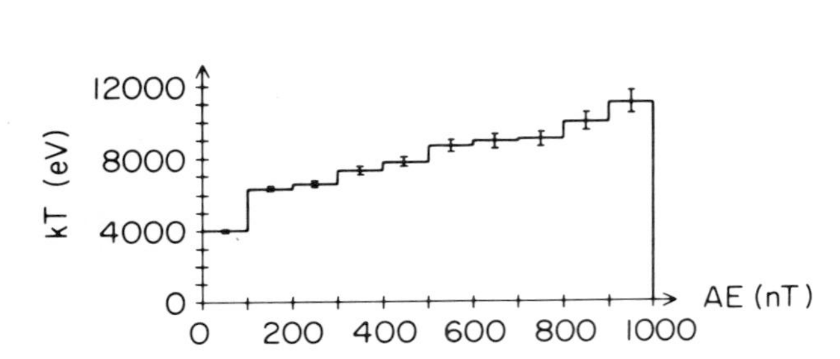

Fig. 10. Continuation of Figure 9 for the bulk speed, density, and temperature in the plasma sheet (after Huang and Frank, 1986).

motion is sketched in Figure 11. This poleward leap should be readily identifiable in global images of the auroral oval.

Examples of the latitudinal motions of the boundaries of the auroral oval are given by Craven and Frank (1987). An example of a sequence of global auroral images as taken during a typical magnetic substorm during 1436 to 1702 UT on November 4, 1981, is shown in Figure 12. These images were taken at 12-min intervals in the visible emissions of atomic oxygen, 557.7 nm, with DE 1. The imaging instrumentation on this spacecraft is the only such optical device to successfully gain global images of the earth's visible aurora with the sunlit atmosphere in its field-of-view. Without this capability this sequence of images of a single substorm is not possible. This capability is achieved in part with super-polished surfaces in an off-axis catoptric telescope (Frank et al., 1981a). The sunlit atmosphere (electronically suppressed) is positioned on the left-hand side of each image in Figure 12. Luminosities are color coded: dark blue is ~3 kR (kilorayleighs) and white is ~30 kR. Local dawn is located in the upper section of the auroral oval, local evening in the lower section. A small intensification of auroral luminosities is first noted at ~1448 UT (second image, first row) and the expansive phase of a second, more intense substorm begins at ~1525 UT (fifth image; first image, second row). Intense emissions are subsequently seen in the late evening sector. The luminosities in this region begin to subside by ~1637 UT (eleventh image). Equatorward of these bright arcs and in the local morning sector, a more diffuse region of luminosities appears at ~1537 UT (sixth image). This region progressively increases in size and extends toward the morning sector. This zone of luminosities is the signature of the injection of electrons deeply into the magnetosphere and their subsequent eastward gradient drift.

The locations of the poleward and equatorward boundaries of the auroral oval for various local times during this substorm are summarized in Figure 13 (Craven and Frank, 1987). The boundaries are given for the 2-kR intensity contours. There is no evidence of a poleward leap of the poleward boundary of the auroral oval. At magnetic local midnight the average latitudinal speed of the poleward boundary is ~230 m s^{-1}

LATE STAGES OF SUBSTORM EXPANSIVE PHASE
— NEAR-EARTH NEUTRAL LINE MODEL

TAILWARD MOTION OF PLASMOID

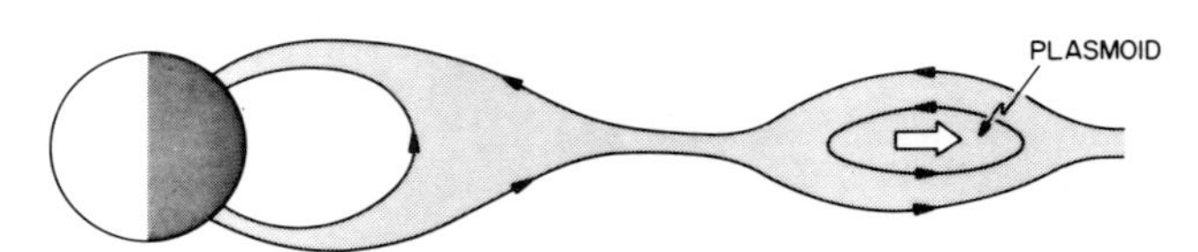

PLASMA SHEET RECOVERY

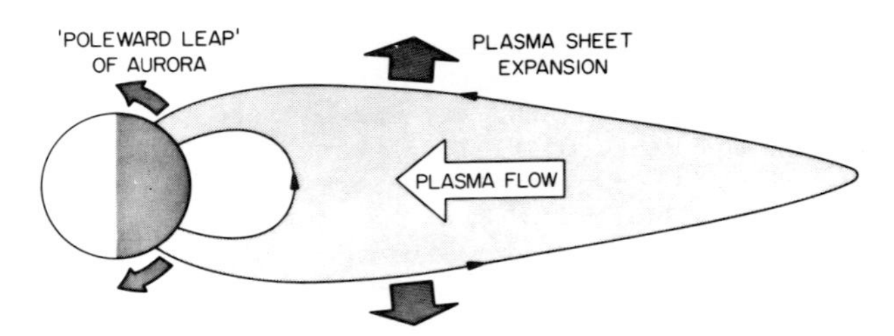

Fig. 11. Sketch of the anticipated poleward leap of the auroral oval during recovery of the plasma sheet and following the ejection of a near-earth plasmoid into the distant magnetotail.

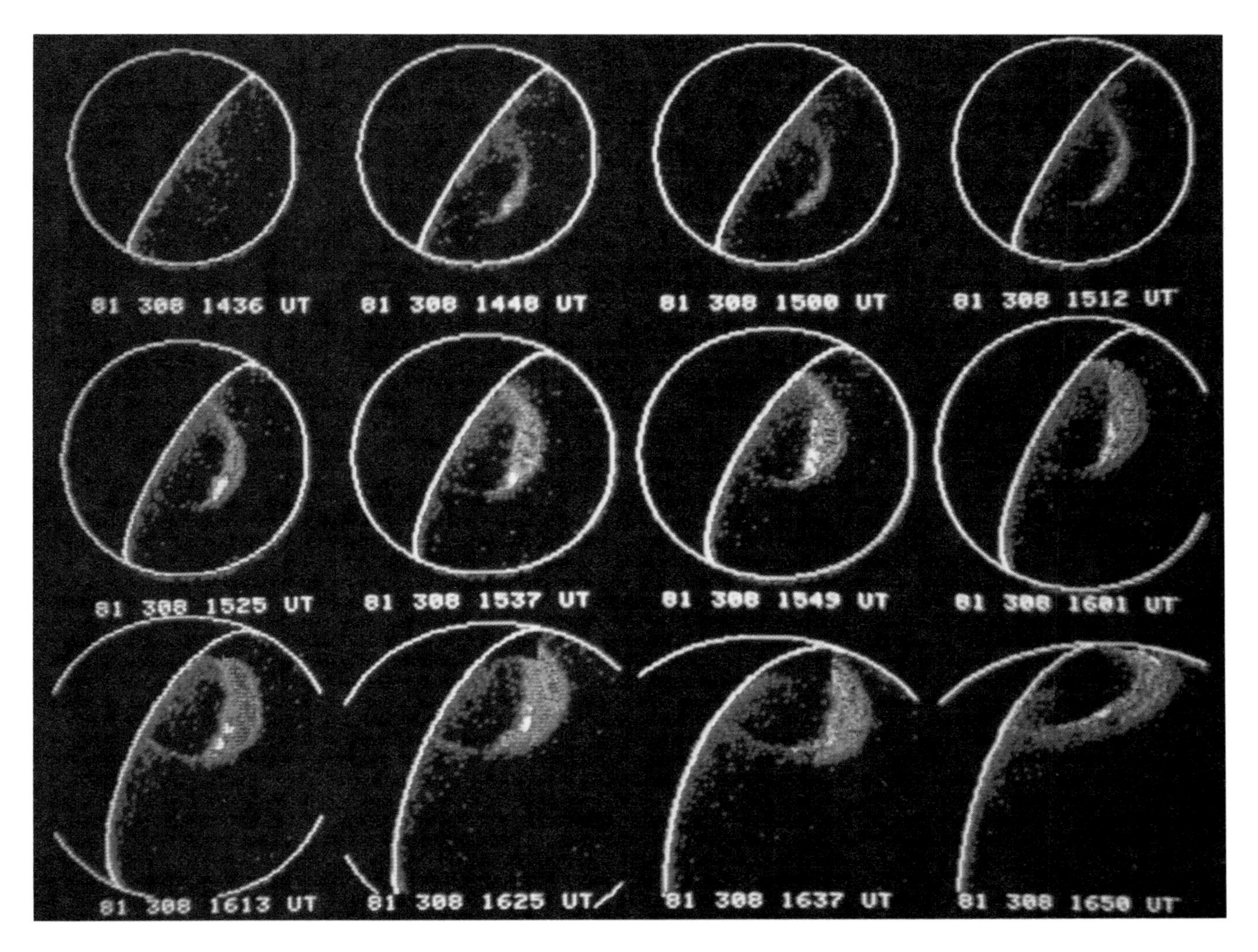

Fig. 12. Sequence of 12 consecutive false-color auroral images of the 557.7-nm emissions of atomic oxygen for the time interval 1436-1702 UT on November 4, 1981. Universal Time for the beginning of the 12-min telemetering period for each image is given. Initial onset of weak substorm activity begins with the second image of the northern auroral oval, followed during the fifth image by onset of more intense activity (after Craven and Frank, 1987).

(poleward) during ~1500 to 1555 UT and ~80 m s^{-1} (equatorward) during ~1555 to 1710 UT. Rostoker (1986) argued similarly against the existence of the poleward leap on the basis of the temporal variations of the electrojet during substorms. Hones et al. (1987) were unable to find evidence of the poleward leap in the ultraviolet images of an auroral substorm from the Viking spacecraft. Thus the anticipated poleward leap associated with the substorm recovery phase and the rapid refilling of the plasma sheet are not supported by results from global auroral imaging.

The boundary layer model does not give a poleward leap of the aurora (cf. Figure 8). This model predicts that the intense poleward arcs in the evening sector (white zones, Figure 12) are associated with intensification of field-aligned currents in the plasma sheet boundary layers, along with increased energization of electrons in the auroral acceleration region. This part of the substorm current wedge is sketched in Figure 14. A distant spacecraft located near this field-aligned current in the plasma sheet boundary layer observes a magnetic field that is dependent upon the spacecraft's position relative to this current (Eastman et al., 1987). Occasionally this position is such that the magnetic field exhibits a southward component which, to-

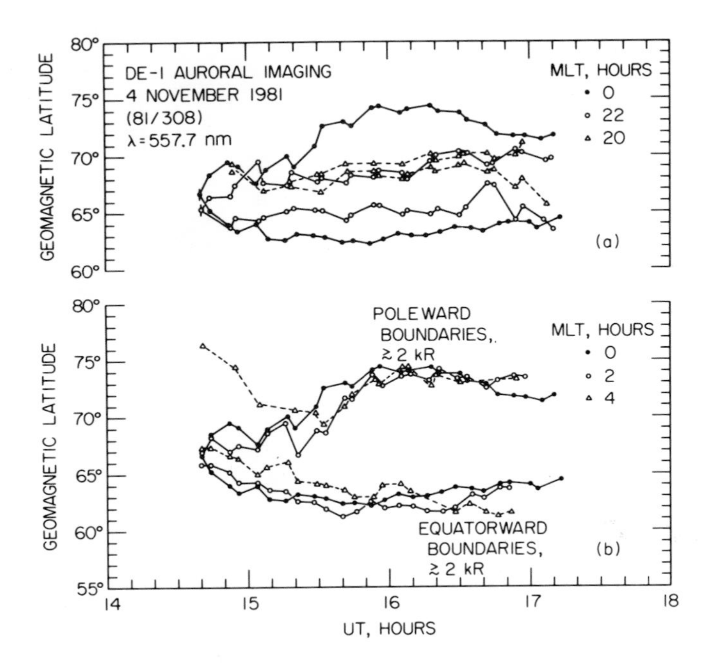

Fig. 13. Continuation of Figure 12. Latitudes of the poleward and equatorward boundaries of the auroral oval for column emission rates ~2 kR at 557.7 nm from atomic oxygen. These latitudes are shown for several magnetic local times (after Craven and Frank, 1987).

gether with tailward flow of ionospheric plasmas, can mimic the signature of a plasmoid.

Variations of Magnetotail Energy During an Auroral Substorm

Consider the northern auroral zone as viewed with DE 1 and shown in Figure 15. Also given in this figure is a contour defining the poleward boundary of the auroral oval for an intensity of ~1 kR. The polar cap is taken as the area within this contour. Magnetic field lines threading the polar cap extend into the northern magnetotail lobe. The total magnetic flux from the polar cap is readily determined by numerical integration over the irregular-shaped area defined in Figure 15. If the static and dynamic pressures in the solar wind are also simultaneously determined, then the total energy of the flaring magnetotail may be estimated (Coroniti and Kennel, 1972). A time series of images such as that shown in Figure 15 prior to and during a substorm can evidently provide insight into the gain and loss of magnetotail energy. Previous research indicates that the polar cap area expands prior to the onset of a substorm and subsequently diminishes after onset. The reader is referred to the statistical study by Meng and Makita (1986) and the references cited therein. The polar cap area is found to expand when the interplanetary magnetic field turns southward. Similarly the radius of the magnetotail is observed to increase after such southward turnings as noted by Baker et al. (1987) from an analysis of distant magnetotail observations with ISEE 3.

Continuous imaging of the auroral oval over time periods comparable to those for substorms is clearly advantageous in determining the variations of magnetic energy in the magnetotail and their association with auroral activity and solar wind parameters. An example is shown in Figure 16 for a series of images spanning a modest auroral substorm on November 10, 1981 (L. A. Frank et al., unpublished manuscript, 1987). The images were provided by the instrumentation onboard DE 1. The onset of the auroral substorm occurs at ~1540 UT as determined by a rapid brightening of an auroral arc in the late evening sector of the northern auroral oval. For this auroral substorm the luminosities are decreasing by ~1615 UT. Prior to the substorm onset the auroral oval is dim and relatively featureless for a period of 2 hours or more. Such an isolated substorm, and the availability of the solar wind parameters and interplanetary magnetic field vectors, provides an excellent opportunity to observe the response of the polar cap to a southward turning of the interplanetary field. This southward turning occurs at ~1353 UT (see Figure 16).

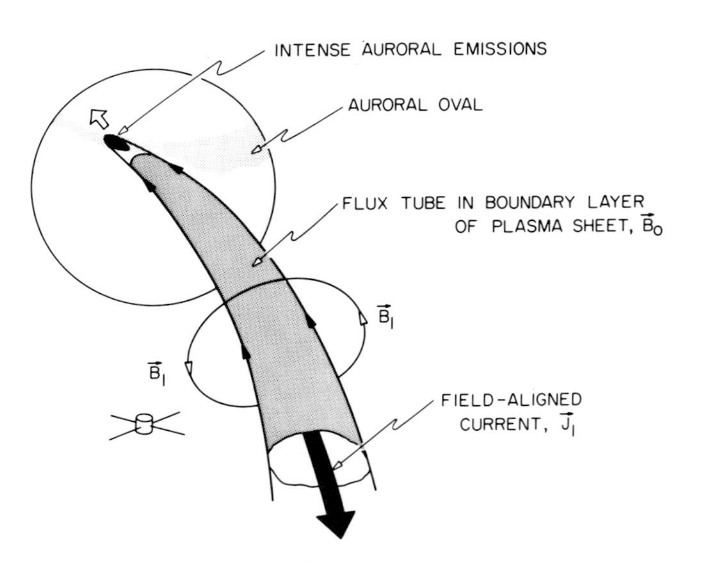

Fig. 14. Sketch of the magnetic field perturbations in the vicinity of an intense field-aligned current in the plasma sheet boundary layer during the expansion phase of an auroral substorm.

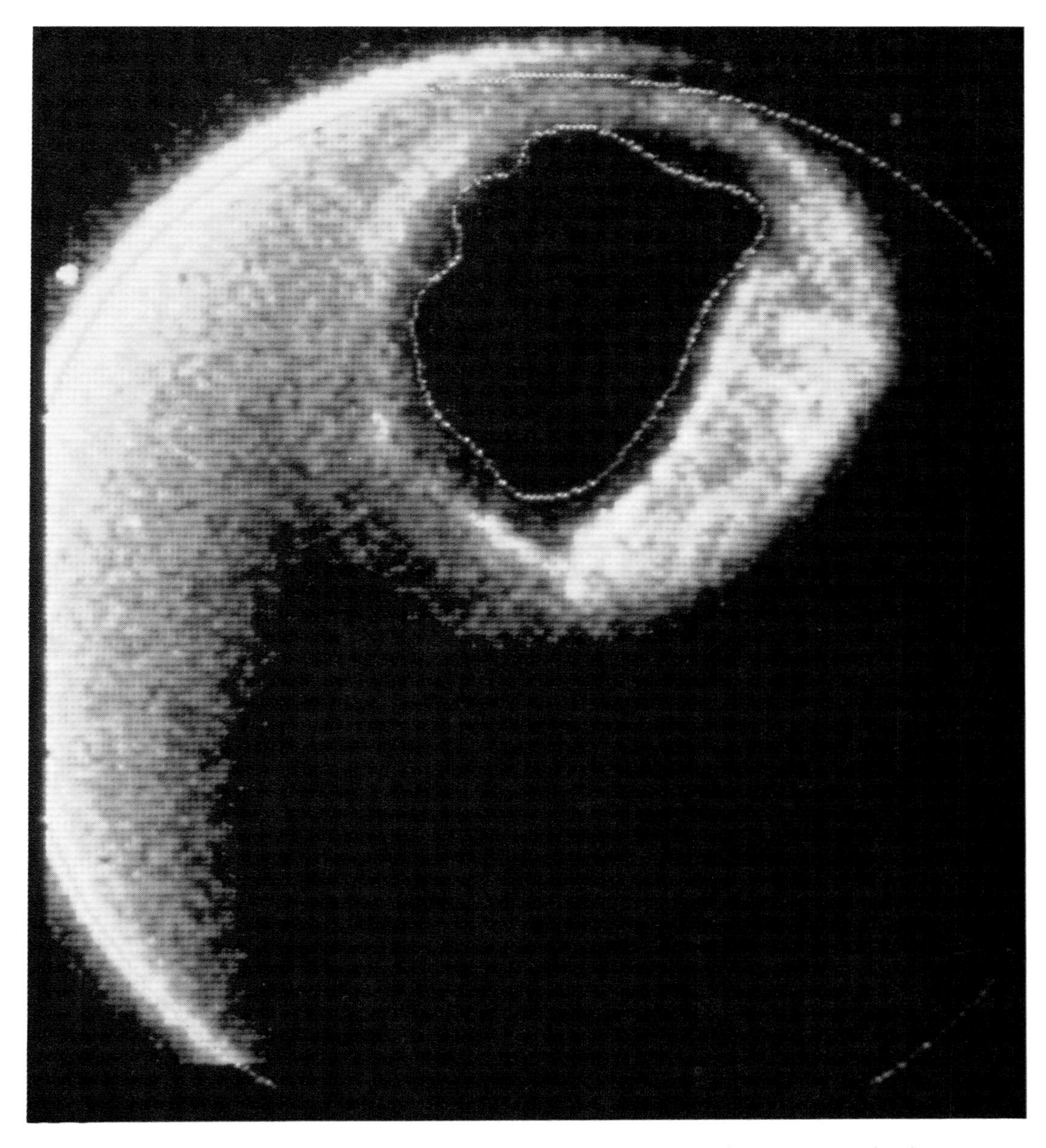

Fig. 15. Earth's northern auroral oval as observed in atomic oxygen emissions primarily at 130.4 nm with Dynamics Explorer 1 at 1242 UT on November 11, 1981. The boundary of the polar cap is shown. This area is used to calculate the magnetic energy of the flaring magnetotail.

The magnetic flux through the polar cap as a function of time is shown in the top panel of this figure. Each measurement corresponds to the area determined from a single image. Prior to the southward turning of the interplanetary magnetic field, no persistent fluctuation of polar cap magnetic flux is observed. Coincident with the southward turning, the magnetic flux increases for a period of ~80 min. The magnetotail energy, as determined with the Coroniti-Kennel relationships, increases by a factor of ~2 during this period. The rate of increase of magnetic energy in the magnetotail is ~ $+2 \times 10^{19}$ erg s^{-1}. The static pressure due to solar wind electrons is neglected in our calculation of energy. Approximately 20 min prior to the onset of the substorm the energy input ceases. During the expansion phase of the substorm at ~1540 to 1600 UT, the rate of energy loss from the magnetotail is ~ -2×10^{19} erg s^{-1}, or similar in

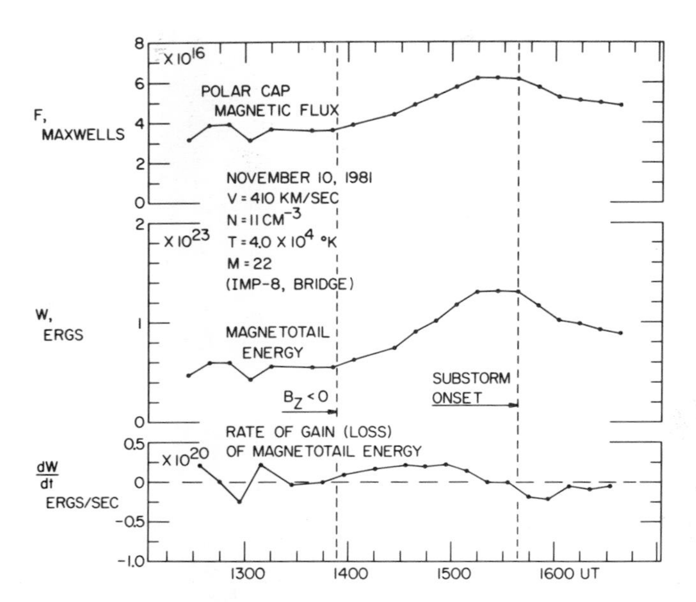

Fig. 16. The polar cap magnetic flux, the magnetic energy of the flaring magnetotail, and the rate of change of magnetotail energy during a modest, isolated substorm on November 10, 1981.

magnitude to the energy acquisition rate prior to the auroral substorm. No fluctuation of the solar wind parameters or of the interplanetary magnetic field can be unambiguously identified with substorm onset. Thus the magnetotail is clearly driven by the solar wind during the period preceding the substorm onset, but the release of energy during the substorm appears to be an unloading of energy without any coincident stimulus from the solar wind.

It is notable for global models of magnetospheric energy transport that the polar cap area provides an instantaneous measure of magnetotail energy. With the assumption of the Coroniti-Kennel relationships, the estimated amount of magnetotail energy lost during the substorm discussed above is found to be $\sim 3 \times 10^{22}$ erg. For global models of the subsequent energy transport, there are four major sinks: (1) particle precipitation into the upper atmosphere, (2) particle injection into the ring current, (3) joule heating in the atmosphere, and (4) loss of hot plasmas to the solar wind (cf. Baker et al., 1986). The first two sinks can be quantitatively evaluated with global auroral imaging and the magnetic Dst(H) index, respectively. On the other hand, global joule heating requires considerable modeling of ionospheric currents and electric fields (cf. Kamide et al., 1986; Craven et al., 1984). A quantitative assessment of the hot plasma loss to the solar wind during a substorm is not available.

The Theta Aurora

A future successful model of the plasma transport and the magnetic topology within the earth's magnetosphere should account for the character of polar cap phenomena during periods of northward-directed interplanetary magnetic fields. A brief summary of the responses of the polar caps for northward-directed fields is given here. Two major states for the convection fields and luminosities over the polar caps are identified at this date: (1) highly disordered convection and a complex system of polar arcs and (2) orderly anti-sunward convection that is bisected by a relatively narrow zone of sunward convection and luminosities extending from the night sector of the auroral oval to the local day sector, i.e., the theta aurora. Observations of the disordered, active polar cap are discussed in the literature (cf. Burke et al., 1982; Hardy et al., 1982), but the nature of this polar cap activity prevents a readily achievable interpretation in terms of magnetospheric dynamics. Instead this discussion centers on the second state of the polar cap, i.e., the theta aurora, because its relative topographic simplicity affords more opportunity for interpretation and may also provide insight into the disordered state. An extensive description of the theta aurora and associated fields and plasmas is given by Frank et al. (1986).

An image of a theta aurora over the southern polar cap is shown in Figure 17. This image provides a convenient reference frame for our discussion. The transpolar arc extends continuously from the night sector of the auroral oval to the day sector, giving the striking appearance of the Greek letter θ. This spatial distribution of luminosities was first detected in global auroral images taken at ultraviolet wavelengths (Frank et al., 1982). It should be possible, but observationally difficult, to detect the presence of a theta aurora with a ground-based array of all-sky cameras in Antarctica during the winter months. Over the northern polar cap, the transpolar arc moves in the direction of the y-component of the magnetic field. It is not certain as to which direction the transpolar arc moves over the southern polar cap in response to the interplanetary magnetic field vector. This uncertainty may be resolved with simultaneous imaging of the northern and southern polar caps with Viking and DE 1, respectively. Plasma convection due to $\mathbf{E} \times \mathbf{B}$ drift is anti-sunward over the polar cap with the exception of a narrow zone of sunward convection that coincides with the position of the transpolar arc. The potential difference across the transpolar arc region is typically several kilovolts.

The composition of the hot plasmas on magnetic field lines threading the transpolar arc, a mixture of H^+, He^{++}, and O^+ ions, is similar to that found within the plasma sheet and its boundary

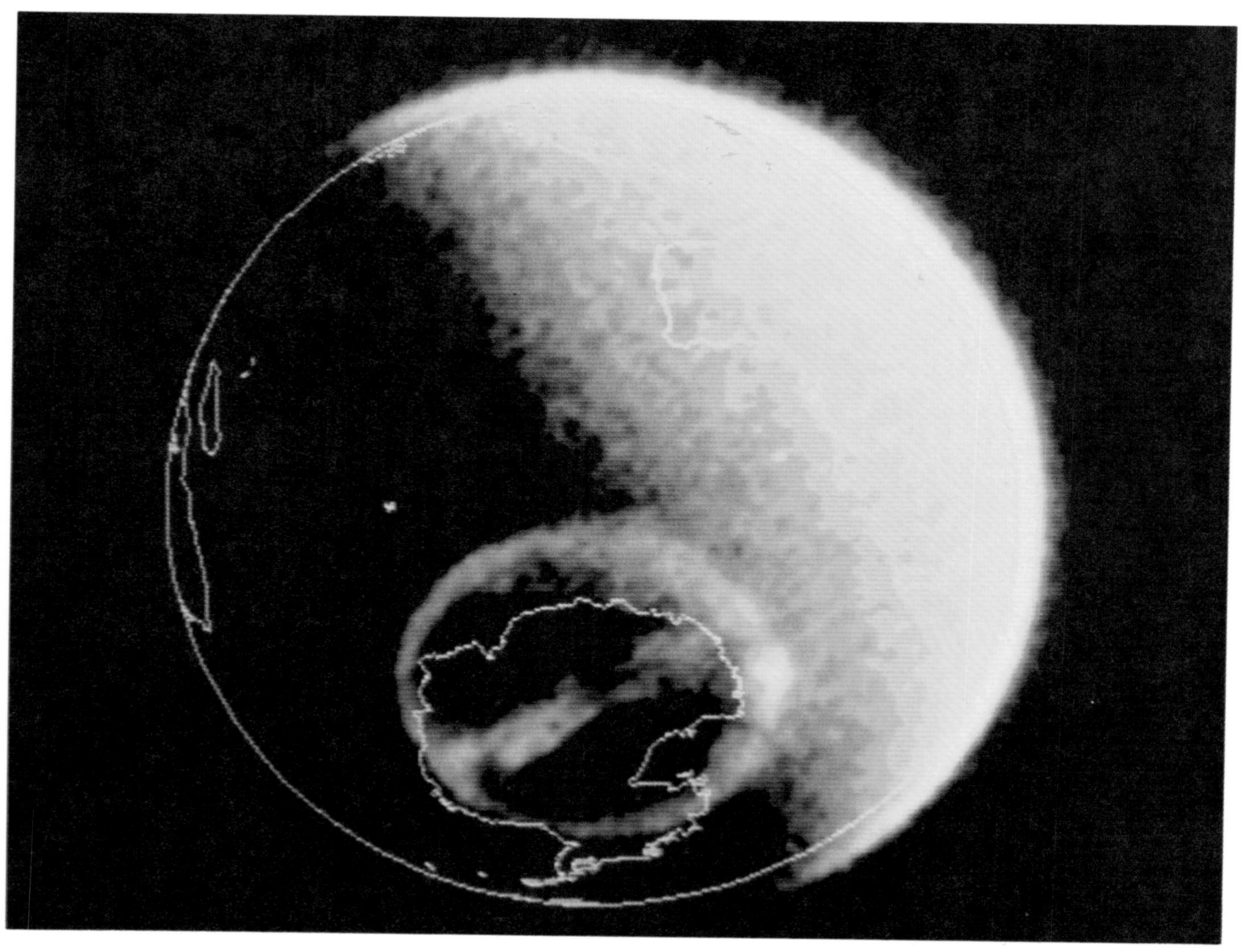

Fig. 17. A theta aurora observed over the southern hemisphere with DE 1 at 0022 UT on May 11, 1983 (after Frank et al., 1985).

layers (Peterson and Shelley, 1984; Frank et al., 1986). Thus, the plasmas over the transpolar arc are not directly transported from the polar magnetopause into the upper atmosphere. Over the other regions of the polar cap the density ratio, He^{++}/H^{+}, is considerably higher, an observation that lends further support to the interpretation that the transpolar arc is associated with a magnetospheric plasma region other than the magnetotail lobe. The fortuitous occurrence of a solar electron event coincident with observations of a theta aurora with the two Dynamics Explorer spacecraft allows a determination of whether or not the magnetic field lines threading the transpolar arc are connected to those in the interplanetary medium, i.e., open field lines.

These magnetic field lines associated with the transpolar arc are found to be closed in contrast to the open field lines threading the remainder of the polar cap. The magnetic field lines threading the poleward arcs of the auroral oval, which are mapped into the plasma sheet boundary layer, are also closed. The similarities of phenomena associated with the two luminosity regions, the transpolar arc and the poleward zone of the auroral oval, extend to the presence of field-aligned current sheets, field-aligned acceleration of electrons associated with the auroral luminosities, broadband electrostatic noise, funnel-shaped auroral hiss, and thermal electron density decreases (Frank et al., 1986). In fact the transpolar arc and the poleward zone of the auroral oval are so similar that the transpolar arc cannot be uniquely distinguished from a poleward arc in the auroral oval with only fields and plasma measurements.

The behavior of the polar cap with the dimming or disappearance of the transpolar arc remains unexplained. The polar cap reverts to its other state for northward interplanetary magnetic fields: highly disordered electric fields and a complex distribution of polar arcs and glows. Yet the hot ion plasmas associated with the previously existing transpolar arc are observed at times to remain in a narrow zone above the polar cap. When the transpolar arc subsequently brightens, the ordered convection pattern returns. This author believes that a well-founded insight into this dynamical behavior of the polar cap will provide considerable progress in understanding plasma entry and transport in the polar magnetosphere and the magnetotail lobes.

The direction of the interplanetary magnetic field is important in determining the spatial distribution of stresses at the magnetopause boundary layers. The asymmetric stresses associated with the y-component of the interplanetary field are expected to exert a torque on the magnetosphere about the earth-sun direction (Cowley, 1981). At least to distances out to the moon's orbit, the magnetotail does not respond to this torque with a large angular twisting of the plasma sheet or its boundary layer (Hardy et al., 1979). Thus, the transpolar arc cannot be due to a major reorientation of the plasma sheet. Akasofu and Roederer (1984) and Lyons (1985) proposed that the transpolar arc of a theta aurora is due to major distortion of the magnetotail plasma regimes from the superposition of the interplanetary magnetic field. Lyons (1985) extended his arguments to include convection electric fields over the polar cap. Reiff and Burch (1985) invoked several magnetospheric convection cells in order to provide the four-cell auroral and polar cap convection pattern that is observed with the appearance of the theta aurora. Frank et al. (1982) noted the similarities of plasmas over the transpolar arc with those over poleward arcs of the auroral oval and proposed that the magnetotail lobes are bifurcated with plasmas from the plasma sheet boundary layer. Such bifurcation is sketched in Figure 18. Huang and coworkers (1987) reported the occasional existence of filamentary plasma structures in the magnetotail lobes that are supportive of the proposed bifurcation of the magnetotail lobes. These filamentary structures may be also associated with polar arcs other than the transpolar arc of the theta aurora.

Acknowledgments. This research was supported in part by NASA under contract NAS5-28700 and grants NAG5-295, NAG5-483, and NGL-16-001-002 and by ONR under grant N00014-85-K-0404.

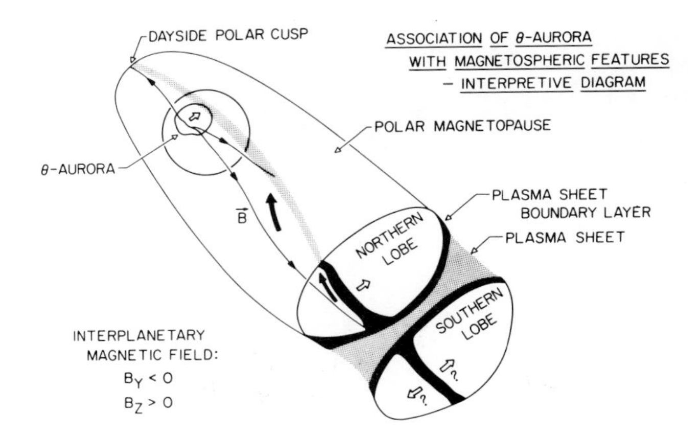

Fig. 18. An interpretive diagram of the bifurcation of the magnetotail lobes during the occurrence of a theta aurora.

References

Akasofu, S.-I., and M. Roederer, Dependence of the polar cap geometry on the IMF, Planet. Space Sci., 32, 111, 1984.

Baker, D. N., L. F. Bargatze, and R. D. Zwickl, Magnetospheric response to the IMF: Substorms, J. Geomag. Geoelectr., 38, 1047, 1986.

Baker, D. N., R. C. Anderson, R. D. Zwickl, and J. A. Slavin, Average plasma and magnetic field variations in the distant magnetotail associated with near-earth substorm effects, J. Geophys. Res., 92, 71, 1987.

Birn, J., T. G. Forbes, E. W. Hones, Jr., and S. J. Bame, On the velocity distribution of ion jets during substorm recovery, J. Geophys. Res., 86, 9001, 1981.

Burke, W. J., M. S. Gussenhoven, M. C. Kelley, D. A. Hardy, and F. J. Rich, Electric and magnetic field characteristics of discrete arcs in the polar cap, J. Geophys. Res., 87, 2431, 1982.

Coroniti, F. V., and C. F. Kennel, Changes in magnetospheric configuration during the substorm growth phase, J. Geophys. Res., 77, 3361, 1972.

Cowley, S.W.H., Magnetospheric asymmetries associated with the y-component of the IMF, Planet. Space Sci., 29, 79, 1981.

Craven, J. D., and L. A. Frank, Latitudinal motions of the aurora during substorms, J. Geophys. Res., 92, 4565, 1987.

Craven, J. D., Y. Kamide, L. A. Frank, S.-I. Akasofu, and M. Sugiura, Distribution of aurora and ionospheric currents observed simultaneously on a global scale, in Magnetospheric Currents, Geophys. Monogr. Ser., vol. 28, edited by T. A. Potemra, p. 137, AGU, Washington, D.C., 1984.

DeCoster, R. J., and L. A. Frank, Observations pertaining to the dynamics of the plasma sheet, J. Geophys. Res., 84, 5099, 1979.

Eastman, T. E., and L. A. Frank, Boundary layers of the earth's outer magnetosphere, in Magnetic Reconnection in Space and Laboratory Plasmas, Geophys. Monogr. Ser., vol. 30, edited by E. W. Hones, Jr., p. 249, AGU, Washington, D.C., 1984.

Eastman, T. E., L. A. Frank, W. K. Peterson, and W. Lennartsson, The plasma sheet boundary layer, J. Geophys. Res., 89, 1553, 1984.

Eastman, T. E., L. A. Frank, and C. Y. Huang, The boundary layers as the primary transport regions of the earth's magnetotail, J. Geophys. Res., 90, 9541, 1985.

Eastman, T. E., G. Rostoker, L. A. Frank, C. Y. Huang, and D. G. Mitchell, Boundary layer dynamics in the description of magnetospheric substorms, J. Geophys. Res., in press, 1987.

Elphic, R. C., J. D. Craven, L. A. Frank, and M. Sugiura, A study of field-aligned currents observed at high and low altitudes in the nightside magnetosphere, Physica Scripta, in press, 1987.

Elphic, R. C., P. A. Mutch, and C. T. Russell, Observations of field-aligned currents at the plasma sheet boundary: An ISEE 1 and 2 survey, Geophys. Res. Lett., 12, 631, 1985.

Frank, L. A., Hot plasmas in the earth's magnetosphere, in Physics of Solar Planetary Environments, vol. II, edited by D. J. Williams, p. 685, Washington, D.C., 1976.

Frank, L. A., Plasmas in the earth's magnetotail, in Space Plasma Simulations, edited by M. Ashour-Abdalla and D. A. Dutton, p. 211, D. Reidel Publ. Co., Dordrecht, Holland, 1985; also Space Sci. Rev., 42, 211, 1985.

Frank, L. A., K. L. Ackerson, and R. P. Lepping, On hot tenuous plasmas, fireballs, and boundary layers in the earth's magnetotail, J. Geophys. Res., 81, 5859, 1976.

Frank, L. A., J. D. Craven, K. L. Ackerson, M. R. English, R. H. Eather, and R. L. Carovillano, Global auroral imaging instrumentation for the Dynamics Explorer Mission, Space Sci. Instrum., 5, 369, 1981a.

Frank, L. A., R. L. McPherron, R. J. DeCoster, B. G. Burek, K. L. Ackerson, and C. T. Russell, Field-aligned currents in the earth's magnetotail, J. Geophys. Res., 86, 687, 1981b.

Frank, L. A., J. D. Craven, J. L. Burch, and J. D. Winningham, Polar views of the earth's aurora with Dynamics Explorer, Geophys. Res. Lett., 9, 1001, 1982.

Frank, L. A., C. Y. Huang, and T. E. Eastman, Currents in the earth's magnetotail, in Magnetospheric Currents, Geophys. Monogr. Ser., vol. 28, edited by T. A. Potemra, p. 147, AGU, Washington, D.C., 1984.

Frank, L. A., J. D. Craven, and R. L. Rairden, Images of the earth's aurora and geocorona from the Dynamics Explorer Mission, Adv. Space Res., 5, 53, 1985.

Frank, L. A., J. D. Craven, D. A. Gurnett, S. D. Shawhan, D. R. Weimer, J. L. Burch, J. D. Winningham, C. R. Chappell, J. H. Waite, R. A. Heelis, N. C. Maynard, M. Sugiura, W. K. Peterson, and E. G. Shelley, The theta aurora, J. Geophys. Res., 91, 3177, 1986.

Hardy, D. A., P. H. Reiff, and W. J. Burke, Response of magnetotail plasma at lunar distance to changes in the interplanetary magnetic field, the solar wind plasma, and substorm activity, J. Geophys. Res., 84, 1382, 1979.

Hardy, D. A., W. J. Burke, and M. S. Gussenhoven, DMSP optical and electron measurements in the vicinity of polar cap arcs, J. Geophys. Res., 87, 2413, 1982.

Hones, E. W., Jr., Transient phenomena in the magnetotail and their relation to substorms, Space Sci. Rev., 23, 393, 1979.

Hones, E. W., Jr. (editor), Plasma sheet behavior during substorms, in Magnetic Reconnection in Space and Laboratory Plasmas, Geophys. Monogr. Ser., vol. 30, p. 178, AGU, Washington, D.C., 1984.

Hones, E. W., Jr., The poleward leap of the auroral electrojet as seen in auroral images, J. Geophys. Res., 90, 5333, 1985.

Hones, E. W., Jr., Reply, J. Geophys. Res., 91, 5881, 1986.

Hones, E. W., Jr., A. T. Y. Lui, S. J. Bame, and S. Singer, Prolonged tailward flow of plasma in the thinned plasma sheet observed at $r \simeq 18\ R_E$ during substorms, J. Geophys. Res., 79, 1385, 1974.

Hones, E. W., Jr., C. D. Anger, J. Birn, J. S. Murphree, and L. L. Cogger, A study of a magnetospheric substorm recorded by the Viking auroral imager, Geophys. Res. Lett., 14, 411, 1987.

Huang, C. Y., and L. A. Frank, A statistical study of the central plasma sheet: Implications for substorm models, Geophys. Res. Lett., 13, 652, 1986.

Huang, C. Y., L. A. Frank, W. K. Peterson, D. J. Williams, W. Lennartsson, D. G. Mitchell, R. C. Elphic, and C. T. Russell, Filamentary structures in the magnetotail lobes, J. Geophys. Res., 92, 2349, 1987.

Kamide, Y., J. D. Craven, L. A. Frank, B.-H. Ahn, and S.-I. Akasofu, Modeling substorm current systems using conductivity distributions inferred from DE auroral images, J. Geophys. Res., 91, 11,235, 1986.

Lui, A. T. Y., E. W. Hones, Jr., F. Yasuhara, S.-I. Akasofu, and S. J. Bame, Magnetotail plasma flow during plasma sheet expansions: Vela 5 and 6 and IMP 6 observations, J. Geophys. Res., 82, 1235, 1977.

Lyons, L. R., A simple model for polar cap convection patterns and generation of θ auroras, J. Geophys. Res., 90, 1561, 1985.

Lyons, L. R., and T. W. Speiser, Evidence for current sheet acceleration in the geomagnetic tail, J. Geophys. Res., 87, 2276, 1982.

McComas, D. J., C. T. Russell, R. C. Elphic, and S. J. Bame, The near-earth cross-tail current sheet: Detailed ISEE 1 and 2 case studies, J. Geophys. Res., 91, 4287, 1986.

McPherron, R. L., Magnetospheric substorms, Rev. Geophys. Space Phys., 17, 657, 1979.

Meng, C.-I., and K. Makita, Dynamic variations of the polar cap, in Solar Wind-Magnetosphere Coupling, edited by Y. Kamide and J. A. Slavin, p. 605, Terra Scientific Publ. Co., Tokyo, 1986.

Mitchell, D. G., F. Kutchko, D. J. Williams, T. E. Eastman, L. A. Frank, and C. T. Russell, An extended study of the low-latitude boundary layer on the dawn and dusk flanks of the magnetosphere, J. Geophys. Res., 92, 7394, 1987.

Ness, N. F., The earth's magnetic tail, J. Geophys. Res., 70, 2989, 1965.

Peterson, W. K., and E. G. Shelley, Origin of the plasma in a cross-polar cap auroral feature (theta aurora), J. Geophys. Res., 89, 6729, 1984.

Potemra, T. A., Current systems in the earth's magnetosphere, Rev. Geophys. Space Phys., 17, 640, 1979.

Reiff, P. H., and J. L. Burch, IMF B_y-dependent plasma flow and Birkeland currents in the dayside magnetosphere 2. Global model for northward and southward IMF, J. Geophys. Res., 90, 1595, 1985.

Rostoker, G., Comment on "The poleward leap of the auroral electrojet as seen in auroral images" by Edward W. Hones, Jr., J. Geophys. Res., 91, 5879, 1986.

Rostoker, G., and T. E. Eastman, A boundary layer model for magnetospheric substorms, J. Geophys. Res., in press, 1987.

Russell, C. T., and R. L. McPherron, The magnetotail and substorms, Space Sci. Rev., 15, 205, 1973.

Slavin, J. A., E. J. Smith, D. G. Sibeck, D. N. Baker, R. D. Zwickl, and S.-I. Akasofu, An ISEE 3 study of average and substorm conditions in the distant magnetotail, J. Geophys. Res., 90, 10,875, 1985.

Spitzer, L., Jr., Equations of motion for an ideal plasma, Ap. J., 116, 299, 1952.

Tsyganenko, N. A., and A. V. Usmanov, Determination of the magnetospheric current system parameters and development of experimental geomagnetic field models based on data from IMP and HEOS satellites, Planet. Space Sci., 30, 985, 1982.

Williams, D. J., Energetic ion beams at the edge of the plasma sheet: ISEE 1 observations plus a simple explanatory model, J. Geophys. Res., 86, 5507, 1981.

PLASMA SHEET THEORIES

T. W. Speiser

Department of Astrophysical, Planetary, and Atmospheric Sciences
University of Colorado, Boulder, Colorado 80309

and

NOAA, SEL, Boulder, Colorado 80303

Abstract. Theories relevant to the geomagnetic tail plasma sheet are reviewed. The topics discussed include: kinetic instabilities, simulations, current sheet particle acceleration, and plasmoid and plasma sheet boundary layer pictures. Kinetic treatment is often appropriate for these topics since the tail plasma is collisionless. Fluid calculations are appropriate when stochastic processes dominate and for studies where long wavelengths are appropriate. Fluid theories are reviewed in an accompanying review. Fluid treatments of tearing modes and reconnection require a finite resistivity in the diffusion region. Thus, although anomalous resistivity can be guessed or in some cases calculated, ideally the kinetic treatment is often preferred. Particle motion and acceleration in the current sheet can give rise to beam-like distributions in the plasma sheet boundary layer. Studies of current sheet particle motion have also been used as the basis for kinetic tail equilibria models. Furthermore, quite recently current sheet particle motion is used directly in Coroniti's explosive tail reconnection model. Inertial effects from the equilibrium models provide the dissipation necessary for reconnection. Sketches of the current near-earth neutral line/plasmoid and plasma sheet boundary layer pictures are given, and a possible combination picture is discussed.

Introduction

The plasma sheet (and associated boundary layers) in the geomagnetic tail is a major region for transport of energy and particles into the inner magnetosphere. As observations and stability calculations are summarized in companion papers (Frank, 1987; Birn, 1987), we will briefly review here kinetic instabilities and simulations and models based on single particle motion, i.e., medium scale models. The current plasmoid and plasma sheet boundary layer pictures are sketched, and a synthesis model is discussed.

Dungey (1953, 1958) started much of the whole field of reconnection and neutral point stability following ideas of Giovanelli and Hoyle. Using fluid theory, Dungey showed that a magnetic neutral point in a magnetofluid is unstable, and with a small perturbation will quickly collapse, with the current density and inverse thickness going to infinity in a finite time. It was a fortuitous situation wherein the system was described by a finite set of equations with a finite set of unknowns. The basic approximations were neglect of pressure gradients and resistivity. Dungey argued that pressure gradients might change the time constants but would not stop the discharge, whereas a finite conductivity would stop the collapse to a finite thickness with a finite current density. Under this situation, a finite electric field is left across the region, and plasma flows across the separatrices. As the plasma is frozen to the field in the external region, it is as if a field line moves across the separatrices, but this requires the field line to be broken as it passes through the neutral point and to be reconnected or rejoined with its partner. A measure of the reconnection rate is given by the plasma flow across the separatrices or by the electric field left across the region (Vasyliunas, 1975).

Kinetic Equilibria

In one of the first kinetic equilibrium studies of a current sheet topology, Harris (1962) solved the combined Vlasov-Maxwell equations. Since the Vlasov (or collisionless Boltzmann) equation is identical to the Liouville theorem ($Df/Dt = 0$) in phase space, a distribution function built up from constants of the motion for individual particle orbits will satisfy the Vlasov equation. Harris utilized this fact and started with a distribution function of the form

$$f_s = (M_s/2\pi\theta_s)^{3/2} N_s$$

$$\exp\{-(M_s/2\theta_s)[\alpha_1^2 + (\alpha_2 - V_s)^2 + \alpha_3^2]\} \quad (1)$$

where s is an index for species, M_s, θ_s, and N_s are the species mass, thermal energy, and density, V_s is the bulk speed, and α_1, α_2, and α_3 are the constants of motion. (These are combinations of the total energy and y and z momentum components. For this one-dimensional model, the dimension of variation is taken to be x, with $\hat{e}_x$ the direction perpendicular to the current sheet.) Using this distribution function along with the first and second moments to get the charge and current density, the Poisson equation and Ampere's law then give the coupled differential equations for the electrostatic (Φ) and magnetic (A_y) potentials. One solution is for $\Phi = 0$ and then $A_y \sim \log \cosh(x/h)$, where $h = cL_D/V$ is the scale for the variation. (c is the speed of light, L_D is the Debye length, and V is the streaming velocity in the y-direction.) The magnetic field thus varies as $\tanh(x/h)$ and the plasma density as $\cosh^{-2}(x/h)$, results which are widely quoted in the literature as well as used as a starting point for stability studies. Harris noted that his result showed that the magnetic pressure at infinity was just balanced by the plasma pressure at the center of the current sheet. In addition, using his solutions, it can be shown that the total pressure, gas plus magnetic, is constant everywhere. Finally, Harris showed that the fields could be found in other systems moving relative to his by a Lorentz transformation, and that his solution becomes the Bennet pinch solution in cylindrical coordinates.

Galeev (1982) showed for a weakly two-dimensional current sheet that the simplest generalization of the Harris sheet includes a weak normal magnetic field component, with the additional magnetic field tension being balanced by a weak pressure gradient.

Stability and Tearing Modes

In the tearing mode instability, a current sheet tears or induces filamentation as parallel current elements attract. For the tearing mode, the classic MHD result is due to Furth et al. (1963). There it was found that the growth rate, $\gamma \sim \eta_c^{3/5} k^{-2/5}$, where η_c is the resistivity and k is the wave number, favoring growth at longer wavelengths. Furth (1964) also discussed the collisionless tearing mode from the standpoint of the Vlasov equation. He started with the one-dimensional equilibrium solution, as discussed above, and considered perturbations of the form $f(z) \exp(\gamma t + ikx)$ where z is the dimension of variation perpendicular to the sheet and x is the dimension along the sheet in the direction of the equilibrium field. (We changed Furth's notation to correspond to the usual solar magnetospheric coordinates in the geomagnetic tail.) For flow velocities (V) that are large compared to the thermal speed (V_t), Furth estimated the collisionless growth rate as $\gamma \sim (4\pi k N_o he^2/MC^2)^{1/2} V$, where N_o is the maximum unperturbed plasma density, e and M the ion charge and mass, h the scale for the z-variation (as in the Harris sheet), and V the bulk velocity. Furth then noted that this result is equivalent to replacing the MHD collisional resistivity η_c by $\eta^* = \eta_c + \eta_i$, where η_i is an effective resistivity due to the ion mass and $\eta_i \propto \gamma$ is the collisionless growth rate. To quote Furth: "The extra term corresponds to the dj/dt term in Ohm's Law. We can use all of the results of section A (MHD), simply replacing η by η^*." Thus, in the collisionless case, current-carrier inertia takes the place of resistivity in fluid calculations.

Coppi et al. (1966), using semi-intuitive arguments, analyzed the stability of the Harris sheet. They found a growth rate of the order $(V_{the}/de)(r_{Le}/h)^2$, where V_{the} is the electron thermal speed, r_{Le} is the mean electron larmor radius, h, as before, is the sheet thickness, and $de \simeq (r_{Le}h)^{1/2}$ is the effective slab thickness centered on the neutral sheet, where macroscopic energy is transferred to resonant electrons. The result is for long wavelengths such that $k^2h^2 < 1$ and agrees with the more formal result of Laval et al. (1965). Using values from IMP-1 of 2h = 600 km, B_o = 16 nT, $\theta_i \simeq 1$ keV, Copi et al. (1966) found a growth time of order 5 to 15 s for electron thermal energies from 1 to 0.01 keV. They also estimated that electrons could be energized to about 10 keV in times of the order of the growth time.

A very powerful method to investigate stability of a system utilizes a variational approach on a function δW where $\delta W_{min} \geq 0$ implies stability and $\delta W_{min} < 0$ implies instability (e.g., Schindler, 1972, 1974; Schindler and Birn, 1978; Galeev, 1982). As ΔW represents perturbed system potential energy and δK represents the kinetic energy, and $\delta W + \delta K = 0$, if δW is negative, δK must be positive; i.e., the system kinetic energy grows and the system is unstable. If we can find the minimum of δW and show that it is positive, then the system is stable. Conversely, if the minimum is negative, then the system is unstable.

In many cases, the function δW may be separated into two parts: $\delta W = \delta F + \delta G$, where δF always has the same form

$$\delta F = \frac{1}{2} \int [(\nabla A_1)^2 - 4\pi A_1^2 d^2P_0/dA_0^2] d^2r \qquad (2)$$

and δG depends on which model is being considered. Here, $P_0(A_0)$ is the equilibrium pressure, A_0 is the equilibrium y-component of the vector potential (considering field lines in the x-z plane), and A_1 is the first-order perturbation. Schindler and Birn (1978) showed that for a large class of equilibria, $\delta G \geq 0$ for all A_1. So a sufficient criterion for stability is $\delta F_{min} \geq 0$.

Grad (1964) showed that the minimum of the expression $F = [B^2/8\pi - P]d^2r$ for ideal MHD and under constant pressure variations gives a sufficient criterion for instability. Schindler and Birn (1978) showed that the first variation of this function gives rise to the equilibrium condition $-\nabla^2 A_0 = 4\pi\, dP_0(A_0)/dA_0$, and the second variation yields equation (2) above. Schindler and Birn (1978) identified F with the free energy of the system. If $\delta F \geq 0$, the system is stable since all neighboring states have higher free energy than the equilibrium state. If $\delta F < 0$, then there are some (one?) neighboring states that have less free energy than the equilibrium and the system is potentially unstable, but the full δW, including δG, must be investigated. For various models, δG has the following forms: ideal MHD (Bernstein et al., 1958)

$$\delta G = \int \gamma P[\nabla \cdot (A_1 \nabla A_0/|\nabla A_0|^2)]^2 d^2r , \qquad (3)$$

incompressible, resistive MHD with resistivity frozen in (Tasso, 1975)

$$\delta G = \int (d^2P_0/dA_0^2)A_1{}^{-2}d^2r , \qquad (4)$$

Vlasov theory (Schindler et al., 1973; Schindler, 1974)

$$\delta G = \int \left|\partial\rho_0/\partial\phi_0\right| \{\langle\Psi_1\rangle - [\langle\Psi_1\rangle]^2\}d^2r, \qquad (5)$$

where $\partial\rho_0/\partial\Phi_0 = \Sigma\, e^2 \int (\partial F_0/\partial H_0)d^3v$; $F_0(H_0,P_0)$ is the particle distribution function for a given species as a function of the constants of the motion [H_0 (the Hamiltonian) and P_{y0} (the generalized momentum conjugate to the y-coordinate)], $\partial F_0/\partial H_0$ is assumed to be negative; $e\Psi_1$ is the perturbed Hamiltonian, $e\Psi_1 = -(e/c)v_y A_1 + e\Phi_1$; Σ sums over the particle species. The various averages have different meanings; see Schindler and Birn (1978) for details.

Utilizing these concepts, Schindler and Birn (1978) showed that for the magnetospheric tail, free energy is in fact available and that for the realistic case of a diverging tail, $\delta F_{min} < 0$ always, while the tail would be stable if it converged (cross-sectional area reduces with increasing distance away from the earth). If $P(x,0)$ and B_x vary monotonically with x, then the stability condition can be simply written as: $B_z > 0$ for all z implies F-stability; whereas if B_z changes sign, then F-instability is implied. (B_z would change sign for example if $B_z > 0$ in the plasma sheet and yet with increasing z, the tail diverges, so after some z, B_z would become negative.)

In order to find a growth rate, the dynamic equations have to be studied as indicated above for the Harris sheet perturbation by Coppi et al. (1966). For sufficiently weak B_z where both electrons and ions are non-adiabatic in the neutral sheet, Schindler et al. (1973) found the growth rate for the electron tearing mode of order $\gamma_e \sim V_{Te}a_e{}^{3/2}\, L_z{}^{-5/2}$, where V_{Te} is the electron thermal velocity.

For an intermediate B_z (non-adiabatic ions, adiabatic electrons), one finds the growth rate (τ_i) for the ion tearing mode (Schindler, 1974) as

$$\tau_i{}^{-1} = \gamma_i = V_{Ti}a_i{}^{3/2}\, L_z{}^{-5/2} .$$

The adiabatic electrons can lead to a stabilizing effect (Galeev and Zelenyi, 1975, 1976); however, according to the earlier theories, for a range of B_z values, there still exists a considerable range where the ion tearing mode may take place. But more recently, Lembege and Pellat (1982) found that compressibility of the adiabatic electron gas does indeed stabilize. Coroniti (1985) estimated that instability begins for wavelengths greater than about 100 R_E, and this is so long compared to observed scales in the tail as to be unreasonable. On the other hand, Goldstein and Schindler (1982) obtained a tearing unstable mode for more reasonable wavelengths of order R_E, but the assumed tail plasma parameters were not very realistic.

Taking a different tack of investigating the effect on tearing of initially anisotropic distribution functions, Chen and Palmadesso (1984) (see also Chen and Lee, 1985; Chen et al., 1984) showed that the tearing growth rates can be substantially enhanced for ion temperature anisotropies $T_\perp/T_\parallel > 1$. Conversely, anisotropies with $T_\perp/T_\parallel < 1$ strongly stabilizes the mode. Other potentially destabilizing conditions, such as electron pitch angle scattering, are discussed in the excellent reviews by Galeev (1982) and Coroniti (1985).

Finally, Baker et al. (1982) studied the role of heavy ions in the initiation of the tearing mode. Regions of enhanced ionospheric O^+, for example, can lead to locally enhanced tearing growth rates.

Explosive Tearing Reconnection

Past the onset of instability, Galeev (1982) (Galeev et al., 1978) estimated that the amplitude of the perturbation $b_1(t)$ increases explosively as: $b_1(t) = b_1(0)/(1 - t/\tau_R)$, where τ_R is the characteristic time of an explosive growth.

Note that $b_1(t)$ is proportional to the initial perturbation $b_1(0)$. This solution therefore is reminiscent of Dungey's (1953) discharge, the starting point for ideas about reconnection (see also Forbes and Speiser, 1979; Forbes, 1982). For typical tail parameters, Galeev (1982) estimated $\tau_R \sim 100$ s, with highest possible energies of accelerated protons of order 1 MeV, and the acceleration time for these protons of ~100 s. Galeev also showed that the ion spectrum goes like a power law (as observed) and that observed inverse velocity dispersion can be produced by the model as well. (The lower-energy ions are produced at an earlier time by the explosively growing electric field.)

Zelenyi et al. (1984) used the above results of the explosively growing tearing mode and simulated particle dynamics in this model. They then compared the simulation with observations of proton spectra in the tail by Sarris et al. (1981). The calculations are in reasonable agreement with the observations, but the observations were made during quiet to moderately disturbed times, while the model should, of course, apply to the beginning of a substorm.

Coroniti (1985) has developed a model incorporating the nonlinear explosive phase plus saturation. In Coroniti's picture, enhanced erosion of dayside flux launches a fast rarefaction wave tailward and tailward convection is stimulated. After about 20 min the magnetosphere reaches a stressed state with an earthward displaced magnetosphere, reduced B_z in the plasma sheet, and initiation of slow earthward flow in the plasma sheet. A crisis point is reached: the stressed plasma sheet is pulled tailward by the enhanced solar wind stress on the tail and earthward by the fast rarefaction wave stresses trying to return flux to the dayside. Coroniti concluded that reconnection starts out in the tail at an initially slow rate and then evolves to a rapid rate associated with substorm expansion phase onset. In the final saturation phase, steady collisionless reconnection flow takes place at a rate comparable to the maximum Petschek rate. In Coroniti's model, inertial conductivity (see section on "Current Sheet Particle Motion") is responsible for the explosive behavior. The particle lifetime in the system diminishes as the vector potential increases, and as the total current is fixed by the external solution, the electric field has to increase faster than linearly to compensate for the decreasing conductivity and current layer thickness. After a period of about 1 hour of slow reconnection, the explosive reconnection time scale is found to be of order 5-10 min, in agreement with substorm onset times, and cross tail potentials of order 1 MV are predicted during explosive phase.

Kinetic Simulations

Simulations in the earth's magnetosphere and tail fall into several categories: MHD, kinetic, mixed kinetic and fluid, and laboratory simulations. Additionally, simulations may be local, dealing with one small part, or global in nature--attempts to mimic the entire solar wind-magnetosphere interaction. Here a few of the computer simulations that are mostly based on kinetics will be reviewed. For reviews of MHD models, see Coroniti (1987) and Birn (1985, 1987). For a review of laboratory simulations, see Stenzel and Gekelman (1984), Baum and Bratenahl (1980), and Podgorny (1981).

An ideal kinetic simulation would contain full electron and ion dynamics, over a volume large enough to model say a significant portion of the geomagnetic tail. Unfortunately, this is not possible at this time. Katanuma and Kamimura (1980) have included complete electron and ion dynamics but in their treatment they are limited to scales of the order of several Debye lengths, i.e., approximately hundreds of meters, small compared with relevant lengths in the tail. Terasawa (1981) presented results of a hybrid simulation where the electrons were treated as a fluid but full ion dynamics were considered. Terasawa basically confirmed the nonlinear theory of Galeev et al. (1978). After a linear tearing mode phase, the nonlinear explosive phase is found, depending on the initial perturbation (as with GCA), but also depending on the system scale size. The explosive mode was found for the larger system ($L_x/\rho \sim 51$) but not for one half as large. The nonlinear growth rate depended on the equilibrium drift velocity V_o and varied as $V_o^{3/2}$, consistent with Galeev et al. (1978), but a factor of 3 smaller in magnitude. The energetics were very interesting. Flow velocities were reduced inside the magnetic island in the explosive phase as compared to the linear phase, consistent also with a reversed electric field inside the island. Acceleration of particles by the rapidly growing electric field is significant in the vicinity of the x-line, but since the density is lower there as compared to the magnetic island where deceleration occurs, the deceleration effect dominates considering the system as a whole. In a fully electron and ion kinetic study, Leboeuf et al. (1982) essentially confirmed the explosive growth, but they assumed an ion-to-electron mass ratio of 10, which according to Lembege and Pellat (1982) might have caused an important underestimation of the electron compressibility.

Swift (1982) also used a particle code that included ion dynamics. He concluded that the dawn-dusk electric field results in plasma sheet thinning and that the electric field along with a lobe particle population was necessary for the maintenance of the extended magnetotail. Birn and Schindler (1985), however, pointed out that Swift's system was not in equilibrium initially, and they questioned the addition of a constant E_y everywhere. For Birn and Schindler's MHD simulation, only a small fraction of the lobe E_y penetrates into the interior of the plasma sheet. In an update of his first model, Swift (1983) showed that the revisions allow two-dimensional simulations for more extended time periods and under more realistic conditions. He also found that when the magnetic field starts to collapse, most of the energy goes into earthward streaming of plasma sheet particles (see following section). The previous criticism of Birn and Schindler may still apply, but Swift noted that he obtained an equilibrium convection electric field proportional to B_xB_z which is nearly the equilibrium condition obtained from particle orbits (following section) (Lyons and Speiser, 1985; Hill, 1975).

Current Sheet Particle Motion

Motion in Quasi-Steady Fields

Following Dungey's (1953) original ideas concerning neutral point acceleration, particle motion in current sheets has been studied by many authors (e.g., Parker, 1957; Speiser, 1965, 1967, 1968, 1970; Schindler, 1965; Alexeev and Kropotkin, 1970; Sonnerup, 1971; Eastwood, 1972, 1974; Pudovkin and Tsyganenko, 1973; Cowley, 1973, 1980, 1982, 1984; Jaeger and Speiser, 1974; Swift, 1977; Stern, 1977; Wagner et al., 1979; Lyons and Speiser, 1982; Speiser and Lyons, 1984; Lyons, 1984). In the absence of stochastic processes, Liouville's theorem says that the particle distribution function is constant along a particle orbit. Thus, in this approximation, particle orbit solutions are equivalent to solutions of the Vlasov equation. The simplest model of a current sheet involves a sheared magnetic field which changes direction by 180° on passing through a current sheet of finite width. The field strength goes to zero everywhere at the sheet center. There may or may not be an electric field within the current sheet perpendicular to the field but tangential to the sheet. If there is an electric field, particles become trapped in the sheet, execute damped oscillations about the sheet, and become accelerated by the electric field (Speiser, 1965, 1967). In this configuration, plasma **E** x **B** drifts into the current sheet from both sides, becomes accelerated in the interior, and satisfies the definition of reconnection from the fluid point of view (Dungey, 1961; Alfvén, 1968; Vasyliunas, 1975). Such a simple model would not be too interesting if applied to the geomagnetic tail as all accelerated particles would get shot out of the dawn (electrons) and dusk (ions) sides of the tail into the sheath region. However, the Alfvén (1968) consistency condition ($\Phi_A = B_{x0}^2/\mu_0 n_0 e$) allows the prediction of the overall voltage drop as a function of the field strength and plasma density, and was therefore one of the earliest predictions of a reconnection rate (more strictly speaking, annihilation rate for this simple model). This model may also be applicable in a limited region around a neutral point or neutral line.

In the next approximation, we add a constant, weak normal magnetic field to the current sheet. The particle orbits now include a gyration about the weak normal field in addition to the oscillation about the sheet (Speiser, 1965, 1967, 1968, 1970; Lyons and Speiser, 1982; Speiser and Lyons, 1984). The particles, however, are turned 90° by the gyromotion to the accelerating electric field and when turned this much, they pop out of the sheet with small pitch angles. (The Lorentz force causing the oscillation about the sheet changes sign, thus ejecting the particles from the sheet.) The Alfvén consistency condition becomes generalized as the Eastwood (1972) condition, now a function of the normal field ($\Phi_E = B_{x0}B_zL/(2\mu_0 n_0 m)^{1/2}$, where B_{x0} is the external, tangential field component, B_z the normal field component, n_0 the external plasma density, L the current sheet width, and m the ion mass; see Eastwood (1972) for the connection between Φ_E and Φ_A. The Eastwood consistency condition is close to, but not quite, the prediction we would like to have of the reconnection rate, as the relevant value of B_z is unknown. Of course, we can take typical observed values for the geomagnetic tail and find the predicted electric field, for example. Values of the cross tail potential drop come out to be of order 100 kV, about what is expected. In one sense, this model is simpler than the first model, for the electric field can be transformed away by moving with speed E_y/B_z toward the earth (Speiser, 1965). This moving frame where **E** = 0 is sometimes referred to as the DeHoffman-Teller frame. The particle interaction with the current sheet can be likened to specular reflection off a moving mirror (Cowley, 1980). Particles initially moving away from the earth (into the mirror) gain twice the transformation velocity. Those moving toward the earth with the transformation velocity are not energized, i.e., do not interact with the current sheet. Current sheet oscillation is expected when B_z is weak enough so that the usual adiabatic theory (μ = const) is violated. [When B_z is sufficiently strong, particles merely follow field lines through the sheet (Eastwood, 1972; Stern 1977; Wagner et al., 1979), but the moving mirror picture also works for the adiabatic case (Chapman and Cowley, 1984)]. However, because of the new oscillatory character of the orbits about the current sheet, new adiabatic invariants come into play (Schindler, 1965; Speiser, 1970; Sonnerup, 1971). When this model is valid, electron drift is on the average in the opposite direction, supplying the requisite cross tail current. For larger B_z, such that the electrons remain adiabatic (μ = const), the ions would undergo the current sheet oscillation and would supply most of the cross tail current, and the electron drift would be in the **E** x B_z direction, i.e., toward the earth. Speiser (1970), in fact, showed that for this simple, constant B_z model, there is a **j**,**E** relationship even though this is single particle motion with no collisions. The average--say ion--drift is roughly the E_y/B_z transformation velocity and the negative of that for the (non-adiabatic) electrons. Therefore, the net current density is just $j = n_e \bar{V}_D$, where $\bar{V}_D$ is the average velocity difference between the ions and electrons, or $j \approx n_e (E_y/B_z)\hat{e}y$, giving our **j,E** relation. Cowley (1978) showed that for a one-dimensional current sheet, trapped particles, whether adiabatic or not, produce no net current. However, untrapped particles can give rise to a net current. For our use here, net currents can be produced if these accelerated particles do not return (or return to some other location where conditions have changed). The effective (or inertial) conductivity is therefore $\sigma_i \approx n_e/B_z$ or $\sigma_i \approx n_e^2/m\omega_g$, where ω_g is the gyrofrequency about the weak B_z. Thus, this gyrofrequency takes the place of a collision frequency in the conductivity expression. Energy dissipation is from the electric field into accelerated particles and not into stochastic, heat producing processes (although these may be important at other stages). It is this inertial concept that Coroniti (1985) uses for his explosive reconnection model (see above).

For the simple sheet with constant B_z, Cowley (1978) was in fact able to find general constants of the motion for arbitrary $B_x(z)$, including a possible $E_z(z)$. Cowley's study showed that particles are confined to a flux tube during this interaction (his Figures 1 and 2) whether they remain adiabatic or not. He also concluded that for self-consistency in such one-dimensional models, there must be a class of untrapped, i.e., transitory particles. If (magnetically or electrically) trapped particles are to maintain the current, then one-dimensional models are not appropriate.

The Eastwood eqiulibrium condition has been generalized by Hill (1975) to include the effects of initial streaming and pressure anistropy. These effects will of course change the simple **j,E** relation above. Lyons and Speiser (1985) showed how the Eastwood-Hill results (except for the anisotropy terms) arise from the results of single particle motion. The consistency in Hill's more general form is just the stress balance condition which reduces to the marginal firehose condition just outside the current sheet.

Lyons and Speiser (1982) followed many ion orbits numerically in a model of the tail current sheet corresponding to this second model (E_y = const, B_z = const). For various initial pitch angles, particles can be ejected from the current sheet over a wide range of pitch angles although there is some tendency for initially very low-energy particles to come out field-aligned, and otherwise for pitch angles to be preserved. [In the adiabatic (μ = const) limit, particles obviously preserve pitch angle, and Alexeev and Kropotkin (1970) have shown that pitch angle is also preserved in the limit of a very thin current sheet, or equivalently in the limit of very large initial particle energies.] For nominal fields, the energy increase is markedly higher for the particles ejected with small pitch angles than for those ejected with large pitch angles. Assuming an initial mantle or boundary layer distribution, the calculated orbits allow a mapping to be made to the energized, streaming distribution on the edge of the plasma sheet. Comparison of this modeled distribution with observations resulted in good agreement. When B_z is weak enough, analytic solutions may be applied and initial distributions may be analytically mapped into energized, streaming distributions in the plasma sheet boundary layer (Speiser and Lyons, 1984).

For this (B_z = const) model, Cowley (1982) showed how one can simply map an initial distribution into an accelerated, streaming distribution by reflecting the initial distribution about the plane $V_\parallel = V_F$, where V_F is the transformation velocity, E_y/B_z.

Using a model slightly more complicated than the B_z = const model, Jaeger and Speiser (1974) again numerically followed ion orbits. The model had a varying B_z, falling off with distance as observed and as would be expected, for example, with a neutral line model. However, the rate of fall-off was like that imposed by a dipole field (r^{-3}) and therefore was somewhat unrealistic. Again (using Liouville's theorem), initial distributions were mapped into accelerated, streaming distributions. The initial distributions were from fluid models in the magnetosheath, incorporating reasonable solar wind compositional values. Final (post-acceleration) energies gave nearly--but not exactly--equal energy per unit charge, i.e., what would be expected for various particles falling through the same potential drop. This was surprising since the previous B_z = const model predicted iso-velocity energization. These results and those of Speiser (1967) show that heavy ions should be found in streaming distributions in the plasma sheet boundary layer inside the proton layer. Taking alpha particles as an example, the reason for this is that for the same energy, alpha particles will have a larger gyroradius about the weak normal field, B_z. Therefore, in a region where the proton gyroradius is about the tail width, proton beams toward the earth will be formed, but alphas will not get turned sufficiently and will thus be shot out of the sides of the tail. If we move the interaction region closer to the earth (where B_z is larger in this model), alphas will now be turned toward the earth and ejected along field lines that are necessarily inside (i.e., closer to the mid-plane) the energetic-proton streaming layer. Möbius et al. (1980) have indeed found evidence of alpha particle layers within proton layers. However, it should be noted that any model (such as Sarris and Axford, 1979) which has alphas moving more slowly than protons will tend to produce this type of separation via **E x B** drifts. One problem with the Jaeger and Speiser (1974) model was that the intensities were much larger than the observations. Lyons and Speiser

(1982) used a more realistic initial distribution, and comparisons with observations were more favorable.

As previously discussed for the B_z = const model, B_z may be weak as seen by one species, yet strong as seen by another. For a tail field of 20 nT and B_z of 1 nT, ions, for example, with moderate energies, essentially violate the condition μ = const and thus participate in current sheet oscillation. Electrons on the other hand would be expected to remain adiabatic. Lyons (1984) followed many electron orbits in the B_z = const model. He found that most electrons incident on the current sheet become trapped in it and they gain energy from the cross tail electric field. Even for nominal parameters as above, slight violations of the adiabatic condition are important for electrons. Their energy gain goes as $[B_z(x)]^{2/3}$. Some small fraction of electrons within the current sheet may undergo pitch angle scattering and thus end up in the atmospheric loss cone.

If initial streaming (V_{x0}) of the particles is included, the current-electric field relationship becomes

$$j_y = \left|\frac{2nq}{\pi}(V_{x0} - E_y/B_z)\right| = \left|\sigma_i E_y\left(1 - \frac{V_{x0}}{E_y/B_z}\right)\right|, \quad (6)$$

which is Lyons and Speiser's (1985) equation (6). Thus, this looks like a form of Ohm's law, but it should be emphasized that it comes from a study of the individual particle orbits, and not from the generalized Ohms' law. Specifically, the average drift velocity component in the y-direction, $\bar{V}_y$, is needed. Using the above frame where there is a finite E_y, we see that $\mathbf{E} \cdot \mathbf{j}$ is positive, i.e., an energy sink that goes into accelerated particles. It is an inertial effect and as such is affected by frame transformations, unlike a true (stochastic) resistivity. In magnetotail dynamics a resistivity is important to decouple the plasma motion from the magnetic field motion or, equivalently, from the $\mathbf{E} \times \mathbf{B}/B^2$ motion. That is, we are interested in finding a slippage between plasma and field. If we define $\mathbf{V_s}$ as such a slippage velocity, then

$$\mathbf{V_s} = \bar{\mathbf{V}}_\mathbf{p} - \mathbf{E} \times \mathbf{B}/B^2 \neq 0, \quad (7)$$

where $\bar{\mathbf{V}}_\mathbf{p}$ is the average particle velocity in the sheet; it has the following x and y components (Lyons and Speiser, 1985):

$$\bar{V}_x = E_y/B_z, \; \bar{V}_y = \frac{2}{\pi}\left|V_{x0} - E_y/B_z\right| . \quad (8)$$

Plugging equation (8) into (7), we find that the slippage velocity $\mathbf{V_s}$ is just

$$\mathbf{V_s} = \bar{\mathbf{V}}_\mathbf{y}. \quad (9)$$

Thus, inertial resistivity does supply a slippage between plasma and field motion, but it is of an entirely different form than that derived from a resistive term in the generalized Ohm's law ($\mathbf{V_s}(\eta) = \eta\, \mathbf{j} \times \mathbf{B}/B^2$).

Motion Near Neutral Points or Lines

There have been a few studies made of particle motion near magnetic neutral lines or points (e.g., Aström, 1956; Rusbridge, 1971; Stern, 1979). As indicated in the previous section, numerical studies of particle orbits with a model with B_z varying did produce somewhat different results from the B_z = const model (Jaeger and Speiser, 1974). Aström (1956) categorized the types of electron trajectories in the vicinity of an x-type neutral line and found orbits resembling trochoids, meandering, and figure-8 types. He also investigated the stability of orbit types if they were perturbed out of the initial plane. Rusbridge (1971) was interested in the change of a particle's magnetic moment on interaction with an x-type neutral line. He found that the maximum change was by a factor of about 5.75, and the key factor was the particle's phase angle at the point where it ceased to be adiabatic. Stern (1979) analyzed equations of motion near o-type neutral lines in some limiting cases and found that particles become decoupled from the field line motion and then could undergo a runaway acceleration mode along the neutral line. Furthermore, Stern concluded that the acceleration process is more efficient along o-type than x-type neutral lines, as there is a kind of stability there. However, Vasyliunas (1980) showed that the line integral of the electric field along an o-line is severely limited to Φ_A, the Alfvén potential, twice the average magnetic energy per paticle. o-lines, thus, cannot account for the energetic component of the charged particle population in the magnetotail.

Martin (1986a) studied particle motion in an x-type magnetic neutral line and electric field along the line. He showed that the system is basically non-integrable but can be studied qualitatively with a two-dimensional effective potential (see also Stern, 1979). For a thin current sheet, the motion is relatively simple and particles can be significantly energized. For a thick current sheet, the motion can be chaotic (many transitions are made and the motion becomes randomized) for larger pitch angle particles, and particles can lose energy. For the chaotic motion, Martin (1986b) estimated an effective conductivity (minimum value) $\sigma_{ch} \sim 10^{-4}$ mho m^{-1}. ($\sigma_{ch} = ne^2/m\lambda$; m = particle mass; e = charge, n = number density; λ = Lyapunov exponent; λ is determined numerically and is dependent on the model fields, scale lengths, etc., and it is a measure of the time scale for particle orbits to become uncorrelated.) Martin also found that all particles entering the neutral line region become chaotic. Chen and

Palmadesso (1985) found a similar regime of stochastic orbits for a tail-like magnetic field, but they also found three types of orbits which remain distinct in the absence of noise. The distribution will be strongly non-Maxwellian and may well have implications for the plasma dynamics.

Motion in Time-Dependent Fields

In the preceding section, it has been assumed that $E_y \approx$ const in space and time. Obviously, there are times such as during substorms when the tail fields and plasmas suffer marked changes. In the reconnection sense, at the onset of reconnection, one would expect time changes in the fields to be large until saturation or quasi-steady state is reached (if indeed that ever happens) (see e.g., Galeev et al., 1978; Schindler, 1974; Coroniti et al., 1977). Even when the fields are varying in time and space, the current sheet orbits discussed above may at times be valid if the variations are on a scale large compared to the particle scale lengths in the current sheet or if the variations are slow compared to the time spent by the particles in the current sheet interaction. Nevertheless, we would not expect cross tail potentials to typically exceed say ~200 kV; thus we would be hard pressed to explain 0.5-MeV particles with the previous current sheet acceleration models. Pellinen and Heikkila (1978) followed particle orbits in such time-dependent fields and found that the most efficient accelerating mechanism is a two-step process with initial acceleration along a neutral line followed by batatron acceleration. Energies into the MeV range can possibly be achieved. Galeev (1982) in fact showed that the explosive phase of the tearing instability in the magnetospheric tail also generates particle bursts with a power law spectrum, and for typical tail parameters, he estimated maximum proton energies of ~1 MeV, acceleration time of ~10 s, and a power law of ~-5, $[J(E_p) \alpha E_p^{-5}]$, which compares favorably with observations. Inverse velocity dispersion can also be qualitatively explained, as the lower-energy portion is generated earlier by the explosively growing electric field. The induction electric field is also generated near the x-line and directed from dawn to dusk leading to harder protons (electrons) on the evening (morning) side of the tail. Electron acceleration would take place but with smaller intensities as compared with the ions since their acceleration region would be reduced in size.

Plasmoid and Plasma Sheet Boundary Layer Models

The plasmoid model suggests that at the onset of a substorm, reconnection or tearing is initiated in the central plasma sheet close to the earth. Subsequently a plasmoid or magnetic bubble is formed which then propagates downtail. This picture is an elaboration of the earlier ideas on reconnection of Dungey's (1953, 1958, 1961). Such reconnection and plasmoid formation would seem to follow naturally from tearing mode studies (Schindler, 1972, 1974; Schindler and Birn, 1978; Galeev, 1982; Coroniti, 1985). See also Schindler and Ness (1972), Russell (1972), Vasyliunas (1976), and Hones (1979). Simulations also often show the appearance of x-lines, o-lines, plasmoids, etc. (see e.g., Birn, 1987). Observations have been reported in the distant tail that are consistent with plasmoid ejection following substorm onset in the near-earth tail (Hones et al., 1982, 1986). In the near-earth tail it is often found that there are large normal magnetic field components at the center of the neutral sheet and these normal components are correlated with substorm onsets (Speiser and Forbes, 1981). Sometimes these normal components change sign with subsequent neutral sheet crossings, consistent with the picture of plasmoid propagation. Speiser and Schindler (1981) argued that for normal components outside the range 6 nT > B_z> -3 nT, large volumes of the tail plasma sheet are necessarily involved; thus the tail is affected macroscopically, not just microscopically. Thus, the large (in absolute value) normal components reported by Speiser and Forbes are consistent with the plasmoid model.

In the plasma sheet boundary layer (PSBL) picture (Eastman and Frank, 1984; Eastman et al., 1985; Frank, 1985, 1987), the PSBL is considered to be the dominant region of transport of plasma, energy, and momentum in the tail. The PSBL is thought to be topologically connected to a distant (Dungey-like) neutral point in one scenario [Frank's (1985) F region]. The distant neutral point (or line) coupled with the dawn-dusk electric field could thus provide the ion beams (above) found in the PSBL, using the simple current sheet acceleration mechanisms as outlined above.

Two questions which need to be resolved for both of these pictures are as follows: (1) For the PSBL, what initiates a substorm? (2) How does the observation that southernmost quiet arcs typically brighten first at substorm onset fit into the plasmoid model? [Lyons (1987) has pointed out that this observation, coupled with a mapping of quiet arcs into the PSBL, would seem to imply that neutral point dynamics (plasmoids) play no role in substorm initiation.]

A possible resolution may be to have a combined neutral point-PSBL model. That is, we can imagine a distant neutral line coupled with current sheet particle motion producing a quasi-steady plasma sheet boundary layer. As tail stresses increase, a tearing instability, or Coroniti's near-earth reconnection, begins. Initially the electric field and the local plasma sheet deformation occurs at only a very slow rate (say for about an hour). Then the explosive tearing begins at about the time the pre-existing PSBL has been squeezed--or is topologically connected--into the near-earth neutral line. Thus, the substorm onset would occur with the onset of Coroniti's explosive phase, and the PSBL would be connected along the separatrices attached to the explosively growing neutral line. Southern quiet arcs associated with the inner edge of the PSBL would brighten first in accordance with observations. The inner neutral line and associated current sheet would continue to feed particles into the now active PSBL. The inner PSBL would become disconnected from the outer PSBL and distant neutral line, at least until the inner neutral line retreated tailward or disappeared.

Summary

1. Collisionless linear tearing mode analysis (Schindler et al., 1973) has progressed to the point that it looks like several unstable regimes exist. Electron magnetization stabilization in finite B_z fields can apparently be overcome by pitch angle scattering, by anisotropic temperature distributions, and/or by electrical coupling to the ionosphere.

2. Explosive nonlinear tearing (Galeev et al., 1978) has reasonable onset times, megavolt potentials, and has been verified by Terasawa's ion simulation and also by a fully kinetic (Leboeuf et al., 1982) simulation. Some uncertainties still exist.

3. Coroniti's (1985) explosive model uniquely depends upon particle inertia directly from the current sheet dynamics. Reasonable onset times and megavolt potentials are deduced. Electron Hall and mirror dynamics are ignored.

4. Simple quasi-steady current sheet particle motion leads to consistency conditions between the fields and particles and to an inertial Ohm's law. Motion near neutral points or lines introduces a new chaotic regime. Few studies have been made in time-dependent fields.

5. Perhaps plasmoid and plasma sheet boundary layer models can be combined to explain the brightening of quiet auroral arcs at the time of substorm onset.

Acknowledgments. This study was supported by grants ATM-8219436 (Johns Hopkins University, APL Task G203) and ATM-8318203 from the National Science Foundation. I appreciate useful discussions with D. J. Williams, K. Schindler, R. Martin, and P. Dusenbery.

References

Alexeev, I. I., and A. P. Kropotkin, Interaction of energetic particles with the neutral layer of the magnetospheric tail, Geomag. Aeron., 10, 615, 1970.

Alfvén, H., Some properties of magnetospheric neutral surfaces, J. Geophys. Res., 73, 4379, 1968.

Aström, E., Electron orbits in hyperbolic magnetic fields, Tellus, 8, 260, 1956.

Baker, D. N., E. W. Hones, Jr., D. T. Young, and J. Birn, The possible role of ionospheric oxygen in the initiation and development of plasma sheet instabilities, Geophys. Res. Lett., 9, 1337, 1982.

Baum, P., and A. Bratenahl, Magnetic reconnection experiments, in Advances in Electronics and Electron Physics, vol. 54, edited by L. Marton and C. Marton, p. 1, Academic Press, New York, 1980.

Bernstein, I. B., E. A. Frieman, M. D. Kruskal, and R. M. Kulsrud, An energy principle for hydromagnetic stability problems, Proc. Roy. Soc., A244, 17, 1958.

Birn, J., Computer simulation of reconnection in planetary magnetospheres, in Unstable Current Systems and Plasma Instabilities in Astrophysics, edited by M. R. Kundu and G. D. Holman, p. 167, D. Reidel Publ. Co., Dordrecht, Holland, 1985.

Birn, J., and K. Schindler, Large-scale instabilities and dynamics of the magnetotail plasma sheet, this volume, 1987.

Birn, J., and K. Schindler, Computer modeling of magnetotail convection, J. Geophys. Res., 90, 3441, 1985.

Chapman, S. C., and S.W.H. Cowley, Acceleration of lathium test ions in the quiet-time geomagnetic tail, J. Geophys. Res., 89, 7357, 1984.

Chen, J., and P. J. Palmadesso, Tearing instability in an anistropic neutral sheet, Phys. Fluids, 27, 1198, 1984.

Chen, J., and P. J. Palmadesso, Chaos and nonlinear dynamics of single-particle orbits in a magnetotail-like magnetic field, J. Geophys. Res., 91, 1499, 1985.

Chen, J., and Y. C. Lee, Collisionless tearing instability in a non-Maxwellian neutral sheet: An integro-differential formulation, Phys. Fluids, 28, 2137, 1985.

Chen, J., P. J. Palmadesso, J. A. Fedder, and J. G. Lyon, Fast Collisionless tearing in an anisotropic neutral sheet, Geophys. Res. Lett., 11, 12, 1984.

Coppi, B., G. Laval, and R. Pellat, Dynamics of the geomagnetic tail, Phys. Rev. Lett., 16, 1207, 1966.

Coroniti, F. V., Explosive tail reconnection: The growth and expansion phases of magnetospheric substorms, J. Geophys. Res., 90, 7427, 1985.

Coroniti, F. V., MHD aspects of magnetotail dynamics, this volume, 1987.

Coroniti, F. V., F. L. Scarf, L. A. Frank, and R. P. Lepping, Microstructure of a magnetotail fireball, Geophys. Res. Lett., 4, 219, 1977.

Cowley, S.W.H., A self-consistent model of a simple magnetic neutral sheet system surrounded by a cold, collisionless plasma, Cosmic Electrodyn., 3, 448, 1973.

Cowley, S.W.H., A note on the motion of charged particles in one-dimensional magnetic current sheets, Planet. Space Sci., 26, 539, 1978.

Cowley, S.W.H., Plasma populations in a simple open model magnetosphere, Space Sci. Rev., 26, 217, 1980.

Cowley, S.W.H., The causes of convection in the earth's magnetosphere: A review of the developments during the IMS, Rev. Geophys. Space Phys., 20, 531, 1982.
Cowley, S.W.H., The distant geomagnetic tail in theory and observation, in Magnetic Reconnection in Space and Laboratory Plasmas, Geophys. Monogr. Ser., vol. 30, edited by E. W. Hones, Jr., p. 228, AGU, Washington, D.C., 1984.
Dungey, J. W., Conditions for the occurrence of electrical discharges in astrophysical systems, Phil. Mag., 44, 725, 1953.
Dungey, J. W., Cosmic Electrodyn., Cambridge University Press, Cambridge, Massachusetts, 1958.
Dungey, J. W., Interplanetary magnetic field and auroral zones, Phys. Rev. Lett., 6, 47, 1961.
Eastman, T. E., and L. A. Frank, Boundary layers of the earth's outer magnetosphere, Magnetic Reconnection in Space and Laboratory Plasmas, Geophys. Monogr. Ser., vol. 30, edited by E. W. Hones, Jr., p. 249, AGU, Washington, D.C., 1984.
Eastman, T. E., L. A. Frank, and C. Y. Huang, The boundary layers as the primary transport regions of the earth's magnetotail, J. Geophys. Res., 90, 9541, 1985.
Eastwood, J. W., Consistency of fields and particle motion in the Speiser model of the current sheet, Planet. Space Sci., 20, 1555, 1972.
Eastwood, J. W., The warm current sheet model and its implication on the temporal behavior of the geomagnetic tail, Planet. Space Sci., 22, 1641, 1974.
Forbes, T. G., Implosion of a uniform current sheet in a low-beta plasma, J. Plasma Phys., 27, 491, 1982.
Forbes, T. G., and T. W. Speiser, Temporal evolution of magnetic reconnection in the vicinity of a magnetic neutral line, J. Plasma Phys., 21, 107, 1979.
Frank, L. A., Plasmas in the earth's magnetotail, Space Sci. Rev., 42, 211, 1985.
Frank, L. A., Dynamics of the magnetotail - Recent observations, this volume, 1987.
Furth, H. P., Instabilities due to finite resistivity or finite current-carrier mass, in Advanced Plasma Theory, edited by M. N. Rosenbluth, p. 159, Academic Press, New York, 1964.
Furth, H. P., J. Killeen, and M. N. Rosenbluth, Finite resistivity instabilities of a sheet pinch, Phys. Fluids, 6, 459, 1963.
Galeev, A. A., Magnetospheric tail dynamics, in Magnetospheric Plasma Physics, Development in Earth & Planet. Sciences Series, No. 4, edited by A. Nishida, p. 143, D. Reidel Publ. Co., Boston, Massachusetts, 1982.
Galeev, A. A., and L. M. Zelenyi, Metastable states of diffuse neutral sheet and the substorm explosive phase, JETP Lett., 22, 170, 1975.
Galeev, A. A., and L. M. Zelenyi, Tearing instabilities in plasma configurations, Sov. Phys. JETP, 43, 1113, 1976.
Galeev, A. A., F. V. Coroniti, and M. Ashour-Abdalla, Explosive tearing mode reconnection in the magnetospheric tail, Geophys. Res. Lett., 5, 707, 1978.
Goldstein, H., and K. Schindler, Large-scale, collision-free instability of two-dimensional plasma sheets, Phys. Rev. Lett., 48, 1468, 1982.
Grad, H., Some new variational properties of hydromagnetic equilibria, Phys. Fluids, 7, 1283, 1964.
Harris, E. G., On a plasma sheath separating regions of oppositely directed magnetic field, Il Nuovo Cimento, 23, 115, 1962.
Hill, T. W., Magnetic merging in a collisionless plasma, J. Geophys. Res., 80, 4689, 1975.
Hones, E. W., Jr., Plasma flow in the magnetotail and its implications for substorm theories, in Dynamics of the Magnetosphere, edited by S.-I. Akasofu, p. 545, D. Reidel Publ. Co., Hingham, Massachusetts, 1979.
Hones, E. W., Jr., J. Birn, S. J. Bame, G. Paschmann, and C. T. Russell, On the three-dimensional magnetic structure of the plasmoid created in the magnetotail at substorm onset, Geophys. Res. Lett., 9, 203, 1982.
Hones, E. W., Jr., R. D. Zwickl, T. A. Fritz, and S. J. Bame, Structural and dynamical aspects of the distant magnetotail determined from ISEE 3 plasma measurements, Planet. Space Sci., 34, 889, 1986.
Jaeger, E. G., and T. W. Speiser, Energy and pitch angle distributions for auroral ions using the current sheet model, Astrophys. Space Sci., 28, 129, 1974.
Katanuma, I., and T. Kamimura, Simulation studies of the collisionless tearing instabilities, Phys. Fluids, 23, 2500, 1980.
Laval, G. R., R. Pellat, and M. Vuillemin, in Proceedings of the Second International Conference on Plasma Physics and Thermonuclear Fusion, vol. 2, p. 259, International Atomic Energy Agency, Vienna, 1965.
Leboeuf, J. N., T. Tajima, and J. M. Dawson, Dynamic magnetic X-points, Phys. Fluids, 25, 784, 1982.
Lembege, B., and R. Pellat, Stability of a thick two-dimensional quasi-neutral sheet, Phys. Fluids, 25, 1995, 1982.
Lyons, L. R., Electron energization in the geomagnetic tail current sheet, J. Geophys. Res., 89, 5479, 1984.
Lyons, L. R., Processes associated with the plasma sheet boundary layer, Physica Scripta, in press, 1987.
Lyons, L. R., and T. W. Speiser, Evidence for current sheet acceleration in the geomagnetic tail, J. Geophys. Res., 87, 2276, 1982.
Lyons, L. R., and T. W. Speiser, Ohm's law for a current sheet, J. Geophys. Res., 90, 8543, 1985.
Martin, R. F., Jr., The effect of plasma sheet

thickness on ion acceleration near a magnetic neutral line, in Ion Acceleration in the Magnetosphere and Ionosphere, Geophys. Monogr. Ser., vol. 38, edited by T. Chang, p. 141, AGU, Washington, D.C., 1986a.

Martin, R. F., Jr., Chaotic particle dynamics near a two-dimensional magnetic neutral point with application to the geomagnetic tail, J. Geophys. Res., 91, 11,985, 1986b.

Möbius, E., F. M. Ipavich, N. Scholer, G. Gloeckler, D. Hovestadt, and B. Klecker, Observations of a nonthermal ion layer at the plasma sheet boundary during substorm recovery, J. Geophys. Res., 85, 5143, 1980.

Parker, E. N., Newtonian development of the dynamical properties of ionized gases of low density, Phys. Rev., 107, 924, 1957.

Pellinen, R. J., and W. J. Heikkila, Energization of charged particles to high energies by an induced substorm electric field within the magnetotail, J. Geophys. Res., 83, 1544, 1978.

Podgorny, I. M., Active experiments in space, laboratory experiments, and numerical simulation, Report 11p-65I, Academy of Sciences USSR Space Research Institute to IAGA Assembly, Edinburgh, 1981.

Pudovkin, M. I., and N. A. Tsyganenko, Particle motions and currents in the neutral sheet of the magnetospheric tail, Planet. Space Sci., 21, 2027, 1973,

Rusbridge, M. G., Non-adiabatic charged particle motion near a magnetic field zero line, Plasma Phys., 13, 977, 1971.

Russell, C. T., The configuration of the magnetosphere, in Critical Problems of Magnetospheric Physics, edited by E. R. Dyer, p. 1, IUCSTP Secretariat, Washington, D.C., 1972.

Sarris, E. T., and W. I. Axford, Energetic protons near the plasma sheet boundary, Nature, 277, 460, 1979.

Sarris, E. T., S. M. Krimigis, A.T.Y. Lui, K. L. Ackerson, and L. A. Frank, Relationship between energetic particle and plasmas in the distant plasma sheet, Geophys. Res. Lett., 8, 349, 1981.

Schindler, K., Adiabatic orbits in discontinuous fields, J. Math. Phys., 6, 313, 1965.

Schindler, K., A self-consistent theory of the tail of the magnetosphere, in Earth's Magnetospheric Processes, edited by M. McCormac, p. 200, D. Reidel Publ. Co., Dordrecht, Holland, 1972.

Schindler, K., A theory of the substorm mechanism, J. Geophys. Res., 79, 2803, 1974.

Schindler, K., and J. Birn, Magnetospheric Physics, Phys. Repts., 47, 109, 1978.

Schindler, K., and N. F. Ness, Internal structure of the geomagnetic neutral sheet, J. Geophys. Res., 77, 91, 1972.

Schindler, K., D. P. Pfirsch, and H. Wobig, Stability of two-dimensional collision-free plasmas, Plasma Phys., 15, 1165, 1973.

Sonnerup, B.U.Ö., Adiabatic particle orbits in a magnetic null sheet, J. Geophys. Res., 76, 8211, 1971.

Speiser, T. W., Particle trajectories in model current sheets, 1, Analytical solutions, J. Geophys. Res., 70, 4219, 1965.

Speiser, T. W., Particle trajectories in model current sheets, 2, Applications to auroras using a geomagnetic tail model, J. Geophys. Res., 72, 3919, 1967.

Speiser, T. W., On the uncoupling of parallel and perpendicular motion in a neutral sheet, J. Geophys. Res., 73, 1112, 1968.

Speiser, T. W., Conductivity without collisions or noise, Planet. Space Sci., 18, 613, 1970.

Speiser, T. W., and K. Schindler, Magnetospheric substorm models: Comparison with neutral sheet magnetic field observations, Astrophys. Space Sci., 77, 443, 1981.

Speiser, T. W., and T. G. Forbes, Explorer 34 magnetic field measurements near the tail current sheet and auroral activity, Astrophys. Space Sci., 77, 409, 1981.

Speiser, T. W., and L. R. Lyons, Comparison of analytical approximation for particle motion in a current sheet with precise numerical calculations, J. Geophys. Res., 89, 147, 1984.

Stenzel, R. L., and W. Gekelman, Laboratory experiments on magnetic field line reconnection, in Proceedings of the 1982-84 MIT Symposium on Physics of Space Plasmas, edited by J. Belcher, H. Bridge, T. Chang, B. Coppi, and J. Jasperse, MIT Press, Cambridge, Massachusetts, 1984.

Stern, D. P., Adiabatic particle motion in a nearly drift free magnetic field: Application to the geomagnetic tail, Rep. X-602-77-89, Goddard Space Flight Center, Greenbelt, Maryland, 1977.

Stern, D. P., The role of O-type neutral lines in magnetic merging during substorms and solar flares, J. Geophys. Res., 84, 63, 1979.

Swift, D. W., The effect of the neutral sheet on magnetospheric plasma, J. Geophys. Res., 82, 1288, 1977.

Swift, D. W., Numerical simulation of the interaction of the plasma sheet with the lobes of the earth's magnetotail, J. Geophys. Res., 87, 2287, 1982.

Swift, D. W., A two-dimensional simulation of the interaction of the plasma sheet with the lobes of the earth's magnetotail, J. Geophys. Res., 88, 125, 1983.

Tasso, H., Energy principle for two-dimensional resistive instabilities, Plasma Phys., 17, 1131, 1975.

Terasawa, T., Numerical study of explosive tearing mode instabiity in one-component plasmas, J. Geophys. Res., 86, 9007, 1981.

Vasyliunas, V. M., Theoretical models of magnetic field merging, 1, Rev. Geophys. Space Phys., 13, 303, 1975.

Vasyliunas, V. M., An overview of magnetospheric dynamics, in Magnetospheric Particles and Fields, edited by B. M. McCormac, p. 99, D. Reidel Publ. Co., Dordrecht, Holland, 1976.

Vasyliunas, V. M., Upper limit on the electric field along a magnetic O line, J. Geophys. Res., 85, 4616, 1980.

Wagner, J. S., J. R. Kan, and S.-I. Akasofu, Particle dynamics in the plasma sheet, J. Geophys. Res., 84, 891, 1979.

Zelenyi, L. M., A. S. Lipatov, D. G. Lominadze, and A. L. Taktakishvili, The dynamics of the energetic proton bursts in the course of the magnetic field topology reconstruction in the earth's magnetotail, Planet. Space Sci., 32, 313, 1984.

ION-CYCLOTRON WAVE HEATING OF HEAVY IONS IN THE EQUATORIAL MAGNETOSPHERE: A NUMERICAL SIMULATION STUDY

M. W. Chen

Department of Physics, University of California at Los Angeles, Los Angeles, California 90024

T. Hada

Institute of Geophysics and Planetary Physics, University of California at Los Angeles
Los Angeles, California 90024

M. Ashour-Abdalla

Department of Physics, Institute of Geophysics and Planetary Physics
University of California at Los Angeles, Los Angeles, California 90024

Abstract. The heating of cold H^+ ions and heavy ions by electromagnetic ion-cyclotron waves (ICWs) in the ring current region of the equatorial magnetosphere is studied using a 1-2/2 dimensional hybrid numerical simulation code. In this study, we consider a plasma consisting of electrons, hot H^+ ions, cold H^+ ions, and cold heavy ions in which the ICWs are driven by the temperature anisotropy ($T_\perp > T_\parallel$) of the hot protons. We found for large-amplitude ICWs that the cold H^+ ions are preferentially heated over the heavy ions although the cold H^+ ions are not initially resonant with the ICW. We propose that the cold H^+ ions are heated by a three-step process. First, the large-amplitude ICW undergoes a resonant decay instability analogous to the decay instability of a large-amplitude Alfvén wave. The ICW will decay into daughter electromagnetic waves and an acoustic wave. Second, the cold H^+ ions are heated in the parallel direction by Landau trapping with the generated acoustic waves. Finally, the increase of the parallel velocity of the cold H^+ allows more cold H^+ to be resonant with the ICW. Thus, the cold H^+ can heat in the perpendicular direction also.

Introduction

GEOS observations show that He^+ ions of ionospheric origin can heat up to suprathermal energies (~100 eV) in the equatorial ring current at L ~ 6-7 R_E (Young et al., 1981). Left-handed polarized (LHP) ICWs and hot anisotropic H^+ ions are detected in conjunction with the energized He^+ ions suggesting that the He^+ ions are heated due to wave-particle interactions with the ICWs. Linear theories (Gendrin et al., 1984) and numerical simulations (Omura et al., 1985) which used plasma parameters very similar to the GEOS observations show that resonant interactions between heavy ions and ICWs, which are driven by anisotropic hot protons that presumably originate from the plasma sheet, can lead to quasi-linear heating of heavy ions. The numerical simulations of Omura et al. (1985) also show the preferential heating of He^+ ions over cold H^+ ions.

Recent Dynamics Explorer 1 (DE 1) observations of a region of the ring current closer to the earth (L ~ 4-5 R_E), however, show that there is preferential heating of cold H^+ over He^+ ions (Olsen et al., 1987). Olsen et al. (1987) report that the cold hydrogen ion distribution frequently exhibits evidence of heating in the high-energy tail (10-100 eV), whereas heavy ions such as He^+ and O^+ show only occasional evidence of heating. From linear theory, it is unlikely that the cold H^+ ions will be initially resonant with the ICWs. The wave frequency of the ICW on the H^+ branch is considerably less than the H^+ gyrofrequency which implies that the parallel velocities of the cold H^+ ions are far from the ICW resonance velocity. Clearly, the mechanism for heating the cold H^+ is different than that for the heating of the cold heavy ions.

In this paper, we will show by numerical simulations that in addition to quasi-linear heating of heavy ions, heating of the cold H^+ can occur for large-amplitude ICWs. We propose the following mechanism for the heating of the cold H^+ ions. First, the large-amplitude ICW undergoes a resonant decay instability analogous to the decay instability of a large-amplitude

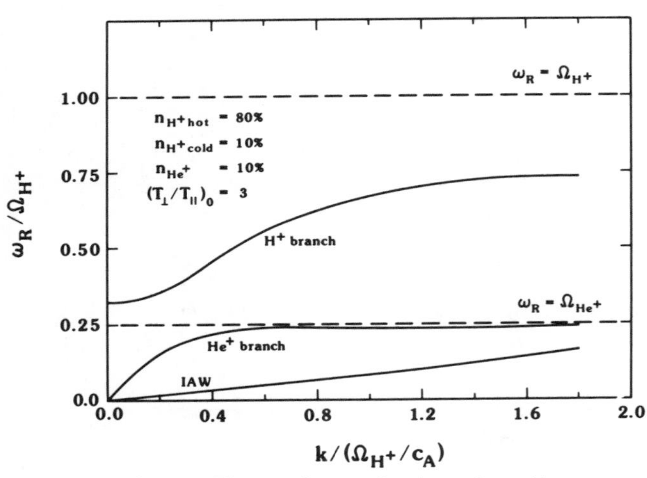

Fig. 1. Linear dispersion relation when the heavy ion species is He^+. The plasma parameters are the same as run 1 of Table 1.

Alfvén wave (Goldstein, 1978). The ICW will decay into electromagnetic daughter waves and an acoustic wave. Second, the cold H^+ ions are heated by Landau trapping with the generated acoustic waves. The increase of the parallel velocity of the cold H^+ ions allows more cold H^+ to be resonant with the ICW. This results in heating in the perpendicular direction as well. We will briefly review the linear theory necessary for the discussion of our simulation results. We will discuss the simulation model and its results, followed by our conclusions.

Linear Analysis

This section is a brief review of the linear analysis of the ICW. Previous work has shown that LHP ICWs are excited by hot anisotropic ($T_\perp > T_\parallel$) ions (Cornwall, 1964; Kennel and Petschek, 1966, and others) due to a resonant cyclotron interaction. For a given temperature anisotropy, the ICW is unstable at $0 < \omega < \left((T_\perp - T_\parallel)/T_\perp\right)\Omega_{H^+}$, where $T_\parallel$ and $T_\perp$ are the parallel and perpendicular temperatures, respectively (Kennel and Petschek, 1966). We will treat only the case of ICWs propagating parallel to a uniform magnetic field, B_o, since ion Landau damping substantially reduces the growth rate for oblique propagation.

When heavy ions, denoted by the subscript σ, are introduced, the ICW branch splits into two branches: a heavy ion branch at $0 < \omega < \Omega_\sigma$ and an H^+ branch at $\omega_{cut} < \omega < \Omega_{H^+}$, where ω_{cut} is the LHP cutoff frequency (Stix, 1962). The stop band, $\Omega_\sigma < \omega < \omega_{cut}$, is the frequency range in which no LHP normal modes are allowed to exist (Brice, 1965). The growth rate, $\gamma = \mathrm{Im}(\omega)$ can be positive on both branches (Gendrin et al., 1984). The relative magnitudes of the maximum growth rates of γ_σ and γ_{H^+} depend on the plasma parameters. In our simulation (see Table 1), $\gamma_\sigma < \gamma_{H^+}$ by about 3 orders of magnitude; whereas the linear analysis of the parameters of the Gendrin et al. (1984) study showed that $\gamma_\sigma \sim \gamma_{H^+}$.

A plot of the dispersion relation of the ICW when the heavy ion species is He^+ is shown in Figure 1 (the plasma parameters are summarized in run 1 of Table 1). The plot shows the He^+ branch which has an asymptote at Ω_{He^+} and the H^+ branch which has an asymptote at Ω_{H^+}. In addition, there is an ion-acoustic mode, $\omega = c_s k$, where $c_s = 0.09\ c_A$. A large-amplitude ICW can decay into daughter electromagnetic waves and an acoustic wave through a resonant decay instability process. The linear analysis shows the possibility that the ICW decays into several acoustic wave modes, but only one mode is shown in Figure 1. The phase velocity of this mode is very close to the value calculated from $c_s = \sqrt{(\gamma_s(T_e/m_H))} \simeq 0.13$, where γ_s is the ratio of specific heats ($\gamma_s = 5/3$, $\beta_e = 10^{-2}$).

A plot of the normalized growth rate versus the wave number with the same plasma parameters is shown in Figure 2. The plot shows the range of wave numbers in which there is positive growth of the waves for both the He^+ and the H^+ branches. The wave modes 7, 8, and 9 for the system length chosen in the simulation are labeled on the growth curve corresponding to the H^+ branch. From this linear analysis, we expect these modes to be dominant and we compare them later to the dominant modes found in the simulation.

Quasi-linear heating of heavy ions is possible if the bulk velocity of the heavy ion, $v_{\parallel\sigma}$, is resonant with the ICWs, that is, if it satisfies the following condition, $v_R - v_{tr} < v_{\parallel\sigma} < v_R + v_{tr}$, where $v_R = |\omega - \Omega_\sigma|/k$ is the resonance velocity, $v_{tr} = \omega_{tr}/k = \sqrt{(B_w/B_o)(v_\perp\Omega_\sigma/k)}$ is the trapping velocity, B_w is the magnetic field amplitude, and ω and k correspond to the frequency and wave number of the fastest growing mode of the ICW.

TABLE 1. Initial Parameters for Runs 1 and 2

Run	m_σ/m_{H^+}	n_{H^+hot}/n_t	n_{H^+cold}/n_t	n_σ/n_t	$\beta_{\parallel H^+hot}$	$\beta_{\parallel H^+cold}$	$\beta_{\parallel\sigma}$	β_e
1	4	80%	10%	10%	0.2	2×10^{-5}	8×10^{-5}	0.01
2	16	80%	10%	10%	0.2	2×10^{-5}	3.2×10^{-4}	0.01

Simulation Method

We use a hybrid code (Chodura, 1975; Sgro and Nielson, 1976) in which the ions and electrons are treated as particles and a massless fluid, respectively. The hybrid code allows us to use much larger time steps compared with full particle codes because the electron spatial and temporal scales have been eliminated. Also since we are interested in the ion kinetic behavior and we consider only the parallel propagating waves, the hybrid code is appropriate for our problem. The code is one dimensional in space while including all three velocity components. We define a Cartesian coordinate system specifying the spatial dimension as x. A uniform static magnetic field, B_o, is assumed to point along x. The code uses periodic boundary conditions which treat the system as infinite and homogeneous. Although this is an idealized system, it gives us insight on a local model of the heating of the ions.

The motion of the ions is determined by integrating the equations of motion

$$\frac{d\mathbf{v}}{dt} = \frac{q_i}{m_i}(\mathbf{E} + \mathbf{v} \times \mathbf{B}) , \quad (1)$$

$$\frac{dx}{dt} = v_x , \quad (2)$$

where q_i and m_i are the charge and the mass of the ion, respectively. The electric field is obtained by evaluating the electron momentum equation

$$\mathbf{E} = -\mathbf{u}_e \times \mathbf{B} - \frac{1}{en}\nabla P_e , \quad (3)$$

where $\mathbf{u}_e = \mathbf{u}_i - \mathbf{J}/en$ and $\mathbf{u}_i$ and p_e are the electron bulk velocity, the ion bulk velocity, and the electron pressure, respectively. We assume the quasi-neutrality condition. The current density, J, is obtained from Maxwell's equations

$$\mathbf{J} = \frac{1}{\mu_o}\nabla \times \mathbf{B} , \quad (4)$$

while the magnetic field is advanced by

$$\nabla \times \mathbf{E} = -\frac{\partial \mathbf{B}}{\partial t} . \quad (5)$$

In the code, time is normalized by the inverse of the H^+ gyrofrequency, $\Omega_{H^+}^{-1}$, velocities are normalized by the Alfvén speed, c_A, and mass is normalized by m_{H^+}. In our simulation runs, we used 128 grid points with a mesh size of $dx = 0.5\ c_A/\Omega_{H^+}$. The time step is limited by the gyrofrequency of the lightest ion, the H^+ ion. We used a time step of $0.1\ \Omega_{H^+}^{-1}$ throughout the simluation runs.

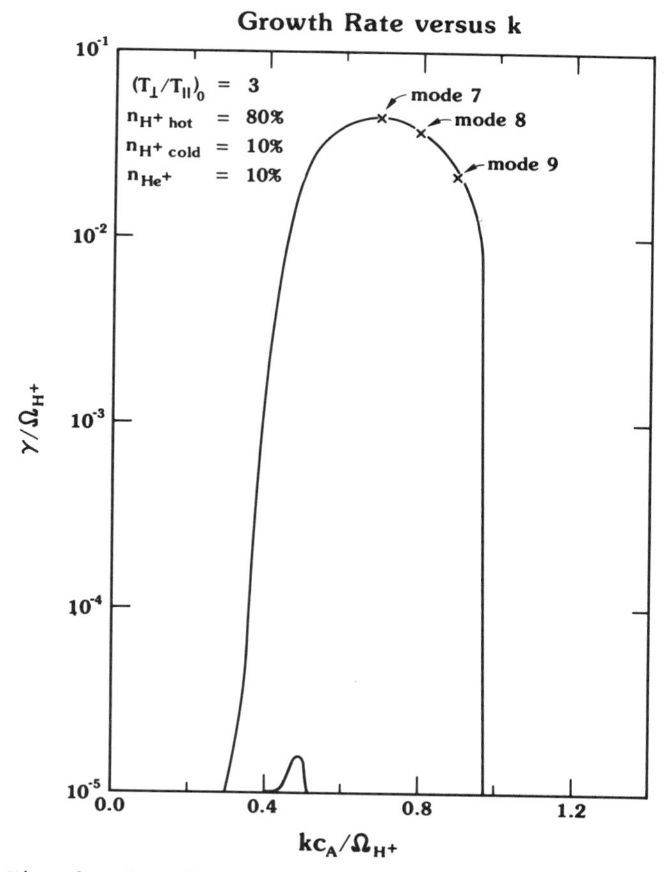

Fig. 2. Growth rate versus wave number for plasma parameters listed under run 1 of Table 1. The wave modes 7, 8, and 9 for the system length of the simulation are labeled by an "x" on the growth curve corresponding to the H^+ branch.

Simulation Results and Discussion

The plasma parameters used in the simulation runs are summarized in Table 1 where σ denotes the heavy ion species and n_t is the total ion density. For these runs we used a high density concentration of hot anisotropic H^+ ions (80%) which is much greater than typically observed by spacecraft in the ring current region (~10%) (Young et al., 1981). In addition to the runs listed in Table 1, we made one run using the same parameters as run 1 except that the percentage density of the hot H^+, the cold H^+, and the He^+ components are 10%, 80%, and 10%, respectively. In this case we found that when the concentration of hot H^+ is low, the same mechanism operates for the heating of the cold H^+ although it takes longer before it occurs. In all the runs, we considered hot H^+ with a bi-Maxwellian velocity distribution, cold H^+, and a cold heavy ion species. Initially, the cold ion species were

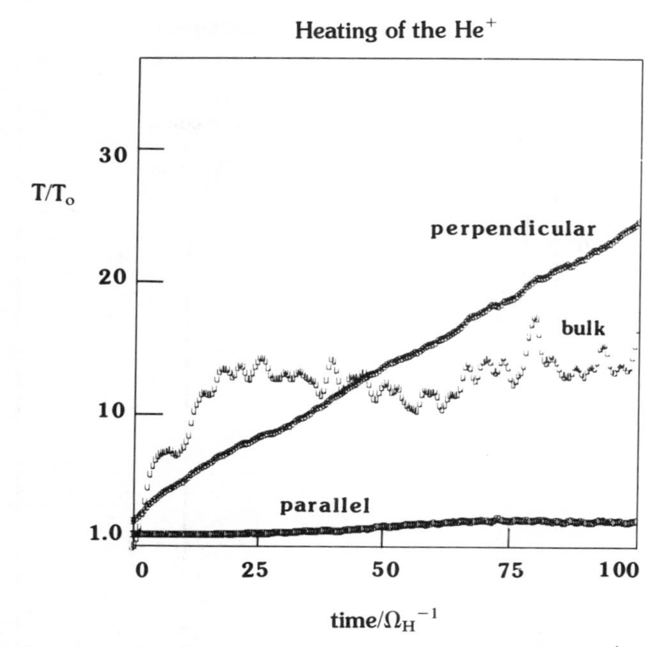

Fig. 3. Time history plot of the heating of He^+ ions for run 1. The parallel, perpendicular, and bulk temperatures are normalized by the initial parallel temperature of the cold H^+ ion.

isotropic, whereas the hot protons had an anisotropy of $T_\perp/T_\parallel = 3$ which provided the free energy to drive the growth of the ICW. In the simulation, there were 10,000 particles of each species.

Time history plots of parallel, perpendicular, and bulk temperatures $\left(T_\parallel = m(\overline{v_\parallel - \bar{v}_\parallel})^2, T_\perp = m(\overline{v_\perp - \bar{v}_\perp})^2, T_{bulk} = m(\bar{v}_\parallel{}^2 + \bar{v}_\perp{}^2),\right.$ where the bar denotes an average over a cell length$\left.\right)$ of run 1 and 2 of the He^+ ions and the O^+ ions, normalized by the corresponding initial temperatures, are shown respectively in Figures 3 and 4. The heating of these heavy ion species is primarily explained by the quasi-linear heating model (Tanaka, 1985; Omura et al., 1985). There is more heating of He^+ ions than O^+ ions by roughly a factor of 12 in the perpendicular direction and essentially no parallel heating of O^+ ions at $t = 100\ \Omega_{H^+}{}^{-1}$. In all the runs, there is always preferential heating in the perpendicular direction. This occurs because the wave electric field which scatters the heavy ions has a larger perpendicular component than parallel component, $\delta E_\parallel < \delta E_\perp$. The parallel component of the wave electric field remains small, since the electrons are cold ($B_e = 0.01$).

The smaller O^+ ion heating rate agrees with the linear theory. The O^+ gyrofrequency, $\Omega_{O^+} = 1/16\ \Omega_{H^+}$, is considerably smaller than Ω_{He^+}, thus fewer O^+ ions will be resonate with the growing wave, whose typical frequency is $\omega \sim 1/2\ \Omega_{H^+}$. An exception to the agreement between the quasi-linear theory and the simulation, however, is the heating of the cold protons. From the linear theory we do not initially expect that many cold protons will resonate with the ICWs and get trapped. Yet, we find that there is more heating of the cold protons than the heavy ions.

The time history plots of the cold H^+, Figure 5, show that the rate of heating of the cold H^+ does not resemble the quasi-linear behavior of the heavy ions (Figures 3 and 4). In the perpendicular direction, the cold H^+ ions are heated at a slow rate until about $t = 30\ \Omega_{H^+}{}^{-1}$ at which time the rate of heating suddenly increases and subsequently becomes quasi-linear. Again in the parallel direction, the rate of heating is slow until roughly $t \simeq 30\ \Omega_{H^+}{}^{-1}$ with a subsequent increase of heating until $t \simeq 50\ \Omega_{H^+}{}^{-1}$ when the heating rate saturates. Since the cold protons do not resonate with the ICW in the initial stage of the run ($t \lesssim 30\ \Omega_{H^+}{}^{-1}$), the cold protons are merely sloshed back and forth by the waves. This increases the ion bulk energy ($E_b = E_{b\parallel} + E_{b\perp}$, where $E_{b\parallel} = 1/2\ m\bar{v}_\parallel{}^2$, $E_{b\perp} = 1/2\ m\bar{v}_\perp{}^2$, $v_\parallel$ is the parallel component of the velocity, $v_\perp$ is the perpendicular component of the velocity, and the bar denotes an average taken over a cell length). Around $t \simeq 30\ \Omega_{H^+}{}^{-1}$, however, there is a rapid increase of the ion thermal energy $\left(E_\parallel = 1/2\right.$

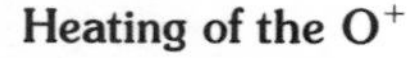

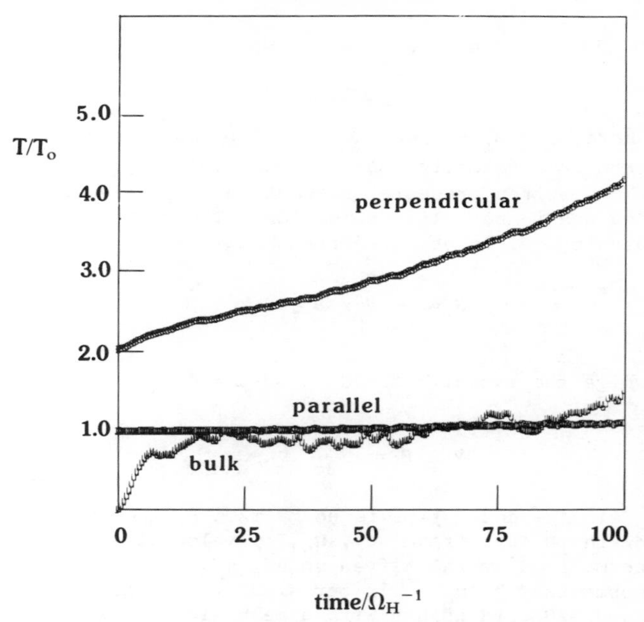

Fig. 4. Time history plot of the heating of O^+ ions for run 2. Note that there is very little heating of the O^+ ions as compared with the He^+ ions in Figure 1.

$m(\overline{v_\parallel - \bar{v}_\parallel})^2$ and $E_\perp = 1/2\ m(\overline{v_\perp - \bar{v}_\perp})^2)$. As the ion thermal energy increases, the bulk energy decreases.

The heating of the cold H^+ can be explained in terms of a nonlinear resonant decay instability. Goldstein (1978) showed that finite large-amplitude LHP Alfvén waves will spontaneously decay into two daughter Alfvén waves and an acoustic wave. Analogously, the finite amplitude ICW will decay into two daughter waves and an acoustic wave. The acoustic wave will produce a perturbed electrostatic electric field in the parallel direction. If the cold H^+ ions are Landau resonant with the acoustic waves, they will be heated. Recent numerical simulations (Terasawa et al., 1986) have also confirmed ion heating due to the resonant decay instability process.

The matching conditions for the resonant decay instability can be found easily by vector addition of (ω_o, k_o), the frequency and wave number of the fastest growing mode of the ICW, and $(\omega_{AC} \pm k_{AC})$, the frequency and wave number of the fastest growing mode of the acoustic wave is $k_{AC} = 2\ k_o$. The matching conditions for the frequency and wave number of the daughter electromagnetic waves are

$$\omega = \omega_o \pm \omega_{AC}$$

$$k = k_o \pm k_{AC} = k_o \pm 2\ k_o\ .$$

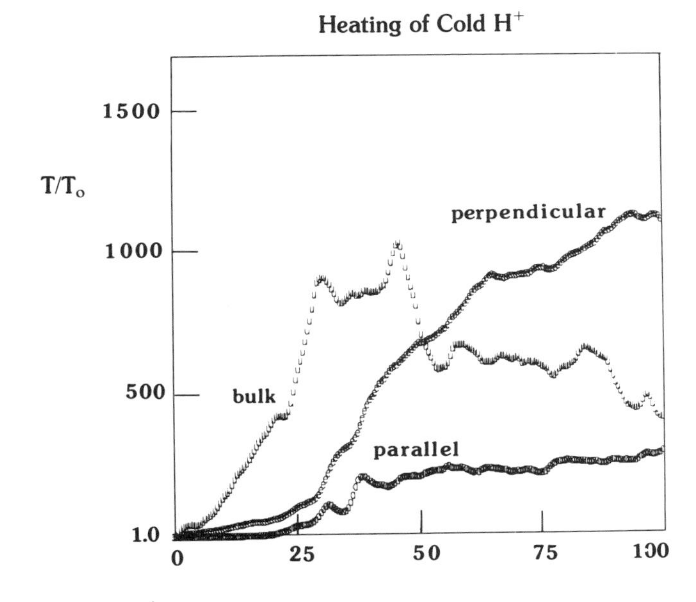

Fig. 5. Time history plot of the heating of cold H^+ ions for run 1.

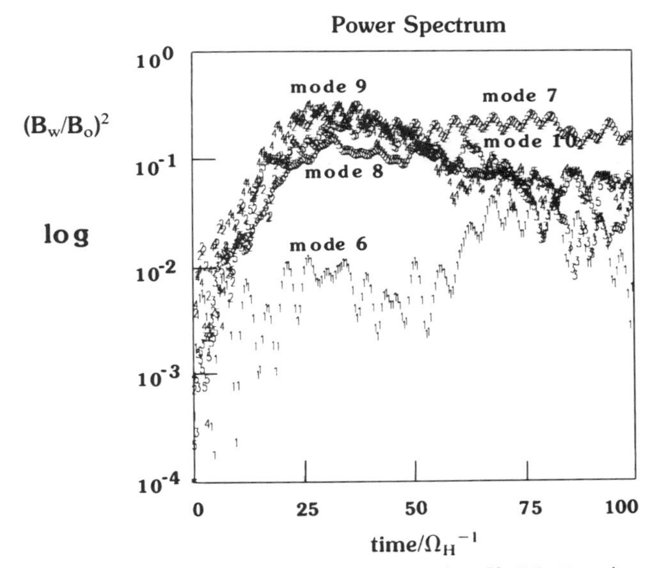

Fig. 6. Power spectrum of magnetic field Fourier transformed in k-space $\left((B_w(k,t)/B_o\right)^2$ versus time for the most dominant modes of the system. The curves with "2," "3," "4," and "5," correspond to modes 7, 8, 9, and 10, respectively.

The plot of the power spectrum of the wave magnetic field Fourier transformed in k-space $[(B_w(k,t))/B_o]^2$ versus time in Figure 6 shows that the ICWs are indeed of large amplitude. The dominant wave modes 7, 8, 9, and 10 with a maximum power of $(B_w/B_o)^2 \sim 0.3$ are shown. Wave modes 7, 8, and 9 are expected to be dominant from the linear analysis (Figure 2). From the plot it can also be seen that the modes decay with time.

The acoustic wave generated from the resonant decay instability of the large-amplitude ICW will produce perturbations in density and v_x. If there are only electromagnetic ion-cyclotron waves in the system, then there should be no density perturbations. Figure 7 is a time evolution plot of density perturbations over the system length for run 1. At the beginning of the simulation, we see no density perturbations. At $t \sim 20\ \Omega_{H^+}^{-1}$, the density perturbations become distinct. This indicates the presence of a longitudinal wave in the system which could possibly be an acoustic wave. To confirm that the longitudinal waves are acoustic waves, the matching conditions for the resonant decay instability must be shown to be satisfied.

Evidence that the matching conditions for the resonant decay instability are satisfied can be seen in Figure 8. Figure 8a shows the power spectrum of the y-component of the wave magnetic field Fourier transformed in both k-space and ω-space versus frequency at mode 8 which is one of the dominant decaying modes. The two inner peaks are near the helium gyrofrequency which is

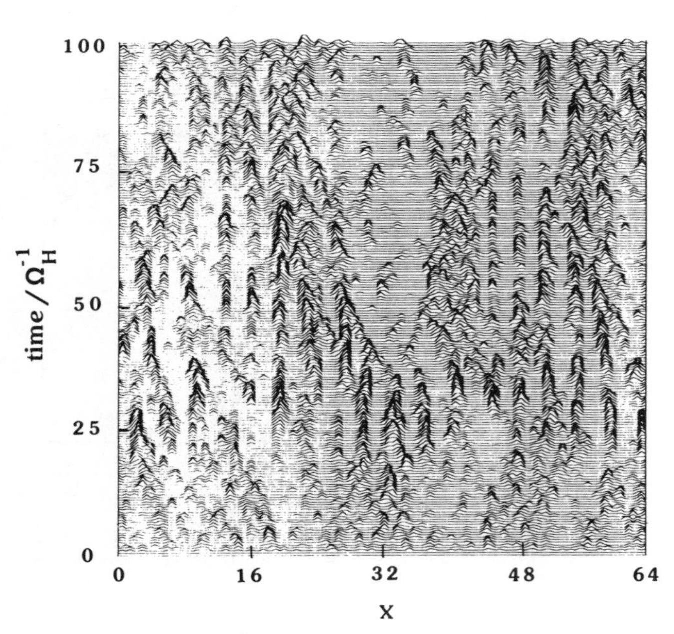

Fig. 7. Time evolution of the density perturbations over the system length.

the asymptote of the He^+ branch of the linear dispersion relation. The two outer peaks are near the frequency ω_o which corresponds to the frequency on the H^+ branch of the linear dispersion relation. Figure 8a shows the ICW at ω_o and k_o. Figure 8b shows the density perturbations Fourier transformed for mode 16 which corresponds to 2 k_o. The peak is near the acoustic frequency expected. Thus, Figure 8b shows the acoustic wave at ω_{AC} and $k_{AC} = 2\ k_o$. Figure 8c shows $B_y(\omega,k)$ versus frequency for mode 24 which corresponds to 3 k_o. The peaks occur near the frequencies $\pm(\omega_o + \omega_{AC})$ as required by the matching conditions. The trapping frequency of the He^+, $\omega_{tr} \sim 0.6$, should be added to the matching condition to account for the frequency shift due to particle trapping. There is some error because mode 8 is not the only decaying dominant mode (Figure 6) and because the acoustic mode is not well resolved. Taking these factors into account, the power spectrum shows that the matching conditions hold in the simulation.

The acoustic wave produces a parallel electrostatic electric field. In the simulation, we find that $E_\parallel \sim 0.01$. We expect that the cold H^+ can become Landau resonant with the acoustic wave. Heating by Landau resonance requires the parallel velocity of a particle to be near the phase velocity of the wave. A particle will be Landau resonant with an electrostatic wave if the parallel velocity of the particle satisfies the following condition:

$$v_{RL} - v_{trL} < v_\parallel < v_{RL} + v_{trL} ,$$

where

$$v_{RL} = \frac{\omega}{k}$$

and

$$v_{trL} = \frac{\omega_{trL}}{k} = \left(\frac{eE_\parallel}{mk}\right)^{1/2} ,$$

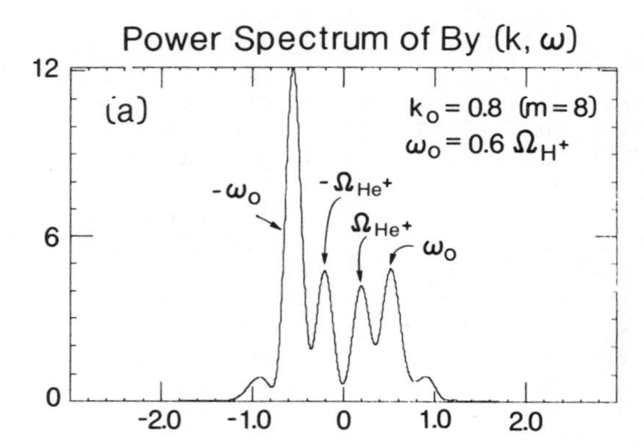

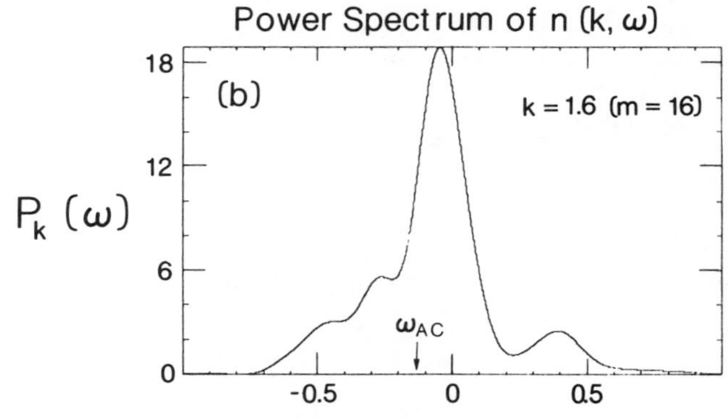

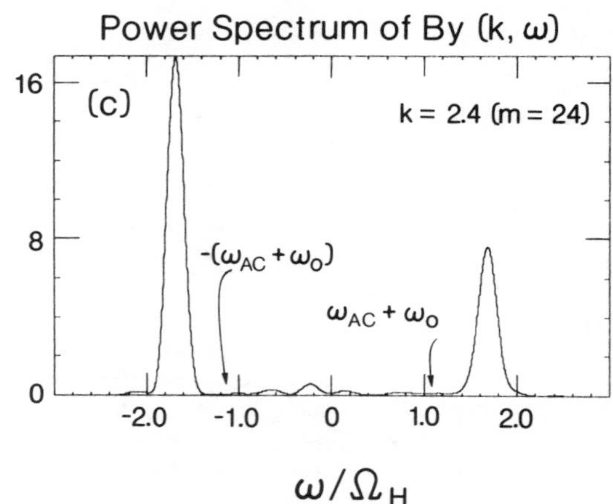

Fig. 8. Power spectrum versus frequency of the y-component of the wave magnetic field, $B_y(k,\omega)$, at modes 8(a) and 24(c) and the density perturbations, $n(k,\omega)$ at mode 16(b).

where $E_\parallel$ is the amplitude of the parallel electric field and ω and k correspond to the frequency and wave number of the wave. From the simulation, the trapping velocity is $v_{trL} \approx 0.09$ which is almost an order of magnitude larger than the initial thermal velocity of the cold H^+, $v_{th} = 0.01$. Thus, we expect that most of the cold H^+ ions will have a parallel velocity within the trapping region.

The acoustic wave is associated with density perturbations. If we look at the cold H^+ phase-space plots, v_x versus x, in Figure 9a at $t = 40\ \Omega_{H^+}^{-1}$ we see modulations of the cold H^+ ions. The modulations have become phase mixed due to trapping of the cold H^+ ions by the acoustic wave. The time in which the phase space begins to become phase mixed is at roughly $t \approx 30\ \Omega_{H^+}^{-1}$ which coincides with the time scale of the sudden increase in the heating rate of the cold H^+ ions (Figure 5).

The phase-space plots, v_x versus x, for the He^+ ions at $t = 40\ \Omega_{H^+}^{-1}$ are shown in Figure 9b. The phase-space plot for the He^+ ions shows strong modulations but no phase mixing at $t = 40\ \Omega_{H^+}^{-1}$. This is in contrast with the phase plots of the H^+ ions (Figure 9a) where the ions are Landau trapped with the acoustic wave. The increase of $v_\parallel$ now allows more cold H^+ to be resonant with the ICW. Thus, the cold H^+ ions can get heated in the perpendicular direction also as can be seen in the simulation (Figure 5).

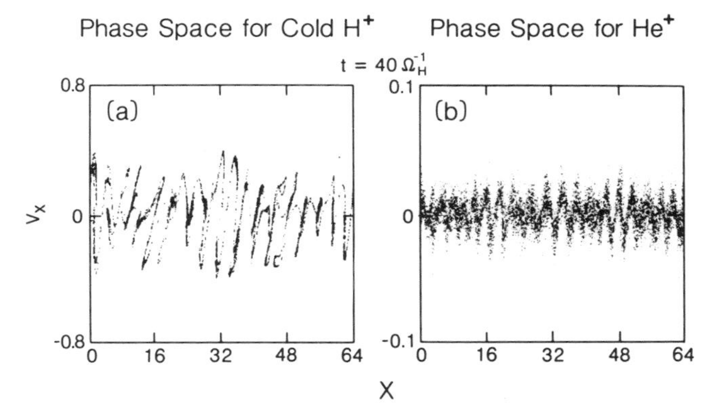

Fig. 9. Phase space plots, v_x versus x, for cold H^+ (a) and He^+ (b) ions at $t = 40\ \Omega_{H^+}^{-1}$ in run 1. The phase-space plot of cold H^+ shows the phase mixing of the ion modulations. The modulations in the phase-space plot for He^+ ions are not phase mixed.

Conclusions

The quasi-linear heating of heavy ions by ICWs in the ring current region at L ~ 4-5 R_E was studied by using a hybrid simulation code. In addition to quasi-linear heating of the heavy ions by ICW resonance, we found heating of cold H^+ ions even though they were not initially resonant with the ICWs. The cold H^+ can get heated by Landau resonance with the acoustic waves generated from the resonant decay instability of the large-amplitude ICW. The simulations showed the existence of these acoustic waves and the decay instability process as well as the phase mixing of the cold H^+ in phase space. The time scale for the heating of the cold H^+ ions to become significantly large matches the time scale for the modulations of the cold H^+ ions in phase space to phase mix as we would expect for the Landau resonance mechanism. The subsequent increase of $v_\parallel$ of the cold H^+ ions allows more cold H^+ to become resonant with the ICW so that they heat in the perpendicular direction as well. This explains one mechanism for the preferential heating of cold H^+ ions over He^+ ions in this region as was observed by the DE 1 satellite (Olsen et al., 1987). From this study, we anticipate observable signatures indicating the presence of acoustic waves in this region.

Acknowledgments. The authors would like to thank Richard Sydora for the helpful discussions on resonant decay instabilities. This work was supported by NASA Marshall Space Flight Center grant NAGW 8-074, NASA Solar Terrestrial Theory Program grant NAGW-78, Air Force Geophysics Laboratory contract F-196-28-85-K-0027, and San Diego Supercomputer.

References

Brice, N., Fundamentals of very low frequency emission generation mechanisms, J. Geophys. Res., 70, 61, 1965.

Chodura, R. A., A hybrid fluid-particle model of ion heating in high Mach number shock waves, Nucl. Fusion, 15, 55, 1975.

Cornwall, J. M., Cyclotron instabilities and electromagnetic emission generation mechanisms, J. Geophys. Res., 69, 4515, 1964.

Gendrin, R., M. Ashour-Abdalla, Y. Omura, and Kevin Quest, Linear analysis of ion cyclotron interaction in a multicomponent plasma, J. Geophys. Res., 89, 9119, 1984.

Goldstein, M. L., An instability of finite amplitude circularly polarized Alfvén waves, Astrophys. J., 219, 700, 1978.

Kennel, C. F., and H. E. Petschek, Limit on stably trapped particle fluxes, J. Geophys. Res., 71, 1, 1966.

Olsen, R. C., S. D. Shawhan, D. L. Gallagher, J. L. Green, C. R. Chappell, and R. R. Anderson, Plasma observations at the earth's magnetic equator, J. Geophys. Res., 92, 2385, 1987.

Omura, Y., M. Ashour-Abdalla, R. Gendrin, and K. Quest, Heating of the thermal helium in the equatorial magnetospere: A simulation study, J. Geophys. Res., 90, 8281, 1985.

Sgro, A. G., and C. W. Nielson, Hybrid model of

ion dynamics and magnetic field diffusion during pinch implosions, Phys. Fluids, 19, 126, 1976.
Stix, T., The Theory of Plasma Waves, McGraw-Hill, New York, 1962.
Tanaka, M., Simulations of heavy ion heating by electromagnetic ion cyclotron waves driven by proton temperature anisotropies, J. Geophys. Res., 90, 6459, 1985.
Terasawa, T., M. Hoshino, J. I. Sakai, and T. Hada, Decay instability of finite-amplitude circularly polarized Alfvén waves: A numerical simulation of stimulated Brillouin scattering, J. Geophys. Res., 91, 4171, 1986.
Young, D. T., S. Perraut, A. Roux, C. de Villedary, R. Gendrin, A. Korth, G. Kremser, and D. Jones, Wave-particle interactions near Ω_{He^+} observed on GEOS 1 and 2; Propagation of ion-cyclotron waves in He^+-rich plasma, J. Geophys. Res., 86, 6755, 1981.

THE MOTION OF THE WTS AS A FUNCTION OF ELECTRON TEMPERATURE ANISOTROPY IN THE PLASMA SHEET

P. L. Rothwell

Air Force Geophysics Laboratory, Hanscom AFB, Bedford, Massachusetts 01731

M. B. Silevitch

Center for Electromagnetic Research, Northeastern University, Boston, Massachusetts 02115

L. P. Block

Department of Physics, Royal Institute of Technology, Stockholm, Sweden S 100 44

Abstract. Our previous studies on the motion of the westward traveling surge and the simultaneous generation of Pi 2 pulsations during substorms. Now it is found that the requirement that the Pi 2 pulsations be damped leads to a relationship between the direction of surge motion and the electron temperature anisotropy in the plasma sheet. For example, embryonic surge structures that are quite narrow ($\lesssim$50-100 km) in the north-south dimension require substantial electron heating parallel to the magnetic field in order for poleward surge motion to take place. Otherwise, these structures will move westward.

Introduction

In Rothwell et al. (1984), referred to herein as paper 1, the ionospheric response of the westward traveling surge (WTS) to energetic electron precipitation was determined. We found that the conductivity gradients along the surge's poleward and westward boundaries propagated as ionization waves with a speed determined by the electron precipitation energy and flux. In a subsequent article, Rothwell et al. (1986), referred to herein as paper 2, it was found that the WTS could act like a resonant ionospheric cavity for Alfvén waves propagating along the attached magnetic field lines. A feedback instability driven by the east-west electric field of the substorm current wedge was shown to develop which caused the generation of Pi 2 waves (≈40-160 s). Higher-energy electrons turned on higher-order modes for a fixed surge size and, therefore, created more complex pulse structures. The damping of the Pi 2 pulsations was found to depend on the zero-order ionization level (N_0) inside the surge region. The value of N_0 depends directly on the energy flux of the precipitating electrons as well as on their flux intensity. Therefore, a major result of the previous two papers is that both the surge motion and the damping of Pi 2 pulsations critically depend on the relation between the precipitating flux and energy, i.e., on field-aligned potential drops. Fridman and Lemaire (1980) calculated the precipitating flux for a bi-Maxwellian plasma passing through a potential drop. They found that for a wide range of potentials, the precipitating current closely followed Ohm's law. We will use this important relation in deriving the physical consequences of damped Pi 2 pulsations.

First, we briefly review the dynamics of the surge motion and the generation of Pi 2 pulsations with emphasis on the underlying unity between the two phenomena. We then look at the conditions for surge motion and damped Pi 2 pulsations and relate them to the work of Fridman and Lemaire (1980). Finally, a correlation is made between plasma sheet conditions and the surge motion.

We assume the current system inside the surge as shown in Figure 1 (Inhester et al., 1981; Baumjohann, 1983). The substorm current wedge is created by an external electric field, E_0. This field drives a westward Pedersen current and a poleward Hall current. The Hall current is closed off into the magnetosphere by precipitating electrons along the poleward WTS boundary and by equatorward Pedersen currents in the ionosphere. Closure is governed by the parameter α. The fraction of the Hall current closed off by the magnetospheric currents is called the closure parameter, α. The Pedersen current arises from polarization charges along the poleward boundary. In paper 1 we found that WTS motion is controlled by: (1) the energy and intensity of the precipi-

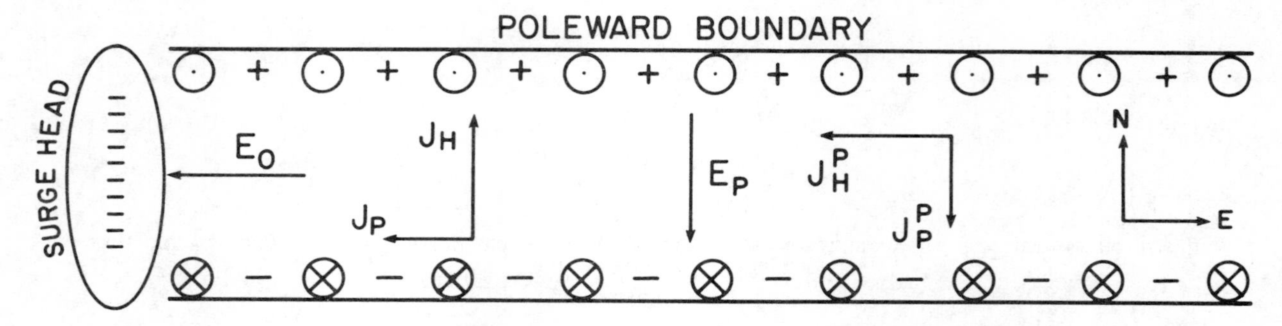

Fig. 1. Inhester-Baumjohann current model for the WTS.

tating electrons, (2) the electron ion recombination rate, and (3) the degree of current closure on the poleward boundary of the surge.

On the basis of work by Rostoker and Samson (1981) and Samson and Rostoker (1983), there is clearly an observed relationship between the WTS and the generation of Pi 2 pulsations during substorm onsets. It is natural, therefore, to look for ways by which this could be done using the current system shown in Figure 1. It was determined in paper 2 that the north-south current in the WTS could produce standing ionospheric waves due to the reflection from conductivity gradients along the boundaries. The attached magnetic field lines, acting as an equivalent transmission line (Sato, 1978), serve as an ac port to the surge. Under the right conditions this port can resonate with the ionospheric cavity and Pi 2 pulsations will be generated from the ionosphere into the magnetosphere.

We will now demonstrate the underlying unity between papers 1 and 2. The continuity equation for the ionization density is given by

$$\frac{\partial N}{\partial t} = Q\, j_{\parallel}/e - \sigma_r N^2, \tag{1}$$

where N is the ionization density in the ionosphere, Q is the production efficiency for ionization (Rees, 1963), $j_{\parallel}/e$ is the downward electron flux that is creating the ionization, e is electric charge, and σ_r is the electron ion recombination rate. The coordinate system is such that x points north, y points west, and z is toward the zenith. In paper 1, $j_{\parallel}$ is related to the closure of the ionospheric Hall current along the surge boundaries. When this closure condition is included, then equation (1) becomes a wave equation. Inside the surge region no currents are assumed to close into the magnetosphere and this constraint does not exist. Therefore, in the interior the zero-order ionization density is considered constant, being the equilibrium value determined by the downward electron flux and the electron ion recombination rate. The downward electron flux is presumed to be balanced by upward moving lower-energy electrons in the surge interior yielding a zero net field-aligned current. In this view the precipitating flux on the surge boundaries is not necessarily equal to the flux inside the surge.

Motion of the WTS

The zero-order equation for equation (1) is given by equation (17) in paper 1. If we momentarily ignore electron-ion recombination effects, then the velocity of the poleward boundary is given by

$$V_x = QHV_d\alpha \tag{2}$$

where H is the ionospheric height over which Q is significant (Rees, 1963) and V_d (≈ 0.25 km s^{-1}) is the **E** x **B** drift in the ionosphere. An expression for the poleward surge velocity which includes electron-ion recombinatin effects is given by

$$V = V_x - G\,\Sigma_{Ho}^{\;2}\, E_o\alpha/j_{\parallel}(0); \quad G = \sigma_r B/eH . \tag{3}$$

See equation (22) in paper 1. Σ_{Ho} is the height-integrated zero-order Hall conductivity inside the surge region, B is the magnetic field at the ionosphere and $j_{\parallel}(0)$ is the upward field-aligned current density at the poleward surge boundary. Using $\Sigma_{Ho} = eHN_o/B$ where $N_o = [Qj_{\parallel}(<)/e\sigma_r]^{1/2}$ and $j_{\parallel}(<)$ is the upward current density in the surge interior, we find that equation (3) can be simplified as

$$V = V_x[1 - j_{\parallel}(<)Q_i/j_{\parallel}(0)Q_b] . \tag{4}$$

Now $Q_i j_{\parallel}(<)$ and $Q_b j_{\parallel}(0)$ are the ionization rates in the surge interior and along the poleward boundary while the Q's are the energy-dependent ionization efficiencies (Rees, 1963). If the ionization rate along the poleward surge boundary is larger than that required to sustain N_o in the surge interior, then the surge moves poleward. If it is less the surge moves equatorward and the surge is stationary if they are equal. Therefore, our model predicts an auroral brightening along the poleward surge boundary coincident with the poleward leaps has been observed.

Pi 2 Pulsations

It was shown in paper 2 that the first-order approximation to equation (1) leads to the following expressions for the Pi 2 frequencies and growth rates:

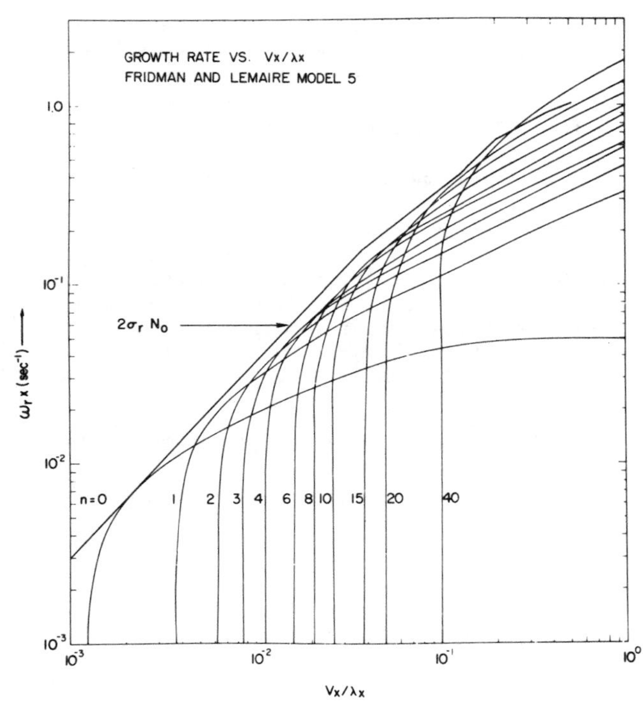

Fig. 2. Plot of the growth rates of Pi 2 pulsations as derived in paper 2 (Rothwell et al., 1986). The top curve, that is closely parallel to the growth curves, is the Pi 2 damping rate using model 5 of Fridman and Lemaire (1980). This calculation was made with $\Sigma_{op}Z_o$ = 10, V_A/L = 0.005, and λ_x = 500 km.

$$\omega_r = \frac{2\pi V_x}{\lambda_x(1 + X^2)} \qquad (5)$$

$$\omega_i = \omega_r X - 2\sigma_r N_o, \qquad (6)$$

where $X = \Sigma_{po}Z_o \cot(\omega_r L/V_A - n\pi)$. The zero-order Pedersen conductivity is given by Σ_{po}, and Z_o ($\approx$1 Ω) is the equivalent transmission line impedance for an Alfven wave guided by the magnetic field lines (Sato, 1978). The length of the field line between the ionozphere and the equator is denoted by L, V_A is the Alfven speed along the field lines ($\approx$1000 km s^{-1}), and n is the frequency mode. Now ω_r and ω_i are the real and imaginary parts of the frequency. A positive value ω_i implies continuous wave growth, while a negative value implies damping. Note that our approach of breaking the electron precipitation flux into a zero-order dc component and a first-order ac component differs from the treatment of Lysak (1986). Figure 2 shows the growth and decay curves as defined by equation (6). A new mode is turned on whenever X passes through zero. Notice that the envelope of the growth curves closely follows a straight line. This line is the focus of the maximum growth rates for each value of V_A/λ_x, where λ_x is the north-south surge dimension. This is found by inserting equation (5) into equation (6) and setting the derivative with respect to X = 0. One finds that the maximum growth rate occurs at X = 1 where both the frequency and growth rate are equal to $\pi V_x/\lambda_x$. Therefore, the condition for damped Pi 2 pulsations is clearly

$$2\sigma_r N_o \geq \pi V_x/\lambda_x \qquad (7)$$

Now if one inserts equation (2) for V_x into equation (7) one finds

$$\lambda_x/\alpha \geq \pi QHV_d/(2\sigma_r N_o) = (\lambda_x/\alpha)_t \ . \qquad (8)$$

The subscript t denotes the threshold value of λ_x/α as defined by variables immediately to the left. This is the minimum required value for this ratio in order to have marginal damping of the Pi 2 pulsations. The results of Fridman and Lemaire (1980) are now used to relate $j_\parallel$(<) and Φ (the field-aligned potential drop) for a given set of plasma sheet parameters. One finds that between 5 and 50 kV, Ohm's law is obeyed to a good approximation. The maximum ionization efficiency (Q) as a function of incident electron energy can also be fitted from the results of Rees (1963). Noting the energy dependence of QH from paper 1 we find that to a very good approximation the threshold value of equation (8) is independent of the precipitation energy. (An implicit assumption is that the precipitation energy inside the surge is the same as that along the poleward boundary. Relaxing this assumption would multiply equation (9) below by the ratio of incident energies raised to a power close to 1.) With this assumption, however, equation (8) is dependent only on the coefficient of the power law fit to the Fridman and Lemaire curves, A_1. Equation (8) now becomes

$$(\lambda_x/\alpha)_t = 0.106/\sqrt{A_1} \text{ km} \ . \qquad (9)$$

Equation (9) gives threshold values between 56 and 456 km as a function of $T_\perp/T_\parallel$. $T_\perp$ and $T_\parallel$ are the perpendicular and parallel electron temperatures in the plasma sheet relative to the local magnetic field. Note that we have assumed that all the ionization arises from the energetic electrons. Relaxing this assumption would lower the threshold further.

For fixed α equation (9) places a lower limit on λ_x for damped Pi 2 pulsations to occur. What happens if the initial surge size is below this limit and positive growing pulsations are stimulated? The period of the mode with the fastest growth rate was found above to be $2\lambda_x/V_x$. The speed of the poleward boundary is given by equations (2) and (4) so that the surge can expand very rapidly during one pulsation period. In the limiting case of equation (2), for example, the surge triples in size. Therefore, growing pulsations are finite in size and are quenched by the zero-order surge expansion. This feature vali-

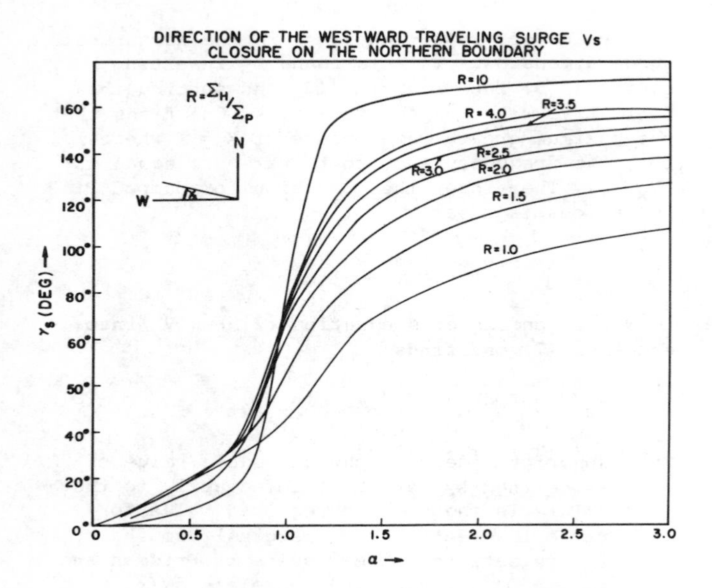

Fig. 3. Direction of the WTS as a function of the closure parameter α and the ratio, R, of the Hall to Pedersen conductivities.

dates the linear approximation used in deriving the Pi 2 pulsations.

In order for the Pi 2 pulsations to be finite they must be excited near the marginal damping limit as given by equation (9). This limit, therefore, is a reasonable physical constraint on our model. Figure 3 is taken from paper 1 and shows that the direction of the WTS is very dependent on α, with $\alpha \leq 0.8$ implying westward motion. The above results show that if λ_x is less than $(\lambda_x/\alpha)_t$ then α must be less than 1 in order to have finite Pi 2 waves. This case is most likely to occur in the early stages of substorm breakup when the surge is in its narrower, embryonic stages. Rapid westward motion (≈30 km s^{-1}) has been observed, for example, in the leading branch of the WTS at the beginning of breakup (Opgenoorth et al., 1983; Yahnin, 1983) consistent with this interpretation.

One can also use these results to develop scenarios for the evolution of the WTS motion. For example, assume the temperature ratio in the plasma sheet stays constant throughout the substorm and a surge initially develops with λ_x below threshold. Then as mentioned α is less than 1 and the surge starts to move northwestward; see Figure 3. The surge north-south dimension, λ_x, increases due to the poleward component of motion and α may then eventually reach ≥ 1 which implies subsequent poleward or even eastward motion as seen from Figure 3.

However, it is probably unrealistic to expect the plasma sheet temperature ratio, $T_\perp/T_\parallel$, to stay constant throughout the substorm. We, therefore, recalculated the Fridman and Lemaire (1980) model for a wider range of plasma sheet electron temperatures and inserted the results into equation (9). It was found that the threshold value decreases as $T_\perp/T_\parallel$ decreases for fixed $T_\perp$. In physical terms, at substorm onsets preferential plasma sheet heating parallel to the magnetic field, B, allows poleward motion of embryonic (i.e., narrower) surge regions. On the other hand, perpendicular heating (larger $T_\perp/T_\parallel$) implies the surge must move northwestward until λ_x reaches the appropriate threshold value. Therefore, the initial motion of embryonic surge regions can be a measure of electron temperature anisotropy in the plasma sheet.

We found that a reasonable fit to $(\lambda_x/\alpha)_t$ is given by ($n_s = 0.3$ cm^{-3})

$$(\lambda_x/\alpha)_t = 122.8 + 26.9 \ln(T_\perp/T_\parallel) \text{ km}$$

$$0.02 \leq T_\perp/T_\parallel \leq 0.5 \qquad (10)$$

and

$$(\lambda_x/\alpha)_t = 131.8(T_\perp/T_\parallel)^{0.53} \text{ km}$$

$$1.0 \leq T_\perp/T_\parallel \leq 11. \qquad (11)$$

Note that the threshold scales inversely as the square root of the plasma sheet number density, n_s.

Discussion

The requirement that Pi 2 pulsations be damped provides a connection between surge motion and the electron temperature anisotropy in the plasma sheet. It has also been shown that positive growing solutions are eventually damped due to the rapid poleward motion of the surge. The linear theory presented here is, therefore, consistent with the generation of finite amplitude Pi 2 pulsations as observed.

Poleward motion of narrower surge regions implies an initial heating and/or Fermi acceleration of plasma sheet electrons parallel to the magnetic field. Fermi acceleration could arise at higher latitudes due to the initial dipolar relaxation of the more extended field lines and the **E** x B earthward convection. During the latter stages perpendicular heating arises in part from betatron acceleration due to the magnetic field lines becoming more dipolar. Based on the above model, during substorm onsets one expects rapid poleward expansion of the surge followed by predominantly westward motion.

References

Baumjohann, W., Ionospheric and field-aligned current systems in the auroral zone: A concise review, Adv. Space Res., 2, 55, 1983.

Fridman, M., and J. Lemaire, Relationship between

auroral electron fluxes and field-aligned electric potential difference, J. Geophys. Res., 85, 664, 1980.
Inhester, B., W. Baumjohann, R. A. Greenwald, and E. Nielsen, Two-dimensional observations of ground magnetic fields associated with auroral zone currents, 3, Auroral zone currents during passage of a westward traveling surge, J. Geophys., 49, 155, 1981.
Lysak, R. L., Coupling of the dynamic ionosphere to auroral flux tubes, J. Geophys. Res., 91, 7047, 1986.
Opgenoorth, H. J., R. J. Pellinen, W. Baumjohann, E. Nielsen, G. Marklund, and L. Eliasson, Three-dimensional current flow and particle precipitation in a westward traveling surge (observed during the Barium-GEOS rocket experiment), J. Geophys. Res., 88, 3138, 1963.
Rees, M. H., Auroral ionization and excitation by incident electrons, Planet. Space Sci., 11, 1209, 1963.
Rostoker, G., and J. C. Samson, Polarization characteristics of Pi 2 pulsations and implications for their source mechanisms: Location of source regions with respect to auroral electrojets, Planet. Space Sci., 29, 225, 1981.
Rothwell, Paul L., Michael B. Silevitch, and Lars P. Block, A model for the propagation of the westward traveling surge, J. Geophys. Res., 89, 8941, 1984.
Rothwell, Paul L., Michael B. Silevitch, and Lars P. Block, Pi 2 pulsations and the westward traveling surge, J. Geophys. Res., 91, 6921, 1986.
Samson, J. C., and G. Rostoker, Polarization characteristics of Pi 2 pulsations and implications for their source mechanisms: Influence of the westward traveling surge, Planet. Space Sci., 31, 435, 1983.
Sato, T., A Theory of quiet auoral arcs, J. Geophys. Res., 83, 1042, 1978.
Yahnin, A. G., A. Sergeev, R. J. Pellinen, W. Baumjohann, K. A. Kaila, H. Ranta, J. Kangas, and O. M. Raspopov, Substorm time sequence and microstructure on 11 November 1976, J. Geophys., 53, 182, 1983.

A SIMULATION STUDY OF BROADBAND ELECTROSTATIC NOISE IN THE PRESENCE OF IONOSPHERIC ELECTRONS

D. Schriver

Department of Physics, University of California, Los Angeles, California 90024

M. Ashour-Abdalla

Department of Physics and Institute of Geophysics and Planetary Physics
University of California, Los Angeles, California 90024

Abstract. Ion beams have been observed flowing along magnetic field lines in the earth's plasma sheet boundary layer and are believed to generate intense electrostatic wave activity known as broadband electrostatic noise. Cold electrons of ionospheric origin have also been observed in this same region and it has been shown that the addition of these cold electrons modifies substantially the plasma wave dispersion properties. With cold electrons present, four instabilities can be excited: (1) ion acoustic, (2) electron acoustic, (3) beam resonant, and (4) Buneman two stream. These instabilities can generate waves with large growth rates at frequencies consistent with broadband electro-static noise. Using computer simulations, we examine the wave energy frequency spectrum generated by these instabilities, as well as cold electron heating and nonlinear effects on the ion beam.

Introduction

The plasma sheet boundary layer (PSBL) in the earth's magnetotail is a highly dynamic region of space wherein fast plasma flows and field-aligned beams, plasma inhomogeneities, and entry of ionospheric plasma occur. Strong ac and dc electric fields are also observed in the PSBL and the most commonly seen waves are known as broadband electrostatic noise (BEN). BEN was first observed in the magnetotail by Scarf et al. (1974) and studied in detail by Gurnett et al. (1976) and Cattel et al. (1986b). BEN is most intense in the PSBL but has also been observed with less intensity in the lobe and central plasma sheet regions as well and is characterized by its wide frequency range, from 10 Hz to several kilohertz and its intensity of the order of millivolts per meter.

The cause of BEN is not completely understood, although as suggested by Cattel and Mozer (1986), it is probably a combination of instabilities that generate the entire frequency spectrum. The peak in wave power is at low frequencies (<50 Hz) and is near the lower hybrid frequency (Cattel and Mozer, 1986). Ashour-Abdalla and Thorne (1978) proposed a theory for BEN generated by current-driven ion cyclotron waves, but the instability excites frequencies that are too low to account for the entire BEN spectrum. Another possibility is the lower hybrid drift instability (Huba et al., 1978) driven by density gradients. This instability, however, is not broadbanded enough to explain the entire BEN spectrum, which is observed to extend up to the electron plasma frequency near 10 kHz (Gurnett et al., 1976). Ion beams are commonly observed in the PSBL and can drive high-frequency instabilities as first discussed by Grabbe and Eastman (1984) and are a likely candidate for the generation of the upper frequency portion of BEN.

Ion beams are a common feature of the PSBL and were first observed by Frank et al. (1976) and studied in detail by Eastman et al. (1984). These beams are field-aligned and stream earthward, probably formed by either current sheet acceleration (Speiser, 1965) or as a consequence of reconnection (Schindler and Birn, 1987). The ion beams have variable drift kinetic energies from 1 to >40 keV and have temperatures of the order of a few hundred electron volts (Eastman et al., 1984). Counterstreaming beams, which are probably magnetically reflected earthward flowing beams (Eastman et al., 1984), are also sometimes observed as well as cooler (<100 eV) tailward flowing beams which are likely to be ionospheric ions accelerated out from the auroral zone region into the PSBL (Sharp et al., 1981). Simultaneous observations of ion beams and BEN led Grabbe and Eastman (1984) to the conclusion that ion beams serve as a free energy source in generating at least a portion of the BEN spectrum.

Linear theory analysis that included an ion beam drifting relative to a background of hot plasma sheet ions and electrons showed that instabilities are driven that have a wide range in frequency with an upper cutoff near the electron plasma frequency (Grabbe and Eastman, 1984; Omidi, 1985; Akimoto and Omidi, 1986; Dusenbery and Lyons, 1985; Ashour-Abdalla and Okuda, 1986a). The instabilities driven are the ion-acoustic and ion-ion instabilities which are broadbanded, but have two difficulties in explaining BEN. The first is that the upper frequency cutoff is about 0.1 ω_{pe} which is too low to account for observations that show the upper cutoff of BEN at or above ω_{pe}. The other difficulty is that the more commonly observed earthward (and counterstreaming) ion beams have temperatures that are too high to excite the ion-ion and ion-acoustic instabilities and only the cooler, less frequently observed ionospheric tailward flowing beams can be unstable. Since BEN is often observed only with earthward flowing or counterstreaming warm beams, other instabilities must be active to generate BEN.

Recently there have been observations that suggest the existence of cold electrons in the PSBL (Etcheto and Saint-Marc, 1985; Sojka, private communication, 1986). These cold electrons are most likely of ionoshperic origin and have temperatures that range from 1 to 130 electron volts. Including cold electrons into a PSBL model, as first done by Grabbe (1985), changes the wave dispersion properties such that new instabilities are excited that in the absence of cold electrons would not be present. These are the electron acoustic, Buneman, and beam resonant instabilities (Schriver and Ashour-Abdalla, 1987). These instabilities are better suited to explain BEN because they address both of the earlier problems of the ion-ion and ion-acoustic instabilities. The frequency range is extended up closer to ω_{pe}($\omega \sim 0.5\ \omega_{pe}$) and much warmer beams can be unstable. In fact, with cold electrons included, observed beam temperatures ($T_b \geq$ 500 eV) can be used to generate broadbanded frequency instabilities.

Simulation studies of ion beam instabilities in relation to BEN have been performed by Ashour-Abdalla and Okuda (1986a,b). The first study (Ashour-Abdalla and Okuda, 1986a) examined the ion-ion and ion-acoustic instabilities (no cold electrons) and showed that the instabilities are broadbanded, with frequencies up to ω_{pe}, but require cold beams to be excited. The ion-ion instability was saturated by ion beam heating, while the ion-acoustic instability was saturated by a plateau formation in the hot electron velocity distribution near the sound speed. The addition of cold electrons into the plasma model was considered by Ashour-Abdalla and Okuda (1986b), and they were able to excite the electron acoustic instability. A narrow-banded frequency spectrum was excited, and the instability was saturated by cold electron heating. The beam resonant instability, which can also be excited and is more broadbanded, was not studied by Ashour-Abdalla and Okuda (1986b) and is the main focus of the simulation study in this work.

This paper will concentrate on the generation of the high-frequency portion of BEN by ion beam instabilities with ionospheric cold electrons included. The linear theory, which gives mode identification and unstable frequency characteristics, will be reviewed in the next section. Simulation results describing nonlinear saturation and the wave power spectrum generated by these instabilities will be discussed, followed by the conclusions.

Linear Theory

We consider a uniform, infinite, homogeneous plasma consisting of four species: (1) ion beam, (2) hot background ions, (3) hot background electrons, and (4) cold background electrons. Each species is modeled by a Maxwellian distribution, and for the observed PSBL conditions, two instabilities can be excited. These are the beam resonant and electron acoustic instabilities. These instabilities are excited for warm beams (T_b > 100 eV) and are present only if cold electrons are included.

The plasma model for the PSBL is illustrated in Figure 1. The velocity distribution for each species is shown, indicated by f_α where α is the species subscripts as follows: b for beam ions, e or h for hot background electrons, i for hot background ions, and c for cold background electrons. The conditions on the beam speed, U, for when either instability is excited is indicated on Figure 1, and it is seen that the electron acoustic instability is excited at lower beam drift speeds. Both instabilities are driven unstable by resonance with the positive slope of the beam. The beam resonant instability is the high beam temperature (kinetic) limit of the Buneman two-stream instability, which is an interaction between the ion beam and the cold electron background. The electron acoustic mode is a normal mode of any plasma that has two electron species of different temperatures (Watanabe and Taniuti, 1977) and is driven unstable by resonance with the positive slope of the ion beam (Ashour-Abdalla and Okuda, 1986b). The background ions play a relatively small role in the dispersion and growth rates of these instabilities and merely preserve charge neutrality. This is a consequence of the ion beam preferring to interact with the more mobile, oppositely-charged cold electrons. In the absence of cold electrons, the ion beam-cold electron two-stream (Buneman) instability is replaced by the ion beam-ion background ion-ion fluid instability.

The frequency characteristics of each instability is illustrated in Figure 2. Plotted is real frequency versus effective beam drift

With Cold Electrons - High Beam Temperature

(a) Beam Resonant

$$\frac{V_h\sqrt{\frac{n_c}{n_h}}}{(1+k^2\lambda_h^2+T_h n_b/T_b n_h)^{1/2}} \lesssim U\cos\theta \lesssim V_h\sqrt{\frac{n_c}{n_h}} + V_b$$

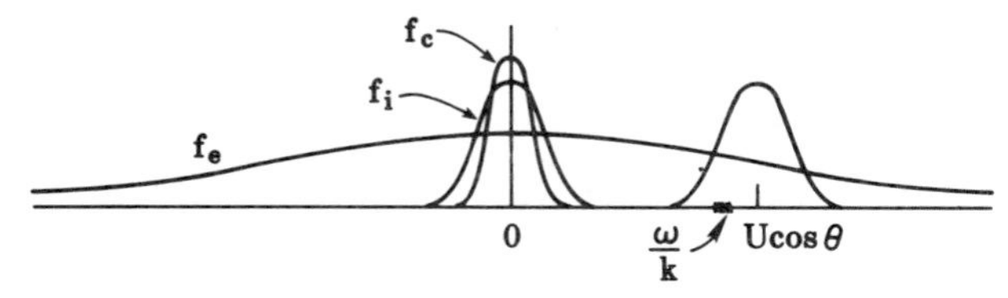

(b) Electron Acoustic $V_i < U\cos\theta < \frac{V_h\sqrt{\frac{n_c}{n_h}}}{(1+k^2\lambda_h^2+T_h n_b/T_b n_h)^{1/2}}$

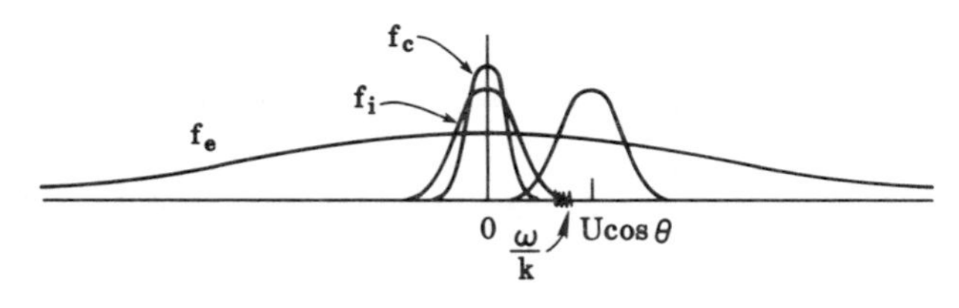

Fig. 1. The velocity distributions for each species are shown for the model of the plasma sheet boundary layer used. The two instabilities are illustrated for the four species plasma considered: (1) warm ion beam, (2) hot background electrons f_h, (3) hot background ions f_i, and (4) cold background electrons f_c. The conditions on beam drift, $U\cos\theta$, are shown for each instability, where θ is the angle of wave propagation. The electron acoustic instability has the lower velocity threshold.

speed $U\cos\theta$ (where θ is the angle between k and U, and U is normalized to the hot electron thermal velocity V_h). The shaded region indicates where positive growth rates occur ($\gamma > 0$), and it is seen that the electron acoustic instability, although having a lower drift velocity threshold, is narrow banded in frequency. This is because the electron acoustic mode frequency asymptotically approaches the cold electron plasma frequency ($\omega \rightarrow 0.4\ \omega_{pe}$) as k increases, and since the mode is unstable only at large k, where the frequency changes very little, only a narrow-banded spectrum is unstable. The beam resonant instability on the other hand can be very broadbanded, as seen in Figure 2, because $\omega \sim kV\cos\theta$. This instability has a frequency range over many orders of magnitude and has the same range as that of BEN.

As the temperature of the cold electrons is increased, both instabilities weaken until the temperature reaches that of the hot electrons, where neither instability (or any other for the given beam temperature of 500 eV) is present. This is seen in Figure 3, which shows the dependence of maximum growth rate (γ_m) versus beam drift speed $U\cos\theta$ as the cold electron temperature is varied. The ratio of hot to cold electron temperature is used (T_h/T_c), and as this parameter decreases, or as T_c increases, the growth rates come down. The electron acoustic instability is most sensitive to T_c, with the beam resonant instability being affected only when $T_h/T_c < 250$. This dependence on cold electron temperature will become important when looking at wave saturation from the computer simulations in the next section.

Regions of Wave Growth (High Beam Temperature) Frequency vs. Beam Drift

Electron Acoustic
Beam Resonant
Plasma Frequency
$\gamma > 0$
$\frac{\omega_r}{\omega_{pe}}$: 10.0, 1.0, 0.1, 0.01, 0.001
$\frac{U\cos\theta}{V_h}$: 0.0, 0.2, 0.4, 0.6

Fig. 2. Real frequency is shown versus beam drift speed ($U\cos\theta$) with the unstable region ($\gamma > 0$) shaded. The beam resonant instability can be broadbanded but the electron acoustic instability is narrow banded. Frequency is normalized to the total electron plasma frequency, while beam drift is normalized to hot electron thermal velocity V_h. Here $T_b = T_h = T_i = 500\ T_c$ and $n_c = 0.25\ n_h$, $n_b = 1.5\ n_i$.

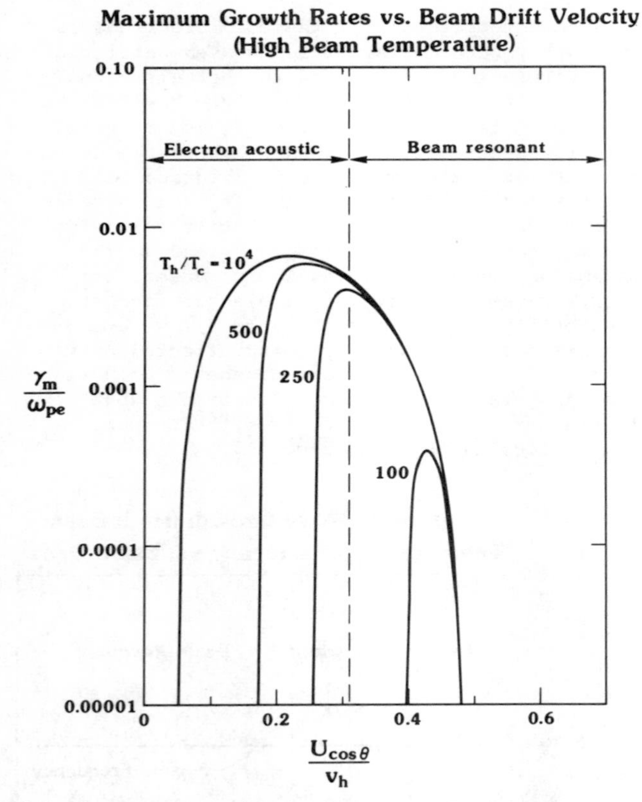

Fig. 3. Maximum growth rate is plotted versus $U\cos\theta$ for four different values of the cold electron temperature. As T_c increases (T_h/T_c decreases), the growth rates decrease with the electron acoustic instability affected the most. As $T_c \to T_h$ both instabilities vanish. Density parameters are the same as in the previous figure.

Simulation Results

To study nonlinear effects, such as wave saturation, how the various particle distributions are affected by the waves, and what type of wave power spectrum is created, computer simulations have been done. A one-dimensional, electrostatic, full particle code is used with periodic boundary conditions. The grid spacing is scaled to the cold electron Debye length ($\Delta \simeq \lambda_c$) and the length of the system is $L = 1024\Delta$. With periodic boundary conditions, the wave number is restricted to $k = 2\pi\, n/L$ ($n = 0, 1, 2...$). Because of limited computer resources and the fact that both ions and electrons are treated dynamically as full particles, a reduced mass ratio is used where $M_i/M_e = 400$. The total number of particles is 200,000 (100,000 ions and 100,000 electrons), and the initial conditions are set up such that the electron acoustic and beam resonant instabilities can be excited by simply varying the beam drift speed U. Each of the species is loaded at $t = 0$ as a Maxwellian distribution with $T_h = T_b = T_i = 500\ T_c$. The densities of each species are $n_c = 0.25\ n_h$ and $n_b = 1.50\ n_i$. For each run, $\theta = 0°$, so that only parallel propagation is considered and only one instability can be excited per run depending on U.

The wave power spectrum generated over each simulation run for a number of different beam speeds is shown in Figure 4. Wave energy is summed up over all k modes and plotted versus frequency for four different beam speeds. At the lowest beam speed ($U/V_h = 0.20$), the wave spectrum is narrow banded. This is consistent with the electron acoustic instability which has the lowest drift velocity threshold and is narrow banded as predicted by linear theory. As the beam speed is raised ($U/V_h > 0.60$), the wave spectrum broadens out, characteristic of the beam resonant instability. The wave power peaks at the lower frequency end ($\sim 0.1\ \omega_{pe}$) and steadily decreases up to the electron plasma frequency. This is consistent with observations of BEN which show the wave power peaked at low frequencies (<100 Hz) and a steady decrease in power with

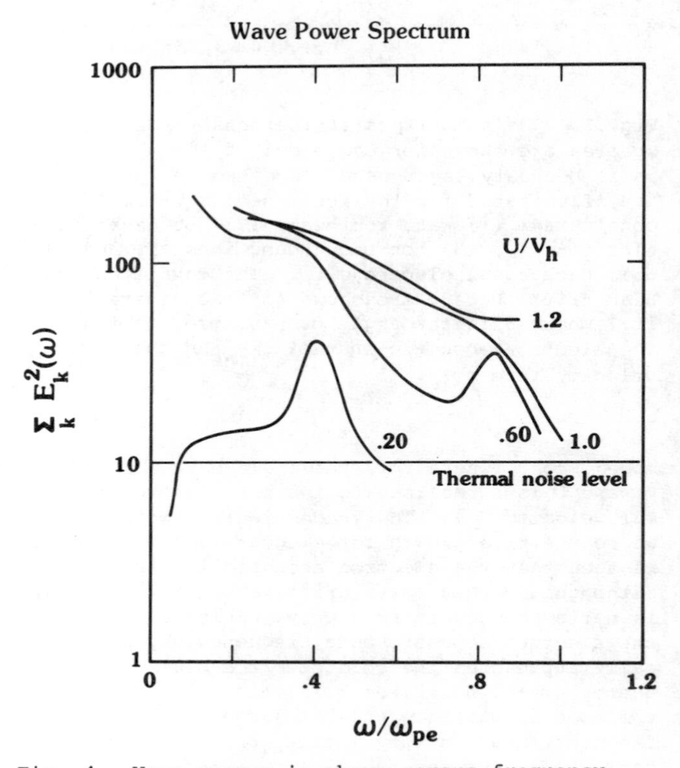

Fig. 4. Wave energy is shown versus frequency for various simulation runs with four different drift speeds U. The wave energy is summed over all k modes to give the continuous spectrum. Here $\theta = 0°$ with other parameters given in the section on "Simulation Results."

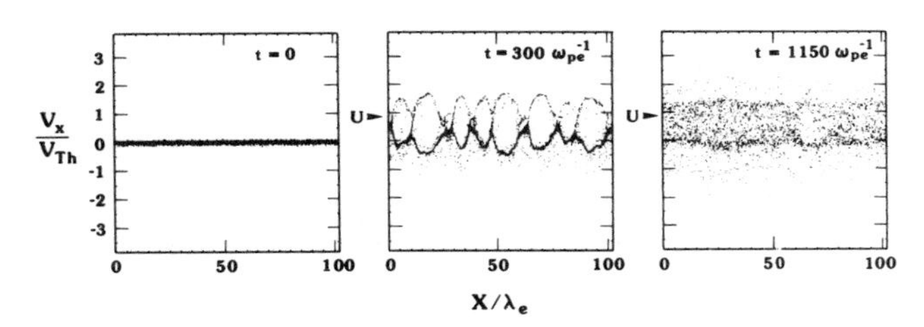

Fig. 5. Cold electron phase space is shown for three different times in a particular simulation run. Plotted in each panel is V_x versus x, where x is the distance along the simulation (x) axis. For this case $U/V_h = 0.8$ and $\theta = 0°$. Other parameters are given in the text.

increasing frequency up to ω_{pe} (Gurnett et al., 1976; Cattel et al., 1986b; Roeder et al., 1986). The wave spectra generated by these beam instabilities does not go low enough to explain the entire BEN spectrum, but nicely duplicates a large portion of the high-frequency end. The very low frequencies observed in BEN are likely to be caused by another instability, with the entire spectrum generated by a superposition of wave modes as suggested by Cattel and Mozer (1986). The low frequencies are probably generated by a lower hybrid instability (Huba et al., 1978), with the high frequencies generated by ion beams.

Both instabilities, beam resonant and electron acoustic, are saturated by cold electron heating. The electron acoustic instability has already been simulated by Ashour-Abdalla and Okuda (1986b), and they also saw cold electron heating as the saturation mechanism. Since the electron acoustic instability has already been simulated and the beam resonant instability is more broad-banded, thus having more application to BEN, the saturation mechanism for the beam resonant instability will be examined in detail here. From linear theory, as the cold electron temperature increases, the beam resonant instability growth rates tend to decrease (see Figure 3). Although the beam resonant instability (and electron acoustic) relies on the positive slope of the ion beam for wave growth, the ion beam remains relatively unaffected by the waves. This is because the cold electrons are easily and quickly heated, saturating the instability before the more massive ion beam is affected.

The evolution of the cold electrons can be followed by examining their phase space at different times. This is seen in Figure 5, which shows cold electron phase space (velocity versus box length) at three different times. The parameters are set up in this case such that the beam resonant instability is excited. At t = 0, the cold electrons are gathered around $V_x = 0$ with a small thermal spread. At $t = 300\ \omega_{pe}^{-1}$ (middle panel), the unstable waves have reached an appreciable amplitude and the cold electrons are strongly (nonlinearly) phase trapped by the large-amplitude waves. The number of trapping vortices indicates the wave mode dominant of that time (mode 8). Near the end of the run, at $t = 1150\ \omega_{pe}^{-1}$), the waves have saturated and the cold electrons are thermalized with a temperature increase of about 250 times their original temperature. This is very close to the value at which linear theory predicts that there will be no unstable wave modes (Schriver and Ashour-Abdalla, 1987). The other species' (hot background and beam) velocity distributions are relatively unaffected at the end of the run, which means that nearly all the wave power has gone into heating the cold electrons.

A look at the cold electron velocity distribution shows that at the end of the run, not only has heating occurred, but the distribution of the now heated cold electrons has aquired a net positive drift velocity. This is seen in Figure 6. The distribution at t = 0 is illustrated by the dashed line and is sharply peaked, indicative of the low initial temperature. At the end of the run ($t = 1250\ \omega_{pe}^{-1}$), the cold electrons have been heated drastically and have a net drift velocity just less than the beam speed U (shown on the diagram). This results from the phase

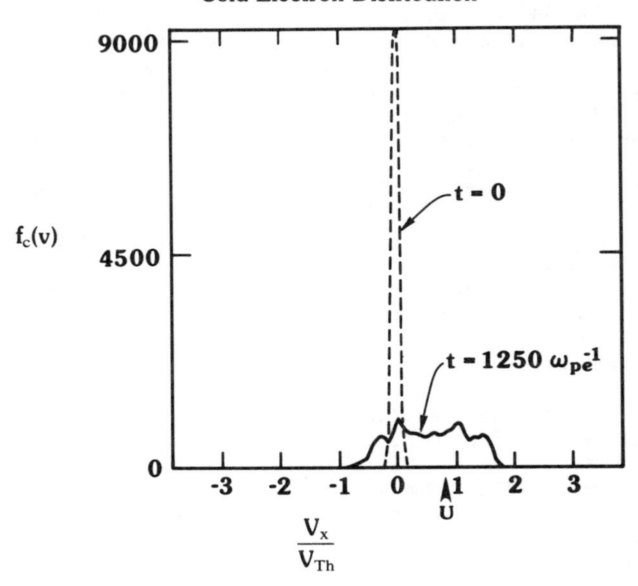

Fig. 6. The cold electron velocity distribution, f_c, is shown at two different times during the simulation run. The dashed line is at t = 0 and the solid line illustrating the cold electron heating is near the end of the run at $t = 1250\ \omega_{pe}^{-1}$. The parameters for this run are the same as in the previous figure.

velocity of the waves being slightly below the beam speed so that when the cold electrons are trapped by the waves, they are pulled up to the phase velocity (as well as heated) and the distribution as a whole ends up with a drift velocity equal to the wave phase velocity. The cold electrons are, in effect, dragged along by the ion beam. In each of the simulation runs, for variable drift speed U, the results are essentially the same; the cold electrons heat, saturating the instability, and pick up a net drift velocity. Only the amount of heating varies slightly from case to case, satisfying the marginal stability criteria, whereby no mode is unstable.

Conclusions

Ion beams are a souce of free energy that can drive a broadbanded instability using PSBL parameters and realistic beam temperatures. The presence of ionospheric cold electrons is vital in allowing high temperature ion beams to excite the beam resonant instability and are important in saturation of the growing waves.

The wave frequency spectrum excited by the beam resonant instability is broadbanded, with peaks at low frequencies ($\lesssim 0.1\ \omega_{pe}$) and upper cutoff near ω_{pe}. This is consistent with BEN observations at high frequencies but does not go to low enough frequencies to entirely explain BEN. The beam resonant instability combined with a low-frequency instability can account for the whole BEN spectrum. Since density gradients are observed in the PSBL (Cattel et al., 1986a), the low-frequency waves can possibly be driven by the lower hybrid drift instability (Huba et al., 1978). It appears then that at least two free energy sources are responsible for BEN, and it is possible there are more. This supports the suggestion by Cattel and Mozer (1986) that the entire BEN spectrum is caused by a superposition of wave modes. Future work should try to combine the various free energy sources into one model, whereby the entire BEN spectrum could be generated self consistently.

In the initial value problem where no new cold electrons are input into the system, particle transport is affected by BEN mainly through heating of the cold electrons. The hot background plasma and the ion beam are affected very little by BEN. With the ion beam hardly affected, it can travel long distances along magnetic field lines. This is consistent with observations that show earthward flowing beams traveling the length of the tail to be magnetically reflected and forming counterstreaming beams (Eastman et al., 1984). Because the cold electrons are rapidly heated, however, they must be constantly replenished to be observed and this would allow a steady level of wave excitation to occur in the PSBL; if the wave level is maintained in this way, the ion beam eventually will be heated since the wave growth is due to the positive slope of the ion beam. Simulations being done now suggest that this ion beam heating would be slow (compared to the cold electron heating), occurring roughly for $t > 20{,}000\ \omega_{pe}^{-1}$, and thus the beam would heat but still travel long distances, which is consistent with the observations of Eastman et al. (1984) that indicate some ion heating occurs.

The mechanism that sends the cold electrons into the tail is not well understood and needs to be studied further. The cold electrons, once heated, should form a secondary electron beam, as seen in the simulations. This would be difficult to observe, though, since the hot electron background has a wide thermal spread and would mask the secondary drifting electrons. Also, no cold electrons would be expected in the plasma sheet proper, since upon entering the boundary layer, they are quickly heated and by the time they could diffuse into the central plasma sheet, they would be quite warm. More observations of cold electrons are needed to determine the density and temperatures in the PSBL, as well as how they travel from the ionosphere into the tail. Future theoretical studies should concentrate on how a constant source of cold electrons may affect the generation of BEN.

The PSBL is a highly dynamic region of space where earthward streaming ion beams, tailward flowing ionospheric plasma, density and magnetic field gradients, and intense electric field activity are observed. Although the causes and consequences of BEN are falling into place, the acceleration mechanisms for the ion beams and ionospheric plasma into the boundary layer are not completely understood. Since the PSBL represents a meeting place for ionospheric plasma and solar wind plasma, more work needs to be done to understand why and how the PSBL is such an important area for plasma transport in the magnetosphere.

Acknowledgments. This work was supported by National Science Foundation grant ATM 85-13215, NASA STTP grant NAGW-78, and U.S. Air Force contract F19628-85-K-0027. Computing resourses were provided by the San Diego Supercomputer Center.

References

Akimoto, K., and N. Omidi, The generation of broadband electrostatic noise by an ion beam in the magnetotail, Geophys. Res. Lett., 13, 97, 1986.

Ashour-Abdalla, M., and H. Okuda, Theory and simulations of broadband electrostatic noise in the geomagnetic tail, J. Geophys. Res., 91, 6833, 1986a.

Ashour-Abdalla, M., and H. Okuda, Electron acoustic instabilities in the geomagnetic tail, Geophys. Res. Lett., 13, 366, 1986b.

Ashour-Abdalla, M., and R. M. Thorne, Towards a unified view of diffuse auroral precipitation, J. Geophys. Res., 83, 4755, 1978.

Cattel, C. A., and F. S. Mozer, Experimental determination of the dominant wave mode in the active near-earth magnetotail, Geophys. Res. Lett., 13, 221, 1986.

Cattel, C. A., F. S. Mozer, E. W. Hones, Jr., R. R. Anderson, and R. D. Sharp, ISEE observations of the plasma sheet boundary, plasma sheet, and neutral sheet, 1. Electric field, magnetic field, plasma and ion composition, J. Geophys. Res., 91, 5663, 1986a.

Cattel, C. A., F. S. Mozer, E. W. Hones, Jr., R. R. Anderson, and R. D. Sharp, ISEE observations of the plasma sheet boundary, plasma sheet, and neutral sheet, 2. Waves, J. Geophys. Res., 91, 5681, 1986b.

Dusenbery, P. B., and L. R. Lyons, The generation of electrostatic noise in the plasma sheet boundary layer, J. Geophys Res., 90, 10,935, 1985.

Eastman, T. E., L. A. Frank, and W. Peterson, The plasma sheet boundary layer, J. Geophys. Res., 89, 1553, 1984.

Etcheto, J., and A. Saint-Marc, Anomalously high plasma densities in the plasma sheet boundary layer, J. Geophys. Res., 90, 5338, 1985.

Frank, L. A., K. L. Ackerson, and R. P. Lepping, On hot tenuous plasmas, fireballs, and boundary layers in the Earth's magnetotail, J. Geophys. Res., 81, 5859, 1976.

Grabbe, C. L., and T. E. Eastman, Generation of broadband electrostatic waves in the magnetotail, J. Geophys. Res., 89, 3865, 1984.

Grabbe, C. L., New results on the generation of broadband electrostatic waves in the magnetotail, Geophys. Res. Lett., 12, 483, 1985.

Gurnett, D. A., L. A. Frank, and R. P. Lepping, Plasma waves in the distant magnetotail, J. Geophys. Res., 81, 6059, 1976.

Huba, J. D., N. T. Gladd, and K. Papadopolous, Lower hybrid drift wave turbulence in the distant magnetotail, J. Geophys. Res., 83, 5217, 1978.

Omidi, N., Broadband electrostatic noise produced by ion beams in the Earth's magnetotail, J. Geophys. Res., 90, 12,330, 1985.

Roeder, J. L., H. C. Koons, O. H. Bauer, G. Haerendel, R. Treumann, R. R. Anderson, D. A. Gurnett, and R. H. Holzworth, A survey of impulsive broadband electrostatic waves in the magnetosphere (abstract), Eos Trans. AGU, 67, 1167, 1986.

Scarf, F., L. A. Frank, K. L. Ackerman, and R. P. Lepping, Plasma wave turbulence at distant crossings of the plasma sheet boundaries and neutral sheet, Geophys. Res. Lett., 1, 189, 1974.

Schindler, K., and J. Birn, On the generation of field-aligned plasma flow at the boundary of the plasma sheet, J. Geophys. Res., 92, 95, 1987.

Schriver, D., and M. Ashour-Abdalla, Generation of high-frequency broadband electrostatic noise: The role of cold electrons, J. Geophys. Res., 92, 5807, 1987.

Sharp, R. D., D. L. Carr, W. K. Peterson, and E. G. Shelley, Ion streams in the magnetotail, J. Geophys. Res., 86, 4639, 1981.

Speiser, T. W., Particle trajectories in model current sheets, 1. Analytical solutions, J. Geophys. Res., 70, 4219, 1965.

Watanabe, K., and T. Taniuti, Electron acoustic mode in a plasma of two temperature electrons, J. Phys. Soc. Japan, 43, 1819, 1977.

CONSEQUENCES OF THE DURATION OF SOLAR ENERGETIC PARTICLE-ASSOCIATED MAGNETIC STORMS ON THE INTENSITY OF GEOMAGNETICALLY TRAPPED PROTONS

W. N. Spjeldvik

Department of Physics, Weber State College, Ogden, Utah 84408

Abstract. The development of a time-dependent model for geomagnetically confined protons has proceeded to the stage where model calculations may be used to simulate and predict the time-variable proton radiation zone. The model has been used to study the diffusive injection of solar energetic protons into the stable trapping region during major magnetic storms. Although essentially all model parameters can be time variable, this paper analyzes the effects of time variations in the radial diffusion coefficient and in the outer zone boundary condition imposed at the approximate outer limit of the stable trapping region (L = 7) beyond which protons are unable to complete a drift trajectory around the world. First, the steady state equilibrium for radiation belt protons was simulated throughout the equatorial trapping region consistent with imposed quiet-time boundary conditions and the quiet-time average radial diffusion coefficient. Then, the fully time-dependent proton transport equation was solved in time, space, and energy equivalent for characteristic major magnetic storm and solar energetic proton events. In this paper, perturbation magnitudes have been kept fixed, but the duration of the events has been varied. It was found that for several tens of kiloelectron volt protons, short-lived events on time scales of hours have almost the same radiation belt consequences as more enduring events. In contrast, radiation belt protons in the megaelectron volt range are strongly dependent on the duration time scales of the arrival of energetic solar protons and on the maintenance of an enhanced radial diffusion coefficient.

Introduction

After the discovery of the earth's radiation zone in the late 1950's (Van Allen, 1959), theoretical efforts were made to understand and eventually physically model the properties and time variations of this region of earthspace. In the early 1960's Van Allen (1962) presented a discourse on the observations and their interpretations. Much of the early Soviet work was summarized by Tverskoy (1965, 1969, 1971), and the European and American contributions were reviewed by Fälthammar (1968), Roederer (1968, 1970), Hess (1968), Schulz and Lanzerotti (1974), Schulz (1975), Walt (1977), Spjeldvik (1979), and others.

Early modelings were made of proton distributions (Nakada and Mead, 1965), proton and helium ion distributions (Cornwall, 1972), and electron distributions (Lyons and Thorne, 1973). The latter authors coupled radial transport theory for electrons with a model for ELF wave-electron interactions to explain the division of trapped electrons into two fairly distinct belts at quiet times. Spjeldvik and Thorne (1975) subsequently found that the consequential energetic electron precipitation into the atmosphere also explained the enhancement and maintenance of the ionospheric D region as observed by ground-based radio wave experiments.

Following the concepts of Cornwall (1968, 1972), Spjeldvik (1977) applied realistic, quiet-time outer zone boundary conditions to model the steady state proton radiation zone, also taking into account charge exchange and proton energy degradation due to Coulomb collisions with exospheric neutral hydrogen atoms. Later, works by Spjeldvik and Fritz (1978a,b) carried the boundary condition driven modeling effect further by including the charge exchange reaction schemes necessary to simulate the quiet-time structure of heavier ion species such as helium and oxygen.

Steady state studies have generally been quite successful in describing the overall structures of the earth's radiation zone and appear to have substantial merit at intermediate energies, hundreds of kiloelectron volts to several megaelectron volts. However, it appears that a time-independent approach may not properly describe the stably trapped ion and electron population either at low or very high radiation belt energies.

Below several tens of kiloelectron volts, the ion charge exchange cross sections for ionic neutralization are typically quite high for most elements, of the order 10^{-15} to 10^{-16} cm^2. In the exospheric environment of the earth, this leads to decay times (negative e-folding times) of tens of minutes to a few hours for many radiation belt ion species. Under such circumstances the particle flux distribution will respond quickly to variations in the effective source rate which can be internal (the high-latitude ionosphere or local acceleration of lower-energy particles) or external (convective or diffusive injection). It seems quite unlikely that such sources will remain steady on time scales of months or years required to establish equilibrium conditions for the higher-energy particles.

At high energies, above several megaelectron volts, the ionic life times versus charge exchange loss and the electronic and ionic Coulomb collision energy degradation times become very long and are typically measured in years to hundreds of years, depending on location in the radiation zone. An effective upper limit to the particle lifetimes is provided by the known secular variations of the geomagnetic field itself (e.g., Schulz and Paulikas, 1972). Substantial fluxes of very energetic particles are probably injected into the trapping region only on rare occasions. Radial and angular redistribution, adiabatic energization during net inward transport (which slows down with lower L shells), long time scale particle losses, and collisional energy degradation eventually change the injected distribution. A steady state may actually never be reached.

Modeling of the earth's radiation zone using a time-dependent approach is therefore necessary in order to understand the structure of the most energetic particles. A second reason stems from the fact that the trapping region is disturbed by magnetic storms which cause both adiabatic and nonadiabatic variations in the trapped particle populations. Even non-energetic particle events can influence these populations; for example, enhancement in solar ultraviolet radiation will heat the upper atmosphere and ionosphere and cause exospheric density increases. These in turn decrease all collisional lifetimes, and so alter the radiation belt structure.

In this paper, the time-dependent proton model is used to study one aspect of the time-variable radiation zone of the earth, its response to solar energetic particle-associated magnetic storms as parameterized by a time-variable radial diffusion coefficient and a time-variable outer zone boundary condition. Emphasis is placed on the consequences of the duration of such events on the trapped proton population.

The Proton Transport Equation

The theoretical framework for diffusive transport of radiation belt protons is well established (e.g., Schulz and Lanzerotti, 1974; Spjeldvik and Rothwell, 1983). Equatorially mirroring protons are transported across the magnetic field lines by radial diffusion caused by fluctuations in the geoelectric and geomagnetic fields on time scales comparable with the azimuthal drift times. They are also subject to losses, either by transport out of the trapping region (diffusion is both inward and outward, but with an overall net inward transport) or by losses within the trapping region. The internal losses are due to charge exchange with exospheric neutral hydrogen atoms which results in proton transformation to fast neutral hydrogen. Energy degradation by proton collisions with plasmaspheric electrons and with exospheric neutrals generally act as a loss of the higher-energy protons although it produces (and becomes a source of) lower-energy protons. Following Cornwall (1972) the time-dependent proton transport equation may be written in the form:

$$\frac{df}{dt} = L^2 \frac{d}{dL} \left(D_{LL} L^{-2} \frac{df}{dL}\right) - \Lambda f + G\mu^{-1/2}\frac{df}{d\mu} + S,$$

where

- t = time
- f = $f(L,\mu,t)$ = distribution function
- L = L shell parameter
- D_{LL} = radial diffusion coefficient
- G = Coulomb energy degradation factor
- μ = first-adiabatic invariant
- Λ = σ_{10} V[H] = charge exchange factor
- σ_{10} = charge exchange cross section
- V = proton speed
- [H] = exospheric hydrogen number density
- S = internal proton source function.

The present study disregards any internal proton source function ($S = 0$) and thus is not applicable to $E > 30$ MeV protons at $L < 1.2$ where cosmic ray albedo neutron decay (CRAND) could be significant. Numerical values of Λ, G, [H], and the density of the average plasmasphere are taken from the compilation of Spjeldvik (1977). The radial diffusion coefficient is taken from the work of Cornwall (1972) and is analytically given by

$$D_{LL} = D_{om}L^{10} + D_{oe}L^{10}/[L^4 + (\mu/\mu_o Z)^2],$$

where the first term is due to geomagnetic fluctuations assuming a power spectral density following a ν^{-2} relation (ν being fluctuation frequency), and the second term is due to geoelectric fluctuations on substorm time scales (cf. Cornwall, 1972). Z is the net ionic charge state number ($Z = 1$ for protons), and $\mu_o = 1$ MeV/G. The sub-coefficients D_{om} and D_{oe} are taken to be 2×10^{-10} and 10^{-5} per day, respectively, for the quiet-time pre-event situation.

Time-Dependent Parameters

In this study, both the radial transport coefficients and the outer radiation zone boundary condition are considered time variable. The internal parameters (atmospheric density, plasmaspheric free electron concentration, and magnetic topology) are considered fixed. The geometry of the geomagnetic field does change during the main phase of magnetic storms, and this leads to adiabatic effects (e.g., Söraas ond Davis, 1968) during the field distortion. Such adiabatic effects are not explicitly included in this work and would superimpose on the results given herein; they would, however, quickly disappear as the magnetic field configuration returns to normal. The L shell coordinates used here may be considered to move with the field distortion and therefore not represent fixed radial distance from the earth. This work is primarily concerned with the principal long-term (after-) effects of the solar energetic as modeled by a time-dependent diffusion approximation.

Time dependence of D_{LL} is parametrically taken into account by

$$D_{LL}(t) = D_{LL}(t = 0) \times F_D(t),$$

where

$$F_D(t) = F_o \qquad \text{for } t < t_1$$

$$F_D(t) = \exp[(\log F_o) + (\log F_{pd} - \log F_o)(t - t_1)/(t_2 - t_1)] \qquad \text{for } t_1 \leqq t \leqq t_2$$

$$F_D(t) = F_{pd} \qquad \text{for } t_2 \leqq t \leqq t_3$$

$$F_D(t) = \exp[(\log F_{pd}) - (\log F_{pd} - \log F_o)(t - t_3)/(t_4 - t_3)] \qquad \text{for } t_3 \leqq t \leqq t_4$$

$$F_D(t) = F_o \qquad \text{for } t > t_4.$$

The function $F_D(t)$ thus describes a pulse in the overall magnitude of the radial diffusion coefficient with exponential rise at times t_1 to t_2, constant at an elevated level at times t_2 to t_3, exponential decline to the pre-event level at times t_3 to t_4, and constant at times beyond t_4. $F_o = 1$ and F_{pd} describe the peak enhancement of D_{LL}.

A similar parametric function is invoked for the outer zone proton boundary condition. It is thought that emission of energetic protons from the sun during certain solar events is highly variable from event to event. Some events are of short duration (a few hours), while others last for days. Empirically, arrival of solar energetic protons shows a roughly exponential rise (over a time scale of hours) up to a peak (or a series of peaks), which may or may not be sustained in time, and eventually is followed by a roughly exponential decline to the pre-event background flux intensity. These protons will therefore maintain a presence in earthspace into the effective geomagnetic cut-off distance (which varies somewhat with energy). Fluctuations in the geomagnetic field will cause some of these protons to diffusively leak in onto drift orbits around the earth. In the absence of B-field variations this leakage is probably near zero, but with increasing fluctuation magnitudes, the diffusive entry will become more substantial.

Magnetic field variations and the arrival of solar energetic particles may not necessarily always coincide. It is perhaps possible to have arrival of energetic protons (from the sun, from Jupiter, or from other extraterrestrial sources) without a magnetic storm and storms without the arrival of energetic protons, but in this parameterized simulation, we shall consider the two correlated. Such correlation will give optimum diffusive penetration of energetic protons into the radiation zone.

The time dependence of the outer zone boundary conditions are therefore parameterized in a similar fashion as the D_{LL} variations

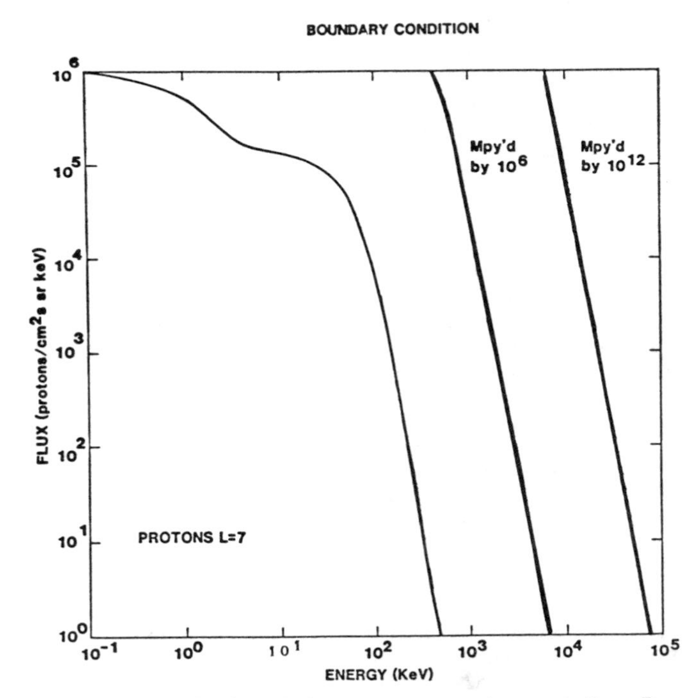

Fig. 1. Radiation belt proton spectrum at L = 7 applied as quiet-time boundary condition on radiation belt proton flux simulations. The spectrum was constructed using quiet-time observations from the ATS 6 spacecraft in the kiloelectron volt energy range and extrapolated assuming a power law spectral continuation through the mega-electron volt energy range (e.g., Spjeldvik, 1977). For details about the ATS 6 proton data, see Fritz and Wilken (1976) and references therein.

$$f(L_{max},\mu,t) = f(L_{max},\mu,0) \times E_B(t),$$

where $F_B(t)$ has the same mathematical form as $F_D(t)$ except that F_{pd} is replaced by F_{pb}, the peak boundary condition enhancement.

Empirically D_{LL} had been found to increase by 2 orders of magnitude or more during some magnetic storms (e.g., Spjeldvik and Fritz, (1981) and is known to be correlated with the Kp index (i.e., Lanzerotti et al., 1978). Consequently, here we parameterize: $F_{pd} = 100$.

The fluxes of solar energetic protons can increase by many orders of magnitude during event periods. The spectral shapes can also change, although data from the GOES 2 spacecraft at the geostationary orbit often indicate an overall flux enhancement without much spectral change. For simplicity, we shall assume that to be the case and apply the peak enhancement factor: $F_{pb} = 1000$, independent of proton energy. This would make the results applicable to large events but not to rare extreme ones and not to events with highly variable spectral configuration.

The quiet-time (pre-event) proton boundary condition used here is shown in Figure 1. These data were obtained with the ATS 6 spacecraft (i.e., Fritz and Wilken, 1976) and have previously been applied in the study of the steady state proton population (Spjeldvik, 1977). The parameterized time-variable boundary condition thus scales with this spectrum.

It remains to specify the parameterized duration of the simulated events. Three cases are considered: short duration, medium duration, and long duration events. The characteristic times (in hours) are specified in Table 1. The short duration event is assumed to be a single peak (rise immediately followed by decline), the medium duration event is taken to last for 1 day (exclusive of rise and decline time), and the long duration event is assumed to last for 3 days. The longer duration events perhaps constitute a superposition of several closely timed events, such as the situation seen in the August 1972 solar particle event(s) and magnetic storm(s). However, for illustrative purposes we take the sustained peak period as constant fluxes.

Results

A numerical solution to the time dependent proton transport equation was developed using a finite difference approximation to the differentials. Numerical resolution in L shell was L = 0.05 L shell intervals with 15 grid points per

TABLE 1. Time Table (hours)

Event case	t_1	t_2	t_3	t_4	t_3-t_2
Short (case 1)	1	5	5	25	0
Medium (case 2)	1	5	29	40	24
Long (case 3)	1	5	77	105	72

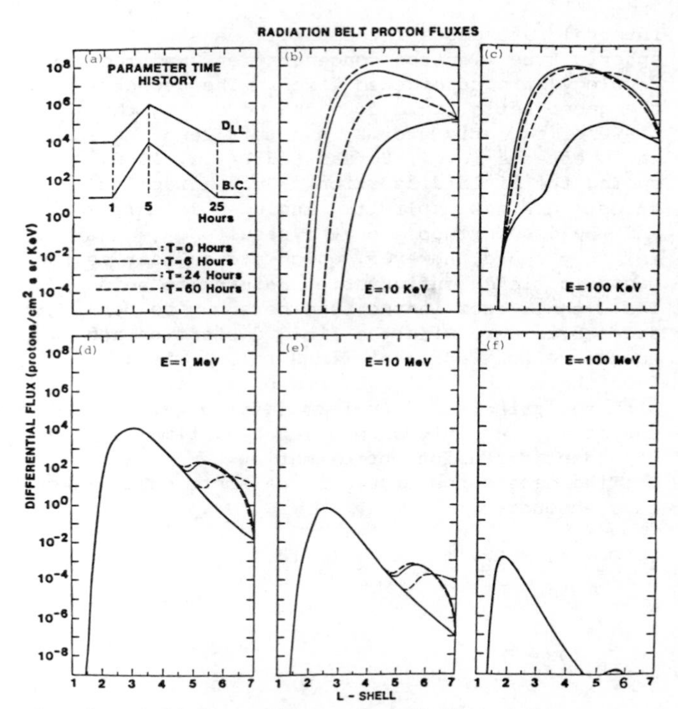

Fig. 2. Radial profiles of radiation belt protons computed for case 1 simulations. Panel (a) shows the imposed parameter variations, and the remaining panels show the proton profiles.

decade in the first-adiabatic invariant μ. The time stepping utilized Δt = 30 min. The solution was obtained by a recursive application of the tri-diagonal matrix inversion algorithm which insures an unconditionally stable solution at each recursive step in μ and t.

Case 1 - Short Duration Event

Figure 2 illustrates the evolution of radiation belt protons at energies ranging from 10 keV to 100 MeV. The computed proton distribution functions were converted to differential fluxes at fixed energy and time, and they are displayed as radial profiles in this figure. Panel (a) contains a cartoon showing the imposed parameter time variations, and panels (b)-(f) show the evolving proton flux profiles. In this event, the higher-energy protons are almost excluded from the trapping region. At E = 100 MeV the protons penetrate barely into L = 5, and a similar situation is found at E = 10 MeV and at E = 1 MeV. In contrast, the kiloelectron volt proton population readily penetrates deeply into the trapping region, to below L = 3. The physical reason for this stems from the energy dependence of the combined (geoelectric and geomagnetic fluctuations) D_{LL} values which are higher in the kiloelectron volt range than in the megaelectron volt range (e.g., Spjeldvik, 1979). Following the storm-time diffusive proton injection, this figure also illustrates that the E = 10 keV proton population quickly relaxes toward pre-event

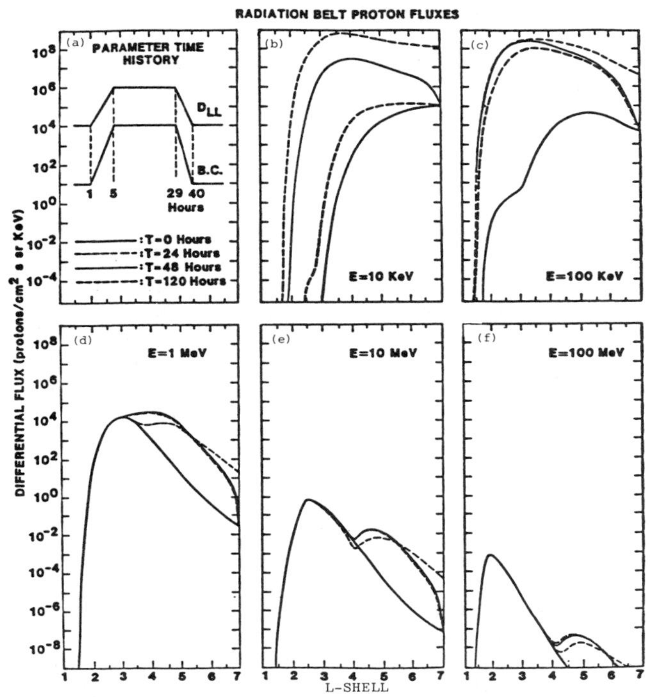

Fig. 3. Radial profiles of radiation belt protons computed for case 2 numerical simulations. Panel (a) shows the imposed parameter variations, and the remaining panels show the computed proton profiles.

conditions. This is because the loss time scales are very short at those energies. The relaxation times are much longer at E > 100 keV, a result qualitatively expected from the known energy variation of the proton charge exchange cross section σ_{10}. The features most apparent therefore can be summarized as follows: (1) Low-energy protons (tens of kiloelectron volts) are quickly injected deeply into the radiation belt region but are also lost quickly (tens of hours) when the source disappears. (2) Medium energy protons (hundreds of kiloelectron volts) are quickly injected deeply into the radiation belt region but remain there for an extended period (tens of days) when the enhancement of the boundary source subsides. (3) High-energy protons (megaelectron volts) are diffusively injected more slowly and thus effectively do not penetrate as deeply as the lower-energy protons, but they persist for longer periods (months or years, depending on location), their lifetimes in the outer zone being primarily diffusively limited by transport loss across the outer boundary.

Case 2 - Medium Duration Events

The event corresponding to a sustained parameter enhancement is depicted in panel (a) of Figure 3, and (b)-(f) panels show the proton radial profiles at different times at E = 10 keV, 100 keV, 1 MeV, 10 MeV, and 100 MeV energies. Qualitatively similar to the results in Figure 2, one nevertheless notices quantitative differences. At E = 100 MeV the proton flux enhancement penetrates to L ~ 4, an effect also seen at E = 10 MeV and at E = 1 MeV. Because the megaelectron volt protons generally have their peak in the L shell profile at lower L shells with higher energies, the relative penetration of the new protons toward this location appears more dramatic toward lower energies in the megaelectron volt range. In the kiloelectron volt range the longer event duration produces only marginally different effects, suggesting that the lower energy proton distribution almost saturates diffusively early in the event. At E = 10 keV and E = 100 keV a somewhat deeper penetration is seen with somewhat higher flux intensities, but the relative effect is much less than that seen in the megaelectron volt range. The noteworthy features are: (1) The lower-energy protons (tens of kiloelectron volts) are only marginally influenced by the sustenance of the boundary source beyond a few hours duration (except the maintenance of the elevated flux level). (2) The intermediate energy protons (hundreds of kiloelectron volts) are similarly in an apparent diffusive saturation state after a few hours. (3) The high-energy protons (megaelectron volts) penetrate significantly deeper into the trapping region as the event persists.

Case 3 - Long Duration Event

Results from the 3-day event simulation are shown in Figure 4. Again, panel (a) depicts the parameter time evolution utilized, while panels (b)-(f) show the computed proton profiles. The outstanding feature (in comparison with Figures 2 and 3) is the substantially further penetration of the megaelectron volt proton fluxes. At E = 100 MeV the new fluxes are reaching L = 3, a feature also seen at E = 10 MeV. In the latter case the transient peak in the E = 10 MeV proton radial profile seen at L ~ 3.5 for t > 48 hours is actually higher than the equilibrium controlled (pre-event) peak at L ~ 2.6. At E = 1 MeV there is penetration beyond the pre-event maximum with substantially enhanced proton fluxes peaking at L ~ 2.5 compared to the equilibrium peak at L ~ 3. The lower-energy proton fluxes, depicted at E = 10 keV and E = 100 keV hardly appear affected by the event duration. This is of course expected on physical grounds and it is consistent with the expection that arises from the study of case 2 above. One therefore concludes that (1) lower-energy (kiloelectron volt) protons are largely unaffected by extended event periods beyond the first few hours (except by sustained compensation for losses), (2) radiation belt protons around 1 MeV display spectacularly increased simulated flux intensities in sustained events that last for several days, and (3) the high-energy protons in the megaelectron volt range penetrate to lower L shells with increasing event duration times.

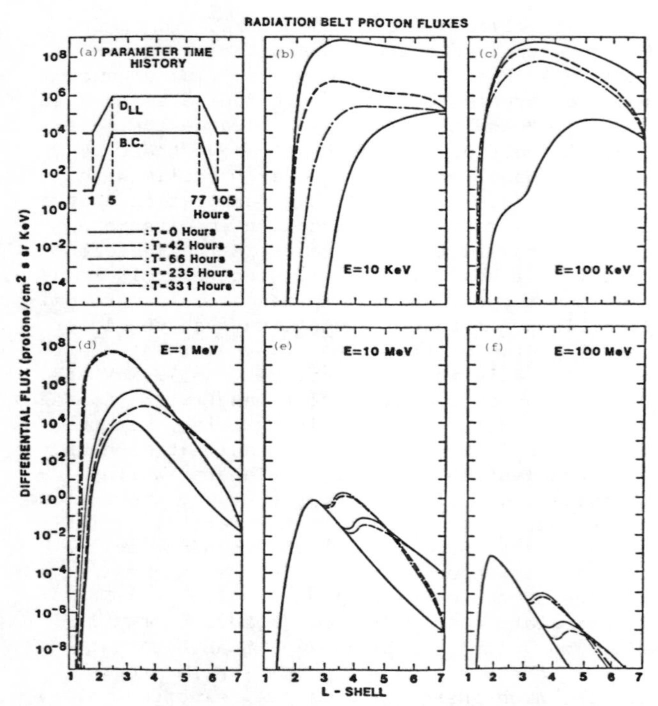

Fig. 4. Radial profiles of radiation belt protons computed for case 3 numerical proton flux simulations. Panel (a) shows the imposed parameter variations, and the remaining panels show the flux profiles.

Summary and Conclusions

The effects of solar energetic protons at kiloelectron volt and megaelectron volt energies on the trapped fluxes of radiation belt protons have been assessed from a diffusion point of view. The solar protons are thought to be ejected from the sun during solar flares and local acceleration events and may undergo acceleration in traveling shock waves in interplanetary space. These protons form a dynamic outer boundary condition on trapped radiation belt fluxes. When there are higher fluxes outside the stable trapping region than just inside the boundary, geomagnetic and geoelectric fluctuations cause a net diffusion flux to enter the trapping region and give rise to a diffusive injection event. At the outer boundary this diffusion flux is

$$\psi_b = -(D_{LL} L^{-2} \, df/dL) \quad \text{at} \quad L = L_{max}.$$

As time goes on (and the enhanced boundary condition is maintained) the protons penetrate into the radiation belts with effective penetration velocities depending on energy, the kiloelectron volt protons faster and the megaelectron volt protons slower. When the enhancement of the solar proton flux diminishes, protons in the outer zone start to diffuse outward (in the net diffusion sense), and this leads to a steepening of the radial profiles toward the pre-event configuration. It requires only a few hours of solar proton event time to inject kiloelectron volt protons to lower L shells ($L \sim 3$) while it requires several days to diffusively inject megaelectron volt protons to these locations.

There is probably an effective lower limit to the validity of the results from this study. Below a few tens of kiloelectron volts, the influence of large-scale geoelectric fields (induction and charge separation) may compete with the magnetic azimuthal drift. Also, for large values of the D_{LL} temporal variations may occur over a shorter time scale than the proton drift period. Thus, the diffusion approximation may not be strictly valid for these low-energy particles.

The intent of the present work is to show the principal proton flux features from a parameterized study. The results are not suitable for a quantitative comparison with actual space observations for a particular event period. However, the present results point to qualitative features which one could look for in the data. In particular, when displayed as distribution functions versus L shell at constant proton magnetic moment, there should be a positive df/dL value near the outer boundary traveling inward in the beginning of the event, and a negative df/dL value near the outer boundary also traveling inward in the end of the event. With limited data (and for another ion species), Spjeldvik and Fritz (1981) qualitatively verified the latter feature.

Future work in this area might be directed toward using actual proton time-dependent boundary conditions (for example, as observed by geostationary spacecraft) together with estimated $D_{LL}(t)$ values (Lanzerotti et al., 1978), and the predicted radiation belt structure should be tested against in situ observations near the geomagnetic equatorial plane.

References

Cornwall, J. M., Diffusion processes influenced by conjugate point wave phenomena, Radio Sci., 3, 740, 1968.

Cornwall, J. M., Radial diffusion of ionized helium and protons: A probe for magnetospheric dynamics, J. Geophys. Res., 77, 1756, 1972.

Fälthammar, C. G., Radial diffusion by violation of the third adiabatic invariant, in Earth's Particles and Fields, edited by B. M. McCormac, p. 157, Reinholt, New York, 1968.

Fritz, T. A., and B. Wilken, Substorm generated fluxes of heavy ions at the geostationary orbit, in Magnetospheric Particles and Fields, edited by B. M. McCormac, p. 171, D. Reidel Publ. Co., Dordrecht, Holland, 1976.

Hess, W. N., The Radiation Belts and the Magnetosphere, Blaisdell Publ. Co., Waltham, Massachusetts, 1968.

Lanzerotti, L. J., D. C. Webb, and C. W. Arthur, Geomagnetic field fluctuations at synchronous orbit, 2. Radial diffusion, J. Geophys. Res., 83, 3866, 1978.

Lyons, L. R., and R. M. Thorne, Equilibrium structure of radiation belt electrons, J. Geophys. Res., 78, 2142, 1973.

Nakada, M. P., and G. D. Mead, Diffusion of protons in the outer belt, J. Geophys. Res., 70, 4777, 1965.

Roederer, J. G., Experimental evidence on radial diffusion of geomagnetically trapped particles, in Earth's Particles and Fields, edited by B. M. McCormac, p. 143, Reinholt, New York, 1968.

Roederer, J. G., Dynamics of Geomagnetically Trapped Radiation, Springer-Verlag, New York, 1970.

Schulz, M., Geomagnetically trapped radiation, Space Sci. Rev., 17, 481, 1975.

Schulz, M., and G. A. Paulikas, Secular magnetic variation and the inner proton belt, J. Geophys. Res., 77, 744, 1972.

Schulz, M., and L. J. Lanzerotti, Particle Diffusion in the Radiation Belts, Springer-Verlag, New York, 1974.

Söraas, F., and L. R. Davis, Temporal variations of the 10 keV to 1700 keV trapped protons observed on satellite Explorer 26 during the first half of 1965, NASA TMX-63320, Goddard Space Flight Center, Maryland, 1968.

Spjeldvik, W. N., Equilibrium structure of equatorially mirroring radiation belt protons, J. Geophys. Res., 82, 2801, 1977.

Spjeldvik, W. N., Expected charge states of energetic ions in the magnetosphere, Space Sci. Rev., 23, 499, 1979.

Spjeldvik, W. N., and R. M. Throne, The cause of storm after effects in the middle latitude D-region, J. Atm. Terr. Phys., 37, 777, 1975.

Spjeldvik, W. N., and T. A. Fritz, Energetic ionized helium in the quiet-time radition belts: Theory and comparison with observation, J. Geophys. Res., 83, 654, 1978a.

Spjeldvik, W. N., and T. A. Fritz, Theory for charge states of energetic oxygen ions in the earth's radiation belts, J. Geophys. Res., 83, 1583, 1978b.

Spjeldvik, W. N., and T. A. Fritz, Observations of energetic helium ions in the earth's radiation belts during a sequence of geomagnetic storms, J. Geophys. Res., 86, 2317, 1981.

Spjeldvik, W. N., and P. L. Rothwell, The earth's radiation belts, Environmental Research Paper No. 854, AFGL-TR-83-0240, Air Force Geophysics Laboratory, Massachusetts, 1983.

Tverskoy, B. A., Transport and acceleration of particles in the earth's magnetosphere, Geomagn. Aeronomy, 6, 617, 1965.

Tverskoy, B. A., Main mechanisms in the formation of the radiation belts of the earth, Rev. Geophys. Space Phys., 7, 219, 1969.

Tverskoy, B. A., Dynamics of the Radiation Belts of the Earth, NASA Technical Translation, TTF-635, Washington, D.C., 1971.

Van Allen, J. A., The geomagnetically trapped corpuscular radiation, J. Geophys. Res., 64, 1683, 1959.

Van Allen, J. A., Dynamics, composition, and origin of the geomagnetically trapped corpuscular radiation, Trans. Int. Astron. Union, XIB, 99, 1962.

Walt, M., History of artifical radiation belts, in Trapped Radiation Handbook, DNA-2524H, edited by J. B. Cladis, G. T. Davidson, and L. L. Newkirk, Lockheed Palo Alto Research Laboratory, January 1977.

MAGNETOHYDRODYNAMIC BOUNDARY CONDITIONS FOR GLOBAL MODELS

T. G. Forbes

Space Science Center, Institute for the Study of Earth, Oceans, and Space
University of New Hampshire, Durham, New Hampshire 03824

Abstract. Boundary conditions in the ionosphere and the upstream solar wind are important in determining the dynamics of global magnetohydrodynamic models of the magnetosphere. It is generally recognized that the orientation of the magnetic field in the upstream solar wind strongly modulates the rate of energy input into the magnetosphere by magnetic reconnection. However, other aspects of the upstream boundary conditions may determine whether the reconnection occurs in a patchy manner, as in flux transfer events, or in a global manner, as in the Paschmann et al. (1979) events. Ionospheric boundary conditions should also affect the reconnection process. For example, ionospheric line-tying can cause x-line motion in the outer magnetosphere. If it is assumed that auroras occur on field lines mapping to x-lines, then auroral motions are different than the local convective motion of the plasma in which they occur. Global magnetohydrodynamic models which incorporate both magnetospheric reconnection and ionospheric convection could be used to investigate the effect of reconnection and convection upon dayside and nightside auroral motions during the course of a magnetic substorm.

1. Introduction

The use of the magnetohydrodynamic (MHD) approximation for collisionless plasmas like those occurring in the magnetosphere-ionosphere system has always been problematical. Even in the double-adiabatic, two-fluid form of the MHD equations, it is still necessary to make somewhat ad hoc assumptions about transport coefficients for resistive and viscous dissipation and thermal conduction (Chew et al., 1956; Coroniti, 1985). Unlike kinetic theory, the transport coefficients in MHD are not necessarily determined self-consistently yet, despite this limitation, interest in the MHD equations has remained strong. There are several reasons for this. First, from a qualitative point of view, most of the convective dynamics of the magnetosphere and ionosphere are encompassed in MHD. Second, the MHD equations constitute a nonlinear system; consequently, they are superior to any linearized system, whether kinetic or otherwise, in their ability to describe the nonlinear behavior typical in nature. Third, and perhaps most important, they can often be solved in many situations where the full nonlinear kinetic equations cannot.

Efforts to construct numerical MHD models began a few years ago (e.g., Birn, 1980; Wu et al., 1981; Lyon et al., 1981; Leboeuf et al., 1981; Brecht et al., 1982; also see the review by Walker, 1983), but to a large extent these models are still preliminary, as even within an MHD context, the construction of a global model is a major task. Perhaps, the greatest difficulty facing global modelers is the representation of thin boundary layers (e.g., the magnetopause) within a large-scale numerical model. However, there are other problems as well. For example, no model has yet attempted a truly global scale, i.e., one that includes the region where the very distant tail has completely merged back into the solar wind. Yet another problem area is boundary conditions, since realistic ionospheric boundary conditions are yet to be incorporated into these models.

All of the above problem areas will probably be removed during the next decade or so as efforts continue to improve the numerical models. In this paper we will consider just one of these areas, namely, boundary conditions. We consider first the upstream solar wind boundary conditions (section 2) and then the ionospheric boundary conditions (section 3).

2. Solar Wind Boundary Conditions

As numerical models of the magnetosphere and ionosphere become increasingly more complex, difficulties with the boundary conditions are becoming more evident. This leads naturally to a consideration of what are the mathematical requirements for a physically meaningful initial boundary value problem (IBVP). Unfortunately, IBVPs for the nonlinear MHD systems in two and three dimensions are not easy to analyze, and to

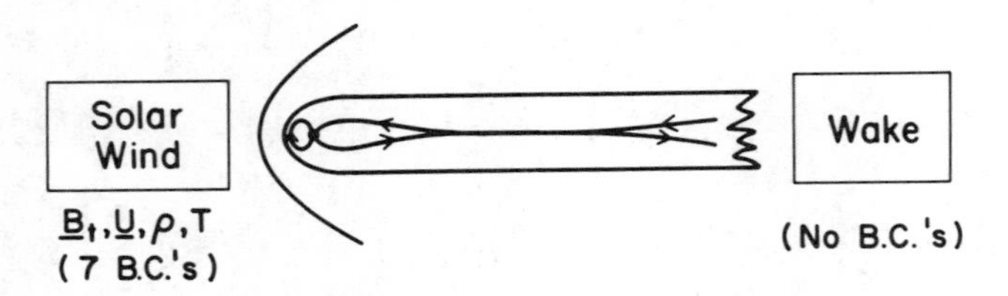

Fig. 1. Schematic drawing illustrating boundary conditions needed for time-dependent, three-dimensional MHD models of the global magnetosphere.

date no rigorous analysis exists to determine whether a given IBVP is physically meaningful.

Historically, the question of what constitutes a physically meaningful IBVP was embodied in the concept of well-posedness introduced by Hadamard in the 19th century (Payne, 1973). Roughly, an IBVP is well posed in the Hadamard sense if a unique solution exists and if small variations in the initial or boundary conditions lead to correspondingly small variations in the solution (Chester, 1971). This definition was widely accepted as a necessary condition for IBVPs because it ensured determinism which was thought to be essential for a physically meaningful solution. However, it is now recognized that there is a large class of physical problems, such as those involving turbulence or catastrophic behavior, which is not well posed in the Hadamard sense (Vemuri and Karplus, 1981).

The boundary conditions for a nonlinear IBVP in MHD can reasonably be considered overspecified if no solutions exist and underspecified if an infinite number of solutions exists. Intermediate to these two extremes, cases may occur where the addition of an extra boundary condition simply alters the character of the allowable solutions rather than eliminating them. For example, in ideal Poiseuille flow the addition of a no-slip condition to the wall condition introduces a boundary layer discontinuity, so that the previous continuous solution is replaced by a discontinuous one. Thus, the additional no-slip condition overspecifies the IBVP with respect to continuous solutions but not with respect to discontinuous solutions.

2.1 Allowable Number of Boundary Conditions

For ideal MHD over- or underspecification with respect to continuous solutions can be determined using the method of characteristics (Chester, 1971). Specifically, one can show that the number of locally independent boundary conditions which may be imposed at a given point on a boundary at a given time is equal to the number of characteristics propagating from the boundary and into the solution domain (Forbes and Priest, 1987). The characteristics of the MHD equations are the ray paths of the various wave modes (i.e., slow, fast, Alfvén, and entropy). For example, the two characteristics associated with the Alfvén wave are defined by $d\mathbf{X}/dt = \mathbf{u} \pm \mathbf{c}_a$, where $\mathbf{u}$ is the flow velocity, $\mathbf{c}_a$ is the Alfvén velocity, and $\mathbf{X}(t)$ is a ray path. For a three-dimensional global model, the upstream solar wind inflow boundary is always supermagnetosonic with respect to the fast-mode wave speed, and so seven variables can be specified (e.g., $\mathbf{B}_t$, $\mathbf{u}$, p, and ρ, where B_t is the magnetic field component tangential to the boundary, $\mathbf{u}$ is the velocity, p is the pressure, and ρ is the density) corresponding to the seven characteristic speeds, $u_n \pm c_f$, $u_n \pm c_a$, $u_n \pm c_s$, and u_n, where u_n is the bulk flow speed normal to the boundary, and c_f, c_a, and c_s are the corresponding fast-mode, Alfvén, and slow-mode wave speeds, respectively. Independent specification of all eight variables would mathematically overspecify the model with respect to the ideal MHD system. In the very distant wake of the magnetotail the flow is again supermagnetosonic, but the plasma flow direction with respect to the boundary is reversed. So here no boundary conditions can be specified independently of the initial conditions and the other boundary conditions (cf. Figure 1).

2.2 Different Regimes of Reconnection

A new family of steady state reconnection solutions was recently found by Priest and Forbes (1986) by expanding the incompressible MHD equations in powers of the Alfvén Mach number. Following Petschek (1964) and Vasyliunas (1975), solutions were sought by linearizing about a uniform horizontal field $B_e\,\hat{\mathbf{x}}$ in the rectangular domain $(-1 \leq x \leq 1,\ -1 \leq y \leq 1)$ by putting

$$\mathbf{B} = B_e\,\hat{\mathbf{x}} + \mathbf{B}_1 + \dots$$

$$\mathbf{V} = V_{1y}\,\hat{\mathbf{y}} + \mathbf{V}_2 + \dots$$

with $V_{1y} = M_e V_{Ae}$, where M_e and V_{Ae} are the external Alfvén Mach number and velocity at the input boundary $y = 1$. The region $y > 0$ consists of magnetic field lines which are oppositely directed from those in the region $y < 0$. The boundary conditions for the solution in the region $0 < y < 1$ are $B_{1x} = 0$ at the top $(y = 1)$, $\partial B_{1y}/\partial x = 0$ (free floating) at the sides $(|x| = 1)$, and $B_{1y} = f(x)$ at $y = 0$, where $f(x)$ is a function which gives the appropriate conditions at the shocks in terms of B_N. The resulting solution is

$$B_{1x} = \sum_{n=0}^{\infty} a_n\,[b - \cos(m\pi x)]\sinh[m\pi(1-y)],$$

$$B_{1y} = \sum_{n=0}^{\infty} a_n\,\sin(m\pi x)\cosh[m\pi(1-y)],$$

where $m = n + 1/2$ and,

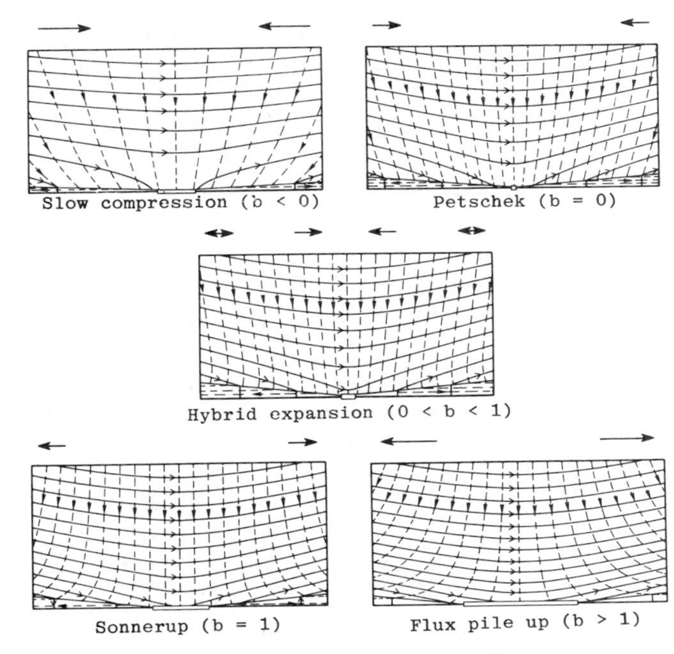

Fig. 2. Magnetic field lines (solid) and streamlines (dashed) for different regimes of reconnection. Here the external magnetic Reynolds number is 500. The external Alfvén Mach number, M_e, for the slow-mode compression (b < 0) and Petschek (b = 0) cases is 0.043 and 0.091, respectively. For each of the other three cases M_e = 0.10 (after Priest and Forbes, 1986).

$$a_n = \frac{4B_N \sin(m\ L/L_e)}{(L/L_e) m^2 \pi^2 \cosh(m\pi)}.$$

The parameter b labels different reconnection solutions as shown in Figure 2. When b = 0, there are no first-order currents or pressure gradients in the inflow, and we have a Petschek solution with a weak fast-mode expansion (Petschek, 1964). Setting b = 1 gives a Sonnerup-like solution. As the plasma comes in toward the reconnection point it experiences a slow-mode expansion with the streamlines diverging, the magnetic field strength remaining constant on the axis, the pressure decreasing, and the current density (positive) increasing (Sonnerup, 1970).

The importance of the parameter b is that it determines the value of the x-component of the velocity at the corner (x,y) = (1,1). To lowest order this is

$$v_{2x}(1,1) = M_e v_{Ae}\ a_o \pi(b-2/\pi)/(2B_e),$$

so that when $b < 2/\pi$, the flow at the corners is directed in toward the numerical box, and when $b > 2/\pi$, it is directed outward.

When b < 0, the flow on the upper boundary (y = 1) is highly converging and there is a family of slow-mode compressions. The inflowing plasma is compressed as it comes in, and the current density (negative) increases in magnitude. At the same time the magnetic field strength falls.

When b > 1, there is a flux-pile-up regime of slow-mode expansions with long diffusion regions. These are created by a strongly diverging flow on the upper boundary. As the plasma comes in, the streamlines diverge, the magnetic field strength increases, the pressure falls, and the current density increases.

For 0 < b < 1, a hybrid regime exists in which there is a fast-mode expansion on the y-axis and a slow-mode expansion at the edge ($|x|$ = 1) of the box. The pressure falls and the current density (positive) rises as the plasma flows in; however, the magnetic field strength falls along the y-axis but rises along the edges.

Examples of these regimes are given in Figure 2 for varying values of the parameter b and the Alfvén Mach number (M_e) at the inflow boundary. All but the upper left panel are for roughly the same value of the reconnection rate (M_e), so these may be directly compared with each other. The upper left panel is for a much slower reconnection rate. The lengths of the diffusion regions, but not the widths, are indicated by rectangular boxes since the widths are too small to be shown to scale. If b increases and M_e is held constant, the diffusion region becomes longer and the flow diverges more and more. In the limit when b is very large the expansion procedure breaks down, but it is natural to suspect that one recovers the stagnation-point solution (Parker, 1963; Sonnerup and Priest, 1975) in which the field lines are straight and the flow is diverging.

When the flow is driven from outside, any of the above regimes are possible in principle, depending on the detailed boundary conditions. When, however, the reconnection results from the nonlinear development of some local instability such as the tearing mode, something like the Petschek mechanism is much more likely to result. The reason is that it is the only solution for which the slow characteristics are purely outgoing, taking information away from the site where reconnection is being initiated (Soward and Priest, 1977). In other words it sucks the plasma through the reconnection site by a weak, fast-mode expansion.

2.3 Stability of Flux-Pile-Up Regime

It is the impinging solar wind which drives reconnection at the dayside magnetopause. Even after the solar wind is slowed by its passage through the bow shock, it is still moving toward the magnetopause at a rate faster than consistent with a Petschek-like solution (i.e., M_a > 0.1, where M_a is the Alfvén Mach number). Hence, we expect the flux-pile-up regime of reconnection to occur when the interplanetary magnetic field is

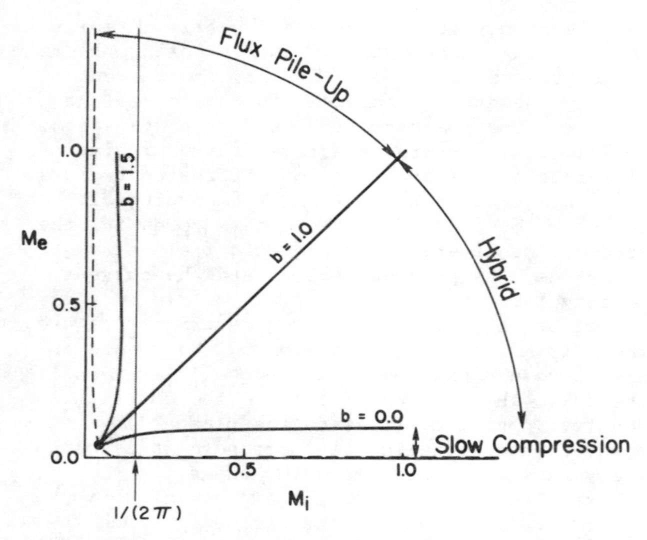

Fig. 3. The region in the M_e,M_i plane where instability of the diffusion region current sheet might occur. The quantities M_e and M_i are the external and internal Alfvén Mach numbers of the plasma flowing in towards the x-line. The external Mach numbers are measured far away from the diffusion region, while the internal Mach numbers are measured just outside the diffusion region at the inflow boundary (after Forbes and Priest, 1987).

favorably oriented for reconnection (i.e., southward).

As previously mentioned, the stagnation-point flow represents a limit of the flux-pile-up regime, and it has a diffusion region current sheet whose length is infinite. A preliminary analysis by Sonnerup and Sakai (1981) indicates that stagnation-point flow may be unstable in some circumstances. Their results suggest that the diffusion region current sheet is susceptible to the tearing mode instability when the speed of the plasma flowing toward the stagnation point is small compared to the Alfvén speed. Or conversely, the current sheet is increasingly more stable as the inflow speed increases.

If we now assume this result holds in general for all reconnection regimes, we can determine a sufficient condition for stability. If the flow and fields in the diffusion region always act to stabilize rather than destabilize, then the current sheet there will certainly be stable if it satisfies the ordinary tearing mode stability criterion, namely,

$$2\pi\ell > L\alpha_c ,$$

where ℓ and L are the half width and half length of the current sheet. The quantity α_c is a factor on the order of unity and its exact value depends on the functional form of the current distribution in the sheet (Furth et al., 1963; Priest, 1985).

If we now apply this stability criteria to the diffusion region at the x-line we obtain

$$M_i > (2\pi)^{-1}$$

as a sufficient condition for stability of the diffusion region current sheet to the tearing mode. The quantity M_i is the inflow Alfvén Mach number just outside the diffusion region, and it is equal to the ratio of the width, ℓ, to the length, L, of the diffusion region current sheet. Figure 3 shows the region in the M_e,M_i plane where the steady state solutions of Priest and Forbes (1986) might be unstable. It is clear that the most susceptible solutions are those lying in the flux-pile-up regime near the $M_i = 0$ axis.

Figure 4 shows results from a numerical experiment by Lee and Fu (1986) which imply that the flux-pile-up regime is indeed unstable whenever the length of the central current sheet is sufficiently long. In their experiment the ratio of the length L to width ℓ at which instability occurs is about 12. This is consistent with our

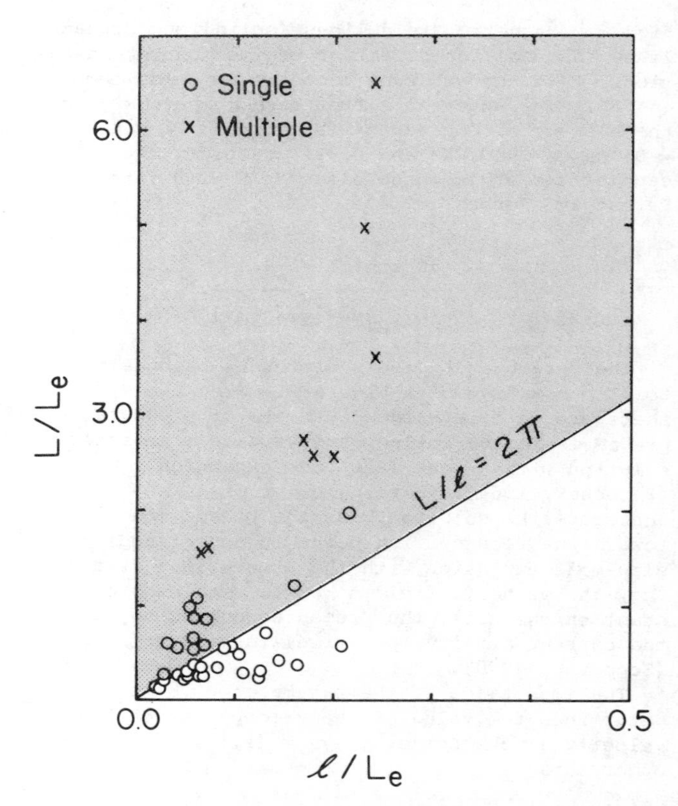

Fig. 4. The length of the diffusion region as a function of its width in the numerical experiments of Lee and Fu (1986). The points designated "multiple" indicate the diffusion regions which developed multiple x-lines as a result of tearing. The shaded area indicates where instability might occur according to simple tearing mode theory.

sufficient condition that this ratio must at least exceed 2π, that is, about 6. Thus, one is led to the conclusion that dayside reconnection is likely to occur in a non-steady state or patchy manner at least some of the time. Perhaps observations of flux-transfer events can be interpreted as evidence for an unstable flux-pile-up regime of reconnection (Russell and Elphic, 1978; see also the review by Galeev et al., 1986).

Since the upstream solar wind configuration varies considerably, occasionally there may be times when the upstream boundary conditions result in a stable regime. Perhaps these conditions might account for the global type of quasi-steady dayside reconnection events reported by Paschmann et al. (1979) and Paschmann (1984). A global MHD model could be used to investigate this possibility.

3. Ionospheric Boundary Conditions

It is not really clear at present what the appropriate boundary conditions are at the ionosphere within the context of an MHD model. However, it seems reasonable to expect that one key aspect is ionospheric line-tying of magnetic field lines mapping from the magnetosphere to the ionosphere. In the following three sections we first define line-tying (section 3.1), and then discuss its implications for the reconnection in the geomagnetic tail (section 3.2) and the motion of auroras during magnetic substorms (section 3.3).

3.1 Line-Tying

Here we define a line-tying boundary as a boundary where the field line velocity **U**, given by

$$\mathbf{U} = \mathbf{u} + (\mathbf{j} \times \mathbf{B})/(\sigma B^2)$$

is zero. Here **u** is the plasma velocity, **j** is the current density, **B** is the magnetic field, and σ is the electrical conductivity. Note that this is not equivalent to the frozen-flux condition since frozen flux requires $\mathbf{U} = \mathbf{u}$ rather than $\mathbf{U} = 0$.

The solar photosphere is a good example of a line-tying boundary which satisfies the above definition. The photosphere effectively anchors coronal field lines threading it because coronal plasma is too tenuous to affect convection of dense photospheric plasma. Although the conductivity of the photosphere is poor compared to the corona, both conductivities are so large that the relative conductivity is of little significance (i.e., the second term for **U** on the right-hand side of the above equation is essentially zero). Instead, it is the comparatively large difference in the inertia of the two plasmas that prevents coronal disturbances, such as flares, from inducing plasma motions in the photosphere. Thus, this kind of line-tying is often referred to as inertial line-tying.

Compared to the photosphere, the terrestrial ionosphere is a relatively poor line-tying boundary due to its relatively low mass. Typically the time required for ionospheric convection to be externally induced in response to an imposed force from the magnetosphere is about 5 min (Coroniti and Kennel, 1973). However, if the external force is only applied at a localized region, it may take 20-30 min before the whole ionosphere can globally respond (Holzer and Reid, 1975; Reid and Holzer, 1975). Hence, ionospheric line-tying, coupled with travel time effects, can be significant for the rapid changes in convection associated with magnetospheric substorms.

3.2 Auroral Motion

To demonstrate how line-tying might be involved in magnetospheric substorms, we consider the field configuration shown in Figure 5. Here the surface $x = 0$ corresponds to the ionosphere and the region $x < 0$ to the geomagnetic tail. It is further assumed that the auroras occur near the ionospheric footpoints ($0,z_a$ and $0,-z_a$) of the field lines mapping to the x-line ($-x_o,0$). In practice the actual location of the aurora should probably be identified with the region where the electric field parallel to the magnetic field is the strongest (Lyons and Williams, 1984). Intuitively, one expects this region to be close to the footpoints of the field lines mapping to the x-line. However, it is possible that the movement of this region might sometimes be quite different than the x-line footpoints. For the sake of simplicity we will ignore this possibility in the following discussion.

In two dimensions we can express the vector magnetic field **B** as

$$\mathbf{B} = \nabla \times \mathbf{A} = \nabla A \times \hat{e}_y,$$

where the condition A = constant defines a field line. In two dimensions we can also write Faraday's equation in terms of A as

$$E = -\partial A/\partial t - \nabla\phi(y),$$

or

$$E = -\partial A/\partial t + k,$$

where k is a constant. Now if we assume that **B** is constant in time at the ionosphere, then $E(0,z) = E_i$ = constant, since $\nabla \times \mathbf{E} = 0$ if **B** is constant. Hence,

$$E(x_o,0) = E_i - \partial A/\partial t,$$

$$E(x_o,0) = E_i - (\partial A/\partial z_a)(z_a/\partial t),$$

or

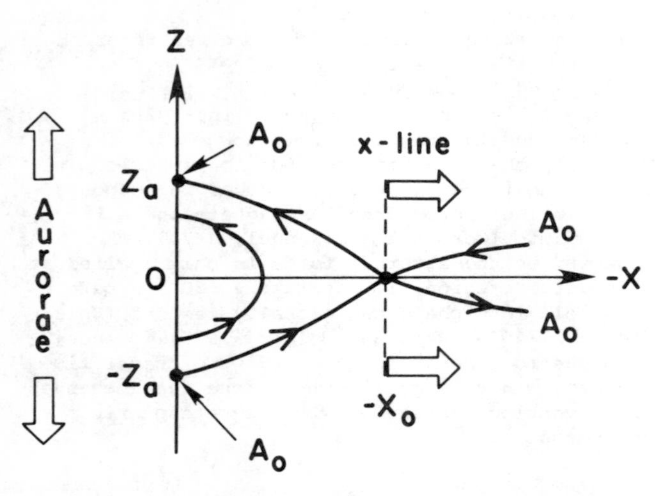

Fig. 5. Topological sketch of reconnection in a two-dimensional magnetotail model with a line-tying boundary at the ionosphere (z-axis). As field lines reconnect, the x-line moves outward and the footpoints of the separatrices move away from the origin.

$$E_o \equiv E(x_o, 0) = E_i + B_x(0, z_a)\dot{z}_a \; .$$

By definition, E_o is the rate of magnetic reconnection in the tail. If the reconnection is steady state, then $E_o = E_i$ and $\dot{z}_a = 0$. Hence, in steady state the auroras are stationary, and the ionospheric convective electric field E_o equals the magnetospheric electric field. If the ionospheric convective electric field is small, so that $E_i << E_o$, then the auroras move poleward at the speed $E_o/B_x(0, \pm z_a)$. These auroral motions are not due to bulk motions of the plasma; instead they are due to the reformation of the x-line onto new field lines which are at progressively greater distances from the equator (0,0).

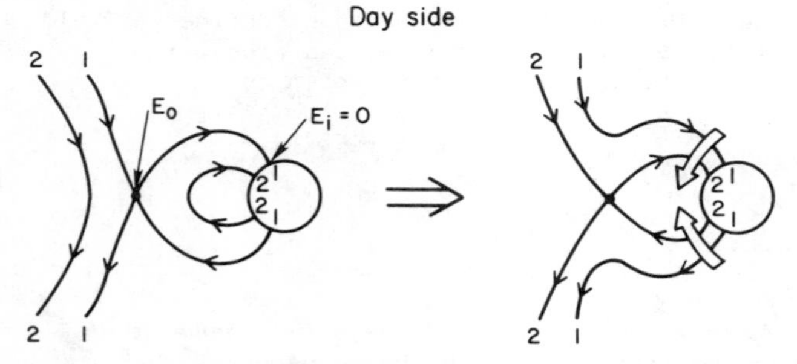

Fig. 6. Erosion of the dayside magnetopause due to the onset of reconnection. The erosion occurs when the electric field, E_o, at the dayside x-line exceeds the electric field, E_i, at the ionospheric footpoints of the field lines mapping to the x-line. If auroras are associated with these footpoints, they will appear to move equatorward as shown on the right.

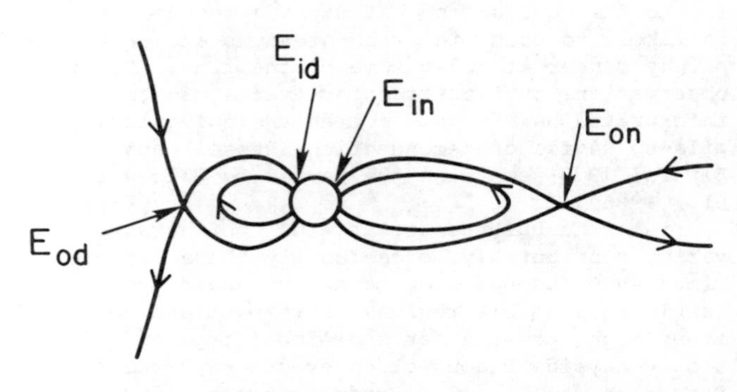

Fig. 7. Magnetic field configuration for a two-dimensional analogue model of the magnetosphere. The apparent and actual motions of aurora associated with x-line separatrices depend on the relative values of the electric fields at the x-lines (E_{od} and E_{on}) and the ionosphere (E_{id} and E_{in}).

On the other hand, if the ionospheric convective electric field is so large that $E_i >> E_o$, then the auroral motion is entirely due to bulk motion of the plasma generated by ionospheric convection.

The relation between reconnection at the dayside magnetopause and ionospheric convection has already been extensively investigated by Coroniti and Kennel (1973), Holzer and Reid (1975), and Reid and Holzer (1975). The latter authors used a circuit analogy to model the erosion of the dayside magnetopause before substorm onset (Aubrey et al., 1970, 1971). This erosion process in the Holzer and Reid model occurs when the reconnection rate, E_o at the magnetopause, exceeds the ionospheric convective electric field E_i as illustrated in Figure 6. As the magnetopause erodes, the footpoints of the field lines mapping to the x-line (i.e., the auroras) move equatorward until E_i becomes equal to or greater than E_i. In the Holzer and Reid model E_i actually overshoots E_o as the relative values of the two electric fields oscillate back and forth.

Numerically it should be possible to construct MHD models of the global magnetosphere which incorporate this kind of reconnection-induced auroral motion. Such models would at least provide a qualitative picture of how the reconnection electric fields at the magnetopause and magnetotail x-lines interact with the convective electric fields at the dayside and nightside ionospheres. As an illustrative example of how such a model might work, we consider the schematic shown in Figure 7. For simplicity we consider only a two-dimensional model and replace the three-dimensional motion of field lines slipping around the earth by letting field lines slip through the two-dimensional earth. The combined apparent and convective velocities of the iono-

TABLE 1. A Scenario for Auroral Motions During a Substorm

Substorm Phase	Electric Fields				Auroral Motions	
	E_{od}	E_{id}	E_{in}	E_{on}	$\dot{S}_{id}$ (dayside)	$\dot{S}_{in}$ (nightside)
$B_z\downarrow$	1	0	0	0	1 (equatorward)	0
Growth	1	1	0	0	0	0
Growth	1	1	1	0	0	1 (equatorward)
Onset	1	1	1	1	0	0
Expansion	1	1	1	2	0	-1 (poleward)
$B_z\uparrow$	0	1	1	1	-1 (poleward)	0
Recovery	0	0	1	1	0	0
Recovery	0	0	0	1	0	-1 (poleward)
Quiet	0	0	0	0	0	0

spheric footpoints of field lines mapping to the x-lines (i.e., the separatrices) can then be written as

$$\dot{S}_{id} = (E_{od} - E_{id})/B_{rd}$$

$$\dot{S}_{in} = (E_{in} - E_{on})/B_{rn} ,$$

where $\dot{S}_{id}$ and $\dot{S}_{in}$ are the speeds of the dayside and nightside ionospheric footpoints of the separatrices, E_{od} and E_{on} are the dayside and nightside convective electric fields at the footpoints, and B_{rd} and B_{rn} are the corresponding dayside and nightside radial magnetic field components.

Now let us consider how $\dot{S}_{id}$ and $\dot{S}_{in}$ vary in response to electric field variations initiated by a southward turning of the interplanetary magnetic field. Let us also assume for simplicity that $B_{rd} = B_{rn} = 1$ and $E = 0$, 1, or 2 in some appropriate dimensionless units. Prior to the southward turning of the interplanetary magnetic field, we assume that both the dayside and nightside x-lines exist but that the reconnection rates there are negligible. Thus, at the moment of the southward turning, a significant electric field exists only at the dayside x-line. Consequently, the footpoints of the dayside separatrices begin to move equatorward and any auroras associated with them also move equatorward (cf. Table 1). Now after a short period of perhaps a few Alfvén travel times, the dayside ionosphere will begin to convect in response to the dayside reconnection. When the electric field E_{id} at the dayside ionosphere becomes equal to the reconnection rate E_{od}, the equatorward motion of the dayside aurora halts. But while this is happening, the induced ionospheric convection electric field spreads over the whole polar cap. When the electric field reaches the nightside, any preexisting, nightside auroras will move equatorward. As the reconnection electric field penetrates into the geomagnetic tail, the electric field at the nightside x-line begins to increase. At this point we might imagine either that the enhanced reconnection in the tail occurs at an old, pre-existing x-line or that it occurs at a newly formed near-earth x-line. In the latter case there would be an additional equatorward jump of the auroras just at onset. After onset, the equatorward motion stops, but it may be replaced by a poleward motion if the electric field in the tail overshoots the ionospheric electric field E_{in}. Such an overshoot seems especially likely if a new tail x-line forms due to instability or loss of mechanical equilibrium.

Table 1 summarizes the above hypothetical scenario for the auroral motions and extends it to include a recovery phase triggered by a northward turning of the interplanetary magnetic field. We reemphasize that the auroral motions are of two different types. At the dayside (nightside) ionosphere, the equatorward (poleward) motions are apparent motions caused by the topological reconfiguration of the x-line, while the poleward (equatorward) motions are actual bulk plasma motions caused by ionospheric convection. The two different kinds of motion could, in principle, be observationally distinguished by using ground-based Doppler radar in conjuction with optical imaging.

At the moment there is considerable controversy about the timing of the observed auroral motions with respect to the various substorm phases. For example, Hones (1985, 1986) has argued for the existence of a poleward leap of the aurora during the recovery phase, while Rostoker (1986) has argued that such a poleward leap is only a feature of the expansion phase; see also Sergeev and Yahnin (1979). In our scenario, poleward leaps occur during both phases. Whether these leaps would occur as we have conjectured is uncertain, but hopefully, future global MHD models will ascertain this.

With three-dimensional models it should be possible to include the parallel electric fields which we expect to be important in determining the actual auroral locations. Parallel electric fields in MHD are given by $\eta j_\parallel$, or $(\eta \cdot j)_\parallel$, where η is the electrical resistivity and $j_\parallel$ is the field-aligned current density. Hence, the electrical resistivity coefficient, or tensor, contains much of the physics involved in generating the parallel electric field. In a collisionless plasma, it is very difficult to determine a

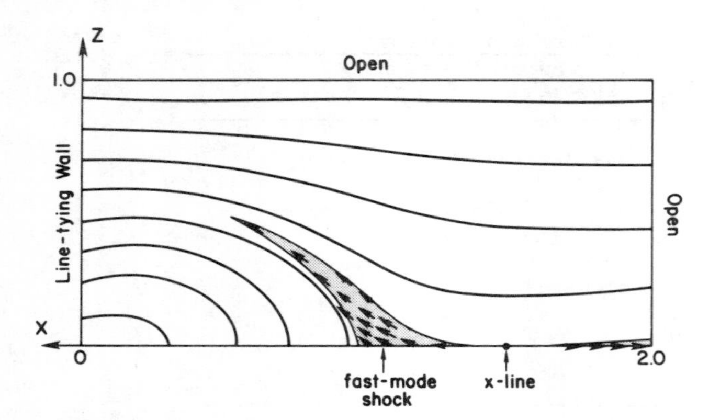

Fig. 8. Formation of a plasma sheet boundary layer in the numerical experiment of Forbes and Priest (1983). At the open boundaries, plasma is free to move in or out of the numerical domain. Arrows and shading indicate the region where the flow speed exceeds 0.3 times the external Alfvén speed (i.e., the Alfvén speed of the tail lobe). Because the outflow from the x-line has a Mach number of about 2 with respect to the local fast-mode wave speed, the jet hitting the wall at left forms a fast-mode shock. This shock is a permanent flow feature. Both the x-line and the shock move to the right (more negative x) as the region of closed field lines at left increases with time (after Forbes and Priest, 1983).

transport coefficient like electrical resistivity in a self-consistent way, since it may be a function of the global solution rather than just the local fluid properties. But given the present-day lack of any collisionless global model, an MHD model should at least provide instructional insight into how parallel electric fields interact with the convection dynamics.

3.3 Plasma Sheet Boundary Layer

A persistent feature of substorm flows in the near-earth magnetotail is their confinement to the outer edges of the plasma sheet (Forbes et al., 1981; Andrews et al., 1981; Williams, 1981). These flows are essentially parallel to the magnetic field and they form a boundary layer whose existence was predicted by Cowley (1980) before its actual discovery. Cowley realized that the inner magnetosphere would act as an obstacle to free outflow from the x-line and that most of the plasma sheet earthward of the x-line would be at rest. However, at the outer edges of the plasma sheet, particles on newly reconnected field lines would not yet have had time to be brought to rest, and so an earthward flow would exist.

Cowley's prediction of the boundary layer was based on a collisionless, cold temperature formulation, but recently Soward and Priest (1986) and Schindler and Birn (1987) have demonstrated that a plasma sheet boundary layer is also a general feature of MHD reconnection models.

One of the earliest MHD numerical experiments to demonstrate the formation of field-aligned flows along the separatrices (i.e., a plasma sheet boundary layer) was done by Sato et al. (1978) in which they compared driven reconnection using free outflow boundary conditions (i.e., open conditions) to driven reconnection using blocked outflow boundary conditions (i.e., wall conditions). In the former case the outflow from the x-line resembled the usual Petschek picture of two oppositely directed outflow jets, but in the latter case the reconnection resulted in four outflow jets, each aligned along one of the four separatrices forming the x-lines.

A similar bifurcation of the two standard Petschek jets into four field-aligned jets was also found by Biskamp (1986). Soward and Priest (1986) have also shown that the cause of the bifurcation is the boundary conditions. Yet another example of boundary layer formation is the numerical experiment on line-tied reconnection carried out by Forbes and Priest (1983). The outflow boundary layer occurring in this experiment is shown in Figure 8. At the left boundary ($x = 0$) a perfectly conducting wall blocks the outflow and also line-ties the magnetic field threading it, while at the right boundary ($x = -2$) free outflow is implemented by using open boundary conditions. Because the outflow from the x-line is supermagnetosonic with respect to the fast-mode wave speed, a fast shock forms in the jet directed toward the left boundary (Forbes, 1986). Upstream of the shock the jet is similar to its counterpart at the right of the x-line, but downstream of the shock the jet bifurcates to form a field-aligned boundary layer flow enclosing a region of relatively stagnant flow. As the reconnection proceeds, the x-line and the fast shock shift to the right as more and more field lines are closed.

It is interesting to speculate whether a fast-mode shock ever occurs in the magnetotail. From the MHD point of view, such a shock should form when the earthward-directed reconnection jet is supermagnetosonic with respect to the fast-mode wave speed. In compressible Petschek theory this occurs when the β parameter of the plasma flowing in toward the x-line is less than 0.8 (Soward and Priest, 1982). If the effective inflow region is the tail lobe, then this condition is well met since in the lobes, $\beta \lesssim 10^{-2}$. However, if the effective inflow region is the plasma mantle, then β might be only slightly less than 1. Even though a fast shock might be present, it may be difficult to detect because it is relatively weak (compression factor <2) and small in spatial extent ($\lesssim 10^4$ km).

In MHD models, the plasma sheet boundary layer is due to the blockage of the outflow from the x-line either by a wall (Sato, 1978; Forbes and Priest, 1983) or by a region of increasing magnetic field strength (Schindler and Birn, 1987). In the collisionless cold plasma models (Cowley, 1980; Williams, 1981), the plasma sheet boundary layer arises because particles accelerated in the vicinity of the x-line are returned to the tail by mirroring in the inner magnetosphere. By the time the mirrored particles return to the tail, the field lines on which they are located have moved in toward the center of the plasma sheet, and the bulk flow, which is the integral first moment of both mirrored and un-mirrored particles, is nearly zero. Thus, a significant bulk flow exists only near the edges of the plasma sheet where mirrored particles are absent.

In both MHD and collisionless cold plasma models, the existence of the plasma sheet boundary layer depends on the presence of a downstream obstacle, namely the earth's inner magnetosphere. Although similar in spirit to collisionless models, MHD models of the plasma sheet boundary layer do not explicitly incorporate nonlocal effects of free-streaming particles. If such particles are dominant, then the plasma sheet boundary layer may be an example of a thin boundary layer where non-MHD techniques are needed.

4. Summary

MHD provides a convenient formalism for considering the global aspects of the magnetosphere/ionosphere system; however, since the magnetospheric plasma is collisionless, MHD models can never be entirely satisfactory. Nevertheless, there are questions the MHD models can successfully address. To illustrate this we have examined how the upstream solar wind boundary conditions and the ionospheric boundary conditions might affect reconnection dynamics.

Based on a recent generalization of Petschek's theory by Priest and Forbes (1986), we have argued that different upstream conditions in the solar wind might give rise to different types of magnetic reconnection at the magnetopause. Normally, the reconnection there is expected to be of the unstable flux-pile-up type. This type could produce patchy reconnection events which are similar to the ubiquitous flux-transfer events. Occasionally, the upstream solar wind conditions might give rise to a stable type of reconnection and this might explain the quasi-steady, large-scale reconnection events reported by Paschmann et al. (1979).

Ionospheric boundary conditions have been examined within the context of line-tying. Although the ionosphere only weakly line-ties the magnetospheric field lines threading it, line-tying is still potentially significant since it plays a role in apparent motions of the aurora induced by reconnection in the outer magnetosphere (e.g., the poleward leap during expansion and/or recovery phases of substorms). Global MHD models could, in principle, determine how magnetospheric reconnection, ionospheric convection, and parallel electric fields all couple together to produce the observed auroral motions.

Finally, MHD models based on reconnection could also be used to study the dynamics of the plasma sheet boundary layer. This layer is a natural result of the obstruction of the earthward-directed rconnection jet by the inner magnetosphere.

Acknowledgment. This work was supported by NSF grant ATM-8507035 and NASA grant NAGW-76 to the University of New Hamsphire.

References

Andrews, M. K., P. W. Daly, and E. Keppler, Ion jetting at the plasma sheet boundary; simultaneous observations of incident and reflected particles, Geophys. Res. Lett., 8, 987, 1981.

Aubry, M. P., C. T. Russell, and M. G. Kivelson, On the inward motion of the magnetopause before a substorm, J. Geophys. Res., 75, 7018, 1970.

Aubry, M. P., M. G. Kivelson, and C. T. Russell, Motion and structure of the magnetopause, J. Geophys. Res., 76, 1673, 1971.

Birn, J., Computer studies of the dynamic evolution of the geomagnetic tail, J. Geophys. Res., 85, 1214, 1980.

Biskamp, D., Magnetic reconnection via current sheets, Phys. Fluids, 29, 1520, 1986.

Brecht, S. H., J. G. Lyon, J. A. Fedder, and K. Hain, A time-dependent three-dimensional simulation of the earth's magnetosphere: Reconnection events, J. Geophys. Res., 87, 6098, 1982.

Chester, C. R., Techniques in Partial Differential Equations, p. 73, McGraw-Hill, New York, 1971.

Chew, G. F., M. L. Goldberger, and F. E. Low, The Boltzmann equation and the one-fluid hydromagnetic equations in the absence of particle collisions, Pro. Roy. Soc. (London), A236, 112, 1956.

Coroniti, F. V., Space plasma turbulent dissipation: Reality or myth?, in Space Plasma Simulations, edited by M. Ashour-Abdalla and D. A. Dutton, p. 399, D. Reidel Publ. Co., Dordrecht, 1985.

Coroniti, F. B., and C. F. Kennel, Can the ionosphere regulate magnetospheric convection?, J. Geophys. Res., 78, 2837, 1973.

Cowley, S.W.H., Plasma populations in a simple open model magnetosphere, Space Sci. Rev., 25, 217, 1980.

Forbes, T. G., Fast-shock formation in line-tied magnetic reconnection models of solar flares, Astrophys. J., 305, 553, 1986.

Forbes, T. G., E. W. Hones, Jr., S. J. Bame, J. R. Asbridge, G. Paschmann, N. Sckopke, and C. T. Russell, Evidence for the tailward retreat of a magnetotail during substorm recovery, Geophys. Res. Lett., 8, 261, 1981.

Forbes, T. G., and E. R. Priest, A numerical experiment relevant to line-tied reconnection in two-ribbon flares, Solar Phys., 84, 169, 1983.

Forbes, T. G., and E. R. Priest, A comparison of analytical and numerical models for steadily-driven magnetic reconnection, Rev. of Geophys., in press, 1987.

Furth, H. P., J. Killeen, and M. N. Rosenbluth, Finite-resistivity instabilities of a sheet pinch, Phys. Fluids, 6, 459, 1963.

Galeev, A. A., M. M. Kuznetsova, and L. M. Zeleny, Magnetopause stability threshold for patchy reconnection, Space Sci. Rev., 44, 1, 1986.

Holzer, T. E., and G. C. Reid, The response of the dayside magnetopause-ionosphere system to time-varying field line reconnection at the magnetopause, 1, Theoretical model, J. Geophys. Res., 80, 2041, 1975.

Hones, E. W., Jr., The poleward leap of the auroral electrojet as seen in auroral images, J. Geophys. Res., 90, 5333, 1985.

Hones, E. W., Jr., Reply to comment by Gordon Rostoker, J. Geophys. Res., 91, 5881, 1986.

Leboeuf, J. N., T. Tajima, C. F. Kennel, and J. M. Dawson, Global simulations of the three-dimensional magnetosphere, Geophys. Res. Lett., 8, 257, 1981.

Lee, L. C., and Z.F. Fu, A simulation study of magnetic reconnection: Transition from a fast-mode to a slow-mode expansion, J. Geophys. Res., 91, 4551, 1986.

Lyon, J. G., S. H. Brecht, J. D. Huba, J. A. Fedder, and P. J. Palmadesso, Computer simulation of a geomagnetic substorm, Phys. Rev. Lett., 46, 1038, 1981.

Lyons, L. R., and D. J. Williams (Eds.), Quantitative Aspects of Magnetospheric Physics, p. 99, D. Reidel Publ. Co., Dordrecht, 1984.

Parker, E. N., The solar-flare phenomenon and the theory of reconnection and annihiliation of magnetic fields, Astrophys. J. Suppl. Ser., 177, 1963.

Paschmann, G., Plasma and particle observations at the magnetopause: Implications for reconnection, in Magnetic Reconnection in Space and Laboratory Plasmas, Geophys. Monogr. Ser., vol. 30, edited by E. W. Hones, Jr., p. 114, AGU, Washington, D.C., 1984.

Paschmann, G., B.U.Ö. Sonnerup, I. Papamastorakis, N. Sckopke, G. Haerendel, S. J. Bame, J. R. Asbridge, J. T. Gosling, C. T. Russell, and R. C. Elphic, Plasma acceleration at the earth's magnetopause: Evidence for reconnection, Nature, 282, 243, 1979.

Payne, L. E., Some general remarks on improperly posed problems for partial differential equations, in Symposium on Non-Well Posed Problems and Logarithmic Convexity, Lecture Notes in Mathematics, vol. 316, edited by R. J. Knops, p. 1., Springer-Verlag, New York, 1973.

Petschek, H. E., Magnetic field annihilation, in AAS-NASA Symposium on the Physics of Solar Flares, NASA SP-50, edited by W. N. Hess, p. 425, 1964.

Priest, E. R., The magnetohydrodynamics of current sheets, Rep. Prog. Phys., 48, 955, 1985.

Priest, E. R., and T. G. Forbes, New models for fast steady-state magnetic reconnection, J. Geophys. Res., 91, 5579, 1986.

Reid, G. C., and T. E. Holzer, The response of the dayside magnetosphere-ionosphere system to time-varying field line reconnection at the magnetopause, 2, Erosion event of March 27, 1968, J. Geophys. Res., 80, 2050, 1975.

Rostoker, G. R., Comment on "The poleward leap of the auroral electrojet as seen in auroral images" by Edward W. Hones, Jr., J. Geophys. Res., 91, 5879, 1986.

Russell, C. T., and R. C. Elphic, ISEE observations of flux transfer events at the dayside magnetopause, Geophys. Res. Lett., 6, 33, 1978.

Sato, T., T. Hayashi, and T. Tamao, Confinement and jetting of plasmas by magnetic reconnection, Phys. Rev. Lett., 41, 1548, 1978.

Schindler, K., and J. Birn, On the generation of field-aligned plasma flow at the boundary of the plasma sheet, J. Geophys. Res., 92, 95, 1987.

Sergeev, V. A., and A. G. Yahnin, The features of auroral bulge expansion, Planet. Space Sci., 27, 1429, 1979.

Sonnerup, B.U.Ö., Magnetic-field reconnection in a highly conducting incompressible fluid, J. Plasma Phys., 4, 161, 1970.

Sonnerup, B.U.Ö., and E. R. Priest, Resistive MHD stagnation-point flows at a current sheet, J. Plasma Phys., 14, 283, 1975.

Sonnerup, B.U.Ö., and J.-I. Sakai, Stability of a current sheet with resistive MHD stagnation-point flows (abstract), Eos Trans. AGU, 62, 353, 1981.

Soward, A. M., and E. R. Priest, Magnetic field line reconnection, Phil. Trans. Roy. Soc. of London, 284, 369, 1977.

Soward, A. M., and E. R. Priest, Fast magnetic field-line reconnection in a compressible fluid, part 1. Coplanar field lines, J. Plasma Phys., 28, 335, 1982.

Soward, A. M., and E. R. Priest, Magnetic field line reconnection with jets, J. Plasma Phys., 35, 333, 1986.

Vasyliunas, V. M., Theoretical models of magnetic field line merging, 1, Rev. Geophys. Space Phys., 13, 303, 1975.

Vemuri, V., and W. J. Karplus, Digital Computer Treatment of Partial Differential Equations, p. 73, Prentice-Hall, Englewood Cliffs, 1981.

Walker, R. J., Modeling planetary magnetospheres, Rev. Geophys., 21, 495, 1983.

Williams, D. J., Energetic ion beams at the edge of the plasma sheet: ISEE-1 observations plus a simple explanatory model, J. Geophys. Res., 86, 5507, 1981.

Wu, C. C., R. J. Walker, and J. M. Dawson, A three-dimensional MHD model of the earth's magnetosphere, Geophys. Res. Lett., 8, 523, 1981.

MAKING CONNECTIONS BETWEEN GLOBAL MAGNETOSPHERIC MODELS AND SIMULATIONS OF MICROSCOPIC PLASMA PROCESSES

R. L. Lysak

School of Physics and Astronomy, University of Minnesota, Minneapolis, Minnesota 55455

Abstract. Magnetospheric phenomena occur over a wide range of spatial and temporal scales and include a wide variety of physical processes, making the construction of a comprehensive magnetospheric model very difficult if not impossible. This paper will outline some of the problems involved in constructing such models and some potential paths for their solution. An attainable goal of attempting to meld together various types of models will be described, along with some early efforts along these lines.

Introduction

The magnetosphere, ionosphere, and neutral atmosphere constitute a large, complex system in which a wide variety of physical phenomena, including chemical reactions, ionization phenomena, hydrodynamic and magnetohydrodynamic motions, and plasma instabilities and turbulence, occur. No single model on any computer existing or planned to date can hope to cover all of these phenomena. Computations must take place using a finite number of grid points or elements which evolve according to a finite set of equations. Thus, phenomena occurring on spatial scales smaller than the grid size cannot be modeled. Similarly, larger-scale phenomena which may play an important and time varying role on the boundary conditions are outside the scope of the model. In addition, introducing new physical effects requires increasing the set of evolution equations which increases the demands on memory and computation time.

Given these limitations, it is not surprising that a comprehensive model of the magnetosphere-ionosphere-atmosphere system does not yet exist. As this meeting has demonstrated, many varied approaches have been attempted in order to address various problems within the total system. At the risk of oversimplifying the present picture, one might say that two general approaches to the problem exist, which might be called the empirical approach and the first principles approach. In the empirical approach, the modeler attempts to accumulate a large base of data and, using certain notions about the important physics of the problem, parameterizes the data statistically as a function of appropriate input parameters. In this way an attempt can be made to predict the behavior of the dependent variables for future observations given the input. The first principles approach, on the other hand, begins with a set of preconceived notions about the important physical properties of the system and models the important processes, often with a set of differential equations. Again given certain input, these equations can be solved numerically and the diagnostics generated compared with relevant observations.

The difference between these approaches is one of emphasis rather than one of substance. It would not be possible for an empirical model to simply collect and parameterize data without some guidance from theoretical concepts. Indeed, the success of an empirical model in predicting future observations depends on whether the data are organized by the parameters that truly describe the physical system being modeled. Similarly, a first principles model places the emphasis on discovering the consequences of a few simple physical ideas, but actual calculations require the inclusion of initial and boundary values which often must be determined empirically. Thus, in both types of models both empirical data and theoretical ideas are used, but in the empirical model the emphasis is more on predicting in some detail the behavior in future observations, while the first principles model is perhaps less concerned with actually predicting the details of a particular event and more interested in determining which physical processes are important in a particular type of situation.

Within these large classifications, various sorts of models exist depending on what physical processes and what sort of spatial and temporal scales are of interest to the modeler. The remainder of this paper will concentrate on models of the first principles type, in particular, models of magnetohydrodynamic and plasma phenomena occurring in the outer magnetosphere. Even within this restriction, one finds no one model can describe the entirety of phenomena because of the wide variety of spatial and temporal scales and the finite nature of computing resources. One approach to come to grips with this problem will be described.

Spatial Scales of Magnetospheric Phenomena

The magnetosphere is often described as a very large plasma physics laboratory in which one advantage over conventional plasma laboratories is the absence of boundaries. This statement is of course not strictly true since there are in fact various boundaries, e.g., the ionosphere, the magnetopause, and the plasma sheet boundary layers. A more precise statement is that the boundaries in the magnetospheric plasma are very far apart in terms of the fundamental length scales of the plasma, such as the Debye length and the gyroradius. In terms of typical magnetospheric parameters, the Debye length $\lambda_D \simeq 7$ m $\sqrt{[T(eV)/n(cm^{-3})]}$ and the gyroradius $\rho_i \simeq 1$ m $\sqrt{[T(eV)(m/m_p)]}/B(G)$ are much smaller than the macroscopic scale sizes of the magnetosphere ($\sim$10 R_E) so that an auroral field line, for example, contains millions of Debye lengths. Similarly the time scales are quite disparate, with the electron plasma frequency $\omega_p \simeq 9$ kHz $\sqrt{n(cm^{-3})}$ and the ion gyrofrequency $\Omega_i \simeq 1.5$ kHz $B(G)/(m/m_p)$ being quite fast compared to the time scales of tens of seconds to minutes typical of the transit time of particles or Alfvén waves along an auroral field line.

These distance and time scales are of importance since they govern the stability of the most common type of model for the microscopic problems in plasma physics, namely, the explicit particle-in-cell simulation method (e.g., Birdsall and Langdon, 1985). This method involves following a large number of superparticles as they interact with the self-consistent fields produced by the charge and current densities of the particles themselves. This type of simulation allows one to study plasma waves and instabilities in great detail since diagnostic information on both the particle distribution functions and the wave spectrum for every allowed wavelength in the system can be obtained as the system evolves. This method has been used successfully in modeling such phenomena as auroral double layer formation (e.g., Goertz and Joyce, 1975; Sato and Okuda, 1980; Barnes et al., 1985), ion acceleration in the auroral zone (Okuda and Ashour-Abdalla, 1981, 1983; Hudson and Roth, 1986), active experiments in space (Roth et al., 1983; Hudson and Roth, 1984), and interactions of the solar wind with man-made and natural comets (e.g., Lui et al., 1986; Madlund et al., 1986), to name only a few areas.

Particle simulations cannot be easily extended to large-scale systems due to stability considerations. For an explicit, electrostatic, one-dimensional particle code to be stable, the time step must satisfy $\omega_p \Delta t < 2$, where Δt is the time step. Similarly, finite cell size instabilities can take place if $\Delta x > \pi \lambda_D$, effectively limiting the total spatial size of the system. Electromagnetic codes have an additional restriction that $\Delta x/\Delta t > c$, the so-called Courant condition. Extending the system to three dimensions in velocity is not difficult, but things get worse in more spatial dimensions, since not only does the number of grid points increase, but the number of particles required also increases. Most particle simulations in the space physics literature are therefore limited to a few thousands of Debye lengths in scale in one dimension and a few hundreds of Debye lengths in two dimensions. Three-dimensional particle simulations are at present only in a very early stage of development. Therefore, a particle simulation can at best only hope to model a region of the magnetosphere of perhaps 1000 km in size. One consequence of this is that the system is quite dependent on the boundary conditions, which must be imposed rather arbitrarily in regions of space where no physical boundaries exist. Similar restrictions apply to models that solve the Vlasov equation directly (Singh and Schunk, 1982; Smith, 1982) which are, in practice, limited to one dimension and systems of a few hundred Debye lengths.

Therefore, global scale modeling of the magnetosphere cannot be done on the particle level. To model large-scale phenomena, the detailed information on the distribution functions must be lumped into moments such as the density, fluid velocity, and temperature for each species. At this level the magnetohydrodynamics (MHD) description becomes appropriate. The MHD equations take the form of partial differential equations which can then be stepped forward in time. The grid size of an MHD model is limited mainly by the requirement that spatial scales of interest are well resolved, since modes will be damped due to numerical diffusion unless the wave number satisfies $k\Delta x \ll 1$. The time step in this model is determined by a Courant condition, $\Delta t < \Delta x/V$, where V is a characteristic velocity (e.g., fluid velocity or Alfvén speed). Since the total number of cells is limited by available computer resources, the spatial resolution in existing models of the entire magnetosphere is limited, (e.g., Lyon et al., 1986; Ogino et al., 1986). This has the consequence that reconnection rates in the tail are determined by the numerical diffusion and not by any physical effect. Since different numerical methods have different degrees of diffusion, results using different methods may yield quite different results (Brackbill, 1985).

One way of improving the spatial resolution in an MHD model is to restrict the domain of the model to a limited region of space. Such mesoscale models have been applied to various regions of the magnetosphere such as the tail reconnection region (e.g., Birn et al., 1986), the polar wind region (e.g., Schunk, 1977), and the auroral acceleration region (e.g., Lysak and Dum, 1983; Mitchell and Palmadesso, 1983; Lysak, 1985, 1986). These models have the advantage of higher spatial resolution at the cost of a more limited simulation domain; therefore, this type of model is dependent on boundary conditions which must be applied in a more arbitrary way than in the global models, in which the solar wind constitutes a boundary which is conceptually straightforward. All of these models, however, being based on the MHD approximation, are unable to self-consistently model microscopic processes such as wave-particle interactions since detailed information on particle distributions is lost.

The difficulty of modeling phenomena occurring on a wide range of scales suggests that some means of communicating information between models of various types must be found. Phenomena occurring on scales smaller than the grid size of the model must somehow be included in the dynamics of the system being considered. Conversely, information on phenomena occurring on scales larger than the system size must be incoporated into the boundary conditions of the model. A way of proceeding based on multiple scale analysis used in the analysis of MHD turbulence is considered in the next section.

Multiple Spatial Scale Analysis

The problem of relating phenomena at a wide variety of spatial scales has long been central to the study of magnetohydrodynamic turbulence (e.g., Kraichnan, 1967; Batchelor, 1970; Kraichnan and Montgomery, 1980). It is well known that steady state spectra of turbulence exhibit a power law behavior k^{-p}, where the spectral index p depends on the dimensionality of the turbulence. Such power law spectra extend to a maximum wave number that depends on the Reynolds number of the system. To fully describe the evolution of turbulence in the high Reynolds number (i.e., low dissipation) limit appropriate for most space plasmas, it is therefore necessary to use a very large number of wave numbers, resulting in numerical problems similar to those described above. The approach taken in MHD turbulence theory has been to model the small-scale interactions by means of effective transport coefficients, such as the so-called eddy viscosity.

In hydrodynamics such coefficients are introduced by the following procedure (e.g., Kraichnan, 1976). The coupling between models of different wavelengths enters in through the nonlinear term in the fluid momentum equation

$$\frac{\partial \mathbf{v}}{\partial t} = -\mathbf{v} \cdot \nabla \mathbf{v} - \nabla p. \qquad (1)$$

If the spectrum is such that it can be divided into separate regions in which $kL_c < 1$ and those in which $kL_c > 1$ where L_c is some scale length, the equation of motion can be averaged over the scale L_c producing an average momentum equation

$$\frac{\partial \langle\mathbf{v}\rangle}{\partial t} = -\langle\mathbf{v}\rangle \cdot \nabla\langle\mathbf{v}\rangle - \langle\delta\mathbf{v} \cdot \nabla\delta\mathbf{v}\rangle - \nabla\langle p\rangle, \qquad (2)$$

where $\delta\mathbf{v}$ represents the fluctuating part of the velocity, $\mathbf{v} = \langle\mathbf{v}\rangle + \delta\mathbf{v}$. In the case of isotropic spectra, the averaging of the nonlinear term can, to lowest order, be taken to have the simple form

$$-\langle\delta\mathbf{v} \cdot \nabla\delta\mathbf{v}\rangle = \chi\nabla^2\langle\mathbf{v}\rangle, \qquad (3)$$

where the coefficient χ is referred to as the eddy viscosity. The eddy viscosity can be calculated if the spectrum is assumed (e.g., Pouquet et al., 1976) and under some restrictions can be calculated more self-consistently (Montgomery and Hatori, 1984; Montgomery and Chen, 1984). In MHD, the magnetic induction equation

$$\frac{\partial \mathbf{B}}{\partial t} = \nabla \times (\mathbf{v} \times \mathbf{B}) \qquad (4)$$

can be averaged to yield

$$\frac{\partial \langle\mathbf{B}\rangle}{\partial t} = \nabla \times (\langle\mathbf{v}\rangle \times \langle\mathbf{B}\rangle) + \nabla \times \langle\delta\mathbf{v} \times \delta\mathbf{B}\rangle, \qquad (5)$$

but in this case, in the presence of a large-scale background field the averaged nonlinear term has the form

$$\langle\delta\mathbf{v} \times \delta\mathbf{B}\rangle = \alpha\langle\mathbf{B}\rangle - \beta\nabla \times \langle\mathbf{B}\rangle, \qquad (6)$$

where the first term is the so-called alpha effect which can be responsible for dynamo action and magnetic field generation (e.g., Moffatt, 1978) and the second term, the beta effect, gives the anomalous resistivity. Due to the assumption of separated spatial scales, these models are not strictly rigorous in the usual situation of a smooth spectrum; however, use of such assumed transport coefficients does allow some useful results to be obtained.

The self-consistent calculation of these transport coefficients is complicated even in situations with a great deal of symmetry. Inclusion of plasma effects such as finite ion gyroradii complicates matters further. At these scales plasma instabilities rather than the fluid nonlinearities can provide the major nonlinear effects, and, for weak turbulence, the quasi-linear theory has been used to estimate such effective transport coefficients (Papadopoulos, 1977). In space plasmas, quasi-linear theory has been used to account for parallel electric fields in the aurora (e.g., Hudson et al., 1978) and the electric fields along the neutral line during reconnection (e.g., Huba et al., 1978). Quasi-linear results have been used to estimate transport coefficients in mesoscale MHD models of reconnection (Hayashi and Sato, 1978) and auroral current flows (Lysak and Dum, 1983). As in the case of MHD turblence, the quasi-linear theory is not completely rigorous, particularly in situations where strong turbulence effects, such as double layer formation, are important. However, the quasi-linear theory in practice gives results

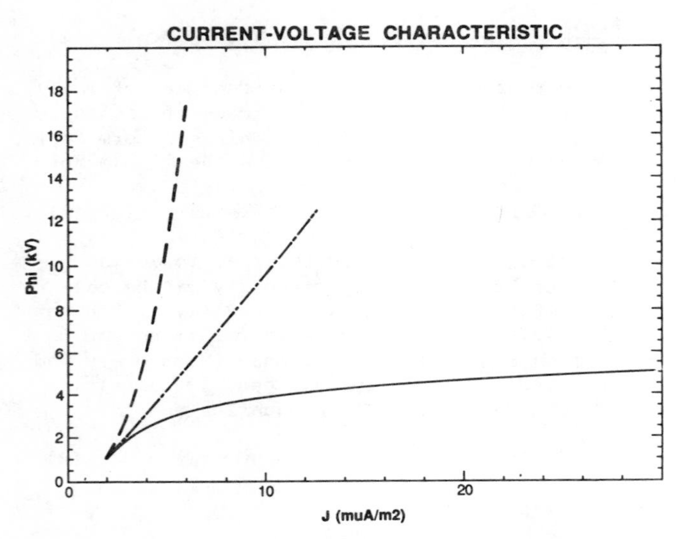

Fig. 1. Current-voltage relationship for the auroral circuit for three different microscopic models: double layer model (solid line), quasi-linear theory (dot-dash line), and quasi-linear theory modified by resonance broadening (dashed line). Details can be found in Lysak and Hudson (1987).

which are a good first approximation for the parallel electric field.

Use of Microscopic Simulation to Obtain Transport Coefficients

A possible way of using simulations in order to improve on quasi-linear theory in the context of the auroral parallel electric field problem was recently discussed by Lysak and Hudson (1987). In order to estimate the parallel electric field due to the formation of double layers, an estimate of the average electric field due to the double layers was found. This model was used in the MHD model of auroral current flows (Lysak and Dum, 1983) in order to assess the global effects of this electric field. The results are compared both with quasi-linear theory and with quasi-linear theory modified by resonance broadening effects (Dum and Dupree, 1970) in Figure 1 (Lysak and Hudson, 1987). As can be seen, the different models have different consequences with regard to the exact shape of the current-voltage curve for the auroral zone, although all the models give qualitative agreement with typical auroral zone observations of potential drops of kilovolts being associated with field-aligned currents of a few microamps per square meter. Athough this model represents only a crude characterization of the simulation results, future work will concentrate on performing particle simulations under varying parameters in order to construct a more complete local current-voltage relation.

The essential ingredient in constructing transport coefficients based on microscopic simulations is to determine what the transport is as a function of the fluid variables. Whether the properties of kinetic effects such as double layers can, in fact, be reduced to such a simple parameterization is not at all certain and a subject of future research. The fact remains that even an approximate characterization of the kinetic effects can improve the understanding of what the macroscopic consequences of microscopic processes might be, especially since many of the details of the microphysics may be averaged out at the macroscopic level.

Use of larger-scale models can influence the development of the microscopic model by indicating the proper boundary conditions to use. One might consider the microscopic model to be essentially one cell of the larger scale model; thus the appropriate boundary conditions could be simply the quantities on each side of a cell at some given time in the larger model. Since the time scale of the microscopic simulation is usually much faster than that of a larger scale model, these boundary conditions would be constant, or perhaps slowly varying in time. One possible approach is to consider the larger scale model as an external circuit that closes the current which flows through the simulation (e.g., Birdsall and Langdon, 1985). Some models with external circuits of varying degrees of complexity have already appeared in the literature (Sato and Okuda, 1980; Smith, 1982).

Conclusions

Any attempt to construct a comprehensive model of the magnetosphere as well as other parts of the total terrestrial system will necessarily involve a variety of models covering different physical processes and different spatial and temporal scales. To a great extent, the success of the overall program will depend on how information gathered by the use of one class of models can be incorporated into others. This paper has suggested that one way to include plasma kinetic effects into large-scale MHD models is to parameterize the results of microscopic plasma simulations in terms of nonlinear transport coefficients that depend upon the MHD variables. Conversely, the results of large-scale models should be used qualitatively, if not quantitatively, to determine the appropriate initial and boundary conditions in the microscopic simulation. Existing efforts along these lines have been mainly at the qualitative level, in which the general characteristics of one type of model have been used in the other. Future efforts should attempt to make this correspondence more quantitative by, for example, doing a set of particle simulations with the express goal in mind of constructing a form for transport coefficients to use in an MHD model. Coordinated use of different types of models in this way will help compensate for the present limitations of each type of model individually.

Acknowledgments. Since this paper was not meant to be a comprehensive review of all types of models, I wish to apologize to all the modelers whom I did not reference in this paper. I wish to thank Yan Song, Mary Hudson, and Bill Lotko for useful discussions on this topic, and Tom Moore and Hunter Waite for bringing together the variety of modelers present at this meeting. This work was supported in part in NSF grants ATM-8451168 and ATM-8508949.

References

Barnes, C., M. K. Hudson, and W. Lotko, Weak double layers in ion acoustic turbulence, Phys. Fluids, 28, 1055, 1985.

Batchelor, G. K., Theory of Homogeneous Turbulence, Cambridge University Press, Cambridge, 1970.

Birdsall, C. K., and A. B. Langdon, Plasma Physics via Computer Simulation, McGraw Hill, New York, 1985.

Birn, J., E. W. Hones, and K. Schindler, Field-aligned plasma flow in MHD simulations of magnetotail reconnection and the formation of boundary layers, J. Geophys. Res., 91, 11,116, 1986.

Brackbill, J. U., Fluid modeling of magnetized plasmas, in Space Plasma Simulations, edited by M. Ashour-Abdalla and D. A. Dutton, p. 153, Reidel, Boston, 1985.

Dum, C. T., and T. H. Dupree, Nonlinear stabilization of high frequency instabilities in a magnetic field, Phys. Fluids, 13, 2064, 1970.

Goertz, C. K., and G. Joyce, Numerical simulation of the plasma double layer, Astrophys. Space Sci., 32, 165, 1975.

Hayashi, T., and T. Sato, Magnetic reconnection: Acceleration, heating, and shock formation, J. Geophys. Res., 83, 217, 1978.

Huba, J. D., N. T. Gladd, and K. Papadopoulos, Lower hybrid drift wave turbulence in the distant magnetotail, J. Geophys. Res., 83, 5217, 1978.

Hudson, M. K., and I. Roth, Thermal fluctuations from an artificial ion beam injection into the ionosphere, J. Geophys. Res., 89, 9812, 1984.

Hudson, M. K., and I. Roth, Ion heating in the cusp, in Ion Acceleration in the Magnetosphere and Ionosphere, Geophys. Monogr. Ser., vol. 38, edited by T. Chang, p. 271, AGU, Washington, D.C., 1986.

Hudson, M. K., R. L. Lysak, and F. S. Mozer, Magnetic field-aligned potential drops due to electrostatic ion cyclotron turbulence, Geophys. Res. Lett., 5, 143, 1978.

Kraichnan, R. H., Inertial ranges in two-dimensional turbulence, Phys. Fluids, 10, 1417, 1967.

Kraichnan, R. H., Eddy viscosity in two and three dimensions, J. Atmos. Sci., 33, 1521, 1976.

Kraichnan, R. H., and D. Montgomery, Two-dimensional turbulence, Rept. Prog. Phys., 43, 547, 1980.

Lui, A.T.Y., C. C. Goodrich, A. Mankofsky, and K. Papadopoulos, Early time interaction of lithium ions with the solar wind in the AMPTE mission, J. Geophys. Res., 91, 1333, 1986.

Lyon, J. G., J. A. Fedder, and J. D. Huba, The effect of different resistivity models on magnetotail dynamics, J. Geophys. Res., 91, 8057, 1986.

Lysak, R. L., Auroral electrodynamics with current and voltage generators, J. Geophys. Res., 90, 4178, 1985.

Lysak, R. L., Coupling of the dynamic ionosphere to auroral flux tubes, J. Geophys. Res., 91, 7047, 1986.

Lysak, R. L., and C. T. Dum, Dynamics of magnetosphere-ionosphere coupling including turbulent transport, J. Geophys. Res., 88, 365, 1983.

Lysak, R. L., and M. K. Hudson, Effect of double layers on magnetosphere-ionosphere coupling, Laser and Particle Beams, in press, 1987.

Madlund, C. D., S. P. Gary, and D. Windke, Ion/ion instabilities and the development of shell-like ion distributions (abstract), Eos Trans. AGU, 67, 1144, 1986.

Mitchell, H. G., and P. J. Palmadesso, A dynamic model for the auroral field line plasma in the presence of field-aligned current, J. Geophys. Res., 88, 2131, 1983.

Moffatt, H. K., Magnetic Field Generation in Electrically Conducting Fluids, Cambridge University Press, Cambridge, 1978.

Montgomery, D., and T. Hatori, Analytical estimates of turbulent MHD transport coefficients, Plasma Phys. and Controlled Fusion, 26, 717, 1984.

Montgomery, D., and H. Chen, Turbulent amplification and large-scale magnetic fields, Plasma Phys. and Controlled Fusion, 26, 1199, 1984.

Ogino, T., R. J. Walker, M. Ashour-Abdalla, and J. M. Dawson, An MHD simulation of the effects of the interplanetary magnetic field B_y component on the interaction of the solar wind with the earth's magnetosphere during southward interplanetary magnetic field, J. Geophys. Res., 91, 10,029, 1986.

Okuda, H., and M. Ashour-Abdalla, Formation of a conical distribution and intense ion heating in the presence of hydrogen cyclotron waves, Geophys. Res. Lett., 8, 811, 1981.

Okuda, H., and M. Ashour-Abdalla, Acceleration of hydrogen ions and conic formation along auroral field lines, J. Geophys. Res., 88, 889, 1983.

Papadopoulos, K., A review of anomalous resistivity for the ionosphere, Rev. Geophys. Space Phys., 15, 113, 1977.

Pouquet, A., U. Frisch, and J. Leurat, Strong MHD helical turbulence and the nonlinear dynamo effect, J. Fluid Mech., 77, 321, 1976.

Roth, I., C. W. Carlson, M. K. Hudson, and R. L.

Lysak, Simulation of beam excited minor species gyroharmonics in the Porcupine experiment, J. Geophys. Res., 88, 8115, 1983.

Sato, T., and H. Okuda, Numerical simulation of ion acoustic double layers, J. Geophys. Res., 85, 3357, 1980.

Schunk, R. W., Mathematical structure of transport equations for multispecies flows, Rev. Geophys. Space Phys., 15, 429, 1977.

Singh, N., and R. W. Schunk, Dynamical features of moving double layers, J. Geophys. Res., 87, 3561, 1982.

Smith, R. A., Vlasov simulation of plasma double layers, Physica Scripta, 25, 413, 1982.

CONFERENCE PARTICIPANTS

FIRST HUNTSVILLE WORKSHOP ON
MAGNETOSPHERE/IONOSPHERE PLASMA MODELS
GUNTERSVILLE STATE PARK, ALABAMA
OCTOBER 13-16, 1986

Roger R. Anderson
Maha Ashour-Abdalla
A. R. Barakat
Robert Bingham
Donald Brautingam
J. L. Burch
Don Carpenter
Michael O. Chandler
Tom Chang
C. R. Chappell
P. K. Chaturvedi
Margaret W. Chen
John B. Cladis
Richard H. Comfort
Paul D. Craven
G. B. Crew
Dwight Decker
Pierette Decreau
Odile de la Beaujardiere
Alex Dessler
Paul B. Dusenbery
Tim Eastman
Joel A. Fedder
Terry G. Forbes
L. A. Frank
Dennis L. Gallagher
Supriya B. Ganguli
Stephanie Gannaway-Osborn
Barbara Giles
Ed Gillis
George Gloeckler
T. Gombosi
Donald A. Gurnett
M. Susan Gussenhoven
Rod Heelis
Tom Holzer
James L. Horwitz
Charles F. Kennel
Timothy L. Killee
William Lotko
Richard Lundin
John Lyon
L. R. Lyons
Bob Lysak
Richard F. Martin, Jr.
Carl E. McIlwain
R. J. Moffett
Thomas E. Moore
Julie J. Moses
Richard Olsen
Nojan Omidi
Peter J. Palmadesso
W. K. Peterson
Craig J. Pollock
David L. Reasoner
Robert H. Redus
John M. Retterer
Timothy Reyes
Phil Richards
Raymond G. Roble
David J. Rogers
Paul L. Rothwell
Per Even Sandholt
Rudolph Schmidt
J. David Schriver
Robert W. Schunk
N. Singh
George L. Siscoe
Frank Six
Yan Song
Theodore Speiser
Walter N. Spjeldvik
Kenneth Swinney
Roy B. Torbert
Douglas Torr
Vytenis M. Vasyliunas
J. Hunter Waite
Raymond J. Walker
Daniel Weimer
Elden C. Whipple
Robert M. Winglee
J. David Winningham
R. A. Wolf
Andrew W. Yau